INTRODUCTION TO CHEMICAL ENGINEERING THERMODYNAMICS

McGraw-Hill Chemical Engineering Series

Building the Literature of a Profession

Fifteen prominent chemical engineers first met in New York more than 60 years ago to plan a continuing literature for their rapidly growing profession. From industry came such pioneer practitioners as Leo H. Baekeland, Arthur D. Little, Charles L. Reese, John V. N. Dorr, M. C. Whitaker, and R. S. McBride. From the universities came such eminent educators as William H. Walker, Alfred H. White, D. D. Jackson, J. H. James, Warren K. Lewis, and Harry A. Curtis. H. C. Parmelee, then editor of *Chemical and Metallurgical Engineering,* served as chairman and was joined subsequently by S. D. Kirkpatrick as consulting editor.

After several meetings, this committee submitted its report to the McGraw-Hill Book Company in September 1925. In the report were detailed specifications for a correlated series of more than a dozen texts and reference books which have since become the McGraw-Hill Series in Chemical Engineering and which became the cornerstone of the chemical engineering curriculum.

From this beginning there has evolved a series of texts surpassing by far the scope and longevity envisioned by the founding Editorial Board. The McGraw-Hill Series in Chemical Engineering stands as a unique historical record of the development of chemical engineering education and practice. In the series one finds the milestones of the subject's evolution: industrial chemistry, stoichiometry, unit operations and processes, thermodynamics, kinetics, and transfer operations.

Chemical engineering is a dynamic profession, and its literature continues to evolve. McGraw-Hill, with its editor, B.J. Clark and its consulting editors, remains committed to a publishing policy that will serve, and indeed lead, the needs of the chemical engineering profession during the years to come.

The Series

INTRODUCTION TO CHEMICAL ENGINEERING THERMODYNAMICS

Fifth Edition

J. M. Smith
Professor Emeritus of Chemical Engineering
University of California, Davis

H. C. Van Ness
Institute Professor Emeritus of Chemical Engineering
Rensselaer Polytechnic Institute

M. M. Abbott
Professor of Chemical Engineering
Rensselaer Polytechnic Institute

The McGraw-Hill Companies, Inc.

New York   St. Louis   San Francisco   Auckland   Bogotá   Caracas   Lisbon
London   Madrid   Mexico City   Milan   Montreal   New Delhi
San Juan   Singapore   Sydney   Tokyo   Toronto

McGraw-Hill

A Division of The McGraw·Hill Companies

This book was set in Computer Modern Roman by Publication Services, Inc.
The editors were B. J. Clark, Kiran Kimbell, and John M. Morriss;
the production supervisor was Annette Mayeski.
The cover was designed by Joan Greenfield.
Project supervision was done by Publication Services, Inc.
R. R. Donnelley & Sons Company was printer and binder.

Library of Congress Catalog Card Number: 95-81742

CONTENTS

LIST OF SYMBOLS

A	Area
A	Helmholtz energy
A	Parameter in Eq. (4.4)
a	Acceleration
a	Molar area of an adsorbed phase
a	Parameter in cubic equations of state
a_i	Activity of pure species i
$\bar{a}_i$	Partial parameter, cubic equations of state
$\hat{a}_i$	Activity of species i in solution
B	Second virial coefficient, density expansion
B	Parameter in Eq. (4.4)
B'	Second virial coefficient, pressure expansion
B^0, B^1	Functions in generalized second-virial-coefficient correlation
B_{ij}	Interaction second virial coefficient
b	Parameter in cubic equations of state
$\bar{b}_i$	Partial parameter, cubic equations of state
C	Third virial coefficient, density expansion
C	Parameter in Eq. (4.4)
C'	Third virial coefficient, pressure expansion
C_{ijk}	Interaction third virial coefficient
C_P	Molar or specific heat capacity at constant pressure
C_V	Molar or specific heat capacity at constant volume
C_P^E	Excess heat capacity at constant pressure
C_P°	Standard-state heat capacity at constant pressure
ΔC_P°	Standard heat-capacity change of reaction
$\langle C_P \rangle_H$	Mean heat capacity for enthalpy calculations
$\langle C_P \rangle_S$	Mean heat capacity for entropy calculations
$\langle C_P^\circ \rangle_H$	Mean standard heat capacity for enthalpy calculations
$\langle C_P^\circ \rangle_S$	Mean standard heat capacity for entropy calculations
c	Speed of sound
D	Fourth virial coefficient, density expansion
D	Parameter in Eq. (4.4)

D'	Fourth virial coefficient, pressure expansion
E_i	Energy level
E_K	Kinetic energy
E_P	Gravitational potential energy
F	Degrees of freedom (phase rule)
F	Force
f_i	Fugacity of pure species i
f_i°	Standard-sate fugacity
$\hat{f}_i$	Fugacity of species i in solution
G	Molar or specific Gibbs energy $\equiv H - TS$
G_i°	Standard-state Gibbs energy of species i
$\bar{G}_i$	Partial Gibbs energy of species i in solution
G^E	Excess Gibbs energy $\equiv G - G^{id}$
G^R	Residual Gibbs energy $\equiv G - G^{ig}$
ΔG	Gibbs energy change of mixing
ΔG°	Standard Gibbs-energy change of reaction
ΔG_f°	Standard Gibbs-energy change of formation
g	Local acceleration of gravity
g_c	Dimensional constant $= 32.1740 (\text{lb}_\text{m})(\text{ft})(\text{lb}_\text{f})^{-1}(\text{s})^{-2}$
g_i	Degeneracy
H	Molar or specific enthalpy $\equiv U + PV$
$\bar{H}_i$	Partial enthalpy of species i in solution
H^E	Excess enthalpy $\equiv H - H^{id}$
H^R	Residual enthalpy $\equiv H - H^{ig}$
$(H^R)^0, (H^R)^1$	Functions in generalized residual-enthalpy correlation
ΔH	Enthalpy change ("heat") of mixing
$\widetilde{\Delta H}$	Heat of solution
ΔH°	Standard enthalpy change of reaction
ΔH_0°	Standard heat of reaction at reference temperature T_0
ΔH_f°	Standard enthalpy change of formation
h	Planck's constant
I	First ionization potential
K_j	Equilibrium constant for chemical reaction j
K_i	Vapor/liquid equilibrium constant for species i
k	Boltzmann's constant
k_i	Henry's constant for species i in solution
k_{ij}	Equation-of-state interaction parameter
$\mathcal{L}$	Molar fraction of system that is liquid
$\mathbf{M}$	Mach number
M	Molar mass (molecular weight)
M	Molar or specific value of extensive thermodynamic property
$\bar{M}_i$	Partial property of species i in solution
M^E	Excess property $\equiv M - M^{id}$
M^R	Residual property $\equiv M - M^{ig}$

ΔM	Property change of mixing
$\Delta M°$	Standard property change of reaction
$\Delta M_f°$	Standard property change of formation
m	Mass
$\dot{m}$	Mass flow rate
N	Number of chemical species (phase rule)
N_A	Avogadro's number
n	Number of moles
$\tilde{n}$	Moles of solvent per mole of solute
n_i	Number of moles of species i
P	Absolute pressure
$P°$	Pressure in the standard state
P_c	Critical pressure
P_r	Reduced pressure
P_0	Reference pressure
p_i	Partial pressure of species i
P_i^{sat}	Saturation vapor pressure of species i
Q	Heat
$\dot{Q}$	Rate of heat transfer
q	Electric charge
R	Universal gas constant
r	Intermolecular separation
r	Number of independent chemical reactions (phase rule)
S	Molar or specific entropy
$\bar{S}_i$	Partial entropy of species i in solution
S^E	Excess entropy $\equiv S - S^{id}$
S^R	Residual entropy $\equiv S - S^{ig}$
$(S^R)^0, (S^R)^1$	Functions in generalized residual entropy correlation
$\dot{S}_G$	Rate of entropy generation in control volume
$S_{G,\text{total}}$	Total entropy generation per unit amount of fluid
$\dot{S}_{G,\text{total}}$	Total rate of entropy generation
ΔS	Entropy change of mixing
$\Delta S°$	Standard entropy change of reaction
$\Delta S_f°$	Standard entropy change of formation
T	Absolute temperature in kelvins or rankines
T_c	Critical temperature
T_n	Normal-boiling-point temperature
T_r	Reduced temperature
T_0	Reference temperature
T_σ	Absolute temperature of surroundings
T_i^{sat}	Saturation temperature of species i
t	Temperature in °C or (°F)
U	Molar or specific internal energy
$\mathcal{U}$	Intermolecular pair-potential function

u	Velocity
V	Molar or specific volume
$\mathcal{V}$	Molar fraction of system that is vapor
$\bar{V}_i$	Partial volume of species i in solution
V_c	Critical volume
V_r	Reduced volume
V^E	Excess volume $\equiv V - V^{id}$
V^R	Residual volume $\equiv V - V^{ig}$
ΔV	Volume change of mixing
W	Work
$\dot{W}$	Work rate (power)
W_{ideal}	Ideal work
$\dot{W}_{\text{ideal}}$	Ideal work rate
W_{lost}	Lost work
$\dot{W}_{\text{lost}}$	Lost work rate
W_s	Shaft work for flow process
$\dot{W}_s$	Shaft power for flow process
x_i	Mole fraction of species i in general or in a liquid phase
x^v	Quality
y_i	Mole fraction of species i in a vapor phase
Z	Compressibility factor $\equiv PV/RT$
Z_c	Critical compressibility factor $\equiv P_c V_c/RT_c$
Z^0, Z^1	Functions in generalized compressibility-factor correlation
$\mathcal{Z}$	Partition function
z	Adsorbed phase compressibility factor; defined by Eq. (14.48)
z	Elevation above a datum level
z_i	Overall mole fraction or mole fraction in a solid phase

Superscripts

E	Denotes excess thermodynamic property
av	Denotes phase transition from adsorbed phase to vapor
id	Denotes value for an ideal solution
ig	Denotes value for an ideal gas
l	Denotes liquid phase
lv	Denotes phase transition from liquid to vapor
R	Denotes residual thermodynamic property
s	Denotes solid phase
sl	Denotes phase transition from solid to liquid
t	Denotes a total value of an extensive thermodynamic property
v	Denotes vapor phase
∞	Denotes a value at infinite dilution

Subscripts

C	Denotes a value for a cold heat reservoir
c	Denotes a value for the critical state
cv	Denotes a control volume
fs	Denotes flowing streams
H	Denotes a value for a hot heat reservoir
r	Denotes a reduced value

Greek letters

α	Polarizability
α,β	As superscripts, identify phases
β	Volume expansivity
Γ_i	Integration constant
γ	Ratio of heat capacities C_P/C_V
γ_i	Activity coefficient of species i in solution
ϵ	Well depth in intermolecular potential function
ϵ_0	Electric permittivity of vacuum
ε	Reaction coordinate
η	Efficiency
κ	Isothermal compressibility
Π	Spreading pressure in adsorbed phase
π	Number of phases (phase rule)
μ	Dipole moment
μ_i	Chemical potential of species i
ν_i	Stoichiometric number of species i
ρ	Molar density
ρ_c	Critical density
ρ_r	Reduced density
σ	Molecular collision diameter
τ	Time
τ	Temperature ratio $\equiv T/T_0$ $[\equiv 1 - T_r$ in Eq. (6.56)]
Φ_i	Ratio of fugacity coefficients, defined by Eq. (12.2)
ϕ_i	Fugacity coefficient of pure species i
$\hat{\phi}_i$	Fugacity coefficient of species i in solution
ϕ^0, ϕ^1	Functions in generalized fugacity-coefficient correlation
ω	Acentric factor

Notes

$^{\circ}$	As a superscript, denotes the standard state
$^{-}$	Overbar denotes a partial property
$\cdot$	Overdot denotes a time rate
$^{\wedge}$	Circumflex denotes a property in solution
Δ	Difference operator

PREFACE

The purpose of this text is to present thermodynamics from a chemical engineering viewpoint. The laws of thermodynamics are universal, but are most effectively taught in the context of the discipline of student commitment. This is the justification for a separate text for chemical engineers, just as it has been for the previous four editions, which have been in print for 47 years.

In writing this text, we have sought to maintain the rigor characteristic of sound thermodynamic analysis, while at the same time providing a treatment that may be readily understood by the average undergraduate. The material includes much that is of an introductory nature, but the development is carried far enough to allow application to significant problems in chemical-engineering practice. Indeed, the content is more than adequate for an academic-year undergraduate course, and is sufficiently comprehensive to make the book a useful reference both in graduate courses and for professional practice.

For a student new to this subject a demanding task of discovery lies ahead. New ideas, terms, and symbols appear at a bewildering rate. The challenge, ever present, is to think topics through to the point of understanding, to acquire the capacity to reason, and to apply this fundamental body of knowledge to the solution of practical problems. Moreover, knowledge gained here is enlarged and refined as the educational experience continues.

The first two chapters of the book present basic definitions and a development of the first law as it applies to nonflow and simple steady-flow processes. Chapters 3 and 4 treat the pressure/volume/temperature behavior of fluids and certain heat effects, allowing early application of the first law to important engineering problems. The second law and some of its applications are considered in Chap. 5. A treatment of the thermodynamic properties of pure fluids in Chap. 6 leads to application in Chap. 7 of the first and second laws to flow processes in general and in Chaps. 8 and 9 to power production and refrigeration processes. The remainder of the book, dealing with fluid mixtures, treats topics in the unique domain of chemical-engineering thermodynamics. Chapters 10 and 11 provide a comprehensive exposition of the thermodynamic properties of fluid mixtures, and of their uses in vapor/liquid equilibrium and in mixing processes. Chapter 12 is devoted to a detailed treatment of vapor/liquid equilibrium for systems at modest pressures. The application of equations of state in thermodynamic calculations, particularly in vapor/liquid equilibrium, is discussed in Chap. 13, and additional topics related to phase equilibria are treated in Chap. 14. Chemical-reaction equilibrium is covered at length in Chap. 15. Finally, Chap. 16 deals with the thermodynamic analysis of real processes, affording a review of much of the practical subject matter of thermodynamics.

We gratefully acknowledge the contributions of Professor Charles Muckenfuss, Debra L. Sauke, and Eugene N. Dorsi, whose efforts produced computer programs for calculation of the thermodynamic properties of steam and ultimately the steam tables of Appendix F. We thank those who offered detailed and constructive criticism of the fourth edition: Philip T. Eubank, Texas A&M University; Dana E. Knox, New Jersey Institute of Technology; Joseph C. Mullins, Clemson University, and Bruce E. Poling, University of Toledo. With respect to the present edition, we appreciate the efforts of Alan L. Myers, University of Pennsylvania, and Keith E. Gubbins, Cornell University, who reviewed parts of the manuscript, and of John J. Hwalek, University of Maine, who contributed expert advice regarding the Mathcad® solutions of Appendix D.

J. M. Smith
H. C. Van Ness
M. M. Abbott

INTRODUCTION TO CHEMICAL
ENGINEERING THERMODYNAMICS

CHAPTER 1

INTRODUCTION

1.1 The Scope of Thermodynamics

The science of thermodynamics was born in the nineteenth century of the need to describe the operation of steam engines and to set forth the limits of what they can accomplish. Thus the name itself denotes power developed from heat, and its initial applications were to heat engines, of which the steam engine is an example. However, the principles observed to be valid for engines were soon generalized into postulates now known as the first and second laws of thermodynamics. These laws have no proof in the mathematical sense; their validity lies in the absence of contrary experience. Thus thermodynamics shares with mechanics and electromagnetism a basis in laws.

These laws lead through mathematical deduction to a network of equations which find application in all branches of science and engineering. The chemical engineer must cope with a wide variety of problems. Among them are the calculation of heat and work requirements for physical and chemical processes, and the determination of equilibrium conditions for chemical reactions and for the transfer of chemical species between phases.

Thermodynamic considerations do not establish the *rates* of chemical or physical processes. Rates depend on both driving force and resistance. Although driving forces are thermodynamic variables, resistances are not. Neither can thermodynamics, a macroscopic-property formulation, reveal the microscopic (molecular) mechanisms of physical or chemical processes. On the other hand, knowledge of the microscopic behavior of matter can be useful in the calculation of thermodynamic properties. Property values are essential to the practical application of thermodynamics. The chemical engineer must deal with many chemical species and their mixtures, and experimental data are often unavailable. Fortunately, generalized correlations developed from a limited data base provide estimates in the absence of data.

Table 1.1: Prefixes for SI Units

Multiple	Prefix	Symbol
10^{-9}	nano	n
10^{-6}	micro	μ
10^{-3}	milli	m
10^{-2}	centi	c
10^{3}	kilo	k
10^{6}	mega	M
10^{9}	giga	G

The application of thermodynamics to any real problem starts with the identification of a particular body of matter as the focus of attention. This body of matter is called the *system*, and its thermodynamic state is defined by a few measurable macroscopic properties. These depend on the fundamental *dimensions* of science, of which length, time, mass, temperature, and amount of substance are of interest here.

1.2 Dimensions and Units

The *fundamental* dimensions are *primitives*, recognized through our sensory perceptions and not definable in terms of anything simpler. Their use, however, requires the definition of arbitrary scales of measure, divided into specific *units* of size. Primary units have been set by international agreement, and are codified as the International System of Units (abbreviated SI, for Système International).

The *second*, symbol s, the SI unit of time, is the duration of 9,192,631,770 cycles of radiation associated with a specified transition of the cesium atom. The *meter*, symbol m, is the fundamental unit of length, defined as the distance light travels in a vacuum during 1/299,792,458 of a second. The *kilogram*, symbol kg, is the mass of a platinum/iridium cylinder kept at the International Bureau of Weights and Measures at Sèvres, France. The unit of temperature is the *kelvin*, symbol K, equal to 1/273.16 of the thermodynamic temperature of the triple point of water. A more detailed discussion of temperature, the characteristic dimension of thermodynamics, is given in Sec. 1.4. The measure of the amount of substance is the *mole*, symbol mol, defined as the amount of substance represented by as many elementary entities (e.g., molecules) as there are atoms in 0.012 kg of carbon-12. This is equivalent to the "gram mole" commonly used by chemists.

Multiples and decimal fractions of SI units are designated by prefixes. Those in common use are listed in Table 1.1. Thus, the centimeter is given as 1 cm $= 10^{-2}$ m and 10^{3} g $=$ 1 kg.

Other systems of units, such as the English engineering system, use units that are related to SI units by fixed conversion factors. Thus, the foot (ft) is defined as

0.3048 m, the pound *mass* (lb_m) as 0.45359237 kg, and the pound mole (lb mol) as 453.59237 mol.

1.3 Force

The SI unit of force is the *newton*, symbol N, derived from Newton's second law, which expresses force F as the product of mass m and acceleration a:

$$F = ma$$

The newton is defined as the force which when applied to a mass of 1 kg produces an acceleration of 1 m s^{-2}; thus the newton is a *derived* unit representing 1 kg m s^{-2}.

In the English engineering system of units, force is treated as an additional independent dimension along with length, time, and mass. The pound *force* (lb_f) is defined as that force which accelerates 1 pound *mass* 32.1740 feet per second per second. Newton's law must here include a dimensional proportionality constant if it is to be reconciled with this definition. Thus, we write

$$F = \frac{1}{g_c}ma$$

whence[1]

$$1(lb_f) = \frac{1}{g_c} \times 1(lb_m) \times 32.1740(ft)(s)^{-2}$$

and

$$g_c = 32.1740(lb_m)(ft)(lb_f)^{-1}(s)^{-2}$$

The pound *force* is equivalent to 4.4482216 N.

Since force and mass are different concepts, a pound *force* and a pound *mass* are different quantities, and their units do not cancel one another. When an equation contains both units, (lb_f) and (lb_m), the dimensional constant g_c must also appear in the equation to make it dimensionally correct.

Weight properly refers to the force of gravity on a body, and is therefore correctly expressed in newtons or in pounds *force*. Unfortunately, standards of mass are often called "weights", and the use of a balance to compare masses is called "weighing". Thus, one must discern from the context whether force or mass is meant when the word "weight" is used in a casual or informal way.

Example 1.1   An astronaut weighs 730 N in Houston, Texas, where the local acceleration of gravity is $g = 9.792$ m s^{-2}. What are the astronaut's mass and weight on the moon, where $g = 1.67$ m s^{-2}?

SOLUTION   Letting $a = g$, we write Newton's law as

$$F = mg$$

---

[1]Where non-SI units (e.g., English units) are employed, parentheses enclose the abbrevations of all units.

whence

$$m = \frac{F}{g} = \frac{730 \text{ N}}{9.792 \text{ m s}^{-2}} = 74.55 \text{ N m}^{-1} \text{ s}^2$$

Since the newton N has the units kg m s^{-2},

$$m = 74.55 \text{ kg}$$

This *mass* of the astronaut is independent of location, but *weight* depends on the local acceleration of gravity. Thus on the moon the astronaut's weight is

$$F(\text{moon}) = mg(\text{moon}) = 74.55 \text{ kg} \times 1.67 \text{ m s}^{-2}$$

or

$$F(\text{moon}) = 124.5 \text{ kg m s}^{-2} = 124.5 \text{ N}$$

Use of the English engineering system of units requires conversion of the astronaut's weight to (lb$_f$) and the values of g to (ft)(s)$^{-2}$. With 1 N equivalent to 0.224809(lb$_f$) and 1 m to 3.28084(ft), we have

$$\text{Weight of astronaut in Houston} = 164.1(\text{lb}_f)$$

$$g(\text{Houston}) = 32.13 \qquad \text{and} \qquad g(\text{moon}) = 5.48(\text{ft})(\text{s})^{-2}$$

Newton's law then gives

$$m = \frac{Fg_c}{g} = \frac{164.1(\text{lb}_f) \times 32.1740(\text{lb}_m)(\text{ft})(\text{lb}_f)^{-1}(\text{s})^{-2}}{32.13(\text{ft})(\text{s})^{-2}}$$

or

$$m = 164.3(\text{lb}_m)$$

Thus the astronaut's mass in (lb$_m$) and weight in (lb$_f$) in Houston are *numerically* almost the same, but on the moon this is not the case:

$$F(\text{moon}) = \frac{mg(\text{moon})}{g_c} = \frac{(164.3)(5.48)}{32.1740} = 28.0(\text{lb}_f)$$

1.4 Temperature

Temperature is commonly measured with liquid-in-glass thermometers, wherein the liquid expands when heated. Thus a uniform tube, partially filled with mercury, alcohol, or some other fluid, can indicate degree of "hotness" simply by the length of the fluid column. However, numerical values are assigned to the various degrees of hotness by arbitrary definition.

For the Celsius scale, the ice point (freezing point of water saturated with air at standard atmospheric pressure) is zero, and the steam point (boiling point of pure water at standard atmospheric pressure) is 100. We may give a thermometer a numerical scale by immersing it in an ice bath and making a mark for zero at the fluid level, and then immersing it in boiling water and making a mark for 100 at this greater fluid level. The distance between the two marks is divided into 100 equal spaces called *degrees*. Other spaces of equal size may be marked off below zero and above 100 to extend the range of the thermometer.

All thermometers, regardless of fluid, provide the same reading at zero and 100 if they are calibrated by the method described, but at other points the readings do not usually correspond, because fluids vary in their expansion characteristics. An arbitrary choice of fluid could be made, and for many purposes this would be entirely satisfactory. However, as will be shown, the temperature scale of the SI system, with its kelvin unit, symbol K, is based on the ideal gas as thermometric fluid. Since the definition of this scale depends on the properties of gases, its detailed discussion is delayed until Chap. 3. We note, however, that as an absolute scale, it depends on the concept of a lower limit of temperature.

Kelvin temperatures are given the symbol T; Celsius temperatures, given the symbol t, are defined in relation to Kelvin temperatures by

$$t°C = T \text{ K} - 273.15$$

The unit of Celsius temperature is the degree Celsius, °C, equal in size to the kelvin. However, temperatures on the Celsius scale are 273.15 degrees lower than on the Kelvin scale. This means that the lower limit of temperature, called absolute zero on the Kelvin scale, occurs at -273.15°C.

In practice it is the *International Temperature Scale of 1990* (ITS-90) which is used for calibration of scientific and industrial instruments.[2] This scale has been so chosen that temperatures measured on it closely approximate ideal-gas temperatures; the differences are within the limits of present accuracy of measurement. The ITS-90 is based on assigned values of temperature for a number of reproducible equilibrium states (*fixed points*) and on *standard instruments* calibrated at these temperatures. Interpolation between the fixed-point temperatures is provided by formulas that establish the relation between readings of the standard instruments and values on ITS-90. The fixed points are specified phase-equilibrium states of pure substances. The platinum-resistance thermometer is an example of a standard instrument; it is used for temperatures from -259.35°C (the triple point of hydrogen) to 961.78°C (the freezing point of silver).

In addition to the Kelvin and Celsius scales two others are still used by engineers in the United States: the Rankine scale and the Fahrenheit scale. The Rankine scale is directly related to the Kelvin scale by

$$T(\text{R}) = 1.8\,T \text{ K}$$

and is an absolute scale.

The Fahrenheit scale is related to the Rankine scale by an equation analogous to the relation between the Celsius and Kelvin scales.

$$t(°\text{F}) = T(\text{R}) - 459.67$$

Thus the lower limit of temperature on the Fahrenheit scale is $-459.67(°\text{F})$. The relation between the Fahrenheit and Celsius scales is given by

$$t(°\text{F}) = 1.8\,t°C + 32$$

---

[2]The English-language text of the definition of ITS-90 is given by H. Preston-Thomas, *Metrologia*, vol. 27, pp. 3–10, 1990.

The ice point is therefore 32(°F) and the normal boiling point of water is 212(°F).

The Celsius degree and the kelvin represent the same temperature *interval*, as do the Fahrenheit degree and the rankine. The relationships among the four temperature scales are shown in Fig. 1.1. In thermodynamics, it is absolute temperature that is implied by an unqualified reference to temperature.

1.5 Defined Quantities; Volume

In the international system of units force is defined through Newton's law. Additional defined quantities find frequent use in applications of thermodynamics. Some, like volume, are so common as to need almost no discussion. Others, requiring detailed explanation, are treated in the following sections.

Volume is a quantity representing the product of three lengths. The volume of a substance, like its mass, depends on the amount of material considered. Specific or molar volume V, on the other hand, is defined as volume per unit mass or per mole, and is therefore independent of the total amount of material considered. Density ρ is the reciprocal of specific or molar volume.

1.6 Pressure

The pressure P exerted by a fluid on a surface is defined as the normal force exerted by the fluid per unit area of the surface. If force is measured in N and area in m^2, the unit is the newton per square meter or N m^{-2}, called the pascal, symbol Pa,

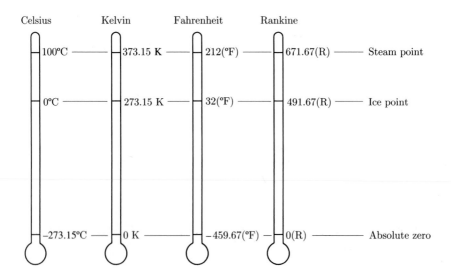

Figure 1.1: Relations among temperature scales.

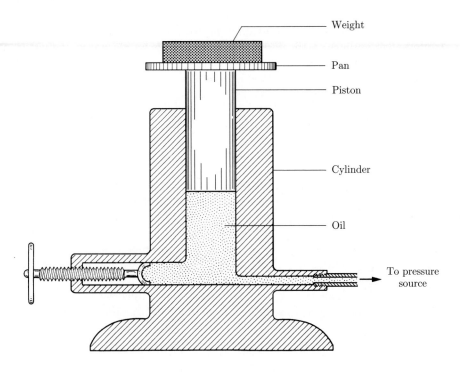

Figure 1.2: Dead-weight gauge.

the basic SI unit of pressure. In the English engineering system the most common unit is the pound *force* per square inch (psi).

The primary standard for the measurement of pressure derives from its definition. A known force is balanced by a fluid pressure acting on a known area; whence $P \equiv F/A$. The apparatus providing this direct pressure measurement is the dead-weight gauge. A simple design is shown in Fig. 1.2. The piston is carefully fitted to the cylinder so that the clearance is small. Weights are placed on the pan until the pressure of the oil, which tends to make the piston rise, is just balanced by the force of gravity on the piston and all that it supports. With the force of gravity given by Newton's law, the pressure of the oil is

$$P = \frac{F}{A} = \frac{mg}{A}$$

where m is the mass of the piston, pan, and weights, g is the local acceleration of gravity, and A is the cross-sectional area of the piston. Gauges in common use, such as Bourdon gauges, are calibrated by comparison with dead-weight gauges.

Since a vertical column of a given fluid under the influence of gravity exerts a pressure at its base in direct proportion to its height, pressure is also expressed as the equivalent height of a fluid column. This is the basis for the use of manometers for pressure measurement. Conversion of height to force per unit area follows from Newton's law applied to the force of gravity acting on the mass of fluid in the column. The mass is given by

$$m = Ah\rho$$

where A is the cross-sectional area of the column, h is its height, and ρ is the fluid density. Therefore

$$P = \frac{F}{A} = \frac{mg}{A} = \frac{Ah\rho g}{A} = h\rho g$$

The pressure to which a fluid height corresponds is determined by the density of the fluid (which depends on its identity and temperature) and the local acceleration of gravity. Thus the (torr) is the pressure equivalent of 1 millimeter of mercury at $0°C$ in a standard gravitational field, and is equal to 133.322 Pa.

Another unit of pressure is the standard atmosphere (atm), the approximate average pressure exerted by the earth's atmosphere at sea level, defined as 101,325 Pa, 101.325 kPa, or 0.101325 MPa. The bar, an SI unit defined as 10^5 Pa, is equal to 0.986923(atm).

Most pressure gauges give readings which are the difference between the pressure of interest and the pressure of the surrounding atmosphere. These readings are known as *gauge* pressures, and can be converted to *absolute* pressures by addition of the barometric pressure. Absolute pressures must be used in thermodynamic calculations.

Example 1.2 A dead-weight gauge with a 1-cm-diameter piston is used to measure pressures very accurately. In a particular instance a mass of 6.14 kg (including piston and pan) brings it into balance. If the local acceleration of gravity is 9.82 m s^{-2}, what is the *gauge* pressure being measured? If the barometric pressure is 748(torr), what is the *absolute* pressure?

SOLUTION   The force exerted by gravity on the piston, pan, and weights is

$$F = mg = (6.14)(9.82) = 60.295 \text{ N}$$

$$\text{Gauge pressure} = \frac{F}{A} = \frac{60.295}{(1/4)(\pi)(1)^2} = 76.77 \text{ N cm}^{-2}$$

The absolute pressure is therefore

$$P = 76.77 + (748)(0.013332) = 86.74 \text{ N cm}^{-2}$$
$$\text{torr}$$

or

$$P = 867.4 \text{ kPa}$$

Example 1.3 At $27°C$ the reading on a manometer filled with mercury is 60.5 cm. The local acceleration of gravity is 9.784 m s^{-2}. To what pressure does this height of mercury correspond?

SOLUTION   Recall the equation in the preceding text:

$$P = h\rho g$$

At 27°C the density of mercury is 13.53 g cm^{-3}. Then

$$P = 60.5 \text{ cm} \times 13.53 \text{ g cm}^{-3} \times 9.784 \text{ m s}^{-2} = 8{,}009 \text{ g m s}^{-2} \text{ cm}^{-2}$$

or

$$P = 8.009 \text{ kg m s}^{-2} \text{ cm}^{-2} = 8.009 \text{ N cm}^{-2} = 80.09 \text{ kPa} = 0.8009 \text{ bar}$$

1.7 Work

Work W is performed whenever a force acts through a distance. By definition, the quantity of work is given by the equation

$$dW = F \, dl \tag{1.1}$$

where F is the component of force acting along the line of the displacement dl. When integrated, this equation yields the work of a finite process. By convention, work is regarded as positive when the displacement is in the same direction as the applied force and negative when they are in opposite directions.

The work which accompanies a change in volume of a fluid is often encountered in thermodynamics. Consider the compression or expansion of a fluid in a cylinder caused by the movement of a piston. The force exerted by the piston on the fluid is equal to the product of the piston area and the pressure of the fluid. The displacement of the piston is equal to the volume change of the fluid divided by the area of the piston. Equation (1.1) therefore becomes

$$dW = -PA \, d\frac{V}{A}$$

or, since A is constant,

$$dW = -P \, dV \tag{1.2}$$

Integrating,

$$W = -\int_{V_1}^{V_2} P \, dV \tag{1.3}$$

The minus sign is included in these equations so that they conform to the sign convention adopted for work. When the piston moves into the cylinder so as to compress the fluid, the applied force and its displacement are in the same direction; the work is therefore positive. The minus sign is required because the volume change is negative. For an expansion process, the applied force and its displacement are in opposite directions. The volume change in this case is positive, and the minus sign is required to make the work negative.

Equation (1.3) expresses the work done by a finite compression or expansion process.[3] Consider the compression of a gas with initial volume V_1 at pressure P_1 to volume V_2 at pressure P_2 along the path shown in Fig. 1.3 from point 1 to point 2. This path relates the pressure at any point during the process to the volume. The

---

[3]However, as explained in Sec. 2.9, it may be applied only in special circumstances.

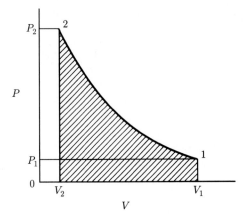

Figure 1.3: *PV* diagram.

work required for the process is given by Eq. (1.3) and is represented on Fig. 1.3 by the area under the curve. The SI unit of work is the newton-meter or joule, symbol J. In the English engineering system the unit often used is the foot-pound *force* (ft lb$_f$).

1.8 Energy

The general principle of conservation of energy was established about 1850. The germ of this principle as it applies to mechanics was implicit in the work of Galileo (1564–1642) and Isaac Newton (1642–1726). Indeed, it follows directly from Newton's second law of motion once work is defined as the product of force and displacement. When a body of mass m, acted upon by a force F, is displaced a distance dl during a differential interval of time dt, the work done is given by Eq. (1.1). In combination with Newton's second law this equation becomes

$$dW = ma\,dl$$

By definition the acceleration is $a = du/dt$, where u is the velocity of the body. Thus

$$dW = m\frac{du}{dt}dl$$

which may be written

$$dW = m\frac{dl}{dt}du$$

Since the definition of velocity is $u = dl/dt$, the expression for work becomes

$$dW = mu\,du$$

This equation may now be integrated for a finite change in velocity from u_1 to u_2:

$$W = m \int_{u_1}^{u_2} u \, du = m \left(\frac{u_2^2}{2} - \frac{u_1^2}{2} \right)$$

or

$$W = \frac{mu_2^2}{2} - \frac{mu_1^2}{2} = \Delta \left(\frac{mu^2}{2} \right) \tag{1.4}$$

Each of the quantities $\frac{1}{2}mu^2$ in Eq. (1.4) is a *kinetic energy*, a term introduced by Lord Kelvin[4] in 1856. Thus, by definition,

$$E_K \equiv \frac{1}{2}mu^2 \tag{1.5}$$

Equation (1.4) shows that the work done *on* a body in accelerating it from an initial velocity u_1 to a final velocity u_2 is equal to the change in kinetic energy of the body. Conversely, if a moving body is decelerated by the action of a resisting force, the work done *by* the body is equal to its change in kinetic energy. In the SI system of units with mass in kg and velocity in m s^{-1}, kinetic energy E_K has the units of kg m^2 s^{-2}. Since the newton is the composite unit kg m s^{-2}, E_K is measured in newton-meters or joules. In accord with Eq. (1.4), this is the unit of work.

In the English engineering system, kinetic energy is expressed as $\frac{1}{2}mu^2/g_c$, where g_c has the value 32.1740 and the units $(\text{lb}_\text{m})(\text{ft})(\text{lb}_\text{f})^{-1}(\text{s})^{-2}$. Thus the unit of kinetic energy in this system is

$$E_K = \frac{mu^2}{2g_c} = \frac{(\text{lb}_\text{m})(\text{ft})^2(\text{s})^{-2}}{(\text{lb}_\text{m})(\text{ft})(\text{lb}_\text{f})^{-1}(\text{s})^{-2}} = (\text{ft lb}_\text{f})$$

Dimensional consistency here requires the inclusion of g_c .

If a body of mass m is raised from an initial elevation z_1 to a final elevation z_2, an upward force at least equal to the weight of the body must be exerted on it, and this force must move through the distance $z_2 - z_1$. Since the weight of the body is the force of gravity on it, the minimum force required is given by Newton's law as

$$F = ma = mg$$

where g is the local acceleration of gravity. The minimum work required to raise the body is the product of this force and the change in elevation:

$$W = F(z_2 - z_1) = mg(z_2 - z_1)$$

or

$$W = mz_2g - mz_1g = \Delta(mzg) \tag{1.6}$$

---

[4]Lord Kelvin, or William Thomson (1824–1907), was an English physicist who, along with the German physicist Rudolf Clausius (1822–1888), laid the foundations for the modern science of thermodynamics.

We see from Eq. (1.6) that the work done *on* the body in raising it is equal to the change in the quantity mzg. Conversely, if the body is lowered against a resisting force equal to its weight, the work done *by* the body is equal to the change in the quantity mzg. Equation (1.6) is similar in form to Eq. (1.4), and both show that the work done is equal to the change in a quantity which describes the condition of the body in relation to its surroundings. In each case the work performed can be recovered by carrying out the reverse process and returning the body to its initial condition. This observation leads naturally to the thought that, if the work done on a body in accelerating it or in elevating it can be subsequently recovered, then the body by virtue of its velocity or elevation contains the ability or capacity to do the work. This concept proved so useful in rigid-body mechanics that the capacity of a body for doing work was given the name *energy*, a word derived from the Greek and meaning "in work". Hence the work of accelerating a body is said to produce a change in its *kinetic energy*, or

$$W = \Delta E_K = \Delta \left(\frac{mu^2}{2} \right)$$

and the work done on a body in elevating it is said to produce a change in its *potential energy*, or

$$W = \Delta E_P = \Delta(mzg)$$

Thus potential energy[5] is defined as

$$E_P \equiv mzg \qquad\qquad (1.7)$$

In the SI system of units with mass in kg, elevation in m, and the acceleration of gravity in m s^{-2}, potential energy has the units of kg m^2 s^{-2}. This is the newton-meter or joule, the unit of work, in agreement with Eq. (1.6).

In the English engineering system, potential energy is expressed as mzg/g_c. Thus the unit of potential energy in this system is

$$E_P = \frac{mzg}{g_c} = \frac{(\text{lb}_\text{m})(\text{ft})(\text{ft})(\text{s})^{-2}}{(\text{lb}_\text{m})(\text{ft})(\text{lb}_\text{f})^{-1}(\text{s})^{-2}} = (\text{ft lb}_\text{f})$$

Again, g_c must be included for dimensional consistency.

In any examination of physical processes, an attempt is made to find or to define quantities which remain constant regardless of the changes which occur. One such quantity, early recognized in the development of mechanics, is mass. The great utility of the law of conservation of mass suggests that further conservation principles should be of comparable value. Thus the development of the concept of energy logically led to the principle of its conservation in mechanical processes. If a body is given energy when it is elevated, then the body should conserve or retain this energy until it performs the work of which it is capable. An elevated body, allowed to fall freely, should gain in kinetic energy what it loses in potential energy

---

[5]This term was first proposed in 1853 by the Scottish engineer William Rankine (1820–1872).

so that its capacity for doing work remains unchanged. For a freely falling body, we should be able to write

$$\Delta E_K + \Delta E_P = 0$$

or

$$\frac{mu_2^2}{2} - \frac{mu_1^2}{2} + mz_2g - mz_1g = 0$$

The validity of this equation has been confirmed by countless experiments. Success in application to freely falling bodies led to the generalization of the principle of energy conservation to apply to all *purely mechanical processes*. Ample experimental evidence to justify this generalization was readily obtained.

Other forms of mechanical energy besides kinetic and gravitational potential energy are possible. The most obvious is potential energy of configuration. When a spring is compressed, work is done by an external force. Since the spring can later perform this work against a resisting force, the spring possesses capacity for doing work. This is potential energy of configuration. Energy of the same form exists in a stretched rubber band or in a bar of metal deformed in the elastic region.

The generality of the principle of conservation of energy in mechanics is increased if we look upon work itself as a form of energy. This is clearly permissible, because both kinetic- and potential-energy changes are equal to the work done in producing them [Eqs. (1.4) and (1.6)]. However, work is energy in transit and is never regarded as residing in a body. When work is done and does not appear simultaneously as work elsewhere, it is converted into another form of energy.

The body or assemblage on which attention is focused is called the *system*. All else is called the *surroundings*. When work is done, it is done by the surroundings on the system, or vice versa, and energy is transferred from the surroundings to the system, or the reverse. It is only during this transfer that the form of energy known as work exists. In contrast, kinetic and potential energy reside with the system. Their values, however, are measured with reference to the surroundings, i.e., kinetic energy depends on velocity with respect to the surroundings, and potential energy depends on elevation with respect to a datum level. *Changes* in kinetic and potential energy do not depend on these reference conditions, provided they are fixed.

Example 1.4   An elevator with a mass of 2,500 kg rests at a level 10 m above the base of an elevator shaft. It is raised to 100 m above the base of the shaft, where the cable holding it breaks. The elevator falls freely to the base of the shaft and strikes a strong spring. The spring is designed to bring the elevator to rest and, by means of a catch arrangement, to hold the elevator at the position of maximum spring compression. Assuming the entire process to be frictionless, and taking $g = 9.8$ m s^{-2}, calculate:

   (*a*) The potential energy of the elevator in its initial position relative to the base of the shaft.

   (*b*) The work done in raising the elevator.

   (*c*) The potential energy of the elevator in its highest position relative to the base of the shaft.

(d) The velocity and kinetic energy of the elevator just before it strikes the spring.

(e) The potential energy of the compressed spring.

(f) The energy of the system consisting of the elevator and spring (1) at the start of the process, (2) when the elevator reaches its maximum height, (3) just before the elevator strikes the spring, and (4) after the elevator has come to rest.

SOLUTION   Let subscript 1 designate the initial conditions; subscript 2, conditions when the elevator is at its highest position; and subscript 3, conditions just before the elevator strikes the spring.

(a) By Eq. (1.7),

$$E_{P_1} = mz_1g = (2{,}500)(10)(9.8) = 245{,}000 \text{ J}$$

(b) By Eq. (1.1),

$$W = \int_{z_1}^{z_2} F\,dl = \int_{z_1}^{z_2} mg\,dl = mg(z_2 - z_1)$$

whence

$$W = (2{,}500)(9.8)(100 - 10) = 2{,}205{,}000 \text{ J}$$

(c) By Eq. (1.7),

$$E_{P_2} = mz_2g = (2{,}500)(100)(9.8) = 2{,}450{,}000 \text{ J}$$

Note that $W = E_{P_2} - E_{P_1}$.

(d) From the principle of conservation of mechanical energy, one may write that the sum of the kinetic- and potential-energy changes during the process from conditions 2 to 3 is zero; that is,

$$\Delta E_{K_{2\rightarrow3}} + \Delta E_{P_{2\rightarrow3}} = 0$$

or

$$E_{K_3} - E_{K_2} + E_{P_3} - E_{P_2} = 0$$

However, E_{K_2} and E_{P_3} are zero. Therefore

$$E_{K_3} = E_{P_2} = 2{,}450{,}000 \text{ J}$$

Since $E_{K_3} = \frac{1}{2}mu_3^2$,

$$u_3^2 = \frac{2E_{K_3}}{m} = \frac{(2)(2{,}450{,}000)}{2{,}500}$$

whence

$$u_3 = 44.27 \text{ m s}^{-1}$$

(e) Since the changes in the potential energy of the spring and the kinetic energy of the elevator must sum to zero,

$$\Delta E_P(\text{spring}) + \Delta E_K(\text{elevator}) = 0$$

The initial potential energy of the spring and the final kinetic energy of the elevator are zero; therefore, the final potential energy of the spring must equal the kinetic energy of the elevator just before it strikes the spring. Thus the final potential energy of the spring is 2,450,000 J.

(*f*) If the elevator and the spring together are taken as the system, the initial energy of the system is the potential energy of the elevator, or 245,000 J. The total energy of the system can change only if work is transferred between it and the surroundings. As the elevator is raised, work is done on the system by the surroundings in the amount of 2,205,000 J. Thus the energy of the system when the elevator reaches its maximum height is 245,000 + 2,205,000 = 2,450,000 J. Subsequent changes occur entirely within the system, with no work transfer between the system and surroundings. Hence the total energy of the system remains constant at 2,450,000 J. It merely changes from potential energy of position (elevation) of the elevator to kinetic energy of the elevator to potential energy of configuration of the spring.

This example illustrates application of the law of conservation of mechanical energy. However, the entire process is assumed to occur without friction; the results obtained are exact only for such an idealized process.

During the period of development of the law of conservation of mechanical energy, heat was not generally recognized as a form of energy, but was considered an indestructible fluid called *caloric*. This concept was firmly entrenched, and for many years no connection was made between heat resulting from friction and the established forms of energy. The law of conservation of energy was therefore limited in application to frictionless mechanical processes. No such limitation is necessary; heat like work is now regarded as energy in transit, a concept that gained acceptance during the years following 1850, largely on account of the classic experiments of J. P. Joule (1818–1889), a brewer of Manchester, England. These experiments are considered in detail in Chap. 2, but first we examine some of the characteristics of heat.

1.9 Heat

We know from experience that a hot object brought in contact with a cold object becomes cooler, whereas the cold object becomes warmer. A reasonable view is that something is transferred from the hot object to the cold one, and we call that something heat Q.[6] Thus we say that heat always flows from a higher temperature to a lower one. This leads to the concept of temperature as the driving force for the transfer of energy as heat. More precisely, the rate of heat transfer from one body to another is proportional to the temperature difference between the two bodies; when there is no temperature difference, there is no net transfer of heat. In the thermodynamic sense, heat is never regarded as being stored within a body. Like work, it exists only as energy in transit from one body to another, or between a

---

[6]An equally reasonable view would have been to regard "cool" as something transferred from the cold object to the hot one.

system and its surroundings. When energy in the form of heat is added to a body, it is stored not as heat but as kinetic and potential energy of the atoms and molecules making up the body.

In spite of the transient nature of heat, it is often viewed in relation to its effect on the body from which or to which it is transferred. As a matter of fact, until about 1930 the definitions of units of heat were based on the temperature changes of a unit mass of water. Thus the *calorie* was long defined as that quantity of heat which when transferred to one gram of water raised its temperature one degree Celsius. Likewise, the *British thermal unit*, or (Btu), was defined as the quantity of heat which when transferred to one pound *mass* of water raised its temperature one degree Fahrenheit. Although these definitions provide a "feel" for the size of heat units, they depend on experiments made with water and are thus subject to change as measurements become more accurate. The calorie and (Btu) are now recognized as units of energy, and are defined in relation to the joule, the SI unit of energy, equal to 1 N m. This is the mechanical work done when a force of one newton acts through a distance of one meter. All other energy units are defined as multiples of the joule. The foot-pound *force*, for example, is equivalent to 1.3558179 J, the calorie to 4.1840 J, and the (Btu) to 1055.04 J. The SI unit of power is the watt, symbol W, defined as an energy rate of one joule per second.

Table A.1 of App. A provides an extensive list of conversion factors for energy as well as for other units.

PROBLEMS

1.1. What is the value of g_c and what are its units in a system in which the second, the foot, and the pound *mass* are defined as in Sec. 1.2, and the unit of force is the *poundal*, defined as the force required to give $1(\text{lb}_m)$ an acceleration of $1(\text{ft})(\text{s})^{-2}$?

1.2. Pressures up to 3,500 bar are measured with a dead-weight gauge. The piston diameter is 0.30 cm. What is the approximate mass in kg of the weights required?

1.3. Pressures up to 3,500(atm) are measured with a dead-weight gauge. The piston diameter is 0.15(in). What is the approximate mass in (lb_m) of the weights required?

1.4. The reading on a mercury manometer at 25°C (open to the atmosphere at one end) is 43.62 cm. The local acceleration of gravity is 9.806 m s^{-2}. Atmospheric pressure is 101.45 kPa. What is the absolute pressure in kPa being measured? The density of mercury at 25°C is 13.534 g cm^{-3}.

1.5. The reading on a mercury manometer at 70(°F) (open to the atmosphere at one end) is 27.36(in). The local acceleration of gravity is 32.187(ft)(s)$^{-2}$. Atmospheric pressure is 30.06(in Hg). What is the absolute pressure in (psia) being measured? The density of mercury at 70(°F) is 13.543 g cm^{-3}.

1.6. The first accurate measurements of the properties of high-pressure gases were made by E. H. Amagat in France between 1869 and 1893. Before developing the dead-weight gauge, he worked in a mine shaft, and used a mercury manometer for measurements of pressure to more than 400 bar. Estimate the height of manometer required.

1.7. An instrument to measure the acceleration of gravity on Mars is constructed of a spring from which is suspended a mass of 0.38 kg. At a place on earth where the local acceleration of gravity is 9.80 m s^{-2}, the spring extends 1.03 cm. When the instrument package is landed on Mars, it radios the information that the spring is extended 0.38 cm. What is the Martian acceleration of gravity?

1.8. A group of engineers has landed on the moon, and they wish to determine the mass of some rocks. They have a spring scale calibrated to read pounds *mass* at a location where the acceleration of gravity is 32.192(ft)(s)$^{-2}$. One of the moon rocks gives a reading of 13.37 on this scale. What is its mass? What is its weight on the moon? Take g(moon) = 5.32(ft)(s)$^{-2}$.

1.9. A gas is confined in a 1.5(ft)-diameter cylinder by a piston, on which rests a weight. The mass of the piston and weight together is 300(lb$_m$). The local acceleration of gravity is 32.158(ft)(s)$^{-2}$, and atmospheric pressure is 29.84(in Hg).

(*a*) What is the force in (lb$_f$) exerted on the gas by the atmosphere, the piston, and the weight, assuming no friction between the piston and cylinder?

(*b*) What is the pressure of the gas in (psia)?

(*c*) If the gas in the cylinder is heated, it expands, pushing the piston and weight upward. If the piston and weight are raised 2(ft), what is the work done by the gas in (ft lb$_f$)? What is the change in potential energy of the piston and weight?

1.10. A gas is confined in a 0.47 m-diameter cylinder by a piston, on which rests a weight. The mass of the piston and weight together is 150 kg. The local acceleration of gravity is 9.813 m s^{-2}, and atmospheric pressure is 101.57 kPa.

(*a*) What is the force in newtons exerted on the gas by the atmosphere, the piston, and the weight, assuming no friction between the piston and cylinder?

(*b*) What is the pressure of the gas in kPa?

(*c*) If the gas in the cylinder is heated, it expands, pushing the piston and weight upward. If the piston and weight are raised 0.83 m, what is the work done by the gas in kJ? What is the change in potential energy of the piston and weight?

1.11. Verify that the SI unit of kinetic and potential energy is the joule.

1.12. An automobile having a mass of 1,400 kg is traveling at 30 m s^{-1}. What is its kinetic energy in kJ? How much work must be done to bring it to a stop?

1.13. The turbines in a hydroelectric plant are fed by water falling from a 40-m height. Assuming 93% efficiency for conversion of potential to electrical energy, and 9% loss of the resulting power in transmission, what is the mass-flow rate of water required to power a 150-watt light bulb?

CHAPTER 2

THE FIRST LAW AND OTHER BASIC CONCEPTS

2.1 Joule's Experiments

During the years 1840–1878, J. P. Joule[1] carried out careful experiments on the nature of heat and work. These experiments are fundamental to an understanding of the first law of thermodynamics and the modern concept of energy.

In their essential elements Joule's experiments were simple enough, but he took elaborate precautions to insure accuracy. In his most famous series of measurements, he placed known amounts of water in an insulated container and agitated the water with a rotating stirrer. The amounts of work done on the water by the stirrer were accurately measured, and the temperature changes of the water were carefully noted. He found that a fixed amount of work was required per unit mass of water for every degree of temperature rise caused by the stirring, and that the original temperature of the water could be restored by the transfer of heat through simple contact with a cooler object. Thus Joule was able to show conclusively that a quantitative relationship exists between work and heat and, therefore, that heat is a form of energy.

2.2 Internal Energy

In experiments such as those conducted by Joule, energy is added to the water as work, but is extracted from the water as heat. The question arises as to what happens to this energy between the time it is added to the water as work and the

---

[1] An account of Joule's celebrated experiments is given by H. J. Steffens, *James Prescott Joule and the Concept of Energy*, Chap. IV, Neale Watson Academic Publications, Inc., New York, 1979.

time it is extracted as heat. Logic suggests that this energy is contained in the water in another form, which is defined as *internal energy*.

The internal energy of a substance does not include energy that it may possess as a result of its macroscopic position or movement. Rather it refers to the energy of the molecules making up the substance, which are in ceaseless motion and possess kinetic energy of translation; except for monatomic molecules, they also possess kinetic energy of rotation and of internal vibration. The addition of heat to a substance increases this molecular activity, and thus causes an increase in its internal energy. Work done on the substance can have the same effect, as was shown by Joule.

The internal energy of a substance also includes the potential energy resulting from intermolecular forces. On a submolecular scale there is energy associated with the electrons and nuclei of atoms, and bond energy resulting from the forces holding atoms together as molecules. Although absolute values of internal energy are unknown, this is not a disadvantage in thermodynamic analysis, because only *changes* in internal energy are required.

The designation of this form of energy as *internal* distinguishes it from kinetic and potential energy which the substance may possess as a result of its macroscopic position or motion, and which can be thought of as *external* forms of energy.

2.3   The First Law of Thermodynamics

The recognition of heat and internal energy as forms of energy suggests a generalization of the law of conservation of mechanical energy (Sec. 1.8) to apply to heat and internal energy as well as to work and external potential and kinetic energy. Indeed, the generalization can be extended to still other forms, such as surface energy, electrical energy, and magnetic energy. This generalization was at first no more than a postulate, but without exception all observations of ordinary processes support it.[2] Hence it has achieved the stature of a law of nature, and is known as the first law of thermodynamics. One formal statement is:

> *Although energy assumes many forms, the total quantity of energy is constant, and when energy disappears in one form it appears simultaneously in other forms.*

In application of the first law to a given process, the sphere of influence of the process is divided into two parts, the *system* and its *surroundings*. The part in which the process occurs is taken as the system; everything with which the system interacts is taken as the surroundings. The system may be of any size depending on the particular conditions, and its boundaries may be real or imaginary, rigid or flexible. Frequently a system is made up of a single substance; in other cases it may be complex. In any event, the equations of thermodynamics are written with

---

[2]For nuclear-reaction processes, the Einstein equation applies, $E = mc^2$, where c is the velocity of light. Here, mass is transformed into energy, and the laws of conservation of mass and energy combine to state that mass and energy together are conserved.

reference to some well-defined system. This focuses attention on the particular process of interest and on the equipment and material directly involved in the process. However, the first law applies to the system *and* surroundings, and not to the system alone. In its most basic form, the first law may be written

$$\Delta(\text{Energy of the system}) + \Delta(\text{Energy of surroundings}) = 0 \qquad (2.1)$$

where the difference operator "Δ" signifies finite changes in the quantities enclosed in parentheses. The changes in the system may be in its internal energy, in its potential or kinetic energy, and in the potential or kinetic energy of its finite parts. Since attention is focused on the system, the nature of the energy changes in the surroundings is not of interest.

In the thermodynamic sense, heat and work refer to energy *in transit across the boundary* which divides the system from its surroundings. These forms of energy are not stored, and are never *contained in* a body or system. Energy is stored in its potential, kinetic, and internal forms; these reside with material objects and exist because of the position, configuration, and motion of matter.

If the boundary of a system does not permit the transport of matter between the system and its surroundings, the system is said to be *closed*, and its mass is necessarily constant. In this event there can be no *transport* of internal energy across the boundary of the system. All energy exchange between a system and its surroundings is as heat and work, and the total energy change of the surroundings equals the net energy transferred to or from it as heat and work. The second term of Eq. (2.1) may therefore be replaced by

$$\Delta(\text{Energy of surroundings}) = \pm Q \pm W$$

The choice of signs used with Q and W depends on which direction of transport is regarded as positive.

The first term of Eq. (2.1) may be expanded to show energy changes in various forms. If the mass of the system is constant and if only internal-, kinetic-, and potential-energy changes are involved,

$$\Delta(\text{Energy of the system}) = \Delta U^t + \Delta E_K + \Delta E_P$$

where U^t is the *total* internal energy of the system. Equation (2.1) is now written

$$\Delta U^t + \Delta E_K + \Delta E_P = \pm Q \pm W$$

The modern sign convention for both heat Q and work W makes the numerical values of both quantities positive for transport across the boundary from the surroundings and into the system. With this understanding, we write[3]

$$\Delta U^t + \Delta E_K + \Delta E_P = Q + W \qquad (2.2)$$

---

[3]This is the convention recommended by the International Union of Pure and Applied Chemistry. However, the original choice of sign for work and the one used in earlier editions of this text was the opposite of that adopted here; the right-hand side of Eq. (2.2) was then written $Q - W$.

Equation (2.2) indicates that the total energy change of a constant-mass system equals the net energy transported into it as heat and work.

Closed systems often undergo processes that cause no change in their external potential or kinetic energy, but only a change in internal energy. For such processes, Eq. (2.2) reduces to

$$\Delta U^t = Q + W \qquad (2.3)$$

Equation (2.3) applies to processes involving *finite* changes in the internal energy of the system. For *differential* changes it is written

$$dU^t = dQ + dW \qquad (2.4)$$

This equation, like Eq. (2.3), applies to closed systems which undergo changes in *internal* energy only. The system must of course be clearly defined, as illustrated in the examples of this and later chapters.

The units used in Eqs. (2.2) through (2.4) must be the same for all terms. In the SI system the energy unit is the joule. Other energy units still in use are the calorie, the (ft lb$_f$), and the (Btu).

2.4 Thermodynamic State and State Functions

The notation of Eqs. (2.2) and (2.3) suggests that the terms on the left sides are different in kind from the ones on the right. Those on the left represent *changes* in the characteristics of the system which determine its macroscopic condition. The kinetic- and potential-energy terms account for changes in what we have termed external forms of energy. The internal-energy term reflects changes at the molecular or microscopic level, i.e., in the internal state or the *thermodynamic state* of the system. It is this state that is reflected by its *thermodynamic properties*, among which are temperature, pressure, and density. We know from experience that for a homogeneous pure substance fixing two of these properties automatically fixes all the others, and thus determines its thermodynamic state. For example, nitrogen gas at a temperature of 300 K and a pressure of 10^5 kPa (1 bar) has a fixed specific volume or density and a fixed internal energy. Indeed, it has an established set of thermodynamic properties. If this gas is heated or cooled, compressed or expanded, and then returned to its initial conditions, it is found to have exactly the same set of properties as before. These properties do not depend on the past history of the substance nor on the means by which it reaches a given state. They depend only on present conditions, however reached. Such quantities are known as *state functions.* When two of them are fixed or held at fixed values for a homogeneous pure substance,[4] the *thermodynamic state* of the substance is fully determined. This means that a state function, such as internal energy or density, is a property that

---

[4]For systems more complex than a simple homogeneous pure substance, the number of properties or state functions that must be arbitrarily specified in order to define the state of the system may be different from two. The method of determining this number is the subject of Sec. 2.8.

always has a value; it may therefore be expressed mathematically as a function of other thermodynamic properties, such as temperature and pressure, and its values may be identified with points on a graph.

On the other hand, the terms on the right sides of Eqs. (2.2) and (2.3), representing heat and work, are not properties; they account for the energy changes that occur in the surroundings and appear only when changes occur in a system. They depend on the nature of the process causing the change, and are associated with areas rather than points on a graph, as suggested by Fig. 1.3. Although time is not a thermodynamic coordinate, we note that the passage of time is inevitable whenever heat is transferred or work is accomplished.

The differential of a state function represents an infinitesimal *change* in its value. The integration of such a differential results in a finite difference between two of its values. For example,

$$\int_{P_1}^{P_2} dP = P_2 - P_1 = \Delta P \qquad \text{and} \qquad \int_{V_1}^{V_2} dV = V_2 - V_1 = \Delta V$$

The differentials of heat and work are not *changes*, but are infinitesimal *amounts*. When integrated, these differentials give not finite changes, but finite amounts. Thus

$$\int dQ = Q \qquad \text{and} \qquad \int dW = W$$

For closed systems which undergo the same change in state by different processes, experiment shows that the amounts of heat and work required differ from one process to another, but that *the sum $Q + W$ is the same for all such processes.* This gives experimental justification to the statement that internal energy is a state function. Equation (2.3) yields the same value of ΔU^t regardless of the process, provided only that the change in the system is always from the same initial to the same final state.

The internal energy of a system, like its volume, depends on the quantity of material involved; such properties are said to be *extensive*. In contrast, temperature and pressure, the principal thermodynamic coordinates for homogeneous fluids, are independent of the quantity of material making up the system, and are known as *intensive* properties.

The first-law equations may be written for systems containing any quantity of material; the values of Q, W, and the energy terms then refer to the entire system. More often, however, we write the equations of thermodynamics for a representative unit amount of material, either a unit mass or a mole. The volume and internal energy are then on a unit basis; they become intensive properties, independent of the quantity of material actually present. Thus, although the total volume V^t and total internal energy U^t of an arbitrary quantity of material are extensive properties, specific and molar volume V (or density) and specific and molar internal energy U are intensive. Writing Eqs. (2.3) and (2.4) for a representative unit amount of the system puts all of the terms on a unit basis, but this does *not* make Q and W into thermodynamic properties or state functions. Multiplication of a quantity on a unit basis (e.g., V or U) by the mass (or number of moles) of the system gives the total quantity (e.g., $V^t = mV$ or $U^t = mU$).

Internal energy (through the enthalpy, defined in Sec. 2.5) is useful for the calculation of heat and work quantities for such equipment as heat exchangers, evaporators, distillation columns, pumps, compressors, turbines, engines, etc., precisely because it *is* a state function. The tabulation of all possible Q's and W's for all possible processes is impossible. But the intensive state functions, such as specific volume and specific internal energy, are properties of matter. They can be measured and their values tabulated as functions of temperature and pressure for a particular substance for future use in the calculation of Q and W for any process involving that substance. We treat the measurement, correlation, and use of these state functions in later chapters.

Example 2.1   Water flows over a waterfall 100 m in height. Consider 1 kg of the water, and assume that no energy is exchanged between the 1 kg and its surroundings.

(*a*) What is the potential energy of the water at the top of the falls with respect to the base of the falls?

(*b*) What is the kinetic energy of the water just before it strikes bottom?

(*c*) After the 1 kg of water enters the river below the falls, what change has occurred in its state?

SOLUTION   Taking the 1 kg of water as the system, and noting that it exchanges no energy with its surroundings, we may set Q and W equal to zero and write Eq. (2.2) as

$$\Delta U + \Delta E_K + \Delta E_P = 0$$

This equation applies to each part of the process.

(*a*) From Eq. (1.7),

$$E_P = mzg = 1 \text{ kg} \times 100 \text{ m} \times 9.8066 \text{ m s}^{-2}$$

where g has been taken as the standard value. This gives

$$E_P = 980.66 \text{ N m} \qquad \text{or} \qquad 980.66 \text{ J}$$

(*b*) During the free fall of the water no mechanism exists for the conversion of potential or kinetic energy into internal energy. Thus ΔU must be zero, and

$$\Delta E_P + \Delta E_K = E_{K_2} - E_{K_1} + E_{P_2} - E_{P_1} = 0$$

For practical purposes we may take $E_{K_1} = E_{P_2} = 0$. Then

$$E_{K_2} = E_{P_1} = 980.66 \text{ J}$$

(*c*) As the 1 kg of water strikes bottom and mixes with other falling water to form a river, the resulting turbulence has the effect of converting kinetic energy into internal energy. During this process, ΔE_P is essentially zero, and Eq. (2.2) becomes

$$\Delta U + \Delta E_K = 0 \qquad \text{or} \qquad \Delta U = E_{K_2} - E_{K_3}$$

However, the river velocity is assumed small, and therefore E_{K_3} is negligible. Thus

$$\Delta U = E_{K_2} = 980.66 \text{ J}$$

The overall result of the process is the conversion of potential energy of the water into internal energy of the water. This change in internal energy is manifested by a temperature rise of the water. Since energy in the amount of 4,184 J kg^{-1} is required for a temperature rise of 1°C in water, the temperature increase is $980.66/4,184 = 0.234$°C, if there is no heat transfer with the surroundings.

Example 2.2 A gas is confined in a cylinder by a piston. The initial pressure of the gas is 7 bar, and the volume is 0.10 m^3. The piston is held in place by latches in the cylinder wall. The whole apparatus is placed in a total vacuum. What is the energy change of the apparatus if the retaining latches are removed so that the gas suddenly expands to double its initial volume? The piston is again held by latches at the end of the process.

SOLUTION Since the question concerns the entire apparatus, the system is taken as the gas, piston, and cylinder. No work is done during the process, because no force external to the system moves, and no heat is transferred through the vacuum surrounding the apparatus. Hence Q and W are zero, and the total energy of the system remains unchanged. Without further information we can say nothing about the distribution of energy among the parts of the system. This may well be different than the initial distribution.

Example 2.3 If the process described in Example 2.2 is repeated, not in a vacuum but in air at standard atmospheric pressure of 101.3 kPa, what is the energy change of the apparatus? Assume the rate of heat exchange between the apparatus and the surrounding air to be slow compared with the rate at which the process occurs.

SOLUTION The system is chosen exactly as before, but in this case work is done by the system in pushing back the atmosphere. This work is given by the product of the force exerted by the atmospheric pressure on the piston and the displacement of the piston. If the area of the piston is A, the force is $F = P_{atm}A$. The displacement of the piston is equal to the volume change of the gas divided by the area of the piston, or $\Delta l = \Delta V^t/A$. The work done by the system on the surroundings, according to Eq. (1.1), is then

$$\text{Work done } by \text{ system} = F\,\Delta l = P_{atm}\,\Delta V^t$$
$$= (101.3)(0.2 - 0.1) = 10.13 \text{ kPa m}^3$$

Since W is work done *on* the system, it is the negative of this result:

$$W = -10.13 \text{ kN m} = -10.13 \text{ kJ}$$

Heat transfer between the system and surroundings is also possible in this case, but the problem is worked for the instant after the process has occurred and before appreciable heat transfer has had time to take place. Thus Q is assumed to be zero in Eq. (2.2), giving

$$\Delta(\text{Energy of the system}) = Q + W = 0 - 10.13 = -10.13 \text{ kJ}$$

The total energy of the system has *decreased* by an amount equal to the work done on the surroundings.

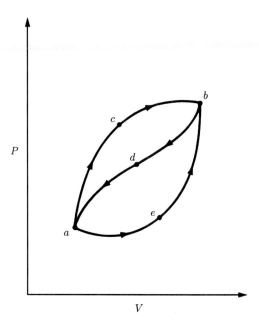

Figure 2.1: Diagram for Example 2.4.

Example 2.4 When a system is taken from state a to state b in Fig. 2.1 along path acb, 100 J of heat flows into the system and the system does 40 J of work. How much heat flows into the system along path aeb if the work done by the system is 20 J? The system returns from b to a along path bda. If the work done on the system is 30 J, does the system absorb or liberate heat? How much?

SOLUTION   We presume that the system changes only in its internal energy and that Eq. (2.3) is applicable. For path acb,

$$\Delta U_{ab}^t = Q_{acb} + W_{acb} = 100 - 40 = 60 \text{ J}$$

This is the internal energy change for the state change from a to b by *any* path. Thus for path aeb,

$$\Delta U_{ab}^t = 60 = Q_{aeb} + W_{aeb} = Q_{aeb} - 20$$

whence

$$Q_{aeb} = 80 \text{ J}$$

For path bda,

$$\Delta U_{ba}^t = -\Delta U_{ab}^t = -60 = Q_{bda} + W_{bda} = Q_{bda} + 30$$

Thus

$$Q_{bda} = -60 - 30 = -90 \text{ J}$$

Heat is therefore liberated from the system.

2.5 Enthalpy

A number of thermodynamic properties related to internal energy are *defined* because of their usefulness in the application of thermodynamics to practical processes. Enthalpy (en-thal'-py) is introduced in this section, and others are treated later. Enthalpy is explicitly defined for any system by the mathematical expression

$$H^t \equiv U^t + PV^t$$

where U^t = total internal energy
$\quad\ P$ = absolute pressure
$\quad\ V^t$ = total volume

For a unit mass or a mole this becomes

$$\boxed{H \equiv U + PV} \tag{2.5}$$

The units of all terms of this equation must be the same. The product PV has the units of energy, as does U; therefore H also has units of energy. In the SI system the basic unit of pressure is the pascal or N m^{-2} and, for volume, the m^3. Thus the PV product has the unit N m or joule. In the English engineering system a common unit for the PV product is the (ft lb$_f$), which arises when pressure is in (lb$_f$)(ft)$^{-2}$ with volume in (ft)^3. This result is usually converted to (Btu) through division by 778.16 for use in Eq. (2.5), because the common English engineering unit for U and H is the (Btu).

Since U, P, and V are all state functions, H as defined by Eq. (2.5) must also be a state function. In differential form Eq. (2.5) may be written

$$dH = dU + d(PV) \tag{2.6}$$

This equation applies whenever a differential change occurs in the system. Integration of Eq. (2.6) gives

$$\Delta H = \Delta U + \Delta(PV) \tag{2.7}$$

an equation applicable whenever a finite change occurs in the system. Equations (2.5) through (2.7) apply to a unit mass of substance or to a mole. Like volume and internal energy, enthalpy is an extensive property; specific or molar enthalpy is of course intensive.

Enthalpy is useful as a thermodynamic property because the $U + PV$ group appears frequently, particularly in problems involving flow processes. The calculation of a numerical value for H is carried out in the following example.

> **Example 2.5**   Calculate ΔU and ΔH for 1 kg of water when it is vaporized at the constant temperature of 100°C and the constant pressure of 101.33 kPa. The specific volumes of liquid and vapor water at these conditions are 0.00104 and 1.673 m^3 kg^{-1}. For this change, heat in the amount of 2,256.9 kJ is added to the water.

SOLUTION   The kilogram of water is taken as the system, because it alone is of interest. We imagine the fluid contained in a cylinder by a frictionless piston which exerts a constant pressure of 101.33 kPa. As heat is added, the water expands from its initial to its final volume, doing work on the piston. By Eq. (1.3),

$$W = -P\,\Delta V = -101.33 \text{ kPa} \times (1.673 - 0.001) \text{ m}^3$$

whence

$$W = -169.4 \text{ kPa m}^3 = -169.4 \text{ kN m}^{-2}\text{ m}^3 = -169.4 \text{ kJ}$$

Since $Q = 2{,}256.9$ kJ, Eq. (2.3) gives

$$\Delta U = Q + W = 2{,}256.9 - 169.4 = 2{,}087.5 \text{ kJ}$$

With P constant, Eq. (2.7) becomes

$$\Delta H = \Delta U + P\,\Delta V$$

But $P\Delta V = -W$. Therefore

$$\Delta H = \Delta U - W = Q = 2{,}256.9 \text{ kJ}$$

2.6   The Steady-State Steady-Flow Process

The application of Eqs. (2.3) and (2.4) is restricted to nonflow (constant-mass) processes in which only internal-energy changes occur. Far more important industrially are processes in which fluid flows at a steady rate through equipment. We consider here only the special case of a steady-state steady-flow process, for which conditions and flow rates at all points along the flow path are constant with time. This means that there can be no accumulation of material or energy at any point. Such processes require application of the general first-law expression, Eq. (2.2), but we first put it in more convenient form.

Figure 2.2 represents a steady-state steady-flow process in which a fluid, either liquid or gas, flows through equipment from a single entrance to a single exit. The mass-flow rate is therefore constant over the entire flow path. At section 1, the entrance to the apparatus, conditions in the fluid are denoted by subscript 1. At this point the fluid has an elevation above an arbitrary datum level of z_1, an average velocity u_1, a specific volume V_1, a pressure P_1, an internal energy U_1, etc. Similarly, the conditions in the fluid at section 2, the exit of the apparatus, are denoted by subscript 2.

The system is taken as a unit mass of the fluid, and we consider the overall changes which occur in this unit mass of fluid as it flows through the apparatus from section 1 to section 2. The energy of the unit mass may change in all three of the forms taken into account by Eq. (2.2), that is, potential, kinetic, and internal. The kinetic-energy change of a unit mass of fluid between sections 1 and 2 follows from Eq. (1.5):

$$\Delta E_K = \tfrac{1}{2}u_2^2 - \tfrac{1}{2}u_1^2 = \tfrac{1}{2}\Delta u^2$$

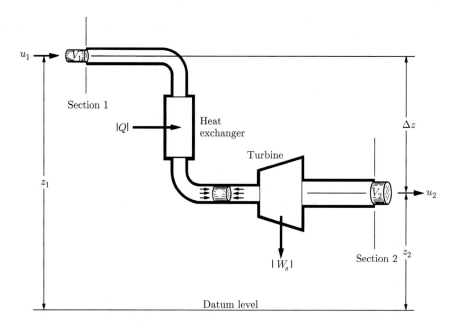

Figure 2.2: Steady-state steady-flow process.

In this equation u represents the average velocity of the flowing fluid, defined as the volumetric flow rate divided by the cross-sectional area.[5] The potential-energy change of a unit mass of fluid between sections 1 and 2 is based on Eq. (1.7):

$$\Delta E_P = z_2 g - z_1 g = g\,\Delta z$$

Equation (2.2) now becomes

$$\Delta U + \frac{\Delta u^2}{2} + g\,\Delta z = Q + W \qquad (2.8)$$

where Q and W represent *all* the heat and work added per unit mass of fluid flowing through the apparatus.

It might appear that W is just the shaft work W_s indicated in Fig. 2.2, but this is not the case. The term shaft work means work done by or on the fluid flowing through a piece of equipment and transmitted by a shaft which protrudes from the equipment and which rotates or reciprocates. Therefore, W_s represents the work which is interchanged between the system and its surroundings through this shaft.

---

[5]This is an approximation appropriate to fully developed turbulent flow, normally the only case for which the kinetic-energy term is significant.

In addition to W_s there is work exchanged between the unit mass of fluid taken as the system and the fluid on either side of it. The unit mass of fluid regarded as the system may be imagined as enclosed by flexible diaphragms and to flow through the apparatus as a fluid cylinder whose dimensions respond to changes in cross-sectional area, temperature, and pressure. As illustrated in Fig. 2.2, a drawing of this cylinder at any point along its path shows pressure forces at its ends exerted by the adjacent fluid. These forces move with the system and do work. The force on the upstream side of the cylinder does work *on* the system. The force on the downstream side is in the opposite direction and results in work done *by* the system. From section 1 to section 2 these two pressure forces follow exactly the same path and vary in exactly the same manner. Hence, the net work which they produce between these two sections is zero. However, the terms representing work done by these pressure forces as the fluid enters and leaves the apparatus do not, in general, cancel.

In Fig. 2.2 the unit mass of fluid is shown just before it enters the apparatus. This cylinder of fluid has a volume equal to its specific volume V_1 at the conditions existing at section 1. If its cross-sectional area is A_1, its length is V_1/A_1. The force exerted on its upstream face is $P_1 A_1$, and the work done by this force in pushing the cylinder into the apparatus is

$$W_1 = P_1 A_1 \frac{V_1}{A_1} = P_1 V_1$$

This represents work done *on* the system by the surroundings. At section 2 work is done *by* the system on the surroundings as the fluid cylinder emerges from the apparatus. This work is therefore given by

$$W_2 = -P_2 A_2 \frac{V_2}{A_2} = -P_2 V_2$$

Since W in Eq. (2.8) represents *all* the work done *on* the unit mass of fluid, it is equal to the sum of the shaft work and the work done at the entrance and exit; that is,

$$W = W_s + P_1 V_1 - P_2 V_2$$

In combination with this result, Eq. (2.8) becomes

$$\Delta U + \frac{\Delta u^2}{2} + g\,\Delta z = Q + W_s + P_1 V_1 - P_2 V_2$$

or

$$\Delta U + \Delta(PV) + \frac{\Delta u^2}{2} + g\,\Delta z = Q + W_s$$

But by Eq. (2.7),

$$\Delta U + \Delta(PV) = \Delta H$$

Therefore,

$$\boxed{\Delta H + \frac{\Delta u^2}{2} + g\,\Delta z = Q + W_s} \tag{2.9a}$$

This equation is the mathematical expression of the first law for a steady-state steady-flow process between a single entrance and a single exit. Equations of greater generality are developed in Chap. 7. All terms in Eq. (2.9a) are expressions for energy per unit mass of fluid; in the SI system of units, all would be expressed in joules or in some multiple of the joule. For the English engineering system of units, this equation must be re-expressed to include the dimensional constant g_c in the kinetic- and potential-energy terms.

$$\Delta H + \frac{\Delta u^2}{2g_c} + \frac{g}{g_c}\Delta z = Q + W_s \qquad (2.9b)$$

Here, the usual unit for ΔH and Q is the (Btu), whereas kinetic energy, potential energy, and work are usually expressed as (ft lb$_f$). Therefore the factor $778.16(\text{ft lb}_f)(\text{Btu})^{-1}$ must be used with the appropriate terms to put them all in consistent units of either (ft lb$_f$) or (Btu).

For many of the applications considered in thermodynamics, the kinetic- and potential-energy terms are very small compared with the others and may be neglected. In such a case Eqs. (2.9a) and (2.9b) reduce to

$$\Delta H = Q + W_s \qquad (2.10)$$

This expression of the first law for a steady-state steady-flow process is analogous to Eq. (2.3) for a nonflow process. Here, however, the enthalpy rather than the internal energy is the thermodynamic property of importance.

Equations (2.9) and (2.10) are useful for the solution of many steady-state steady-flow problems. For most such applications values of the enthalpy must be available. Since H is a state function and a property of matter, its values depend only on point conditions; once determined, they may be tabulated for subsequent use whenever the same sets of conditions are encountered again. Thus Eq. (2.9) may be applied to laboratory processes designed specifically for the measurement of enthalpy data.

One such process employs a flow calorimeter. A simple example of this device is illustrated schematically in Fig. 2.3. Its essential feature is an electric heater immersed in a flowing fluid. The apparatus is designed so that the kinetic- and potential-energy changes of the fluid from section 1 to section 2 (Fig. 2.3) are negligible. This requires merely that the two sections be at the same elevation and that the velocities be small. Furthermore, no shaft work is accomplished between sections 1 and 2. Hence Eq. (2.9a) reduces to

$$\Delta H = H_2 - H_1 = Q$$

Heat is added to the fluid from the electric resistance heater; the rate of energy input is determined from the resistance of the heater and the current passing through it. The entire apparatus is well insulated. In practice a number of details need attention, but in principle the operation of the flow calorimeter is simple. Measurements

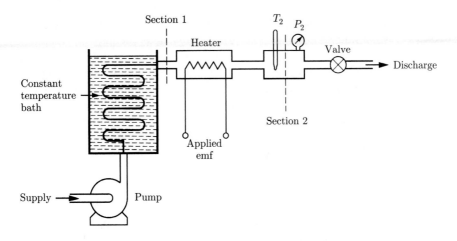

Figure 2.3: Flow calorimeter.

of the rate of heat input and the rate of flow of the fluid allow calculation of values of ΔH between sections 1 and 2.

As an example, consider the measurement of enthalpies of H_2O, both as liquid and as vapor. Liquid water is supplied to the apparatus by the pump. The constant-temperature bath might be filled with a mixture of crushed ice and water to maintain a temperature of $0°C$. The coil which carries the test fluid (in this case water) through the constant-temperature bath is made long enough so that the fluid emerges essentially at the bath temperature of $0°C$. Thus the fluid at section 1 is always liquid water at $0°C$. The temperature and pressure at section 2 are measured by suitable instruments. Values of the enthalpy of H_2O for various conditions at section 2 may be calculated by the equation

$$H_2 = H_1 + Q$$

where Q is the heat added by the resistance heater per unit mass of water flowing.

Clearly, H_2 depends not only on Q but also on H_1. The conditions at section 1 are always the same, i.e., liquid water at $0°C$, except that the pressure varies from run to run. However, pressure has a negligible effect on the properties of liquids unless very high pressures are reached, and for practical purposes H_1 may be considered a constant. Absolute values of enthalpy, like absolute values of internal energy, are unknown. An arbitrary value may therefore be assigned to H_1 as the *basis* for all other enthalpy values. If we set $H_1 = 0$ for liquid water at $0°C$, then the values of H_2 are given by

$$H_2 = H_1 + Q = 0 + Q = Q$$

These results may be tabulated along with the corresponding conditions of T and P existing at section 2 for a large number of runs. In addition, specific-volume measurements may be made for these same conditions, and these may be tabulated. Corresponding values of the internal energy of water may be calculated by Eq. (2.5), $U = H - PV$, and these numbers too may be tabulated. In this way tables of thermodynamic properties may be compiled over the entire useful range of conditions. The most widely used such tabulation is for H_2O and is known as the *steam tables.*[6]

The enthalpy may be taken as zero for some other state than liquid at $0°C$. The choice is arbitrary. The equations of thermodynamics, such as Eq. (2.9), apply to *changes* of state, for which the enthalpy *differences* are independent of where the origin of values is placed. However, once an arbitrary zero point is selected for the enthalpy, an arbitrary choice cannot be made for the internal energy, for values of internal energy are then calculable from the enthalpy by Eq. (2.5).

Example 2.6 For the flow calorimeter just discussed, the following data are taken with water as the test fluid:

$$\text{Flow rate} = 4.15 \text{ g s}^{-1}$$

$$t_1 = 0°C \qquad t_2 = 300°C \qquad P_2 = 3 \text{ bar}$$

$$\text{Rate of heat addition from resistance heater} = 12{,}740 \text{ W}$$

It is observed that the water is completely vaporized in the process. Calculate the enthalpy of steam at $300°C$ and 3 bar based on $H = 0$ for liquid water at $0°C$.

SOLUTION   If Δz and Δu^2 are negligible and if W_s and H_1 are zero, then $H_2 = Q$, and

$$H_2 = \frac{12{,}740 \text{ J s}^{-1}}{4.15 \text{ g s}^{-1}} = 3{,}070 \text{ J g}^{-1}$$

Example 2.7 Air at 1 bar and $25°C$ enters a compressor at low velocity, discharges at 3 bar, and enters a nozzle in which it expands to a final velocity of 600 m s^{-1} at the initial conditions of pressure and temperature. If the work of compression is 240 kJ per kilogram of air, how much heat must be removed during compression?

SOLUTION   Since the air returns to its initial conditions of T and P, the overall process produces no change in enthalpy of the air. Moreover, the potential-energy change of the air is presumed negligible. Neglecting also the initial kinetic energy of the air, we write Eq. (2.9a) as

$$Q = \frac{u_2^2}{2} - W_s$$

The kinetic-energy term is evaluated as follows:

$$\tfrac{1}{2}u_2^2 = \tfrac{1}{2}(600)^2 = 180{,}000 \text{ m}^2 \text{ s}^{-2}$$

or

$$\tfrac{1}{2}u_2^2 = 180{,}000 \text{ N m kg}^{-1} = 180 \text{ kJ kg}^{-1}$$

---

[6]Steam tables are given in App. F. Tables for various other substances are found in the literature. A discussion of compilations of thermodynamic properties appears in Chap. 6.

Then

$$Q = 180 - 240 = -60 \text{ kJ kg}^{-1}$$

Thus, heat must be removed in the amount of 60 kJ for each kilogram of air compressed.

Example 2.8   Water at 200($^\circ$F) is pumped from a storage tank at the rate of 50(gal)(min)$^{-1}$. The motor for the pump supplies work at the rate of 2(HP). The water goes through a heat exchanger, giving up heat at the rate of 40,000(Btu)(min)$^{-1}$, and is delivered to a second storage tank at an elevation 50(ft) above the first tank. What is the temperature of the water delivered to the second tank?

SOLUTION   This is a steady-state steady-flow process for which Eq. (2.9b) applies. The initial and final velocities of water in the storage tanks are negligible, and the term $\Delta u^2/2g_c$ may be omitted. The remaining terms are expressed in units of (Btu)(lb$_m$)$^{-1}$ through use of appropriate conversion factors. At 200($^\circ$F) the density of water is 60.1(lb$_m$)(ft)$^{-3}$, and 1(ft)^3 is equivalent to 7.48(gal); thus the mass flow rate is

$$(50)(60.1/7.48) = 402(\text{lb}_m)(\text{min})^{-1}$$

from which we obtain

$$Q = -40,000/402 = -99.50(\text{Btu})(\text{lb}_m)^{-1}$$

Since 1(HP) is equivalent to 42.41(Btu)(min)$^{-1}$, the shaft work is

$$W_s = (2)(42.41)/(402) = 0.21(\text{Btu})(\text{lb}_m)^{-1}$$

If the local acceleration of gravity is taken as the standard value of 32.174(ft)(s)$^{-2}$, the potential-energy term becomes

$$\frac{g}{g_c}\Delta z = \left(\frac{32.174}{32.174}\right)\frac{(50)}{(778.16)} = 0.06(\text{Btu})(\text{lb}_m)^{-1}$$

Equation (2.9b) now yields ΔH:

$$\Delta H = Q + W_s - \frac{g}{g_c}\Delta z = -99.50 + 0.21 - 0.06 = -99.35(\text{Btu})(\text{lb}_m)^{-1}$$

The enthalpy of water at 200($^\circ$F) is given in the steam tables as 168.09(Btu)(lb$_m$)$^{-1}$. Thus

$$\Delta H = H_2 - H_1 = H_2 - 168.09 = -99.35$$

and

$$H_2 = 168.09 - 99.35 = 68.74(\text{Btu})(\text{lb}_m)^{-1}$$

The temperature of water having this enthalpy is found from the steam tables to be

$$t = 100.74(^\circ\text{F})$$

In this example W_s and $(g/g_c)\Delta z$ are small compared with Q, and for practical purposes they could be neglected.

2.7   Equilibrium

Equilibrium is a word denoting a static condition, the absence of change. In thermodynamics it is taken to mean not only the absence of change but the absence of any *tendency* toward change on a macroscopic scale. Thus a system at equilibrium is one which exists under such conditions that there is no tendency for a change in state to occur. Since any tendency toward change is caused by a driving force of one kind or another, the absence of such a tendency indicates also the absence of any driving force. Hence a system at equilibrium may be described as one in which all forces are in exact balance. Whether a change actually occurs in a system not at equilibrium depends on resistance as well as on driving force. Many systems undergo no measurable change even under the influence of large driving forces, because the resistance is very large.

Different kinds of driving forces tend to bring about different kinds of change. Mechanical forces such as pressure on a piston tend to cause energy transfer as work; temperature differences tend to cause the flow of heat; chemical potentials tend to cause substances to react chemically or to be transferred from one phase to another. At equilibrium all such forces are in balance. We often deal with systems at partial equilibrium. In many applications of thermodynamics, chemical reactions are of no concern. For example, a mixture of hydrogen and oxygen at ordinary conditions is not in chemical equilibrium, because of the large driving force for the formation of water. In the absence of chemical reaction, this system may well be in thermal and mechanical equilibrium, and purely physical processes may be analyzed without regard to the possible chemical reaction.

2.8   The Phase Rule

As mentioned earlier, the state of a pure homogeneous fluid is fixed whenever two intensive thermodynamic properties are set at definite values. In contrast, when *two* phases are in equilibrium, the state of the system is fixed when only a single such property is specified. For example, a mixture of steam and liquid water in equilibrium at 101.33 kPa can exist only at 100°C. It is impossible to change the temperature without also changing the pressure if vapor and liquid are to continue to exist in equilibrium.

The number of independent variables that must be arbitrarily fixed to establish the *intensive* state of *any* system, i.e., the *degrees of freedom* F of the system, is given by the celebrated phase rule of J. Willard Gibbs,[7] who deduced it by theoretical reasoning in 1875. It is presented here without proof in the form applicable to nonreacting systems:[8]

$$\boxed{F = 2 - \pi + N} \tag{2.11}$$

---

[7]Josiah Willard Gibbs (1839–1903), American mathematical physicist.

[8]The justification of the phase rule for nonreacting systems is given in Sec. 12.2, and the phase rule for reacting systems is considered in Sec. 15.8.

where π = number of phases, and N = number of chemical species.

The intensive state of a system at equilibrium is established when its temperature, pressure, and the compositions of all phases are fixed. These are therefore phase-rule variables, but they are not all independent. The phase rule gives the number of variables from this set which must be abitrarily specified to fix all remaining phase-rule variables.

A *phase* is a homogeneous region of matter. A gas or a mixture of gases, a liquid or a liquid solution, and a solid crystal are examples of phases. A phase need not be continuous; examples of discontinuous phases are a gas dispersed as bubbles in a liquid, a liquid dispersed as droplets in another liquid with which it is immiscible, and a crystalline solid dispersed in either a gas or liquid. In each case a dispersed phase is distributed throughout a continuous phase. An abrupt change in properties always occurs at the boundary between phases. Various phases can coexist, but they *must be in equilibrium* for the phase rule to apply. An example of a system at equilibrium which is made up of three phases is a boiling saturated solution of a salt in water with excess salt crystals present. The three phases are crystalline salt, the saturated aqueous solution, and the vapor generated by boiling.

The phase-rule variables are *intensive* properties, which are independent of the extent of the system and of the individual phases. Thus the phase rule gives the same information for a large system as for a small one and for different relative amounts of the phases present. Moreover, the only compositions that are phase-rule variables are those of the individual phases. Overall or total compositions are not phase-rule variables when more than one phase is present.

The minimum number of degrees of freedom for any system is zero. When $F = 0$, the system is invariant, and Eq. (2.11) becomes $\pi = 2 + N$. This value of π is the maximum number of phases which can coexist at equilibrium for a system containing N chemical species. When $N = 1$, this number is 3, characteristic of a triple point. For example, the triple point of water, where liquid, vapor, and the common form of ice exist together in equilibrium, occurs at $0.01°C$ and 0.0061 bar. Any change from these conditions causes at least one phase to disappear.

Example 2.9 How many degrees of freedom has each of the following systems?

(*a*) Liquid water in equilibrium with its vapor.

(*b*) Liquid water in equilibrium with a mixture of water vapor and nitrogen.

(*c*) A liquid solution of alcohol in water in equilibrium with its vapor.

SOLUTION (*a*) The system contains a single chemical species. There are two phases (one liquid and one vapor). Thus

$$F = 2 - \pi + N = 2 - 2 + 1 = 1$$

This result is in agreement with the well-known fact that for a given pressure water has but one boiling point. Temperature or pressure, but not both, may be specified for a system consisting of water in equilibrium with its vapor.

(*b*) In this case two chemical species are present. Again there are two phases. Thus

$$F = 2 - \pi + N = 2 - 2 + 2 = 2$$

We see from this example that the addition of an inert gas to a system of water in equilibrium with its vapor changes the characteristics of the system. Now temperature and pressure may be independently varied, but once they are fixed the system described can exist in equilibrium only at a particular composition of the vapor phase. (If nitrogen is taken to be negligibly soluble in water, we need not consider the composition of the liquid phase.)

(*c*) Here $N = 2$, and $\pi = 2$. Thus

$$F = 2 - \pi + N = 2 - 2 + 2 = 2$$

The phase-rule variables are temperature, pressure, and the phase compositions. The composition variables are either the mass or mole fractions of the species in a phase, and they must sum to unity for each phase. Thus fixing the mole fraction of the water in the liquid phase automatically fixes the mole fraction of the alcohol. These two compositions cannot both be arbitrarily specified.

2.9   The Reversible Process

The development of thermodynamics is facilitated by the introduction of a special kind of nonflow process characterized as *reversible*. A process is reversible *when its direction can be reversed at any point by an infinitesimal change in external conditions.*

To indicate the nature of reversible processes, we examine the simple expansion of a gas in a piston/cylinder arrangement. The apparatus is shown in Fig. 2.4, and is imagined to exist in an evacuated space. The gas trapped inside the cylinder is chosen as the system; all else is the surroundings. Expansion processes result when mass is removed from the piston. To make the process as simple as possible, we assume that the piston slides within the cylinder without friction and that the piston and cylinder neither absorb nor transmit heat. Moreover, because the density of the gas in the cylinder is low and because the mass of gas is small, we ignore the effects of gravity on the contents of the cylinder. This means that gravity-induced pressure gradients in the gas are considered very small relative to its pressure and that changes in potential energy of the gas are taken as negligible in comparison with the potential-energy changes of the piston assembly.

The piston in Fig. 2.4 confines the gas at a pressure just sufficient to balance the weight of the piston and all that it supports. This is a condition of equilibrium, for the system has no tendency to change. Mass must be removed from the piston if it is to rise. We imagine first that a mass m is suddenly slid from the piston to a shelf (at the same level). The piston assembly accelerates upward, reaching its maximum velocity at the point where the upward force on the piston just balances its weight. Its momentum then carries it to a higher level, where it reverses direction. If the piston were held in this position of maximum elevation, its potential-energy increase would very nearly equal the work done by the gas during the initial stroke. However,

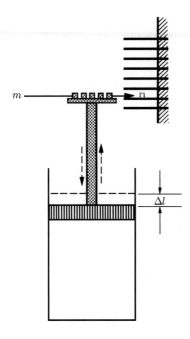

Figure 2.4: Expansion of a gas.

when unconstrained, the piston assembly oscillates, with decreasing amplitude, ultimately coming to rest at a new equilibrium position at a level above its initial position.

The oscillations of the piston assembly are damped out because the viscous nature of the gas gradually converts gross directed motion of the molecules into chaotic molecular motion. This *dissipative* process transforms some of the work initially done by the gas in accelerating the piston back into internal energy of the gas. Once the process is initiated, no *infinitesimal* change in external conditions can reverse its direction; the process is *irreversible*.

All processes carried out in finite time with real substances are accompanied in some degree by dissipative effects of one kind or another, and all are therefore irreversible. However, we can *imagine* processes that are free of dissipative effects. For the expansion process of Fig. 2.4, such effects have their origin in the sudden removal of a finite mass from the piston. The resulting imbalance of forces acting on the piston causes its acceleration, and leads to its subsequent oscillation. The sudden removal of smaller mass increments reduces but does not eliminate this dissipative effect. Even the removal of an infinitesimal mass leads to piston oscillations of infinitesimal amplitude and a consequent dissipative effect. However, one may *imagine* a process wherein small mass increments are removed one after another at a rate such that the piston's rise is continuous, with oscillation only at the end of the process.

The limiting case of removal of a succession of infinitesimal masses from the piston is approximated when the mass m in Fig. 2.4 is replaced by a pile of powder, blown in a very fine stream from the piston. During this process, the piston rises at a uniform but very slow rate, and the powder collects in storage at ever higher levels. The system is never more than differentially displaced either from internal equilibrium or from equilibrium with its surroundings. If the removal of powder from the piston is stopped and the direction of transfer of powder is reversed, the process reverses direction and proceeds backwards along its original path. Both the system and its surroundings are ultimately restored to their initial conditions. The original process is *reversible*.

Without the assumption of a frictionless piston, we cannot imagine a reversible process. If the piston sticks because of friction, a finite mass must be removed before the piston breaks free. Thus the equilibrium condition necessary to reversibility is not maintained. Moreover, friction between two sliding parts is a mechanism for the dissipation of mechanical energy into internal energy.

Our discussion has centered on a single nonflow process, the expansion of a gas in a cylinder. The opposite process, compression of a gas in a cylinder, is described in exactly the same way. There are, however, many processes which are driven by other than mechanical forces. For example, heat flow occurs when a temperature difference exists, electricity flows under the influence of an electromotive force, and chemical reactions occur because a chemical potential exists. In general, a process is reversible when the net force driving it is only differential in size. Thus heat is transferred reversibly when it flows from a finite object at temperature T to another such object at temperature $T - dT$.

The concept of a reversible chemical reaction may be illustrated by the decomposition of calcium carbonate, which when heated forms calcium oxide and carbon dioxide gas. At equilibrium, this system exerts a definite decomposition pressure of CO_2 for a given temperature. When the pressure falls below this value, $CaCO_3$ decomposes. Assume that a cylinder is fitted with a frictionless piston and contains $CaCO_3$, CaO, and CO_2 in equilibrium. It is immersed in a constant-temperature bath, as shown in Fig. 2.5, with the temperature adjusted to a value such that the decomposition pressure is just sufficient to balance the weight on the piston. The system is in mechanical equilibrium, the temperature of the system is equal to that of the bath, and the chemical reaction is held in balance by the pressure of the CO_2. Any change of conditions, however slight, upsets the equilibrium and causes the reaction to proceed in one direction or the other. If the weight is differentially increased, the CO_2 pressure rises differentially, and CO_2 combines with CaO to form $CaCO_3$, allowing the weight to fall slowly. The heat given off by this reaction raises the temperature in the cylinder, and heat flows to the bath. Decreasing the weight differentially sets off the opposite chain of events. The same results are obtained if the temperature of the bath is raised or lowered. If the temperature of the bath is raised differentially, heat flows into the cylinder and calcium carbonate decomposes. The CO_2 generated causes the pressure to rise differentially, which in turn raises the piston and weight. This continues until the $CaCO_3$ is completely de-

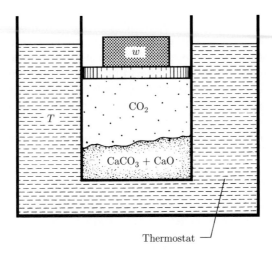

Figure 2.5: Reversibility of a chemical reaction.

composed. The process is reversible, for the system is never more than differentially displaced from equilibrium, and only a differential lowering of the temperature of the bath causes the system to return to its initial state.

Chemical reactions can sometimes be carried out in an electrolytic cell, and in this case they can be held in balance by an applied potential difference. If such a cell consists of two electrodes, one of zinc and the other of platinum, immersed in an aqueous solution of hydrochloric acid, the reaction that occurs is

$$\text{Zn} + 2\text{HCl} \rightleftharpoons \text{H}_2 + \text{ZnCl}_2$$

The cell is held under fixed conditions of temperature and pressure, and the electrodes are connected externally to a potentiometer. If the electromotive force produced by the cell is exactly balanced by the potential difference of the potentiometer, the reaction is held in equilibrium. The reaction may be made to proceed in the forward direction by a slight decrease in the opposing potential difference, and it may be reversed by a corresponding increase in the potential difference above the emf of the cell.

In summary, a reversible process is frictionless; it is never more than differentially removed from equilibrium, and therefore traverses a succession of equilibrium states; the driving forces are differential in magnitude; its direction can be reversed at any point by a differential change in external conditions, causing the process to retrace its path, leading to restoration of the initial state of the system and its surroundings.

In Sec. 1.7 we derived an equation for the work of compression or expansion of a gas caused by the differential displacement of a piston in a cylinder:

$$dW = -P\,dV \tag{1.2}$$

The work done on the *system* is given by this equation only when certain character-istics of the reversible process are realized. The first requirement is that the system be no more than infinitesimally displaced from a state of *internal* equilibrium char-acterized by uniformity of temperature and pressure. The system then always has an identifiable set of properties, including pressure P. The second requirement is that the system be no more than infinitesimally displaced from mechanical equi-librium with its surroundings. In this event, the internal pressure P is never more than minutely out of balance with the external force, and we may make the substi-tution $F = PA$ that transforms Eq. (1.1) into Eq. (1.2). Processes for which these requirements are met are said to be *mechanically reversible*. For such processes, Eq. (1.3) correctly yields the work done on the system:

$$W = - \int_{V_1}^{V_2} P\,dV \qquad (1.3)$$

The reversible process is ideal in that it can never be fully realized; it represents a limit to the performance of actual processes. In thermodynamics, the calculation of work is usually made for reversible processes, because of their tractability to mathematical analysis. The choice is between these calculations and no calculations at all. Results for reversible processes in combination with appropriate *efficiencies* yield reasonable approximations of the work for actual processes.

> **Example 2.10**   A horizontal piston/cylinder arrangement is placed in a constant-temperature bath. The piston slides in the cylinder with negligible friction, and an external force holds it in place against an initial gas pressure of 14 bar. The initial gas volume is $V_1^t = 0.03 \text{ m}^3$, where the superscript denotes a *total* rather than a molar volume. The external force on the piston is reduced gradually, allowing the gas to expand until its volume doubles. Experiment shows that under these conditions the volume of the gas is related to its pressure in such a way that the product PV^t is constant. Calculate the work done by the gas in moving the external force.
>
> How much work would be done if the external force were suddenly reduced to half its initial value instead of being gradually reduced?
>
> SOLUTION   The process, carried out as first described, is mechanically reversible, and Eq. (1.3) is applicable. If $PV^t = k$, then $P = k/V^t$, and
>
> $$W = -k \int_{V_1^t}^{V_2^t} \frac{dV^t}{V^t} = -k \ln \frac{V_2^t}{V_1^t}$$
>
> But
>
> $$V_1^t = 0.03 \text{ m}^3 \qquad V_2^t = 0.06 \text{ m}^3$$
>
> and
>
> $$k = PV^t = P_1 V_1^t = (14 \times 10^5)(0.03) = 42{,}000 \text{ J}$$
>
> Therefore
>
> $$W = -42{,}000 \ln 2 = -29{,}112 \text{ J}$$

The final pressure is

$$P_2 = \frac{k}{V_2^t} = \frac{42{,}000}{0.06} = 700{,}000 \text{ Pa} \qquad \text{or} \qquad 7 \text{ bar}$$

In the second case, after half the initial force has been removed, the gas undergoes a sudden expansion against a constant force equivalent to a pressure of 7 bar. Eventually the system returns to an equilibrium condition identical with the final state attained in the reversible process. Thus ΔV^t is the same as before, and the net work accomplished equals the equivalent external pressure times the volume change, or

$$W = -(7 \times 10^5)(0.06 - 0.03) = -21{,}000 \text{ J}$$

This process is clearly irreversible, and compared with the reversible process is said to have an efficiency of

$$\frac{21{,}000}{29{,}112} = 0.721 \qquad \text{or} \qquad 72.1\%$$

Example 2.11 The piston/cylinder arrangement shown in Fig. 2.6 contains nitrogen gas trapped below the piston at a pressure of 7 bar. The piston is held in place by latches. The space above the piston is evacuated. A pan is attached to the piston rod and a mass m of 45 kg is fastened to the pan. The piston, piston rod, and pan together have a mass of 23 kg. The latches holding the piston are released, allowing the piston to rise rapidly until it strikes the top of the cylinder. The distance moved by the piston is 0.5 m. The local acceleration of gravity is 9.8 m s^{-2}. Discuss the energy changes that occur because of this process.

SOLUTION This example serves to illustrate some of the difficulties encountered when irreversible nonflow processes are analyzed. We take the gas alone as the system. According to the basic definition, the work done by the gas on the surroundings is equal to $\int P'dV^t$, where P' is the pressure exerted on the face of the piston by the gas. Because the expansion is very rapid, pressure gradients exist in the gas, and neither P' nor the integral can be evaluated. However, we can avoid the calculation of work by returning to Eq. (2.1). The total energy change of the system (the gas) is its internal-energy change. For $Q = 0$, the energy changes of the surroundings consist of potential-energy changes of the piston, rod, pan, and mass m and of internal-energy changes of the piston, rod, and cylinder. Therefore, Eq. (2.1) may be written

$$\Delta U_{\text{sys}} + (\Delta U_{\text{surr}} + \Delta E_{P\,\text{surr}}) = 0$$

The potential-energy term is

$$\Delta E_{P\,\text{surr}} = (45 + 23)(9.8)(0.5) = 333.2 \text{ N m}$$

Therefore

$$\Delta U_{\text{sys}} + \Delta U_{\text{surr}} = -333.2 \text{ N m} = -333.2 \text{ J}$$

and one cannot determine the individual internal-energy changes which occur in the piston/cylinder assembly.

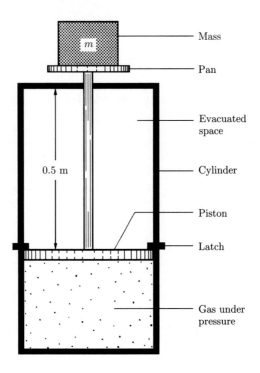

Mass

Pan

Evacuated space

0.5 m   Cylinder

Piston

Latch

Gas under pressure

Figure 2.6: Diagram for Example 2.11.

2.10   Constant-V and Constant-P Processes

Throughout this book plain uppercase letters such as V and U are used to represent the *specific* or *molar* values of extensive thermodynamic properties. The superscript t is added to denote the total properties of a system: V^t, U^t, etc. An alternative for a system of specified mass m or number of moles n is simply to multiply each specific or molar property by the appropriate symbol to indicate explicitly the amount of substance in the system: mU or nU, mV or nV, etc. For a closed system of n moles, Eq. (2.4) then becomes

$$d(nU) = dQ + dW \qquad (2.12)$$

where Q and W always represent *total* heat and work, whatever the value of n.

The work of a mechanically reversible, nonflow process is given by

$$dW = -P\,d(nV) \qquad (2.13)$$

If such a process occurs at constant volume, the work is zero; if it occurs at constant pressure, then integration yields

$$W = -nP\,\Delta V$$

When combined with Eq. (2.13), Eq. (2.12) becomes

$$d(nU) = dQ - Pd(nV) \tag{2.14}$$

This is the general first-law equation for a mechanically reversible, nonflow process. If in addition the process occurs at constant volume, then the work is zero, and

$$dQ = d(nU) \qquad (\text{const } V) \tag{2.15}$$

Integration yields

$$Q = n\,\Delta U \qquad (\text{const } V) \tag{2.16}$$

Thus for a mechanically reversible, constant-volume, nonflow process, the heat transferred is equal to the internal-energy change of the system.

Equation (2.5), which defines the enthalpy, may be written

$$nH = nU + P(nV)$$

For an infinitesimal constant-pressure change of state,

$$d(nH) = d(nU) + Pd(nV)$$

Combining this with Eq. (2.14) gives

$$dQ = d(nH) \qquad (\text{const } P) \tag{2.17}$$

Integration yields

$$Q = n\,\Delta H \qquad (\text{const } P) \tag{2.18}$$

Thus for a mechanically reversible, constant-pressure, nonflow process, the heat transferred equals the enthalpy change of the system. Comparison of the last two equations with Eqs. (2.15) and (2.16) shows that the enthalpy plays a role in constant-pressure processes analogous to the internal energy in constant-volume processes.

2.11   Heat Capacity

We remarked earlier that heat is often viewed in relation to its effect on the object to which or from which it is transferred. This is the origin of the idea that a body has a capacity for heat. The smaller the temperature change in a body caused by the transfer of a given quantity of heat, the greater its capacity. Indeed, a *heat capacity* might be defined as

$$C = \frac{dQ}{dT}$$

The difficulty with this is that it makes C, like Q, a process-dependent quantity rather than a state function. However, it does suggest the possibility that more than one useful heat capacity might be defined.

In fact two heat capacities are in common use for homogeneous fluids; although their names belie the fact, both are state functions, defined unambigously in relation to other state functions:

Heat capacity at constant volume

$$C_V \equiv \left(\frac{\partial U}{\partial T}\right)_V \qquad\qquad (2.19)$$

Heat capacity at constant pressure

$$C_P \equiv \left(\frac{\partial H}{\partial T}\right)_P \qquad\qquad (2.20)$$

These definitions accommodate both molar heat capacities and specific heat capacities (usually called specific heats), depending on whether U and H are molar or specific properties.

Although the definitions of C_V and C_P make no reference to any process, each allows an especially simple description of a particular process. Thus, if we have a constant-volume process, Eq. (2.19) may be written

$$dU = C_V\,dT \qquad (\text{const } V) \qquad\qquad (2.21)$$

Integration yields

$$\Delta U = \int_{T_1}^{T_2} C_V\,dT \qquad (\text{const } V) \qquad\qquad (2.22)$$

For a mechanically reversible, constant-volume process, this result may be combined with Eq. (2.16) to give

$$Q = n\,\Delta U = n \int_{T_1}^{T_2} C_V\,dT \qquad (\text{const } V) \qquad\qquad (2.23)$$

Consider now the case in which the volume varies during the process, but is the same at the end as at the beginning. Such a process cannot rightly be called one of constant volume, even though $V_2 = V_1$ and $\Delta V = 0$. However, changes in state functions or properties are independent of path and are, therefore, the same for all processes which lead from the same initial to the same final conditions. Hence, property changes for this case may be calculated from the equations for a truly constant-volume process leading from the same initial to the same final conditions. For such processes Eq. (2.22) gives $\Delta U = \int C_V\,dT$, because U, C_V, and T are all state functions or properties. On the other hand, Q does depend on path, and Eq. (2.23) is a valid expression for Q only for a *constant-volume* process. For the same reason, W is in general zero only for a constant-volume process. This discussion illustrates the reason for the careful distinction made between state functions and heat and work. The principle that state functions are independent of the process is an important and useful concept. Thus for the calculation of property changes an actual process may be replaced by any other process which accomplishes the same change in state. Such an alternative process may be selected, for example, because of its simplicity.

For a constant-pressure process, Eq. (2.20) may be written

$$dH = C_P dT \qquad (\text{const } P) \qquad (2.24)$$

whence

$$\Delta H = \int_{T_1}^{T_2} C_P dT \qquad (\text{const } P) \qquad (2.25)$$

For a mechanically reversible, constant-pressure process, this result may be combined with Eq. (2.18) to give

$$Q = n\,\Delta H = n \int_{T_1}^{T_2} C_P dT \qquad (\text{const } P) \qquad (2.26)$$

Since H, C_P, and T are all state functions, Eq. (2.25) applies to any process for which $P_2 = P_1$ whether or not it is actually carried out at constant pressure. However, only for the mechanically reversible, constant-pressure process can heat and work can be calculated by the equations $Q = n\,\Delta H$, $Q = n \int C_P dT$, and $W = -Pn\,\Delta V$.

Example 2.12   An ideal gas is one for which PV/T is a constant, regardless of the changes it undergoes. Such a gas has a volume of 0.02271 m^3 mol^{-1} at 0°C and 1 bar. In the following problem, air may be considered an ideal gas with the constant heat capacities

$$C_V = (5/2)R \qquad \text{and} \qquad C_P = (7/2)R$$

where $R = 8.314$ J mol^{-1} K^{-1}. Thus

$$C_V = 20.785 \qquad \text{and} \qquad C_P = 29.099 \text{ J mol}^{-1} \text{ K}^{-1}$$

The initial conditions of the air are 1 bar and 25°C. It is compressed to 5 bar and 25°C by two different mechanically reversible processes. Calculate the heat and work requirements and ΔU and ΔH of the air for each path:

(*a*) Cooling at constant pressure followed by heating at constant volume.

(*b*) Heating at constant volume followed by cooling at constant pressure.

SOLUTION   In each case we take the system as 1 mol of air contained in an imaginary piston/cylinder arrangement. Since the processes considered are mechanically reversible, the piston is imagined to move in the cylinder without friction. The initial volume of air is

$$V_1 = (0.02271) \left(\frac{298.15}{273.15} \right) = 0.02479 \text{ m}^3$$

The final volume is

$$V_2 = V_1 \frac{P_1}{P_2} = (0.02479) \left(\frac{1}{5} \right) = 0.004958 \text{ m}^3$$

(a) In this case during the first step the air is cooled at the constant pressure of 1 bar until the final volume of 0.004958 m^3 is reached. During the second step the volume is held constant at this value while the air is heated to its final state. The temperature of the air at the end of the cooling step is

$$T = (298.15)\left(\frac{0.004958}{0.02479}\right) = 59.63 \text{ K}$$

For this step the pressure is constant. By Eq. (2.26),

$$Q = \Delta H = C_P\,\Delta T = (29.099)(59.63 - 298.15) = -6{,}941 \text{ J}$$

Since $\Delta U = \Delta H - \Delta(PV) = \Delta H - P\,\Delta V$, then

$$\Delta U = -6{,}941 - (1 \times 10^5)(0.004958 - 0.02479) = -4{,}958 \text{ J}$$

In the second step the air is heated at constant volume. By Eq. (2.23),

$$\Delta U = Q = C_V\,\Delta T = (20.785)(298.15 - 59.63) = 4{,}958 \text{ J}$$

The complete process represents the sum of its steps. Hence

$$Q = -6{,}941 + 4{,}958 = -1{,}983 \text{ J}$$

and

$$\Delta U = -4{,}958 + 4{,}958 = 0$$

Since the first law applies to the entire process, $\Delta U = Q + W$, and therefore

$$0 = -1{,}983 + W$$

Whence

$$W = 1{,}983 \text{ J}$$

Equation (2.7), $\Delta H = \Delta U + \Delta(PV)$, also applies to the entire process. But $T_1 = T_2$, and therefore $P_1 V_1 = P_2 V_2$. Hence $\Delta(PV) = 0$, and

$$\Delta H = \Delta U = 0$$

(b) Two different steps are used in this case to reach the same final state of the air. In the first step the air is heated at a constant volume equal to its initial value until the final pressure of 5 bar is reached. During the second step the air is cooled at the constant pressure of 5 bar to its final state. The air temperature at the end of the first step is

$$T = (298.15)(5/1) = 1{,}490.75 \text{ K}$$

For this step the volume is constant, and

$$Q = \Delta U = C_V\,\Delta T = (20.785)(1{,}490.75 - 298.15) = 24{,}788 \text{ J}$$

For the second step pressure is constant, and

$$Q = \Delta H = C_P\,\Delta T = (29.099)(298.15 - 1{,}490.75) = -34{,}703 \text{ J}$$

Also

$$\Delta U = \Delta H - \Delta(PV) = \Delta H - P \, \Delta V$$
$$\Delta U = -34{,}703 - (5 \times 10^5)(0.004958 - 0.02479) = -24{,}788 \text{ J}$$

For the two steps combined,

$$Q = 24{,}788 - 34{,}703 = -9{,}915 \text{ J}$$
$$\Delta U = 24{,}788 - 24{,}788 = 0$$
$$W = \Delta U - Q = 0 - (-9{,}915) = 9{,}915 \text{ J}$$

and as before

$$\Delta H = \Delta U = 0$$

The property changes ΔU and ΔH calculated for the given change in state are the same for both paths. On the other hand the answers to parts (*a*) and (*b*) show that Q and W depend on the path.

Example 2.13 Calculate the internal-energy and enthalpy changes that occur when air is changed from an initial state of 40(°F) and 10(atm), where its molar volume is $36.49(\text{ft})^3(\text{lb mole})^{-1}$ to a final state of 140(°F) and 1(atm). Assume for air that PV/T is constant and that $C_V= 5$ and $C_P= 7(\text{Btu})(\text{lb mole})^{-1}(°\text{F})^{-1}$.

SOLUTION Since property changes are independent of the process that brings them about, we can base calculations on a simple two-step, mechanically reversible process in which 1(lb mole) of air is (*a*) cooled at constant volume to the final pressure, and (*b*) heated at constant pressure to the final temperature. The absolute temperatures here are on the Rankine scale:

$$T_1 = 40 + 459.67 = 499.67(\text{R})$$
$$T_2 = 140 + 459.67 = 599.67(\text{R})$$

Since $PV = kT$, the ratio T/P is constant for step (*a*). The intermediate temperature between the two steps is therefore

$$T' = (499.67)(1/10) = 49.97(\text{R})$$

and the temperature changes for the two steps are

$$\Delta T_a = 49.97 - 499.67 = -449.70(\text{R})$$

and

$$\Delta T_b = 599.67 - 49.97 = 549.70(\text{R})$$

For step (*a*), Eq. (2.22) becomes

$$\Delta U_a = C_V \, \Delta T_a$$

whence

$$\Delta U_a = (5)(-449.70) = -2{,}248.5(\text{Btu})(\text{lb mole})^{-1}$$

For step (*b*), Eq. (2.25) becomes

$$\Delta H_b = C_P \, \Delta T_b$$

whence

$$\Delta H_b = (7)(549.70) = 3{,}847.9(\text{Btu})(\text{lb mole})^{-1}$$

For step (a), Eq. (2.7) becomes

$$\Delta H_a = \Delta U_a + V\,\Delta P_a$$

Whence

$$\Delta H_a = -2{,}248.5 + 36.49(1 - 10)(2.7195) = -3{,}141.6(\text{Btu})$$

The factor 2.7195 converts the PV product from $(\text{atm})(\text{ft})^3$, which is an energy unit, into (Btu). For step (b), Eq. (2.7) becomes

$$\Delta U_b = \Delta H_b - P\,\Delta V_b$$

The final volume of the air is given by

$$V_2 = V_1 \frac{P_1 T_2}{P_2 T_1}$$

from which we find that $V_2 = 437.93(\text{ft})^3$. Therefore

$$\Delta U_b = 3{,}847.9 - (1)(437.93 - 36.49)(2.7195) = 2{,}756.2(\text{Btu})$$

For the two steps together,

$$\Delta U = -2{,}248.5 + 2{,}756.2 = 507.7(\text{Btu})$$

and

$$\Delta H = -3{,}141.6 + 3{,}847.9 = 706.3(\text{Btu})$$

PROBLEMS

2.1. A nonconducting container filled with 25 kg of water at $20°C$ is fitted with a stirrer, which is made to turn by gravity acting on a weight of mass 35 kg. The weight falls slowly through a distance of 5 m in driving the stirrer. Assuming that all work done on the weight is transferred to the water and that the local acceleration of gravity is 9.8 m s^{-2}, determine:

(a) The amount of work done on the water.

(b) The internal-energy change of the water.

(c) The final temperature of the water, for which $C_P = 4.18$ kJ kg^{-1} °C^{-1}.

(d) The amount of heat that must be removed from the water to return it to its initial temperature.

(e) The total energy change of the universe because of (1) the process of lowering the weight, (2) the process of cooling the water back to its initial temperature, and (3) both processes together.

2.2. Rework Prob. 2.1 for an insulated container that changes in temperature along with the water and has a heat capacity equivalent to 5 kg of water. Work the problem in two ways:

(*a*) Take the water and container as the system.

(*b*) Take the water alone as the system.

2.3. Comment on the feasibility of cooling your kitchen in the summer by opening the door to the electrically powered refrigerator.

2.4. Liquid water at $180°C$ and $1,002.7$ kPa has an internal energy (on an arbitrary scale) of 762.0 kJ kg^{-1} and a specific volume of 1.128 cm^3 g^{-1}.

(*a*) What is its enthalpy?

(*b*) The water is brought to the vapor state at $300°C$ and $1,500$ kPa, where its internal energy is $2,784.4$ kJ kg^{-1} and its specific volume is 169.7 cm^3 g^{-1}. Calculate ΔU and ΔH for the process.

2.5. If a tank containing 20 kg of water at $20°C$ is fitted with a stirrer that delivers work to the water at the rate of 0.25 kW, how long does it take for the temperature of the water to rise to $30°C$ if no heat is lost from the water to its surroundings? For water, $C_P = 4.18$ kJ kg^{-1} $°C^{-1}$.

2.6. With respect to a mass of 1 kg,

(*a*) How much change in elevation must it undergo to change its potential energy by 1 kJ?

(*b*) Starting from rest, to what velocity must it accelerate so that its kinetic energy is 1 kJ?

(*c*) What conclusions are indicated by these results?

2.7. Heat in the amount of 7.5 kJ is added to a closed system while its internal energy decreases by 12 kJ. How much energy is transferred as work? For a process causing the same change of state but for which the work is zero, how much heat is transferred?

2.8. A steel casting weighing 2 kg has an initial temperature of $500°C$; 40 kg of water initially at $25°C$ is contained in a perfectly insulated steel tank weighing 5 kg. The casting is immersed in the water and the system is allowed to come to equilibrium. What is its final temperature? Ignore any effect of expansion or contraction, and assume constant specific heats of 4.18 kJ kg^{-1} K^{-1} for water and 0.50 kJ kg^{-1} K^{-1} for steel.

2.9. Nitrogen flows at steady state through a horizontal, insulated pipe with inside diameter of 1.5(in) [3.81 cm]. A pressure drop results from flow through a partially opened valve. Just upstream from the valve the pressure is 100(psia) [689.5 kPa], the temperature is 120(°F) [$48.9°C$], and the average velocity is 20(ft)(s)$^{-1}$ [6.09 m s^{-1}]. If the pressure just downstream from the valve is 20(psia) [137.9 kPa], what is the temperture? Assume for nitrogen that PV/T is constant, $C_V = (5/2)R$, and $C_P = (7/2)R$. (Values for R are given in App. A.)

2.10. Liquid water at 80(°F) ($26.67°C$) flows in a straight horizontal pipe in which there is no exchange of either heat or work with the surroundings. Its velocity is 40(ft)(s)$^{-1}$ [12.19 m s^{-1}] in a pipe with an internal diameter of 1(in) [2.54 cm] until it flows into a section where the pipe diameter abruptly increases. What is the temperature change of the water if the downstream diameter is 1.5(in) [3.81 cm]? If it is 3(in) [7.62 cm]? What is the maximum temperature change for an enlargement in the pipe?

2.11. Water flows through a horizontal coil heated from the outside by high-temperature flue gases. As it passes through the coil the water changes state from liquid at 200 kPa and 80°C to vapor at 100 kPa and 125°C. Its entering velocity is 3 m s^{-1} and its exit velocity is 200 m s^{-1}. Determine the heat transferred through the coil per unit mass of water. Enthalpies of the inlet and outlet streams are:

Inlet: 334.9 kJ kg^{-1}; Outlet: 2,726.5 kJ kg^{-1}

2.12. Steam flows at steady state through a converging, insulated nozzle, 25 cm long and with an inlet diameter of 5 cm. At the nozzle entrance (state 1), the temperature and pressure are 325°C and 700 kPa, and the velocity is 30 m s^{-1}. At the nozzle exit (state 2), the steam temperature and pressure are 240°C and 350 kPa. Property values are:

$H_1 = 3,112.5$ kJ kg^{-1} $V_1 = 388.61$ cm^3 g^{-1}

$H_2 = 2,945.7$ kJ kg^{-1} $V_2 = 667.75$ cm^3 g^{-1}

What is the velocity of the steam at the nozzle exit, and what is the exit diameter?

2.13. A system comprised of chloroform, 1,4-dioxane, and ethanol exists as a two-phase vapor/liquid system at 50°C and 55 kPa. It is found, after the addition of some pure ethanol, that the system can be returned to two-phase equilibrium at the initial T and P. In what respect has the system changed, and in what respect has it not changed?

2.14. For the system described in Pb. 2.13:

 (a) How many phase-rule variables in addition to T and P must be chosen so as to fix the compositions of both phases?

 (b) If the temperature and pressure are to remain the same, can the *overall* composition of the system be changed (by adding or removing material) without affecting the compositions of the liquid and vapor phases?

2.15. In the following take $C_V = 20.8$ and $C_P = 29.1$ J mol^{-1} °C^{-1} for nitrogen gas:

 (a) Three moles of nitrogen at 30°C, contained in a rigid vessel, is heated to 250°C. How much heat is required if the vessel has a negligible heat capacity? If the vessel weighs 100 kg and has a heat capacity of 0.5 kJ kg^{-1} °C^{-1}, how much heat is required?

 (b) Four moles of nitrogen at 200°C is contained in a piston/cylinder arrangement. How much heat must be extracted from this system, which is kept at constant pressure, to cool it to 40°C if the heat capacity of the piston and cylinder is neglected?

2.16. In the following take $C_V = 5$ and $C_P = 7$(Btu)(lb mole)$^{-1}$(°F)$^{-1}$ for nitrogen gas:

 (a) Three pound moles of nitrogen at 70(°F) contained in a rigid vessel, is heated to 350(°F). How much heat is required if the vessel has a negligible heat capacity? If it weighs 200(lb$_m$) and has a heat capacity of 0.12(Btu)(lb$_m$)$^{-1}$(°F)$^{-1}$, how much heat is required?

(b) Four pound moles of nitrogen at 400(°F) is contained in a piston/cylinder arrangement. How much heat must be extracted from this system, which is kept at constant pressure, to cool it to 150(°F) if the heat capacity of the piston and cylinder is neglected?

2.17. Find the equation for the work of a reversible, isothermal compression of 1 mol of gas in a piston/cylinder assembly if the molar volume of the gas is given by

$$V = \frac{RT}{P} + b$$

where b and R are positive constants.

2.18. Steam at 200(psia) and 600(°F) [state 1] enters a turbine through a 3-inch-diameter pipe with a velocity of 10(ft)(s)$^{-1}$. The exhaust from the turbine is carried through a 10-inch-diameter pipe and is at 5(psia) and 200(°F) [state 2]. What is the power output of the turbine?

$H_1 = 1322.6(\text{Btu})(\text{lb}_\text{m})^{-1}$ $V_1 = 3.058(\text{ft})^3(\text{lb}_\text{m})^{-1}$

$H_2 = 1148.6(\text{Btu})(\text{lb}_\text{m})^{-1}$ $V_2 = 78.14(\text{ft})^3(\text{lb}_\text{m})^{-1}$

2.19. Carbon dioxide gas enters a water-cooled compressor at the initial conditions $P_1 = 15(\text{psia})$ and $T_1 = 50(°F)$ and is discharged at the final conditions $P_2 = 520(\text{psia})$ and $T_2 = 200(°F)$. The entering CO_2 flows through a 4-inch-diameter pipe with a velocity of 20(ft)(s)$^{-1}$, and is discharged through a 1-inch-diameter pipe. The shaft work supplied to the compressor is 5,360(Btu)(mol)$^{-1}$. What is the heat-transfer rate from the compressor in (Btu)(hr)$^{-1}$?

$H_1 = 307(\text{Btu})(\text{lb}_\text{m})^{-1}$ $V_1 = 9.25(\text{ft})^3(\text{lb}_\text{m})^{-1}$

$H_2 = 330(\text{Btu})(\text{lb}_\text{m})^{-1}$ $V_2 = 0.28(\text{ft})^3(\text{lb}_\text{m})^{-1}$

2.20. One kilogram of air is heated reversibly at constant pressure from an initial state of 300 K and 1 bar until its volume triples. Calculate W, Q, ΔU, and ΔH for the process. Assume that air obeys the relation $PV/T = 83.14$ bar cm^3 mol^{-1} K^{-1} and that $C_P = 29$ J mol^{-1} K^{-1}.

2.21. The conditions of a gas change in a steady-flow process from 20°C and 1,000 kPa to 60°C and 100 kPa. Devise a reversible nonflow process (any number of steps) for accomplishing this change of state, and calculate ΔU and ΔH for the process on the basis of 1 mol of gas. Assume for the gas that PV/T is constant, $C_V = (5/2)R$, and $C_P = (7/2)R$.

2.22. Show that W and Q for an *arbitrary* mechanically reversible nonflow process are given by

$$W = \int V\,dP - \Delta(PV)$$

$$Q = \Delta H - \int V\,dP$$

CHAPTER 3

VOLUMETRIC PROPERTIES OF PURE FLUIDS

3.1 *PVT* Behavior of Pure Substances

Thermodynamic properties, such as internal energy and enthalpy, from which one calculates the heat and work requirements of industrial processes, are often evaluated from volumetric data. Moreover, pressure/volume/temperature (PVT) relations are themselves important for such purposes as the metering of fluids and the sizing of vessels and pipelines. Therefore, the PVT behavior of pure fluids is described in this chapter.

Homogeneous fluids are normally divided into two classes, liquids and gases. However, the distinction cannot always be sharply drawn, because the two phases become indistinguishable at what is called the *critical point*. Measurements of the vapor pressure of a pure solid at temperatures up to its triple point and measurements of the vapor pressure of the pure liquid at temperatures above the triple point lead to a pressure-vs.-temperature curve such as the one made up of lines 1-2 and 2-C in Fig. 3.1. The third line (2-3) shown on this graph gives the solid/liquid equilibrium relationship. These three curves represent the conditions of P and T required for the coexistance of two phases and thus are boundaries for the single-phase regions. Line 1-2, the sublimation curve, separates the solid and gas regions; line 2-3, the fusion curve, separates the solid and liquid regions; line 2-C, the vaporization curve, separates the liquid and gas regions. The three curves meet at the triple point, where all three phases coexist in equilibrium. According to the phase rule [Eq. (2.11)], the triple point is invariant. If the system exists along any of the two-phase lines of Fig. 3.1, it is univariant, whereas in the single-phase regions it is divariant.

The vaporization curve 2-C terminates at point C, the critical point. The coordinates of this point are the critical pressure P_c and the critical temperature T_c,

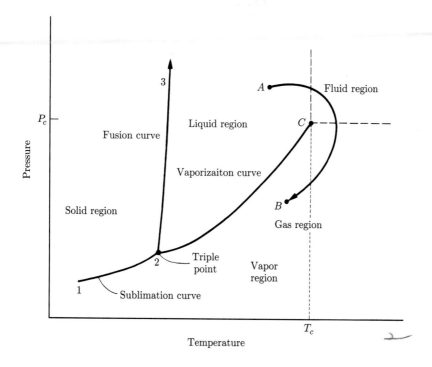

Figure 3.1: *PT* diagram for a pure substance.

the highest temperature and pressure at which a pure chemical species can exist in vapor/liquid equilibrium. The fluid region, existing at higher temperatures and pressures, is marked off by dashed lines, which do not represent phase transitions, but rather are limits fixed by the meanings accorded the *words* liquid and gas. A phase is generally considered a liquid if it can be vaporized by reduction in pressure at constant temperature. A phase is considered a gas if it can be condensed by reduction of temperature at constant pressure. Since the fluid region fits neither of these definitions, it is neither a gas nor a liquid. The gas region is sometimes divided into two parts, as shown by the dotted line of Fig. 1.3. A gas to the left of this line, which can be condensed either by compression at constant temperature or by cooling at constant pressure, is called a vapor.

Because of the existence of the critical point, a path can be drawn from the liquid region to the gas region that does not cross a phase boundary; e.g., the path from A to B in Fig. 3.1. This path represents a gradual transition from the liquid to the gas region. On the other hand, a path from A to B that crosses the phase boundary 2-C includes a vaporization step, where an abrupt change of properties occurs.

Figure 3.1 does not provide any information about volume; it merely displays the phase boundaries on a *PT* diagram. These boundaries appear on a *PV* diagram as areas, representing regions where two phases, solid/liquid, solid/vapor, and liquid/vapor, coexist in equilibrium at given temperature and pressure. The molar (or specific) volume, however, depends on the relative amounts of the phases present. The triple point of Fig. 3.1 here becomes a horizontal line, where the three phases coexist at a single temperature and pressure.

In Fig. 3.3 we superimpose four isotherms on that part of the PV diagram of Fig. 3.2 representing the liquid, liquid/vapor, and vapor regions. The line labeled $T > T_c$ is an isotherm for a temperature greater than the critical. As seen from Fig. 3.1, such isotherms do not cross a phase boundary and are therefore smooth. The lines labeled T_1 and T_2 are for subcritical temperatures, and consist of three distinct segments. The horizontal segments represent the phase change between liquid and vapor. The constant pressure at which this occurs for a given temperature is the saturation or vapor pressure, and is given by the point on Fig. 3.1 where the isotherm crosses the vaporization curve. Points along the horizontal lines of Fig. 3.3 represent all possible mixtures of liquid and vapor in equilibrium, ranging from 100 percent liquid at the left end to 100 percent vapor at the right end. The locus of these end points is the dome-shaped curve labeled BCD, the left half of which (from B to C) represents *saturated liquid*, and the right half (from C to D) *saturated vapor*. The two-phase region lies under the dome BCD, while the liquid and gas regions lie to its left and right. The isotherms in the liquid region are very steep, because liquid volumes change little with large changes in pressure. The horizontal segments of the isotherms in the two-phase region become progressively shorter at higher temperatures, being ultimately reduced to a point at C. Thus, the critical isotherm, labeled T_c, exhibits a horizontal inflection at the critical point C at the top of the dome. Here the liquid and vapor phases cannot be distinguished from one another, because their properties are the same.

Some insight into the nature of the critical point is gained from a description of the changes that occur when a pure substance is heated in a sealed upright tube of constant volume. Such changes follow vertical lines on Fig. 3.3. They are also shown on the PT diagram of Fig. 3.4, where the vaporization curve of Fig. 3.1 appears as a solid line. The dashed lines are constant-volume paths in the single-phase regions only. If the tube is filled with either liquid or gas, the heating process produces changes described by these lines, for example by the change from E to F (liquid region) and by the change from G to H (vapor region). The corresponding vertical lines on Fig. 3.3 lie to the left and to the right of BCD.

If the tube is only partially filled with liquid (the remainder being vapor in equilibrium with the liquid), heating at first causes changes described by the vapor-pressure curve (solid line) of Fig. 3.4. If the meniscus is originally near the top of the tube, the liquid expands upon heating until it completely fills the tube. One such process is represented by the path from (J, K) to Q; it then follows the line of constant molar volume V_2^l with continued heating. If the meniscus separating the two phases is initially somewhat lower in the tube, liquid vaporizes, the meniscus recedes to the bottom of the tube, and disappears as the last drop of liquid vaporizes. For example, on Fig. 3.4, one such path is from (J, K) to N; it then follows the line of constant molar volume V_2^v upon further heating. The

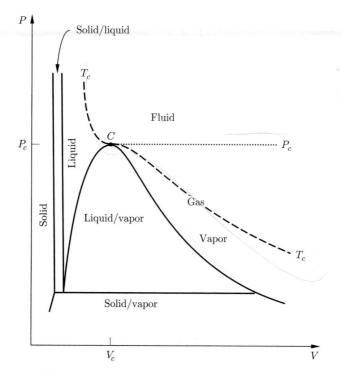

Figure 3.2: *PV* diagram for a pure substance.

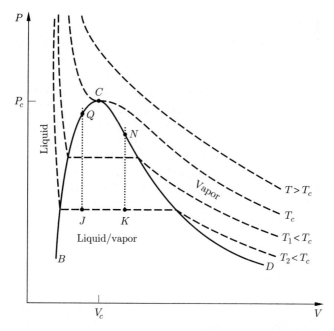

Figure 3.3: *PV* diagram for the liquid, liquid/vapor, and vapor regions of a pure fluid.

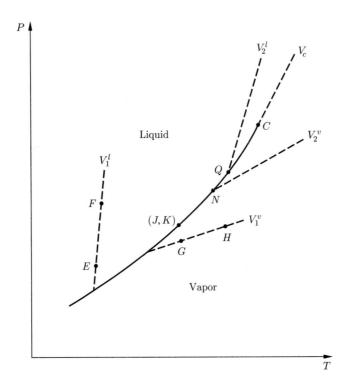

Figure 3.4: *PT* diagram for a pure fluid showing the vapor-pressure curve and constant-volume lines in the single-phase regions.

two paths are also shown by the dashed lines of Fig. 3.3, the first passing through points J and Q, and the second, through K and N.

A unique filling of the tube, with a particular intermediate meniscus level, causes the path of the heating process to coincide with the vapor-pressure curve of Fig. 3.4 all the way to its end at the critical point C. On Fig. 3.3 the path is a vertical line passing through the critical point. Physically, heating does not produce much change in the level of the meniscus. As the critical point is approached, the meniscus becomes indistinct, then hazy, and finally disappears as the system changes from two phases (as represented by the vapor-pressure curve) to a single phase (as represented by the region above C). Further heating produces changes represented in Fig. 3.4 by a path along V_c, the line of constant molar volume corresponding to the critical volume of the fluid.

For the regions of the diagram where a single phase exists, Fig. 3.3 implies a relation connecting P, V, and T which may be expressed by the functional equation:

$$f(P, V, T) = 0$$

This means that an *equation of state* exists relating pressure, molar or specific volume, and temperature for any pure homogeneous fluid in equilibrium states.

The simplest equation of state is for an ideal gas, $PV = RT$, a relation which has approximate validity for the low-pressure gas region of Fig. 3.3 and which is discussed in detail in Sec. 3.3.

An equation of state may be solved for any one of the three quantities P, V, or T as a function of the other two. For example if V is considered a function of T and P, then $V = V(T, P)$, and

$$dV = \left(\frac{\partial V}{\partial T}\right)_P dT + \left(\frac{\partial V}{\partial P}\right)_T dP \tag{3.1}$$

The partial derivatives in this equation have definite physical meanings and are measurable quantities. They are related to two properties that are commonly tabulated for liquids:

1. *Volume expansivity*

$$\beta \equiv \frac{1}{V}\left(\frac{\partial V}{\partial T}\right)_P \tag{3.2}$$

2. *Isothermal compressibility*

$$\kappa \equiv -\frac{1}{V}\left(\frac{\partial V}{\partial P}\right)_T \tag{3.3}$$

Combination of Eqs. (3.1) through (3.3) provides the equation

$$\frac{dV}{V} = \beta \, dT - \kappa \, dP \tag{3.4}$$

The isotherms for the liquid phase on the left side of Fig. 3.3 are very steep and closely spaced. Thus both $(\partial V/\partial P)_T$ and $(\partial V/\partial T)_P$ and hence both β and κ are small. This characteristic behavior of liquids (outside the region of the critical point) suggests an idealization, commonly employed in fluid mechanics and known as the *incompressible fluid*, for which β and κ are both zero. No real fluid is in fact incompressible, but the idealization is nevertheless useful, because it often provides a sufficiently realistic model of liquid behavior for practical purposes. There is no PVT equation of state for an incompressible fluid, because V is independent of T and P.

For real liquids β and κ are weak functions of temperature and pressure. Thus for small changes in T and P little error is introduced if we regard them as constant. Then Eq. (3.4) may be integrated to give

$$\ln \frac{V_2}{V_1} = \beta(T_2 - T_1) - \kappa(P_2 - P_1) \tag{3.5}$$

This is a less restrictive approximation than the assumption of an incompressible fluid.

Example 3.1   For acetone at $20°C$ and 1 bar,

$$\beta = 1.487 \times 10^{-3} \, °C^{-1}$$
$$\kappa = 62 \times 10^{-6} \, bar^{-1}$$
$$V = 1.287 \, cm^3 \, g^{-1}$$

Find:

(a) The value of $(\partial P / \partial T)_V$.

(b) The pressure generated when acetone is heated at constant volume from $20°C$ and 1 bar to $30°C$.

(c) The volume change when acetone is changed from $20°C$ and 1 bar to $0°C$ and 10 bar.

SOLUTION   (a) The derivative $(\partial P / \partial T)_V$ is determined by application of Eq. (3.4) to the case for which V is constant and $dV = 0$:

$$\beta \, dT - \kappa \, dP = 0 \qquad (const \; V)$$

or

$$\left(\frac{\partial P}{\partial T} \right)_V = \frac{\beta}{\kappa} = \frac{1.487 \times 10^{-3}}{62 \times 10^{-6}} = 24 \, bar \, °C^{-1}$$

(b) If β and κ are assumed constant in the $10°C$ temperature interval, then the equation derived in (a) may be written ($V = const$)

$$\Delta P = \frac{\beta}{\kappa} \Delta T = (24)(10) = 240 \, bar$$

and

$$P_2 = P_1 + \Delta P = 1 + 240 = 241 \, bar$$

(c) Direct substitution into Eq. (3.5) gives

$$\ln \frac{V_2}{V_1} = (1.487 \times 10^{-3})(-20) - (62 \times 10^{-6})(9) = -0.0303$$

Whence

$$\frac{V_2}{V_1} = 0.9702$$

and

$$V_2 = (0.9702)(1.287) = 1.249 \, cm^3 \, g^{-1}$$

which gives

$$\Delta V = V_2 - V_1 = 1.249 - 1.287 = -0.038 \, cm^3 \, g^{-1}$$

3.2 Virial Equations

Figure 3.2 indicates the complexity of the PVT behavior of a pure substance and suggests the difficulty of its description by an equation. However, for the gas region by itself relatively simple equations often suffice. For an isotherm such as T_1 we note from Fig. 3.3 that as P increases V decreases. The PV product for a gas or vapor should therefore be much more nearly constant than either of its members, and hence more easily represented. For example, PV along an isotherm may be expresed by a power series expansion in P:

$$PV = a + bP + cP^2 + \cdots$$

If we let $b = aB'$, $c = aC'$, etc., this equation becomes

$$PV = a(1 + B'P + C'P^2 + D'P^3 + \cdots) \tag{3.6}$$

where a, B', C', etc., are constants for a given temperature and a given chemical species.

In principle, the right-hand side of Eq. (3.6) is an infinite series. However, in practice a finite number of terms is used. In fact, PVT data show that at low pressures truncation after two terms provides satisfactory results. In general, the greater the pressure range, the larger the number of terms required.

Parameters B', C', etc., are functions of temperature and the identity of the chemical species; parameter a, however, is the same function of temperature for all species. This is shown by data taken for various gases at a specific constant temperature (fixed by use of a reproducible state such as the triple point of water or the normal boiling point of water). The results plotted as PV vs. P in Fig. 3.5 have the same limiting value of PV for all gases as $P \to 0$. In the limit as $P \to 0$, Eq. (3.6) becomes

$$\lim_{P \to 0} (PV) \equiv (PV)^* = a$$

Thus, a is the same for all gases and depends on temperature only:

$$(PV)^* = a = f(T)$$

It is this property of gases that makes them valuable in thermometry, for the limiting values of $(PV)^*$ are used to establish a temperature scale which is independent of the identity of the gas used as thermometric fluid. One need only fix the form of the functional relationship $f(T)$ and define a quantitative scale; both steps are completely arbitrary. The simplest procedure, and the one adopted internationally, is:

1. Fix the functional relationship so that $(PV)^*$ is directly proportional to T,

$$(PV)^* = a = RT \tag{3.7}$$

   where R is the proportionality constant.

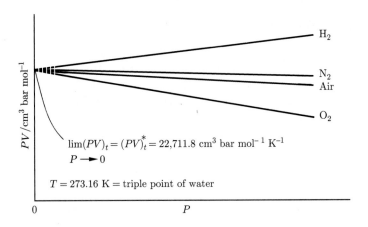

Figure 3.5: The limit of PV as $P \to 0$ is independent of the gas.

2. Assign a value of 273.16 K to the temperature of the triple point of water,

$$(PV)_t^* = R \times 273.16 \text{ K} \tag{3.8}$$

where the subscript t denotes the value at the triple point of water.

Division of Eq. (3.7) by Eq. (3.8) gives

$$\frac{(PV)^*}{(PV)_t^*} = \frac{T/\text{K}}{273.16 \text{ K}}$$

or

$$\boxed{T/\text{K} = 273.16 \frac{(PV)^*}{(PV)_t^*}} \tag{3.9}$$

Equation (3.9) establishes the Kelvin temperature scale throughout the temperature range for which limiting values of PV as $P \to 0$ [values of $(PV)^*$] are experimentally accessible.

The state of a gas at the limiting condition where $P \to 0$ deserves some discussion. As the pressure on a gas is decreased, the individual molecules become more and more widely separated. The volume of the molecules themselves becomes a smaller and smaller fraction of the total volume occupied by the gas. Furthermore, the forces of attraction between molecules become ever smaller because of the increasing distances between them. In the limit, as the pressure approaches zero, the molecules are separated by infinite distances. Their volumes become negligible compared with the total volume of the gas, and the intermolecular forces approach zero. At these conditions all gases are said to be ideal, and the temperature scale established by Eq. (3.9) is known as the ideal-gas temperature scale. The proportionality constant R in Eq. (3.7) is called the *universal gas constant*. Its numerical value is determined by means of Eq. (3.8) from experimental PVT data:

$$R = \frac{(PV)_t^*}{273.16 \text{ K}}$$

Since PVT data cannot in fact be taken at zero pressure, data taken at finite pressures are extrapolated to the zero-pressure state. The currently accepted value of $(PV)_t^*$ is 22,711.8 cm^3 bar mol^{-1}. Figure 3.5 shows how this determination is made. It leads to the following value of R:

$$R = \frac{22{,}711.8 \text{ cm}^3 \text{ bar mol}^{-1}}{273.16 \text{ K}} = 83.1447 \text{ cm}^3 \text{ bar mol}^{-1} \text{ K}^{-1}$$

Through the use of conversion factors, R may be expressed in various units. Commonly used values are given by Table A.2 of App. A.

With the establishment of the ideal-gas temperature scale, the constant a in Eq. (3.6) may be replaced by RT, in accord with Eq. (3.7). Thus Eq. (3.6) becomes

$$Z \equiv \frac{PV}{RT} = 1 + B'P + C'P^2 + D'P^3 + \cdots \tag{3.10}$$

where the dimensionless ratio PV/RT is called the *compressibility factor* and is given the symbol Z. An alternative expression for Z, also in common use, is

$$Z = 1 + \frac{B}{V} + \frac{C}{V^2} + \frac{D}{V^3} + \cdots \tag{3.11}$$

Both of these equations are known as *virial expansions*, and the parameters B', C', D', etc., and B, C, D, etc., are called *virial coefficients*. Parameters B' and B are second virial coefficients; C' and C are third virial coefficients; etc. For a given gas the virial coefficients are functions of temperature only.

The two sets of coefficients in Eqs. (3.10) and (3.11) are related as follows:

$$B' = \frac{B}{RT}$$

$$C' = \frac{C - B^2}{(RT)^2}$$

$$D' = \frac{D - 3BC + 2B^3}{(RT)^3}$$

$$\text{etc.}$$

The first step in the derivation of these relations is elimination of P on the right-hand side of Eq. (3.10) through use of Eq. (3.11). The resulting equation is a power series in $1/V$ which is compared term by term with Eq. (3.11). This comparison provides the equations relating the two sets of virial coefficients. They hold exactly only for the two virial expansions as infinite series. For the truncated forms of the virial equations treated in Sec. 3.4, these relations are only approximate.

Many other equations of state have been proposed for gases, but the virial equations are the only ones having a firm basis in theory. The methods of statistical mechanics allow derivation of the virial equations and provide physical significance to the virial coefficients. Thus, for the expansion in $1/V$, the term B/V arises on account of interactions between pairs of molecules (see Sec. 3.9); the C/V^2 term, on account of three-body interactions; etc. Since two-body interactions are many times more common than three-body interactions, and three-body interactions are many times more numerous than four-body interactions, etc., the contributions to Z of the successively higher-ordered terms decrease rapidly.

3.3   The Ideal Gas

Since the terms B/V, C/V^2, etc., of the virial expansion [Eq. (3.11)] arise on account of molecular interactions, the virial coefficients B, C, etc., would be zero if no such interactions existed. The virial expansion would then reduce to

$$Z = 1 \qquad \text{or} \qquad PV = RT$$

For a real gas, molecular interactions *do* exist, and exert an influence on the observed behavior of the gas. As the pressure of a real gas is reduced at constant temperature, V increases and the contributions of the terms B/V, C/V^2, etc., decrease. For a pressure approaching zero, Z approaches unity, not because of any change in the virial coefficients, but because V becomes infinite. Thus in the limit as the pressure approaches zero, the equation of state assumes the same simple form as for the hypothetical case of $B = C = \cdots = 0$; that is

$$Z = 1 \qquad \text{or} \qquad PV = RT$$

We know from the phase rule that the internal energy of a real gas is a function of pressure as well as of temperature. This pressure dependency arises as a result of forces between the molecules. If such forces did not exist, no energy would be required to alter the average intermolecular distance, and therefore no energy would be required to bring about volume and pressure changes in a gas at constant temperature. We conclude that in the absence of molecular interactions, the internal energy of a gas depends on temperature only. These considerations of the behavior of a hypothetical gas in which no molecular forces exist and of a real gas in the limit as pressure approaches zero lead to the definition of an *ideal gas* as one whose macroscopic behavior is characterized by:

1. The equation of state,

$$\boxed{PV = RT} \tag{3.12}$$

2. An internal energy that is a function of temperature only, and as a result of Eq. (2.19) a heat capacity C_V which is also a function of temperature only.

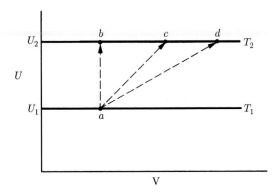

Figure 3.6: Internal energy changes for an ideal gas.

The ideal gas is a model fluid that is useful because it is described by simple equations that are frequently applicable as good approximations for actual gases. In engineering calculations, gases at pressures up to a few bars may often be considered ideal. The remainder of this section is therefore devoted to the development of thermodynamic relationships for ideal gases.

The Constant-Volume (Isochoric) Process

The equations which apply to a mechanically reversible constant-volume process were developed in Sec. 2.10. No simplification results for an ideal gas. Thus for one mole

$$dU = dQ = C_V dT \qquad (3.13)$$

For a finite change,

$$\Delta U = Q = \int C_V dT \qquad (3.14)$$

Since both the internal energy and C_V of an ideal gas are functions of temperature only, ΔU for an ideal gas may *always* be calculated by $\int C_V dT$, regardless of the kind of process causing the change. This is demonstrated in Fig. 3.6, which shows a graph of internal energy as a function of molar volume with temperature as parameter. Since U is independent of V at constant temperature, a plot of U vs. V at constant temperature is a horizontal line. For different temperatures, U has different values, with a separate line for each temperature. Two such lines are shown in Fig. 3.6, one for temperature T_1 and one for a higher temperature T_2. The dashed line connecting points a and b represents a constant-volume process for which the temperature increases from T_1 to T_2 and the internal energy changes by $\Delta U = U_2 - U_1$. This change in internal energy is given by Eq. (3.14) as $\Delta U = \int C_V dT$. The dashed lines connecting points a and c and points a and d represent other processes not occurring at constant volume but which also lead from an initial temperature T_1 to a final temperature T_2. The graph clearly shows that the change in U for these processes is the same as for the constant-volume process, and it is therefore given by the same equation, namely, $\Delta U = \int C_V dT$. However, ΔU is *not* equal to Q for these processes, because Q depends not only on T_1 and T_2 but also on the path of the process.

The Constant-Pressure (Isobaric) Process

The equations which apply to a mechanically reversible, constant-pressure nonflow process were developed in Sec. 2.10. For one mole,

$$dH = dQ = C_P dT \qquad (3.15)$$

and

$$\Delta H = Q = \int C_P dT \qquad (3.16)$$

Because the internal energy of an ideal gas is a function of temperature only, both enthalpy and C_P also depend on temperature alone. This is evident from the definition $H = U + PV$, or $H = U + RT$ for an ideal gas, and from Eq. (2.20). Just as $\Delta U = \int C_V dT$ for any process involving an ideal gas, so $\Delta H = \int C_P dT$ not only for constant-pressure processes but for *all* finite processes.

The expressions for dU and dH and the definition of enthalpy imply a simple relationship between C_P and C_V for an ideal gas; since

$$dH = dU + R\, dT \qquad \text{(ideal gas)}$$

then from Eqs. (3.13) and (3.15),

$$C_P dT = C_V dT + R\, dT$$

and

$$\boxed{C_P = C_V + R} \qquad (3.17)$$

This equation does *not* imply that C_P and C_V are themselves constant for an ideal gas, but only that they vary with temperature in such a way that their difference is equal to the constant R.

The Constant-Temperature (Isothermal) Process

The internal energy of an ideal gas cannot change in an isothermal process. Thus for one mole of an ideal gas in any nonflow process,

$$dU = dQ + dW = 0$$

and

$$Q = -W$$

For a mechanically reversible nonflow process and with $P = RT/V$, we have immediately that

$$Q = -W = \int P\, dV = \int RT \frac{dV}{V}$$

Integration at constant temperature from the initial volume V_1 to the final volume V_2 gives

$$Q = -W = RT \ln \frac{V_2}{V_1} \qquad (3.18)$$

Since $P_1/P_2 = V_2/V_1$ for the isothermal process, Eq. (3.18) may also be written

$$Q = -W = RT \ln \frac{P_1}{P_2} \qquad (3.19)$$

The Reversible Adiabatic Process

An adiabatic process is one for which there is no heat transfer between the system and its surroundings; that is, $dQ = 0$. Therefore, application of the first law to one mole of an ideal gas in mechanically reversible nonflow processes gives

$$dU = dW = -P \, dV$$

Since the change in internal energy for any process involving an ideal gas is given by Eq. (3.13), this becomes

$$C_V \, dT = -P \, dV$$

Substituting RT/V for P and rearranging, we get

$$\frac{dT}{T} = -\frac{R}{C_V} \frac{dV}{V} \qquad (3.20)$$

If the ratio of heat capacitites C_P/C_V is designated by γ, then in view of Eq. (3.17),

$$\gamma = \frac{C_V + R}{C_V} = 1 + \frac{R}{C_V}$$

or

$$\frac{R}{C_V} = \gamma - 1 \qquad (3.21)$$

Substitution in Eq. (3.20) gives

$$\frac{dT}{T} = -(\gamma - 1) \frac{dV}{V}$$

If γ is constant,[1] integration yields

$$\ln \frac{T_2}{T_1} = -(\gamma - 1) \ln \frac{V_2}{V_1}$$

---

[1] The assumption that γ is constant for an ideal gas is equivalent to the assumption that the heat capacities themselves are constant. This is the only way that the ratio $C_P/C_V \equiv \gamma$ and the difference $C_P - C_V = R$ can *both* be constant. However, since both C_P and C_V increase with temperature, their ratio γ is less sensitive to temperature than the heat capacities themselves.

or

$$\frac{T_2}{T_1} = \left(\frac{V_1}{V_2}\right)^{\gamma-1} \qquad (3.22)$$

This equation relates temperature and volume for a mechanically reversible adiabatic process involving an ideal gas with constant heat capacities. The analogous relationships between temperature and pressure and between pressure and volume can be obtained from Eq. (3.22) and the ideal-gas equation. Since $P_1V_1/T_1 = P_2V_2/T_2$, we may eliminate V_1/V_2 from Eq. (3.22), obtaining

$$\frac{T_2}{T_1} = \left(\frac{P_2}{P_1}\right)^{(\gamma-1)/\gamma} \qquad (3.23)$$

A comparison of Eqs. (3.22) and (3.23) shows that

$$\left(\frac{V_1}{V_2}\right)^{\gamma-1} = \left(\frac{P_2}{P_1}\right)^{(\gamma-1)/\gamma}$$

or

$$P_1V_1^\gamma = P_2V_2^\gamma = PV^\gamma = \text{const} \qquad (3.24)$$

The work of an adiabatic process may be obtained from the relation

$$dW = dU = C_V dT \qquad (3.25)$$

If C_V is constant, integration gives

$$W = \Delta U = C_V \,\Delta T \qquad (3.26)$$

Alternative forms of Eq. (3.26) are obtained if C_V is eliminated by Eq. (3.21):

$$W = C_V \,\Delta T = \frac{R \,\Delta T}{\gamma - 1} = \frac{RT_2 - RT_1}{\gamma - 1}$$

Since $RT_1 = P_1V_1$ and $RT_2 = P_2V_2$, this expression may also be written

$$W = \frac{P_2V_2 - P_1V_1}{\gamma - 1} \qquad (3.27)$$

Equations (3.25), (3.26), and (3.27) are general for an adiabatic process, whether reversible or not. However, V_2 is usually not known, and is eliminated from Eq. (3.27) by Eq. (3.24), valid only for mechanically reversible processes. This leads to the expression

$$W = \frac{P_1V_1}{\gamma - 1}\left[\left(\frac{P_2}{P_1}\right)^{(\gamma-1)/\gamma} - 1\right] = \frac{RT_1}{\gamma - 1}\left[\left(\frac{P_2}{P_1}\right)^{(\gamma-1)/\gamma} - 1\right] \qquad (3.28)$$

The same result is obtained when the relation between P and V given by Eq. (3.24) is used for integration of the expression $W = -\int P\,dV$.

Equations (3.22) through (3.28) are for ideal gases with constant heat capacities. Equations (3.22), (3.23), (3.24), and (3.28) also require the process to be mechanically reversible. Processes which are adiabatic but not mechanically reversible are *not* described by these equations.

When applied to real gases, Eqs. (3.22) through (3.28) often yield satisfactory approximations, provided the deviations from ideality are relatively small. For monatomic gases, $\gamma = 1.67$; approximate values of γ are 1.4 for diatomic gases and 1.3 for simple polyatomic gases such as CO_2, SO_2, NH_3, and CH_4.

The Polytropic Process

The *general* equations applying to one mole of an ideal gas undergoing a mechanically reversible nonflow process are:

$$dU = dQ + dW \qquad \Delta U = Q + W \qquad \text{(first law)}$$

$$dW = -P\,dV \qquad W = -\int P\,dV$$

$$dU = C_V\,dT \qquad \Delta U = \int C_V\,dT$$

$$dH = C_P\,dT \qquad \Delta H = \int C_P\,dT$$

Values for Q cannot be determined directly, but are obtained from the first law, with the work obtained from the integral $\int P\,dV$. However, evaluation of this integral requires specification of a P vs. V relation. The *polytropic process* is usually defined as one for which this relation is given by

$$PV^\delta = K$$

where K is a constant for any given process. With this relation between P and V, Eqs. (3.22), (3.23), and (3.24) can be rewritten with γ replaced by δ, and evaluation of $\int P\,dV$ yields Eq. (3.28) with γ replaced by δ:

$$W = \frac{RT_1}{\delta - 1}\left[\left(\frac{P_2}{P_1}\right)^{(\delta-1)/\delta} - 1\right] \tag{3.29}$$

Moreover, for constant heat capacities, the first law solved for Q yields

$$Q = \frac{(\delta - \gamma)RT_1}{(\delta - 1)(\gamma - 1)}\left[\left(\frac{P_2}{P_1}\right)^{(\delta-1)/\delta} - 1\right] \tag{3.30}$$

For particular values of δ, Eqs. (3.29) and (3.30) reduce to the following cases for an ideal gas:

$$\delta = 0 \qquad \text{Isobaric}$$
$$\delta = 1 \qquad \text{Isothermal}$$
$$\delta = \gamma \qquad \text{Adiabatic}$$
$$\delta = \infty \qquad \text{Isochoric}$$

The equations developed in this section have been *derived* for mechanically reversible, nonflow processes involving ideal gases. However, those equations which relate state functions only are valid for ideal gases regardless of the process. They apply equally to reversible and irreversible flow and nonflow processes, because changes in state functions depend only on the initial and final states of the system. On the other hand, an equation for Q or W is specific to the process considered in its derivation.

The work of an *irreversible* process is calculated by a two-step procedure. First, W is determined for a mechanically reversible process that accomplishes the same change of state. Second, this result is multiplied or divided by an efficiency to give the actual work. If the process produces work, the value for the reversible process is numerically too large and must be multiplied by an efficiency. If the process requires work, the value for the reversible process is too small and must be divided by an efficiency.

Applications of the concepts and equations developed in this section are illustrated in the examples that follow. In particular, the work of irreversible processes is treated in Example 3.3.

Example 3.2   Air is compressed from an initial condition of 1 bar and 25°C to a final state of 5 bar and 25°C by three different mechanically reversible processes:

(a) Heating at constant volume followed by cooling at constant pressure.

(b) Isothermal compression.

(c) Adiabatic compression followed by cooling at constant volume.

At these conditions, air may be considered an ideal gas with the constant heat capacities, $C_V = (5/2)R$ and $C_P = (7/2)R$.

Calculate the work required, heat transferred, and the changes in internal energy and enthalpy of the air for each process.

SOLUTION   In each case the system is taken as 1 mol of air, contained in an imaginary frictionless piston/cylinder arrangement. For $R = 8.314$ J mol^{-1} K^{-1},

$$C_V = 20.785 \quad \text{and} \quad C_P = 29.099 \text{ J mol}^{-1} \text{ K}^{-1}$$

The initial and final conditions of the air are identical with those of Example 2.12. It was shown there that

$$V_1 = 0.02479 \quad \text{and} \quad V_2 = 0.004958 \text{ m}^3$$

(a) This part of the problem is idential with part (b) of Example 2.12. However, it may now be solved in a simpler manner. The temperature at the end of the constant-volume heating step was calculated in Example 2.12 as 1,490.75 K. Also for this step $W = 0$ and therefore

$$Q = \Delta U = C_V \, \Delta T = 24{,}788 \text{ J}$$

Moreover,

$$\Delta H = C_P \, \Delta T = (29.099)(1{,}490.75 - 298.15) = 34{,}703 \text{ J}$$

For the second step at constant pressure, Eq. (3.16) yields

$$Q = \Delta H = C_P \, \Delta T = (29.099)(298.15 - 1{,}490.75) = -34{,}703 \text{ J}$$

$$\Delta U = C_V \, \Delta T = (20.785)(298.15 - 1{,}490.75) = -24{,}788 \text{ J}$$

and

$$W = \Delta U - Q = -24{,}788 - (-34{,}703) = 9{,}915 \text{ J}$$

For the entire process,

$$\begin{aligned}
\Delta U &= 24{,}788 - 24{,}788 = 0 \\
\Delta H &= 34{,}703 - 34{,}703 = 0 \\
Q &= 24{,}788 - 34{,}703 = -9{,}915 \text{ J} \\
W &= 9{,}915 + 0 = 9{,}915 \text{ J}
\end{aligned}$$

(b) For the isothermal compression of an ideal gas,

$$\Delta U = \Delta H = 0$$

Equation (3.19) gives

$$Q = -W = RT \ln \frac{P_1}{P_2} = (8.314)(298.15) \ln \frac{1}{5} = -3{,}990 \text{ J}$$

(c) The initial adiabatic compression of the air takes it to its final volume of 0.004958 m^3. The temperature and pressure at this point are given by Eqs. (3.22) and (3.24):

$$T_2 = T_1 \left(\frac{V_1}{V_2}\right)^{\gamma-1} = (298.15) \left(\frac{0.02479}{0.004958}\right)^{0.4} = 567.57 \text{ K}$$

and

$$P_2 = P_1 \left(\frac{V_1}{V_2}\right)^{\gamma} = (1) \left(\frac{0.02479}{0.004958}\right)^{1.4} = 9.52 \text{ bar}$$

For this step $Q = 0$. Hence

$$\Delta U = W = C_V \, \Delta T = (20.785)(567.57 - 298.15) = 5{,}600 \text{ J}$$

and

$$\Delta H = C_P \, \Delta T = (29.099)(567.57 - 298.15) = 7{,}840 \text{ J}$$

For the second step $\Delta V = 0$; therefore

$$Q = \Delta U = C_V \, \Delta T = (20.785)(298.15 - 567.57) = -5{,}600 \text{ J}$$

and

$$\Delta H = C_P \, \Delta T = (29.099)(298.15 - 567.57) = -7{,}840 \text{ J}$$

For the entire process,

$$\begin{aligned}
\Delta U &= 5{,}600 - 5{,}600 = 0 \\
\Delta H &= 7{,}840 - 7{,}840 = 0 \\
Q &= 0 - 5{,}600 = -5{,}600 \text{ J} \\
W &= 5{,}600 + 0 = 5{,}600 \text{ J}
\end{aligned}$$

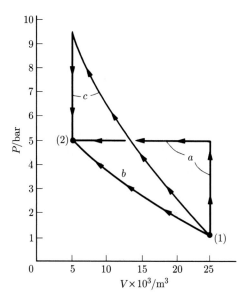

Figure 3.7: Diagram for Example 3.2.

Figure 3.7 shows these processes sketched on a PV diagram.

A comparison of the answers to the three parts of this problem shows that the property changes ΔU and ΔH are the same regardless of the path for which they are calculated. On the other hand, Q and W depend on the path.

The work for each of these mechanically reversible processes can also be calculated by $W = -\int P\,dV$. The value of this integral is proportional to the area below the curve on the PV diagram representing the process. The relative sizes of these areas correspond to the numerical values of W.

.

Example 3.3   An ideal gas undergoes the following sequence of mechanically reversible processes:

(a) From an initial state of 70°C and 1 bar, it is compressed adiabatically to 150°C.

(b) It is then cooled from 150 to 70°C at constant pressure.

(c) Finally, it is expanded isothermally to its original state.

Calculate W, Q, ΔU, and ΔH for each of the three processes and for the entire cycle. Take $C_V = (3/2)R$ and $C_P = (5/2)R$.

If these processes are carried out *irreversibly* but so as to accomplish exactly the the same *changes of state* (i.e., the same changes in P, T, U, and H), then the values of Q and W are different. Calculate values of Q and W for an efficiency of 80 percent for each step.

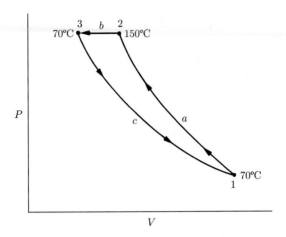

Figure 3.8: Diagram for Example 3.3.

SOLUTION From the given information, we have

$$C_V = (3/2)(8.314) = 12.471 \text{ J mol}^{-1} \text{ K}^{-1}$$

and

$$C_P = (5/2)(8.314) = 20.785 \text{ J mol}^{-1} \text{ K}^{-1}$$

The cycle is represented on a PV diagram in Fig. 3.8. Consider first the mechanically reversible operation of the cycle, and take as a basis 1 mol of gas.

(a) For an ideal gas undergoing adiabatic compression,

$$\Delta U = W = C_V \, \Delta T = (12.471)(150 - 70) = 998 \text{ J}$$

$$\Delta H = C_P \, \Delta T = (20.785)(150 - 70) = 1{,}663 \text{ J}$$

and

$$Q = 0$$

Pressure P_2 can be found from Eq. (3.23)

$$P_2 = P_1 \left(\frac{T_2}{T_1}\right)^{\gamma/(\gamma-1)} = (1) \left(\frac{150 + 273.15}{70 + 273.15}\right)^{2.5} = 1.689 \text{ bar}$$

(b) Equation (3.16) is applicable to the constant-pressure process:

$$\Delta H = Q = C_P \, \Delta T = (20.785)(70 - 150) = -1{,}663 \text{ J}$$

Also

$$\Delta U = C_V \, \Delta T = (12.471)(70 - 150) = -998 \text{ J}$$

By the first law,

$$W = \Delta U - Q = -998 - (-1{,}663) = 665 \text{ J}$$

(*c*) For ideal gases ΔU and ΔH are zero for an isothermal process. Since $P_3 = P_2$, Eq. (3.19) gives

$$Q = -W = RT \ln \frac{P_2}{P_1} = (8.314)(343.15) \ln \frac{1.689}{1} = 1{,}495 \text{ J}$$

For the entire process

$$Q = 0 - 1{,}663 + 1{,}495 = -168 \text{ J}$$
$$W = 998 + 665 - 1{,}495 = 168 \text{ J}$$
$$\Delta U = 998 - 998 + 0 = 0$$
$$\Delta H = 1{,}663 - 1{,}663 + 0 = 0$$

The property changes ΔU and ΔH both are zero for the entire cycle, because the initial and final states are identical. Note also that $Q = -W$ for the cycle. This follows from the first law with $\Delta U = 0$.

If the same changes of state are carried out by irreversible processes, the property changes for the steps are identical with those already calculated. However, the values of Q and W are different.

(*a*) This step can no longer be adiabatic. For mechanically reversible, adiabatic compression, $W = 998$ J. If the process is 80 percent efficient compared with this, then

$$W = \frac{998}{0.80} = 1{,}248 \text{ J}$$

Since ΔU is still 998 J, by the first law,

$$Q = \Delta U - W = 998 - 1{,}248 = -250 \text{ J}$$

(*b*) The work for the mechanically reversible cooling process is 665 J. For the irreversible process,

$$W = \frac{665}{0.80} = 831 \text{ J}$$

and

$$Q = \Delta U - W = -998 - 831 = -1{,}829 \text{ J}$$

(*c*) As work is done *by* the system in this step, the irreversible work is numerically less than the reversible work:

$$W = (0.80)(-1{,}495) = -1{,}196 \text{ J}$$

and

$$Q = \Delta U - W = 0 + 1{,}196 = 1{,}196 \text{ J}$$

For the entire cycle, ΔU and ΔH are again zero, but

$$Q = -250 - 1{,}829 + 1{,}198 = -883 \text{ J}$$

and

$$W = 1{,}248 + 831 - 1{,}196 = 883 \text{ J}$$

A summary of these results is given in the accompanying table. All values are in joules.

	Mechanically reversible				Irreversible			
	ΔU	ΔH	Q	W	ΔU	ΔH	Q	W
Step a	998	1,663	0	998	998	1,663	-250	1,248
Step b	-998	$-1,663$	$-1,663$	665	-998	$-1,663$	$-1,829$	831
Step c	0	0	1,495	$-1,495$	0	0	1,196	$-1,196$
Cycle	0	0	-168	168	0	0	-883	883

The cycle is one which requires work and produces an equal amount of heat. The striking feature of the comparison shown in the table is that the total work required when the cycle consists of three irreversible steps is more than five times the total work required when the steps are mechanically reversible, even though each irreversible step is assumed 80 percent efficient.

Example 3.4 A 0.4-kg mass of nitrogen at $27°C$ is held in a vertical cylinder by a frictionless piston. The weight of the piston makes the pressure of the nitrogen 0.35 bar higher than that of the surrounding atmosphere, which is at 1 bar and $27°C$. Thus the nitrogen is initially at a pressure of 1.35 bar, and is in mechanical and thermal equilibrium with its surroundings. Consider the following sequence of processes:

(a) The apparatus is immersed in an ice/water bath and is allowed to come to equilibrium.

(b) A variable force is slowly applied to the piston so that the nitrogen is compressed reversibly at the constant temperature of $0°C$ until the gas volume reaches one-half that at the end of step (a). At this point the piston is held in place by latches.

(c) The apparatus is removed from the ice/water bath and comes to thermal equilibrium in the surrounding atmosphere at $27°C$.

(d) The latches are removed, and the apparatus is allowed to return to complete equilibrium with its surroundings.

Sketch the entire cycle on a PV diagram, and calculate Q, W, ΔU^t, and ΔH^t for the nitrogen for each step of the cycle. Nitrogen may be considered an ideal gas for which $C_V = (5/2)R$ and $C_P = (7/2)R$.

SOLUTION   At the end of the cycle the nitrogen returns to its initial conditions of $27°C$ and 1.35 bar. The steps making up the cycle are

$$(a) \qquad 27°C, 1.35 \text{ bar} \xrightarrow{\text{const P}} 0°C, 1.35 \text{ bar}$$

$$(b) \qquad 0°C, V_2 \xrightarrow{\text{const T}} 0°C, V_3 = \tfrac{1}{2}V_2$$

$$(c) \qquad 0°C, V_3 \xrightarrow{\text{const V}} 27°C, V_4 = V_3$$

$$(d) \qquad 27°C, V_4 \xrightarrow{T_4 = T_1} 27°C, 1.35 \text{ bar}$$

(a) In this step, represented by the horizontal line marked a in Fig. 3.9, the nitrogen

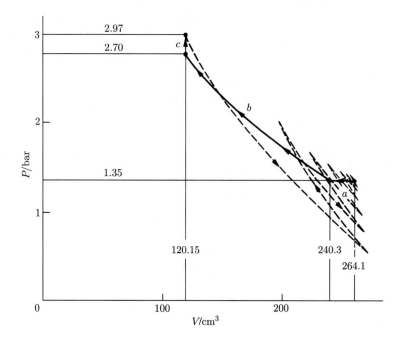

Figure 3.9: Diagram for Example 3.4.

is cooled at constant pressure. The process is mechanically reversible, even though the heat transfer occurs irreversibly as the result of a finite temperature difference. Thus for the mass m of nitrogen

$$W_a = -m \int P\,dV = -mP\,\Delta V = -\frac{mR\,\Delta T}{M}$$

With $R = 8.314$ J mol^{-1} K^{-1}, $m = 400$ g, and the molar mass (molecular weight) $M = 28$, we have

$$W_a = -\frac{(400)(8.314)(0 - 27)}{28} = 3{,}207 \text{ J}$$

and

$$Q_a = m\,\Delta H_a = mC_P\,\Delta T = (400)(7/2)(8.314/28)(0 - 27) = -11{,}224 \text{ J}$$

From the first law,

$$m\,\Delta U_a = Q_a + W_a = -11{,}224 + 3{,}207 = -8{,}017 \text{ J}$$

The internal-energy change may also be evaluated from Eq. (3.14):

$$m\,\Delta U_a = mC_V\,\Delta T = (400)(5/2)(8.314/28)(0 - 27) = -8{,}017 \text{ J}$$

(*b*) The process carried out here is an isothermal compression shown by curve *b* in Fig. 3.9. Neither the internal energy nor the enthalpy changes at constant temperature:

$$\Delta U_b = \Delta H_b = 0$$

and under conditions of mechanical reversibility,

$$Q_b = -W_b = \frac{mRT}{M} \ln \frac{V_3}{V_2} = \frac{(400)(8.314)(273.15)}{28} \ln \frac{1}{2} = -22{,}487 \text{ J}$$

(*c*) For this constant-volume process, $W_c = 0$ and, according to Eq. (3.14),

$$Q_c = m \, \Delta U_c = mC_V \, \Delta T = (400)(5/2)(8.314/28)(27 - 0) = 8{,}017 \text{ J}$$

In addition,

$$m \, \Delta H_c = mC_P \, \Delta T = (400)(7/2)(8.314/28)(27 - 0) = 11{,}224 \text{ J}$$

(*d*) The first three steps of the cycle can be sketched on a *PV* diagram without difficulty, because their paths are known. For the final step this is not possible, because the process is irreversible. When the latches holding the frictionless piston are removed, the piston moves rapidly upward, and owing to its inertia goes beyond its equilibrium position. This initial expansion is nearly equivalent to a reversible, adiabatic process, because little turbulence results from a single stroke of the piston and because heat transfer is slow. The subsequent oscillations of the piston as it gradually reaches its final equilibrium position are the primary source of the irreversibility. This process goes on for a considerable time during which heat transfer occurs in an amount sufficient to return the nitrogen to its initial temperature of 27°C at a pressure of 1.35 bar. It is not possible to specify the exact path of an irreversible process. However, the dashed lines in Fig. 3.9 indicate roughly the form that it takes.

Since the process is irreversible, the work done cannot be obtained from the integral $\int P \, dV$. Indeed, it is not possible to calculate W from the given information. During the initial expansion of the gas, the work is approximately that of a mechanically reversible adiabatic expansion. This work transfers energy from the gas to the surroundings, where it pushes back the atmosphere and increases the potential energy of the of the piston. If the piston were held at its position of maximum travel, the major part of the irreversibility would be avoided, and the work could be calculated to a good approximation by the equations for a reversible adiabatic expansion. However, as the process actually occurs, the oscillating piston causes turbulence or stirring in both the gas and the atmosphere, and there is no way to know the extent of either. This makes impossible the calculation of either Q or W.

Unlike work and heat, the property changes of the system for step *d* can be computed, since they depend solely on the initial and final states, and these are known. The internal energy and enthalpy of an ideal gas are functions of temperature only. Therefore, ΔU and ΔH are zero, because the initial and final temperatures are both 27°C. The first law applies to irreversible as well as to reversible processes, and for step *d* it becomes

$$\Delta U_d = Q_d + W_d = 0$$

or

$$Q_d = -W_d$$

Although neither Q_d nor W_d can be calculated, they clearly are equal. Step d results in net potential-energy increase because of elevation of the piston and atmosphere and a compensating decrease in the internal energy of the surrounding atmosphere.

Example 3.5 Air flows at a steady rate through a horizontal insulated pipe which contains a partly closed valve. The conditions of the air upstream from the valve are $20°C$ and 6 bar, and the downstream pressure is 3 bar. The line leaving the valve is enough larger than the entrance line so that the kinetic-energy change of the air as it flows through the valve is negligible. If air is regarded as an ideal gas, what is the temperature of the air some distance downstream from the valve?

SOLUTION Flow through a partly closed valve is known as a *throttling process*. Since flow is at a steady rate, Eq. (2.9) applies. The line is insulated, making Q small; moreover, the potential-energy and kinetic-energy changes are negligible. Since no shaft work is accomplished, $W_s = 0$. Hence, Eq. (2.9) reduces to

$$\Delta H = 0$$

Thus, for an ideal gas,

$$\Delta H = \int_{T_1}^{T_2} C_P dT = 0$$

whence

$$T_2 = T_1$$

The result that $\Delta H = 0$ is general for a throttling process, because the assumptions of negligible heat transfer and potential- and kinetic-energy changes are usually valid. If the fluid is an ideal gas, no temperature change occurs. The throttling process is inherently irreversible, but this is immaterial to the calculation: $\Delta H = \int C_P dT$ is valid for an ideal gas whatever the process.

3.4 Application of the Virial Equations

The two forms of the virial expansion given by Eqs. (3.10) and (3.11) are infinite series. For engineering purposes their use is practical only where convergence is very rapid, that is, where no more than two or three terms are required to yield reasonably close approximations to the values of the series. This is realized for gases and vapors at low to moderate pressures.

Figure 3.10 shows a compressibility-factor graph for methane. Values of the compressibility factor Z (as calculated from PVT data for methane by the defining equation $Z = PV/RT$) are plotted against pressure for various constant temperatures. The resulting isotherms show graphically what the virial expansion in P is intended to represent analytically. All isotherms originate at the value $Z = 1$ for $P = 0$. In addition the isotherms are nearly straight lines at low pressures. Thus

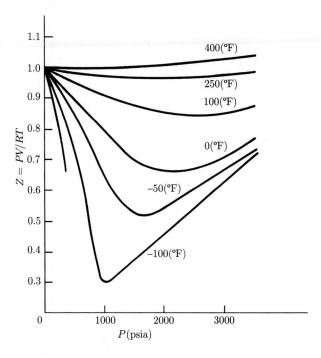

Figure 3.10: Compressibility-factor graph for methane.

the tangent to an isotherm at $P = 0$ is a good approximation of the isotherm for a finite pressure range. Differentiation of Eq. (3.10) for a given temperature gives

$$\frac{dZ}{dP} = B' + 2C'P + 3D'P^2 + \cdots$$

from which

$$\left(\frac{dZ}{dP}\right)_{P=0} = B'$$

Thus the equation of the tangent line is

$$Z = 1 + B'P$$

a result also given by truncating Eq. (3.10) to two terms. Alternatively, we may use the approximate relation $B' = B/RT$ to express the equation for Z in terms of the coefficient B:

$$\boxed{Z = \frac{PV}{RT} = 1 + \frac{BP}{RT}} \qquad (3.31)$$

Since Eq. (3.11) may also be truncated to two terms for application at low pressures,

$$Z = \frac{PV}{RT} = 1 + \frac{B}{V} \qquad (3.32)$$

a question arises as to which equation provides the better representation of low-pressure PVT data. Experience shows that Eq. (3.31) is at least as accurate as Eq. (3.32), and is much more convenient for use in most applications. Thus when the virial equation is truncated to two terms, Eq. (3.31) is preferred. This equation satisfactorily represents the PVT behavior of most vapors at subcritical temperatures up to a pressure of about 15 bar. At higher temperatures it is appropriate for gases over an increasing pressure range as the temperature increases. Values of B, the second virial coefficient, depend on the nature of the gas and on temperature. Experimental values are available for a number of gases. Moreover, estimation of second virial coefficients is possible where no data are available, as discussed in Sec. 3.6. For pressures above the range of applicability of Eq. (3.31) but below about 50 bar, the virial equation truncated to three terms usually provides excellent results. In this case Eq. (3.11), the expansion in $1/V$, is far superior to Eq. (3.10). Thus when the virial equation is truncated to three terms, the appropriate form is

$$\boxed{Z = \frac{PV}{RT} = 1 + \frac{B}{V} + \frac{C}{V^2}} \qquad (3.33)$$

This equation is explicit in pressure, but cubic in volume. Solution for V is easily done by an iterative scheme with a calculator.

Values of C, like those of B, depend on the gas and on the temperature. However, much less is known about third virial coefficients than about second virial coefficients, though data for a number of gases can be found in the literature. Since virial coefficients beyond the third are rarely known and since the virial expansion with more than three terms becomes unwieldy, virial equations of more than three terms are rarely used.

Figure 3.11 illustrates the effect of temperature on the virial coefficients B and C for nitrogen; although numerical values are different for other gases, the trends are similar. The curve of Fig. 3.11 suggests that B increases monotonically with T; however, at temperatures higher than shown B reaches a maximum and then slowly decreases. The effect of T on C is more difficult to establish experimentally, but its main features are clear: C is negative at low temperatures, passes through a maximum at a temperature near the critical, and thereafter decreases slowly with increasing T.

Example 3.6 Reported values for the virial coefficients of isopropanol vapor at 200°C are

$$B = -388 \text{ cm}^3 \text{ mol}^{-1}$$
$$C = -26{,}000 \text{ cm}^6 \text{ mol}^{-2}$$

Calculate V and Z for isopropanol vapor at 200°C and 10 bar by:

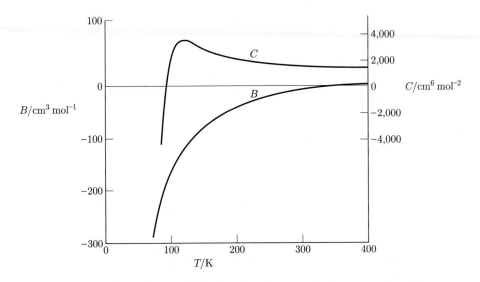

Figure 3.11: Density-series virial coefficients B and C for nitrogen.

(*a*) The ideal-gas equation.

(*b*) Equation (3.31).

(*c*) Equation (3.33).

SOLUTION   The absolute temperature is $T = 473.15$ K, and the appropriate value of the gas constant is $R = 83.14$ cm^3 bar mol^{-1} K^{-1}.

(*a*) By the ideal-gas equation,

$$V = \frac{RT}{P} = \frac{(83.14)(473.15)}{10} = 3{,}934 \text{ cm}^3 \text{ mol}^{-1}$$

and of course $Z = 1$.

(*b*) Solving Eq. (3.31) for V, we find

$$V = \frac{RT}{P} + B = 3{,}934 - 388 = 3{,}546 \text{ cm}^3 \text{ mol}^{-1}$$

Whence

$$Z = \frac{PV}{RT} = \frac{V}{RT/P} = \frac{3{,}546}{3{,}934} = 0.9014$$

(*c*) To facilitate iteration, we write Eq. (3.33) as

$$V_{i+1} = \frac{RT}{P}\left(1 + \frac{B}{V_i} + \frac{C}{V_i^2}\right)$$

where subscript i denotes the iteration number. For the first iteration, $i = 0$, and

$$V_1 = \frac{RT}{P}\left(1 + \frac{B}{V_0} + \frac{C}{V_0^2}\right)$$

where V_0 is an initial estimate of the molar volume. For this we use the ideal-gas value, which gives

$$V_1 = 3{,}934\left(1 - \frac{388}{3{,}934} - \frac{26{,}000}{(3{,}934)^2}\right) = 3{,}539$$

The second iteration depends on this result:

$$V_2 = \frac{RT}{P}\left(1 + \frac{B}{V_1} + \frac{C}{V_1^2}\right)$$

whence

$$V_2 = 3{,}934\left(1 + \frac{388}{3{,}539} - \frac{26{,}000}{(3{,}539)^2}\right) = 3{,}495$$

Iteration continues until the difference $V_{i+1} - V_i$ is insignificant, and leads after five iterations to the final value,

$$V = 3{,}488 \text{ cm}^3 \text{ mol}^{-1}$$

from which $Z = 0.8866$. In comparison with this result, the ideal-gas value is 13 percent too high and Eq. (3.31) gives a value 1.7 percent too high.

3.5 Cubic Equations of State

For an accurate description of the PVT behavior of fluids over wide ranges of temperature and pressure, an equation of state more comprehensive than the virial equation is required. Such an equation must be sufficiently general to apply to liquids as well as to gases and vapors. Yet it must not be so complex as to present excessive numerical or analytical difficulties in application.

Polynomial equations that are cubic in molar volume offer a compromise between generality and simplicity that is suitable to many purposes. Cubic equations are in fact the simplest equations capable of representing both liquid and vapor behavior. The first practical cubic equation of state was proposed by J. D. van der Waals[2] in 1873:

$$P = \frac{RT}{V - b} - \frac{a}{V^2} \tag{3.34}$$

Here, a and b are positive constants; when they are zero, the ideal-gas equation is recovered.

Given values of a and b for a particular fluid, one can calculate P as a function of V for various values of T. Figure 3.12 is a schematic PV diagram showing

---

[2]Johannes Diderik van der Waals (1837–1923), Dutch physicist who won the 1910 Nobel Prize for physics.

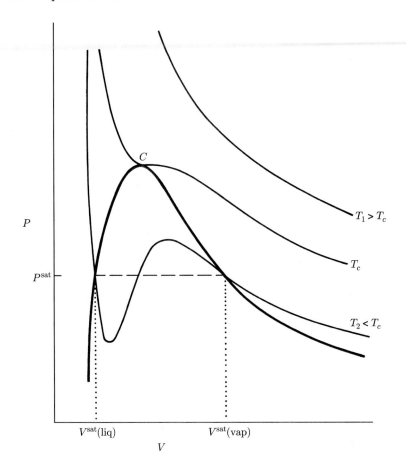

Figure 3.12: Isotherms as given by a cubic equation of state.

three such isotherms. Superimposed is the curve representing states of saturated liquid and saturated vapor. For the isotherm $T_1 > T_c$, pressure is a monotonically decreasing function with increasing molar volume. The critical isotherm (labeled T_c) contains the horizontal inflection at C characteristic of the critical point. For the isotherm $T_2 < T_c$, the pressure decreases rapidly in the liquid region with increasing V; after crossing the saturated-liquid line, it goes through a minimum, rises to a maximum, and then decreases, crossing the saturated-vapor line and continuing into the vapor region. Experimental isotherms do not exhibit this smooth transition from the liquid to the vapor region; rather, they contain a horizontal segment within the two-phase region where saturated liquid and saturated vapor coexist in varying proportions at the saturation or vapor pressure. This behavior, shown by the dashed line in Fig. 3.12, is nonanalytic, and we accept as inevitable the unrealistic behavior of equations of state in the two-phase region.

Actually, the *PV* behavior predicted in this region by proper cubic equations of state is not wholly fictitious. When the pressure is decreased on a saturated liquid devoid of vapor-nucleation sites in a carefully controlled experiment vaporization does not occur, and the liquid phase persists alone to pressures well below its vapor pressure. Similarly, raising the pressure on a saturated vapor in a suitable experiment does not cause condensation, and the vapor persists alone to pressures well above the vapor pressure. These non-equilibrium or metastable states of superheated liquid and subcooled vapor are approximated by those portions of the PV isotherm which lie in the two-phase region adjacent to the saturated-liquid and saturated-vapor states.

The modern development of cubic equations of state started in 1949 with publication of the Redlich/Kwong equation:[3]

$$P = \frac{RT}{V - b} - \frac{a}{T^{1/2}V(V + b)} \tag{3.35}$$

This equation, like other cubic equations of state, has three volume roots, of which two may be complex. Physically meaningful values of V are always real, positive, and greater than the constant b. With reference to Fig. 3.12, we see that when $T > T_c$, solution for V at any positive value of P yields only one real positive root. When $T = T_c$, this is also true, except at the critical pressure, where there are three roots, all equal to V_c. For $T < T_c$, only one real positive root exists at high pressures, but for a range of lower pressures three real positive roots exist. Here, the middle root is of no significance; the smallest root is a liquid or liquid-like volume, and the largest root is a vapor or vapor-like volume. The volumes of saturated liquid and saturated vapor are given by the smallest and largest roots when P is the saturation or vapor pressure.

Although one may solve explicitly for the roots of a cubic equation of state, in practice iterative procedures are often used.[4] These are practical only when they converge on the desired root. Complete assurance in this regard cannot be given, but the following schemes are usually effective for the Redlich/Kwong equation.

Vapor Volumes

Equation (3.35) is multiplied through by $(V - b)/P$ to give

$$V - b = \frac{RT}{P} - \frac{a(V - b)}{T^{1/2}PV(V + b)} \tag{3.36}$$

For iteration, we write

$$V_{i+1} = \frac{RT}{P} + b - \frac{a(V_i - b)}{T^{1/2}PV_i(V_i + b)} \tag{3.37}$$

---

[3]Otto Redlich and J. N. S. Kwong, *Chem. Rev.*, vol. 44, pp. 233–244, 1949.

[4]Such iterative procedures are built into computer software packages for technical calculations. With Mathcad®, for example, one can solve routinely for V in equations such as (3.35) with little thought as to how it is done.

The ideal-gas equation provides a suitable initial value, $V_0 = RT/P$.

Liquid Volumes

Equation (3.35) is put into standard polynomial form:

$$V^3 - \frac{RT}{P}V^2 - \left(b^2 + \frac{bRT}{P} - \frac{a}{PT^{1/2}}\right)V - \frac{ab}{PT^{1/2}} = 0$$

An iteration scheme results when this is written

$$V_{i+1} = \frac{1}{c}\left(V_i^3 - \frac{RT}{P}V_i^2 - \frac{ab}{PT^{1/2}}\right) \tag{3.38}$$

where

$$c = b^2 + \frac{bRT}{P} - \frac{a}{PT^{1/2}} \tag{3.39}$$

For an initial value, take $V_0 = b$.

The constants in an equation of state may of course be evaluated by a fit to available PVT data. For simple cubic equations of state, however, suitable estimates come from the critical constants T_c and P_c. Since the critical isotherm exhibits a horizontal inflection at the critical point, we may impose the mathematical conditions:

$$\left(\frac{\partial P}{\partial V}\right)_{T;cr} = \left(\frac{\partial^2 P}{\partial V^2}\right)_{T;cr} = 0$$

where the subscript cr denotes the critical point. Differentiation of Eq. (3.34) or Eq. (3.35) yields expressions for both derivatives, which may be equated to zero for $P = P_c$, $T = T_c$, and $V = V_c$. The equation of state may itself be written for the critical conditions, providing three equations in the five constants P_c, V_c, T_c, a, and b. Of the several ways to treat these equations, the most suitable is elimination of V_c to yield expressions relating a and b to P_c and T_c. The reason is that P_c and T_c are usually more accurately known than V_c. The expressions that result are

The van der Waals equation

$$a = \frac{27R^2T_c^2}{64P_c} \qquad b = \frac{RT_c}{8P_c}$$

The Redlich/Kwong equation

$$a = \frac{0.42748R^2T_c^{2.5}}{P_c} \tag{3.40}$$

$$b = \frac{0.08664RT_c}{P_c} \tag{3.41}$$

Although these equations may not yield the best possible values, they give values that are reasonable and which can almost always be determined, because critical temperatures and pressures (in contrast to extensive PVT data) are often known, or can be reliably estimated. A list of values of T_c and P_c is provided in App. B.

Since the introduction of the Redlich/Kwong equation, scores of cubic equations of state have been proposed. They are all special cases of the *generic cubic equation of state*:

$$P = \frac{RT}{V - b} - \frac{\theta(V - \eta)}{(V - b)(V^2 + \delta V + \epsilon)} \tag{3.42}$$

Here, b, θ, δ, ϵ, and η are parameters which in general depend on temperature and (for mixtures) composition. Although Eq. (3.42) appears to possess great flexibility, it has inherent limitations because of its cubic form.[5] The Redlich/Kwong equation is obtained from Eq. (3.42) with the identifications: $\theta = a/T^{1/2}$, $\eta = \delta = b$, and $\epsilon = 0$. Other common forms of Eq. (3.42) result when different identifications are made; these are most commonly used in connection with vapor/liquid equilibrium (Chapt. 13).

Equations of greater overall accuracy are necessarily more complex, as is illustrated by the Benedict/Webb/Rubin equation:

$$P = \frac{RT}{V} + \frac{B_0 RT - A_0 - C_0/T^2}{V^2} + \frac{bRT - a}{V^3}$$

$$+ \frac{a\alpha}{V^6} + \frac{c}{V^3 T^2}\left(1 + \frac{\gamma}{V^2}\right)\exp\frac{-\gamma}{V^2} \tag{3.43}$$

where A_0, B_0, C_0, a, b, c, α, and γ are all constant for a given fluid. This equation and its modifications, despite their complexity, are used in the petroleum and natural-gas industries for light hydrocarbons and a few other commonly encountered gases.

Example 3.7   Given that the vapor pressure of methyl chloride at 60°C is 13.76 bar, use the Redlich/Kwong equation to estimate the molar volumes of saturated vapor and saturated liquid at these conditions.

SOLUTION   We evaluate the constants a and b by Eqs. (3.40) and (3.41) with values of T_c and P_c taken from App. B:

$$a = \frac{(0.42748)(83.14)^2(416.3)^{2.5}}{66.80} = 1.5641 \times 10^8 \text{ cm}^6 \text{ bar mol}^{-2} \text{ K}^{1/2}$$

and

$$b = \frac{(0.08664)(83.14)(416.3)}{66.80} = 44.891 \text{ cm}^3 \text{ mol}^{-1}$$

For evaluation of the molar volume of saturated vapor, we substitute known values into Eq. (3.37); this gives

$$V_{i+1} = 2{,}057.83 - \frac{622{,}768}{V_i}\left(\frac{V_i - 44.891}{V_i + 44.891}\right)$$

[5]M. M. Abbott, *AIChE J.*, vol. 19, pp. 596–601, 1973; *Adv. in Chem. Series 182*, K. C. Chao and R. L. Robinson, Jr., eds., pp. 47–70, Am. Chem. Soc., Washington, D.C., 1979.

Iteration starts with $V_i = V_0 = RT/P \approx 2{,}000$ cm^3 mol^{-1}, and continues to convergence on the value

$$V = 1{,}713 \text{ cm}^3 \text{ mol}^{-1}$$

The experimental result is $1{,}635.6$ cm^3 mol^{-1}.

For evaluation of the molar volume of saturated liquid, we substitute known values into Eqs. (3.38) and (3.39); the resulting equation is

$$V_{i+1} = \frac{V_i^3 - 2{,}012.94 V_i^2 - 2.79567 \times 10^7}{-530{,}390}$$

Iteration starts with $V_i = V_0 = b \approx 45$ cm^3 mol^{-1}, and continues to convergence on the value

$$V = 71.34 \text{ cm}^3 \text{ mol}^{-1}$$

The experimental result is 60.37 cm^3 mol^{-1}.

Roots of the Redlich/Kwong equation are much more easily found with a software package such as Mathcad® or Maple®, in which iteration is an integral part of the equation-solving routine. Starting values or bounds are required, and must be appropriate to the particular root of interest. The Mathcad® program for solving Example 3.7 is given in App. D.2.

3.6 Generalized Correlations for Gases

An alternative form of the Redlich/Kwong equation is obtained by multiplication of Eq. (3.35) by V/RT:

$$Z = \frac{1}{1-h} - \frac{a}{bRT^{1.5}}\left(\frac{h}{1+h}\right)$$

where

$$h \equiv \frac{b}{V} = \frac{b}{ZRT/P} = \frac{bP}{ZRT}$$

Elimination of a and b in these equation by Eqs. (3.40) and (3.41) gives

$$Z = \frac{1}{1-h} - \frac{4.9340}{T_r^{1.5}}\left(\frac{h}{1+h}\right) \tag{3.44a}$$

$$h \equiv \frac{0.08664 P_r}{Z T_r} \tag{3.44b}$$

where $T_r \equiv T/T_c$ and $P_r \equiv P/P_c$ are called *reduced temperature* and *reduced pressure*. This pair of equations is arranged for convenient iterative solution for the compressibility factor Z for any gas at any conditions T_r and P_r. For an initial value of $Z = 1$, h is calculated by Eq. (3.44b). With this value of h, Eq. (3.44a) yields a new value of Z for substitution into Eq. (3.44b). This procedure is continued until a new iteration produces a change in Z less than some small preset tolerance. The procedure does not converge for liquids.

Equations of state which express Z as a function of T_r and P_r are said to be *generalized*, because of their general applicability to all gases. Any equation of state can be put into this form, thus providing a generalized correlation for the properties of fluids. Such a correlation has the advantage of allowing the estimation of property values from very limited information. One needs only the critical temperature and critical pressure of the fluid. This is the basis for the two-parameter *theorem of corresponding states*:

> All fluids, when compared at the same reduced temperature and reduced pressure, have approximately the same compressibility factor, and all deviate from ideal-gas behavior to about the same degree.

Although use of an equation based on the two-parameter theorem of corresponding states provides far better results in general than the ideal-gas equation, significant deviations from experiment still exist for all but the *simple fluids* argon, krypton, and xenon. Appreciable improvement results from introduction of a third corresponding-states parameter, characteristic of molecular structure; the most popular such parameter is the *acentric factor* ω, introduced by K. S. Pitzer and coworkers.[6]

The acentric factor for a pure chemical species is defined with reference to its vapor pressure. Since the logarithm of the vapor pressure of a pure fluid is approximately linear in the reciprocal of absolute temperature, we may write

$$\frac{d\log P_r^{\text{sat}}}{d(1/T_r)} = a$$

where P_r^{sat} is the reduced vapor presssure, T_r is the reduced temperature, and a is the slope of a plot of $\log P_r^{\text{sat}}$ vs. $1/T_r$. Note that "log" denotes a logarithm to the base 10. If the two-parameter theorem of corresponding states were generally valid, the slope a would be the same for all pure fluids. This is observed not to be true; each fluid has its own characteristic value of a, which could in principle serve as a third corresponding-states parameter. However, Pitzer noted that all vapor-pressure data for the simple fluids (Ar, Kr, Xe) lie on the same line when plotted as $\log P_r^{\text{sat}}$ vs. $1/T_r$ and that the line passes through $\log P_r^{\text{sat}} = -1.0$ at $T_r = 0.7$. This is illustrated in Fig. 3.13. Data for other fluids define other lines whose locations can be fixed in relation to the line for the simple fluids (SF) by the difference:

$$\log P_r^{\text{sat}}(\text{SF}) - \log P_r^{\text{sat}}$$

The acentric factor is defined as this difference evaluated at $T_r = 0.7$:

$$\omega \equiv -1.0 - \log(P_r^{\text{sat}})_{T_r=0.7} \tag{3.45}$$

---

[6]Fully described in K. S. Pitzer, *Thermodynamics*, 3d ed., App. 3, McGraw-Hill, New York, 1995.

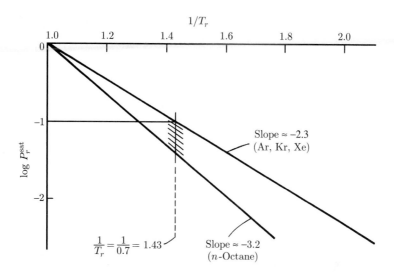

Figure 3.13: Approximate temperature dependence of the reduced vapor pressure.

Therefore ω can be determined for any fluid from T_c, P_c, and a single vapor-pressure measurement made at $T_r = 0.7$. Values of ω and the critical constants T_c, P_c, and V_c for a number of fluids are listed in App. B.

The definition of ω makes its value zero for argon, krypton, and xenon, and experimental data yield compressibility factors for all three fluids that are correlated by the same curves when Z is represented as a function of T_r and P_r. This is the basic premise of the three-parameter theorem of corresponding states:

> *All fluids having the same value of ω, when compared at the same T_r and P_r, have the same value of Z, and all deviate from ideal-gas behavior to about the same degree.*

The correlation for Z developed by Pitzer and coworkers takes the form

$$Z = Z^0 + \omega Z^1 \tag{3.46}$$

where Z^0 and Z^1 are functions of both T_r and P_r. When $\omega = 0$, as is the case for the simple fluids, the second term disappears, and Z^0 becomes identical with Z. Thus a generalized correlation for Z as a function of T_r and P_r based on data for just argon, krypton, and xenon provides the relationship $Z^0 = F^0(T_r, P_r)$. By itself, this represents a *two*-parameter corresponding-states correlation for Z. Since the second term of Eq. (3.46) is a relatively small correction to this correlation, its omission does not introduce large errors, and a correlation for Z^0 may be used alone for quick but less precise estimates of Z than are obtained from a three-parameter correlation.

Equation (3.46) is a simple linear relation between Z and ω for given values of T_r and P_r. Experimental data for Z for non-simple fluids plotted vs. ω at

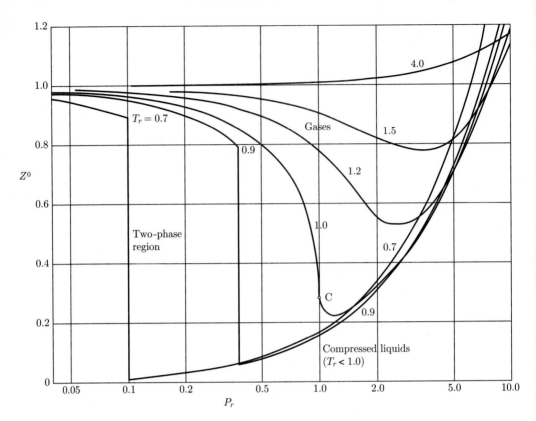

Figure 3.14: The Lee/Kesler correlation for $Z^0 = F^0(T_r, P_r)$.

constant T_r and P_r do indeed yield approximately straight lines, and their slopes provide values for Z^1 from which the generalized function $Z^1 = F^1(T_r, P_r)$ can be constructed.

Of the generalized three-parameter correlations available, the one developed by Lee and Kesler[7] has found greatest favor. Although its development is based on a modified form of the Benedict/Webb/Rubin equation of state, it takes the form of tables which present values of Z^0 and Z^1 as functions of T_r and P_r. These are given in App. E as Tables E.1 through E.4. The nature of the correlation is indicated by Fig. 3.14, a plot of Z^0 vs. P_r for six isotherms.

The Lee/Kesler correlation provides reliable results for gases which are non-polar or only slightly polar; for these, errors of no more than 2 or 3 percent are indicated. When applied to highly polar gases or to gases that associate, larger

[7]B. I. Lee and M. G. Kesler, *AIChE J.*, vol. 21, pp. 510–527, 1975.

errors can be expected. The quantum gases (e.g., hydrogen, helium, and neon) do not conform to the same corresponding-states behavior as normal fluids. Their treatment by the usual correlations is sometimes accommodated by use of *effective* critical constants.[8]

The tabular nature of the generalized compressibility-factor correlation is a disadvantage, but the complexity of the functions Z^0 and Z^1 precludes their accurate representation by simple equations. However, we can give approximate analytical expression to these functions for a limited range of pressures. The basis for this is Eq. (3.31), the simplest form of the virial equation, which may be written

$$Z = 1 + \frac{BP}{RT} = 1 + \left(\frac{BP_c}{RT_c}\right)\frac{P_r}{T_r} \tag{3.47}$$

Thus, Pitzer and coworkers proposed a second correlation, which expresses the quantity BP_c/RT_c as

$$\frac{BP_c}{RT_c} = B^0 + \omega B^1 \tag{3.48}$$

Combining Eqs. (3.47) and (3.48) gives

$$Z = 1 + B^0\frac{P_r}{T_r} + \omega B^1\frac{P_r}{T_r}$$

Comparison of this equation with Eq. (3.46) provides the following identifications:

$$Z^0 = 1 + B^0\frac{P_r}{T_r} \tag{3.49}$$

and

$$Z^1 = B^1\frac{P_r}{T_r}$$

Second virial coefficients are functions of temperature only, and similarly B^0 and B^1 are functions of reduced temperature only. They are well represented by the following equations:

$$B^0 = 0.083 - \frac{0.422}{T_r^{1.6}} \tag{3.50}$$

$$B^1 = 0.139 - \frac{0.172}{T_r^{4.2}} \tag{3.51}$$

The simplest form of the virial equation has validity only at low to moderate pressures where Z is linear in pressure. The generalized virial-coefficient correlation is therefore useful only at low to moderate reduced pressures where Z^0 and Z^1 are at least approximately linear functions of reduced pressure. In Fig. 3.15 we present a graph which compares the linear relation of Z^0 to P_r as given by Eqs. (3.49) and

[8]J. M. Prausnitz, R. N. Lichtenthaler, and E. G. de Azevedo, *Molecular Thermodynamics of Fluid-Phase Equilibria*, pp. 165–168, Prentice-Hall, Englewood Cliffs, NJ, 1986.

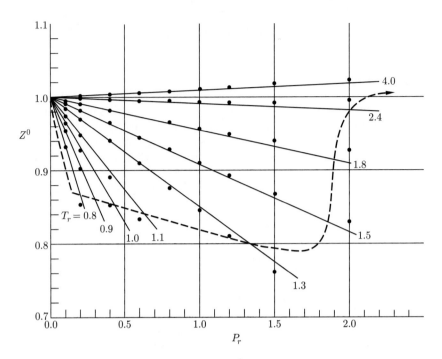

Figure 3.15: Comparison of correlations for Z^0. The straight lines represent the virial-coefficient correlation; the points, the Lee/Kesler correlation. In the region above the dashed line the two correlations differ by less than 2 %.

(3.50) with values of Z^0 from the Lee/Kesler compressibility-factor correlation. The range of values of T_r and P_r within which the correlations differ by less than two percent lies above the dashed line drawn on the figure. For reduced temperatures greater than $T_r \approx 3$, there appears to be no limitation on the pressure. For lower values of T_r the allowable pressure range decreases with decreasing temperature. A point is reached, however, at $T_r \approx 0.7$ where the pressure range is limited by the saturation pressure. This is indicated by the left-most segment of the dashed line. The minor contributions of Z^1 to the correlations are here neglected. In view of the uncertainty associated with any generalized correlation, deviations of no more than 2 percent in Z^0 are not significant.

The relative simplicity of the generalized virial-coefficient correlation does much to recommend it. Moreover, the temperatures and pressures of most chemical-processing operations lie within the region where it does not deviate by a significant amount from the compressibility-factor correlation. Like the parent correlation, it is most accurate for nonpolar species and least accurate for highly polar and associating molecules.

Example 3.8 Determine the molar volume of *n*-butane at 510 K and 25 bar by each of the following:

(*a*) The ideal-gas equation.

(b) The generalized compressibility-factor correlation.

(c) The generalized virial-coefficient correlation.

SOLUTION   (a) By the ideal-gas equation,

$$V = \frac{RT}{P} = \frac{(83.14)(510)}{25} = 1{,}696.1 \text{ cm}^3 \text{ mol}^{-1}$$

(b) Taking values of T_c and P_c from App. B, we find

$$T_r = \frac{510}{425.1} = 1.200 \qquad P_r = \frac{25}{37.96} = 0.659$$

Interpolation in Tables E.1 and E.2 then provides

$$Z^0 = 0.865 \qquad Z^1 = 0.038$$

Thus, by Eq. (3.46) with $\omega = 0.200$,

$$Z = Z^0 + \omega Z^1 = 0.865 + (0.200)(0.038) = 0.873$$

and

$$V = \frac{ZRT}{P} = \frac{(0.873)(83.14)(510)}{25} = 1{,}480.7 \text{ cm}^3 \text{ mol}^{-1}$$

If we take $Z = Z^0 = 0.865$, in accord with the two-parameter corresponding states correlation, then $V = 1{,}467.1 \text{ cm}^3 \text{ mol}^{-1}$, which is less than 1 percent lower than the value given by the three-parameter correlation.

(c) Values of B^0 and B^1 are given by Eqs. (3.50) and (3.51):

$$B^0 = -0.232 \qquad B^1 = 0.059$$

By Eq. (3.48),

$$\frac{BP_c}{RT_c} = B^0 + \omega B^1 = -0.232 + (0.200)(0.059) = -0.220$$

Then by Eq. (3.47),

$$Z = 1 + (-0.220)\frac{0.659}{1.200} = 0.879$$

from which we find $V = 1{,}489.1 \text{ cm}^3 \text{ mol}^{-1}$, a value less than 1 percent higher than that given by the compressibility-factor correlation. For comparison, the experimental value is 1,480.7.

Example 3.9 What pressure is generated when 1(lb mol) of methane is stored in a volume of 2(ft)^3 at 122($^\circ$F)? Base calculations on each of the following:

(a) The ideal-gas equation.

(b) The Redlich/Kwong equation.

(c) A generalized correlation.

SOLUTION   (*a*) By the ideal-gas equation,

$$P = \frac{RT}{V} = \frac{(0.7302)(122 + 459.67)}{2} = 212.4(\text{atm})$$

(*b*) For the Redlich/Kwong equation, values of *a* and *b* come from Eqs. (3.40) and (3.41):

$$a = \frac{(0.42748)(0.7302)^2(343.1)^{2.5}}{45.4} = 10{,}945.4(\text{atm})(\text{ft})^6(\text{R})^{1/2}$$

and

$$b = \frac{(0.08664)(0.7302)(343.1)}{45.4} = 0.4781(\text{ft})^3$$

where values of T_c and P_c from App. B have been converted to (R) and (atm). Substitution of known values into Eq. (3.35) now gives

$$P = \frac{(0.7302)(581.67)}{2 - 0.4781} - \frac{10{,}945.4}{(581.67)^{1/2}(2)(2 + 0.4781)} = 187.5(\text{atm})$$

(*c*) Since the pressure here is high, the generalized compressibility-factor correlation is the proper choice. In the absence of a known value for P_r, an iterative procedure is based on the following equation:

$$P = \frac{ZRT}{V} = \frac{Z(0.7302)(581.67)}{2} = 212.4Z$$

Since $P = P_c P_r = 45.4P_r$, this equation becomes

$$Z = \frac{45.4P_r}{212.4} = 0.2138P_r$$

or

$$P_r = \frac{Z}{0.2138}$$

One now assumes a starting value for Z, say $Z = 1$. This gives $P_r = 4.68$, and allows a new value of Z to be calculated by Eq. (3.46) from values interpolated in Tables E.3 and E.4 at the reduced temperature of $T_r = 581.67/343.1 = 1.695$. With this new value of Z, a new value of P_r is calculated, and the procedure continues until no significant change occurs from one step to the next. The final value of Z so found is 0.889 at $P_r = 4.14$. This may be confirmed by substitution into Eq. (3.46) of values for Z^0 and Z^1 from Tables E.3 and E.4 interpolated at $P_r = 4.14$ and $T_r = 1.695$. Since $\omega = 0.012$, we have

$$Z = Z^0 + \omega Z^1 = 0.887 + (0.012)(0.258) = 0.890$$

and

$$P = \frac{ZRT}{V} = \frac{(0.890)(0.7302)(581.67)}{2} = 189.0(\text{atm})$$

Since the acentric factor is small, the two- and three-parameter compressibility-factor correlations are little different. Both the Redlich/Kwong equation and the generalized compressibility-factor correlation give answers very close to the experimental value of 185(atm). The ideal-gas equation yields a result that is high by 14.6 percent.

Example 3.10 A mass of 500 g of gaseous ammonia is contained in a 30,000-cm³ vessel immersed in a constant-temperature bath at 65°C. Calculate the pressure of the gas by each of the following:

(a) The ideal-gas equation.

(b) The Redlich/Kwong equation.

(c) A generalized correlation.

SOLUTION The molar volume of ammonia in the vessel is given by

$$V = \frac{V^t}{n} = \frac{V^t}{m/M}$$

where n is the number of moles, m is the mass of ammonia in the vessel of total volume V^t, and M is the molar mass of ammonia. Thus

$$V = \frac{30,000}{500/17.02} = 1,021.2 \text{ cm}^3 \text{ mol}^{-1}$$

(a) By the ideal-gas equation,

$$P = \frac{RT}{V} = \frac{(83.14)(65 + 273.15)}{1,021.2} = 27.53 \text{ bar}$$

(b) For application of the Redlich/Kwong equation, we first evaluate a and b by Eqs. (3.40) and (3.41):

$$a = \frac{(0.42748)(83.14)^2(405.7)^{2.5}}{112.8} = 8.684 \times 10^7 \text{ bar cm}^6 \text{ K}^{1/2}$$

and

$$b = \frac{(0.08664)(83.14)(405.7)}{112.8} = 25.91 \text{ cm}^3$$

where values of T_c and P_c are from App. B. Substitution of known values into Eq. (3.36) now gives

$$P = \frac{(83.14)(338.15)}{1,021.2 - 25.9} - \frac{8.684 \times 10^7}{(338.15)^{1/2}(1,021.2)(1,021.2 + 25.9)} = 23.84 \text{ bar}$$

(c) Since the reduced pressure is low ($\simeq 0.2$), we use the generalized virial-coefficient correlation. For a reduced temperature of $T_r = 338.15/405.7 = 0.834$, values of B^0 and B^1 as given by Eqs. (3.50) and (3.51) are

$$B^0 = -0.482 \qquad B^1 = -0.232$$

Substitution into Eq. (3.48) with $\omega = 0.253$ yields

$$\frac{BP_c}{RT_c} = -0.482 + (0.253)(-0.232) = -0.541$$

and

$$B = \frac{-0.541 RT_c}{P_c} = \frac{-(0.541)(83.14)(405.7)}{112.8} = -161.8 \text{ cm}^3 \text{ mol}^{-1}$$

Solving Eq. (3.31) for P, we obtain

$$P = \frac{RT}{V - B} = \frac{(83.14)(338.15)}{1{,}021.2 + 161.8} = 23.76 \text{ bar}$$

An iterative solution is not necessary, because B is independent of pressure. The calculated P corresponds to a reduced pressure of $P_r = 23.76/112.8 = 0.211$, and reference to Fig. 3.15 confirms the suitability of the generalized virial-coefficient correlation.

Experimental data indicate that the pressure is 23.82 bar at the given conditions. Thus the ideal-gas equation yields an answer that is high by about 15 percent, whereas the other two methods give answers in substantial agreement with experiment, even though ammonia is a polar molecule.

3.7 Generalized Correlations for Liquids

Although the molar volumes of liquids can be calculated by means of generalized cubic equations of state, the results are often not of high accuracy. However, the Lee/Kesler correlation is equally suitable for both liquids and gases. Figure 3.14 illustrates curves for both phases, and the values of Tables E.1 through E.4 cover both phases. We note again, however, that this correlation is most suitable for nonpolar and slightly polar fluids.

In addition, generalized equations are available for the estimation of molar volumes of *saturated* liquids. The simplest equation, proposed by Rackett,[9] is an example:

$$V^{\text{sat}} = V_c Z_c^{(1-T_r)^{0.2857}} \tag{3.52}$$

The only data required are the critical constants, given in App. B. Results are usually accurate to 1 or 2 percent.

Lydersen, Greenkorn, and Hougen[10] developed a general method for estimation of liquid volumes, based on the principle of corresponding states. It applies to liquids just as the two-parameter compressibility-factor correlation applies to gases, but is based on a correlation of reduced density as a function of reduced temperature and pressure. Reduced density is defined as

$$\rho_r \equiv \frac{\rho}{\rho_c} = \frac{V_c}{V} \tag{3.53}$$

where ρ_c is the density at the critical point. The generalized correlation is shown in Fig. 3.16. This figure may be used directly with Eq. (3.53) for determination of

---

[9]H. G. Rackett, *J. Chem. Eng. Data*, vol. 15, pp. 514–517, 1970; see also C. F. Spencer and S. B. Adler, *ibid.*, vol. 23, pp. 82–89, 1978, for a review of available equations.

[10]A. L. Lydersen, R. A. Greenkorn, and O. A. Hougen, "Generalized Thermodynamic Properties of Pure Fluids," *Univ. Wisconsin, Eng. Expt. Sta. Rept. 4*, 1955.

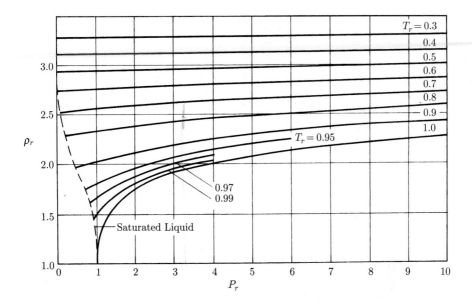

Figure 3.16: Generalized density correlation for liquids.

liquid volumes if the value of the critical volume is known. A better procedure is to make use of a single known liquid volume (state 1) by the identity,

$$V_2 = V_1 \frac{\rho_{r_1}}{\rho_{r_2}} \tag{3.54}$$

where $\qquad V_2 =$ required volume
$\qquad\qquad V_1 =$ known volume
$\qquad \rho_{r_1},\, \rho_{r_2} =$ reduced densities read from Fig. 3.16

This method gives good results and requires only experimental data that are usually available. Figure 3.16 makes clear the increasing effects of both temperature and pressure on liquid density as the critical point is approached.

Correlations for the molar densities as functions of temperature are given for many pure liquids by Daubert and coworkers.[11]

Example 3.11 For ammonia at 310 K, estimate the density of

(*a*) The saturated liquid.

(*b*) The density of the liquid at 100 bar.

SOLUTION (*a*) We apply the Rackett equation at the reduced temperature,

$$T_r = \frac{310}{405.7} = 0.7641$$

---

[11]T. E. Daubert, R. P. Danner, H. M. Sibul, and C. C. Stebbins, *Physical and Thermodynamic Properties of Pure Chemicals: Data Compilation*, Taylor & Francis, Bristol, PA, extant 1995.

With $V_c = 72.47$ and $Z_c = 0.242$ (from App. B), we get

$$V^{\text{sat}} = V_c Z_c^{(1-T_r)^{0.2857}} = (72.47)(0.242)^{(0.2359)^{0.2857}} = 28.33 \text{ cm}^3 \text{ mol}^{-1}$$

This compares with the experimental value of $29.14 \text{ cm}^3 \text{ mol}^{-1}$, and is in error by 2.7 percent.

(b) The reduced conditions are

$$T_r = 0.764 \qquad P_r = \frac{100}{112.8} = 0.887$$

From Fig. 3.16, we have $\rho_r = 2.38$. Substituting this value along with V_c into Eq. (3.53) gives

$$V = \frac{V_c}{\rho_r} = \frac{72.47}{2.38} = 30.45 \text{ cm}^3 \text{ mol}^{-1}$$

In comparison with the experimental value of $28.6 \text{ cm}^3 \text{ mol}^{-1}$, this result is in error by 6.5 percent.

If we start with the experimental value of $29.14 \text{ cm}^3 \text{ mol}^{-1}$ for saturated liquid at 310 K, Eq. (3.54) may be used. For the saturated liquid at $T_r = 0.764$, we find from Fig. 3.16 that $\rho_{r_1} = 2.34$. Substitution of known values into Eq. (3.54) gives

$$V_2 = V_1 \frac{\rho_{r_1}}{\rho_{r_2}} = (29.14)\left(\frac{2.34}{2.38}\right) = 28.65 \text{ cm}^3 \text{ mol}^{-1}$$

This result is in essential agreement with the experimental value.

Direct application of the Lee/Kesler correlation with values of Z^0 and Z^1 interpolated from Tables E.1 and E.2 leads to a value of $33.87 \text{ cm}^3 \text{ mol}^{-1}$, which is significantly in error, no doubt owing to the highly polar nature of ammonia.

3.8   Molecular Theory of Fluids

Classical thermodynamics is a deductive science, in which the general features of macroscopic-system behavior follow from a few laws and postulates. However, the practical application of thermodynamics requires values for the properties of individual chemical species and their mixtures. These may be presented either as numerical data (e.g., the steam tables for water) or as correlating equations (e.g., the PVT equation of state and expressions for the temperature dependence of ideal-gas heat capacities).

The usual source of property values is experiment. For example, the ideal-gas equation of state evolved as a statement of observed volumetric behavior of gases at low pressures. Similarly, the rule of thumb that $C_P \approx 29 \text{ J mol}^{-1} \text{ K}^{-1}$ for diatomic gases at normal temperatures is based on experimental observation. However, macroscopic experiments provide no insight into why substances exhibit their observed property values. The basis for insight is a microscopic view of matter.

A central dogma of modern physics is that matter is *particulate*. The quest for the ultimate elementary particles is still in progress, but for engineering purposes we may adopt the following picture: ordinary matter consists of molecules; molecules consist of atoms; and atoms consist of a positively charged nucleus (comprising neutrons and protons), surrounded by negatively charged electrons. Atoms and molecules with equal numbers of electrons and protons have no net charge and are neutral.

Molecules are small and light: typical linear dimensions are 10^{-10} to 10^{-8} m, and typical masses are 10^{-27} to 10^{-25} kg. Hence the number of molecules in a macroscopic system is enormous. For example, one mole of matter contains 6.022×10^{23} molecules (Avogadro's number). Because of these features—smallness, lightness, and numerical abundance—the proper description of behavior at the molecular level and its extrapolation to a macroscopic scale require the special methods of quantum mechanics and statistical mechanics. We pursue neither of these topics here. Instead, we present material useful for relating molecular concepts to observed thermodynamic properties.

Intermolecular Forces and the Pair-Potential Function

We noted in Sec. 3.3 that although an ideal gas is characterized by the absence of molecular interactions, it still possesses internal energy. This energy is associated with the individual molecules, and results from their motion. Real gases and other fluids are comprised of molecules that have not only the energy of individual molecules, but also energy that is shared among them because of intermolecular forces. This *intermolecular potential energy* is associated with *collections* of molecules, and is the form of energy that reflects the existence of such forces. Well established is the fact that two molecules attract each other when they are far apart and repel one another when close together. Electromagnetic in origin, intermolecular forces represent interactions among the charge distributions of neighboring molecules.

Figure 3.17 is a sketch of the intermolecular potential energy $\mathcal{U}$ for an isolated pair of spherically symmetric neutral molecules, for which $\mathcal{U}$ depends only on the distance between the molecular centers, i.e., on the intermolecular separation r. (More generally, $\mathcal{U}$ is also a function of the relative orientations of the two molecules.) The *intermolecular force* F is proportional to the r-derivative of $\mathcal{U}$:

$$F(r) = -\frac{d\mathcal{U}(r)}{dr}$$

By convention, a positive F represents an intermolecular repulsion, and a negative F an intermolecular attraction. Hence (see Fig. 3.17) molecules repel each other at small separations, and attract each other at modest-to-large separations.

An algebraic expression for the *pair-potential function* $\mathcal{U}$ is one of the tools of the trade of the molecular scientist or engineer. The methods of statistical mechanics provide for its relation to both thermodynamic and transport properties. Shown in Fig. 3.17 are specific values for $\mathcal{U}$ and r that may appear as species-dependent parameters in a pair-potential function.

The *hard-core diameter d* is a measure of the center-to-center separation for which $\mathcal{U}$, and hence F, becomes infinite. It is not subject to precise determination, but plays the role of a modeling parameter in some expressions for $\mathcal{U}$. The *collision*

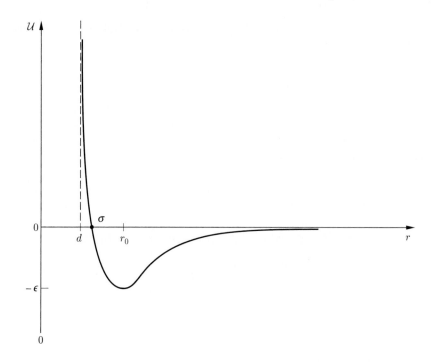

Figure 3.17: The intermolecular potential energy $\mathcal{U}$ for a pair of structureless, neutral molecules. (Not to scale.)

diameter σ is defined as the separation for which $\mathcal{U} = 0$. The *equilibrium separation* r_0 is the separation for which $\mathcal{U}$ attains its minimum value of $-\epsilon$. At $r = r_0$, the net intermolecular force is zero. Quantity ϵ is called the *well depth*. For a particular class of chemical species (e.g., noble gases, or cyclic alkanes), each of these special quantities increases with increasing molecular size. Typical ranges of values for σ and ϵ are $\sigma \approx 3$ to 8×10^{-10} m and $\epsilon \approx 0.1$ to 1.0×10^{-20} J. Commonly, r_0 is about 10 to 15% greater than σ.

Scores of expressions have been proposed for $\mathcal{U}$.[12] All are essentially empirical, although their functional forms often have some basis in theory. The most widely used is the *Lennard-Jones* (LJ) *12/6 pair-potential function*:

$$\mathcal{U}(\text{LJ}) = 4\epsilon\left[\left(\frac{\sigma}{r}\right)^{12} - \left(\frac{\sigma}{r}\right)^{6}\right] \tag{3.55}$$

Equation (3.55) provides semiquantitative representations of thermodynamic and transport properties for nonpolar substances of relatively simple molecular structure. In Eq. (3.55), the r^{-12} term is supposed to represent bimolecular repulsions,

---

[12]For a compilation of expressions for $\mathcal{U}$ see G. C. Maitland, M. Rigby, E. B. Smith, and W. A. Wakeham, *Intermolecular Forces: Their Origin and Determination*, App. 1, Clarendon Press, Oxford, 1981.

Table 3.1: Bimolecular potential energy as a function of separation for the Lennard-Jones 12/6 potential.

r/σ	$\mathcal{U}/\epsilon$
0.0000	∞
0.6279	1000
0.7521	100
0.8797	10
0.9691	1
1.0000	0
1.0267	-0.5
1.1225	-1
1.8413	-0.1
2.7133	-0.01
3.9841	-0.001
∞	0

and the r^{-6} term bimolecular attractions. Although the attraction term has significant theoretical justification, the r^{-12} dependence for repulsions is chosen primarily for mathematical convenience. Table 3.1 shows some values of the dimensionless potential energy $\mathcal{U}/\epsilon$ as a function of the dimensionless separation r/σ, as implied by Eq. (3.55). Clearly, the effects of intermolecular forces are significant only over a modest range of separations. Even though they operate over an infinite range ($\mathcal{U}$ and F together approach zero only in the limit as $r \to \infty$), both $\mathcal{U}$ and F for neutral molecular pairs are numerically negligible for separations greater than about 10 molecular diameters.

Contributions to the Intermolecular Potential Energy [13]

We now consider briefly the *origins* of intermolecular forces. An essential concept here is that *a molecule is a distribution of charges*: a collection of positively charged nuclei, surrounded by a cloud of negatively charged electrons. Intermolecular repulsions at sufficiently small separations therefore result from the overlap of electron clouds of interacting molecules, giving rise to a Coulombic repulsion. At still smaller separations the positively charged nuclei "see" each other, again promoting repulsion.

The origins of intermolecular attractions are less obvious. Here, several mechanisms can contribute. First, consider the electrostatic interaction of two rigid charge distributions A and B. By Coulomb's law, the electrostatic potential energy

---

[13] A comprehensive discussion of these contributions to $\mathcal{U}(r)$ is given by J. O. Hirschfelder, C. F. Curtiss, and R. B. Bird, *Molecular Theory of Gases and Liquids*, pp. 25–30, 209, 983–1020, John Wiley and Sons, New York, 1954.

of interaction $\mathcal{U}(\text{el})$ is

$$\mathcal{U}(\text{el}) = \frac{1}{4\pi\epsilon_0} \sum_i^A \sum_j^B \frac{q_i\, q_j}{r_{ij}} \tag{3.56}$$

Here, q_i is a charge in distribution A, q_j is a charge in distribution B, and r_{ij} is the separation between charges q_i and q_j. Quantity ϵ_0 is the *electric permittivity of vacuum*, a physical constant[14] equal to 8.85419×10^{-12} C V^{-1} m^{-1}. (The unit of electric charge is the coulomb C, and the unit of electric potential difference is the volt V.) The sums are taken over all charges in the distributions.

Equation (3.56) is exact, but awkward to use as it stands. Application is facilitated when the charge separation r_{ij} is replaced by the center-of-mass separation r of the two distributions. This is done by series-expansion techniques; the result is written in symbolic form as

$$\mathcal{U}(\text{el}) = \frac{1}{4\pi\epsilon_0} \left(\frac{\beta_1}{r} + \frac{\beta_2}{r^2} + \frac{\beta_3}{r^3} + \cdots \right)$$

The coefficients $\beta_1, \beta_2, \beta_3, \ldots$ depend upon details of the charge distributions, and upon the relative orientations of the distributions. For neutral molecules, coefficients β_1 and β_2 are zero. Since the β_3/r^3 term involves orientations of the distributions, which change continually, the contributions from all orientations must be averaged. A statistical averaging procedure ("Boltzmann averaging"), when applied to the β_3/r^3 term, yields the following approximate expression for $\mathcal{U}(\text{el})$ for two neutral rigid charge distributions:[15]

$$\mathcal{U}(\text{el}) = -\frac{2}{3}\frac{\mu_A^2\mu_B^2}{kT(4\pi\epsilon_0)^2}\frac{1}{r^6} \tag{3.57}$$

Quantity k is *Boltzmann's constant*, equal to 1.381×10^{-23} J K^{-1}; μ_A and μ_B are the permanent *dipole moments* for charge distributions A and B associated with the molecules. This contribution to the pair-potential function vanishes only when one of the permanent dipole moments is zero.

The averaging procedure which leads to Eq. (3.57) produces several remarkable results:

1. Even though the distributions are electrically neutral, there is a net *attraction* between them.

2. The original r^{-3} dependence of $\mathcal{U}(\text{el})$ becomes on averaging an r^{-6} dependence.

3. As given by Eq. (3.57) $\mathcal{U}(\text{el})$ varies with T^{-1}. Hence the magnitude of the permanent-dipole interaction decreases with increasing temperature.

---

[14]Unrelated to the well depth ϵ of Fig. 3.17.

[15]This result is only valid for modest dipole moments. For a discussion, see T. M. Reed and K. E. Gubbins, *Applied Statistical Mechanics*, sec. 5-7, McGraw-Hill, New York, 1973.

Equation (3.57) is the simplest example of a direct electrostatic potential for two neutral molecules; here, the dipole moment emerges as an important physical property. Dipole moments are measures of the net separation of charge within a molecule. For a spherically symmetric neutral charge distribution (e.g., an atom of argon), μ is zero. For a molecule in which the charge $+|q|$ is separated from charge $-|q|$ by distance l, the dipole moment is

$$\mu = |q|\, l$$

Hence μ has dimensions of charge $\times$ length; its SI unit is the coulomb(C)·meter(m). However, values are usually reported in *debyes* (D); 1 D $= 3.3357 \times 10^{-30}$ C m. A molecule with a nonzero dipole moment is called *polar*. Water ($\mu = 1.9$ D), acetone ($\mu = 2.9$ D), and acetonitrile ($\mu = 4.0$ D) are strongly polar molecules. Carbon monoxide ($\mu = 0.1$ D), propylene ($\mu = 0.4$ D), and toluene ($\mu = 0.4$ D) are slightly polar. Carbon dioxide, neon, nitrogen, and n-octane are nonpolar ($\mu = 0$).

The dipole moments just discussed are *permanent* dipole moments, intrinsic properties of a molecule. A net separation of charge may also be *induced* in any molecule by application of an external electric field. The *induced dipole moment* μ(ind) so created is approximately proportional to the strength of the applied field. Thus, for a molecule A,

$$\mu_A(\text{ind}) = \alpha_A E$$

where E is the applied field strength and α_A is the *polarizability* of A. If the source of the electric field is a permanent dipole in a neighboring molecule B, then the contribution to $\mathcal{U}$ from the permanent dipole/induced dipole interaction is

$$\mathcal{U} = -\frac{\mu_B^2 \alpha_A}{(4\pi\epsilon_0)^2}\frac{1}{r^6}$$

If molecules A and B are both polar (μ_A, $\mu_B \neq 0$), then the complete expression for the potential energy of induction $\mathcal{U}$(ind) is[16]

$$\mathcal{U}(\text{ind}) = -\frac{(\mu_A^2 \alpha_B + \mu_B^2 \alpha_A)}{(4\pi\epsilon_0)^2}\frac{1}{r^6} \tag{3.58}$$

The polarizability α, like the permanent dipole moment μ, is an intrinsic property of a molecule. SI units for α are C m^2 V^{-1}, but values are usually reported for the quantity $\hat{\alpha} \equiv \alpha/4\pi\epsilon_0$, in cm^3. The volumetric units for $\hat{\alpha}$ suggest a possible connection between polarizability and molecular volume. Typically, $\hat{\alpha}$ increases with molecular volume: very roughly, $\hat{\alpha} \approx 0.05\,\sigma^3$, where σ is the molecular collision diameter. Hence $\hat{\alpha}$ normally falls in the range of about 1 to 25×10^{-24} cm^3.

We have so far discussed two types of dipole (and dipole moment): *permanent* and *induced*. Both can be rationalized and treated by the methods of classical electrostatics, and both produce a contribution to $\mathcal{U}$ proportional to r^{-6}. There

---

[16]See Reed and Gubbins, *op. cit.*

is yet a third kind of dipole, an *instantaneous dipole*, whose calculation requires the methods of quantum mechanics. However, its existence can be rationalized on semi-classical grounds. If we picture a molecule A as nuclei with orbiting (i.e., *moving*) electrons, then we can imagine that a snapshot might show an instantaneous but temporary net separation of molecular charge. This is manifested as an instantaneous dipole, which induces a dipole in a neighboring molecule B. Interaction of the dipoles results in the intermolecular *dispersion force*, with corresponding dispersion potential $\mathcal{U}(\text{disp})$ given for large separations as

$$\mathcal{U}(\text{disp}) = -\frac{3}{2}\left(\frac{I_A I_B}{I_A + I_B}\right)\frac{\alpha_A \alpha_B}{(4\pi\epsilon_0)^2}\frac{1}{r^6} \tag{3.59}$$

Here, I is the *first ionization potential*, the energy required to remove one electron from a neutral molecule. Typically, I is of magnitude 1 to 4×10^{-18} J. All molecules possess nonzero ionization potentials and polarizabilities; hence all molecular pairs experience the dispersion interaction.

The dispersion potential $\mathcal{U}(\text{disp})$, like $\mathcal{U}(\text{el})$ and $\mathcal{U}(\text{ind})$, varies as r^{-6}. When molecules A and B are of the same kind, we may take these three special results as providing justification for the r^{-6} attraction term in empirical intermolecular potential functions such as the Lennard-Jones 12/6 potential, Eq. (3.55). For identical molecules A and B, $\mu_A = \mu_B = \mu$, and Eqs. (3.57), (3.58), and (3.59) produce the expressions

$$\mathcal{U}(\text{el}) = -\frac{2}{3}\frac{\mu^4}{kT(4\pi\epsilon_0)^2}\frac{1}{r^6} \tag{3.60}$$

$$\mathcal{U}(\text{ind}) = -\frac{2\mu^2\alpha}{(4\pi\epsilon_0)^2}\frac{1}{r^6} \tag{3.61}$$

$$\mathcal{U}(\text{disp}) = -\frac{3}{4}\frac{\alpha^2 I}{(4\pi\epsilon_0)^2}\frac{1}{r^6} \tag{3.62}$$

These equations can be used to estimate the contributions of direct-electrostatic, induction, and dispersion forces to the intermolecular potential for pairs of identical molecules. Thus, if we write

$$\mathcal{U}(\text{long range}) = -\frac{C_6}{r^6}$$

then

$$C_6 = \frac{1}{(4\pi\epsilon_0)^2}\left(\frac{2}{3}\frac{\mu^4}{kT} + 2\mu^2\alpha + \frac{3}{4}\alpha^2 I\right) \tag{3.63}$$

Quantity C_6 is a measure of the strength of long-range intermolecular attractions. Fractional contributions of the three mechanisms to long-range forces are

$$f(\text{el}) = \mathcal{U}(\text{el})/\Sigma$$
$$f(\text{ind}) = \mathcal{U}(\text{ind})/\Sigma$$
$$f(\text{disp}) = \mathcal{U}(\text{disp})/\Sigma$$

where

$$\Sigma \equiv \mathcal{U}(\text{el}) + \mathcal{U}(\text{ind}) + \mathcal{U}(\text{disp})$$

and the $\mathcal{U}$'s are given by Eqs. (3.60), (3.61), and (3.62).

Values of C_6 calculated by Eq. (3.63) and the fractional contributions made by electrostatic, induction, and dispersion interactions to $\mathcal{U}$ are summarized in Table 3.2 for 15 polar substances, illustrating concepts just developed. Also shown are values of μ, $\hat{\alpha}$, and I for each species, and, in the last column, the ratio of the direct electrostatic and dispersion contributions:

$$\frac{f(\text{el})}{f(\text{disp})} \equiv \frac{\mathcal{U}(\text{el})}{\mathcal{U}(\text{disp})} = \frac{8}{9} \frac{\mu^4}{\alpha^2 I k T}$$

The dimensionless ratio $f(\text{el})/f(\text{disp})$ is a measure of the effective polarity of a species. We note the following:

1. In all cases, the *magnitude* of the dispersion interaction is substantial, even when $f(\text{disp})$ is small. These interactions can rarely be ignored.

2. The fractional contribution $f(\text{ind})$ of induction interactions is generally small, never exceeding about 7%.

3. Contributions from permanent dipoles at near-ambient temperatures, through $\mathcal{U}(\text{el})$ and $\mathcal{U}(\text{ind})$, are small (less than about 5% of the total) for values of μ less than 1 D. Hence substances such as propylene (C_3H_6) and toluene ($C_6H_5CH_3$) are commonly classified as nonpolar, even though they have significant dipole moments.

3.9 Second Virial Coefficients from Potential Functions

The molar volume of a fluid is determined by the behavior of its constituent molecules, and is therefore influenced by the forces acting between molecules. One would suppose, for example, that a gas becomes more dense as the attractive forces between molecules become stronger. The reference point is an ideal gas, for which the intermolecular forces are zero. The nature and strength of intermolecular forces in an actual gas therefore determine the departure of its molar volume from that of an ideal gas.

In the virial equation as given by Eq. (3.11), the first term on the right is unity, and by itself provides the ideal-gas value for Z. The remaining terms provide corrections to the ideal-gas value, and of these the term B/V is the most important. As the two-body-interaction term, it is evidently related to the pair-potential function discussed in the preceding section. For spherically symmetric intermolecular force fields, statistical mechanics provides an exact expression relating the second virial coefficient B to the pair-potential function $\mathcal{U}(r)$:[17]

$$B = -2\pi N_A \int_0^\infty \left(e^{-\mathcal{U}(r)/kT} - 1 \right) r^2 dr \qquad (3.64)$$

[17]D. A. McQuarrie, *Statistical Mechanics*, p. 228, Harper and Row, New York, 1976.

Table 3.2: Long-range attractions for polar molecules at 298 K.

Compound	μ/D	$\hat{\alpha}/10^{-24}$ cm^3	$I/10^{-18}$ J	$C_6/10^{-78}$ J m^6	$f(\text{el})$	$f(\text{ind})$	$f(\text{disp})$	$f(\text{el})/f(\text{disp})$
CO	0.1	2.0	2.2	6.6	2.45×10^{-5}	0.001	0.999	2.45×10^{-5}
C$_3$H$_6$	0.4	6.0	1.6	43.4	9.55×10^{-4}	0.004	0.995	9.60×10^{-4}
C$_6$H$_5$CH$_3$	0.4	12.3	1.4	159.3	2.60×10^{-4}	0.003	0.997	2.61×10^{-4}
HI	0.5	5.5	1.7	38.9	0.003	0.007	0.990	0.0026
HBr	0.8	3.6	1.9	19.6	0.034	0.024	0.942	0.0359
CHCl$_3$	1.0	9.0	1.8	112.8	0.014	0.016	0.970	0.0148
HCl	1.1	2.6	2.1	13.7	0.174	0.046	0.780	0.223
(C$_2$H$_5$)$_2$O	1.2	8.7	1.5	91.0	0.037	0.028	0.935	0.0394
NH$_3$	1.5	2.4	1.6	16.2	0.506	0.067	0.427	1.19
HF	1.9	0.8	2.5	22.9	0.922	0.025	0.053	17.6
H$_2$O	1.9	1.5	2.0	25.6	0.826	0.042	0.132	6.26
C$_5$H$_5$N	2.5	9.4	1.5	174.4	0.363	0.067	0.570	0.637
(CH$_3$)$_2$CO	2.9	6.4	1.6	174.5	0.656	0.062	0.282	2.33
HCN	3.2	2.6	2.2	186.3	0.912	0.029	0.059	15.2
CH$_3$CN	4.0	4.5	2.0	459.5	0.903	0.031	0.066	13.7

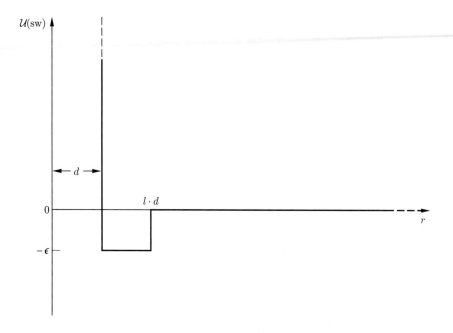

Figure 3.18: The square-well potential $\mathcal{U}(\text{sw})$ for a pair of molecules.

Quantity N_A is Avogadro's number, and $k = R/N_A$ is Boltzmann's constant. By assumption, $\mathcal{U}(r)$ depends only on the intermolecular separation r between pairs of molecules. Given an expression for the pair-potential function $\mathcal{U}(r)$, one determines $B(T)$ by evaluation of the integral in Eq. (3.64). For realistic potential functions (see Fig. 3.17), the integration must generally be done numerically or by series techniques. However, for *rectilinear* potential functions—those in which $\mathcal{U}(r)$ is defined by a collection of straight-line segments—one can obtain closed-form analytical expressions for $B(T)$.

The simplest realistic rectilinear potential function is the *square-well potential* $\mathcal{U}(\text{sw})$, shown in Fig. 3.18. It consists of four segments, producing the following piecewise contributions to $\mathcal{U}$:

$$\left.\begin{aligned}
\mathcal{U}(\text{sw}) &= \infty && \text{for} && r \leq d \\
\mathcal{U}(\text{sw}) &= -\epsilon && \text{for} && d \leq r \leq l \cdot d \\
\mathcal{U}(\text{sw}) &= 0 && \text{for} && l \cdot d \leq r
\end{aligned}\right\} \qquad (3.65)$$

Here, $d = \sigma$, and the hard-core and collision diameters are identical; ϵ is the well depth, and l is a constant which defines the width of the well. Comparison of Fig. 3.18 with Fig. 3.17 shows that $\mathcal{U}(\text{sw})$ mimics many of the features of the "true" intermolecular potential energy, for which repulsions prevail for sufficiently small separations, and attractions dominate for intermediate separations. For sufficiently large separations, $\mathcal{U}$ becomes negligible.

With $\mathcal{U}$ given by Eq. (3.65), evaluation of B by Eq. (3.64) is a straightforward exercise in integration. The result is

$$B(\text{sw}) = \frac{2}{3}\pi N_A d^3 \left[1 - (l^3 - 1)\left(e^{\epsilon/kT} - 1\right) \right] \tag{3.66}$$

where the first term in the square brackets (i.e., 1) arises from the repulsion part of the potential and the remaining term from the attraction part. Equation (3.66) therefore provides the following insight into the behavior of the second virial coefficient:

1. The sign and magnitude of B are determined by the relative contributions of attractions and repulsions.

2. At low temperatures, attractions dominate, producing negative values of B. The stronger the attractions (as determined by the magnitudes of ϵ and l), the more negative is B at fixed T.

3. At high temperatures, repulsions dominate, producing positive values of B. In the (hypothetical) limit of infinite temperature, B approaches the value

$$\lim_{T \to \infty} B(\text{sw}) = \frac{2}{3}\pi N_A d^3 = 4v_m$$

   where v_m is the molar molecular volume, the volume occupied by a mole of hard spheres of diameter d.

4. At the *Boyle temperature* T_B, the contributions of attractions are exactly balanced by those of repulsions, and B is zero. For the square-well potential, according to Eq. (3.66),

$$T_B(\text{sw}) = \frac{\epsilon/k}{\ln\left(\dfrac{l^3}{l^3 - 1}\right)}$$

Hence, the stronger the attractions, the higher is the Boyle temperature.

Although Eq. (3.66) is based on an intermolecular potential function that is in detail unrealistic, it nevertheless often provides an excellent fit of second-virial-coefficient data. An example is provided by argon, for which reliable data for B are available over a wide temperature range, from about 85 to 1,000 K.[18] The correlation of these data by Eq. (3.66) as shown in Fig. 3.19 results from the parameter values $\epsilon/k = 95.2$ K, $l = 1.69$, and $d = 3.07 \times 10^{-8}$ cm. This empirical success depends at least in part on the availability of three adjustable parameters, and is no more than a limited validation of the square-well potential. Use of this

[18] J. H. Dymond and E. B. Smith, *The Virial Coefficients of Pure Gases and Mixtures*, pp. 1–10, Clarendon Press, Oxford, 1980.

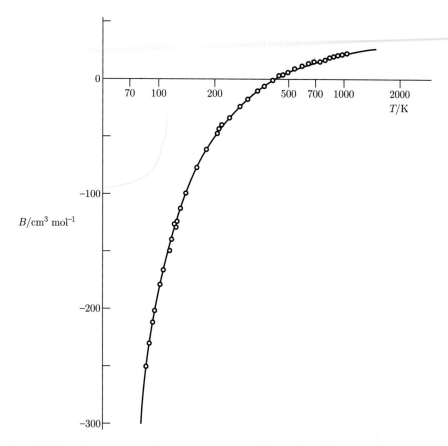

Figure 3.19: Correlation of second-virial-coefficient data for argon by the square-well potential. Circles are data; curve is given by Eq. (3.66).

potential does illustrate by a very simple calculation how the second virial coefficient (and hence the volume of a gas) may be related to molecular parameters.

Such calculations carried out for more-realistic potential functions, though of greater complexity, also lead to values for molecular parameters. For example, values of σ and ϵ for the Lennard-Jones potential [Eq. (3.55)] have been regressed from experimental volumetric data for many gases.[19] This is an essential source of values, because theory does not provide the means for their prediction. One must remember, however, that potential functions are empirical *models*, and as such are approximations. The deficiency of a model is reflected by the values of

---

[19]Values from several potential functions are given by R. L. Rowley, *Statistical Mechanics for Thermophysical Property Calculations*, App. 5, PTR Prentice Hall, Englewood Cliffs, NJ, 1994.

the molecular parameters regressed from the model, and they must therefore be regarded as *effective* values of the parameters. Different effective values result from the use of different potential functions.

Since transport properties, such as viscosity and diffusivity, are also related to these same potential functions, data for the transport properties, particularly viscosity, are also used to provide values for ϵ and σ. We note, however, that any deficiency in the potential function influences the calculation of values in different ways, and the same molecular parameters obtained from different data sources are rarely in exact agreement.

PROBLEMS

3.1. An incompressible fluid is contained in an insulated cylinder fitted with a frictionless piston. Can energy as work be transferred to the fluid? What is the change in internal energy of the fluid when the pressure is increased from P_1 to P_2?

3.2. Express the volume expansivity and the isothermal compressibility as functions of density ρ and its partial derivatives. For water at $50°C$ and 1 bar, $\kappa = 44.18 \times 10^{-6}$ bar^{-1}. To what pressure must water be compressed at $50°C$ to change its density by 1 percent? Assume that κ is independent of P.

3.3. The Tait equation for liquids is written for an isotherm as

$$V = V_0 \left(1 - \frac{AP}{B + P}\right)$$

where V is molar or specific volume, V_0 is the hypothetical molar or specific volume at zero pressure, and A and B are positive constants. Find an expression for the isothermal compressibility consistent with this equation.

3.4. For liquid water the isothermal compressibility is given by

$$\kappa = \frac{c}{V(P + b)}$$

where c and b are functions of temperature only. If 1 kg of water is compressed isothermally and reversibly from 1 to 500 bar at $60°C$, how much work is required? At $60°C$, $b = 2{,}700$ bar and $c = 0.125$ cm^3 g^{-1}.

3.5. Calculate the reversible work done in compressing 1 ft^3 of mercury at a constant temperature of $32(°F)$ from $1(\text{atm})$ to $3{,}000(\text{atm})$. The isothermal compressibility of mercury at $32(°F)$ is

$$\kappa = 3.9 \times 10^{-6} - 0.1 \times 10^{-9} P$$

where P is in (atm) and κ is in (atm)$^{-1}$.

3.6. Five kilograms of liquid carbon tetrachloride undergo a mechanically reversible, isobaric change of state at 1 bar during which the temperature changes from $0°C$ to $20°C$. Determine ΔV^t, W, Q, ΔH^t, and ΔU^t. The following properties for liquid carbon tetrachloride at 1 bar and $0°C$ may be assumed independent of temperature: $\beta = 1.2 \times 10^{-3}$ K^{-1}, $C_P = 0.84$ kJ kg^{-1} K^{-1}, and $\rho = 1{,}590$ kg m^{-3}.

3.7. A substance for which κ is a constant undergoes an isothermal, mechanically reversible process from initial state (P_1, V_1) to final state (P_2, V_2), where V is molar volume.

  (*a*) Starting with the definition of κ, show that the path of the process is described by

$$V = A\exp(-\kappa P)$$

  where A depends on T only.

  (*b*) Determine an exact expression which gives the isothermal work done on 1 mol of this constant-κ substance.

3.8. One mole of an ideal gas with $C_P = (7/2)R$ and $C_V = (5/2)R$ expands from $P_1 = 8$ bar and $T_1 = 600$ K to $P_2 = 1$ bar by each of the following paths:

  (*a*) Constant volume.

  (*b*) Constant temperature.

  (*c*) Adiabatically.

Assuming mechanical reversibility, calculate W, Q, ΔU, and ΔH for each process. Sketch each path on a single PV diagram.

3.9. An ideal gas, $C_P = (5/2)R$ and $C_V = (3/2)R$, is changed from $P = 1$ bar and $V_1^t = 12$ m^3 to $P_2 = 12$ bar and $V_2^t = 1$ m^3 by the following mechanically reversible processes:

  (*a*) Isothermal compression.

  (*b*) Adiabatic compression followed by cooling at constant pressure.

  (*c*) Adiabatic compression followed by cooling at constant volume.

  (*d*) Heating at constant volume followed by cooling at constant pressure.

  (*e*) Cooling at constant pressure followed by heating at constant volume.

Calculate Q, W, ΔU^t, and ΔH^t for each of these processes, and sketch the paths of all processes on a single PV diagram.

3.10. A rigid, nonconducting tank with a volume of 4 m^3 is divided into two unequal parts by a thin membrane. One side of the membrane, representing 1/3 of the tank, contains nitrogen gas at 6 bar and 100°C, and the other side, representing 2/3 of the tank, is evacuated. The membrane ruptures and the gas fills the tank.

  (*a*) What is the final temperature of the gas? How much work is done? Is the process reversible?

  (*b*) Describe a reversible process by which the gas can be returned to its initial state. How much work is done?

Assume nitrogen an ideal gas for which $C_P = (7/2)R$ and $C_V = (5/2)R$.

3.11. An ideal gas, initially at 30°C and 100 kPa, undergoes the following cyclic processes in a closed system:

  (*a*) In mechanically reversible processes, it is first compressed adiabatically to 500 kPa, then cooled at a constant pressure of 500 kPa to 30°C, and finally expanded isothermally to its original state.

(b) The cycle traverses exactly the same changes of state, but each step is irreversible with an efficiency of 80% compared with the corresponding mechanically reversible process.

Calculate Q, W, ΔU, and ΔH for each step of the process and for the cycle. Take $C_P = (7/2)R$ and $C_V = (5/2)R$.

3.12. One cubic meter of an ideal gas at 600 K and 1,000 kPa expands to five times its initial volume as follows:

(a) By a mechanically reversible, isothermal process.

(b) By a mechanically reversible, adiabatic process.

(c) By an adiabatic, irreversible process in which expansion is against a restraining pressure of 100 kPa.

For each case calculate the final temperature, pressure, and the work done by the gas. $C_P = 21$ J mol^{-1} K^{-1}.

3.13. One mole of air, initially at 150°C and 8 bar, undergoes the following mechanically reversible changes. It expands isothermally to a pressure such that when it is cooled at constant volume to 50°C its final pressure is 3 bar. Assuming air an ideal gas for which $C_P = (7/2)R$ and $C_V = (5/2)R$, calculate W, Q, ΔU, and ΔH.

3.14. An ideal gas flows through a horizontal tube at steady state. No heat is added and no shaft work is done. The cross-sectional area of the tube changes with length, and this causes the velocity to change. Derive an equation relating the temperature to the velocity of the gas. If nitrogen at 150°C flows past one section of the tube at a velocity of 2.5 m s^{-1}, what is its temperature at another section where its velocity is 50 m s^{-1}? $C_P = (7/2)R$.

3.15. One mole of an ideal gas, initially at 30°C and 1 bar, is changed to 130°C and 10 bar by three different mechanically reversible processes:

- The gas is first heated at constant volume until its temperature is 130°C; then it is compressed isothermally until its pressure is 10 bar.

- The gas is first heated at constant pressure until its temperature is 130°C; then it is compressed isothermally to 10 bar.

- The gas is first compressed isothermally to 10 bar; then it is heated at constant pressure to 130°C.

Calculate Q, W, ΔU, and ΔH in each case. Take $C_P = (7/2)R$ and $C_V = (5/2)R$. Alternatively, take $C_P = (5/2)R$ and $C_V = (3/2)R$.

3.16. One mole of an ideal gas, initially at 30°C and 1 bar, undergoes the following mechanically reversible changes. It is compressed isothermally to a point such that when it is heated at constant volume to 120°C its final pressure is 12 bar. Calculate Q, W, ΔU, and ΔH for the process. Take $C_P = (7/2)R$ and $C_V = (5/2)R$.

3.17. A process consists of two steps: (1) One mole of air at $T = 800$ K and $P = 4$ bar is cooled at constant volume to $T = 350$ K. (2) The air is then heated at constant pressure until its temperature reaches 800 K. If this two-step process is replaced by a single isothermal expansion of the air from 800 K and 4 bar to some final pressure P, what is the value of P that makes the work of the two processes the same? Assume mechanical reversibility and treat air as an ideal gas with $C_P = (7/2)R$ and $C_V = (5/2)R$.

3.18. A scheme for finding the internal volume V_B^t of a gas cylinder consists of the following steps. The cylinder is filled with a gas to a low pressure P_1, and connected through a small line and valve to an evacuated reference tank of known volume V_A^t. The valve is opened, and gas flows through the line into the reference tank. After the system returns to its initial temperature, a sensitive pressure transducer provides a value for the pressure change ΔP in the cylinder. Determine the cyclinder volume V_B^t from the following data:

- $V_A^t = 256$ cm^3.
- $\Delta P/P_1 = -0.0639$.

3.19. A closed, nonconducting, horizontal cylinder is fitted with a nonconducting, frictionless, floating piston which divides the cylinder into Sections A and B. The two sections contain equal masses of air, initially at the same conditions, $T_1 = 300$ K and $P_1 = 1$(atm). An electrical heating element in Section A is activated, and the air temperatures slowly increase: T_A in Section A because of heat transfer, and T_B in Section B because of adiabatic compression by the slowly moving piston. Treat air as an ideal gas with $C_P = \frac{7}{2}R$, and let n_A be the number of moles of air in Section A. For the process as described, evaluate one of the following sets of quantities:

(a) T_A, T_B, and Q/n_A, if $P(\text{final}) = 1.25(\text{atm})$.

(b) T_B, Q/n_A, and $P(\text{final})$, if $T_A = 425$ K.

(c) T_A, Q/n_A, and $P(\text{final})$, if $T_B = 325$ K.

(d) T_A, T_B, and $P(\text{final})$, if $Q/n_A = 3$ kJ mol^{-1}.

3.20. Derive an equation for the work of mechanically reversible, isothermal compression of 1 mol of a gas from an initial pressure P_1 to a final pressure P_2 when the equation of state is the virial expansion [Eq. (3.10)] truncated to

$$Z = 1 + B'P$$

How does the result compare with the corresponding equation for an ideal gas?

3.21. Show how Eqs. (3.29) and (3.30) reduce to the appropriate expressions for the four particular values of δ listed following Eq. (3.30).

3.22. For methyl chloride at 100°C the virial coefficients are

$$B = -242.5 \text{ cm}^3 \text{ mol}^{-1} \qquad C = 25{,}200 \text{ cm}^6 \text{ mol}^{-2}$$

Calculate the work of mechanically reversible, isothermal compression of 1 mol of methyl chloride from 1 bar to 55 bar at 100°C. Base calculations on the following forms of the virial equation:

(a) $$Z = 1 + \frac{B}{V} + \frac{C}{V^2}$$

(b) $$Z = 1 + B'P + C'P^2$$

where

$$B' = \frac{B}{RT} \qquad \text{and} \qquad C' = \frac{C - B^2}{(RT)^2}$$

Why don't both equations give exactly the same result?

3.23. Calculate Z and V for ethylene at 25°C and 12 bar by the following equations:

(a) The truncated virial equation [Eq. (3.33)] with the following experimental values of virial coefficients:

$$B = -140 \text{ cm}^3 \text{ mol}^{-1} \qquad C = 7{,}200 \text{ cm}^6 \text{ mol}^{-2}$$

(b) The truncated virial equation [Eq. (3.31)], with a value of B from the generalized Pitzer correlation [Eq. (3.48)].

(c) The Redlich/Kwong equation, with estimates of a and b from Eqs. (3.40) and (3.41).

3.24. Calculate Z and V for ethane at 50°C and 15 bar by the following equations:

(a) The truncated virial equation [Eq. (3.33)] with the following experimental values of virial coefficients:

$$B = -156.7 \text{ cm}^3 \text{ mol}^{-1} \qquad C = 9{,}650 \text{ cm}^6 \text{ mol}^{-2}$$

(b) The truncated virial equation [Eq. (3.31)], with a value of B from the generalized Pitzer correlation [Eq. (3.48)].

(c) The Redlich/Kwong equation, with estimates of a and b from Eqs. (3.40) and (3.41).

3.25. Calculate Z and V for sulfur hexafluoride at 75°C and 15 bar by the following equations:

(a) The truncated virial equation [Eq. (3.33)] with the following experimental values of virial coefficients:

$$B = -194 \text{ cm}^3 \text{ mol}^{-1} \qquad C = 15{,}300 \text{ cm}^6 \text{ mol}^{-2}$$

(b) The truncated virial equation [Eq. (3.31)], with a value of B from the generalized Pitzer correlation [Eq. (3.48)].

(c) The Redlich/Kwong equation, with estimates of a and b from Eqs. (3.40) and (3.41).

For sulfur hexafluoride, $T_c = 318.7$ K, $P_c = 37.6$ bar, $V_c = 198$ cm^3 mol^{-1}, and $\omega = 0.286$.

3.26. Determine Z and V for steam at 250°C and 1,800 kPa by the following:

(a) The truncated virial equation [Eq. (3.33)] with the following experimental values of virial coefficients:

$$B = -152.5 \text{ cm}^3 \text{ mol}^{-1} \qquad C = -5{,}800 \text{ cm}^6 \text{ mol}^{-2}$$

(b) The truncated virial equation [Eq. (3.31)], with a value of B from the generalized Pitzer correlation [Eq. (3.48)].

(c) The steam tables.

3.27 With respect to the virial expansions, Eqs. (3.10) and (3.11), show that

$$B' = \left(\frac{\partial Z}{\partial P}\right)_{T,P=0} \quad \text{and} \quad B = \left(\frac{\partial Z}{\partial \rho}\right)_{T,\rho=0}$$

where $\rho \equiv 1/V$.

3.28 Equation (3.11) when truncated to *four* terms accurately represents the volumetric data for methane gas at $0°C$ with

$$B = -53.4 \text{ cm}^3 \text{ mol}^{-1}$$
$$C = 2{,}620 \text{ cm}^6 \text{ mol}^{-2}$$
$$D = 5{,}000 \text{ cm}^9 \text{ mol}^{-3}$$

(*a*) From this information, prepare a plot of Z vs. P for methane at $0°C$ from 0 to 200 bar.

(*b*) To what pressures do Eqs. (3.31) and (3.32) provide good approximations?

3.29. Calculate the molar volume of saturated liquid and the molar volume of saturated vapor by the Redlich/Kwong equation for one of the following and compare results with values found by suitable generalized correlations.

(*a*) Propane at $40°C$ where $P^{\text{sat}} = 13.71$ bar.

(*b*) Propane at $50°C$ where $P^{\text{sat}} = 17.16$ bar.

(*c*) Propane at $60°C$ where $P^{\text{sat}} = 21.22$ bar.

(*d*) Propane at $70°C$ where $P^{\text{sat}} = 25.94$ bar.

(*e*) n-Butane at $100°C$ where $P^{\text{sat}} = 15.41$ bar.

(*f*) n-Butane at $110°C$ where $P^{\text{sat}} = 18.66$ bar.

(*g*) n-Butane at $120°C$ where $P^{\text{sat}} = 22.38$ bar.

(*h*) n-Butane at $130°C$ where $P^{\text{sat}} = 26.59$ bar.

(*i*) Isobutane at $90°C$ where $P^{\text{sat}} = 16.54$ bar.

(*j*) Isobutane at $100°C$ where $P^{\text{sat}} = 20.03$ bar.

(*k*) Isobutane at $110°C$ where $P^{\text{sat}} = 24.01$ bar.

(*l*) Isobutane at $120°C$ where $P^{\text{sat}} = 28.53$ bar.

(*m*) Chlorine at $60°C$ where $P^{\text{sat}} = 18.21$ bar.

(*n*) Chlorine at $70°C$ where $P^{\text{sat}} = 22.49$ bar.

(*o*) Chlorine at $80°C$ where $P^{\text{sat}} = 27.43$ bar.

(*p*) Chlorine at $90°C$ where $P^{\text{sat}} = 33.08$ bar.

(*q*) Sulfur dioxide at $80°C$ where $P^{\text{sat}} = 18.66$ bar.

(*r*) Sulfur dioxide at $90°C$ where $P^{\text{sat}} = 23.31$ bar.

(*s*) Sulfur dioxide at $100°C$ where $P^{\text{sat}} = 28.74$ bar.

(*t*) Sulfur dioxide at $110°C$ where $P^{\text{sat}} = 35.01$ bar.

3.30. Estimate the following:

(*a*) The volume occupied by 18 kg of ethylene at 55 $°C$ and 35 bar.

(b) The mass of ethylene contained in a 0.25-m^3 cylinder at 50°C and 115 bar.

3.31. To a good approximation, what is the molar volume of ethanol vapor at 480°C and 6,000 kPa? How does this result compare with the ideal-gas value?

3.32. A 0.35-m^3 vessel is used to store liquid propane at its vapor pressure. Safety considerations dictate that at a temperature of 320 K the liquid must occupy no more than 80% of the total volume of the vessel. For these conditions, determine the mass of vapor and the mass of liquid in the vessel. At 320 K the vapor pressure of propane is 16.0 bar.

3.33. A 30-m^3 tank contains 14 m^3 of liquid n-butane in equilibrium with its vapor at 25°C. Estimate of the mass of n-butane vapor in the tank. The vapor pressure of n-butane at the given temperature is 2.43 bar.

3.34. Estimate:

(a) The mass of ethane contained in a 0.15-m^3 vessel at 60°C and 14,000 kPa.

(b) The temperature at which 40 kg of ethane storred in a 0.15-m^3 vessel, exerts a pressure 20,000 kPa.

3.35. To what pressure does one fill a 0.15-m^3 vessel at 25°C in order to store 40 kg of ethylene in it?

3.36. If 15 kg of H$_2$O in a 0.4-m^3 container is heated to 400°C what pressure is developed?

3.37. A 0.35-m^3 vessel holds ethane vapor at 25°C and 2,200 kPa. If it is heated to 220°C what pressure is developed?

3.38. What is the pressure in a 0.5-m^3 vessel when it is charged with 10 kg of carbon dioxide at 30°C?

3.39. A rigid vessel, filled to one-half its volume with liquid nitrogen at its normal boiling point, is allowed to warm to 25°C. What pressure is developed? The molar volume of liquid nitrogen at its normal boiling point is 34.7 cm^3 mol^{-1}.

3.40. The specific volume of isobutane liquid at 300 K and 4 bar is 1.824 cm^3 g^{-1}. Estimate the specific volume at 415 K and 75 bar.

3.41. The density of liquid n-pentane is 0.630 g cm^{-3} at 18°C and 1 bar. Estimate its density at 140°C and 120 bar.

3.42. Estimate the density of liquid ethanol at 180°C and 200 bar.

3.43. Estimate the volume change of vaporization for ammonia at 20°C. At this temperature the vapor pressure of ammonia is 857 kPa.

3.44. PVT data may be taken by the following procedure. A mass m of a substance of molar mass M is introduced into a thermostated vessel of known total volume V^t. The system is allowed to equilibrate, and the temperature T and pressure P are measured.

(a) Approximately what percentage errors are allowable in the measured variables (m, M, V^t, T and P) if the maximum allowable error in the calculated compressibility factor Z is ±1%?

(b) Approximately what percentage errors are allowable in the measured variables if the maximum allowable error in calculated values of the second virial coefficient B is ±1%? Assume that $Z \simeq 0.9$ and that values of B are calculated by Eq. (3.32).

3.45. For a gas described by the Redlich/Kwong equation [Eq. (3.35)] and for a temperature greater than T_c, develop expressions for the two limiting slopes,

$$\lim_{P \to 0} \left(\frac{\partial Z}{\partial P} \right)_T \qquad \lim_{P \to \infty} \left(\frac{\partial Z}{\partial P} \right)_T$$

The expressions should contain the temperature T and the Redlich/Kwong parameters a and/or b. Note that in the limit as $P \to 0$, $V \to \infty$, and that in the limit as $P \to \infty$, $V \to b$.

3.46. One mole of an ideal gas with constant heat capacities undergoes an arbitrary mechanically reversible process. Show that

$$\Delta U = \frac{1}{\gamma - 1} \Delta (PV)$$

3.47. The PVT behavior of a certain gas is described by the equation of state

$$P(V - b) = RT$$

where b is a constant. If in addition C_V is constant, show that

(a) U is a function of T only.

(b) $\gamma = \text{const}$.

(c) For a mechanically reversible process, $P(V - b)^\gamma = \text{const}$.

3.48. A certain gas is described by the equation of state

$$PV = RT + \left(b - \frac{\theta}{RT} \right) P$$

Here, b is a constant and θ is a function of T only. For this gas, determine expressions for the isothermal compressibility κ and the thermal pressure coefficient $(\partial P/\partial T)_V$. These expressions should contain only T, P, θ, $d\theta/dT$, and constants.

3.49. If $140(\text{ft})^3$ of methane gas at $60(°F)$ and $1(\text{atm})$ is equivalent to $1(\text{gal})$ of gasoline as fuel for an automobile engine, what would be volume of the tank required to hold methane at $3,000(\text{psia})$ and $60(°F)$ in an amount equivalent to $10(\text{gal})$ of gasoline?

3.50. The following rectilinear potential is an augmentation of the square-well potential [Eq. (3.65)]:

$$\begin{aligned} \mathcal{U} &= \infty \quad &\text{for} \quad r \le d \\ \mathcal{U} &= \xi \quad &\text{for} \quad d \le r \le k \cdot d \\ \mathcal{U} &= -\epsilon \quad &\text{for} \quad k \cdot d \le r \le l \cdot d \\ \mathcal{U} &= 0 \quad &\text{for} \quad l \cdot d \le r \end{aligned}$$

Here, quantities k, l, ξ, and ϵ are positive constants, with $k < l$. Draw a sketch of this potential, and find an algebraic expression for the second virial coefficient $B(T)$. Demonstrate that $B(T)$ for this model can exhibit a *maximum* with respect to T.

3.51. Table 3.2 applies for *like* molecular pairs. Prepare a similar table for all *unlike* molecular pairs comprising species from the following: methane, n-heptane, chloroform, acetone, and acetonitrile. Discuss the result. Data in addition to values that appear in Table 3.2: For methane, $\mu = 0$, $\hat{\alpha} = 2.6 \times 10^{-24}$ cm^3, $I = 2.1 \times 10^{-18}$ J. For n-heptane, $\mu = 0$, $\hat{\alpha} = 13.6 \times 10^{-24}$ cm^3, $I = 1.7 \times 10^{-18}$ J.

CHAPTER 4

HEAT EFFECTS

Heat transfer is one of the most common operations in the chemical industry. Consider, for example, the manufacture of ethylene glycol (an antifreeze agent) by the oxidation of ethylene to ethylene oxide and its subsequent hydration to glycol. The catalytic oxidation reaction is most effective when carried out at temperatures near 250°C. The reactants, ethylene and air, are therefore heated to this temperature before they enter the reactor. To design the preheater one must know how much heat is transferred. The combustion reactions of ethylene with oxygen in the catalyst bed tend to raise the temperature. However, heat is removed from the reactor, and the temperature does not rise much above 250°C. Higher temperatures promote the production of CO_2, an undesired product. Design of the reactor requires knowledge of the rate of heat transfer, and this depends on the heat effects associated with the chemical reactions. The ethylene oxide product is hydrated to glycol by absorption in water. Heat is evolved not only because of the phase change and dissolution process but also because of the hydration reaction between the dissolved ethylene oxide and water. Finally, the glycol is recovered from water by distillation, a process of vaporization and condensation, which results in the separation of a solution into its components.

All of the important heat effects are illustrated by this relatively simple chemical-manufacturing process. In contrast to *sensible* heat effects, which are characterized by temperature changes, the heat effects of chemical reaction, phase transition, and the formation and separation of solutions are determined from experimental measurements made at constant temperature. In this chapter we apply thermodynamics to the evaluation of most of the heat effects that accompany physical and chemical operations. However, the heat effects of mixing processes, which depend on the thermodynamic properties of mixtures, are treated in Chap. 11.

4.1 Sensible Heat Effects

Heat transfer to a system in which there are no phase transitions, no chemical reactions, and no changes in composition causes the temperature of the system to change. Our purpose here is to develop relations between the quantity of heat transferred and the resulting temperature change.

When the system is a homogeneous substance of constant composition, the phase rule indicates that fixing the values of two intensive properties establishes its state. The molar or specific internal energy of a substance may therefore be expressed as a *function of* two other state variables. Arbitrarily selecting these as temperature and molar or specific volume, we write

$$U = U(T, V)$$

whence

$$dU = \left(\frac{\partial U}{\partial T} \right)_V dT + \left(\frac{\partial U}{\partial V} \right)_T dV$$

As a result of Eq. (2.19) this becomes

$$dU = C_V dT + \left(\frac{\partial U}{\partial V} \right)_T dV$$

The final term may be set equal to zero in two circumstances:

1. For any constant-volume process, regardless of substance.

2. Whenever the internal energy is independent of volume, regardless of the process. This is exactly true for ideal gases and incompressible fluids.

In either case,

$$dU = C_V dT$$

and

$$\Delta U = \int_{T_1}^{T_2} C_V dT \qquad (4.1)$$

For a mechanically reversible constant-volume process, $Q = \Delta U$, and Eq. (3.14) may be rewritten here as

$$Q = \Delta U = \int_{T_1}^{T_2} C_V dT$$

Similarly, we may express the molar or specific enthalpy as a function of temperature and pressure:

$$H = H(T, P)$$

whence

$$dH = \left(\frac{\partial H}{\partial T} \right)_P dT + \left(\frac{\partial H}{\partial P} \right)_T dP$$

As a result of Eq. (2.20) this becomes

$$dH = C_P dT + \left(\frac{\partial H}{\partial P}\right)_T dP$$

Again, two circumstances allow the final term to be set equal to zero:

1. For any constant-pressure process, regardless of the substance.

2. Whenever the enthalpy of the substance is independent of pressure, regardless of the process. This is exactly true for ideal gases and approximately true for low-pressure gases, for solids, and for liquids outside the critical region.

In either case,

$$dH = C_P dT$$

and

$$\Delta H = \int_{T_1}^{T_2} C_P dT \tag{4.2}$$

Moreover, $Q = \Delta H$ for mechanically reversible, constant-pressure, nonflow processes [Eq. (2.18)] and for the transfer of heat in steady-flow exchangers where ΔE_P and ΔE_K are negligible and $W_s = 0$. In either case

$$Q = \Delta H = \int_{T_1}^{T_2} C_P dT \tag{4.3}$$

However, the usual engineering application of this equation is to steady-flow heat transfer.

Evaluation of the integral in Eq. (4.3) requires knowledge of the temperature dependence of the heat capacity. This is usually given by an empirical equation; the two simplest expressions of practical value are

$$\frac{C_P}{R} = \alpha + \beta T + \gamma T^2$$

and

$$\frac{C_P}{R} = a + bT + cT^{-2}$$

where α, β, and γ and a, b, and c are constants characteristic of the particular substance. With the exception of the last term, these equations are of the same form. We therefore combine them to provide a single expression:

$$\frac{C_P}{R} = A + BT + CT^2 + DT^{-2} \tag{4.4}$$

where either C or D is zero, depending on the substance considered. Since the ratio C_P/R is dimensionless, the units of C_P are governed by the choice of R.

As shown in Chap. 6, for gases it is the *ideal-gas heat capacity*, rather than the actual heat capacity, that is used in the evaluation of such thermodynamic

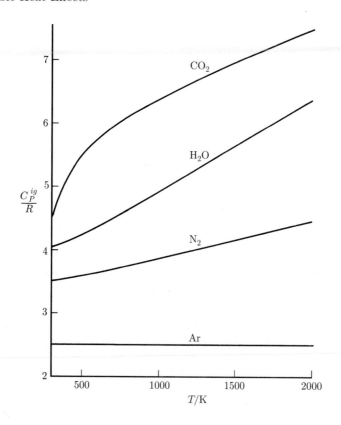

Figure 4.1: Ideal-gas heat capacities of argon, nitrogen, water, and carbon dioxide.

properties as the enthalpy. The reason is that thermodynamic-property evaluation is most conveniently accomplished in two steps: first, calculation of values for a hypothetical *ideal-gas state* wherein ideal-gas heat capacities are used; second, correction of the ideal-gas-state values to the real-gas values. A real gas becomes ideal in the limit as $P \rightarrow 0$; if it were to remain ideal when compressed to finite pressures, its state would remain that of an ideal gas. Gases in their ideal-gas states have properties that reflect their individuality just as do real gases. Ideal-gas heat capacities (designated by C_P^{ig} and C_V^{ig}) are therefore different for different gases; although functions of temperature, they are independent of pressure.

Ideal-gas heat capacities increase smoothly with increasing temperature toward an upper limit, which is reached when all translational, rotational, and vibrational modes of molecular motion are fully excited. The influence of temperature on C_P^{ig} for argon, nitrogen, water, and carbon dioxide is illustrated graphically in Fig. 4.1. Temperature dependence is expressed analytically by equations such as Eq. (4.4), here written

$$\frac{C_P^{ig}}{R} = A + BT + CT^2 + DT^{-2}$$

Values of the parameters are given in Table C.1 of App. C for a number of common organic and inorganic gases. More accurate but more complex equations are found in the literature.[1]

As a result of Eq. (3.17), the two ideal-gas heat capacities are related:

$$\frac{C_V^{ig}}{R} = \frac{C_P^{ig}}{R} - 1 \qquad\qquad (4.5)$$

Thus the temperature dependence of C_V^{ig}/R is readily found from the temperature dependence of C_P^{ig}/R.

The effects of temperature on C_P^{ig} or C_V^{ig} are determined by experiment, most often from spectroscopic data and knowledge of molecular structure by the methods of statistical mechanics. Where experimental data are not available, methods of estimation are employed, as described by Reid, Prausnitz, and Poling.[2]

Although ideal-gas heat capacities are exactly correct for real gases only at zero pressure, real gases seldom depart significantly from ideality at pressures below several bars, and therefore C_P^{ig} and C_V^{ig} are usually good approximations for the heat capacities of real gases at low pressures.

> **Example 4.1** The parameters listed in Table C.1 require use of Kelvin temperatures in Eq. (4.4). Equations of the same form may also be developed for use with temperatures in °C, (R), and (°F), but the parameters are different. The molar heat capacity of methane in the ideal-gas state is given as a function of temperature in Kelvins by
>
> $$\frac{C_P^{ig}}{R} = 1.702 + 9.081 \times 10^{-3}T - 2.164 \times 10^{-6}T^2$$
>
> where the parameters are from Table C.1. Develop an equation for C_P^{ig}/R for temperatures in °C.
>
> SOLUTION   The relation between the two temperature scales is
>
> $$T\text{ K} = t°\text{C} + 273.15$$
>
> Therefore, as a function of t,
>
> $$\frac{C_P^{ig}}{R} = 1.702 + 9.081 \times 10^{-3}(t + 273.15) - 2.164 \times 10^{-6}(t + 273.15)^2$$
>
> or
>
> $$\frac{C_P^{ig}}{R} = 4.021 + 7.899 \times 10^{-3}t - 2.164 \times 10^{-6}t^2$$

---

[1]See F. A. Aly and L. L. Lee, *Fluid Phase Equilibria.*, vol. 6, pp. 169–179, 1981, and its bibliography; see also T. E. Daubert, R. P. Danner, H. M. Sibul, and C. C. Stebbins, *Physical and Thermodynamic Properties of Pure Chemicals: Data Compilation*, Taylor & Francis, Bristol, PA, extant 1995.

[2]R. C. Reid, J. M. Prausnitz, and B. E. Poling, *The Properties of Gases and Liquids*, 4th ed., chap. 6, McGraw-Hill, New York, 1987.

Gas mixtures of constant composition may be treated in exactly the same way as pure gases. An ideal gas, by definition, is a gas whose molecules have no influence on one another. This means that each gas exists in a mixture independent of the others, and that its properties are unaffected by the presence of different molecules. Thus one calculates the ideal-gas heat capacity of a gas mixture by taking the molar average of the heat capacities of the individual species. Consider 1 mol of gas mixture consisting of species A, B, and C, and let y_A, y_B, and y_C represent the mole fractions of these species. The molar heat capacity of the mixture in the ideal-gas state is given by

$$C^{ig}_{P_{\text{mixture}}} = y_A C^{ig}_{P_A} + y_B C^{ig}_{P_B} + y_C C^{ig}_{P_C} \tag{4.6}$$

where $C^{ig}_{P_A}$, $C^{ig}_{P_B}$, and $C^{ig}_{P_C}$ are the molar heat capacities of pure A, B, and C in the ideal-gas state.

As with gases, the heat capacities of solids and liquids are found by experiment. Parameters for the temperature dependence of C_P as expressed by Eq. (4.4) are given for a few solids and liquids in Tables C.2 and C.3 of App. C. Correlations for the heat capacities of many solids and liquids are given by Perry and Green and by Daubert and Danner.[3]

Evaluation of the integral $\int C_P dT$ is accomplished by substitution for C_P, followed by formal integration. For temperature limits of T_0 and T the result is conveniently expressed as follows:

$$\int_{T_0}^{T} \frac{C_P}{R} dT = A T_0 (\tau - 1) + \frac{B}{2} T_0^2 (\tau^2 - 1) + \frac{C}{3} T_0^3 (\tau^3 - 1) + \frac{D}{T_0} \left(\frac{\tau - 1}{\tau} \right) \tag{4.7}$$

where

$$\tau \equiv \frac{T}{T_0}$$

Example 4.2   Calculate the heat required to raise the temperature of 1 mol of methane from 260 to 600°C in a flow process at a pressure sufficiently low that methane may be considered an ideal gas.

SOLUTION   Equation (4.3) in combination with Eq. (4.7) provides the required result. Parameters for C^{ig}_P/R come from Table C.1; temperatures in Kelvins are

$$T_0 = 533.15 \text{ K} \qquad T = 873.15 \text{ K}$$

Then

$$\tau = \frac{873.15}{533.15} = 1.6377$$

and

$$Q = \Delta H = R \int_{533.15}^{873.15} \frac{C^{ig}_P}{R} dT$$

---

[3] R. H. Perry and D. Green, *Perry's Chemical Engineers' Handbook*, 7th ed., Sec. 2, McGraw-Hill, New York, 1996; T. E. Daubert et al., *op. cit.*

$$Q = (8.314) \left[1.702\, T_0(\tau - 1) + \frac{9.081 \times 10^{-3}}{2} T_0^2(\tau^2 - 1) - \frac{2.164 \times 10^{-6}}{3} T_0^3(\tau^3 - 1) \right]$$

$$= 19{,}778 \text{ J}$$

Given T_0 and T, the foregoing example shows that the calculation of Q or ΔH is straightforward. Less direct is the calculation of T, given T_0 and Q or ΔH. Here, an iteration scheme may be useful. Factoring $(\tau - 1)$ from each term on the right-hand side of Eq. (4.7) gives

$$\int_{T_0}^{T} \frac{C_P}{R}\, dT = \left[A T_0 + \frac{B}{2} T_0^2(\tau + 1) + \frac{C}{3} T_0^3(\tau^2 + \tau + 1) + \frac{D}{\tau T_0} \right] (\tau - 1)$$

Since

$$\tau - 1 = \frac{T - T_0}{T_0}$$

this may be written

$$\int_{T_0}^{T} \frac{C_P}{R}\, dT = \left[A + \frac{B}{2} T_0(\tau + 1) + \frac{C}{3} T_0^2(\tau^2 + \tau + 1) + \frac{D}{\tau T_0^2} \right] (T - T_0)$$

We identify the quantity in square brackets as $\langle C_P \rangle_H / R$, where $\langle C_P \rangle_H$ is defined as a *mean heat capacity:*

$$\frac{\langle C_P \rangle_H}{R} = A + \frac{B}{2} T_0(\tau + 1) + \frac{C}{3} T_0^2(\tau^2 + \tau + 1) + \frac{D}{\tau T_0^2} \qquad (4.8)$$

Equation (4.3) may therefore be written

$$\Delta H = \langle C_P \rangle_H (T - T_0) \qquad (4.9)$$

The triangular brackets enclosing C_P identify it as a mean value; subscript "H" denotes a mean value specific to enthalpy calculations, and distinguishes this mean heat capacity from a similar quantity introduced in the next chapter.

Solution of Eq. (4.9) for T gives

$$T = \frac{\Delta H}{\langle C_P \rangle_H} + T_0 \qquad (4.10)$$

A starting value for T (and hence for $\tau = T/T_0$) allows evaluation of $\langle C_P \rangle_H$ by Eq. (4.8). Substitution of this value into Eq. (4.10) provides a new value of T from which to reevaluate $\langle C_P \rangle_H$. Iteration continues in like fashion to convergence on a final value of T.

Thermodynamic calculations frequently require the evaluation of the integral $\int (C_P/R)\, dT$. Convenience therefore suggests that one have at hand a computer routine for this purpose. Equation (4.7) provides a function for evaluation of the integral which for computational purposes we name

ICPH(T0,T;A,B,C,D)

The quantities in parentheses are the variables T_0 and T, followed by the parameters A, B, C, and D. When these quantities are assigned numerical values, the notation represents a value for the integral. Thus, for the evaluation of Q in Example 4.2, we write

```
8.314*ICPH(533.15,873.15;1.702,9.081E-3,-2.164E-6,0.0) ≡ 19,778 J
```

Representative computer programs for evaluation of the integral are given in App. D. To provide added flexibility the programs also evaluate the dimensionless quantity $\langle C_P \rangle_H / R$ as given by Eq. (4.8), another function which for computational purposes is named

$$\text{MCPH(T0,T;A,B,C,D)}$$

A specific numerical value of this function is

```
MCPH(533.15,873.15;1.702,9.081E-3,-2.164E-6,0.0) ≡ 6.9965
```

representing $\langle C_P \rangle_H / R$ for methane in the calculation of Example 4.2. Equation (4.9) then yields

$$\Delta H = (8.314)(6.9965)(873.15 - 533.15) = 19{,}778 \text{ J}$$

Example 4.3 What is the final temperature when heat in the amount of 0.4×10^6(Btu) is added to 25(lb mol) of ammonia initially at 500(°F) in a steady-flow process at approximately 1(atm)?

SOLUTION If ΔH is the enthalpy change for 1(lb mol), $Q = n \, \Delta H$, and

$$\Delta H = \frac{Q}{n} = \frac{0.4 \times 10^6}{25} = 16{,}000 \text{(Btu)(lb mol)}^{-1}$$

The heat-capacity equation requires temperatures in Kelvins; therefore, conversion of all units to the SI system is indicated. Since 1 J mol^{-1} is equivalent to 0.4299(Btu)(lb mol)$^{-1}$, we divide the preceding result by 0.4299 to get

$$\Delta H = 16{,}000/0.4299 = 37{,}218 \text{ J mol}^{-1}$$

With

$$T_0 = \frac{500 + 459.67}{1.8} = 533.15 \text{ K}$$

we may evaluate $\langle C_P^{ig} \rangle_H / R$ for any value of T:

```
MCPH(533.15,T;3.578,3.020E-3,0.0,-0.186E+5)
```
(A)

Iteration between (A) and Eq. (4.10) starts with a value $T \geq T_0$, and converges on the final value,

$$T = 1{,}250 \text{ K} \qquad \text{or} \qquad 1{,}790(°F)$$

4.2 Internal Energy of Ideal Gases: Microscopic View

Thermodynamic properties such as internal energy and enthalpy are manifestations on a gross scale of the positions, motions, and interactions of the countless molecules comprising a macroscopic system. The two theories that relate the behavior of molecules to macroscopic properties are *kinetic theory* and *statistical mechanics*. These theories together represent a very large body of extra-thermodynamic knowledge, well beyond the scope of this text. Our purpose here is to indicate in a very elementary fashion how the energy associated with the individual molecules of an ideal gas relates to the macroscopic internal energy of the ideal-gas state.

A fundamental postulate of quantum theory is that energy is quantized, that energy on the microscopic scale comes in very small discrete units. Thus a macroscopic system has associated with it an enormous number of quanta of energy, which sum to determine its energy level. Quantum theory specifies that the set of energy levels "allowed" to a closed system is determined by its volume. This does not mean that fixing the volume of a system fixes its energy; it just determines the discrete set of energy levels permitted to it. When a system is isolated (constrained from exchanging either mass or energy with its surroundings), it exists at one of the energy levels allowed to it. Each energy level of a system has associated with it a number of *quantum states*. This number g is known as the *degeneracy* of the level. Although an isolated system exists at a particular energy level, it passes over time through all of the g quantum states characterized by this same energy.

A large number of quantum states is accessible to a system of given volume in equilibrium with surroundings at fixed temperature. From the thermodynamic point of view, fixing T and V establishes the state of the system, including its internal energy. However, macroscopic equilibrium does not imply a static situation on the microscopic scale. The ceaseless motion of the molecules and their collisions with the walls of the container result in exchanges of energy with the surroundings, which cause momentary fluctuations in the energy of the system. Thus the internal energy of thermodynamics is an average over the discrete set of energy levels of the allowed quantum states. Statistical mechanics provides the means for arriving at the proper average value.

The fundamental postulate of statistical mechanics for a system of given volume in equilibrium with surroundings at fixed temperature is that the probability of a quantum state depends only on its energy, and all quantum states with the same energy have the same probability. A value for the thermodynamic internal energy is found as the average of the energies of the accessible quantum states, each weighted by its probability. The quantity of fundamental importance to come out of this procedure is the *partition function*:[4]

$$\mathcal{Z} = \sum_i g_i e^{-E_i/kT} \tag{4.11}$$

---

[4]The name implies a sum over states partitioned according to energy levels, and the symbol comes from the German word, *Zustandssumme*, sum-over-states. The other commonly used symbol is Q, which is used in this book to represent heat.

where E_i is the energy of level i, g_i is the degeneracy of the level, and k is Boltzmann's constant. This quantity is a state function, from which all thermodynamic properties may be found once it is known as a function of T and V. In particular, the internal energy is given by

$$U = kT^2 \left(\frac{\partial \ln \mathcal{Z}}{\partial T} \right)_V \tag{4.12}$$

Equations (4.11) and (4.12) are written for the total energy of a macroscopic system. The problem of their application is greatly simplified for ideal gases, which (because of the absence of molecular interactions) do not possess intermolecular potential energy. The internal energy of ideal gases is therefore associated with individual molecules, and results from translation and rotation of each molecule as a whole and from rotations and vibrations internal to the molecule. Energy is also associated with the electrons and nuclei of atoms. For ideal gases all of these forms of energy are usually treated as separable, and this allows the partition function to be factored and written as a product of partition functions, each of which relates to a particular form of molecular energy. It suffices for present purposes to treat the translational kinetic energy as separable from all other contributions. These other contributions together constitute the *intramolecular energy* of a molecule, and they depend on *molecular structure*. We therefore write

$$\mathcal{Z} = \frac{\mathcal{Z}_{\text{translation}} \mathcal{Z}_{\text{intramolecular}}}{N!}$$

whence

$$\ln \mathcal{Z} = \ln \mathcal{Z}_{\text{translation}} + \ln \mathcal{Z}_{\text{intramolecular}} - \ln N! \tag{4.13}$$

In these equations the quantity $N!$ arises because in the summation process the molecules are treated as distinguishable from one another, whereas in fact they are indistinguishable. The translational contribution comes from kinetic theory, whereas the intramolecular contributions derive from quantum mechanics, with the quantized energy levels determined from appropriate spectroscopic measurements.

For one mole of an ideal gas, the equation for the translational partition function is[5]

$$\mathcal{Z}_{\text{translation}} = V^{N_A} \left(\frac{2\pi m k T}{h^2} \right)^{(3/2)N_A}$$

where N_A = Avogadro's number
m = mass of the molecule
h = Planck's constant

In logarithmic form this becomes

$$\ln \mathcal{Z}_{\text{translation}} = N_A \ln V + \frac{3}{2} N_A \ln \frac{2\pi m k T}{h^2}$$

---

[5]D. A. McQuarrie, *Statistical Mechanics*, pp. 81–82, Harper & Row, New York, 1976.

Substitution into Eq. (4.13) gives

$$\ln \mathcal{Z} = N_A \ln V + \frac{3}{2} N_A \ln \frac{2\pi m k T}{h^2} - \ln N_A! + \ln \mathcal{Z}_{\text{intramolecular}} \qquad (4.14)$$

where $\mathcal{Z}_{\text{intramolecular}}$ is evaluated by equations of the form of Eq. (4.11).

Differentiation of Eq. (4.14) with respect to temperature at constant V (and N_A) yields

$$\left(\frac{\partial \ln \mathcal{Z}}{\partial T}\right)_V = \frac{3}{2} \frac{N_A}{T} + \left(\frac{\partial \ln \mathcal{Z}_{\text{intramolecular}}}{\partial T}\right)_V$$

whence by Eq. (4.12) applied to the ideal-gas state

$$U^{ig} = \frac{3}{2} N_A k T + k T^2 \left(\frac{\partial \ln \mathcal{Z}_{\text{intramolecular}}}{\partial T}\right)_V$$

Representing the final term by $\mathcal{F}(T)$ and noting that $N_A k = R$, we rewrite this as

$$U^{ig} = \frac{3}{2} RT + \mathcal{F}(T) \qquad (4.15)$$

where R is the universal gas constant. Since $H^{ig} = U^{ig} + RT$,

$$H^{ig} = \frac{5}{2} RT + \mathcal{F}(T)$$

In view of Eq. (2.20)

$$C_P^{ig} \equiv \left(\frac{\partial H^{ig}}{\partial T}\right)_P = \frac{5}{2} R + \left(\frac{\partial \mathcal{F}(T)}{\partial T}\right)_P \qquad (4.16)$$

The molecules of an ideal *monatomic* gas have no energy of rotation or vibration, and $\mathcal{F}(T)$ in Eq. (4.16) is therefore zero except at high temperatures where electronic energy contributions become important. Thus in Fig. 4.1 the value of C_P^{ig}/R for argon is constant at a value of 5/2. For diatomic and polyatomic gases, $\mathcal{F}(T)$ in Eq. (4.16) contributes importantly at all temperatures of practical importance. The contribution becomes larger the more complex the molecule and increases monotonically with temperature, as is evident from the curves shown in Fig. 4.1 for N_2, H_2O, and CO_2.

4.3 Latent Heats of Pure Substances

When a pure substance is liquefied from the solid state or vaporized from the liquid at constant pressure, no change in temperature occurs; however, the process requires the transfer of a finite amount of heat to the substance. These heat effects are called the latent heat of fusion and the latent heat of vaporization. Similarly, there are heats of transition accompanying the change of a substance from one solid state to another; for example, the heat absorbed when rhombic crystalline sulfur changes to the monoclinic structure at 95°C and 1 bar is 360 J for each gram-atom.

The characteristic feature of all these processes is the coexistence of two phases. According to the phase rule, a two-phase system consisting of a single species is univariant, and its intensive state is determined by the specification of just one intensive property. Thus the latent heat accompanying a phase change is a function of temperature only, and is related to other system properties by an exact thermodynamic equation:

$$\Delta H = T \, \Delta V \frac{dP^{\text{sat}}}{dT}$$
(4.17)

where for a pure species at temperature T,

ΔH = latent heat
ΔV = volume change accompanying the phase change
P^{sat} = vapor pressure

The derivation of this equation, known as the Clapeyron equation, is given in Chap. 6.

When Eq. (4.17) is applied to the vaporization of a pure liquid, dP^{sat}/dT is the slope of the vapor pressure-vs.-temperature curve at the temperature of interest, ΔV is the difference between molar volumes of saturated vapor and saturated liquid, and ΔH is the latent heat of vaporization. Thus values of ΔH may be calculated from vapor-pressure and volumetric data.

Latent heats may also be measured calorimetrically. Experimental values are available at selected temperatures for many substances.[6] Correlations for the latent heats of many compounds as a function of temperature are given by Daubert and Danner.[7] Nevertheless, data are not always available at the temperature of interest, and in many cases the data necessary for application of Eq. (4.17) are also not known. In this event approximate methods are used for estimates of the heat effect accompanying a phase change. Since heats of vaporization are by far the most important from a practical point of view, they have received most attention. One procedure is to use a group-contribution method, known as UNIVAP,[8] which yields ΔH^{lv} values as a function of temperature. Alternative methods serve two purposes:

1. Prediction of the heat of vaporization at the normal boiling point, i.e., at a pressure of 1 standard atmosphere, defined as 101,325 Pa.

2. Estimation of the heat of vaporization at any temperature from the known value at a single temperature.

---

[6]V. Majer and V. Svoboda, *IUPAC Chemical Data Series* No. 32, Blackwell, Oxford, 1985; R. H. Perry and D. Green, *op. cit.*, Sec. 2.

[7]T. E. Daubert et al., *op. cit.*

[8]M. Klüppel, S. Schulz, and P. Ulbig, *Fluid Phase Equilibria*, vol. 102, pp. 1–15, 1994.

A useful method for prediction of the heat of vaporization at the normal boiling point is the equation proposed by Riedel:[9]

$$\frac{\Delta H_n/T_n}{R} = \frac{1.092(\ln P_c - 1.013)}{0.930 - T_{r_n}} \tag{4.18}$$

where T_n = normal boiling point
$\qquad \Delta H_n$ = molar latent heat of vaporization at T_n
$\qquad P_c$ = critical pressure, bar
$\qquad T_{r_n}$ = reduced temperature at T_n

Since $\Delta H_n/T_n$ has the dimensions of the gas constant R, the units of this ratio are governed by the choice of units for R.

Equation (4.18) is surprisingly accurate for an empirical expression; errors rarely exceed 5 percent. Applied to water it gives

$$\Delta H_n/T_n = R\left[\frac{1.092(\ln 220.55 - 1.013)}{0.930 - 0.577}\right] = 13.56R$$

Taking $R = 8.314$ J mol^{-1} K^{-1} and the normal boiling point of water as 100°C or 373.15 K, we get

$$\Delta H_n = (13.56)(8.314)(373.15) = 42{,}065 \text{ J mol}^{-1}$$

This corresponds to 2,334 J g^{-1}, whereas the experimental value is 2,257 J g^{-1}; the error is 3.4 percent.

Estimates of the latent heat of vaporization of a pure liquid at any temperature from the known value at a single temperature may be based on a known experimental value or on a value estimated by Eq. (4.18). The method proposed by Watson[10] has found wide acceptance:

$$\frac{\Delta H_2}{\Delta H_1} = \left(\frac{1 - T_{r_2}}{1 - T_{r_1}}\right)^{0.38} \tag{4.19}$$

This equation is both simple and reliable; its use is illustrated in the following example.

Example 4.4   Given that the latent heat of vaporization of water at 100°C is 2,257 J g^{-1}, estimate the latent heat at 300°C.

SOLUTION   Let ΔH_1 = latent heat at 100°C = 2,257 J g^{-1}
$\qquad\quad \Delta H_2$ = latent heat at 300°C
$\qquad\quad T_{r_1}$ = 373.15/647.1 = 0.577
$\qquad\quad T_{r_2}$ = 573.15/647.1 = 0.886

---

[9]L. Riedel, *Chem. Ing. Tech.*, vol. 26, pp. 679–683, 1954.

[10]K. M. Watson, *Ind. Eng. Chem.*, vol. 35, pp. 398–406, 1943.

Then by Eq. (4.19),

$$\Delta H_2 = (2{,}257) \left(\frac{1 - 0.886}{1 - 0.577} \right)^{0.38} = (2{,}257)(0.270)^{0.38} = 1{,}371 \text{ J g}^{-1}$$

The value given in the steam tables is $1{,}406 \text{ J g}^{-1}$.

4.4 Standard Heat of Reaction

Heat effects so far discussed have been for physical processes. Chemical reactions also are accompanied either by the transfer of heat or by temperature changes during the course of reaction—in some cases by both. These effects are manifestations of the differences in molecular structure, and therefore in energy, of the products and reactants. For example, the reactants in a combustion reaction possess greater energy on account of their structure than do the products, and this energy must either be transferred to the surroundings as heat or produce products at an elevated temperature.

Each of the vast number of possible chemical reactions may be carried out in many different ways, and each reaction carried out in a particular way is accompanied by a particular heat effect. Tabulation of all possible heat effects for all possible reactions is quite impossible. We therefore *calculate* the heat effects for reactions carried out in diverse ways from data for reactions carried out in a standard way. This reduces the required data to a minimum.

The amount of heat required for a specific chemical reaction depends on the temperatures of both the reactants and products. A consistent basis for treatment of reaction heat effects results when the products of reaction and the reactants are all at the *same* temperature.

Consider the flow-calorimeter method for measurement of heats of combustion of fuel gases. The fuel is mixed with air at ambient temperature and ignited. Combustion takes place in a chamber surrounded by a cooling jacket through which water flows. In addition there is a water-jacketed section in which the products of combustion are cooled to the temperature of the reactants. Since no shaft work is produced by the process, and the calorimeter is built so that changes in potential and kinetic energy are negligible, the overall energy balance, Eq. (2.9), reduces to

$$Q = \Delta H$$

Thus the heat Q absorbed by the water is identical to the enthalpy change caused by the combustion reaction, and universal practice is to designate the enthalpy change of reaction ΔH as the *heat of reaction.*

For purposes of data tabulation, we define the *standard* heat of the reaction,

$$aA + bB \rightarrow lL + mM$$

as the enthalpy change when a moles of A and b moles of B in their *standard states at temperature T* react to form l moles of L and m moles of M in their *standard states also at temperature T*. A *standard state* is a particular state of a species at *temperature* T and at specified conditions of pressure, composition, and physical state.

A *standard-state pressure* of 1 standard atmosphere (101,325 Pa) was in use for many years, and older data tabulations are for this pressure. The standard is now 1 bar (10^5 Pa), but for purposes of this chapter, the difference is of negligible consequence. With respect to composition, the standard states used in this chapter are states of the *pure* species. For gases, the physical state is the ideal-gas state and for liquids and solids, the real state at the standard-state pressure and at the system temperature.

In summary, the standard states used in this chapter are:

1. *Gases*: The pure substance in the ideal-gas state at 1 bar.

2. *Liquids and solids*: The actual pure liquid or solid at 1 bar.

Property values in the standard state are denoted by the degree symbol ($°$). For example, $C_P^°$ is the standard-state heat capacity. Since the standard state for gases is the ideal-gas state, $C_P^°$ for gases is identical with C_P^{ig}, and the data of Table C.1 apply to the standard state for gases. All conditions for a standard state are fixed except temperature, which is always the temperature of the system. Standard-state properties are therefore functions of temperature only. The standard state chosen for gases is a hypothetical one, for at 1 bar actual gases are not ideal. However, they seldom deviate much from ideality, and in most instances enthalpies for the real-gas state at 1 bar and the ideal-gas state are little different.

When a heat of reaction is given for a particular reaction, it applies for the stoichiometric coefficients as written. If each stoichiometric coefficient is doubled, the heat of reaction is doubled. For example, the ammonia synthesis reaction may be written

$$\tfrac{1}{2}N_2 + \tfrac{3}{2}H_2 \rightarrow NH_3 \qquad \Delta H_{298}^° = -46{,}110 \text{ J}$$

or
$$N_2 + 3H_2 \rightarrow 2NH_3 \qquad \Delta H_{298}^° = -92{,}220 \text{ J}$$

The symbol $\Delta H_{298}^°$ indicates that the heat of reaction is the *standard* value for a temperature of 298.15 K (25°C).

4.5 Standard Heat of Formation

Tabulation of data for just the *standard* heats of reaction for all of the vast number of possible reactions is impractical. Fortunately, the standard heat of any reaction can be calculated if the *standard heats of formation* of the compounds taking part in the reaction are known. A *formation* reaction is defined as a reaction which forms a single compound *from its constituent elements*. For example, the reaction $C + \tfrac{1}{2}O_2 + 2H_2 \rightarrow CH_3OH$ is the formation reaction for methanol. The reaction $H_2O + SO_3 \rightarrow H_2SO_4$ is *not* a formation reaction, because it forms sulfuric acid not from the elements but from other compounds. Formation reactions are understood to result in the formation of 1 mol of the compound; the heat of formation is therefore based on *1 mol of the compound formed*.

Heats of reaction at any temperature can be calculated from heat-capacity data if the value for one temperature is known; the tabulation of data can therefore be reduced to the compilation of *standard heats of formation at a single temperature.* The usual choice for this temperature is 298.15 K or 25°C. The standard heat of formation of a compound at this temperature is represented by the symbol $\Delta H^\circ_{f_{298}}$. Superscript (°) indicates that it is the standard value, subscript f shows that it is a heat of formation, and the 298 is the approximate absolute temperature in kelvins. Tables of these values for common substances may be found in standard handbooks, but the most extensive compilations available are in specialized reference works.[11] An abridged list of values is given in Table C.4 of App. C.

When chemical equations are combined by addition, the standard heats of reaction may also be added to give the standard heat of the resulting reaction. This is possible because enthalpy is a property, and changes in it are independent of path. In particular, formation equations and standard heats of formation may always be combined to produce any desired equation (not itself a formation equation) and its accompanying standard heat of reaction. Equations written for this purpose often include an indication of the physical state of each reactant and product, i.e., the letter g, l, or s is placed in parentheses after the chemical formula to show whether it is a gas, a liquid, or a solid. This might seem unnecessary since a pure chemical species at a particular temperature and 1 bar can usually exist only in one physical state. However, fictitious states are often assumed for convenience.

Consider the reaction $CO_2(g) + H_2(g) \rightarrow CO(g) + H_2O(g)$ at 25°C. This water-gas-shift reaction is commonly encountered in the chemical industry, though it takes place only at temperatures well above 25°C. However, the data used are for 25°C, and the initial step in any calculation of heat effects concerned with this reaction is to evaluate the standard heat of reaction at 25°C. Since the reaction is actually carried out entirely in the gas phase at high temperature, convenience dictates that the standard states of all products and reactants at 25°C be taken as the ideal-gas state at 1 bar, even though water cannot actually exist as a gas at these conditions. The pertinent formation reactions and their heats of formation from Table C.4 are

$$CO_2(g): \quad C(s) + O_2(g) \rightarrow CO_2(g) \qquad \Delta H^\circ_{f_{298}} = -393{,}509 \text{ J}$$

$$H_2(g): \quad \text{Since hydrogen is an element} \quad \Delta H^\circ_{f_{298}} = 0$$

$$CO(g): \quad C(s) + \tfrac{1}{2}O_2(g) \rightarrow CO(g) \qquad \Delta H^\circ_{f_{298}} = -110{,}525 \text{ J}$$

$$H_2O(g): \quad H_2(g) + \tfrac{1}{2}O_2(g) \rightarrow H_2O(g) \qquad \Delta H^\circ_{f_{298}} = -241{,}818 \text{ J}$$

---

[11]For example, see *TRC Thermodynamic Tables—Hydrocarbons* and *TRC Thermodynamic Tables—Non-hydrocarbons*, serial publications of the Thermodynamics Research Center, Texas A & M Univ. System, College Station, Texas; "The NBS Tables of Chemical Thermodynamic Properties," *J. Physical and Chemical Reference Data*, vol. 11, supp. 2, 1982. See also, T. E. Daubert et al., *op. cit.* Where data are unavailable, estimates based only on molecular structure may be found by the methods of L. Constantinou and R. Gani, *Fluid Phase Equilibria*, vol. 103, pp. 11–22, 1995.

These equations can be written so that their sum gives the desired reaction. This requires that the formation reaction for CO_2 be written in reverse; the heat of reaction is then of opposite sign to the standard heat of formation:

$$CO_2(g) \rightarrow C(s) + O_2(g) \qquad \Delta H^\circ_{298} = 393{,}509 \text{ J}$$

$$C(s) + \tfrac{1}{2}O_2(g) \rightarrow CO(g) \qquad \Delta H^\circ_{298} = -110{,}525 \text{ J}$$

$$H_2(g) + \tfrac{1}{2}O_2(g) \rightarrow H_2O(g) \qquad \Delta H^\circ_{298} = -241{,}818 \text{ J}$$

$$\overline{CO_2(g) + H_2(g) \rightarrow CO(g) + H_2O(g) \quad \Delta H^\circ_{298} = 41{,}166 \text{ J}}$$

The meaning of this result is that the enthalpy of 1 mol of CO plus 1 mol of H_2O is greater than the enthalpy of 1 mol of CO_2 plus 1 mol of H_2 by 41,166 J when each product and reactant is taken as the pure gas at 25°C in the ideal-gas state at 1 bar.

In this example the standard heat of formation of H_2O is available for its hypothetical standard state as a gas at 25°C. One might expect the value of the heat of formation of water to be listed for its actual state as a liquid at 1 bar and 25°C. As a matter of fact, values for both states are given in Table C.4 because they are both frequently used. This is true for many compounds that normally exist as liquids at 25°C and the standard-state pressure. Cases do arise, however, in which a value is given only for the standard state as a liquid or as an ideal gas when what is needed is the other value. Suppose that this were the case for the preceding example and that only the standard heat of formation of liquid H_2O is known. We must now include an equation for the physical change that transforms water from its standard state as a liquid into its standard state as a gas. The enthaply change for this physical process is the difference between the heats of formation of water in its two standard states:

$$-241{,}818 - (-285{,}830) = 44{,}012 \text{ J}$$

This is approximately the latent heat of vaporization of water at 25°C. The sequence of steps is now:

$$CO_2(g) \rightarrow C(s) + O_2(g) \qquad \Delta H^\circ_{298} = 393{,}509 \text{ J}$$

$$C(s) + \tfrac{1}{2}O_2(g) \rightarrow CO(g) \qquad \Delta H^\circ_{298} = -110{,}525 \text{ J}$$

$$H_2(g) + \tfrac{1}{2}O_2(g) \rightarrow H_2O(l) \qquad \Delta H^\circ_{298} = -285{,}830 \text{ J}$$

$$H_2O(l) \rightarrow H_2O(g) \qquad \Delta H^\circ_{298} = 44{,}012 \text{ J}$$

$$\overline{CO_2(g) + H_2(g) \rightarrow CO(g) + H_2O(g) \quad \Delta H^\circ_{298} = 41{,}166 \text{ J}}$$

This result is of course in agreement with the original answer.

Example 4.5   Calculate the standard heat at 25°C for the following reaction:

$$4HCl(g) + O_2(g) \rightarrow 2H_2O(g) + 2Cl_2(g)$$

SOLUTION   Standard heats of formation at 298.15 K from Table C.4 are

$$HCl(g): -92{,}307 \text{ J} \qquad H_2O(g): -241{,}818 \text{ J}$$

The following combination gives the desired result:

$$4HCl(g) \rightarrow 2H_2(g) + 2Cl_2(g) \qquad \Delta H^{\circ}_{298} = (4)(92{,}307)$$
$$2H_2(g) + O_2(g) \rightarrow 2H_2O(g) \qquad \Delta H^{\circ}_{298} = (2)(-241{,}818)$$
$$\overline{4HCl(g) + O_2(g) \rightarrow 2H_2O(g) + 2Cl_2(g) \quad \Delta H^{\circ}_{298} = -114{,}408 \text{ J}}$$

4.6 Standard Heat of Combustion

Only a few *formation* reactions can actually be carried out, and therefore data for these reactions must usually be determined indirectly. One kind of reaction that readily lends itself to experiment is the combustion reaction, and many standard heats of formation come from standard heats of combustion, measured calorimetrically. A combustion reaction is defined as a reaction between an element or compound and oxygen to form specified combustion products. For organic compounds made up of carbon, hydrogen, and oxygen only, the products are carbon dioxide and water, but the state of the water may be either vapor or liquid. Data are always based on 1 mol of the substance burned.

A reaction such as the formation of *n*-butane:

$$4C(s) + 5H_2(g) \rightarrow C_4H_{10}(g)$$

cannot be carried out in practice. However, this equation results from combination of the following combustion reactions:

$$4C(s) + 4O_2(g) \rightarrow 4CO_2(g) \qquad \Delta H^{\circ}_{298} = (4)(-393{,}509)$$
$$5H_2(g) + 2\tfrac{1}{2}O_2(g) \rightarrow 5H_2O(l) \qquad \Delta H^{\circ}_{298} = (5)(-285{,}830)$$
$$4CO_2(g) + 5H_2O(l) \rightarrow C_4H_{10}(g) + 6\tfrac{1}{2}O_2(g) \quad \Delta H^{\circ}_{298} = 2{,}877{,}396$$
$$\overline{4C(s) + 5H_2(g) \rightarrow C_4H_{10}(g) \qquad\qquad \Delta H^{\circ}_{298} = -125{,}790 \text{ J}}$$

This is the value of the standard heat of formation of *n*-butane listed in Table C.4.

4.7 Temperature Dependence of ΔH°

In the foregoing sections, standard heats of reaction are discussed for just a reference temperature of 298.15 K. In this section we treat the calculation of standard heats of reaction at other temperatures from knowledge of the value at the reference temperature.

The general chemical reaction may be written as

$$|\nu_1|A_1 + |\nu_2|A_2 + \cdots \rightarrow |\nu_3|A_3 + |\nu_4|A_4 + \cdots$$

where the $|\nu_i|$ are stoichiometric coefficients and the A_i stand for chemical formulas. The species on the left are reactants; those on the right, products. The sign convention for ν_i is as follows:

$$positive\ (+)\ for\ products \quad and \quad negative\ (-)\ for\ reactants$$

The ν_i with their accompanying signs are called stoichiometric *numbers*. For example, when the ammonia synthesis reaction is written

$$N_2 + 3H_2 \rightarrow 2NH_3$$

then $\qquad\qquad\qquad \nu_{N_2} = -1 \qquad \nu_{H_2} = -3 \qquad \nu_{NH_3} = 2$

This sign convention allows the definition of a standard heat of reaction to be expressed mathematically by the equation

$$\Delta H^\circ \equiv \sum_i \nu_i H_i^\circ \tag{4.20}$$

where H_i° is the enthalpy of species i in its standard state and the summation is over all products and reactants. The standard-state enthalpy of a chemical compound is equal to its heat of formation plus the standard-state enthalpies of its constituent elements. If we arbitrarily set the standard-state enthalpies of all elements equal to zero as the basis of calculation, then the standard-state enthalpy of each compound is its heat of formation. In this event, $H_i^\circ = \Delta H_{f_i}^\circ$ and Eq. (4.20) becomes

$$\Delta H^\circ = \sum_i \nu_i \,\Delta H_{f_i}^\circ \tag{4.21}$$

where the summation is over all products and reactants. This formalizes the procedure described in the preceding section for calculation of standard heats of other reactions from standard heats of formation. Applied to the reaction,

$$4HCl(g) + O_2(g) \rightarrow 2H_2O(g) + 2Cl_2(g)$$

Eq. (4.21) is written
$$\Delta H^\circ = 2\Delta H_{f_{H_2O}}^\circ - 4\Delta H_{f_{HCl}}^\circ$$

With data from Table C.4 for 298.15 K, this becomes

$$\Delta H_{298}^\circ = (2)(-241{,}818) - (4)(-92{,}307) = -114{,}408\ \text{J}$$

in agreement with the result of Example 4.5.

For standard reactions, products and reactants are always at the standard-state pressure of 1 bar. Standard-state enthalpies are therefore functions of temperature only, and by Eq. (2.24),

$$dH_i^\circ = C_{P_i}^\circ\, dT$$

where subscript i identifies a particular product or reactant. Multiplying by ν_i and summing over all products and reactants gives

$$\sum_i \nu_i dH_i^\circ = \sum_i \nu_i C_{P_i}^\circ dT$$

Since ν_i is a constant, it may be placed inside the differential, giving

$$\sum_i d(\nu_i H_i^\circ) = d\sum_i \nu_i H_i^\circ = \sum_i \nu_i C_{P_i}^\circ dT$$

The term $\sum_i \nu_i H_i^\circ$ is the standard heat of reaction, defined by Eq. (4.20). Similarly, we define the standard heat-capacity change of reaction as

$$\Delta C_P^\circ \equiv \sum_i \nu_i C_{P_i}^\circ \tag{4.22}$$

As a result of these definitions, the preceding equation becomes

$$\boxed{d\Delta H^\circ = \Delta C_P^\circ dT} \tag{4.23}$$

This is the fundamental equation relating heats of reaction to temperature.
Integration gives

$$\Delta H^\circ = \Delta H_0^\circ + R \int_{T_0}^{T} \frac{\Delta C_P^\circ}{R} dT \tag{4.24}$$

where ΔH° and ΔH_0° are heats of reaction at temperature T and at reference temperature T_0 respectively. If the temperature dependence of the heat capacity of each product and reactant is given by Eq. (4.4), then the integral is given by the analog of Eq. (4.7):

$$\int_{T_0}^{T} \frac{\Delta C_P^\circ}{R} dT = (\Delta A)T_0(\tau - 1) + \frac{\Delta B}{2}T_0^2(\tau^2 - 1) + \frac{\Delta C}{3}T_0^3(\tau^3 - 1) + \frac{\Delta D}{T_0}\left(\frac{\tau - 1}{\tau}\right) \tag{4.25}$$

where $\tau \equiv T/T_0$ and

$$\Delta A = \sum_i \nu_i A_i$$

with analogous definitions for ΔB, ΔC, and ΔD.

An alternative formulation results when we define a mean heat capacity change of reaction in analogy to Eq. (4.8):

$$\frac{\langle \Delta C_P^\circ \rangle_H}{R} = \Delta A + \frac{\Delta B}{2}T_0(\tau + 1) + \frac{\Delta C}{3}T_0^2(\tau^2 + \tau + 1) + \frac{\Delta D}{\tau T_0^2} \tag{4.26}$$

Equation (4.24) then becomes

$$\Delta H^\circ = \Delta H_0^\circ + \langle \Delta C_P^\circ \rangle_H (T - T_0) \tag{4.27}$$

Equation (4.25) provides a function for evaluation of the integral of interest here that is of exactly the same form as is given by Eq. (4.7). The one comes from the other by simple replacement of C_P by ΔC_P° and of A, etc. by ΔA, etc. The same computer program therefore serves for evaluation of either integral. The only difference is in name:

$$IDCPH(T0,T;DA,DB,DC,DD)$$

where "D" denotes "Δ". In addition, $\langle \Delta C_P^\circ \rangle_H / R$ for computational purposes is named

$$MDCPH(T0,T;DA,DB,DC,DD)$$

Example 4.6   Calculate the standard heat of the methanol-synthesis reaction at 800°C:

$$CO(g) + 2H_2(g) \rightarrow CH_3OH(g)$$

SOLUTION   We apply Eq. (4.21) to this reaction for reference temperature $T_0 = 298.15$ K and with heat-of-formation data from Table C.4:

$$\Delta H_0^\circ = \Delta H_{298}^\circ = -200{,}660 - (-110{,}525) = -90{,}135 \text{ J}$$

Evaluation of the parameters in Eq. (4.25) is based on the following data, taken from Table C.1:

i	ν_i	A	$10^3 B$	$10^6 C$	$10^{-5} D$
CH$_3$OH	1	2.211	12.216	−3.450	0.000
CO	−1	3.376	0.557	0.000	−0.031
H$_2$	−2	3.249	0.422	0.000	0.083

By definition,

$$\Delta A = (1)(2.211) + (-1)(3.376) + (-2)(3.249) = -7.663$$

Similarly, $\Delta B = 10.815 \times 10^{-3}$
$\Delta C = -3.450 \times 10^{-6}$
$\Delta D = -0.135 \times 10^{5}$

Thus the value of the integral of Eq. (4.25) for $T = 1{,}073.15$ K is represented by

$$IDCPH(298.15,1073.15;-7.663,10.815E-3,-3.450E-6,-0.135E+5) \equiv -1{,}615.5 \text{ K}$$

Then by Eq. (4.24),

$$\Delta H^\circ = -90{,}135 + 8.314(-1{,}615.5) = -103{,}566 \text{ J}$$

4.8 Heat Effects of Industrial Reactions

The preceding sections have dealt with the *standard* heat of reaction. Industrial reactions are rarely carried out under standard-state conditions. Furthermore, in actual reactions the reactants may not be present in stoichiometric proportions, the reaction may not go to completion, and the final temperature may differ from the initial temperature. Moreover, inert species may be present, and several reactions may occur simultaneously. Nevertheless, calculations of the heat effects of actual reactions are based on the principles already considered and are best illustrated by example.

Example 4.7 What is the maximum temperature that can be reached by the combustion of methane with 20 percent excess air? Both the methane and the air enter the burner at 25°C.

SOLUTION The reaction is

$$CH_4 + 2O_2 \rightarrow CO_2 + 2H_2O(g)$$

for which

$$\Delta H^\circ_{298} = -393{,}509 + (2)(-241{,}818) - (-74{,}520) = -802{,}625 \text{ J}$$

Since the maximum attainable temperature (often called the theoretical flame temperature) is sought, we assume that the combustion reaction goes to completion adiabatically ($Q = 0$). With the additional assumptions that the kinetic- and potential-energy changes are negligible and that there is no shaft work, the overall energy balance for the process reduces to $\Delta H = 0$. For purposes of calculation of the final temperature, any convenient path between the initial and final states may be used. The path chosen is indicated in the diagram.

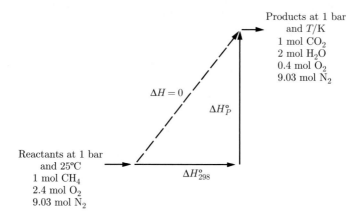

When one mole of methane burned is the basis for all calculations, the following quantities of oxygen and nitrogen are supplied by the entering air:

$$\text{Moles } O_2 \text{ required} = 2.0$$
$$\text{Moles excess } O_2 = (0.2)(2.0) = 0.4$$
$$\text{Moles } N_2 \text{ entering} = (2.4)(79/21) = 9.03$$

The gases leaving the burner contain 1 mol CO_2, 2 mol $H_2O(g)$, 0.4 mol O_2, and 9.03 mol N_2.

Since the enthalpy change must be independent of path,

$$\Delta H^{\circ}_{298} + \Delta H^{\circ}_P = \Delta H = 0 \qquad (A)$$

where all enthalpies are on the basis of 1 mol CH_4 burned. The enthalpy change of the products as they are heated from 298.15 K to T is given by

$$\Delta H^{\circ}_P = \langle C^{\circ}_P \rangle_H (T - 298.15) \qquad (B)$$

where we define $\langle C^{\circ}_P \rangle_H$ as the *total* heat capacity of the product stream:

$$\langle C^{\circ}_P \rangle_H \equiv \sum_i n_i \langle C^{\circ}_{P_i} \rangle_H$$

The simplest procedure here is to sum the mean-heat-capacity equations for the products, each multiplied by its appropriate mole number. Since $C = 0$ for each product gas (Table C.1), Eq. (4.8) yields

$$\langle C^{\circ}_P \rangle_H = \sum_i n_i \langle C^{\circ}_{P_i} \rangle_H = R \left(\sum_i n_i A_i + \frac{\sum_i n_i B_i}{2} T_0 (\tau + 1) + \frac{\sum_i n_i D_i}{\tau T_0^2} \right)$$

Data from Table C.1 are combined as follows:

$$A = \sum_i n_i A_i = (1)(5.457) + (2)(3.470) + (0.4)(3.639) + (9.03)(3.280) = 43.471$$

Similarly,
$$B = \sum_i n_i B_i = 9.502 \times 10^{-3}$$

$$D = \sum_i n_i D_i = -0.645 \times 10^5$$

and $\langle C^{\circ}_P \rangle_H / R$ for the product stream is therefore represented by

```
MCPH(298.15,T;43.471,9.502E-3,0.0,-0.645E+5)
```

Equations (A) and (B) may be combined and solved for T:

$$T = 298.15 - \frac{\Delta H^{\circ}_{298}}{\langle C^{\circ}_P \rangle_H}$$

Because the mean heat capacities depend on T, we evaluate $\langle C^{\circ}_P \rangle_H$ for an assumed value of $T > 298.15$, and substitute the result in the preceding equation. This yields a new value of T for which $\langle C^{\circ}_P \rangle_H$ is reevaluated. The procedure continues to convergence on the final value,

$$T = 2{,}066 \text{ K} \qquad \text{or} \qquad 1{,}793^{\circ}\text{C}$$

Example 4.8   One method for the manufacture of "synthesis gas" (primarily a mixture of CO and H_2) is the catalytic reforming of CH_4 with steam at high temperature and atmospheric pressure:

$$CH_4(g) + H_2O(g) \rightarrow CO(g) + 3H_2(g)$$

The only other reaction which occurs to an appreciable extent is the water-gas-shift reaction:

$$CO(g) + H_2O(g) \rightarrow CO_2(g) + H_2(g)$$

If the reactants are supplied in the ratio, 2 mol steam to 1 mol CH_4, and if heat is supplied to the reactor so that the products reach a temperature of 1,300 K, the CH_4 is completely converted and the product stream contains 17.4 mole percent CO. Assuming the reactants to be preheated to 600 K, calculate the heat requirement for the reactor.

SOLUTION   The standard heats of reaction at $25°C$ for the two reactions are calculated from the data of Table C.4:

$$CH_4(g) + H_2O(g) \rightarrow CO(g) + 3H_2(g) \qquad \Delta H_{298}^\circ = 205{,}813 \text{ J}$$

$$CO(g) + H_2O(g) \rightarrow CO_2(g) + H_2(g) \qquad \Delta H_{298}^\circ = -41{,}166 \text{ J}$$

These two reactions may be added to give a third reaction:

$$CH_4(g) + 2H_2O(g) \rightarrow CO_2(g) + 4H_2(g) \qquad \Delta H_{298}^\circ = 164{,}647 \text{ J}$$

Any pair of these three reactions constitutes an independent set. The third reaction is not independent, since it is obtained by combination of the other two.   The reactions most convenient to work with here are

$$CH_4(g) + H_2O(g) \rightarrow CO(g) + 3H_2(g) \qquad \Delta H_{298}^\circ = 205{,}813 \text{ J} \qquad (A)$$

$$CH_4(g) + 2H_2O(g) \rightarrow CO_2(g) + 4H_2(g) \qquad \Delta H_{298}^\circ = 164{,}647 \text{ J} \qquad (B)$$

We first determine the fraction of CH_4 converted by each of these reactions. As a basis for calculations, let 1 mol CH_4 and 2 mol steam be fed to the reactor. If x mol CH_4 reacts by Eq. (A), then $1 - x$ mol reacts by Eq. (B). On this basis the products of the reaction are

$$
\begin{array}{ll}
\text{CO:} & x \\
\text{H}_2\text{:} & 3x + 4(1 - x) = 4 - x \\
\text{CO}_2\text{:} & 1 - x \\
\text{H}_2\text{O:} & 2 - x - 2(1 - x) = x \\
\hline
\text{Total:} & 5 \text{ mol products}
\end{array}
$$

The mole fraction of CO in the product stream is $x/5 = 0.174$; whence $x = 0.870$. Thus, on the basis chosen, 0.870 mol CH_4 reacts by Eq. (A) and 0.130 mol reacts by Eq. (B). Furthermore, the amount of each species in the product stream is

$$
\begin{array}{l}
\text{Moles CO} = x = 0.87 \\
\text{Moles H}_2 = 4 - x = 3.13 \\
\text{Moles CO}_2 = 1 - x = 0.13 \\
\text{Moles H}_2\text{O} = x = 0.87
\end{array}
$$

We now devise a path, for purposes of calculation, to proceed from reactants at 600 K to products at 1,300 K. Since data are available for the standard heats of reaction at $25°C$, the most convenient path is the one which includes the reactions at 25°C (298.15 K). This is shown schematically in the accompanying diagram. The dashed line represents the actual path for which the enthalpy change is ΔH. Since this enthalpy change is independent of path,

$$\Delta H = \Delta H_R^\circ + \Delta H_{298}^\circ + \Delta H_P^\circ$$

For the calculation of ΔH_{298}°, reactions (A) and (B) must both be taken into account. Since 0.87 mol CH_4 reacts by (A) and 0.13 mol reacts by (B),

$$\Delta H_{298}^\circ = (0.87)(205{,}813) + (0.13)(164{,}647) = 200{,}460 \text{ J}$$

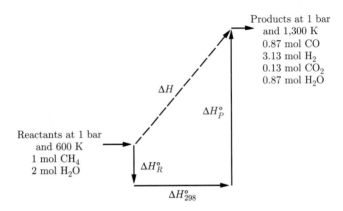

Products at 1 bar
and 1,300 K
0.87 mol CO
3.13 mol H_2
0.13 mol CO_2
0.87 mol H_2O

Reactants at 1 bar
and 600 K
1 mol CH_4
2 mol H_2O

The enthalpy change of the reactants as they are cooled from 600 K to 298.15 K is given by

$$\Delta H_R^\circ = \left(\sum_i n_i \langle C_{P_i}^\circ \rangle_H \right) (298.15 - 600)$$

where the values of $\langle C_{P_i}^\circ \rangle_H / R$ are

CH_4: MCPH(298.15,600;1.702,9.081E-3,-2.164E-6,0.0) $\equiv$ 5.3272
H_2O: MCPH(298.15,600;3.470,1.450E-3,0.0,0.121E+5) $\equiv$ 4.1888

whence

$$\Delta H_R^\circ = (8.314)[(1)(5.3272) + (2)(4.1888)](298.15 - 600) = -34{,}390 \text{ J}$$

The enthalpy change of the products as they are heated from 298.15 to 1,300 K is calculated similarly:

$$\Delta H_P^\circ = \left(\sum_i n_i \langle C_{P_i}^\circ \rangle_H \right) (1{,}300 - 298.15)$$

where $\langle C_{P_i}^\circ \rangle_H / R$ values are

CO: MCPH(298.15,1300;3.376,0.557E-3,0.0,-0.031E+5) $\equiv$ 3.8131
H_2:   MCPH(298.15,1300;3.249,0.422E-3,0.0,0.083E+5) $\equiv$ 3.6076
CO_2: MCPH(298.15,1300;5.457,1.045E-3,0.0,-1.157E+5) $\equiv$ 5.9935
H_2O: MCPH(298.15,1300;3.470,1.450E-3,0.0,0.121E+5) $\equiv$ 4.6599

whence

$$\Delta H_P^{\circ} = (8.314)[(0.87)(3.8131) + (3.13)(3.6076) + (0.13)(5.9935) + (0.87)(4.6599)]$$

$$\times (1{,}300 - 298.15) = 161{,}940 \text{ J}$$

Therefore,

$$\Delta H = -34{,}390 + 200{,}460 + 161{,}940 = 328{,}010 \text{ J}$$

The process is one of steady flow for which W_s, Δz, and $\Delta u^2/2$ are presumed negligible. Thus

$$Q = \Delta H = 328{,}010 \text{ J}$$

This result is on the basis of 1 mol CH_4 fed to the reactor. The factor for converting from J mol^{-1} to (Btu)(lb mol)$^{-1}$ is 0.4299. Therefore on the basis of 1(lb mole) CH_4 fed to the reactor, we have

$$Q = \Delta H = (328{,}010)(0.4299) = 141{,}010 \text{(Btu)}$$

Example 4.9 A boiler is fired with a high-grade fuel oil (consisting only of hydrocarbons) having a standard heat of combustion of $-43{,}515$ J g^{-1} at 25°C with $CO_2(g)$ and $H_2O(l)$ as products. The temperature of the fuel and air entering the combustion chamber is 25°C. The air is assumed dry. The flue gases leave at 300°C, and their average analysis (on a dry basis) is 11.2 percent CO_2, 0.4 percent CO, 6.2 percent O_2, and 82.2 percent N_2. Calculate the fraction of the heat of combustion of the oil that is transferred as heat to the boiler.

SOLUTION   Take as a basis 100 mol dry flue gases, consisting of

CO_2	11.2 mol
CO	0.4 mol
O_2	6.2 mol
N_2	82.2 mol
Total	100.0 mol

This analysis, on a dry basis, does not take into account the H_2O vapor present in the flue gases. The amount of H_2O formed by the combustion reaction is found from an oxygen balance. The O_2 supplied in the air represents 21 mole percent of the air stream. The remaining 79 percent is N_2, which goes through the combustion process unchanged. Thus the 82.2 mol N_2 appearing in 100 mol dry flue gases is supplied with the air, and the O_2 accompanying this N_2 is

$$\text{Moles } O_2 \text{ entering in air} = (82.2)(21/79) = 21.85$$

However,

$$\text{Moles } O_2 \text{ in the dry flue gases} = 11.2 + 0.4/2 + 6.2 = 17.60$$

The difference between these figures is the moles of O_2 that react to form H_2O. Therefore on the basis of 100 mol dry flue gases,

$$\text{Moles } H_2O \text{ formed} = (21.85 - 17.60)(2) = 8.50$$

Moles H_2 in the fuel = moles of water formed = 8.50

The amount of C in the fuel is given by a carbon balance:

Moles C in flue gases = moles C in fuel = $11.2 + 0.4 = 11.60$

These amounts of C and H_2 together give

Mass of fuel burned = $(8.50)(2) + (11.6)(12) = 156.2$ g

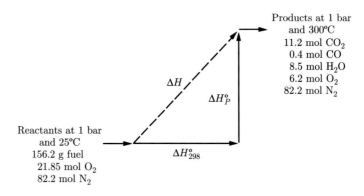

Reactants at 1 bar
and 25°C
156.2 g fuel
21.85 mol O_2
82.2 mol N_2

Products at 1 bar
and 300°C
11.2 mol CO_2
0.4 mol CO
8.5 mol H_2O
6.2 mol O_2
82.2 mol N_2

If this amount of fuel is burned completely to $CO_2(g)$ and $H_2O(l)$ at 25°C, the heat of combustion is

$$\Delta H_{298}^\circ = (-43{,}515)(156.2) = -6{,}797{,}040 \text{ J}$$

However, the reaction actually occurring does not represent complete combustion, and the H_2O is formed as vapor rather than as liquid. The 156.2 g of fuel, consisting of 11.6 moles of C and 8.5 moles of H_2, is represented by the empirical formula $C_{11.6}H_{17}$. Omitting the 6.2 mol O_2 and 82.2 mol N_2 which enter and leave the reactor unchanged, we write the reaction

$$C_{11.6}H_{17}(l) + 15.65O_2(g) \rightarrow 11.2CO_2(g) + 0.4CO(g) + 8.5H_2O(g)$$

This equation is obtained by addition of the following reactions, for each of which the standard heat of reaction at 25°C is known:

$$C_{11.6}H_{17}(l) + 15.85O_2(g) \rightarrow 11.6CO_2(g) + 8.5H_2O(l) \qquad \Delta H_{298}^\circ = -6{,}797{,}040 \text{ J}$$

$$8.5H_2O(l) \rightarrow 8.5H_2O(g) \qquad \begin{aligned}\Delta H_{298}^\circ &= (44{,}012)(8.5) \\ &= 374{,}102 \text{ J}\end{aligned}$$

$$0.4CO_2(g) \rightarrow 0.4CO(g) + 0.2O_2(g) \qquad \begin{aligned}\Delta H_{298}^\circ &= (282{,}984)(0.4) \\ &= 113{,}194 \text{ J}\end{aligned}$$

The sum of these reactions yields the actual reaction, and the sum of the ΔH_{298}° values gives the standard heat of the reaction occurring at 25°C:

$$\Delta H_{298}^\circ = -6{,}309{,}740 \text{ J}$$

The actual process leading from reactants at 25°C to products at 300°C is represented by the dashed line in the accompanying diagram. For purposes of calculating ΔH for this process, we may use any convenient path. The one drawn with solid lines is a logical one, because the enthalpy changes for these steps are easily calculated, and ΔH_{298}° has already been evaluated.

The enthalpy change caused by heating the products of reaction from 25 to 300°C is

$$\Delta H_P^{\circ} = \left(\sum_i n_i \langle C_{P_i}^{\circ} \rangle_H \right) (573.15 - 298.15)$$

where the $\langle C_{P_i}^{\circ} \rangle_H / R$ values are

CO_2: MCPH(298.15,573.15;5.457,1.045E-3,0.0,-1.157E+5) $\equiv$ 5.2352
CO: MCPH(298.15,573.15;3.376,0.557E-3,0.0,-0.031E+5) $\equiv$ 3.6005
H_2O: MCPH(298.15,573.15;3.470,1.450E-3,0.0,0.121E+5) $\equiv$ 4.1725
O_2: MCPH(298.15,573.15;3.639,0.506E-3,0.0,-0.227E+5) $\equiv$ 3.7267
N_2: MCPH(298.15,573.15;3.280,0.593E-3,0.0,0.040E+5) $\equiv$ 3.5618

whence

$$\Delta H_P^{\circ} = (8.314)[(11.2)(5.2352) + (0.4)(3.6005) + (8.5)(4.1725)$$
$$+ (6.2)(3.7267) + (82.2)(3.5618)](573.15 - 298.15)$$
$$= 940{,}660 \text{ J}$$

and

$$\Delta H = \Delta H_{298}^{\circ} + \Delta H_P^{\circ} = -6{,}309{,}740 + 940{,}660 = -5{,}369{,}080 \text{ J}$$

Since the process is one of steady flow for which the shaft work and kinetic- and potential-energy terms in the energy balance [Eq. (2.9)] are zero or negligible, $\Delta H = Q$. Thus, $Q = -5{,}369.08$ kJ, and this amount of heat is transferred to the boiler for every 100 mol dry flue gases formed. This represents

$$\frac{5{,}369{,}080}{6{,}797{,}040}(100) = 79.0 \text{ percent}$$

of the heat of combustion of the fuel.

In the foregoing examples of reactions that occur at approximately 1 bar, we have tacitly assumed that the heat effects of reaction are the same whether gases are mixed or pure, an acceptable procedure for low pressures. For reactions at elevated pressures, this may not be the case, and it may be necessary to account for the effects of pressure and of mixing on the heat of reaction. Suffice it to say at this point that these effects are usually small.

PROBLEMS

4.1. What is the heat required when 10 mol of SO_2 is heated from 200 to 1,100°C at approximately atmospheric pressure in a steady-flow heat exchanger?

4.2. What is the heat required when 12 mol of propane is heated from 250 to 1,200°C at approximately atmospheric pressure in a steady-flow heat exchanger?

4.3. What is the final temperature when heat in the amount of 800 kJ is added to 10 mol of ethylene initially at 200°C in a steady-flow heat exchanger at approximately atmospheric pressure?

4.4. What is the final temperature when heat in the amount of 2,500 kJ is added to 15 mol of 1-butene initially at 260°C in a steady-flow heat exchanger at approximately atmospheric pressure?

4.5. What is the final temperature when heat in the amount of 10^6(Btu) [1.055×10^6 kJ] is added to 40(lb mol) [18.14 kg mol] of ethylene initially at 500(°F) [260°C] in a steady-flow heat exchanger at approximately atmospheric pressure?

4.6. If $250(\text{ft})^3(\text{s})^{-1}$ [7.08 m^3 s^{-1}] of air at 122(°F) [50°C] and approximately atmospheric pressure is preheated for a combustion process to 932(°F) [500°C], what rate of heat transfer is required?

4.7. How much heat is required when 10,000 kg [11.023(tons)] of $CaCO_3$ is heated at atmospheric pressure from 50°C [122(°F)] to 880°C [1,616(°F)]?

4.8. If the heat capacity of a substance is correctly represented by an equation of the form

$$C_P = A + BT + CT^2$$

show that the error resulting when $\langle C_P \rangle_H$ is assumed equal to C_P evaluated at the arithmetic mean of the initial and final temperatures is $C(T_2 - T_1)^2/12$.

4.9. If the heat capacity of a substance is correctly represented by an equation of the form

$$C_P = A + BT + DT^{-2}$$

show that the error resulting when $\langle C_P \rangle_H$ is assumed equal to C_P evaluated at the arithmetic mean of the initial and final temperatures is

$$\frac{D}{T_1 T_2} \left(\frac{T_2 - T_1}{T_2 + T_1} \right)^2$$

4.10. Handbook values for the latent heats of vaporization in J g^{-1} are given in the following table for a number of pure liquids at 25°C and at T_n, the normal boiling point (see App. B for values).

	ΔH^{lv} at 25°C	ΔH^{lv} at T_n
n-Pentane	366.3	357.2
n-Hexane	366.1	336.7
Benzene	433.3	393.9
Toluene	412.3	363.2
Cyclohexane	392.5	358.2

For one of these substances, calculate:

(a) The value of the latent heat at T_n by Eq. (4.19), given the value at 25°C.

(b) The value of the latent heat at T_n by Eq. (4.18).

By what percentages do these values differ from the one listed in the table?

4.11. Handbook values for the latent heats of vaporization in J g^{-1} are given in the following table for several pure liquids at 0°C and at T_n, the normal boiling point (see App. B for values).

	ΔH^{lv} at 0°C	ΔH^{lv} at T_n
Chloroform	270.9	246.9
Methanol	1,189.5	1,099.5
Tetrachloromethane	217.8	194.2

For one of these substances, calculate:

(*a*) The value of the latent heat at T_n by Eq. (4.19), given the value at 0°C.

(*b*) The value of the latent heat at T_n by Eq. (4.18).

By what percentages do these values differ from the one listed in the table?

4.12. Table 9.1 lists the thermodynamic properties of saturated liquid and vapor tetrafluoroethane. Making use of the vapor pressures as a function of temperature and of the saturated-liquid and saturated-vapor volumes, calculate the latent heat of vaporization by Eq. (4.17) at one of the following temperatures and compare the result with the value calculated from the enthalpy values given in the table.

(*a*) 5(°F), (*b*) 30(°F), (*c*) 55(°F), (*d*) 80(°F), (*e*) 105(°F).

4.13. Calculate the heat capacity of a gas sample from the following information: The sample comes to equilibrium in a flask at 25°C and 121.3 kPa. A stopcock is opened briefly, allowing the pressure to drop to 101.3 kPa. With the stopcock closed, the flask warms, returning to 25°C, and the pressure is measured as 104.0 kPa. Determine C_P in J mol^{-1} K^{-1} assuming the gas to be ideal and the expansion of the gas remaining in the flask to be reversible and adiabatic.

4.14. A method for determination of the second virial coefficient of a pure gas is based on the Clapeyron equation through measurements of the latent heat of vaporization ΔH^{lv}, the molar volume of saturated liquid V^l, and the vapor pressure P^{sat}. Determine B in cm^3 mol^{-1} for methyl ethyl ketone (MEK) at 75°C from the following data for this temperature:

- $\Delta H^{lv} = 31,600$ J mol^{-1}.
- $V^l = 96.49$ cm^3 mol^{-1}.
- $\ln P^{sat} = 48.157543 - 5,622.7/T - 4.70504 \ln T$
  where P^{sat} is in kPa and T is in kelvins.

4.15. Estimate the standard heat of formation of liquid ethylbenzene at 25°C.

4.16. A reversible compression of 1 mol of an ideal gas in a piston/cylinder device results in a pressure increase from 1 bar to P_2 and a temperature increase from 400 K to 950 K. The path followed by the gas during compression is given by

$$PV^{1.55} = \text{const}$$

and the molar heat capacity of the gas is given by

$$C_P/R = 3.85 + 0.57 \times 10^{-3}T \qquad [T = K]$$

Determine the heat transferred during the process and the final pressure.

4.17. If the heat of combustion of urea, $(NH_2)_2CO(s)$, at $25°C$ is $631,660$ J mol^{-1} when the products are $CO_2(g)$, $H_2O(l)$, and $N_2(g)$, what is the standard heat of formation of urea at $25°C$?

4.18. Determine the standard heat of each of the following reactions at $25°C$:

(a) $N_2(g) + 3H_2(g) \rightarrow 2NH_3(g)$

(b) $4NH_3(g) + 5O_2(g) \rightarrow 4NO(g) + 6H_2O(g)$

(c) $3NO_2(g) + H_2O(l) \rightarrow 2HNO_3(l) + NO(g)$

(d) $CaC_2(s) + H_2O(l) \rightarrow C_2H_2(g) + CaO(s)$

(e) $2Na(s) + 2H_2O(g) \rightarrow 2NaOH(s) + H_2(g)$

(f) $6NO_2(g) + 8NH_3(g) \rightarrow 7N_2(g) + 12H_2O(g)$

(g) $C_2H_4(g) + \frac{1}{2}O_2(g) \rightarrow \langle(CH_2)_2\rangle O(g)$

(h) $C_2H_2(g) + H_2O(g) \rightarrow \langle(CH_2)_2\rangle O(g)$

(i) $CH_4(g) + 2H_2O(g) \rightarrow CO_2(g) + 4H_2(g)$

(j) $CO_2(g) + 3H_2(g) \rightarrow CH_3OH(g) + H_2O(g)$

(k) $CH_3OH(g) + \frac{1}{2}O_2(g) \rightarrow HCHO(g) + H_2O(g)$

(l) $2H_2S(g) + 3O_2(g) \rightarrow 2H_2O(g) + 2SO_2(g)$

(m) $H_2S(g) + 2H_2O(g) \rightarrow 3H_2(g) + SO_2(g)$

(n) $N_2(g) + O_2(g) \rightarrow 2NO(g)$

(o) $CaCO_3(s) \rightarrow CaO(s) + CO_2(g)$

(p) $SO_3(g) + H_2O(l) \rightarrow H_2SO_4(l)$

(q) $C_2H_4(g) + H_2O(l) \rightarrow C_2H_5OH(l)$

(r) $CH_3CHO(g) + H_2(g) \rightarrow C_2H_5OH(g)$

(s) $C_2H_5OH(l) + O_2(g) \rightarrow CH_3COOH(l) + H_2O(l)$

(t) $C_2H_5CH{:}CH_2(g) \rightarrow CH_2{:}CHCH{:}CH_2(g) + H_2(g)$

(u) $C_4H_{10}(g) \rightarrow CH_2{:}CHCH{:}CH_2(g) + 2H_2(g)$

(v) $C_2H_5CH{:}CH_2(g) + \frac{1}{2}O_2(g) \rightarrow CH_2{:}CHCH{:}CH_2(g) + H_2O(g)$

(w) $4NH_3(g) + 6NO(g) \rightarrow 6H_2O(g) + 5N_2(g)$

(x) $N_2(g) + C_2H_2(g) \rightarrow 2HCN(g)$

(y) $C_6H_5.C_2H_5(g) \rightarrow C_6H_5CH{:}CH_2(g) + H_2(g)$

(z) $C(s) + H_2O(l) \rightarrow H_2(g) + CO(g)$

4.19. What is the standard heat for the reaction of Prob. 4.18(a) at $600°C$?

4.20. What is the standard heat for the reaction of Prob. 4.18(b) at $500°C$?

4.21. What is the standard heat for the reaction of Prob. 4.18(f) at $650°C$?

4.22. What is the standard heat for the reaction of Prob. 4.18(i) at $700°C$?

4.23. What is the standard heat for the reaction of Prob. 4.18(j) at $590(°F)$ [$310°C$]?

4.24. What is the standard heat for the reaction of Prob. 4.18(l) at $770(°F)$ [$410°C$]?

4.25. What is the standard heat for the reaction of Prob. 4.18(m) at 850 K?

4.26. What is the standard heat for the reaction of Prob. 4.18(n) at 1,350 K?

4.27. What is the standard heat for the reaction of Prob. 4.18(o) at 800°C?

4.28. What is the standard heat for the reaction of Prob. 4.18(r) at 450°C?

4.29. What is the standard heat for the reaction of Prob. 4.18(t) 860(°F) [460°C]?

4.30. What is the standard heat for the reaction of Prob. 4.18(u) at 750 K?

4.31. What is the standard heat for the reaction of Prob. 4.18(v) at 900 K?

4.32. What is the standard heat for the reaction of Prob. 4.18(w) at 400°C?

4.33. What is the standard heat for the reaction of Prob. 4.18(x) at 375°C?

4.34. What is the standard heat for the reaction of Prob. 4.18(y) at 1490(°F) [810°C]?

4.35. Develop a general equation for the standard heat of reaction as a function of temperature for one of the reactions given in parts (a), (b), (e), (f), (g), (h), (j), (k), (l), (m), (n), (o), (r), (t), (u), (v), (w), (x), (y), and (z) of Prob. 4.18.

4.36. Hydrocarbon fuels can be produced from methanol by reactions such as the following, which yields 1-hexene:

$$6CH_3OH(g) \rightarrow C_6H_{12}(g) + 6H_2O(g)$$

Compare the standard heat of combustion at 25°C of $6CH_3OH(g)$ with the standard heat of combustion at 25°C of $C_6H_{12}(g)$, reaction products in both cases being $CO_2(g)$ and $H_2O(g)$.

4.37. Calculate the theoretical flame temperature when ethylene at 25°C is burned with

(a) The theoretical amount of air at 25°C.

(b) 25% excess air at 25°C.

(c) 50% excess air at 25°C.

(d) 100% excess air at 25°C.

(e) 50% excess air preheated to 500°C.

4.38. What is the standard heat of combustion of n-pentane gas at 25°C if the combustion products are $H_2O(l)$ and $CO_2(g)$?

4.39. A light fuel oil with an average chemical composition of $C_{10}H_{18}$ is burned with oxygen in a bomb calorimeter. The heat evolved is measured as 43,960 J g^{-1} for the reaction at 25°C. Calculate the standard heat of combustion of the fuel oil at 25°C with $H_2O(g)$ and $CO_2(g)$ as products. Note that the reaction in the bomb occurs at constant volume, produces liquid water as a product, and goes to completion.

4.40. Methane gas is burned completely with 30% excess air at approximately atmospheric pressure. Both the methane and the air enter the furnace at 30°C saturated with water vapor, and the flue gases leave the furnace at 1,500°C. The flue gases then pass through a heat exchanger from which they emerge at 50°C. On the basis of 1 mol of methane, how much heat is lost from the furnace, and how much heat is transferred in the heat exchanger?

4.41. Ammonia gas enters the reactor of a nitric acid plant mixed with 30% more dry air than is required for the complete conversion of the ammonia to nitric oxide and water vapor. If the gases enter the reactor at 75°C [167(°F)], if conversion is 80%, if no side reactions occur, and if the reactor operates adiabatically, what is the temperature of the gases leaving the reactor? Assume ideal gases.

4.42. A gas mixture of methane and steam at atmospheric pressure and 500°C is fed to a reactor, where the following reactions occur:

$$CH_4 + H_2O \rightarrow CO + 3H_2$$

$$CO + H_2O \rightarrow CO_2 + H_2$$

The product stream leaves the reactor at 850°C with the following composition (mole fractions):

$$y_{CO_2} = 0.0275 \qquad y_{CO} = 0.1725 \qquad y_{H_2O} = 0.1725 \qquad y_{H_2} = 0.6275$$

Determine the quantity of heat added to the reactor per mole of product gas.

4.43. A fuel consisting of 75 mole-% methane and 25 mole-% ethane enters a furnace with 80% excess air at 30°C. If 8×10^5 kJ per kg mole of fuel is transferred as heat to boiler tubes, at what temperature does the flue gas leave the furnace? Assume complete combustion of the fuel.

4.44. The gas stream from a sulfur burner consists of 15 mole-% SO_2, 20 mole-% O_2, and 65 mole-% N_2. The gas stream at atmospheric pressure and 400°C enters a catalytic converter where 86% of the SO_2 is further oxidized to SO_3. On the basis of 1 mol of gas entering, how much heat must be removed from the converter so that the product gases leave at 500°C?

4.45. Ethylene gas and steam at 320°C and atmospheric pressure are fed to a chemical-reaction process as an equimolar mixture. The process produces ethanol as represented by the reaction

$$C_2H_4(g) + H_2O(g) \rightarrow C_2H_5OH(l)$$

The liquid ethanol exits the process at 25°C. What is the heat transfer associated with this overall process per mole of ethanol produced?

4.46. Hydrogen is produced by the reaction

$$CO(g) + H_2O(g) \rightarrow CO_2(g) + H_2(g)$$

The feed stream to the reactor is an equimolar mixture of carbon monoxide and steam, and it enters the reactor at 125°C and atmospheric pressure. If 60% of the H_2O is converted to H_2 and if the product stream leaves the reactor at 425°C, how much heat must be transferred from the reactor?

4.47. A direct-fired dryer burns a fuel oil with a net heating value of 19,000(Btu)(lb$_m$)$^{-1}$. [The *net* heating value is obtained when the products of combustion are $CO_2(g)$ and $H_2O(g)$.] The composition of the oil is 85% carbon, 12% hydrogen, 2% nitrogen, and 1% water by weight. The flue gases leave the dryer at 400(°F), and a partial analysis shows that they contain 3 mole-% CO_2 and 11.8 mole-% CO on a dry basis. The fuel, air, and material being dried enter the dryer at 77(°F). If the entering air is saturated with water and if 30% of the net heating value of the oil is allowed for heat losses (including the sensible heat carried out with the dried product), how much water is evaporated in the dryer per (lb$_m$) of oil burned?

4.48. An equimolar mixture of nitrogen and acetylene enters a steady-flow reactor at 25°C and atmospheric pressure. The only reaction occurring is

$$N_2(g) + C_2H_2 \rightarrow 2HCN(g)$$

The product gases leave the reactor at 600°C and contain 24.2 mole-% HCN. How much heat is supplied to the reactor per mole of product gas?

4.49. Chlorine is produced by the reaction

$$4HCl(g) + O_2(g) \rightarrow 2H_2O(g) + 2Cl_2(g)$$

The feed stream to the reactor consists of 60 mole-% HCl, 36 mole-% O_2, and 4 mole-% N_2, and it enters the reactor at 550°C. If the conversion of HCl is 75% and if the process is isothermal, how much heat must be transferred from the reactor per mole of the entering gas mixture?

4.50. A gas consisting only of CO and N_2 is made by passing a mixture of flue gas and air through a bed of incandescent coke (assume pure carbon). The two reactions that occur both go to completion:

$$CO_2 + C \rightarrow 2CO$$

$$2C + O_2 \rightarrow 2CO$$

In a particular instance the flue gas contains 12.8 mole-% CO, 3.7 mole-% CO_2, 5.4 mole-% O_2, and 78.1 mole-% N_2. The flue gas/air mixture is so proportioned that the heats of the two reactions cancel, and the temperature of the coke bed is therefore constant. If this temperature is 875°C, if the feed stream is preheated to 875°C, and if the process is adiabatic, what ratio of moles of flue gas to moles of air is required, and what is the composition of the gas produced?

4.51. A fuel gas consisting of 94 mole-% methane and 6 mole-% nitrogen is burned with 35% excess air in a continuous water heater. Both fuel gas and air enter dry at 77(°F) [25°C]. Water is heated at a rate of 75(lb$_m$)(s)$^{-1}$ [34.0 kg s^{-1}] from 77(°F) [25°C] to 203(°F) [95°C]. The flue gases leave the heater at 410(°F) [210°C]. Of the entering methane, 70% burns to carbon dioxide and 30% burns to carbon monoxide. What volumetric flow rate of fuel gas is required if there are no heat losses to the surroundings?

4.52. A process for the production of 1,3-butadiene results from the catalytic dehydrogenation at atmospheric pressure of 1-butene according to the reaction

$$C_4H_8(g) \rightarrow C_4H_6(g) + H_2(g)$$

To suppress side reactions, the 1-butene feed stream is diluted with steam in the ratio of 10 moles of steam per mole of 1-butene. The reaction is carried out *isothermally* at 525°C, and at this temperature 33% of the 1-butene is converted to 1,3-butadiene. How much heat is transferred to the reactor per mole of entering 1-butene?

CHAPTER 5

THE SECOND LAW OF THERMODYNAMICS

Thermodynamics is concerned with transformations of energy, and the laws of thermodynamics describe the bounds within which these transformations are observed to occur. The first law, stating that energy is conserved in any ordinary process, imposes no restriction on the process direction. Yet, all experience indicates the existence of such a restriction, the concise statement of which constitutes the *second law*.

The differences between the two forms of energy, heat and work, provide some insight into the second law. In an energy balance, both work and heat are included as simple additive terms, implying that one unit of heat, a joule, is equivalent to the same unit of work. Although this is true with respect to an energy balance, experience teaches that there is a difference in quality between heat and work. This experience is summarized by the following facts.

Work is readily transformed into other forms of energy: for example, into potential energy by elevation of a weight, into kinetic energy by acceleration of a mass, into electrical energy by operation of a generator. These processes can be made to approach a conversion efficiency of 100 percent by elimination of friction, a dissipative process that transforms work into heat. Indeed, work is readily transformed completely into heat, as demonstrated by Joule's experiments.

On the other hand, all efforts to devise a process for the continuous conversion of heat completely into work or into mechanical or electrical energy have failed. Regardless of improvements to the devices employed, conversion efficiencies do not exceed about 40 percent. Evidently, heat is a form of energy intrinsically less useful and hence less valuable than an equal quantity of work or mechanical or electrical energy.

Drawing further on our experience, we know that the flow of heat between two bodies always takes place from the hotter to the cooler body, and never in the reverse direction. This fact is of such significance that its restatement serves as an acceptable expression of the second law.

5.1 Statements of the Second Law

The observations just described are results of the restriction imposed by the second law on the directions of actual processes. Many general statements may be made which describe this restriction and, hence, serve as statements of the second law. Two of the most common are:

1. No apparatus can operate in such a way that its *only* effect (in system and surroundings) is to convert heat absorbed by a system completely into work done by the system.

2. No process is possible which consists solely in the transfer of heat from one temperature level to a higher one.

Statement 1 does not say that heat cannot be converted into work; only that the process cannot leave both the system and its surroundings unchanged. Consider a system consisting of an ideal gas in a piston/cylinder assembly expanding reversibly at constant temperature. The work produced can be evaluated from $\int P \, dV$, and for an ideal gas $\Delta U = 0$. Thus, according to the first law, the heat absorbed by the gas from the surroundings is equal to the work produced by the reversible expansion of the gas. At first this might seem a contradiction of statement 1, since in the surroundings the only result is the complete conversion of heat into work. However, the second-law statement requires in addition that there be no change in the system, a requirement which is not met.

This process is limited in another way, because the pressure of the gas soon reaches that of the surroundings, and expansion ceases. Therefore, the continuous production of work from heat by this method is impossible. If the original state of the system is restored in order to comply with the requirements of statement 1, energy from the surroundings in the form of work is needed to compress the gas back to its original pressure. At the same time energy as heat is transferred to the surroundings to maintain constant temperature. This reverse process requires at least the amount of work gained from the expansion; hence no net work is produced. Evidently, statement 1 may be expressed in an alternative way, *viz.*:

1a. It is impossible by a cyclic process to convert the heat absorbed by a system completely into work done by the system.

The word *cyclic* requires that the system be restored periodically to its original state. In the case of a gas in a piston/cylinder assembly the expansion and compression back to the original state constitute a complete cycle. If the process is repeated, it becomes a cyclic process. The restriction to a *cyclic* process in statement 1a amounts to the same limitation as that introduced by the words *only effect* in statement 1.

The second law does not prohibit the production of work from heat, but it does place a limit on the fraction of the heat that may be converted to work in any cyclic process. The partial conversion of heat into work is the basis for nearly all commercial production of power.[1] The development of a quantitative expression for the efficiency of this conversion is the next step in the treatment of the second law.

5.2 Heat Engines

The classical approach to the second law is based on a *macroscopic* viewpoint of properties independent of any knowledge of the structure of matter or behavior of molecules. It arose from study of the *heat engine*, a device or machine that produces work from heat in a cyclic process. An example is a steam power plant in which the working fluid (steam) periodically returns to its original state. In such a power plant the cycle (in simple form) consists of the following steps:

1. Liquid water at approximately ambient temperature is pumped into a boiler at high pressure.

2. Heat from a fuel (heat of combustion of a fossil fuel or heat from a nuclear reaction) is transferred in the boiler to the water, converting it to high-temperature steam at the boiler pressure.

3. Energy is transferred as shaft work from the steam to the surroundings by a device such as a turbine, in which the steam expands to reduced pressure and temperature.

4. Exhaust steam from the turbine is condensed at low temperature and pressure by the transfer of heat to cooling water, thus completing the cycle.

Essential to all heat-engine cycles are the absorption of heat at a high temperature, the rejection of heat at a lower temperature, and the production of work. In the theoretical treatment of heat engines, the two temperature levels which characterize their operation are maintained by *heat reservoirs*, bodies imagined capable of absorbing or rejecting an infinite quantity of heat without temperature change. In operation, the working fluid of a heat engine absorbs heat $|Q_H|$ from a hot reservoir, produces a net amount of work $|W|$, discards heat $|Q_C|$ to a cold reservoir, and returns to its initial state. The first law therefore reduces to

$$|W| = |Q_H| - |Q_C| \tag{5.1}$$

Defining the *thermal efficiency* of the engine as

$$\eta = \frac{\text{net work output}}{\text{heat input}}$$

---

[1] Water and wind power are of course not included.

we get

$$\eta = \frac{|W|}{|Q_H|} = \frac{|Q_H| - |Q_C|}{|Q_H|}$$

or

$$\eta = 1 - \frac{|Q_C|}{|Q_H|} \tag{5.2}$$

Absolute-value signs are used to make the equations independent of the sign conventions for Q and W. We note that for η to be unity (100 percent thermal efficiency) $|Q_C|$ must be zero. No engine has ever been built for which this is true; some heat is always rejected to the cold reservoir. This result of engineering experience is the basis for statements 1 and 1a of the second law.

If a thermal efficiency of 100 percent is not possible for heat engines, what then determines the upper limit? One would certainly expect the thermal efficiency of a heat engine to depend on the degree of reversibility of its operation. Indeed, a heat engine operating in a completely reversible manner is very special, and is called a *Carnot engine*. The characteristics of such an ideal engine were first described by N. L. S. Carnot[2] in 1824. The four steps that make up a *Carnot cycle* are performed in the following order:

1. A system initially in thermal equilibrium with a cold reservoir at temperature T_C undergoes a *reversible* adiabatic process that causes its temperature to rise to that of a hot reservoir at T_H.

2. The system maintains contact with the hot reservoir at T_H, and undergoes a *reversible* isothermal process during which heat $|Q_H|$ is absorbed from the hot reservoir.

3. The system undergoes a *reversible* adiabatic process in the opposite direction of step 1 that brings its temperature back to that of the cold reservoir at T_C.

4. The system maintains contact with the reservoir at T_C, and undergoes a *reversible* isothermal process in the opposite direction of step 2 that returns it to its initial state with rejection of heat $|Q_C|$ to the cold reservoir.

A Carnot engine operates between two heat reservoirs in such a way that all heat absorbed is absorbed at the constant temperature of the hot reservoir and all heat rejected is rejected at the constant temperature of the cold reservoir. Any *reversible* engine operating between two heat reservoirs is a Carnot engine; an engine operating on a different cycle must necessarily transfer heat across finite temperature differences and therefore cannot be reversible.

Since a Carnot engine is reversible, it may be operated in reverse; the Carnot cycle is then traversed in the opposite direction, and it becomes a reversible refrigeration cycle for which the quantities $|Q_H|$, $|Q_C|$, and $|W|$ are the same as for the engine cycle but are reversed in direction.

---

[2]Nicolas Leonard Sadi Carnot (1796–1832), a French engineer.

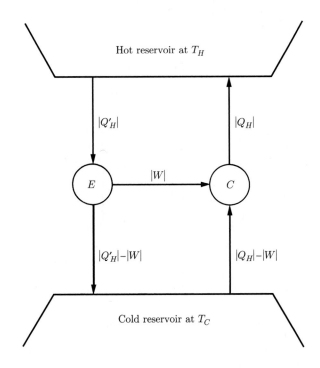

Figure 5.1: Engine E operating a Carnot refrigerator C.

Carnot's theorem states that for two given heat reservoirs no engine can have a higher thermal efficiency than a Carnot engine. Such an engine absorbs heat $|Q_H|$ from a hot reservoir, produces work $|W|$, and discards heat $|Q_C| - |W|$ to a cold reservoir. Assume an engine E *with a greater thermal efficiency* than a Carnot engine for the same heat reservoirs, absorbing heat $|Q'_H|$, producing work $|W|$, and discarding heat $|Q'_H| - |W|$. Then

$$\frac{|W|}{|Q'_H|} > \frac{|W|}{|Q_H|}$$

whence

$$|Q_H| > |Q'_H|$$

Let engine E drive the Carnot engine backward as a Carnot refrigerator, as shown schematically in Fig. 5.1. For the engine/refrigerator combination, the net heat extracted from the cold reservoir is

$$|Q_H| - |W| - (|Q'_H| - |W|) = |Q_H| - |Q'_H|$$

The net heat delivered to the hot reservoir is also $|Q_H| - |Q'_H|$. Thus, the sole result of the engine/refrigerator combination is the transfer of heat from temperature T_C to the higher temperature T_H. Since this is in violation of statement 2 of the second

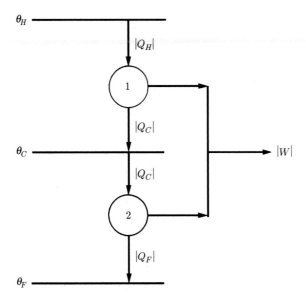

Figure 5.2: Carnot engines 1 and 2 constitute a third Carnot engine.

law, the original premise that engine E has a greater thermal efficiency than the Carnot engine is false, and Carnot's theorem is proved. In similar fashion, one can prove a corollary to Carnot's theorem: All Carnot engines operating between heat reservoirs at the same two temperatures have the same thermal efficiency. These results show that the thermal efficiency of a Carnot engine depends only on the temperature levels T_H and T_C and not upon the working substance of the engine.

5.3 Thermodynamic Temperature Scales

In the preceding discussion we identified temperature levels by the Kelvin scale, established with ideal-gas thermometry. This does not preclude our taking advantage of the opportunity provided by the Carnot engine to establish a *thermodynamic* temperature scale that is truly independent of any material properties. Let θ represent temperature on some empirical scale that unequivocally identifies temperature levels. Consider now two Carnot engines, one operating between a hot reservoir at θ_H and a cold reservoir at temperature θ_C, and a second operating between the reservior at θ_C and a still colder reservoir at θ_F, as shown in Fig. 5.2. The heat rejected by the first engine $|Q_C|$ is absorbed by the second; therefore the two engines working together constitute a third Carnot engine absorbing heat $|Q_H|$ from the reservoir at θ_H and rejecting heat $|Q_F|$ to the reservoir at θ_F. According to Carnot's theorem, the thermal efficiency of the first engine is a function of θ_H and θ_C:

$$\eta = 1 - \frac{|Q_C|}{|Q_H|} = \phi(\theta_H, \theta_C)$$

Rearrangement gives

$$\frac{|Q_H|}{|Q_C|} = \frac{1}{1 - \phi(\theta_H, \theta_C)} = f(\theta_H, \theta_C) \tag{5.3}$$

where f is an unknown function.

For the second and third engines, equations of the same functional form apply:

$$\frac{|Q_C|}{|Q_F|} = f(\theta_C, \theta_F)$$

and

$$\frac{|Q_H|}{|Q_F|} = f(\theta_H, \theta_F)$$

Division of the second of these equations by the first gives

$$\frac{|Q_H|}{|Q_C|} = \frac{f(\theta_H, \theta_F)}{f(\theta_C, \theta_F)}$$

Comparison of this equation with Eq. (5.3) shows that the arbitrary temperature θ_F must cancel from the ratio on the right, leaving

$$\frac{|Q_H|}{|Q_C|} = \frac{\psi(\theta_H)}{\psi(\theta_C)} \tag{5.4}$$

where ψ is another unknown function.

The right-hand side of Eq. (5.4) is the ratio of ψ evaluated at two thermodynamic temperatures; the ψ's are to each other as the absolute values of the heats absorbed and rejected by a Carnot engine operating between reservoirs at these temperatures, quite independent of the properties of any substance. However, Eq. (5.4) still leaves us abritrary choice of the empirical temperature represented by θ; once this choice is made, we must determine the function ψ. If θ is chosen as the Kelvin temperature T, then Eq. (5.4) becomes

$$\frac{|Q_H|}{|Q_C|} = \frac{\psi(T_H)}{\psi(T_C)} \tag{5.5}$$

5.4   Thermodynamic Temperature and the Ideal-Gas Scale

The cycle traversed by an ideal gas serving as the working fluid in a Carnot engine is shown by a PV diagram in Fig. 5.3. It consists of four reversible steps:

1. $a \to b$ Adiabatic compression until the temperature rises from T_C to T_H.

2. $b \to c$ Isothermal expansion to arbitrary point c with absorption of heat $|Q_H|$.

3. $c \to d$ Adiabatic expansion until the temperature decreases to T_C.

4. $d \to a$ Isothermal compression to the initial state with rejection of heat $|Q_C|$.

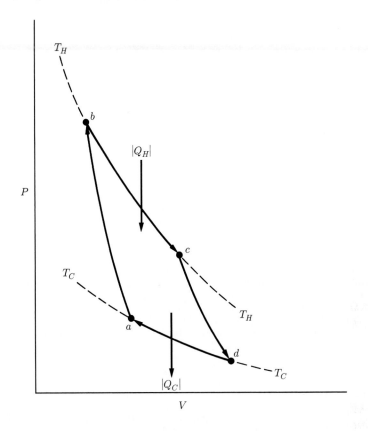

Figure 5.3: *PV* diagram showing Carnot cycle for an ideal gas.

For the isothermal steps $b \to c$ and $d \to a$, Eq. (3.18) yields

$$|Q_H| = RT_H \ln \frac{V_c}{V_b} \qquad \text{and} \qquad |Q_C| = RT_C \ln \frac{V_d}{V_a}$$

Therefore

$$\frac{|Q_H|}{|Q_C|} = \frac{T_H}{T_C} \frac{\ln(V_c/V_b)}{\ln(V_d/V_a)} \qquad\qquad (5.6)$$

For an adiabatic process Eq. (3.20) is written

$$-\frac{C_V}{R} \frac{dT}{T} = \frac{dV}{V}$$

For step $a \to b$, integration gives

$$\int_{T_C}^{T_H} \frac{C_V}{R} \frac{dT}{T} = \ln \frac{V_a}{V_b}$$

Similarly, for step $c \to d$,

$$\int_{T_C}^{T_H} \frac{C_V}{R} \frac{dT}{T} = \ln \frac{V_d}{V_c}$$

Since the left-hand sides of these two equations are the same,

$$\ln \frac{V_a}{V_b} = \ln \frac{V_d}{V_c}$$

This may also be written

$$\ln \frac{V_c}{V_b} = \ln \frac{V_d}{V_a}$$

Equation (5.6) now becomes

$$\boxed{\frac{|Q_H|}{|Q_C|} = \frac{T_H}{T_C}} \tag{5.7}$$

Comparison of this result with Eq. (5.5) yields the simplest possible functional relation for ψ, namely, $\psi(T) = T$. We conclude that the Kelvin temperature scale, based on the properties of ideal gases, is in fact a thermodynamic scale, independent of the characteristics of any particular substance. Substitution of Eq. (5.7) into Eq. (5.2) gives

$$\boxed{\eta = 1 - \frac{T_C}{T_H}} \tag{5.8}$$

Equations (5.7) and (5.8) are known as *Carnot's equations*. In Eq. (5.7) the smallest possible value of $|Q_C|$ is zero; the corresponding value of T_C is the absolute zero of temperature on the Kelvin scale. As mentioned in Sec. 1.4, this occurs at $-273.15°$C. Equation (5.8) shows that the thermal efficiency of a Carnot engine can approach unity only when T_H approaches infinity or T_C approaches zero. Neither of these conditions is realized on earth; all heat engines therefore operate at thermal efficiencies less than unity. The cold reservoirs naturally available are the atmosphere, lakes and rivers, and the oceans, for which $T_C \simeq 300$ K. Hot reservoirs are objects such as furnaces where the temperature is maintained by combustion of fossil fuels and nuclear reactors where the temperature is maintained by fission of radioactive elements. For these practical heat sources, $T_H \simeq 600$ K. With these values,

$$\eta = 1 - \frac{300}{600} = 0.5$$

This is a rough practical limit for the thermal efficiency of a Carnot engine; actual heat engines are irreversible, and their thermal efficiencies rarely exceed 0.35.

> **Example 5.1**   A central power plant, rated at 800,000 kW, generates steam at 585 K and discards heat to a river at 295 K. If the thermal efficiency of the plant is 70 percent of the maximum possible value, how much heat is discarded to the river at rated power?

SOLUTION The maximum possible thermal efficiency is given by Eq. (5.8). Taking T_H as the steam-generation temperature and T_C as the river temperature, we get

$$\eta_{\text{max}} = 1 - \frac{295}{585} = 0.4957$$

The actual thermal efficiency is then

$$\eta = (0.7)(0.4957) = 0.3470$$

$$W = Q_H - Q_C$$

By definition

$$\eta = \frac{|W|}{|Q_H|}$$

Substituting for $|Q_H|$ by Eq. (5.1) gives

$$\eta = \frac{|W|}{|W| + |Q_C|}$$

which may be solved for $|Q_C|$:

$$|Q_C| = \left(\frac{1 - \eta}{\eta}\right) |W|$$

whence

$$|Q_C| = \left(\frac{1 - 0.347}{0.347}\right) (800{,}000) = 1{,}505{,}500 \text{ kW}$$

or

$$|Q_C| = 1{,}505{,}500 \text{ kJ s}^{-1}$$

This amount of heat would raise the temperature of a moderate-size river several degrees Celsius.

5.5 Entropy

Equation (5.7) for a Carnot engine may be written

$$\frac{|Q_H|}{T_H} = \frac{|Q_C|}{T_C}$$

If the heat quantities refer to the engine (rather than to the heat reservoirs), the numerical value of $|Q_H|$ is positive and that of $|Q_C|$ is negative. The equivalent equation written without absolute-value signs is therefore

$$\frac{Q_H}{T_H} = \frac{-Q_C}{T_C}$$

or

$$\frac{Q_H}{T_H} + \frac{Q_C}{T_C} = 0 \tag{5.9}$$

Thus for a complete cycle of a Carnot engine, the two quantities Q/T associated with the absorption and rejection of heat by the working fluid of the engine sum to zero. Since the working fluid of a Carnot engine periodically returns to its initial

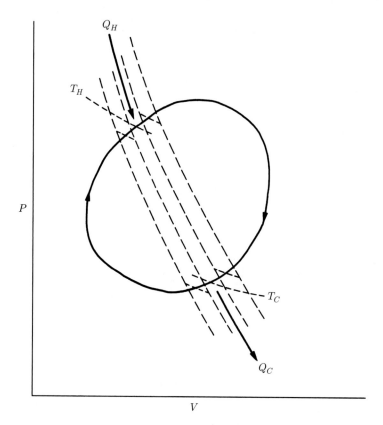

Figure 5.4: An arbitrary cyclic process drawn on a PV diagram.

state, such properties as temperature, pressure, and internal energy return to their initial values even though they vary from one part of the cycle to another. The principal characteristic of a property is that the sum of its changes is zero for any complete cycle. Thus Eq. (5.9) suggests the existence of a property whose changes are here given by the quantities Q/T.

Further insight may be gained through study of a reversible cyclic process as shown by the PV diagram of Fig. 5.4, where the closed curve represents an arbitrary path taken by an arbitrary fluid. We divide the entire enclosed area by a series of reversible adiabatic curves; since such curves cannot intersect (see Pb. 5.1), they may be drawn arbitrarily close to one another. A few of these curves are shown on the figure as long dashed lines. We connect adjacent adiabatic curves by two short reversible isotherms which approximate the curve of the general cycle as closely as possible. The approximation clearly improves as the adiabatic curves are more closely spaced, and by making the separation arbitrarily small, we may approximate the original cycle as closely as we please. Each pair of adjacent adiabatic curves and their isothermal connecting curves represent a Carnot cycle for which Eq. (5.9) applies.

Each cycle has its own pair of isotherms T_H and T_C and associated heat quantities Q_H and Q_C. These are indicated on Fig. 5.4 for a representative cycle. When the adiabatic curves are so closely spaced that the isothermal steps are infinitesimal, the heat quantities become dQ_H and dQ_C, and Eq. (5.9) is written

$$\frac{dQ_H}{T_H} + \frac{dQ_C}{T_C} = 0$$

In this equation T_H and T_C are the absolute temperatures at which the quantities of heat dQ_H and dQ_C are transferred to or from the fluid of the cyclic process. Integration gives the sum of all quantities dQ/T for the entire cycle:

$$\oint \frac{dQ_{\text{rev}}}{T} = 0 \tag{5.10}$$

where the circle in the integral sign signifies that integration is over a complete cycle, and the subscript "rev" indicates that the equation is valid only for reversible cycles.

Thus the quantities dQ_{rev}/T sum to zero for any series of reversible processes that causes a system to undergo a cyclic process. We therefore infer the existence of a property of the system whose differential changes are given by these quantities. The property is called *entropy* (en′-tro-py), and its differential changes are

$$dS^t = \frac{dQ_{\text{rev}}}{T} \tag{5.11}$$

where S^t is the total (rather than molar) entropy of the system. Alternatively,

$$\boxed{dQ_{\text{rev}} = T \, dS^t} \tag{5.12}$$

We represent by points A and B on the PV diagram of Fig. 5.5 two equilibrium states of a particular fluid, and show two arbitrary reversible processes connecting these points along paths ACB and ADB. Integration of Eq. (5.11) for each path gives

$$\Delta S^t = \int_{ACB} \frac{dQ_{\text{rev}}}{T}$$

and

$$\Delta S^t = \int_{ADB} \frac{dQ_{\text{rev}}}{T}$$

where in view of Eq. (5.10) the two integrals must be equal. We therefore conclude that ΔS^t is independent of path and is a property change given by $S_B^t - S_A^t$.

If the fluid is changed from state A to state B by an *irreversible* process, the entropy change must still be $\Delta S^t = S_B^t - S_A^t$, but experiment shows that this result is *not* given by $\int dQ/T$ evaluated for the irreversible process itself, because the calculation of entropy changes by this integral must in general be along reversible paths.

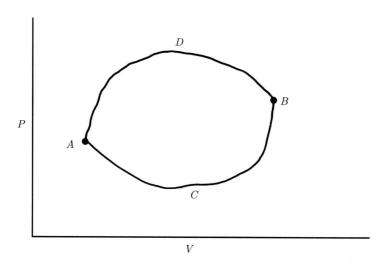

Figure 5.5: Two reversible paths joining equilibrium states A and B.

The entropy change of a *heat reservoir*, however, is always given by Q/T, where Q is the quantity of heat transferred to or from the reservoir at temperature T, whether the transfer is reversible or irreversible. The reason is that the effect of heat transfer on a heat reservoir is the same regardless of the temperature of the source or sink of the heat.

When a process is reversible and adiabatic, $dQ_{\text{rev}} = 0$; then by Eq. (5.11), $dS^t = 0$. Thus the entropy of a system is constant during a reversible adiabatic process, and the process is said to be *isentropic*.

This discussion of entropy can be summarized as follows:

1. The change in entropy of any system undergoing a *reversible* process is found by integration of Eq. (5.11):

$$\Delta S^t = \int \frac{dQ_{\text{rev}}}{T} \qquad (A)$$

2. When a system undergoes an *irreversible* process from one equilibrium state to another, the entropy change of the system ΔS^t must be evaluated by application of Eq. (A) to *an arbitrarily chosen reversible process* that accomplishes the same change of state as the actual process. Integration is *not* carried out for the irreversible path. Since entropy is a state function, the entropy changes of the irreversible and reversible processes are identical.

3. Entropy is useful precisely because it is a state function or *property*. It owes its existence to the second law, from which it arises in much the same way as internal energy does from the first law.

In the special case of a *mechanically reversible* process (Sec. 2.9), the entropy change of the system is correctly evaluated from $\int dQ/T$ applied to the actual process, even though the *heat transfer* between system and surroundings is irreversible. The reason is that it is immaterial, as far as the system is concerned, whether the temperature difference causing the heat transfer is differential (making the process reversible) or finite. The entropy change of a system *caused by the transfer of heat* can always be calculated by $\int dQ/T$, whether the heat transfer is accomplished reversibly or irreversibly. However, when a process is irreversible on account of finite differences in other driving forces, such as pressure, the entropy change is not caused solely by the heat transfer, and for its calculation one must devise a reversible means of accomplishing the same change of state.

This introduction to entropy through a consideration of heat engines is the classical approach, closely following its actual historical development. A complementary approach, based on molecular concepts and statistical mechanics, is considered briefly in Sec. 5.9.

5.6 Entropy Changes of an Ideal Gas

The first law, written for one mole or a unit mass of fluid, is

$$dU = dQ + dW$$

For a reversible process, this becomes

$$dU = dQ_{\text{rev}} - P\,dV$$

By the definition of enthalpy,

$$H = U + PV$$

whence

$$dH = dU + P\,dV + V\,dP$$

Substitution for dU gives

$$dH = dQ_{\text{rev}} - P\,dV + P\,dV + V\,dP$$

or

$$dQ_{\text{rev}} = dH - V\,dP$$

For an ideal gas, we make the substitutions $dH = C_P^{ig}dT$ and $V = RT/P$; then division by T gives

$$\frac{dQ_{\text{rev}}}{T} = C_P^{ig}\frac{dT}{T} - R\frac{dP}{P}$$

As a result of Eq. (5.11), this may be written

$$dS = C_P^{ig}\frac{dT}{T} - R\frac{dP}{P}$$

or

$$\frac{dS}{R} = \frac{C_P^{ig}}{R}\frac{dT}{T} - d\ln P \tag{5.13}$$

where S is the molar entropy of an ideal gas. Integration from an initial state at conditions T_0 and P_0 to a final state at conditions T and P gives

$$\boxed{\frac{\Delta S}{R} = \int_{T_0}^{T} \frac{C_P^{ig}}{R} \frac{dT}{T} - \ln \frac{P}{P_0}}$$
(5.14)

Although *derived* for a reversible process, this equation relates properties only, and is independent of the process causing the change of state. It is therefore a general equation for the calculation of entropy changes of an ideal gas.

Example 5.2   For an ideal gas with constant heat capacities undergoing a reversible adiabatic (and therefore isentropic) process, we found earlier that

$$\frac{T_2}{T_1} = \left(\frac{P_2}{P_1}\right)^{(\gamma-1)/\gamma}$$
(3.23)

Show that this same equation results from application of Eq. (5.14) with $\Delta S = 0$.

SOLUTION   Since C_P^{ig} is constant Eq. (5.14) can be written

$$0 = \ln \frac{T_2}{T_1} - \frac{R}{C_P^{ig}} \ln \frac{P_2}{P_1}$$

whence

$$\frac{T_2}{T_1} = \left(\frac{P_2}{P_1}\right)^{R/C_P^{ig}}$$
(A)

For an ideal gas Eq. (3.17) gives

$$C_P^{ig} = C_V^{ig} + R$$

Upon division by C_P^{ig} this becomes

$$1 = \frac{C_V^{ig}}{C_P^{ig}} + \frac{R}{C_P^{ig}} = \frac{1}{\gamma} + \frac{R}{C_P^{ig}}$$

where $\gamma = C_P^{ig}/C_V^{ig}$. Solving for R/C_P^{ig}, we get

$$\frac{R}{C_P^{ig}} = \frac{\gamma - 1}{\gamma}$$

This transforms Eq. (A) into Eq. (3.23), as required.

Equation (4.4) for the temperature dependence of the molar heat capacity C_P^{ig} allows integration of the first term on the right of Eq. (5.14). The result is conveniently expressed as

$$\int_{T_0}^{T} \frac{C_P^{ig}}{R} \frac{dT}{T} = A \ln \tau + \left[BT_0 + \left(CT_0^2 + \frac{D}{\tau^2 T_0^2} \right) \left(\frac{\tau + 1}{2} \right) \right] (\tau - 1)$$
(5.15)

where

$$\tau \equiv \frac{T}{T_0}$$

Since this integral must often be evaluated, we include in App. D representative computer programs to provide values of the function given by Eq. (5.15). For computational purposes this function is named

ICPS(T0,T;A,B,C,D)

The computer programs also calculate a mean heat capacity defined as

$$\langle C_P^{ig} \rangle_S = \frac{\int_{T_0}^{T} C_P^{ig} dT/T}{\ln(T/T_0)} \tag{5.16}$$

Here, the subscript "S" denotes a mean value specific to entropy calculations. Division of Eq. (5.15) by $\ln(T/T_0)$ or $\ln \tau$ therefore yields

$$\frac{\langle C_P^{ig} \rangle_S}{R} = A + \left[BT_0 + \left(CT_0^2 + \frac{D}{\tau^2 T_0^2} \right) \left(\frac{\tau + 1}{2} \right) \right] \left(\frac{\tau - 1}{\ln \tau} \right) \tag{5.17}$$

For computational purposes this function is named

MCPS(T0,T;A,B,C,D)

Solving for the integral in Eq. (5.16), we get

$$\int_{T_0}^{T} C_P^{ig} \frac{dT}{T} = \langle C_P^{ig} \rangle_S \ln \frac{T}{T_0}$$

and Eq. (5.14) becomes

$$\boxed{\frac{\Delta S}{R} = \frac{\langle C_P^{ig} \rangle_S}{R} \ln \frac{T}{T_0} - \ln \frac{P}{P_0}} \tag{5.18}$$

This form of the equation for entropy changes of an ideal gas may be useful when iterative calculations are required.

Example 5.3 Methane gas at 550 K and 5 bar undergoes a reversible adiabatic expansion to 1 bar. Assuming methane an ideal gas at these conditions, what is its final temperature?

SOLUTION   For this process $\Delta S = 0$, and Eq. (5.18) becomes

$$\frac{\langle C_P^{ig} \rangle_S}{R} \ln \frac{T_2}{T_1} = \ln \frac{P_2}{P_1} = \ln \frac{1}{5} = -1.6094$$

Since $\langle C_P^{ig} \rangle_S$ depends on T_2, we rearrange this equation for iterative solution:

$$\ln \frac{T_2}{T_1} = \frac{-1.6094}{\langle C_P^{ig} \rangle_S/R}$$

whence

$$T_2 = T_1 \exp\left(\frac{-1.6094}{\langle C_P^{ig}\rangle_S/R}\right) \qquad (A)$$

Here, $\langle C_P^{ig}\rangle_S/R$ is given by Eq. (5.17) with constants from Table C.1:

```
MCPS(550,T2;1.702,9.081E-3,-2.164E-6,0.0)
```

With an initial value of $T_2 < 550$, we compute a value of $\langle C_P^{ig}\rangle_S/R$ for substitution into Eq. (A). This yields a new value of T_2 from which to recompute $\langle C_P^{ig}\rangle_S/R$, and the process continues to convergence on a final value of $T_2 = 411.34$ K.

5.7 Mathematical Statement of the Second Law

Consider two heat reservoirs, one at temperature T_H and a second at the lower temperature T_C. Let a quantity of heat $|Q|$ be transferred from the hotter to the cooler reservoir. The entropy change of the reservoir at T_H is

$$\Delta S_H^t = \frac{-|Q|}{T_H}$$

and the entropy change of the reservoir at T_C is

$$\Delta S_C^t = \frac{|Q|}{T_C}$$

These two entropy changes are added to give

$$\Delta S_{\text{total}} = \Delta S_H^t + \Delta S_C^t = \frac{-|Q|}{T_H} + \frac{|Q|}{T_C}$$

or

$$\Delta S_{\text{total}} = |Q|\left(\frac{T_H - T_C}{T_H T_C}\right)$$

Since $T_H > T_C$, the *total* entropy change as a result of this irreversible process is positive. We note also that ΔS_{total} becomes smaller as the difference $T_H - T_C$ gets smaller. When T_H is only infinitesimally higher than T_C, the heat transfer is reversible, and ΔS_{total} approaches zero. Thus for the process of irreversible heat transfer, ΔS_{total} is always positive, approaching zero as the process becomes reversible.

Consider now an adiabatic process wherein no heat transfer occurs. We represent on the PV diagram of Fig. 5.6 an *irreversible*, adiabatic expansion of 1 mole of fluid from an initial equilibrium state at point A to a final equilibrium state at point B. Now suppose the fluid is restored to its initial state by a *reversible* process. If the initial process results in an entropy change of the fluid, then there must be heat transfer during the reversible restoration process such that

$$\Delta S^t = S_A^t - S_B^t = \int_B^A \frac{dQ_{\text{rev}}}{T}$$

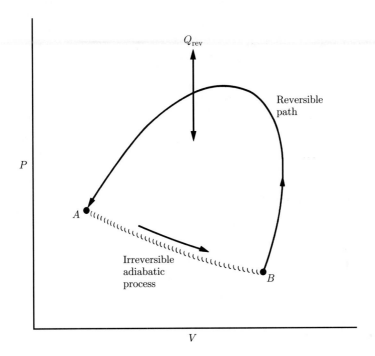

Figure 5.6: Cycle containing an irreversible adiabatic process A to B.

The original irreversible process together with the reversible restoration process constitute a cycle for which $\Delta U = 0$ and for which the work is therefore

$$-W = -W_{\text{irr}} - W_{\text{rev}} = Q_{\text{rev}} = \int_B^A dQ_{\text{rev}}$$

However, according to statement 1a of the second law, Q_{rev} cannot be directed *into* the system, for the cycle would then be a process for the complete conversion of heat into work. Thus, $\int dQ_{\text{rev}}$ is negative, and it follows that $S_A^t - S_B^t$ is also negative; whence $S_B^t > S_A^t$. Since the original irreversible process is adiabatic ($\Delta S_{\text{surr}} = 0$), the total entropy change of the system and surroundings as a result of the process is $\Delta S_{\text{total}} = S_B^t - S_A^t > 0$.

In arriving at this result, our presumption is that the original irreversible process results in an entropy change of the fluid. If we assume that the original process produces no entropy change of the fluid, then we can restore the system to its initial state by a simple reversible adiabatic process. This cycle is accomplished with no heat transfer and therefore with no net work. Thus the system is restored without leaving any change elsewhere, and this implies that the original process is reversible rather than irreversible.

Thus the same result is found for adiabatic processes as for direct heat transfer: ΔS_{total} is always positive, approaching zero as a limit when the process becomes reversible. This same conclusion can be demonstrated for any process whatever, leading to the general equation:

$$\boxed{\Delta S_{\text{total}} \geq 0}$$ (5.19)

This is the mathematical statement of the second law. It affirms that every process proceeds in such a direction that the *total* entropy change associated with it is positive, the limiting value of zero being reached only by a reversible process. No process is possible for which the total entropy decreases.

We return now to a cyclic heat engine that takes in heat $|Q_H|$ from a heat reservoir at T_H, and discards heat $|Q_C|$ to another heat reservoir at temperature T_C. Since the engine operates in cycles, it undergoes no net changes in its properties. The total entropy change of the process is therefore the sum of the entropy changes of the heat reservoirs:

$$\Delta S_{\text{total}} = \frac{-|Q_H|}{T_H} + \frac{|Q_C|}{T_C}$$

The work produced by the engine is

$$|W| = |Q_H| - |Q_C|$$ (5.1)

Elimination of $|Q_C|$ between these two equations and solution for $|W|$ gives

$$|W| = -T_C \, \Delta S_{\text{total}} + |Q_H| \left(1 - \frac{T_C}{T_H} \right)$$

This is the general equation for the work of a heat engine operating between two temperature levels. The minimum work output is zero, resulting when the engine is completely inefficient and the process degenerates into simple irreversible heat transfer between the two heat reservoirs. In this case solution for ΔS_{total} yields the equation obtained at the beginning of this section. The maximum work is obtained when the engine is reversible, in which case $\Delta S_{\text{total}} = 0$, and the equation reduces to the second term on the right, the work of a Carnot engine.

> **Example 5.4**   A steel casting $[C_P = 0.5 \text{ kJ kg}^{-1} \text{ K}^{-1}]$ weighing 40 kg and at a temperature of $450°$C is quenched in 150 kg of oil $[C_P = 2.5 \text{ kJ kg}^{-1} \text{ K}^{-1}]$ at $25°$C. If there are no heat losses, what is the change in entropy of (*a*) the casting, (*b*) the oil, and (*c*) both considered together?
>
> SOLUTION   The final temperature t of the oil and the steel casting is found by an energy balance. Since the change in energy of the oil and steel together must be zero,
>
> $$(40)(0.5)(t - 450) + (150)(2.5)(t - 25) = 0$$
>
> Solution yields $t = 46.52°$C.
>
> (*a*) Change in entropy of the casting:
>
> $$m \, \Delta S = \int \frac{dQ}{T} = m \int \frac{C_P dT}{T} = m C_P \ln \frac{T_2}{T_1}$$
>
> $$m \, \Delta S = (40)(0.5) \ln \frac{273.15 + 46.52}{273.15 + 450} = -16.33 \text{ kJ K}^{-1}$$

(b) Change in entropy of the oil:

$$\Delta S^t = (150)(2.5)\ln\frac{273.15 + 46.52}{273.15 + 25} = 26.13 \text{ kJ K}^{-1}$$

(c) Total entropy change:

$$\Delta S_{\text{total}} = -16.33 + 26.13 = 9.80 \text{ kJ K}^{-1}$$

We note that although the total entropy change is positive, the entropy of the casting has decreased.

Example 5.5 An inventor claims to have devised a process which takes in only saturated steam at 100°C and which by a complicated series of steps makes heat continuously available at a temperature level of 200°C. She claims further that, for every kilogram of steam taken into the process, 2,000 kJ of energy as heat is liberated at the higher temperature level of 200°C. Show whether or not this process is possible. In order to give the inventor the benefit of any doubt, assume cooling water available in unlimited quantity at a temperature of 0°C.

SOLUTION   For any process to be theoretically possible, it must meet the requirements of the first and second laws of thermodynamics. The detailed mechanism need not be known in order to determine whether this is the case; only the overall result is required. If the claims of the inventor satisfy the laws of thermodynamics, means for realizing them are theoretically possible. The determination of a mechanism is then a matter of ingenuity. Otherwise, the process is impossible, and no mechanism for carrying it out can be devised.

In the present instance, a continuous process takes saturated steam into some sort of apparatus, and heat is made continuously available at a temperature level $T' = 200°C$. Since cooling water is available at $T_\sigma = 0°C$, maximum use can be made of the steam by cooling it to this temperature. We therefore assume that the steam is condensed and cooled to 0°C and is discharged from the process at this temperature and at atmospheric pressure. *All* the heat liberated in this operation cannot be made available at temperature level $T' = 200°C$, because this would violate statement 2 of the second law. We must suppose that heat Q_σ is also transferred to the cooling water at $T_\sigma = 0°C$. Moreover, the process must satisfy the first law; thus by Eq. (2.10):

$$\Delta H = Q + W_s$$

where ΔH is the enthalpy change of the steam as it flows through the apparatus and Q is the total heat transfer between the apparatus and its surroundings. Since no shaft work is accomplished by the process, $W_s = 0$. The surroundings consist of the cooling water, which acts as a heat reservoir at the constant temperature of $T_\sigma = 0°C$, and a heat reservoir at $T' = 200°C$ to which heat in the amount of 2,000 kJ is transferred for each kilogram of steam entering the apparatus. The diagram of Fig. 5.7 pictures the overall results of the process.

The values of H and S for saturated steam at 100°C and for liquid water at 0°C are taken from the steam tables. The total heat transfer is

$$Q = Q' + Q_\sigma = -2,000 + Q_\sigma$$

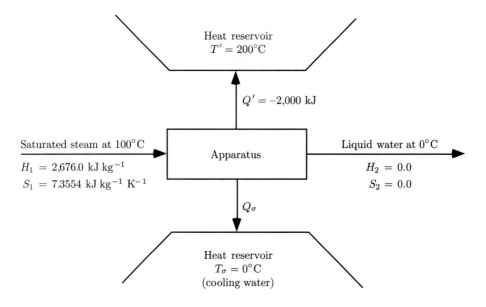

Figure 5.7: Process described in Example 5.5.

Thus on the basis of 1 kg of entering steam, the first law becomes

$$\Delta H = 0.0 - 2{,}676.0 = -2{,}000 + Q_\sigma$$

whence

$$Q_\sigma = -676.0 \text{ kJ}$$

We now examine this result in the light of the second law to determine whether ΔS_{total} is greater than or less than zero for the process.

For 1 kg of steam,

$$\Delta S = 0.0000 - 7.3554 = -7.3554 \text{ kJ K}^{-1}$$

For the heat reservoir at 200°C,

$$\Delta S^t = \frac{2{,}000}{200 + 273.15} = 4.2270 \text{ kJ K}^{-1}$$

For the heat reservoir provided by the cooling water at 0°C,

$$\Delta S^t = \frac{676.0}{0 + 273.15} = 2.4748 \text{ kJ K}^{-1}$$

Thus

$$\Delta S_{\text{total}} = -7.3554 + 4.2270 + 2.4748 = -0.6536 \text{ kJ K}^{-1}$$

Since this result is negative, we conclude that the process as described is impossible, since Eq. (5.19) requires that $\Delta S_{\text{total}} \geq 0$.

This does not mean that all processes of this general nature are impossible, but only that the inventor has claimed too much. Indeed, one can easily calculate the maximum amount of heat which can be transferred to the heat reservoir at 200°C, other conditions remaining the same. This calculation is left as an exercise.

5.8 The Third Law of Thermodynamics

Measurements of heat capacities at very low temperatures provide data for the calculation from Eq. (5.11) of entropy changes down to 0 K. When these calculations are made for different crystalline forms of the same chemical species, the entropy at 0 K appears to be the same for all forms. When the form is non-crystalline, e.g., amorphous or glassy, calculations show that the entropy of the more random form is greater than that of the crystalline form. Such calculations, which are summarized elsewhere,[3] lead to the postulate that *the absolute entropy is zero for all perfect crystalline substances at absolute zero temperature.* While the essential ideas were advanced by Nernst and Planck at the beginning of the twentieth century, more recent studies at very low temperatures have increased our confidence in this postulate, which is now accepted as the third law.

If the entropy is zero at $T = 0$ K, then Eq. (5.11) lends itself to the calculation of absolute entropies. With $T = 0$ as the lower limit of integration, the absolute entropy of a gas at temperature T based on calorimetric data follows from Eq. (5.11) integrated to give

$$ S = \int_0^{T_f} \frac{(C_P)_s}{T} dT + \frac{\Delta H_f}{T_f} + \int_{T_f}^{T_v} \frac{(C_P)_l}{T} dT + \frac{\Delta H_v}{T_v} + \int_{T_v}^{T} \frac{(C_P)_g}{T} dT \qquad (5.20) $$

With respect to this equation,[4] we have supposed that there are no solid-state transitions and thus no heats of transition. The only constant-temperature heat effects are those of fusion at T_f and vaporization at T_v. When a solid-phase transition occurs, a term $\Delta H_t/T_t$ is added.

5.9 Entropy from the Microscopic Viewpoint

We noted in Sec. 4.2 that the internal energy of an ideal gas resides with the individual molecules. This is not true of the entropy, even for molecules that do not interact. The microscopic interpretation of entropy is based on an entirely different concept, as suggested by the following example.

Suppose an insulated container, partitioned into two equal volumes, contains Avogadro's number N_A of ideal-gas molecules in one section and no molecules in the other. When the partition is withdrawn, the molecules quickly distribute themselves uniformly throughout the total volume. The process is an adiabatic expansion that accomplishes no work. Therefore

$$ \Delta U = C_V \Delta T = 0 $$

---

[3]K. S. Pitzer, *Thermodynamics*, 3d ed., chap. 6, McGraw-Hill, New York, 1995.

[4]Evaluation of the first term on the right is not a problem for crystalline substances, because C_P/T remains finite as $T \to 0$.

and the temperature does not change. However, the pressure of the gas decreases by half, and the entropy change as given by Eq. (5.18) is

$$\Delta S = -R\ln\frac{P_2}{P_1} = R\ln 2$$

Since this is the total entropy change, the process is clearly irreversible.

At the molecular level, we note first that the process does not start until the partition is actually removed, and at that instant the molecules occupy only half the space available to them. In this momentary, initial state the molecules are not randomly distributed over the total volume to which they have access, but are crowded into just half the total volume. In this sense they are more ordered than they are in the final state of uniform distribution throughout the entire volume. Thus, the final state can be regarded as a more random, or more disordered, state than the initial state. Generalizing from this example, we are led to the notion that increasing disorder (or decreasing structure) on the molecular level corresponds to increasing entropy.

The means for expressing disorder in a quantitative way was developed by L. Boltzmann and J. W. Gibbs through a quantity Ω, defined as the *number of different ways* that microscopic particles can be distributed among the "states" accessible to them. It is given by the general formula

$$\Omega = \frac{n!}{(n_1!)(n_2!)(n_3!)\cdots} \tag{5.21}$$

where n is the total number of particles, and n_1, n_2, n_3, etc., represent the numbers of particles in "states" 1, 2, 3, etc. The term "state" denotes the condition of the microscopic particles, and we use quotation marks to distinguish this idea of state from the usual thermodynamic meaning as applied to a macroscopic system.

With respect to our example there are but two "states," representing location in one half or the other of the container. The total number of particles is N_A molecules, and initially they are all in a single "state." Thus

$$\Omega_1 = \frac{N_A!}{(N_A!)(0!)} = 1$$

This result confirms that initially the molecules can be distributed between the two accessible "states" in just one way. They are all in a given "state," all in just one half of the container. For an assumed final condition of uniform distribution of the molecules between the two halves of the container, $n_1 = n_2 = N_A/2$, and

$$\Omega_2 = \frac{N_A!}{[(N_A/2)!]^2}$$

This expression gives a very large number for Ω_2, indicating that the molecules can be distributed equally between the two "states" in many different ways. Many other values of Ω_2 are possible, each one of which is associated with a particular *nonuniform* distribution of the molecules between the two halves of the container. The ratio of a particular Ω_2 to the sum of all possible values is the probability of that particular distribution.

The connection established by Boltzmann between entropy S and Ω is given by the equation,

$$S = k \ln \Omega \qquad (5.22)$$

where k is Boltzmann's constant, equal to R/N_A. Integration between states 1 and 2 yields

$$S_2 - S_1 = k \ln \frac{\Omega_2}{\Omega_1}$$

Substituting values for Ω_1 and Ω_2 from our example into this expression gives

$$S_2 - S_1 = k \ln \frac{N_A!}{[(N_A/2)!]^2} = k[\ln N_A! - 2\ln(N_A/2)!]$$

Since N_A is very large, we take advantage of Stirling's formula for the logarithms of factorials of large numbers:

$$\ln X! = X \ln X - X$$

and as a result,

$$S_2 - S_1 = k\left[N_A \ln N_A - N_A - 2\left(\frac{N_A}{2} \ln \frac{N_A}{2} - \frac{N_A}{2} \right) \right]$$

$$= kN_A \ln \frac{N_A}{N_A/2} = kN_A \ln 2 = R \ln 2$$

This is the same value for the entropy change obtained from the classical thermodynamic formula for ideal gases.

The equations of statistical mechanics are derived by application of Eq. (5.21) to a large number of macroscopically identical systems (an ensemble) in all of their many allowed quantum states. The entropy S as given by Eq. (5.22) is then a statistical average value for the ensemble.[5] Ultimately, the result is an equation for the entropy

$$S = k \ln \mathcal{Z} + \frac{U}{T} \qquad (5.23)$$

where $\mathcal{Z}$ is the partition function, given by Eq. (4.11), and U is the internal energy as given by Eq. (4.12).

---

[5]This mathematical development is lengthy but not unduly difficult. An elementary treatment is given by H. C. Van Ness, *Understanding Thermodynamics*, chap. 7, McGraw-Hill, New York, 1969; Dover, New York, 1983. Much more comprehensive are D. A. McQuarrie, *Statistical Mechanics*, Harper & Row, New York, 1976, and R. L. Rowley, *Statistical Mechanics for Thermophysical Property Calculations*, PTR Prentice Hall, Englewood Cliffs, NJ, 1994.

Table 5.1: Absolute Entropies, Ideal-Gas State at 25°C and 1(atm).

$$S^{ig}/\text{J mol}^{-1}\,\text{K}^{-1}$$

	Eq. (5.20)	Eq. (5.24)
CO_2	213.4	213.8
NH_3	192.5	192.9
NO_2	240.6	240.6
CH_4	186.2	186.2
CH_3Cl	234.3	233.5
C_6H_6	269.5	269.9

For an ideal gas, combination of Eqs. (4.14) and (4.15) with Eq. (5.23) yields

$$S^{ig} = kN_A \ln V + \frac{3}{2}kN_A \ln \frac{2\pi mkT}{h^2} - k \ln N_A! + \ln \mathcal{Z}_{\text{intramolecular}} + \frac{3}{2}R + \frac{\mathcal{F}(T)}{T}$$

According to Stirling's formula $\ln N_A! = N_A \ln N_A - N_A$; also $kN_A = R$. Making these substitutions, we get

$$S^{ig} = R \ln V + \frac{3}{2}R \ln \frac{2\pi mkT}{h^2} - R \ln N_A + R + \ln \mathcal{Z}_{\text{intramolecular}} + \frac{3}{2}R + \frac{\mathcal{F}(T)}{T}$$

Upon rearrangement, this may be written

$$S^{ig} = R \ln \left[\left(\frac{2\pi mkT}{h^2} \right)^{3/2} \frac{V e^{5/2}}{N_A} \right] + \ln \mathcal{Z}_{\text{intramolecular}} + \frac{\mathcal{F}(T)}{T} \tag{5.24}$$

Equation (5.24) for the molar entropy of an ideal gas allows calculation of absolute entropies for the ideal-gas state. The data required for evaluation of the last two terms on the right are the bond distances and bond angles in the molecules, and the vibration frequencies associated with the various bonds, as determined from spectroscopic data. The procedure has been very successful in the evaluation of ideal-gas entropies for molecules whose atomic structures are known.

Both the classical and statistical equations [Eqs. (5.20) and (5.24)] yield absolute values of entropy. As is evident from Table 5.1,[6] good agreement between the statistical calculations and those based on calorimetric data are obtained. Results such as these provide impressive evidence for the validity of statistical mechanics and quantum theory. In some instances results based on Eq. (5.24) are considered more reliable because of uncertainties in heat-capacity data or about the crystallinity of the substance near absolute zero. Absolute entropies provide much of the data base for calculation of the equilibrium conversions of chemical reactions, as discussed in Chap. 15.

---

[6]From D. A. McQuarrie, *op. cit.*, p. 138.

PROBLEMS

5.1. Prove that it is impossible for two lines representing reversible, adiabatic processes on a PV diagram to intersect. (*Hint*: Assume that they do intersect, and complete the cycle with a line representing a reversible, isothermal process. Show that performance of this cycle violates the second law.)

5.2. A Carnot engine receives 250 kJ s^{-1} of heat from a heat-source reservoir at 525°C and rejects heat to a heat-sink reservoir at 50°C. What are the power developed and the heat rejected?

5.3. The following heat engines produce power of 95,000 kW. Determine in each case the rates at which heat is absorbed from the hot reservoir and discarded to the cold reservoir.

(*a*) A Carnot engine operates between heat reservoirs at 750 K and 300 K.

(*b*) A practical engine operates between the same heat reservoirs but with a thermal efficiency $\eta = 0.35$.

5.4. A particular power plant operates with a heat-source reservoir at 350°C and a heat-sink reservoir at 30°C. It has a thermal efficiency equal to 55% of the Carnot-engine thermal efficiency for the same temperatures.

(*a*) What is the thermal efficiency of the plant?

(*b*) To what temperature must the heat-source reservoir be raised to increase the thermal efficiency of the plant to 35%? Again η is 55% of the Carnot-engine value.

5.5. Large quantities of liquefied natural gas (LNG) are shipped by ocean tanker. At the unloading port provision is made for vaporization of the LNG so that it may be delivered to pipelines as gas. The LNG arrives in the tanker at atmospheric pressure and 113.7 K, and represents a possible heat sink for use as the cold reservoir of a heat engine. For unloading of LNG as a vapor at the rate of 9,000 m^3 s^{-1}, as measured at 25°C and 1.0133 bar, and assuming the availability of an adequate heat source at 30°C, what is the maximum possible power obtainable and what is the rate of heat transfer from the heat source? Assume that LNG at 25°C and 1.0133 bar is an ideal gas with the molar mass of 17. Also assume that the LNG vaporizes only, absorbing only its latent heat of 512 kJ kg^{-1} at 113.7 K.

5.6. With respect to 1 kg of liquid water:

(*a*) Initially at 0°C, it is heated to 100°C by contact with a heat reservoir at 100°C. What is the entropy change of the water? Of the heat reservoir? What is ΔS_{total}?

(*b*) Initially at 0°C, it is first heated to 50°C by contact with a heat reservoir at 50°C and then to 100°C by contact with a reservoir at 100°C. What is ΔS_{total}?

(*c*) Explain how the water might be heated from 0°C to 100°C so that $\Delta S_{\text{total}} = 0$.

5.7. A rigid vessel of 0.06 m^3 volume contains an ideal gas, $C_V = (5/2)R$, at 500 K and 1 bar.

(*a*) If heat in the amount of 15,000 J is transferred to the gas, determine its entropy change.

(b) If the vessel is fitted with a stirrer that is rotated by a shaft so that work in the amount of 15,000 J is done on the gas, what is the entropy change of the gas if the process is adiabatic? What is ΔS_{total}? What is the irreversible feature of the process?

5.8. An ideal gas, $C_P = (7/2)R$, is heated in a steady-flow heat exchanger from 70°C [158(°F)] to 190°C [374(°F)] by another stream of the same ideal gas which enters at 320°C [608(°F)]. The flow rates of the two streams are the same, and heat losses from the exchanger are negligible.

(a) Calculate the molar entropy changes of the two gas streams for both parallel and countercurrent flow in the exchanger.

(b) What is ΔS_{total} in each case?

(c) Repeat parts (a) and (b) for countercurrent flow if the heating stream enters at 200°C [392(°F)].

5.9. For an ideal gas with constant heat capacities, show that

(a) For a temperature change from T_1 to T_2, ΔS of the gas is greater when the change occurs at constant pressure than when it occurs at constant volume.

(b) For a pressure change from P_1 to P_2, the sign of ΔS for an isothermal change is opposite that for a constant-volume change.

5.10. Imagine that a stream of fluid in steady-state flow serves as a heat source for an infinite set of Carnot engines, each of which absorbs a differential amount of heat from the fluid, causing its temperature to decrease by a differential amount, and each of which rejects a differential amount of heat to a heat reservoir at temperature T_σ. As a result of the operation of the Carnot engines, the temperature of the fluid decreases from T_1 to T_2. Equation (5.8) applies here in differential form, wherein η is defined as

$$\eta \equiv dW/dQ$$

where Q is heat transfer with respect to the flowing fluid. Show that the total work of the Carnot engines is given by

$$W = Q - T_\sigma \Delta S$$

where ΔS and Q both refer to the fluid.

In a particular case the fluid is an ideal gas, $C_P = (7/2)R$, for which $T_1 = 600$ K and $T_2 = 400$ K. If $T_\sigma = 300$ K, what is the value of W in J mol^{-1}? How much heat is discarded to the heat reservoir at T_σ? What is the entropy change of the heat reservoir? What is ΔS_{total}?

5.11. A heat engine operating in outer space may be assumed equivalent to a Carnot engine operating between reservoirs at temperatures T_H and T_C. The only way heat can be discarded from the engine is by radiation, the rate of which is given (approximately) by

$$|\dot{Q}_C| = kAT_C^4$$

where k is a constant and A is the area of the radiator. Prove that, for fixed power output $|\dot{W}|$ and for fixed temperature T_H, the radiator area A is a minimum when the temperature ratio T_C/T_H is 0.75.

5.12. An ideal gas, $C_P = (7/2)R$ and $C_V = (5/2)R$, undergoes a cycle consisting of the following mechanically reversible steps:

- An adiabatic compression from P_1, V_1, T_1 to P_2, V_2, T_2.
- An isobaric expansion from P_2, V_2, T_2 to $P_3 = P_2$, V_3, T_3.
- An adiabatic expansion from P_3, V_3, T_3 to P_4, V_4, T_4.
- A constant-volume process from P_4, V_4, T_4 to P_1, $V_1 = V_4$, T_1.

Sketch this cycle on a PV diagram and determine its thermal efficiency if $T_1 = 200°C$, $T_2 = 500°C$, $T_3 = 1,700°C$, and $T_4 = 700°C$.

5.13. A reversible cycle executed by 1 mol of an ideal gas for which $C_P = (5/2)R$ and $C_V = (3/2)R$ consists of the following:

- Starting at $T_1 = 700$ K and $P_1 = 1.5$ bar, the gas is cooled at constant pressure to $T_2 = 350$ K.
- From 350 K and 1.5 bar, the gas is compressed isothermally to pressure P_2.
- The gas returns to its initial state along a path for which the product PT is constant.

What is the thermal efficiency of the cycle?

5.14. One mole of an ideal gas, $C_P = (7/2)R$ and $C_V = (5/2)R$, is compressed adiabatically in a piston/cylinder device from 2 bar and 25°C to 7 bar. The process is irreversible and requires 35% more work than a reversible, adiabatic compression from the same initial state to the same final pressure. What is the entropy change of the gas?

5.15. One mole of an ideal gas is compressed isothermally but irreversibly at 130°C from 2.5 bar to 6.5 bar in a piston/cylinder device. The work required is 30 percent greater than the work of reversible, isothermal compression. The heat transferred from the gas during compression flows to a heat reservoir at 25°C. Calculate the entropy changes of the gas, the heat reservoir, and ΔS_{total}.

5.16. If 10 mol of SO_2 is heated from 200 to 1,100°C in a steady-flow process at approximately atmospheric pressure, what is its entropy change?

5.17. If 12 mol of propane is heated from 250 to 1,200°C in a steady-flow process at approximately atmospheric pressure, what is its entropy change?

5.18. If heat in the amount of 800 kJ is added to 10 mol of ethylene initially at 200°C in a steady-flow process at approximately atmospheric pressure, what is its entropy change?

5.19. If heat in the amount of 2,500 kJ is added to 15 mol of 1-butene initially at 260°C in a steady-flow process at approximately atmospheric pressure, what is its entropy change?

5.20. If heat in the amount of 10^6 (Btu) [1.055×10^6 kJ] is added to 40(lb mol) [18.14 kg mol] of ethylene initially at 500(°F) [260°C] in a steady-flow process at approximately atmospheric pressure, what is its entropy change?

5.21. A device with no moving parts provides a steady stream of chilled air at $-25°C$ and 1 bar. The feed to the device is compressed air at 25°C and 5 bar. In addition to the stream of chilled air, a second stream of warm air flows from the device at 75°C and 1 bar. Assuming adiabatic operation, what is the ratio of chilled air to warm air that the device produces? Assume that air is an ideal gas for which $C_P = (7/2)R$.

5.22. An inventor has devised a complicated nonflow process in which 1 mol of air is the working fluid. The net effects of the process are claimed to be:

- A change in state of the air from 250°C and 3 bar to 80°C and 1 bar.

- The production of 1,800 J of work.

- The transfer of an undisclosed amount of heat to a heat reservoir at 30°C.

Determine whether the claimed performance of the process is consistent with the second law. Assume that air is an ideal gas for which $C_P = (7/2)R$.

5.23. Consider the heating of a house by a furnace, which serves as a heat-source reservoir at a high temperature T_F. The house acts as a heat-sink reservoir at temperature T, and heat $|Q|$ must be added to the house during a particular time interval to maintain this temperature. Heat $|Q|$ can of course be transferred directly from the furnace to the house, as is the usual practice. However, a third heat reservoir is readily available, namely, the surroundings at temperature T_σ, which can serve as another heat source, thus reducing the amount of heat required from the furnace. Given that $T_F = 810$ K, $T = 295$ K, $T_\sigma = 265$ K, and $|Q| = 1,000$ kJ, determine the minimum amount of heat $|Q_F|$ which must be extracted from the heat-source reservoir (furnace) at T_F. No other sources of energy are available.

5.24. Consider the air conditioning of a house through use of solar energy. At a particular location experiment has shown that solar radiation allows a large tank of pressurized water to be maintained at 175°C. During a particular time interval, heat in the amount of 1,500 kJ must be extracted from the house to maintain its temperature at 24°C when the surroundings temperature is 33°C. Treating the tank of water, the house, and the surroundings as heat reservoirs, determine the minimum amount of heat that must be extracted from the tank of water by any device built to accomplish the required cooling of the house. No other sources of energy are available.

CHAPTER 6

THERMODYNAMIC PROPERTIES OF FLUIDS

The phase rule (Sec. 2.8) tells us that specification of a certain number of intensive properties of a system also fixes the values of all other intensive properties. However, the phase rule provides no information about how values for these other properties may be calculated.

Numerical values for thermodynamic properties are essential to the calculation of heat and work quantities for industrial processes. Consider, for example, the work requirement of a compressor designed to operate adiabatically and to raise the pressure of a gas from P_1 to P_2. This work is given by Eq. (2.9), which becomes

$$W_s = \Delta H = H_2 - H_1$$

when the small kinetic- and potential-energy changes of the gas are neglected. Thus, the shaft work is simply ΔH, the difference between initial and final values of the enthalpy.

Our initial purpose in this chapter is to develop from the first and second laws the fundamental property relations which underlie the mathematical structure of thermodynamics. From these, we derive equations which allow calculation of enthalpy and entropy values from PVT and heat-capacity data. We then discuss the diagrams and tables by which property values are presented for convenient use. Finally, we develop generalized correlations which provide estimates of property values in the absence of complete experimental information.

6.1 Property Relations for Homogeneous Phases

The first law for a closed system of n moles is given by Eq. (2.12):

$$d(nU) = dQ + dW \qquad (2.12)$$

179

For the special case of a reversible process,

$$d(nU) = dQ_{\text{rev}} + dW_{\text{rev}}$$

and by Eqs. (2.13) and (5.12),

$$dW_{\text{rev}} = -P\,d(nV)$$

and

$$dQ_{\text{rev}} = T\,d(nS)$$

These three equations combine to give

$$\boxed{d(nU) = T\,d(nS) - P\,d(nV)} \tag{6.1}$$

where U, S, and V are molar values of the internal energy, entropy, and volume.

This equation, combining the first and second laws, is *derived* for the special case of a reversible process. However, it contains only *properties* of the system. Properties depend on state alone, and not on the kind of process that leads to the state. Therefore, Eq. (6.1) is not restricted in *application* to reversible processes. However, the restrictions placed on the *nature of the system* cannot be relaxed. Thus Eq. (6.1) applies to *any* process in a system of *constant mass* that results in a differential change from one *equilibrium* state to another. The system may consist of a single phase (a homogeneous system), or it may be made up of several phases (a heterogeneous system); it may be chemically inert, or it may undergo chemical reaction. The only requirements are that the system be closed and that the change occur between equilibrium states.

All of the *primary* thermodynamic properties—P, V, T, U, and S—are included in Eq. (6.1). Additional thermodynamic properties arise only by *definition* in relation to these primary properties. In Chap. 2 the enthalpy was defined as a matter of convenience by the equation

$$\boxed{H \equiv U + PV} \tag{2.5}$$

Two additional properties, also defined for convenience, are the *Helmholtz energy*,

$$\boxed{A \equiv U - TS} \tag{6.2}$$

and the *Gibbs energy*,

$$\boxed{G \equiv H - TS} \tag{6.3}$$

Each of these defined properties leads directly to an equation like Eq. (6.1).

Upon multiplication by n, Eq. (2.5) becomes

$$nH = nU + P(nV)$$

Differentiation gives

$$d(nH) = d(nU) + P\,d(nV) + (nV)dP$$

When $d(nU)$ is replaced by Eq. (6.1), this reduces to

$$d(nH) = T\,d(nS) + (nV)dP \tag{6.4}$$

Similarly, from Eq. (6.2),

$$d(nA) = d(nU) - T\,d(nS) - (nS)dT$$

Eliminating $d(nU)$ by Eq. (6.1) gives

$$d(nA) = -P\,d(nV) - (nS)dT \tag{6.5}$$

In analogous fashion, Eq. (6.3) together with Eq. (6.4) yields

$$d(nG) = (nV)dP - (nS)dT \tag{6.6}$$

Equations (6.4) through (6.6) are subject to the same requirements as Eq. (6.1). All are written for the entire mass of any closed system.

Our immediate application of these equations is to one mole (or to a unit mass) of a homogeneous fluid of constant composition. For this case, they simplify to

$$dU = T\,dS - P\,dV \tag{6.7}$$
$$dH = T\,dS + V\,dP \tag{6.8}$$
$$dA = -P\,dV - S\,dT \tag{6.9}$$
$$dG = V\,dP - S\,dT \tag{6.10}$$

These *fundamental property relations* are general equations for a homogeneous fluid of constant composition.

Another set of equations follows from them by application of the criterion of exactness for a differential expression. If $F = F(x, y)$, then the total differential of F is defined as

$$dF = \left(\frac{\partial F}{\partial x}\right)_y dx + \left(\frac{\partial F}{\partial y}\right)_x dy$$

or

$$dF = M\,dx + N\,dy \tag{6.11}$$

where

$$M = \left(\frac{\partial F}{\partial x}\right)_y \qquad \text{and} \qquad N = \left(\frac{\partial F}{\partial y}\right)_x$$

By further differentiation we obtain

$$\left(\frac{\partial M}{\partial y}\right)_x = \frac{\partial^2 F}{\partial y\,\partial x} \qquad \text{and} \qquad \left(\frac{\partial N}{\partial x}\right)_y = \frac{\partial^2 F}{\partial x\,\partial y}$$

Since the order of differentiation in mixed second derivatives is immaterial, these equations give

$$\left(\frac{\partial M}{\partial y}\right)_x = \left(\frac{\partial N}{\partial x}\right)_y \tag{6.12}$$

When F is a function of x and y, the right-hand side of Eq. (6.11) is an *exact differential expression*; since Eq. (6.12) must then be satisfied, it serves as a criterion of exactness.

The thermodynamic properties U, H, A, and G are *known* to be functions of the variables on the right-hand sides of Eqs. (6.7) through (6.10); we may therefore write the relationship expressed by Eq. (6.12) for each of these equations:

$$\left(\frac{\partial T}{\partial V}\right)_S = -\left(\frac{\partial P}{\partial S}\right)_V \tag{6.13}$$

$$\left(\frac{\partial T}{\partial P}\right)_S = \left(\frac{\partial V}{\partial S}\right)_P \tag{6.14}$$

$$\left(\frac{\partial P}{\partial T}\right)_V = \left(\frac{\partial S}{\partial V}\right)_T \tag{6.15}$$

$$\left(\frac{\partial V}{\partial T}\right)_P = -\left(\frac{\partial S}{\partial P}\right)_T \tag{6.16}$$

These are known as *Maxwell's equations*.[1]

Equations (6.7) through (6.10) are the basis not only for derivation of the Maxwell equations but also of a large number of other equations relating thermodynamic properties. We develop here only a few expressions useful for evaluation of thermodynamic properties from experimental data. Their derivation requires application of Eqs. (6.8) and (6.16).

The most useful property relations for the enthalpy and entropy of a homogeneous phase result when these properties are expressed as functions of T and P. What we need to know is how H and S vary with temperature and pressure. This information is contained in the derivatives $(\partial H/\partial T)_P$, $(\partial S/\partial T)_P$, $(\partial H/\partial P)_T$, and $(\partial S/\partial P)_T$.

Consider first the temperature derivatives. As a result of Eq. (2.20), which defines the heat capacity at constant pressure, we have

---

[1] After James Clerk Maxwell (1831–1879), Scottish physicist.

$$\left(\frac{\partial H}{\partial T}\right)_P = C_P \tag{2.20}$$

Another expression for this quantity is obtained by division of Eq. (6.8) by dT and restriction of the result to constant P:

$$\left(\frac{\partial H}{\partial T}\right)_P = T\left(\frac{\partial S}{\partial T}\right)_P$$

Combination of this equation with Eq. (2.20) gives

$$\left(\frac{\partial S}{\partial T}\right)_P = \frac{C_P}{T} \tag{6.17}$$

The pressure derivative of the entropy results directly from Eq. (6.16):

$$\left(\frac{\partial S}{\partial P}\right)_T = -\left(\frac{\partial V}{\partial T}\right)_P \tag{6.18}$$

The correspondong derivative for the enthalpy is found by division of Eq. (6.8) by dP and restriction to constant T:

$$\left(\frac{\partial H}{\partial P}\right)_T = T\left(\frac{\partial S}{\partial P}\right)_T + V$$

As a result of Eq. (6.18) this becomes

$$\left(\frac{\partial H}{\partial P}\right)_T = V - T\left(\frac{\partial V}{\partial T}\right)_P \tag{6.19}$$

Since the functional relations chosen here for H and S are

$$H = H(T, P) \quad \text{and} \quad S = S(T, P)$$

it follows that

$$dH = \left(\frac{\partial H}{\partial T}\right)_P dT + \left(\frac{\partial H}{\partial P}\right)_T dP$$

and

$$dS = \left(\frac{\partial S}{\partial T}\right)_P dT + \left(\frac{\partial S}{\partial P}\right)_T dP$$

Substituting for the partial derivatives in these two equations by Eqs. (2.20) and (6.17) through (6.19), we get

$$\boxed{dH = C_P dT + \left[V - T\left(\frac{\partial V}{\partial T}\right)_P\right] dP} \tag{6.20}$$

and

$$dS = C_P \frac{dT}{T} - \left(\frac{\partial V}{\partial T}\right)_P dP \tag{6.21}$$

These are general equations relating the enthalpy and entropy of homogeneous fluids of constant composition to temperature and pressure.

The coefficients of dT and dP in Eqs. (6.20) and (6.21) are evaluated from heat-capacity and PVT data. As an example of the application of these equations, we note that the PVT behavior of a fluid in the ideal-gas state is expressed by the equations

$$PV^{ig} = RT$$

and

$$\left(\frac{\partial V^{ig}}{\partial T}\right)_P = \frac{R}{P}$$

where V^{ig} is the molar volume of an ideal gas at temperature T and pressure P. Substituting these equations into Eqs. (6.20) and (6.21) reduces them to

$$dH^{ig} = C_P^{ig} dT \tag{6.22}$$

and

$$dS^{ig} = C_P^{ig} \frac{dT}{T} - R\frac{dP}{P} \tag{6.23}$$

where superscript "ig" denotes an ideal-gas value. These equations merely restate for ideal gases equations presented in Secs. 2.11 and 5.6.

Equations (6.18) and (6.19) are expressed in an alternative form by elimination of $(\partial V/\partial T)_P$ in favor of the volume expansivity β by Eq. (3.2):

$$\left(\frac{\partial S}{\partial P}\right)_T = -\beta V \tag{6.24}$$

and

$$\left(\frac{\partial H}{\partial P}\right)_T = (1 - \beta T)V \tag{6.25}$$

The pressure dependence of the internal energy is obtained by differentiation of the equation $U = H - PV$:

$$\left(\frac{\partial U}{\partial P}\right)_T = \left(\frac{\partial H}{\partial P}\right)_T - P\left(\frac{\partial V}{\partial P}\right)_T - V$$

whence by Eqs. (6.25) and (3.3),

$$\left(\frac{\partial U}{\partial P}\right)_T = (\kappa P - \beta T)V \tag{6.26}$$

where κ is the isothermal compressibility. Equations (6.24) through (6.26), which require values of β and κ, are usually applied only to liquids. However, for liquids not near the critical point, the volume itself is small, as are both β and κ. Thus at most conditions pressure has little effect on the entropy, enthalpy, and internal energy of liquids. The important special case of an *incompressible fluid* (Sec. 3.1) is considered in Example 6.2.

When $(\partial V / \partial T)_P$ is replaced in Eqs. (6.20) and (6.21) in favor of the volume expansivity, they become

$$dH = C_P dT + V(1 - \beta T) dP \tag{6.27}$$

and

$$dS = C_P \frac{dT}{T} - \beta V dP \tag{6.28}$$

Since β and V are weak functions of pressure for liquids, they are usually assumed constant at appropriate average values for integration of the final terms of Eqs. (6.27) and (6.28).

Example 6.1 Determine the enthalpy and entropy changes of liquid water for a change of state from 1 bar and 25°C to 1,000 bar and 50°C. The following data for water are available.

$t/°C$	P/bar	$C_P/\text{J mol}^{-1}\,\text{K}^{-1}$	$V/\text{cm}^3\,\text{mol}^{-1}$	β/K^{-1}
25	1	75.305	18.071	256×10^{-6}
25	1,000	......	18.012	366×10^{-6}
50	1	75.314	18.234	458×10^{-6}
50	1,000	......	18.174	568×10^{-6}

SOLUTION For application to the change of state described, Eqs. (6.27) and (6.28) require integration. Since enthalpy and entropy are state functions, the path of integration is arbitrary; the path most suited to the given data is shown in Fig. 6.1. Since the data indicate that C_P is a weak function of T and that both V and β are weak functions of P, integration with arithmetic means is satisfactory. The integrated forms of Eqs. (6.27) and (6.28) that result are

$$\Delta H = \langle C_P \rangle (T_2 - T_1) - \langle V \rangle (1 - \langle \beta \rangle T_2)(P_2 - P_1)$$

and

$$\Delta S = \langle C_P \rangle \ln \frac{T_2}{T_1} - \langle \beta \rangle \langle V \rangle (P_2 - P_1)$$

where for $P = 1$ bar

$$\langle C_P \rangle = \frac{75.305 + 75.314}{2} = 75.310 \text{ J mol}^{-1}\,\text{K}^{-1}$$

and for $t = 50°C$

$$\langle V \rangle = \frac{18.234 + 18.174}{2} = 18.204 \text{ cm}^3\,\text{mol}^{-1}$$

$$\langle \beta \rangle = \frac{458 + 568}{2} \times 10^{-6} = 513 \times 10^{-6} \text{ K}^{-1}$$

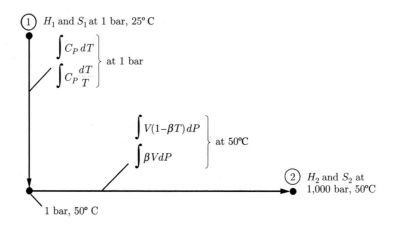

Figure 6.1: Calculational path for Example 6.1.

Substitution of numerical values into the equation for ΔH gives

$$\Delta H = 75.310(323.15 - 298.15)$$

$$+ \frac{(18.204)[1 - (513 \times 10^{-6})(323.15)](1,000 - 1)}{10 \text{ cm}^3 \text{ bar J}^{-1}}$$

$$\Delta H = 1,883 + 1,517 = 3,400 \text{ J mol}^{-1}$$

Similarly for ΔS,

$$\Delta S = 75.310 \ln \frac{323.15}{298.15} - \frac{(513 \times 10^{-6})(18.204)(1,000 - 1)}{10 \text{ cm}^3 \text{ bar J}^{-1}}$$

$$\Delta S = 6.06 - 0.93 = 5.13 \text{ J mol}^{-1} \text{ K}^{-1}$$

Note that the effect of a pressure change of almost 1,000 bar on the enthalpy and entropy of liquid water is less than that of a temperature change of only $25°C$.

Example 6.2   Develop the property relations appropriate to the *incompressible fluid*, a model fluid for which both β and κ are zero (see Sec. 3.1). This is an idealization often employed in fluid mechanics.

SOLUTION   Equations (6.27) and (6.28) written for an incompressible fluid become

$$dH = C_P dT + V \, dP \qquad (A)$$

and

$$dS = C_P \frac{dT}{T} \qquad (B)$$

The enthalpy of an incompressible fluid is therefore a function of both temperature and pressure, whereas the entropy is a function of temperature only, independent of P. We see from Eq. (6.26) that the internal energy as well is a function of temperature only, and is therefore given by the equation

$$dU = C_V dT \qquad (C)$$

By application of Eq. (6.12), the criterion of exactness, to Eq. (A), we find that

$$\left(\frac{\partial C_P}{\partial P}\right)_T = \left(\frac{\partial V}{\partial T}\right)_P$$

However, the definition of β, given by Eq. (3.2), shows that the derivative on the right equals βV, which is zero for an incompressible fluid. This means that C_P is a function of temperature only, independent of P.

The relation of C_P to C_V for an incompressible fluid is of interest. For its derivation we need an equation for S analogous to Eq. (6.28), but in which the independent variables are T and V. In this event, $S = S(T, V)$, and

$$dS = \left(\frac{\partial S}{\partial T}\right)_V dT + \left(\frac{\partial S}{\partial V}\right)_T dV \qquad (D)$$

It remains to evaluate the partial derivatives. Division of Eq. (6.7) by dT and restriction to constant V gives

$$\left(\frac{\partial U}{\partial T}\right)_V = T \left(\frac{\partial S}{\partial T}\right)_V$$

However, $(\partial U/\partial T)_V$ is defined by Eq. (2.19) as C_V. Therefore

$$\left(\frac{\partial S}{\partial T}\right)_V = \frac{C_V}{T} \qquad (E)$$

The volume derivative of the entropy is given directly by Eq. (6.15):

$$\left(\frac{\partial S}{\partial V}\right)_T = \left(\frac{\partial P}{\partial T}\right)_V$$

Equation (3.4) applied to a change of state at constant V becomes

$$\left(\frac{\partial P}{\partial T}\right)_V = \frac{\beta}{\kappa} \qquad (F)$$

By the last two equations,

$$\left(\frac{\partial S}{\partial V}\right)_T = \frac{\beta}{\kappa} \qquad (G)$$

Equations (D), (E), and (G) combine to yield

$$dS = \frac{C_V}{T} dT + \frac{\beta}{\kappa} dV \qquad (H)$$

For a given change of state, Eqs. (6.28) and (H) must give the same value for dS, and they may therefore be equated. The resulting expression, after rearrangement, becomes

$$(C_P - C_V)dT = \beta TV\, dP + \frac{\beta T}{\kappa} dV$$

Upon restriction to constant V, this reduces to

$$C_P - C_V = \beta TV \left(\frac{\partial P}{\partial T}\right)_V$$

Elimination of the derivative by Eq. (F) yields

$$C_P - C_V = \beta T V \left(\frac{\beta}{\kappa} \right)$$

Since $\beta = 0$, the right-hand side of this equation is zero, provided that the indeterminate ratio β/κ is finite. Since this ratio is indeed finite for real fluids, a contrary presumption for the *model* fluid would be irrational. Thus the definition of the incompressible fluid presumes that this ratio is finite, and we conclude for such a fluid that the heat capacities at constant volume and at constant pressure are identical. We may therefore write

$$C_P = C_V = C$$

The fundamental property relations for homogeneous fluids of constant composition given by Eqs. (6.7) through (6.10) show that each of the thermodynamic properties U, H, A, and G is functionally related to a special pair of variables. In particular, Eq. (6.10),

$$dG = V\, dP - S\, dT \qquad\qquad (6.10)$$

expresses the functional relation:

$$G = G(P, T)$$

Thus the special, or *canonical*,[2] variables for the Gibbs energy are temperature and pressure. Since these variables can be directly measured and controlled, the Gibbs energy is a thermodynamic property of great potential utility.

An alternative form of Eq. (6.10), the fundamental property relation, follows from the mathematical identity:

$$d\left(\frac{G}{RT} \right) \equiv \frac{1}{RT} dG - \frac{G}{RT^2} dT$$

Substitution for dG by Eq. (6.10) and for G by Eq. (6.3) gives, after algebraic reduction,

$$\boxed{d\left(\frac{G}{RT} \right) = \frac{V}{RT} dP - \frac{H}{RT^2} dT} \qquad\qquad (6.29)$$

The advantage of this equation is that all terms are dimensionless; moreover, in contrast to Eq. (6.10), the enthalpy rather than the entropy appears on the right-hand side.

Equations such as Eqs. (6.10) and (6.29) are too general for direct practical application, but they are readily applied in restricted form. Thus, from Eq. (6.29) we have immediately that

$$\frac{V}{RT} = \left[\frac{\partial (G/RT)}{\partial P} \right]_T \qquad\qquad (6.30)$$

---

[2] *Canonical* here means that the variables conform to a general rule that is both simple and clear.

and

$$\frac{H}{RT} = -T\left[\frac{\partial(G/RT)}{\partial T}\right]_P \tag{6.31}$$

When G/RT is known as a function of T and P, V/RT and H/RT follow by simple differentiation. The remaining properties are given by defining equations. In particular,

$$\frac{S}{R} = \frac{H}{RT} - \frac{G}{RT}$$

and

$$\frac{U}{RT} = \frac{H}{RT} - \frac{PV}{RT}$$

Thus, when we know how G/RT (or G) is related to its canonical variables, T and P, i.e., when we are given $G/RT = g(T, P)$, we can evaluate all other thermodynamic properties by simple mathematical operations. The Gibbs energy therefore serves as a *generating function* for the other thermodynamic properties, and implicitly represents *complete* property information.

The fundamental property relation most intimately connected with statistical mechanics is Eq. (6.9), which expresses the differential of the Helmholtz energy as a function of its canonical variables T and V:

$$dA = -P\,dV - S\,dT \tag{6.9}$$

We remarked in connection with Eq. (4.11) that the partition function $\mathcal{Z}$ is a state function from which all thermodynamic properties may be found once it is known as a function of T and V. Its relation to the Helmholtz energy follows from Eq. (5.23):

$$S = k\ln\mathcal{Z} + \frac{U}{T} \tag{5.23}$$

which may be written

$$U - TS = -kT\ln\mathcal{Z}$$

Reference to Eq. (6.2) shows that the left-hand side of this equation is by definition the Helmholtz energy A. Therefore

$$\boxed{A = -kT\ln\mathcal{Z}} \tag{6.32}$$

This equation provides a direct link between thermodynamics and statistical mechanics. Since $R = kN_A$, where N_A is Avogadro's number, Eq. (6.32) may be expressed alternatively as

$$\frac{A}{RT} = -\frac{\ln\mathcal{Z}}{N_A}$$

Equations (6.29) through (6.31) for the Gibbs energy have as their counterparts analogous equations for the Helmholtz energy. Derived from Eqs. (6.9) and (6.2), they are

$$d\left(\frac{A}{RT}\right) = -\frac{P}{RT}\,dV - \frac{U}{RT^2}\,dT \tag{6.33}$$

$$\frac{P}{RT} = -\left[\frac{\partial(A/RT)}{\partial V}\right]_T \tag{6.34}$$

$$\frac{U}{RT} = -T\left[\frac{\partial(A/RT)}{\partial T}\right]_V \tag{6.35}$$

The remaining properties come from defining equations:

$$\frac{S}{R} = \frac{U}{RT} - \frac{A}{RT}$$

$$\frac{H}{RT} = \frac{U}{RT} + \frac{PV}{RT}$$

Note that Eq. (4.12) follows immediately from Eqs. (6.35) and (6.32).

Thus, when we know how A/RT (or $\ln \mathcal{Z}$) is related to its canonical variables, T and V, we can evaluate all other thermodynamic properties by simple mathematical operations. The Helmholtz energy and therefore the partition function serve as *generating functions* for the other thermodynamic properties, and implicitly represent complete property information.

Although a powerful tool for the estimation of thermodynamic properties, the application of statistical mechanics requires an enormous number of computations. Making the connection between the microscopic states of matter and its manifest macroscopic properties is best accomplished by molecular simulation, carried out numerically with high-speed computers. In Monte Carlo techniques the generation of a very large number of microscopic replicas of a system containing on the order of one hundred molecules serves to create an ensemble from which by appropriate statistics the partition function of Eq. (4.11) can be deduced. The intermolecular potential-energy function is key to the accurate prediction of the thermodynamic properties of real fluids, and this is a continuing area of research. Molecular simulation is a subject unto itself, and is treated in detail elsewhere.[3]

6.2 Residual Properties

Unfortunately, we have no convenient experimental method for determining numerical values of G or G/RT, and the equations which follow directly from the Gibbs energy are of little practical use. However, the concept of the Gibbs energy as a generating function for other thermodynamic properties carries over to a closely related property for which numerical values *are* readily obtained. Thus we define the *residual* Gibbs energy as

$$G^R \equiv G - G^{ig} \tag{6.36}$$

---

[3]R. L. Rowley, *Statistical Mechanics for Thermophysical Property Calculations*, PTR Prentice Hall, Englewood Cliffs, NJ, 1994. Both Monte Carlo and molecular-dynamics simulations, which have a different basis, are considered.

where G and G^{ig} are the actual and the ideal-gas values of the Gibbs energy at the same temperature and pressure. We can define other residual properties in an analogous way. The residual volume, for example, is

$$V^R \equiv V - V^{ig} \tag{6.37}$$

whence

$$V^R = V - \frac{RT}{P}$$

Since $V = ZRT/P$, the residual volume and the compressibility factor are related:

$$V^R = \frac{RT}{P}(Z - 1) \tag{6.38}$$

We can, in fact, write a general definition for residual properties:

$$\boxed{M^R \equiv M - M^{ig}} \tag{6.39}$$

where M is the molar value of any extensive thermodynamic property, e.g., V, U, H, S, or G.

Equation (6.29), written for the special case of an ideal gas, becomes

$$d\left(\frac{G^{ig}}{RT}\right) = \frac{V^{ig}}{RT}dP - \frac{H^{ig}}{RT^2}dT$$

Subtracting this equation from Eq. (6.29) gives

$$\boxed{d\left(\frac{G^R}{RT}\right) = \frac{V^R}{RT}dP - \frac{H^R}{RT^2}dT} \tag{6.40}$$

This *fundamental property relation* for residual properties applies to fluids of constant composition. From it we get immediately that

$$\frac{V^R}{RT} = \left[\frac{\partial(G^R/RT)}{\partial P}\right]_T \tag{6.41}$$

and

$$\frac{H^R}{RT} = -T\left[\frac{\partial(G^R/RT)}{\partial T}\right]_P \tag{6.42}$$

In addition, the defining equation for the Gibbs energy, $G = H - TS$, written for the special case of an ideal gas is $G^{ig} = H^{ig} - TS^{ig}$; by difference,

$$G^R = H^R - TS^R$$

from which we get the residual entropy:

$$\frac{S^R}{R} = \frac{H^R}{RT} - \frac{G^R}{RT} \tag{6.43}$$

Thus the residual Gibbs energy serves as a generating function for the other residual properties, and here we do have a direct link with experiment. It is provided by Eq. (6.41), written

$$d\left(\frac{G^R}{RT}\right) = \frac{V^R}{RT}dP \qquad (\text{const } T)$$

Integration from zero pressure to arbitrary pressure P gives

$$\frac{G^R}{RT} = \int_0^P \frac{V^R}{RT}dP \qquad (\text{const } T)$$

where at the lower limit we have set G^R/RT equal to zero on the basis that the zero-pressure state is an ideal-gas state. In view of Eq. (6.38), this result becomes

$$\frac{G^R}{RT} = \int_0^P (Z-1)\frac{dP}{P} \qquad (\text{const } T) \qquad (6.44)$$

Differentiation of Eq. (6.44) with respect to temperature in accord with Eq. (6.42) gives

$$\boxed{\frac{H^R}{RT} = -T \int_0^P \left(\frac{\partial Z}{\partial T}\right)_P \frac{dP}{P} \qquad (\text{const } T)} \qquad (6.45)$$

Combining Eqs. (6.44) and (6.45) with Eq. (6.43), we get

$$\boxed{\frac{S^R}{R} = -T \int_0^P \left(\frac{\partial Z}{\partial T}\right)_P \frac{dP}{P} - \int_0^P (Z-1)\frac{dP}{P} \qquad (\text{const } T)} \qquad (6.46)$$

The compressibility factor is defined as $Z = PV/RT$; values of Z and of $(\partial Z/\partial T)_P$ therefore come from experimental PVT data, and the two integrals in Eqs. (6.44) through (6.46) are evaluated by numerical or graphical methods. Alternatively, the two integrals are evaluated analytically when Z is expressed by an equation of state. Thus, given PVT data or an appropriate equation of state, we can evaluate H^R and S^R and hence all other residual properties. It is this direct connection with experiment that makes residual properties essential to the practical application of thermodynamics.

Applied to the enthalpy and entropy, Eq. (6.39) is written

$$H = H^{ig} + H^R \qquad \text{and} \qquad S = S^{ig} + S^R$$

Thus, H and S are found from the corresponding ideal-gas and residual properties by simple addition. General expressions for H^{ig} and S^{ig} are obtained by integration of Eqs. (6.22) and (6.23) from an ideal-gas state at reference conditions T_0 and P_0

to the ideal-gas state at T and P:[4]

$$H^{ig} = H_0^{ig} + \int_{T_0}^{T} C_P^{ig} dT$$

and

$$S^{ig} = S_0^{ig} + \int_{T_0}^{T} C_P^{ig} \frac{dT}{T} - R \ln \frac{P}{P_0}$$

Substitution into the preceding equations gives

$$H = H_0^{ig} + \int_{T_0}^{T} C_P^{ig} dT + H^R \qquad (6.47)$$

and

$$S = S_0^{ig} + \int_{T_0}^{T} C_P^{ig} \frac{dT}{T} - R \ln \frac{P}{P_0} + S^R \qquad (6.48)$$

Recall (Secs. 4.1 and 5.6) that for purposes of computation the integrals in Eqs. (6.47) and (6.48) are represented by

`R*ICPH(T0,T;A,B,C,D)` and `R*ICPS(T0,T;A,B,C,D)`

Equations (6.47) and (6.48) may be expressed alternatively to include the mean heat capacities introduced in Secs. 4.1 and 5.6:

$$H = H_0^{ig} + \langle C_P^{ig} \rangle_H (T - T_0) + H^R \qquad (6.49)$$

and

$$S = S_0^{ig} + \langle C_P^{ig} \rangle_S \ln \frac{T}{T_0} - R \ln \frac{P}{P_0} + S^R \qquad (6.50)$$

where H^R and S^R are given by Eqs. (6.45) and (6.46). Again, for computational purposes, the mean heat capacites are represented by

`R*MCPH(T0,T;A,B,C,D)` and `R*MCPS(T0,T;A,B,C,D)`

Since the equations of thermodynamics which derive from the first and second laws do not permit calculation of absolute values for enthalpy and entropy, and since all we need in practice are relative values, the reference-state conditions T_0 and P_0 are selected for convenience, and values are assigned to H_0^{ig} and S_0^{ig} arbitrarily. The only data needed for application of Eqs. (6.49) and (6.50) are ideal-gas heat capacities and PVT data. Once V, H, and S are known at given conditions of T and P, the other thermodynamic properties follow from defining equations.

---

[4]Thermodynamic properties for organic compounds in the ideal-gas state are given by M. Frenkel, G. J. Kabo, K. N. Marsh, G. N. Roganov, and R. C. Wilhoit, *Thermodynamics of Organic Compounds in the Gas State*, Thermodynamics Research Center, Texas A & M Univ. System, College Station, Texas, 1994.

Table 6.1: Compressibility factors Z for isobutane

P/bar	340 K	350 K	360 K	370 K	380 K
0.1	0.99700	0.99719	0.99737	0.99753	0.99767
0.5	0.98745	0.98830	0.98907	0.98977	0.99040
2	0.95895	0.96206	0.96483	0.96730	0.96953
4	0.92422	0.93069	0.93635	0.94132	0.94574
6	0.88742	0.89816	0.90734	0.91529	0.92223
8	0.84575	0.86218	0.87586	0.88745	0.89743
10	0.79659	0.82117	0.84077	0.85695	0.87061
12	......	0.77310	0.80103	0.82315	0.84134
14	......	......	0.75506	0.78531	0.80923
15.41	......	......	0.71727		

The true worth of the equations for ideal gases is now evident. They are important because they provide a convenient base for the calculation of real-gas properties. Although Eqs. (6.45) and (6.46) as written apply only to gases, residual properties have validity for liquids as well. However, the advantage of Eqs. (6.47) and (6.48) in application to gases is that H^R and S^R, the terms which contain all the complex calculations, are *residuals* that generally are quite small. They have the nature of corrections to the major terms, H^{ig} and S^{ig}. For liquids, this advantage is largely lost, because H^R and S^R must include the large enthalpy and entropy changes of vaporization. Property changes of liquids are usually calculated by integrated forms of Eqs. (6.27) and (6.28), as illustrated in Example 6.1.

Example 6.3   Calculate the enthalpy and entropy of saturated isobutane vapor at 360 K from the following information:

1. Compressibility-factor data (values of Z) for isobutane vapor are given in Table 6.1.

2. The vapor pressure of isobutane at 360 K is 15.41 bar.

3. Set $H_0^{ig} = 18,115.0$ J mol^{-1} and $S_0^{ig} = 295.976$ J mol^{-1} K^{-1} for the ideal-gas reference state at 300 K and 1 bar. [These values are in accord with the bases adopted by R. D. Goodwin and W. M. Haynes, Nat. Bur. Stand. (U.S.), Tech. Note 1051, 1982.]

4. The ideal-gas heat capacity of isobutane vapor in the temperature range of interest is given by

$$C_P^{ig}/R = 1.7765 + 33.037 \times 10^{-3} T \qquad (T/\text{ K})$$

SOLUTION   Calculation of H^R and S^R at 360 K and 15.41 bar by application of Eqs. (6.45) and (6.46) requires the evaluation of two integrals:

$$\int_0^P \left(\frac{\partial Z}{\partial T}\right)_P \frac{dP}{P} \qquad \text{and} \qquad \int_0^P (Z - 1)\frac{dP}{P}$$

Graphical integration requires simple plots of both $(\partial Z/\partial T)_P/P$ and $(Z-1)/P$ vs. P. Values of $(Z-1)/P$ are calculated directly from the given compressibility-factor data at 360 K. The quantity $(\partial Z/\partial T)_P/P$ requires evaluation of the partial derivative $(\partial Z/\partial T)_P$, given by the slope of a plot of Z vs. T at constant pressure. For this purpose, separate plots are made of Z vs. T for each pressure at which compressibility-factor data are given, and a slope is determined at 360 K for each curve (for example, by construction of a tangent line at 360 K). The data for construction of the required plots are shown in Table 6.2.

Table 6.2: Values of the integrands required in Example 6.3

Values in parentheses are by extrapolation.

P/bar	$[(\partial Z/\partial T)_P/P] \times 10^4/\text{K}^{-1}\ \text{bar}^{-1}$	$[-(Z-1)/P] \times 10^2/\text{bar}^{-1}$
0	(1.780)	(2.590)
0.1	1.700	2.470
0.5	1.514	2.186
2	1.293	1.759
4	1.290	1.591
6	1.395	1.544
8	1.560	1.552
10	1.777	1.592
12	2.073	1.658
14	2.432	1.750
15.41	(2.720)	(1.835)

The values of the two integrals are found to be

$$\int_0^P \left(\frac{\partial Z}{\partial T}\right)_P \frac{dP}{P} = 26.37 \times 10^{-4} \text{ K}^{-1}$$

and

$$\int_0^P (Z-1)\frac{dP}{P} = -0.2596$$

Thus by Eq. (6.45)

$$\frac{H^R}{RT} = -(360)(26.37 \times 10^{-4}) = -0.9493$$

and by Eq. (6.46)

$$\frac{S^R}{R} = -0.9493 - (-0.2596) = -0.6897$$

For $R = 8.314$ J mol^{-1} K^{-1},

$$H^R = (-0.9493)(8.314)(360) = -2{,}841.3 \text{ J mol}^{-1}$$

and

$$S^R = (-0.6897)(8.314) = -5.734 \text{ J mol}^{-1} \text{ K}^{-1}$$

Values of the integrals in Eqs. (6.47) and (6.48) are

$$8.314*\text{ICPH}(300,360;1.7765,33.037\text{E}-3,0.0,0.0) \equiv 6{,}324.8 \text{ J mol}^{-1}$$

and

$$8.314*\text{ICPS}(300,360;1.7765,33.037\text{E}-3,0.0,0.0) \equiv 19.174 \text{ J mol}^{-1} \text{ K}^{-1}$$

Substitution of numerical values into Eqs. (6.47) and (6.48) yields

$$H = 18{,}115.0 + 6{,}324.8 - 2{,}841.3 = 21{,}598.5 \text{ J mol}^{-1}$$

and

$$S = 295.976 + 19.174 - 8.314 \ln 15.41 - 5.734 = 286.676 \text{ J mol}^{-1} \text{ K}^{-1}$$

Although calculations have been carried out for just one state, enthalpies and entropies can be evaluated for any number of states, given adequate data. After having completed a set of calculations, one is not irrevocably committed to the particular values of H_0^{ig} and S_0^{ig} initially assigned. The scale of values for either the enthalpy or the entropy can be shifted by addition of a constant to all values. In this way one can give arbitrary values to H and S for some particular state so as to make the scales convenient for one purpose or another. A shift of scale does not affect differences in property values.

The accurate calculation of thermodynamic properties for construction of a table or diagram is an exacting task, seldom required of an engineer. However, engineers do make practical use of thermodynamic properties, and an understanding of the methods used for their calculation leads to an appreciation that some uncertainty is associated with every property value. There are two major reasons for inaccuracy. First, the experimental data are difficult to measure and are subject to error. Moreover, data are frequently incomplete, and are extended by interpolation and extrapolation. Second, even when reliable PVT data are available, a loss of accuracy occurs in the differentiation process required in the calculation of derived properties. This accounts for the fact that data of a high order of accuracy are required to produce enthalpy and entropy values suitable for engineering calculations.

6.3 Two-Phase Systems

The PT diagram of Fig. 3.1 shows curves representing phase boundaries for a pure substance. A phase transition at constant temperature and pressure occurs whenever one of these curves is crossed, and as a result the molar or specific values of the extensive thermodynamic properties change abruptly. Thus the molar or specific volume of a saturated liquid is very different from that for saturated vapor at the same T and P. This is true as well for internal energy, enthalpy, and entropy. The exception is the molar or specific Gibbs energy, which for a pure species does not change during a phase transition such as melting, vaporization, or sublimation. Consider a pure liquid in equilibrium with its vapor in a piston/cylinder arrangement at temperature T and the corresponding vapor pressure P^{sat}. When a differential amount of liquid is caused to evaporate at constant temperature and pressure, Eq. (6.6) applied to the process reduces to $d(nG) = 0$. Since the number of moles n is constant, $dG = 0$, and this requires the molar (or specific) Gibbs energy of the vapor to be identical with that of the liquid. More generally, for two phases α and β of a pure species coexisting at equilibrium,

$$G^\alpha = G^\beta \tag{6.51}$$

where G^α and G^β are the molar Gibbs energies of the individual phases.

The Clapeyron equation, first introduced in Sec. 4.3, follows from this equality. If the temperature of a two-phase system is changed, then the pressure must also change in accord with the relation between vapor pressure and temperature if the two phases continue to coexist in equilibrium. Since Eq. (6.51) holds throughout this change, we have

$$dG^\alpha = dG^\beta$$

Substituting the expressions for dG^α and dG^β given by Eq. (6.10) yields

$$V^\alpha dP^{\text{sat}} - S^\alpha dT = V^\beta dP^{\text{sat}} - S^\beta dT$$

which upon rearrangement becomes

$$\frac{dP^{\text{sat}}}{dT} = \frac{S^\beta - S^\alpha}{V^\beta - V^\alpha} = \frac{\Delta S^{\alpha\beta}}{\Delta V^{\alpha\beta}}$$

The entropy change $\Delta S^{\alpha\beta}$ and the volume change $\Delta V^{\alpha\beta}$ are the changes which occur when a unit amount of a pure chemical species is transferred from phase α to phase β at the equilibrium temperature and pressure. Integration of Eq. (6.8) for this change yields the latent heat of phase transition:

$$\Delta H^{\alpha\beta} = T \Delta S^{\alpha\beta}$$

Thus, $\Delta S^{\alpha\beta} = \Delta H^{\alpha\beta}/T$, and substitution in the preceding equation gives

$$\frac{dP^{\text{sat}}}{dT} = \frac{\Delta H^{\alpha\beta}}{T \Delta V^{\alpha\beta}} \tag{6.52}$$

which is the Clapeyron equation. For the particularly important case of phase transition from liquid l to vapor v, it is written

$$\frac{dP^{\text{sat}}}{dT} = \frac{\Delta H^{lv}}{T \Delta V^{lv}} \tag{6.53}$$

Example 6.4 For vaporization at low pressures, one may introduce reasonable approximations into Eq. (6.53) by assuming that the vapor phase is an ideal gas and that the molar volume of the liquid is negligible compared with the molar volume of the vapor. How do these assumptions alter the Clapeyron equation?

SOLUTION   The assumptions made are expressed by

$$\Delta V^{lv} = V^v = \frac{RT}{P^{\,\text{sat}}}$$

Equation (6.53) then becomes

$$\frac{dP^{\,\text{sat}}}{dT} = \frac{\Delta H^{lv}}{RT^2/P^{\,\text{sat}}}$$

or

$$\frac{dP^{\,\text{sat}}/P^{\,\text{sat}}}{dT/T^2} = \frac{\Delta H^{lv}}{R}$$

or

$$\Delta H^{lv} = -R\frac{d\ln P^{\,\text{sat}}}{d(1/T)}$$

This approximate equation, known as the Clausius/Clapeyron equation, relates the latent heat of vaporization directly to the vapor-pressure curve. Specifically, it shows that ΔH^{lv} is proportional to the slope of a plot of $\ln P^{\,\text{sat}}$ vs. $1/T$. Experimental data for many substances show that such plots produce lines that are nearly straight. According to the Clausius/Clapeyron equation, this implies that ΔH^{lv} is almost constant, virtually independent of T. This is not true; ΔH^{lv} decreases monotonically with increasing temperature from the triple point to the critical point, where it becomes zero. The assumptions on which the Clausius/Clapeyron equation are based have approximate validity only at low pressures.

The Clapeyron equation is an exact thermodynamic relation, providing a vital connection between the properties of different phases. When applied to the calculation of latent heats of vaporization, its use presupposes knowledge of the vapor pressure-vs.-temperature relation. Since thermodynamics imposes no model of material behavior, either in general or for particular species, such relations are empirical. As noted in Example 6.4, a plot of $\ln P^{\,\text{sat}}$ vs. $1/T$ generally yields a line that is nearly straight, i.e.,

$$\ln P^{\,\text{sat}} = A - \frac{B}{T} \tag{6.54}$$

where A and B are constants for a given species. This equation gives a rough approximation of the vapor-pressure relation for the entire temperature range from the triple point to the critical point. Moreover, it provides an excellent basis for interpolation between values that are reasonably spaced.

The Antoine equation, which is more satisfactory for general use, has the form

$$\ln P^{\,\text{sat}} = A - \frac{B}{T + C} \tag{6.55}$$

A principal advantage of this equation is that values of the constants A, B, and C are readily available for a large number of species.[5] Each set of constants is valid for a specified temperature range, and should not be used outside of that range.

---

[5]S. Ohe, *Computer Aided Data Book of Vapor Pressure*, Data Book Publishing Co., Tokyo, 1976; T. Boublik, V. Fried, and E. Hala, *The Vapor Pressures of Pure Substances*, Elsevier, Amsterdam, 1984.

The accurate representation of vapor-pressure data over a wide temperature range requires an equation of greater complexity. The Wagner equation is one of the best available; it expresses the reduced vapor pressure as a function of reduced temperature:

$$\ln P_r^{\text{sat}} = \frac{A\tau + B\tau^{1.5} + C\tau^3 + D\tau^6}{1 - \tau} \tag{6.56}$$

where

$$\tau \equiv 1 - T_r$$

and A, B, C, and D are constants. Values of the constants either for this equation or for Eq. (6.55) are given by Reed, Prausnitz, and Poling[6] for many species.

When a system consists of saturated-liquid and saturated-vapor phases coexisting in equilibrium, the total value of any extensive property of the two-phase system is the sum of the total properties of the phases. Written for the volume, this relation is

$$nV = n^l V^l + n^v V^v$$

where V is the system volume on a molar basis and the total number of moles is $n = n^l + n^v$. Division by n gives

$$V = x^l V^l + x^v V^v$$

where x^l and x^v represent the fractions of the total system that are liquid and vapor. Since $x^l = 1 - x^v$,

$$V = (1 - x^v)V^l + x^v V^v$$

In this equation the properties V, V^l, and V^v may be either molar or unit-mass values. The mass or molar fraction of the system that is vapor x^v is called the *quality*. Analogous equations can be written for the other extensive thermodynamic properties. All of these relations may be summarized by the equation

$$M = (1 - x^v)M^l + x^v M^v \tag{6.57}$$

where M represents V, U, H, S, etc.

6.4 Thermodynamic Diagrams

A thermodynamic diagram represents the temperature, pressure, volume, enthalpy, and entropy of a substance on a single plot. (Sometimes data for all these variables are not included, but the term still applies.) The most common diagrams are: temperature/entropy, pressure/enthalpy (usually $\ln P$ vs. H), and enthalpy/entropy (called a *Mollier* diagram). The designations refer to the variables chosen for the coordinates. Other diagrams are possible, but are seldom used.

[6]R. C. Reid, J. M. Prausnitz, and B. E. Poling, *The Properties of Gases and Liquids*, 4th ed., App. A, McGraw-Hill, 1987.

Figures 6.2 through 6.4 show the general features of the three common diagrams. These figures are based on data for water, but their general character is the
same for all substances. The two-phase states, which fall on lines in the PT diagram
of Fig. 3.1, lie over areas in these diagrams, and the triple point of Fig. 3.1 becomes
a line. When lines of constant quality are shown in the liquid/vapor region, property values for two-phase mixtures are read directly from the diagram. The critical
point is identified by the letter C, and the solid curve passing through this point
represents the states of saturated liquid (to the left of C) and of saturated vapor
(to the right of C). The Mollier diagram (Fig. 6.4) does not usually include volume
data. In the vapor or gas region, lines for constant temperature and constant *superheat* appear. Superheat is a term used to designate the difference between the
actual temperature and the saturation temperature at the same pressure.

Thermodynamic diagrams included in this book are the PH diagram of Fig. 6.5
for methane, the Mollier diagram on the inside of the back cover for steam, and the
PH diagram of Fig. 9.3 for tetrafluoroethane.

Paths of various processes are conveniently traced on a thermodynamic diagram. For example, consider the operation of the boiler in a steam power plant.
The initial state is liquid water at a temperature below its boiling point; the final
state is steam in the superheat region. As the water goes into the boiler and is

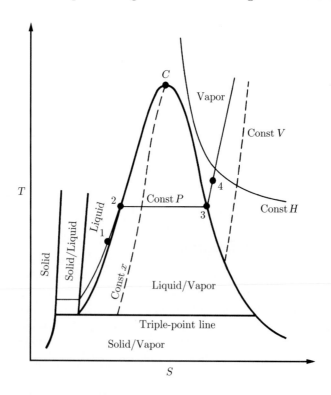

Figure 6.2: TS diagram.

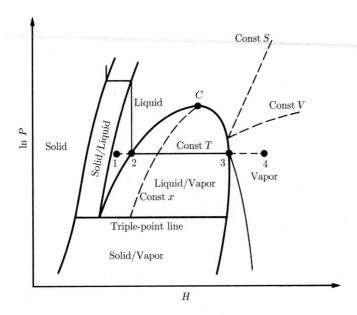

Figure 6.3: PH diagram.

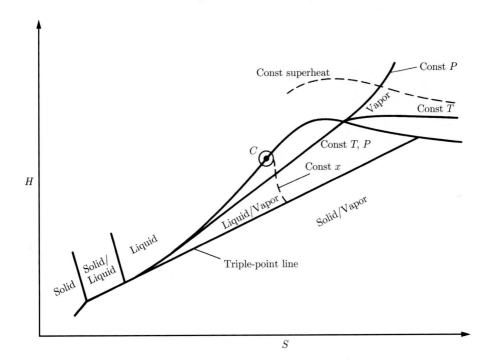

Figure 6.4: Mollier diagram.

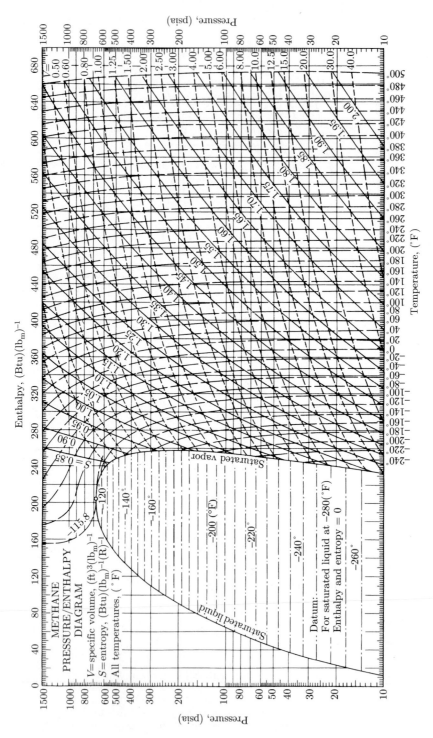

Figure 6.5: *PH* diagram for methane. *(Reproduced by permission of the Shell Development Company, Copyright 1945. Published by C. S. Matthews and C. O. Hurd, Trans. AIChE, vol. 42, pp. 55–78, 1946.)*

heated, its temperature rises at constant pressure (line 1–2 in Figs. 6.2 and 6.3) until saturation is reached. From point 2 to point 3 the water vaporizes, the temperature remaining constant during the process. As more heat is added, the steam becomes superheated along line 3–4. On a pressure/enthalpy diagram (Fig. 6.3) the whole process is represented by a horizontal line corresponding to the boiler pressure. Since the compressibility of a liquid is small for temperatures well below T_c, the properties of liquids change very slowly with pressure. Thus on a TS diagram (Fig. 6.2), the constant-pressure lines in the liquid region lie very close together, and line 1–2 nearly coincides with the saturated-liquid curve.

A reversible adiabatic process is isentropic and is therefore represented on a TS diagram by a vertical line. Hence the path followed by the fluid in reversible adiabatic turbines and compressors is simply a vertical line from the initial pressure to the final pressure. This is also true on the HS or Mollier diagram.

6.5 Tables of Thermodynamic Properties

In many instances thermodynamic properties are reported in tables. The advantage is that, in general, data can be presented more accurately than in diagrams, but the need for interpolation is introduced.

Thermodynamic tables for saturated steam from its normal freezing point to the critical point and for superheated steam over a substantial pressure range, in both SI and in English units, appear in App. F. Values are given at intervals close enough so that linear interpolation is satisfactory. The first table for each system of units presents the equilibrium properties of saturated liquid and saturated vapor at even increments of temperature. The enthalpy and entropy are arbitrarily assigned values of zero for the saturated liquid state at the triple point. The second table is for the gas region, and gives properties of superheated steam at temperatures higher than the saturation temperature for a given pressure. Volume, internal energy, enthalpy, and entropy are tabulated as functions of pressure at various temperatures. The steam tables are the most thorough compilation of properties for any single material. However, tables are available for a number of other substances.[7]

> **Example 6.5** Superheated steam originally at P_1 and T_1 expands through a nozzle to an exhaust pressure P_2. Assuming the process is reversible and adiabatic and that equilibrium is attained, determine the state of the steam at the exit of the nozzle for the following conditions:
>
> (a) $P_1 = 1{,}000$ kPa, $t_1 = 260°$C, and $P_2 = 200$ kPa.
>
> (b) $P_1 = 150$(psia), $t_1 = 500(°$F), and $P_2 = 30$(psia).
>
> SOLUTION Since the process is both reversible and adiabatic, the change in entropy of the steam is zero.

---

[7]Data for many common chemicals are given by R. H. Perry and D. Green, *Perry's Chemical Engineers' Handbook*, 7th ed., Sec. 2, McGraw-Hill, New York, 1996. See also N. B. Vargaftik, *Handbook of Physical Properties of Liquids and Gases*, 2nd ed., Hemisphere Publishing Corp., Washington, DC, 1975. Data for refrigerants appear in the *ASHRAE Handbook: Fundamentals*, American Society of Heating, Refrigerating, and Air-Conditioning Engineers, Inc., Atlanta, 1993.

(*a*) The initial state of the steam is as follows (data from the SI steam tables):

$$t_1 = 260°C$$
$$P_1 = 1,000 \text{ kPa}$$
$$H_1 = 2,965.2 \text{ kJ kg}^{-1}$$
$$S_1 = 6.9680 \text{ kJ kg}^{-1} \text{ K}^{-1}$$

For the final state,

$$P_2 = 200 \text{ kPa}$$
$$S_2 = S_1 = 6.9680 \text{ kJ kg}^{-1} \text{ K}^{-1}$$

Since the entropy of saturated vapor at 200 kPa is greater than S_2, the final state is in the two-phase region. Equation (6.57) applied to the entropy here becomes

$$S = (1 - x^v)S^l + x^v S^v$$

whence

$$6.9680 = 1.5301(1 - x^v) + 7.1268x^v$$

where 1.5301 and 7.1268 are the entropies of saturated liquid and saturated vapor at 200 kPa. Solving, we get

$$x^v = 0.9716$$

On a mass basis, the mixture is 97.16 percent vapor and 2.84 percent liquid. Its enthalpy is obtained by further application of Eq. (6.57):

$$H = (0.0284)(504.7) + (0.9716)(2,706.7) = 2,644.2 \text{ kJ kg}^{-1}$$

(*b*) The initial state of the steam is as follows (data from the steam tables in English units):

$$t_1 = 500(°F)$$
$$P_1 = 150(\text{psia})$$
$$H_1 = 1,274.3(\text{Btu})(\text{lb}_m)^{-1}$$
$$S_1 = 1.6602(\text{Btu})(\text{lb}_m)^{-1}(R)^{-1}$$

In the final state,

$$P_2 = 30(\text{psia})$$
$$S_2 = S_1 = 1.6602(\text{Btu})(\text{lb}_m)^{-1}(R)^{-1}$$

Since the entropy of saturated vapor at 30(psia) is greater than S_2, the final state is in the two-phase region. Equation (6.57) applied to the entropy is written

$$S = (1 - x^v)S^l + x^v S^v$$

whence

$$1.6602 = 0.3682(1 - x^v) + 1.6995x^v$$

where 0.3682 and 1.6995 are the entropies of saturated liquid and saturated vapor at 30(psia). Solving, we get

$$x^v = 0.9705$$

On a mass basis, the mixture is 97.05 percent vapor and 2.95 percent liquid. Its enthalpy follows from another application of Eq. (6.57):

$$H = (0.0295)(218.9) + (0.9705)(1,164.1) = 1,136.2(\text{Btu})(\text{lb}_m)^{-1}$$

6.6 Generalized Property Correlations for Gases

Of the two kinds of data needed for evaluation of thermodynamic properties, heat capacities and PVT data, the latter are most frequently missing. Fortunately, the generalized methods developed in Sec. 3.6 for the compressibility factor are also applicable to residual properties.

Equations (6.45) and (6.46) are put into generalized form by substitution of the relationships

$$P = P_c P_r \qquad\qquad T = T_c T_r$$

$$dP = P_c \, dP_r \qquad\qquad dT = T_c \, dT_r$$

The resulting equations are

$$\frac{H^R}{RT_c} = -T_r^2 \int_0^{P_r} \left(\frac{\partial Z}{\partial T_r} \right)_{P_r} \frac{dP_r}{P_r} \tag{6.58}$$

and

$$\frac{S^R}{R} = -T_r \int_0^{P_r} \left(\frac{\partial Z}{\partial T_r} \right)_{P_r} \frac{dP_r}{P_r} - \int_0^{P_r} (Z - 1) \frac{dP_r}{P_r} \tag{6.59}$$

The terms on the right-hand sides of these equations depend only on the upper limit P_r of the integrals and on the reduced temperature at which they are evaluated. Thus, values of H^R/RT_c and S^R/R may be determined once and for all at any reduced temperature and pressure from generalized compressibility-factor data.

The correlation for Z is based on Eq. (3.46),

$$Z = Z^0 + \omega Z^1$$

Differentiation yields

$$\left(\frac{\partial Z}{\partial T_r} \right)_{P_r} = \left(\frac{\partial Z^0}{\partial T_r} \right)_{P_r} + \omega \left(\frac{\partial Z^1}{\partial T_r} \right)_{P_r}$$

Substitution for Z and $(\partial Z/\partial T_r)_{P_r}$ in Eqs. (6.58) and (6.59) gives

$$\frac{H^R}{RT_c} = -T_r^2 \int_0^{P_r} \left(\frac{\partial Z^0}{\partial T_r} \right)_{P_r} \frac{dP_r}{P_r} - \omega T_r^2 \int_0^{P_r} \left(\frac{\partial Z^1}{\partial T_r} \right)_{P_r} \frac{dP_r}{P_r}$$

and

$$\frac{S^R}{R} = - \int_0^{P_r} \left[T_r \left(\frac{\partial Z^0}{\partial T_r} \right)_{P_r} + Z^0 - 1 \right] \frac{dP_r}{P_r} - \omega \int_0^{P_r} \left[T_r \left(\frac{\partial Z^1}{\partial T_r} \right)_{P_r} + Z^1 \right] \frac{dP_r}{P_r}$$

The first integrals on the right-hand sides of these two equations may be evaluated numerically or graphically for various values of T_r and P_r from the data for Z^0 given in Tables E.1 and E.3, and the integrals which follow ω in each equation may be similarly evaluated from the data for Z^1 given in Tables E.2 and E.4. Alternatively, their evaluation may be based on an equation of state; Lee and Kesler used this means to extend their generalized correlation to residual properties.

If the first terms on the right-hand sides of the preceding equations (including the minus signs) are represented by $(H^R)^0/RT_c$ and $(S^R)^0/R$ and if the terms which follow ω, together with the preceding minus signs, are represented by $(H^R)^1/RT_c$ and $(S^R)^1/R$, then we can write

$$\frac{H^R}{RT_c} = \frac{(H^R)^0}{RT_c} + \omega \frac{(H^R)^1}{RT_c} \tag{6.60}$$

and

$$\frac{S^R}{R} = \frac{(S^R)^0}{R} + \omega \frac{(S^R)^1}{R} \tag{6.61}$$

Calculated values of the quantities $(H^R)^0/RT_c$, $(H^R)^1/RT_c$, $(S^R)^0/R$, and $(S^R)^1/R$ as determined by Lee and Kesler are given as functions of T_r and P_r in Tables E.5 through E.12. These values, together with Eqs. (6.60) and (6.61), allow estimation of residual enthalpies and entropies on the basis of the three-parameter corresponding-states principle as developed by Lee and Kesler (Sec. 3.6). The nature of these correlations is indicated by Fig. 6.6, which shows a plot of $(H^R)^0/RT_c$ vs. P_r for six isotherms.

Tables E.5 and E.6 for $(H^R)^0/RT_c$ and Tables E.9 and E.10 for $(S^R)^0/R$, used alone, provide two-parameter corresponding-states correlations that quickly yield coarse estimates of the residual properties.

Fluids for which the intermolecular potential energy $\mathcal{U}(r)$ is given by the Lennard-Jones equation (Sec. 3.8) are said (as a class) to be *conformal*. More generally, a class of fluids for which $\mathcal{U}(r)$ is of the same functional form is conformal. It is a property of conformal fluids that they obey the two-parameter theorem of corresponding states as stated in Section 3.6. Thus different classes of conformal fluids, distinguished by different functional forms of $\mathcal{U}(r)$, obey different corresponding-states correlations. The purpose of the acentric factor in Pitzer-type correlations (such as Lee/Kesler) is therefore to differentiate between classes of non-polar conformal fluids, primarily on the basis of molecular asymmetry. These classes then obey the three-parameter theorem of corresponding states. An extended set of Lee/Kessler correlations[8] incorporates a fourth parameter to characterize classes of *polar* conformal fluids. Thus an even larger collection of conformal classes of fluids obeys a four-parameter theorem of corresponding states.

As with the generalized compressibility-factor correlation, the complexity of the functions $(H^R)^0/RT_c$, $(H^R)^1/RT_c$, $(S^R)^0/R$, and $(S^R)^1/R$ precludes their general representation by simple equations. However, the correlation for Z based on generalized virial coefficients and valid at low pressures can be extended to the residual properties. The equation relating Z to the functions B^0 and B^1 is derived

[8]R. L. Rowley, *Statistical Mechanics for Thermophysical Property Calculations*, Sec. 11.5 and App. 11, PTR Prentice Hall, Englewood Cliffs, NJ, 1994.

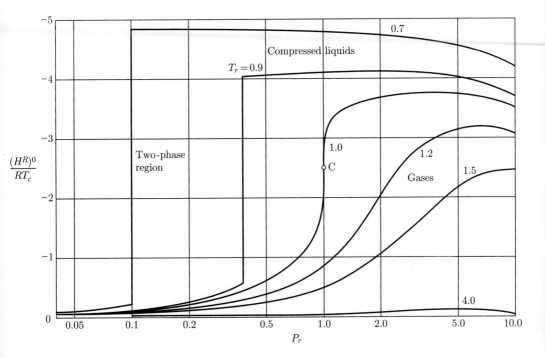

Figure 6.6: The Lee/Kesler correlation for $(H^R)^0/RT_c$ as a function of T_r and P_r.

in Sec. 3.6 from Eqs. (3.47) and (3.48):

$$Z = 1 + B^0 \frac{P_r}{T_r} + \omega B^1 \frac{P_r}{T_r}$$

From this we find

$$\left(\frac{\partial Z}{\partial T_r} \right)_{P_r} = P_r \left(\frac{dB^0/dT_r}{T_r} - \frac{B^0}{T_r^2} \right) + \omega P_r \left(\frac{dB^1/dT_r}{T_r} - \frac{B^1}{T_r^2} \right)$$

Substituting these equations into Eqs. (6.58) and (6.59) gives

$$\frac{H^R}{RT_c} = -T_r \int_0^{P_r} \left[\left(\frac{dB^0}{dT_r} - \frac{B^0}{T_r} \right) + \omega \left(\frac{dB^1}{dT_r} - \frac{B^1}{T_r} \right) \right] dP_r$$

and

$$\frac{S^R}{R} = - \int_0^{P_r} \left(\frac{dB^0}{dT_r} - \omega \frac{dB^1}{dT_r} \right) dP_r$$

Since B^0 and B^1 are functions of temperature only, integration at constant temperature yields

$$\frac{H^R}{RT_c} = P_r \left[B^0 - T_r \frac{dB^0}{dT_r} + \omega \left(B^1 - T_r \frac{dB^1}{dT_r} \right) \right] \tag{6.62}$$

and

$$\frac{S^R}{R} = -P_r \left(\frac{dB^0}{dT_r} + \omega \frac{dB^1}{dT_r} \right) \tag{6.63}$$

The dependence of B^0 and B^1 on reduced temperature is given by Eqs. (3.50) and (3.51). Differentiation of these equations provides expressions for dB^0/dT_r and dB^1/dT_r. Thus the four equations required for application of Eqs. (6.62) and (6.63) are

$$B^0 = 0.083 - \frac{0.422}{T_r^{1.6}} \tag{3.50}$$

$$\frac{dB^0}{dT_r} = \frac{0.675}{T_r^{2.6}} \tag{6.64}$$

$$B^1 = 0.139 - \frac{0.172}{T_r^{4.2}} \tag{3.51}$$

$$\frac{dB^1}{dT_r} = \frac{0.722}{T_r^{5.2}} \tag{6.65}$$

Figure 3.15, drawn specifically for the compressibility-factor correlation, is also used as a guide to the reliability of the correlations of residual properties based on generalized second virial coefficients. However, all residual-property correlations are less precise than the compressibility-factor correlations on which they are based and are, of course, least reliable for strongly polar and associating molecules.

The generalized correlations for H^R and S^R, together with ideal-gas heat capacities, allow calculation of enthalpy and entropy values of gases at any temperature and pressure by Eqs. (6.47) and (6.48). For a change from state 1 to state 2, we write Eq. (6.47) for both states:

$$H_2 = H_0^{ig} + \int_{T_0}^{T_2} C_P^{ig} dT + H_2^R$$

$$H_1 = H_0^{ig} + \int_{T_0}^{T_1} C_P^{ig} dT + H_1^R$$

The enthalpy change for the process, $\Delta H = H_2 - H_1$, is given by the difference between these two equations:

$$\Delta H = \int_{T_1}^{T_2} C_P^{ig} dT + H_2^R - H_1^R \tag{6.66}$$

Similarly, by Eq. (6.48) for the entropy, we get

$$\Delta S = \int_{T_1}^{T_2} C_P^{ig} \frac{dT}{T} - R \ln \frac{P_2}{P_1} + S_2^R - S_1^R \tag{6.67}$$

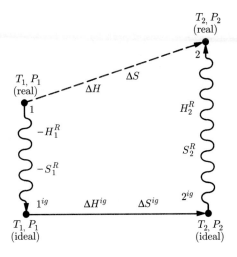

Figure 6.7: Calculational path for property changes ΔH and ΔS.

Again these equations may be written in alternative form:

$$\Delta H = \langle C_P^{ig} \rangle_H (T_2 - T_1) + H_2^R - H_1^R \tag{6.68}$$

and

$$\Delta S = \langle C_P^{ig} \rangle_S \ln \frac{T_2}{T_1} - R \ln \frac{P_2}{P_1} + S_2^R - S_1^R \tag{6.69}$$

Just as we have given names to computational functions useful in evaluation of the integrals in Eqs. (6.66) and (6.67) and the mean heat capacities in Eqs. (6.68) and (6.69), so also do we name functions useful for evaluation of H^R and S^R. Equation (6.62) in combination with Eqs. (3.50), (6.64), (3.51), and (6.65) provides a function for the evaluation of H^R/RT_c which for computational purposes we name

HRB(TR,PR,OMEGA)

A numerical value of H^R is therefore represented by

R*TC*HRB(TR,PR,OMEGA)

Similarly, Eqs. (6.63) through (6.65) provide a function for the evaluation of S^R/R which we name

SRB(TR,PR,OMEGA)

A numerical value of S^R is therefore represented by

R*SRB(TR,PR,OMEGA)

Computer programs for evaluating these functions are given in App. D.

The terms on the right-hand sides of Eqs. (6.66) through (6.69) are readily associated with steps in a *calculational path* leading from an initial to a final state of a system. Thus, in Fig. 6.7, the actual path from state 1 to state 2 (dashed line) is replaced by a three-step calculational path. Step $1 \rightarrow 1^{ig}$ represents a hypothetical process that transforms a real gas into an ideal gas at T_1 and P_1. The enthalpy and entropy changes for this process are

$$H_1^{ig} - H_1 = -H_1^R$$

and

$$S_1^{ig} - S_1 = -S_1^R$$

In step $1^{ig} \rightarrow 2^{ig}$ changes occur in the ideal-gas state from (T_1, P_1) to (T_2, P_2). For this process,

$$\Delta H^{ig} = H_2^{ig} - H_1^{ig} = \int_{T_1}^{T_2} C_P^{ig} dT \tag{6.70}$$

and

$$\Delta S^{ig} = S_2^{ig} - S_1^{ig} = \int_{T_1}^{T_2} C_P^{ig} \frac{dT}{T} - R \ln \frac{P_2}{P_1} \tag{6.71}$$

Finally, step $2^{ig} \rightarrow 2$ is another hypothetical process that transforms the ideal gas back into a real gas at T_2 and P_2. Here,

$$H_2 - H_2^{ig} = H_2^R$$

and

$$S_2 - S_2^{ig} = S_2^R$$

Equations (6.66) and (6.67) result from addition of the enthalpy and entropy changes for the three steps.

> **Example 6.6**   Estimate V, U, H, and S for 1-butene vapor at 200°C and 70 bar if H and S are set equal to zero for saturated liquid at 0°C. Assume that the only data available are
>
> $T_c = 420.0$ K $P_c = 40.43$ bar $\omega = 0.191$
>
> $T_n = 266.9$ K   (normal boiling point)
>
> $C_P^{ig}/R = 1.967 + 31.630 \times 10^{-3} T - 9.837 \times 10^{-6} T^2$ (T/K)

SOLUTION   The volume of 1-butene vapor at 200°C and 70 bar is calculated directly from the equation $V = ZRT/P$, where Z is given by Eq. (3.46) with values of Z^0 and Z^1 interpolated in Tables E.3 and E.4. For the reduced conditions,

$$T_r = \frac{200 + 273.15}{420.0} = 1.127 \qquad P_r = \frac{70}{40.43} = 1.731$$

we find that

$$Z = Z^0 + \omega Z^1 = 0.485 + (0.191)(0.142) = 0.512$$

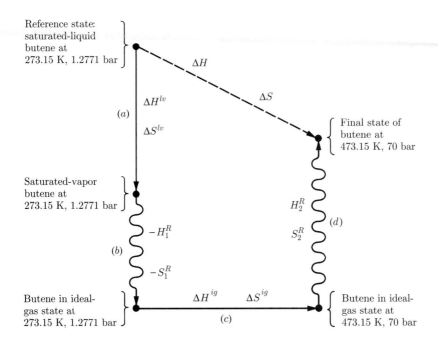

Figure 6.8: Calculational path for Example 6.5.

whence

$$V = \frac{(0.512)(83.14)(473.15)}{70} = 287.8 \text{ cm}^3 \text{ mol}^{-1}$$

For H and S, we use a calculational path like that of Fig. 6.7, leading from an initial state of saturated liquid 1-butene at $0°C$, where H and S are zero, to the final state of interest. In this case, an initial vaporization step is required, and we have the four-step path shown by Fig. 6.8. The steps are:

(a) Vaporization at T_1 and $P_1 = P^{\text{sat}}$.

(b) Transition to the ideal-gas state at (T_1, P_1).

(c) Change to (T_2, P_2) in the ideal-gas state.

(d) Transition to the actual final state at (T_2, P_2).

Step (a): Vaporization of saturated liquid 1-butene at $0°C$. The vapor pressure must be estimated, since it is not given. One method is based on Eq. (6.54):

$$\ln P^{\text{sat}} = A - \frac{B}{T}$$

We know two points on the vapor-pressure curve: the normal boiling point, for which $P^{\text{sat}} = 1.0133$ bar at 266.9 K, and the critical point, for which $P^{\text{sat}} = 40.43$ bar at 420.0 K. For these two points,

$$\ln 1.0133 = A - \frac{B}{266.9}$$

and

$$\ln 40.43 = A - \frac{B}{420.0}$$

Simultaneous solution of these two equations gives

$$A = 10.1260 \qquad \text{and} \qquad B = 2,699.11$$

For 0°C or 273.15 K, we then find that $P^{\text{sat}} = 1.2771$ bar. This result is used in steps (b) and (c). Here, we need an estimate of the latent heat of vaporization. Equation (4.18) provides the value at the normal boiling point, where $T_{r_n} = 266.9/420.0 = 0.636$:

$$\frac{\Delta H_n^{lv}}{RT_n} = \frac{1.092(\ln P_c - 1.013)}{0.930 - T_{r_n}} = \frac{1.092(\ln 40.43 - 1.013)}{0.930 - 0.636} = 9.979$$

whence

$$\Delta H_n^{lv} = (9.979)(8.314)(266.9) = 22,137 \text{ J mol}^{-1}$$

Equation (4.19) now yields the latent heat at 273.15 K, where $T_r = 273.15/420.0 = 0.650$:

$$\frac{\Delta H^{lv}}{\Delta H_n^{lv}} = \left(\frac{1 - T_r}{1 - T_{r_n}}\right)^{0.38}$$

or

$$\Delta H^{lv} = (0.350/0.364)^{0.38}(22,137) = 21,810 \text{ J mol}^{-1}$$

and

$$\Delta S^{lv} = \Delta H^{lv}/T = 21,810/273.15 = 79.84 \text{ J mol}^{-1} \text{ K}^{-1}$$

Step (b): Transformation of saturated-vapor 1-butene into an ideal gas at the initial conditions (T_1, P_1). Since the pressure is relatively low, the values of H_1^R and S_1^R are estimated by Eqs. (6.62) and (6.63). The reduced conditions are

$$T_r = 0.650 \qquad \text{and} \qquad P_r = 0.0316$$

The computational procedure is represented by

$$\text{HRB(0.650,0.0316,0.191)} \equiv -0.0985$$

$$\text{SRB(0.650,0.0316,0.191)} \equiv -0.1063$$

whence

$$H_1^R = (-0.0985)(8.314)(420.0) = -344 \text{ J mol}^{-1}$$

$$S_1^R = (-0.1063)(8.314) = -0.88 \text{ J mol}^{-1} \text{ K}^{-1}$$

As indicated in Fig. 6.8, the property changes for this step are $-H_1^R$ and $-S_1^R$, because the change is from the real to the ideal-gas state.

Step (c): Changes in the ideal-gas state from (273.15 K, 1.2771 bar) to (473.15 K, 70 bar). Here, ΔH^{ig} and ΔS^{ig} are given by Eqs. (6.70) and (6.71), for which we have (Secs. 4.1 and 5.6)

$$8.314*\text{ICPH(273.15,473.15;1.967,31.630E-3,-9.837E-6,0.0)} \equiv 20,564 \text{ J mol}^{-1}$$

and

$$8.314*\text{ICPS}(273.15,473.15;1.967,31.630\text{E-}3,-9.837\text{E-}6,0.0)$$
$$\equiv 55.474 \text{ J mol}^{-1} \text{ K}^{-1}$$

Thus, Eqs. (6.70) and (6.71) yield

$$\Delta H^{ig} = 20{,}564 \text{ J mol}^{-1}$$

and

$$\Delta S^{ig} = 55.474 - 8.314\ln\frac{70}{1.2771} = 22.18 \text{ J mol}^{-1} \text{ K}^{-1}$$

Step (d): Transformation of 1-butene from the ideal-gas state to the real-gas state at T_2 and P_2. The final reduced conditions are

$$T_r = 1.127 \qquad \text{and} \qquad P_r = 1.731$$

At the higher pressure of this step, H_2^R and S_2^R are found by Eqs. (6.60) and (6.61), together with the Lee/Kesler correlation. With interpolated values from Tables E.7, E.8, E.11, and E.12 at $T_r = 1.127$ and $P_r = 1.731$, these equations give

$$\frac{H_2^R}{RT_c} = -2.294 + (0.191)(-0.713) = -2.430$$

and

$$\frac{S_2^R}{R} = -1.566 + (0.191)(-0.726) = -1.705$$

whence

$$H_2^R = (-2.430)(8.314)(420.0) = -8{,}485 \text{ J mol}^{-1}$$
$$S_2^R = (-1.705)(8.314) = -14.18 \text{ J mol}^{-1} \text{ K}^{-1}$$

The sums of the enthalpy and entropy changes for the four steps give the total changes for the process leading from the initial reference state (where H and S are set equal to zero) to the final state:

$$H = \Delta H = 21{,}810 - (-344) + 20{,}564 - 8{,}485 = 34{,}233 \text{ J mol}^{-1}$$

and

$$S = \Delta S = 79.84 - (-0.88) + 22.18 - 14.18 = 88.72 \text{ J mol}^{-1} \text{ K}^{-1}$$

The internal energy is

$$U = H - PV = 34{,}233 - \frac{(70)(287.8)}{10 \text{ cm}^3 \text{ bar J}^{-1}} = 32{,}218 \text{ J mol}^{-1}$$

These results are in far better agreement with experimental values than would have been the case had we assumed 1-butene vapor an ideal gas.

PROBLEMS

6.1. Starting with Eq. (6.8), show that isobars in the vapor region of a Mollier (HS) diagram must have positive slope and positive curvature.

6.2. Making use of the fact that Eq. (6.20) is an exact differential expression, show that

$$(\partial C_P/\partial P)_T = -T(\partial^2 V/\partial T^2)_P$$

What is the result of application of this equation to an ideal gas?

6.3. Estimate the change in enthalpy and entropy when liquid ammonia at 270 K is compressed from its saturation pressure of 381 kPa to 1,200 kPa. For saturated liquid ammonia at 270 K, $V^l = 1.551 \times 10^{-3}$ m^3 kg^{-1}, and $\beta = 2.095 \times 10^{-3}$ K^{-1}.

6.4. Liquid isobutane is throttled through a valve from an initial state of 360 K and 4,000 kPa to a final pressure of 2,000 kPa. Estimate the temperature change and the entropy change of the isobutane. The specific heat of liquid isobutane at 360 K is 2.78 J g^{-1} °C^{-1}. Estimates of V and β may be found from Eq. (3.52).

6.5. Liquid water at 25°C and 1 bar fills a rigid vessel. If heat is added to the water until its temperature reaches 50°C, what pressure is developed? The average value of β between 25 and 50°C is 36.2×10^{-5} K^{-1}. The value of κ at 1 bar and 50°C is 4.42×10^{-5} bar^{-1}, and may be assumed independent of P. The specific volume of liquid water at 25°C is 1.0030 cm^3 g^{-1}.

6.6. Estimate the entropy change of vaporization of benzene at 50°C. The vapor pressure of benzene is given by the equation:

$$\ln P^{\text{sat}}/\text{kPa} = 13.8858 - \frac{2,788.51}{t/°C + 220.79}$$

(*a*) Use Eq. (6.53) with an estimated value of ΔV^{lv}.

(*b*) Use the Clausius/Clapeyron equation of Example 6.4.

6.7. A stream of propane gas is partially liquefied by throttling from 200 bar and 370 K to 1 bar. What fraction of the gas is liquefied in this process? The vapor pressure of propane is given by Eq. (6.56) with parameters: $A = -6.72219$, $B = 1.33236$, $C = -2.13868$, $D = -1.38551$.

6.8. The state of 1(lb$_m$) of steam is changed from saturated vapor at 20(psia) to superheated vapor at 50(psia) and 1,000(°F). What are the enthalpy and entropy changes of the steam? What would the enthalpy and entropy changes be if steam were an ideal gas?

6.9. Very pure liquid water can be subcooled at atmospheric pressure to temperatures well below 0°C. Assume that 1 kg has been cooled as a liquid to −6°C. A small ice crystal (of negligible mass) is added to "seed" the subcooled liquid. If the subsequent change occurs adiabatically at atmospheric pressure, what fraction of the system freezes and what is the final temperature? What is ΔS_{total} for the process, and what is its irreversible feature? The latent heat of fusion of water at 0°C is 333.4 J g^{-1}, and the specific heat of subcooled liquid water is 4.226 J g^{-1} °C^{-1}.

6.10. A two-phase system of liquid water and water vapor in equilibrium at 8,000 kPa consists of equal volumes of liquid and vapor. If the total volume $V^t = 0.15$ m^3, what is the total enthalpy H^t and what is the total entropy S^t?

6.11. A vessel contains 1 kg of H_2O existing as liquid and vapor in equilibrium at 1,000 kPa. If the vapor occupies 70% of the volume of the vessel, determine H and S for the 1 kg of H_2O.

6.12. A pressure vessel contains liquid water and water vapor in equilibrium at 350(°F). The total mass of liquid and vapor is 3(lb_m). If the volume of vapor is 50 times the volume of liquid, what is the total enthalpy of the contents of the vessel?

6.13. Wet steam at 230°C has a density of 0.025 g cm^{-3}. Determine x, H, and S.

6.14. A vessel of 0.15-m^3 volume containing saturated-vapor steam at 150°C is cooled to 30°C. Determine the final volume and mass of *liquid* water in the vessel.

6.15. Wet steam at 1,100 kPa expands at constant enthalpy (as in a throttling process) to 101.33 kPa, where its temperature is 105°C. What is the quality of the steam in its initial state?

6.16. Steam at 2,100 kPa and 260°C expands at constant enthalpy (as in a throttling process) to 125 kPa. What is the temperature of the steam in its final state and what is its entropy change? If steam were an ideal gas, what would be its final temperature and its entropy change?

6.17. Steam at 300(psia) and 500(°F) expands at constant enthalpy (as in a throttling process) to 20(psia). What is the temperature of the steam in its final state and what is its entropy change? If steam were an ideal gas, what would be its final temperature and its entropy change?

6.18. Superheated steam at 500 kPa and 300°C expands isentropically to 50 kPa. What is its final enthalpy?

6.19. A rigid vessel contains 0.014 m^3 of saturated-vapor steam in equilibrium with 0.021 m^3 of saturated-liquid water at 100°C. Heat is transferred to the vessel until one phase just disappears, and a single phase remains. Which phase (liquid or vapor) remains, and what are its temperature and pressure? How much heat is transferred in the process?

6.20. What is the mole fraction of water vapor in air that is saturated with water at 25°C and 101.33 kPa? At 50°C and 101.33 kPa?

6.21. A vessel of 0.25-m^3 capacity is filled with saturated steam at 1,500 kPa. If the vessel is cooled until 25 percent of the steam has condensed, how much heat is transferred and what is the final pressure?

6.22. A vessel of 2-m^3 capacity contains 0.02 m^3 of liquid water and 1.98 m^3 of water vapor at 101.33 kPa. How much heat must be added to the contents of the vessel so that the liquid water is just evaporated?

6.23. A rigid vessel of 0.4-m^3 volume is filled with steam at 800 kPa and 350°C. How much heat must be transferred from the steam to bring its temperature to 200°C?

6.24. One kilogram of steam is contained in a piston/cylinder device at 800 kPa and 200°C.

  (*a*) If it undergoes a mechanically reversible, isothermal expansion to 150 kPa, how much heat does it absorb?

  (*b*) If it undergoes a reversible, adiabatic expansion to 150 kPa, what is its final temperature and how much work is done?

6.25. Steam at 2,000 kPa containing 6% moisture is heated at constant pressure to 575°C. How much heat is required per kilogram?

6.26. Steam at 2,700 kPa and with a quality of 0.90 undergoes a reversible, adiabatic expansion in a nonflow process to 400 kPa. It is then heated at constant volume until it is saturated vapor. Determine Q and W for the process.

6.27. Four kilograms of steam in a piston/cylinder device at 400 kPa and 175°C undergoes a mechanically reversible, isothermal compression to a final pressure such that the steam is just saturated. Determine Q and W for the process.

6.28. One kilogram of water in a piston/cylinder device at 25°C and 1 bar is compressed in a mechanically reversible, isothermal process to 1,500 bar. Determine Q, W, ΔU, ΔH, and ΔS given that $\beta = 250 \times 10^{-6}$ K^{-1} and $\kappa = 45 \times 10^{-6}$ bar^{-1}.

6.29. A piston/cylinder device operating in a cycle with steam as the working fluid executes the following steps:

- Steam at 550 kPa and 200°C is heated at constant volume to a pressure of 800 kPa.

- The steam then expands, reversibly and adiabatically, to the initial temperature of 200°C.

- Finally, the steam is compressed in a mechanically reversible, isothermal process to the initial pressure of 550 kPa.

What is the thermal efficiency of the cycle?

6.30. A piston/cylinder device operating in a cycle with steam as the working fluid executes the following steps:

- Saturated-vapor steam at 300(psia) is heated at constant pressure to 900(°F).

- The steam then expands, reversibly and adiabatically, to the initial temperature of 417.35(°F).

- Finally, the steam is compressed in a mechanically reversible, isothermal process to the initial state.

What is the thermal efficiency of the cycle?

6.31. Steam expands reversibly and adiabatically in a turbine, entering at 4,000 kPa and 400°C.

(*a*) For what discharge pressure is the exit stream a saturated vapor?

(*b*) For what discharge pressure is the exit stream a wet vapor with quality of 0.95?

6.32. A steam turbine, operating reversibly and adiabatically, takes in superheated steam at 2,000 kPa and discharges at 50 kPa.

(*a*) What is the minimum superheat required so that the exhaust contains no moisture?

(*b*) What is the power output of the turbine if it operates under these conditions and the steam rate is 5 kg s^{-1}?

6.33. An operating test of a steam turbine produces the following results. With steam supplied to the turbine at 1,350 kPa and 375°C, the exhaust from the turbine at 10 kPa is saturated vapor. Assuming adiabatic operation and negligible changes in kinetic and potential energies, determine the turbine efficiency, i.e., the ratio of actual work of the turbine to the work of a turbine operating isentropically from the same initial conditions to the same exhaust pressure.

6.34. A steam turbine operates adiabatically with a steam rate of 25 kg s^{-1}. The steam is supplied at 1,300 kPa and 400°C and discharges at 40 kPa and 100°C. Determine the power output of the turbine and the efficiency of its operation in comparison with a turbine that operates *reversibly* and adiabatically from the same initial conditions to the same final pressure.

6.35. Let P_1^{sat} and P_2^{sat} be values of the saturation vapor pressure of a pure liquid at absolute temperatures T_1 and T_2. Justify the following interpolation formula for estimation of the vapor pressure P^{sat} at intermediate temperature T:

$$\ln P^{\text{sat}} = \ln P_1^{\text{sat}} + \frac{T_2(T - T_1)}{T(T_2 - T_1)} \ln \frac{P_2^{\text{sat}}}{P_1^{\text{sat}}}$$

6.36. Assuming the validity of Eq. (6.54), derive *Edmister's formula* for estimation of the acentric factor:

$$\omega = \frac{3}{7} \left(\frac{\theta}{1 - \theta} \right) \log P_c - 1$$

where $\theta \equiv T_n/T_c$, T_n is the normal boiling point, and P_c is in (atm).

6.37. From steam-table data, estimate values for the residual properties V^R, H^R, and S^R for steam at 225°C and 1,600 kPa, and compare with values found by a suitable generalized correlation.

6.38. From data in the steam tables:

(a) Determine numerical values of G^l and G^v for saturated liquid and vapor at 1,000 kPa. Should these be the same?

(b) Determine numerical values of $\Delta H^{lv}/T$ and ΔS^{lv} at 1,000 kPa. Should these be the same?

(c) Find numerical values of V^R, H^R, and S^R for saturated vapor at 1,000 kPa.

(d) Estimate a value for dP^{sat}/dT at 1,000 kPa and apply the Clapeyron equation to evaluate ΔS^{lv} at 1,000 kPa. How well does this result agree with the steam-table value?

Apply appropriate generalized correlations for evaluation of V^R, H^R, and S^R for saturated vapor at 1,000 kPa. How well do these results compare with the values found in (c)?

6.39. From data in the steam tables:

(a) Determine numerical values of G^l and G^v for saturated liquid and vapor at 150(psia). Should these be the same?

(b) Determine numerical values of $\Delta H^{lv}/T$ and ΔS^{lv} at 150(psia). Should these be the same?

(c) Find numerical values of V^R, H^R, and S^R for saturated vapor at 150(psia).

(d) Estimate a value for dP^{sat}/dT at 150(psia) and apply the Clapeyron equation to evaluate ΔS^{lv} at 150(psia). How well does this result agree with the steam-table value?

Apply appropriate generalized correlations for evaluation of V^R, H^R, and S^R for saturated vapor at 150(psia). How well do these results compare with the values found in (c)?

6.40. Estimate V^R, H^R, and S^R for 1,3-butadiene at 500 K and 21 bar by appropriate generalized correlations.

6.41. Estimate V^R, H^R, and S^R for carbon dioxide at 400 K and 200 bar by appropriate generalized correlations.

6.42. Estimate V^R, H^R, and S^R for sulfur dioxide at 450 K and 35 bar by appropriate generalized correlations.

6.43. Steam undergoes a change from an initial state of 450°C and 3,000 kPa to a final state of 140°C and 235 kPa. Determine ΔH and ΔS:

(a) From steam-table data.

(b) By equations for an ideal gas.

(c) By appropriate generalized correlations.

6.44. Propane gas at 1 bar and 35°C is compressed to a final state of 135 bar and 195°C. Estimate the molar volume of the propane in the final state and the enthalpy and entropy changes for the process. In its initial state, propane may be assumed an ideal gas.

6.45. Propane at 70°C and 101.33 kPa is compressed isothermally to 1,500 kPa. Estimate ΔH and ΔS for the process by suitable generalized correlations.

6.46. Estimate the molar volume, enthalpy, and entropy for 1,3-butadiene as a saturated vapor and as a saturated liquid at 380 K. The enthalpy and entropy are set equal to zero for the ideal-gas state at 101.33 kPa and 0°C. The vapor pressure of 1,3-butadiene at 380 K is 1,919.4 kPa.

6.47. Estimate the molar volume, enthalpy, and entropy for n-butane as a saturated vapor and as a saturated liquid at 370 K. The enthalpy and entropy are set equal to zero for the ideal-gas state at 101.33 kPa and 273.15 K. The vapor pressure of n-butane at 370 K is 1,435 kPa.

6.48. Five moles of calcium carbide is combined with 10 mol of liquid water in a closed, rigid, high-pressure vessel of 750-cm^3 capacity. Acetylene gas is produced by the reaction:

$$CaC_2(s) + 2H_2O(l) \rightarrow C_2H_2(g) + Ca(OH)_2(s)$$

Initial conditions are 25°C and 1 bar, and the reaction goes to completion. For a final temperature of 125°C, determine:

(a) The final pressure.

(b) The heat transferred.

At 125°C, the molar volume of Ca(OH)$_2$ is 33.0 cm^3 mol^{-1}. Ignore the effect of any gas present in the tank initially.

6.49. Propylene gas at 127°C and 38 bar is throttled in a steady-state flow process to 1 bar, where it may be assumed an ideal gas. Estimate the final temperature of the propylene and its entropy change.

6.50. Propane gas at 22 bar and 423 K is throttled in a steady-state flow process to 1 bar. Estimate the entropy change of the propane caused by this process. In its final state, propane may be assumed an ideal gas.

6.51. Propane gas at 100°C is compressed isothermally from an initial pressure of 1 bar to a final pressure of 10 bar. Estimate ΔH and ΔS.

6.52. Hydrogen sulfide gas is compressed from an initial state of 400 K and 5 bar to a final state of 600 K and 25 bar. Estimate ΔH and ΔS.

6.53. Carbon dioxide expands at constant enthalpy (as in a throttling process) from 1,600 kPa and 45°C to 101.33 kPa. Estimate ΔS for the process.

6.54. A stream of ethylene gas at 250°C and 3,800 kPa expands isentropically in a turbine to 120 kPa. Determine the temperature of the expanded gas and the work produced if the properties of ethylene are calculated by

(*a*) Equations for an ideal gas.

(*b*) Appropriate generalized correlations.

6.55. A stream of ethane gas at 220°C and 30 bar expands isentropically in a turbine to 2.6 bar. Determine the temperature of the expanded gas and the work produced if the properties of ethane are calculated by

(*a*) Equations for an ideal gas.

(*b*) Appropriate generalized correlations.

6.56. Estimate the final temperature and the work required when 1 mol of *n*-butane is compressed isentropically in a steady-flow process from 1 bar and 50°C to 7.8 bar.

CHAPTER 7

THERMODYNAMICS OF FLOW PROCESSES

Most equipment used in the chemical, petroleum, and related industries is designed for the movement of fluids, and an understanding of fluid flow is essential to a chemical engineer. The underlying discipline is fluid mechanics[1], which is based on the law of mass conservation, the linear-momentum principle (Newton's second law), and the first and second laws of thermodynamics.

The application of thermodynamics to flow processes is also based on mass conservation and on the first and second laws. The addition of the linear-momentum principle makes fluid mechanics a broader field of study. The distinction between *thermodynamics problems* and *fluid-mechanics problems* depends on whether this principle is required for solution. Those problems whose solutions depend only on mass conservation and on the laws of thermodynamics are commonly set apart from the study of fluid mechanics and are treated in courses on thermodynamics. Fluid mechanics then deals with the broad spectrum of problems which *require* application of the momentum principle. This division is arbitrary, but it is traditional and convenient.

The practical application of thermodynamics to flow processes is usually to finite amounts of fluid undergoing finite changes in state. Consider for example the flow of gas through a pipeline. If the states and thermodynamic properties of the gas entering and leaving the pipeline are known, then application of the first

---

[1]Fluid mechanics is treated as an integral part of transport processes by R. B. Bird, W. E. Stewart, and E. N. Lightfoot in *Transport Phenomena*, John Wiley, New York, 1960; by C. O. Bennett and J. E. Myers in *Momentum, Heat, and Mass Transfer*, 2nd ed., McGraw-Hill, New York, 1982; by R. W. Fahien in *Fundamentals of Transport Phenomena*, McGraw-Hill, New York, 1984; and by D. E. Rosner in *Transport Processes in Chemically Reacting Systems*, Butterworths, Boston, 1986.

law establishes the magnitude of the energy exchange with the surroundings of the pipeline. The mechanism of the process, the details of flow, and the state path actually followed by the fluid between entrance and exit are not pertinent to this calculation. On the other hand, if one has only incomplete knowledge of the initial or final state of the gas, then more detailed information about the process is needed before any calculations are made. For example, the exit pressure of the gas may not be specified. In this case, one must apply the momentum principle of fluid mechanics, and this requires an empirical or theoretical expression for the shear stress at the pipe wall.

The fundamental thermodynamic equations generally applicable to flow processes are presented in Sec. 7.1, and in later sections these equations are applied to specific processes.

7.1 Equations of Balance

A mathematical accounting scheme known as an equation of balance may be written for any countable quantity capable of transport. Thus, one can develop balance equations for dollars, for members of human or other species, and more to the point for mass, energy, and entropy. However, no balance equations are written for properties such as temperature, pressure, density, surface tension, and viscosity, which are not countable quantities. Although conservation of mass and of energy are well established principles, countable quantities are not necessarily conserved. Dollars, for example, are printed and burned; humans are born and die; and entropy is generated in any irreversible process. We therefore develop the general equation of balance first, and treat its application to conserved quantities as a special case.

Central to development of equations of balance is the concept of a *control volume*, an arbitrary volume within three-dimensional space identified for analysis. It is enclosed by a bounding *control surface*, which may or may not be identified with material surfaces. The control volume may be closed or it may be open to the transport of matter; it may expand or contract; and in the most general case it may translate or rotate in space. The contents of the control volume constitute the system, and the local part of the universe that interacts with the system forms the surroundings.

We consider here a finite control volume, shown schematically in Fig. 7.1, whose center of mass is stationary. Thus it exhibits no changes in its *gross* kinetic or potential energy. With respect to this control volume we note the following possibilities for countable quantity X:

1. Amounts of X enter and leave the control volume by *transport* across the control surface. Let $\dot{X}_T$ denote the *net* instantaneous rate of transport of X *into* the control volume. The convention here is that transport into the control volume is considered positive, while transport out is negative. Transport constitutes the mode of interaction between the system and its surroundings, and *at the control surface* we can make the essential identification,

$$\dot{X}_T(\text{system}) = -\dot{X}_T(\text{surroundings}) \tag{7.1}$$

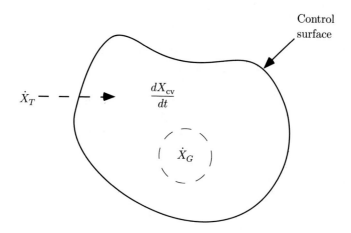

Figure 7.1: Schematic diagram of a control volume.

2. Amounts of X are created and destroyed within the control volume. Let $\dot{X}_G$ denote the *net* instantaneous rate of *generation* of X within the control volume. The implied convention in this case is that creation of X is positive, while destruction is negative.

3. Owing to transport and generation, the amount of X within the control volume can change with time. We therefore represent the instantaneous time rate of change of X within the control volume by dX_{cv}/dt, where subscript "cv" denotes the control volume.

The statement of balance for quantity X is simply

$$\left\{ \begin{array}{c} \text{Net rate of} \\ \text{transport of } X \\ \text{into control} \\ \text{volume} \end{array} \right\} + \left\{ \begin{array}{c} \text{Net rate of} \\ \text{generation of } X \\ \text{within control} \\ \text{volume} \end{array} \right\} = \left\{ \begin{array}{c} \text{Time rate of} \\ \text{change of } X \\ \text{within control} \\ \text{volume} \end{array} \right\}$$

The equivalent *equation of balance* is

$$\boxed{\dot{X}_T + \dot{X}_G = \frac{dX_{cv}}{dt}} \tag{7.2}$$

This equation is the general *rate* form of the equation of balance, applicable at any instant. The two terms on the left are identified as "transport" and "generation" terms; the one on the right is the "accumulation" term. Each can vary with time.

An alternative form of Eq. (7.2), convenient for some applications, is a *differential* equation of balance, obtained by multiplying Eq. (7.2) by dt:

$$\boxed{dX_T + dX_G = dX_{cv}} \tag{7.3}$$

where

$$dX_T \equiv \dot{X}_T \, dt \tag{7.4a}$$
$$dX_G \equiv \dot{X}_G \, dt \tag{7.4b}$$
$$dX_{\text{cv}} \equiv \frac{dX_{\text{cv}}}{dt} dt \tag{7.4c}$$

The *integral* equation of balance follows immediately from Eq. (7.3) by integration from time t_1 to time t_2:

$$\boxed{X_T + X_G = \Delta X_{\text{cv}}} \tag{7.5}$$

where

$$X_T \equiv \int_{t_1}^{t_2} \dot{X}_T \, dt \tag{7.6a}$$

$$X_G \equiv \int_{t_1}^{t_2} \dot{X}_G \, dt \tag{7.6b}$$

$$\Delta X_{\text{cv}} \equiv \int_{t_1}^{t_2} \frac{dX_{\text{cv}}}{dt} dt = X_{\text{cv}}(t_2) - X_{\text{cv}}(t_1) \tag{7.6c}$$

We noted in Sec. 2.4 with respect to energy balances a difference in kind between the terms on the left- and right-hand sides. Here too our notation indicates that the transport and generation terms on the left are different in character from the accumulation terms on the right. Most obviously, the transport and generation terms reflect *causes*, while the accumulation terms represent *effects*.

Quantities $\dot{X}_T$ and $\dot{X}_G$ are *process rates*; when integrated over time they produce *amounts* of X transported into or generated within the control volume [Eqs. (7.6a) and (7.6b)]. The derivative dX_{cv}/dt is an *accumulation rate*; when integrated over time, it yields the *change* in the amount of X within the control volume [Eq. (7.6c)]. The difference operator "Δ" again denotes a change; ΔX is the change in X. Its use with the transport and generation terms is totally inappropriate, because they represent amounts rather than changes.

Conservation of mass and of energy are two of the great generalizations of science. With respect to the equation of balance, when the conservation principle applies to quantity X, then the generation term $\dot{X}_G$ is necessarily zero. Furthermore, conservation also implies that no generation of X can occur in the surroundings. This observation along with Eq. (7.1) leads to the conclusion that the *total* amount of X must be constant. Thus when the conservation law applies to quantity X, we can write

$$\dot{X}_G = \frac{dX_{\text{total}}}{dt} = 0 \tag{7.7}$$

where X_{total} is the amount of X in both system and surroundings.

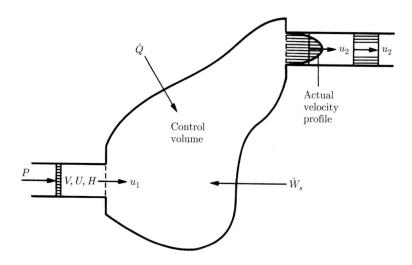

Figure 7.2: Control volume with one entrance and one exit.

The flow processes of interest to chemical engineers usually permit identification of almost the entire control surface with actual material surfaces. Only at specifically provided entrances and exits is the control surface subject to arbitrary location, and here it is universal practice to place the control surface perpendicular to the direction of flow. This allows imposition of two *idealizations* that facilitate the practical application of equations of balance:

1. Flow is presumed unidirectional at any cross section of a conduit where thermodynamic, kinetic, and dynamic properties are assigned or evaluated, namely, at entrances to and exits from the equipment under consideration.

2. At such a cross section these same properties are presumed not to vary in a direction perpendicular to the direction of flow. Thus properties such as velocity, temperature, and density, assigned or evaluated for the cross section, have values which are appropriate averages over the cross section.

These idealizations are pragmatic in nature, and for most practical purposes they introduce negligible error. An example of a control volume with one entrance and one exit is shown in Fig. 7.2. The actual velocity profile shown at the exit is equivalent to the uniform velocity profile indicated to the right that provides the same mass flow rate (idealization 2). In what follows we apply the equation of balance to mass, energy, and entropy with respect to a control volume of the kind shown in Fig. 7.2.

Mass Balance

Since mass is conserved, $\dot{X}_G = 0$, and Eq. (7.2) is written

$$\dot{m}_T = \frac{dm_{cv}}{dt}$$

where $\dot{m}_T$ is the net transport of mass into the control volume by streams flowing in and out at entrances and exits. For convenience in application, we rewrite this equation as

$$\boxed{\frac{dm_{cv}}{dt} + \Delta(\dot{m})_{fs} = 0} \qquad (7.8)$$

where the symbol Δ denotes the difference between exit and entrance streams and the subscript "fs" indicates that the term applies to all flowing streams.

Since the mass flow rate of each stream is given by

$$\dot{m} = \text{mass flow rate} = \rho u A$$

where ρ is the average fluid density, u is its average velocity, and A is the cross-sectional area of the entrance or exit duct, Eq. (7.8) may also be written

$$\frac{dm_{cv}}{dt} + \Delta(\rho u A)_{fs} = 0 \qquad (7.9)$$

In this form the mass-balance equation is often called the *continuity equation*.

The flow process characterized as *steady state* is an important special case for which conditions within the control volume do not change with time. In this case the control volume contains a constant mass of fluid, and the accumulation term of Eq. (7.9) is zero. It therefore reduces to

$$\Delta(\rho u A)_{fs} = 0 \qquad (7.10)$$

The term "steady state" does not imply that flow rates are constant, merely that the inflow of mass is exactly matched by the outflow of mass.

When there is but a single entrance and a single exit stream, as in Fig. 7.2, the mass flow rate $\dot{m}$ is the same for both streams, and Eq. (7.10) becomes

$$\rho_2 u_2 A_2 - \rho_1 u_1 A_1 = 0$$

or

$$\dot{m} = \text{const} = \rho_2 u_2 A_2 = \rho_1 u_1 A_1$$

Since specific volume is the reciprocal of density,

$$\boxed{\dot{m} = \frac{u_1 A_1}{V_1} = \frac{u_2 A_2}{V_2} = \frac{u A}{V}} \qquad (7.11)$$

This form of the continuity equation finds frequent use.

Energy Balance

In Chap. 2 the first law of thermodynamics is applied to closed systems (nonflow processes) and to single-stream, steady-state, steady-flow processes to provide specific equations of energy conservation for these important applications. Our purpose here is to present equations of greater generality written for open systems represented by control volumes. Since energy is conserved, $\dot{X}_G$ is zero, and is omitted from Eq. (7.2).

The transport term is subdivided to reflect the several modes by which energy may cross the control surface. The flowing streams contribute to the transport term by virtue of their internal, potential, and kinetic energies. On the basis of a unit mass, each stream has a total energy $U + \frac{1}{2}u^2 + zg$, where u is the average velocity of the stream, z is its elevation above a datum level, and g is the local acceleration of gravity. Thus, each stream transports energy at the rate $(U + \frac{1}{2}u^2 + zg)\dot{m}$.

Energy is also transported across the control surface as heat $\dot{Q}$ and as work, which may be of several forms. Consider first the work associated with moving the flowing streams into and out of the control volume at entrances and exits. The fluid at any entrance or exit has a set of average properties, P, V, U, H, etc. We imagine that a unit mass of fluid with these properties exists in a conduit adjacent to the entrance or exit, as shown in Fig. 7.2 at the entrance. This unit mass of fluid is pushed into the control volume by additional fluid, here replaced by a piston which exerts the constant pressure P. The work done by this piston in pushing the unit mass into the control volume is PV, and the work rate is $(PV)\dot{m}$. Since Δ denotes the difference between exit and entrance quantities, the net work done *on* the system when all entrance and exit sections are taken into account is $-\Delta[(PV)\dot{m}]_{\text{fs}}$.

Another form of work is the shaft work indicated in Fig. 7.2 by rate $\dot{W}_s$. In addition work may be associated with expansion or contraction of the control volume and there may be stirring work. These forms of work are all included in a rate term represented by $\dot{W}$. The transport term of Eq. (7.2) is therefore written

$$\dot{X}_T = \dot{Q} + \dot{W} - \Delta\left[(PV)\,\dot{m}\right]_{\text{fs}} - \Delta\left[\left(U + \tfrac{1}{2}u^2 + zg\right)\dot{m}\right]_{\text{fs}}$$

Since $U + PV = H$, this is more conveniently expressed as

$$\dot{X}_T = \dot{Q} + \dot{W} - \Delta\left[\left(H + \tfrac{1}{2}u^2 + zg\right)\dot{m}\right]_{\text{fs}} \tag{7.12}$$

The accumulation term is simply the time rate of change of the total internal energy within the control volume, $d(mU)_{\text{cv}}/dt$. In combination with Eq. (7.2) these relations yield the energy-balance equation:

$$\boxed{\frac{d(mU)_{\text{cv}}}{dt} + \Delta\left[\left(H + \tfrac{1}{2}u^2 + zg\right)\dot{m}\right]_{\text{fs}} = \dot{Q} + \dot{W}} \tag{7.13}$$

Although Eq. (7.13) is an energy balance of reasonable generality, it has inherent limitations. In particular, it reflects our original premise that the center of mass of the control volume (Fig. 7.1) is stationary. This means that terms for kinetic- and potential-energy changes of the fluid in the control volume can be omitted. For virtually all applications of interest to chemical engineers, Eq. (7.13) is adequate.[2] For most (but not all) applications, kinetic- and potential-energy changes in the flowing streams are also negligible, and Eq. (7.13) simplifies to

$$\frac{d(mU)_{cv}}{dt} + \Delta(H\dot{m})_{fs} = \dot{Q} + \dot{W} \qquad (7.14)$$

This equation may be applied to a variety of processes of a transient nature, as illustrated in the following examples.

Example 7.1 Consider the filling of an evacuated tank with a gas from a constant-pressure line. What is the relation between the enthalpy of the gas in the entrance line and the internal energy of the gas in the tank? Neglect heat transfer between the gas and the tank. If the gas is ideal and has constant heat capacities, how is the temperature of the gas in the tank related to the temperature in the entrance line?

SOLUTION   If the tank is chosen as the control volume, there is but one opening into the tank and it serves as an entrance, because gas flows into the tank. Since there is no expansion work, stirring work, or shaft work, $\dot{W} = 0$. In the absence of any specific information, we assume that kinetic- and potential-energy changes are negligible. Then Eq. (7.14) becomes

$$\frac{d(mU)_{tank}}{dt} - H'\dot{m}' = 0$$

where the prime (') identifies the entrance stream and the minus sign is required because it *is* an entrance stream. The mass balance is simply

$$\dot{m}' = \frac{dm_{tank}}{dt}$$

Combining these two balance equations yields

$$\frac{d(mU)_{tank}}{dt} - H'\frac{dm_{tank}}{dt} = 0$$

Multiplication by dt and integration over time (noting that H' is constant) gives

$$\Delta(mU)_{tank} - H'\Delta m_{tank} = 0$$

whence

$$m_2 U_2 - m_1 U_1 = H'(m_2 - m_1)$$

Since the mass in the tank initially is zero, $m_1 = 0$, and this reduces to

$$U_2 = H' \qquad (A)$$

---

[2]Another energy balance in common use is known as the *mechanical energy balance*, of which Bernoulli's famous equation is a special case. This equation is not considered in this text, because its proper derivation depends on the momentum principle of fluid mechanics. See R. B. Bird, *Chem. Eng. Edu.*, vol. 27, pp. 102–109, 1993.

a result showing that in the absence of heat transfer the energy of the gas contained within the tank at the end of the process is equal to the enthalpy of the gas added.

If the gas is ideal,

$$H' = U' + P'V' = U' + RT'$$

and Eq. (A) becomes

$$U_2 - U' = RT'$$

For constant heat capacity,

$$U_2 - U' = C_V(T_2 - T')$$

whence

$$C_V(T_2 - T') = RT'$$

or

$$\frac{T_2 - T'}{T'} = \frac{R}{C_V} = \frac{C_P - C_V}{C_V}$$

If C_P/C_V is set equal to γ, this reduces to

$$T_2 = \gamma T'$$

which indicates that the final temperature is independent of the amount of gas admitted to the tank. This result is strongly conditioned by the initial stipulation that heat transfer between the gas and the tank be neglected.

Example 7.2   A 1.5-m^3 tank contains 500 kg of liquid water in equilibrium with pure water vapor, which fills the remainder of the tank. The temperature and pressure are 100°C and 101.33 kPa. From a water line at a constant temperature of 70°C and a constant pressure somewhat above 101.33 kPa, 750 kg of liquid is bled into the tank. If the temperature and pressure in the tank are not to change as a result of the process, how much energy as heat must be transferred to the tank?

SOLUTION   Choose the tank as the control volume. As in Example 7.1, there is no work, and again we assume negligible kinetic- and potential-energy effects. Equation (7.14) therefore is written

$$\frac{d(mU)_{\text{tank}}}{dt} - H'\dot{m}' = \dot{Q}$$

where the prime denotes the state of the inlet stream. The mass balance,

$$\dot{m}' = \frac{dm_{\text{tank}}}{dt}$$

may be combined with the energy balance to yield

$$\frac{d(mU)_{\text{tank}}}{dt} - H'\frac{dm_{\text{tank}}}{dt} = \dot{Q}$$

Multiplication by dt and integration over time (with H' constant) gives

$$Q = \Delta(mU)_{\text{tank}} - H'\Delta m_{\text{tank}}$$

The definition of enthalpy may be applied to the entire contents of the tank to give

$$\Delta(mU)_{\text{tank}} = \Delta(mH)_{\text{tank}} - \Delta(PmV)_{\text{tank}}$$

Since the total tank volume mV and the pressure are constant, $\Delta(PmV)_{\text{tank}} = 0$. Therefore

$$Q = \Delta(mH)_{\text{tank}} - H'\Delta m_{\text{tank}} = (m_2 H_2 - m_1 H_1)_{\text{tank}} - H'\Delta m_{\text{tank}}$$

where Δm_{tank} is the 750 kg of water bled into the tank, and subscripts 1 and 2 refer to conditions in the tank at the beginning and end of the process. At the end of the process the tank still contains saturated liquid and saturated vapor in equilibrium at 100°C and 101.33 kPa. Hence $m_1 H_1$ and $m_2 H_2$ each consist of two terms, one for the liquid phase and one for the vapor phase.

The numerical solution makes use of the following enthalpies taken from the steam tables:

$$H' = 293.0 \text{ kJ kg}^{-1}; \text{ saturated liquid at } 70°\text{C}$$
$$H^l_{\text{tank}} = 419.1 \text{ kJ kg}^{-1}; \text{ saturated liquid at } 100°\text{C}$$
$$H^v_{\text{tank}} = 2{,}676.0 \text{ kJ kg}^{-1}; \text{ saturated vapor at } 100°\text{C}$$

The volume of vapor in the tank initially is 1.5 m³ minus the volume occupied by the 500 kg of liquid water. Thus

$$m^v_1 = \frac{1.5 - (500)(0.001044)}{1.673} = 0.772 \text{ kg}$$

where 0.001044 and 1.673 m³ kg^{-1} are the specific volumes of saturated liquid and saturated vapor at 100°C from the steam tables. Then

$$(m_1 H_1)_{\text{tank}} = m^l_1 H^l_1 + m^v_1 H^v_1 = 500(419.1) + 0.772(2{,}676.0)$$
$$= 211{,}616 \text{ kJ}$$

At the end of the process, the masses of liquid and vapor are determined by the conservation of mass and by the fact that the tank volume is still 1.5 m³. As a result, we may write the two equations:

$$m_2 = 500 + 0.772 + 750 = m^v_2 + m^l_2$$

$$1.5 = 1.673 m^v_2 + 0.001044 m^l_2$$

whence

$$m^l_2 = 1{,}250.65 \text{ kg}$$

$$m^v_2 = 0.116 \text{ kg}$$

Then, since $H^l_2 = H^l_1$ and $H^v_2 = H^v_1$,

$$(m_2 H_2)_{\text{tank}} = 1{,}250.65(419.1) + 0.116(2{,}676.0) = 524{,}458 \text{ kJ}$$

Finally, substituting the values for $(m_1 H_1)_{\text{tank}}$ and $(m_2 H_2)_{\text{tank}}$ in the equation for Q gives

$$Q = 524{,}458 - 211{,}616 - 750(293.0) = 93{,}092 \text{ kJ}$$

Example 7.3 An insulated, electrically heated hot-water heater contains 190 kg of liquid water at 60°C when a power outage occurs. If water is withdrawn from the heater tank at a steady rate of $\dot{m} = 0.2$ kg s^{-1}, how long will it take for the temperature of the water in the tank to drop from 60 to 35°C? Assume that cold water enters the tank at 10°C, and that heat losses from the tank are negligible. Water may be considered an incompressible liquid.

SOLUTION   Here, $\dot{Q} = \dot{W} = 0$. Additionally, we assume perfect mixing of the contents of the tank; this implies that the properties of the water leaving the tank are those of the water in the tank. With the mass flow rate into the tank equal to the mass flow rate out, m_{cv} is constant; moreover, the differences between inlet and outlet kinetic and potential energies can be neglected. Equation (7.14) is therefore written

$$m\frac{dU}{dt} + \dot{m}(H - H_1) = 0 \qquad (A)$$

where unsubscripted quantities refer to the contents of the tank and H_1 is the specific enthalpy of the water entering the tank. For an incompressible liquid with constant heat capacity C (see Example 6.2),

$$\frac{dU}{dt} = C\frac{dT}{dt} \qquad \text{and} \qquad H - H_1 = C(T - T_1)$$

Equation (A) then becomes, on rearrangement,

$$dt = -\frac{m}{\dot{m}}\frac{dT}{T - T_1}$$

Integration from $t = 0$ (where $T = T_0$) to arbitrary time t yields

$$t = -\frac{m}{\dot{m}}\ln\left(\frac{T - T_1}{T_0 - T_1}\right) \qquad (B)$$

Substitution of numerical values into Eq. (B) gives, for the conditions of this problem,

$$t = -\frac{190}{0.2}\ln\left(\frac{35 - 10}{60 - 10}\right) = 658.5 \text{ s}$$

Thus, it takes about 11 minutes for the water temperature in the tank to drop from 60 to 35°C.

Energy Balances for Steady-State Flow Processes

Flow processes for which the accumulation term of Eq. (7.2), dX_{cv}/dt, is zero are said to occur at *steady state*. As discussed with respect to the mass balance, this means that the mass of the system within the control volume is constant; it also means that no changes occur with time in the properties of the fluid within the control volume nor at its entrances and exits. No expansion of the control volume is possible under these circumstances, and the only work of the process is then shaft work. For such a process the general energy balance, Eq. (7.13), becomes

$$\boxed{\Delta\left[\left(H + \tfrac{1}{2}u^2 + zg\right)\dot{m}\right]_{\text{fs}} = \dot{Q} + \dot{W}_s} \qquad (7.15)$$

Although "steady state" does not imply "steady flow,"[3] the usual application of this equation is to steady-state steady-flow processes, because such processes represent the norm in the chemical-process industry.

A further specialization results when there is but one entrance and one exit to the control volume. In this case the same mass flow rate $\dot{m}$ applies to both streams. Equation (7.15) then reduces to

$$\Delta \left(H + \tfrac{1}{2}u^2 + zg\right) \dot{m} = \dot{Q} + \dot{W}_s \qquad (7.16)$$

where subscript "fs" has been omitted in this simple case and Δ denotes the change from entrance to exit. Division by $\dot{m}$ gives

$$\Delta \left(H + \tfrac{1}{2}u^2 + zg\right) = \frac{\dot{Q}}{\dot{m}} + \frac{\dot{W}_s}{\dot{m}} = Q + W_s$$

or

$$\boxed{\Delta H + \frac{\Delta u^2}{2} + g\Delta z = Q + W_s} \qquad (7.17)$$

which is a restatement of Eq. (2.9a). In this equation, each term is based on a unit mass of fluid flowing through the control volume.

The kinetic-energy terms of the various energy balances developed here include the velocity u, which is the bulk-mean velocity as defined by the equation $u = \dot{m}/\rho A$. Fluids flowing in pipes exhibit a velocity profile, as shown in Fig. 7.2, which rises from zero at the wall (the no-slip condition) to a maximum at the center of the pipe. The kinetic energy of a fluid in a pipe depends on the actual velocity profile. For the case of laminar flow, the velocity profile is parabolic, and integration across the pipe shows that the kinetic-energy term should properly be u^2. In fully developed turbulent flow, the more common case in practice, the velocity across the major portion of the pipe is not far from uniform, and the expression $u^2/2$, as used in the energy equations, is more nearly correct.

In all of the equations written here, the energy unit is presumed to be the joule, in accord with the SI system of units. For the English system of units, the kinetic- and potential-energy terms, wherever they appear, require division by the dimensional constant g_c (see Secs. 1.3 and 1.8). However, in many applications, the kinetic- and potential-energy terms are omitted, because they are negligible compared with other terms. Exceptions are applications to nozzles, metering devices, wind tunnels, and hydroelectric power stations.

Entropy Balance

Written for entropy, Eq. (7.2) becomes

$$\dot{S}_T + \dot{S}_G = \frac{dS_{cv}}{dt}$$

---

[3] An example of a steady-state process that is not steady flow is a water heater in which variations in flow rate are exactly compensated by changes in the rate of heat transfer so that temperatures throughout remain constant.

Entropy is transported across the control surface in two ways:

1. Transport by means of heat transfer. If heat flows at the rate $\dot{Q}_j$ across a portion of the control surface at temperature $T_{cs,j}$, the resulting rate of entropy transport is $\dot{Q}_j/T_{cs,j}$. The summation of such terms then gives the net rate of entropy transport into the control volume by this mechanism:

$$\sum_j \frac{\dot{Q}_j}{T_{cs,j}}$$

2. Transport by flowing streams. Each stream carries with it entropy for which the transport rate is $S\dot{m}$. The net rate of transport into the control volume by this mechanism is then

$$-\Delta(S\dot{m})_{fs}$$

The entropy-transport term is therefore

$$\dot{S}_T = \sum_j \frac{\dot{Q}_j}{T_{cs,j}} - \Delta(S\dot{m})_{fs}$$

The accumulation term is the time rate of change of the total entropy of the fluid contained within the control volume $d(mS)_{cv}/dt$. The entropy balance can therefore be written

$$\boxed{\sum_j \frac{\dot{Q}_j}{T_{cs,j}} - \Delta(S\dot{m})_{fs} + \dot{S}_G = \frac{d(mS)_{cv}}{dt}}
\qquad (7.18)$$

According to the second law of thermodynamics, the entropy-generation term is zero for reversible processes and positive for irreversible processes; therefore $\dot{S}_G \geq 0$. An entropy increase reflects irreversibilities *within* the control volume, i.e., *internal* irreversibilities.

The difficulty with Eq. (7.18) is that we can rarely identify the temperature $T_{cs,j}$ at which heat rate $\dot{Q}_j$ crosses the control surface. Rather, the known temperatures are those in the surroundings with which the control volume interacts. We therefore let $T_{\sigma,j}$ represent the temperature in the surroundings associated with heat rate $\dot{Q}_j$, and write the mathematical identity

$$\sum_j \frac{\dot{Q}_j}{T_{cs,j}} \equiv \sum_j \frac{\dot{Q}_j}{T_{\sigma,j}} + \sum_j \frac{\dot{Q}_j}{T_{cs,j}} - \sum_j \frac{\dot{Q}_j}{T_{\sigma,j}}$$

$$= \sum_j \frac{\dot{Q}_j}{T_{\sigma,j}} + \sum_j \dot{Q}_j \left(\frac{T_{\sigma,j} - T_{cs,j}}{T_{\sigma,j} T_{cs,j}} \right)$$

Equation (7.18) may now be written

$$\sum_j \frac{\dot{Q}_j}{T_{\sigma,j}} - \Delta(S\dot{m})_{fs} + \dot{S}_G + \dot{S}'_G = \frac{d(mS)_{cv}}{dt}$$

where by definition

$$\dot{S}'_G \equiv \sum_j \dot{Q}_j \left(\frac{T_{\sigma,j} - T_{cs,j}}{T_{\sigma,j} T_{cs,j}} \right)$$

This quantity is identified as an additional entropy-generation term because it reflects the irreversibilities occuring in the surroundings as the result of heat transfer across finite temperature differences (Sec. 5.7); these are *external thermal irreversibilities*. We note that when $T_{\sigma,j} > T_{cs,j}$ heat is transferred from the surroundings to the control volume. In this event both the temperature difference and its associated heat rate are positive. When $T_{cs,j} > T_{\sigma,j}$, then both the temperature difference and the heat rate are negative. Thus all terms on the right of the preceding equation are positive, and $\dot{S}'_G$ is necessarily positive for heat transfer across finite temperature differences, regardless of direction.

The two quantities $\dot{S}_G$ and $\dot{S}'_G$ account respectively for entropy increases in the control volume and in the surroundings. Therefore together they represent the *total* rate of entropy generation as a result of irreversibilities in any process, and we combine them into a single term:

$$\dot{S}_G + \dot{S}'_G \equiv \dot{S}_{G,\text{total}}$$

The entropy-balance equation can now be written

$$\boxed{\sum_j \frac{\dot{Q}_j}{T_{\sigma,j}} - \Delta(S\dot{m})_{\text{fs}} + \dot{S}_{G,\text{total}} = \frac{d(mS)_{\text{cv}}}{dt}} \tag{7.19}$$

Since the second law requires the entropy-generation term $\dot{S}_{G,\text{total}}$ to be zero or positive, its omission allows the entropy balance to be written as the inequality:

$$\sum_j \frac{\dot{Q}_j}{T_{\sigma,j}} - \Delta(S\dot{m})_{\text{fs}} \leq \frac{d(mS)_{\text{cv}}}{dt}$$

Here, the limiting case of equality applies to a process that is *completely reversible*, which implies:

1. The process is internally reversible within the control volume.

2. Heat transfer external to the control volume is reversible.

The second item means either that the surroundings includes heat reservoirs at temperatures $T_{cs,j}$ or that Carnot engines are interposed between the control surface at temperatures $T_{cs,j}$ and heat reservoirs in the surroundings at temperatures $T_{\sigma,j}$.

For a steady-state flow process the mass and entropy of the fluid in the control volume are constant, and $d(mS)_{\text{cv}}/dt$ is zero. Equation (7.19) then becomes

$$\boxed{\Delta(S\dot{m})_{\text{fs}} = \sum_j \frac{\dot{Q}_j}{T_{\sigma,j}} + \dot{S}_{G,\text{total}}} \tag{7.20}$$

If in addition there is but one entrance and one exit, $\dot{m}$ is the same for both streams, and dividing through by $\dot{m}$ yields

$$\Delta S = \sum_j \frac{Q_j}{T_{\sigma,j}} + S_{G,\text{total}} \qquad (7.21)$$

Each term is here based on a unit amount of fluid flowing through the control volume.

Example 7.4 With reference to Example 5.5, determine the *maximum* amount of heat that can be transferred to the reservoir at 200°C.

SOLUTION   We formulate the problem in the language of the present section, but retain the notation of Example 5.5. The energy balance of Eq. (7.17) is written

$$Q' + Q_\sigma = \Delta H \qquad (A)$$

Similarly, the entropy balance of Eq. (7.21) is

$$\Delta S = \frac{Q'}{T'} + \frac{Q_\sigma}{T_\sigma} + S_{G,\text{total}}$$

The maximum heat rejection to the hot reservoir occurs when the process is completely reversible, in which case $S_{G,\text{total}} = 0$. Then by the last equation,

$$\frac{Q'}{T'} + \frac{Q_\sigma}{T_\sigma} = \Delta S \qquad (B)$$

Combination of Eqs. (A) and (B) and solution for Q' yields

$$Q' = \frac{T'}{T' - T_\sigma}(\Delta H - T_\sigma \Delta S) \qquad (C)$$

Numerical values for ΔH and ΔS are found in Example 5.5:

$$\Delta H = -2{,}676.0 \text{ kJ kg}^{-1} \qquad \Delta S = -7.3554 \text{ kJ kg}^{-1} \text{ K}^{-1}$$

With $T_\sigma = 273.15$ K and $T' = 473.15$ K, Eq. (C) then yields

$$Q' = \frac{473.15}{200}(-2{,}676.0 + 273.15 \times 7.3554) = -1{,}577.7 \text{ kJ kg}^{-1}$$

The corresponding value for Q_σ is found from Eq. (A):

$$Q_\sigma = \Delta H - Q' = -2{,}676.0 + 1{,}577.7 = -1{,}098.3 \text{ kJ kg}^{-1}$$

The value of $Q' = -1{,}577.7$ kJ kg^{-1} is *smaller* in magnitude than the $-2{,}000$ kJ kg^{-1} claimed by the inventor in Example 5.5. As suggested in that example, the inventor's claim implies a negative rate of entropy generation.

Example 7.5   In a steady-state flow process, 1 mol s^{-1} of air at 600 K and 1 atm is continuously mixed with 2 mol s^{-1} of air at 450 K and 1 atm. The product stream is at 400 K and 1 atm. A schematic representation of the process

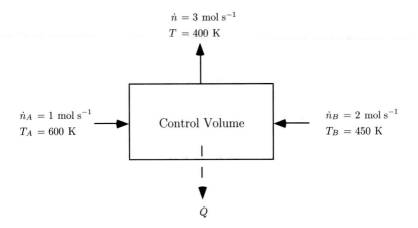

Figure 7.3: Process described in Example 7.5.

is shown in Fig. 7.3. Determine the rate of heat transfer and the rate of entropy generation for the process. Assume that air is an ideal gas with $C_P = (7/2)R$, that the surroundings are at 300 K, and that kinetic- and potential-energy changes are negligible.

SOLUTION   By Eq. (7.15), with $\dot{m}$ replaced by $\dot{n}$,

$$\dot{Q} = \dot{n}H - \dot{n}_A H_A - \dot{n}_B H_B$$

$$= \dot{n}_A(H - H_A) + \dot{n}_B(H - H_B)$$

$$= \dot{n}_A C_P(T - T_A) + \dot{n}_B C_P(T - T_B)$$

$$= 1 \times \frac{7}{2} \times 8.314 \times (400 - 600) + 2 \times \frac{7}{2} \times 8.314 \times (400 - 450)$$

$$= -8{,}729.7 \text{ J s}^{-1}$$

By Eq. (7.20), again with $\dot{m}$ replaced by $\dot{n}$,

$$\dot{S}_{G,\text{total}} = \dot{n}S - \dot{n}_A S_A - \dot{n}_B S_B - \frac{\dot{Q}}{T_\sigma}$$

$$= \dot{n}_A(S - S_A) + \dot{n}_B(S - S_B) - \frac{\dot{Q}}{T_\sigma}$$

$$= \dot{n}_A C_P \ln \frac{T}{T_A} + \dot{n}_B C_P \ln \frac{T}{T_B} - \frac{\dot{Q}}{T_\sigma}$$

$$= 1 \times \frac{7}{2} \times 8.314 \times \ln \frac{400}{600} + 2 \times \frac{7}{2} \times 8.314 \times \ln \frac{400}{450} - \left(\frac{-8{,}729.7}{300}\right)$$

$$= -11.7986 - 6.8547 + 29.0990 = 10.4457 \text{ J K}^{-1}$$

The rate of entropy generation is positive, as it must be for any real process.

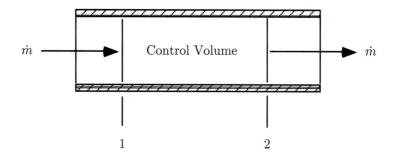

Figure 7.4: Identification of the control volume of Example 7.6.

Example 7.6   Consider the steady-state, adiabatic, irreversible flow of an incompressible liquid in a horizontal pipe of constant cross-sectional area. Show that:

(*a*) The velocity is constant.

(*b*) The temperature increases in the direction of flow.

(*c*) The pressure decreases in the direction of flow.

SOLUTION   (*a*) Figure 7.4 identifies the control volume of interest. By the continuity equation, Eq. (7.11),

$$\frac{u_2 A_2}{V_2} = \frac{u_1 A_1}{V_1}$$

However, $A_2 = A_1$ (constant cross-sectional area) and $V_2 = V_1$ (incompressible fluid). Hence, $u_2 = u_1$.

(*b*) By the entropy balance, Eq. (7.20),

$$\dot{S}_{G,\text{total}} = \dot{m}(S_2 - S_1)$$

For an incompressible liquid with heat capacity C, the entropy change is given by Eq. (*B*) of Example 6.2:

$$S_2 - S_1 = \int_{T_1}^{T_2} C \frac{dT}{T}$$

Thus

$$\dot{m} \int_{T_1}^{T_2} C \frac{dT}{T} = \dot{S}_{G,\text{total}}$$

But $\dot{S}_{G,\text{total}}$ is positive (the flow is irreversible) and hence, by the last equation, $T_2 > T_1$, and temperature increases in the direction of flow.

(*c*) As shown in (*a*), $u_2 = u_1$, and therefore the energy balance, Eq. (7.15) reduces for the stated conditions to

$$H_2 - H_1 = 0$$

For an incompressible liquid, the enthalpy change is given by Eq. (*A*) of Example 6.2:

$$H_2 - H_1 = \int_{T_1}^{T_2} C\,dT + V(P_2 - P_1)$$

Thus

$$V(P_2 - P_1) = -\int_{T_1}^{T_2} C\,dT$$

As shown in (*b*), $T_2 > T_1$; thus by the last equation, $P_2 < P_1$, and pressure decreases in the direction of flow.

Repeating this example for the case of *reversible* adiabatic flow is instructive. In this case $u_2 = u_1$ as before, but $\dot{S}_{G,\text{total}} = 0$. The entropy balance then shows that $T_2 = T_1$, in which case the energy balance yields $P_2 = P_1$. We conclude that the temperature increase of (*b*) and the pressure decrease of (*c*) *originate* from flow irreversibilities, specifically from the irreversibilities associated with fluid friction.

7.2 Duct Flow of Compressible Fluids

Such problems as the sizing of pipes and the shaping of nozzles require application of the momentum principle of fluid mechanics,[4] and therefore do not lie within the province of thermodynamics. However, thermodynamics does provide equations that interrelate the changes occurring in pressure, velocity, cross-sectional area, enthalpy, entropy, and specific volume of a flowing stream. We consider here the adiabatic, steady-state, one-dimensional flow of a compressible fluid in the absence of shaft work and of changes in potential energy. The pertinent thermodynamic equations are first derived; they are then applied to flow in pipes and nozzles.

The appropriate energy balance is Eq. (7.17). With Q, W_s, and Δz all set equal to zero, it reduces to

$$\Delta H + \frac{\Delta u^2}{2} = 0$$

which in differential form becomes

$$dH = -u\,du \tag{7.22}$$

The continuity equation, Eq. (7.11), is also applicable. Since $\dot{m}$ is constant, its differential form is

$$d(uA/V) = 0$$

or

$$\frac{dV}{V} - \frac{du}{u} - \frac{dA}{A} = 0 \tag{7.23}$$

The appropriate fundamental property relation applicable to a unit mass of fluid is Eq. (6.8):

---

[4]See W. L. McCabe, J. C. Smith, and P. Harriott, *Unit Operations of Chemical Engineering*, 5th ed., Sec. 2, McGraw-Hill, New York, 1993; R. H. Perry and D. Green, *Perry's Chemical Engineers' Handbook*, 7th ed., sec. 5, McGraw-Hill, New York, 1996.

$$dH = T\,dS + V\,dP \qquad (6.8)$$

In addition, we may consider the specific volume of the fluid a function of its entropy and pressure: $V = V(S, P)$. Then

$$dV = \left(\frac{\partial V}{\partial S}\right)_P dS + \left(\frac{\partial V}{\partial P}\right)_S dP$$

This equation is put into more convenient form as follows. First, we write the mathematical identity

$$\left(\frac{\partial V}{\partial S}\right)_P = \left(\frac{\partial V}{\partial T}\right)_P \left(\frac{\partial T}{\partial S}\right)_P$$

Substituting for the two partial derivatives on the right by Eqs. (3.2) and (6.17) gives

$$\left(\frac{\partial V}{\partial S}\right)_P = \frac{\beta V T}{C_P}$$

where β is the volume expansivity. The equation derived in physics for the speed of sound c in a fluid is

$$c^2 = -V^2 \left(\frac{\partial P}{\partial V}\right)_S$$

We may therefore write

$$\left(\frac{\partial V}{\partial P}\right)_S = -\frac{V^2}{c^2}$$

Substituting for the two partial derivatives in the equation for dV now yields

$$\frac{dV}{V} = \frac{\beta T}{C_P} dS - \frac{V}{c^2} dP \qquad (7.24)$$

Equations (7.22), (7.23), (6.8), and (7.24) are four expressions relating the six differentials—dH, du, dV, dA, dS, and dP. Thus we may treat dS and dA as the two independent variables and develop equations that express the remaining variables as functions of these two.

First, Eqs. (7.22) and (6.8) are combined:

$$T\,dS + V\,dP = -u\,du \qquad (7.25)$$

or

$$-\frac{du}{u} = \frac{1}{u^2}(T\,dS + V\,dP)$$

Substituting this equation and Eq. (7.24) into Eq. (7.23) gives after rearrangement

$$(1 - \mathbf{M}^2)V\,dP + \left(1 + \frac{\beta u^2}{C_P}\right)T\,dS - \frac{u^2}{A}\,dA = 0 \qquad (7.26)$$

where $\mathbf{M}$ is the Mach number, defined as the ratio of the speed of the fluid in the duct to the speed of sound in the fluid, u/c. Equation (7.26) relates dP to dS and dA.

Equations (7.25) and (7.26) may be combined to eliminate $V\,dP$, giving after rearrangement

$$u\,du - \left(\frac{\dfrac{\beta u^2}{C_P} + \mathbf{M}^2}{1 - \mathbf{M}^2}\right)T\,dS + \left(\frac{1}{1 - \mathbf{M}^2}\right)\frac{u^2}{A}dA = 0 \qquad (7.27)$$

This equation relates du to dS and dA. Combined with Eq. (7.22) it relates dH to dS and dA, and combined with (7.23) it relates dV to these same independent variables.

The differentials in the preceding equations represent changes in the fluid as it traverses a differential length of its path. If this length is dx, then each of the equations of flow may be divided through by dx. Equations (7.26) and (7.27) then become

$$V(1 - \mathbf{M}^2)\frac{dP}{dx} + T\left(1 + \frac{\beta u^2}{C_P}\right)\frac{dS}{dx} - \frac{u^2}{A}\frac{dA}{dx} = 0 \qquad (7.28)$$

and

$$u\frac{du}{dx} - T\left(\frac{\dfrac{\beta u^2}{C_P} + \mathbf{M}^2}{1 - \mathbf{M}^2}\right)\frac{dS}{dx} + \left(\frac{1}{1 - \mathbf{M}^2}\right)\frac{u^2}{A}\frac{dA}{dx} = 0 \qquad (7.29)$$

For adiabatic flow it follows from the second law that the irreversibilities due to fluid friction cause the entropy to increase in the direction of flow, with the limiting value of the rate of increase equal to zero when the flow approaches reversibility. In general, then, we have

$$\frac{dS}{dx} \geq 0$$

Pipe Flow

Consider a compressible fluid in steady-state adiabatic flow in a horizontal pipe of constant cross-sectional area. For this case, $dA/dx = 0$ and Eqs. (7.28) and (7.29) reduce to

$$\frac{dP}{dx} = -\frac{T}{V}\left(\frac{1 + \dfrac{\beta u^2}{C_P}}{1 - \mathbf{M}^2}\right)\frac{dS}{dx}$$

and

$$u\frac{du}{dx} = T\left(\frac{\dfrac{\beta u^2}{C_P} + \mathbf{M}^2}{1 - \mathbf{M}^2}\right)\frac{dS}{dx}$$

For subsonic flow, $\mathbf{M}^2 < 1$, and all quantities on the right-hand sides of these equations are positive; as a result

$$\frac{dP}{dx} < 0 \qquad \text{and} \qquad \frac{du}{dx} > 0$$

Thus the pressure decreases and the velocity increases in the direction of flow. However, the velocity cannot increase indefinitely. If the velocity were to exceed the sonic value, then the above inequalities would reverse. Such a transition is not possible in a pipe of constant cross-sectional area. For subsonic flow, the maximum fluid velocity obtainable in a pipe of constant cross section is the speed of sound, and this value is reached at the *exit* of the pipe. At this point dS/dx reaches its limiting value of zero. Given a discharge pressure low enough for the flow to become sonic, lengthening the pipe does not alter this result; the mass rate of flow decreases so that the sonic velocity is still obtained at the outlet of the lengthened pipe.

The equations for pipe flow indicate that when flow is supersonic the pressure increases and the velocity decreases in the direction of flow. However, such a flow regime is unstable, and when a supersonic stream enters a pipe of constant cross section, a compression shock occurs, the result of which is an abrupt and finite increase in pressure and decrease in velocity to a subsonic value.

Nozzles

The limitations observed for flow in pipes do not extend to properly designed nozzles, which bring about the interchange of internal and kinetic energy of a fluid as the result of a changing cross-sectional area available for flow. The relation between nozzle length and cross-sectional area is not susceptible to thermodynamic analysis, but is a problem in fluid mechanics. In a properly designed nozzle the area changes with length in such a way as to make the flow nearly frictionless. In the limit of reversible flow, the rate of entropy increase approaches zero, and we can set $dS/dx = 0$. In this event Eqs. (7.28) and (7.29) become

$$\frac{dP}{dx} = \frac{u^2}{VA}\left(\frac{1}{1-\mathbf{M}^2}\right)\frac{dA}{dx}$$

and

$$\frac{du}{dx} = -\frac{u}{A}\left(\frac{1}{1-\mathbf{M}^2}\right)\frac{dA}{dx}$$

The characteristics of flow depend on whether the flow is subsonic ($\mathbf{M} < 1$) or supersonic ($\mathbf{M} > 1$). The various cases are summarized in Table 7.1.

Thus, for subsonic flow in a converging nozzle, the velocity increases and the pressure decreases as the cross-sectional area diminishes. The maximum obtainable fluid velocity is the speed of sound, reached at the throat. A further increase in velocity and decrease in pressure would require an increase in cross-sectional area, a diverging section. Because of this, a converging subsonic nozzle can be used to deliver a constant flow rate into a region of variable pressure. Suppose a

Table 7.1: Characteristics of flow for a nozzle.

	Subsonic: **M < 1**		Supersonic: **M > 1**	
	Converging	Diverging	Converging	Diverging
$\dfrac{dA}{dx}$	−	+	−	+
$\dfrac{dP}{dx}$	−	+	+	−
$\dfrac{du}{dx}$	+	−	−	+

compressible fluid enters a converging nozzle at pressure P_1 and discharges from the nozzle into a chamber of variable pressure P_2. As this discharge pressure decreases below P_1, the flow rate and velocity increase. Ultimately, the pressure ratio P_2/P_1 reaches a critical value at which the velocity in the throat is sonic. Further reduction in P_2 has no effect on the conditions in the nozzle. The flow remains constant, and the velocity in the throat is sonic, regardless of the value of P_2/P_1, provided it is always less than the critical value. For steam, the critical value of this ratio is about 0.55 at moderate temperatures and pressures.

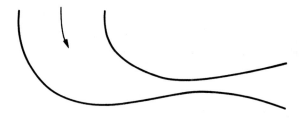

Figure 7.5: Converging/diverging nozzle.

Supersonic velocities are readily attained in the diverging section of a properly designed converging/diverging nozzle (Fig. 7.5). With sonic velocity reached at the throat, a further decrease in pressure requires an increase in cross-sectional area, a diverging section in which the velocity continues to increase. The transition occurs at the throat, where $dA/dx = 0$. The relationships between velocity, area, and pressure in a converging/diverging nozzle are illustrated numerically in Example 7.7.

The speed of sound is attained at the throat of a converging/diverging nozzle only when the pressure at the throat is low enough that the critical value of P_2/P_1 is reached. If insufficient pressure drop is available in the nozzle for the velocity to become sonic, the diverging section of the nozzle acts as a diffuser. That is, after the throat is reached the pressure rises and the velocity decreases; this is the conventional behavior for subsonic flow in diverging sections.

The relation of velocity to pressure in an isentropic nozzle can be expressed analytically if the fluid behaves as an ideal gas. Combination of Eq. (6.8) with (7.22) for isentropic flow gives

$$u \, du = -V \, dP$$

When an ideal gas with constant heat capacities undergoes isentropic expansion, Eq. (3.24) provides a relation between P and V: $PV^\gamma = $ const. Eliminating V in the preceding equation and integrating gives

$$u_2^2 - u_1^2 = -2 \int_{P_1}^{P_2} V \, dP = \frac{2\gamma P_1 V_1}{\gamma - 1} \left[1 - \left(\frac{P_2}{P_1} \right)^{(\gamma-1)/\gamma} \right] \tag{7.30}$$

where conditions at the nozzle entrance are denoted by subscript 1. Equation (7.30) may be solved for the pressure ratio P_2/P_1 at which the sonic velocity is reached. We recall the equation for the speed of sound:

$$c^2 = -V^2 \left(\frac{\partial P}{\partial V} \right)_S$$

Differentiation of the relation $PV^\gamma = $ const with respect to V at constant entropy yields

$$\left(\frac{\partial P}{\partial V} \right)_S = -\frac{\gamma P}{V}$$

We substitute for this derivative in the equation for the speed of sound, applying the result to the downstream section. This leads to

$$c^2 = \gamma P_2 V_2$$

With this value for u_2^2 in Eq. (7.30) and with $u_1 = 0$, solution for the pressure ratio at the throat gives

$$\frac{P_2}{P_1} = \left(\frac{2}{\gamma + 1} \right)^{\gamma/(\gamma-1)} \tag{7.31}$$

Example 7.7 A high-velocity nozzle is designed to operate with steam at 700 kPa and 300°C. At the nozzle inlet the velocity is 30 m s^{-1}. Calculate values of the ratio A/A_1 (where A_1 is the cross-sectional area of the nozzle inlet) for the sections where the pressure is 600, 500, 400, 300, and 200 kPa. Assume that the nozzle operates isentropically.

SOLUTION   The required area ratios are given by Eq. (7.11):

$$\frac{A}{A_1} = \frac{u_1 V}{V_1 u}$$

The velocity u is found from the integrated form of Eq. (7.22)

$$u^2 = u_1^2 - 2(H - H_1)$$

With units for velocity of m s^{-1}, u^2 has the units of m^2 s^{-2}. Units of J kg^{-1} for H are consistent with these,[5] because 1 J = 1 kg m^2 s^{-2}, whence 1 J kg^{-1} = 1 m^2 s^{-2}.

From the steam tables, we have initial values for entropy, enthalpy, and specific volume:

$$S_1 = 7.2997 \text{ kJ kg}^{-1} \text{ K}^{-1}$$
$$H_1 = 3,059.8 \times 10^3 \text{ J kg}^{-1}$$
$$V_1 = 371.39 \text{ cm}^3 \text{ g}^{-1}$$

Thus,

$$\frac{A}{A_1} = \left(\frac{30}{371.39}\right)\frac{V}{u} \qquad (A)$$

and

$$u^2 = 900 - 2(H - 3,059.8 \times 10^3) \qquad (B)$$

Since the expansion process is isentropic, $S = S_1$, and from the steam tables at at 600 kPa,

$$S = 7.2997 \text{ kJ kg}^{-1} \text{ K}^{-1}$$
$$H = 3,020.4 \times 10^3 \text{ J kg}^{-1}$$
$$V = 418.25 \text{ cm}^3 \text{ g}^{-1}$$

From Eq. (B)

$$u = 282.3 \text{ m s}^{-1}$$

and by Eq. (A),

$$\frac{A}{A_1} = \left(\frac{30}{371.39}\right)\left(\frac{418.25}{282.3}\right) = 0.120$$

Area ratios for other pressures are evaluated the same way, and the results are summarized in the accompanying table. The pressure at the throat of the nozzle is about 380 kPa. At lower pressures, the nozzle clearly diverges.

P/kPa	V/cm^3 g^{-1}	u/m s^{-1}	A/A_1
700	371.39	30	1.0
600	418.25	282.3	0.120
500	481.26	411.2	0.095
400	571.23	523.0	0.088
300	711.93	633.0	0.091
200	970.04	752.2	0.104

Example 7.8 Consider again the nozzle of Example 7.7, assuming now that steam behaves as an ideal gas. Calculate:

---

[5]When u is in (ft)(s)$^{-1}$, H in (Btu)(lb$_m$)$^{-1}$ must be multiplied by 778.16(ft lb$_f$)(Btu)$^{-1}$ and by the dimensional constant $g_c = 32.174$(lb$_m$)(ft)(lb$_f$)$^{-1}$(s)$^{-2}$.

(a) The critical pressure ratio and the velocity at the throat.

(b) The discharge pressure if a Mach number of 2.0 is required at the nozzle exhaust.

SOLUTION   (a) The ratio of specific heats for steam is about 1.3. Substituting in Eq. (7.31),

$$\frac{P_2}{P_1} = \left(\frac{2}{1.3+1}\right)^{1.3/(1.3-1)} = 0.55$$

The velocity at the throat, which is equal to the speed of sound, can be found from Eq. (7.30). When P_1 is in Pa (1 Pa = 1 kg m^{-1} s^{-2}) and V_1 is in m^3 kg^{-1}, the product $P_1 V_1$ is in m^2 s^{-2}, the units of velocity squared. For steam as an ideal gas

$$P_1 V_1 = \frac{RT_1}{M} \times 10^{-3} = \frac{(8.314)(573.15)}{18.015} \times 10^{-3} = 264{,}511 \text{ m}^2 \text{ s}^{-2}$$

Substitution in Eq. (7.30) gives

$$u^2_{\text{throat}} = (30)^2 + \frac{(2)(1.3)(264{,}511)}{1.3 - 1}\left[1 - (0.55)^{(1.3-1)/1.3}\right]$$

$$= 900 + 295{,}422 = 296{,}322$$

$$u_{\text{throat}} = 544.4 \text{ m s}^{-1}$$

These results are in good agreement with values obtained in Example 7.7, because steam at these conditions closely approximates an ideal gas.

(b) For a Mach number of 2.0 (based on the velocity of sound at the nozzle throat) the discharge velocity is 1,079.4 m s^{-1}. Substitution of this value in Eq. (7.30) allows calculation of the pressure ratio:

$$(1{,}079.4)^2 = (30)^2 = \frac{(2)(1.3)(264{,}511)}{1.3 - 1}\left[1 - \left(\frac{P_2}{P_1}\right)^{(1.3-1)/1.3}\right]$$

or

$$\left(\frac{P_2}{P_1}\right)^{(1.3-1)/1.3} = 0.492$$

Then

$$P_2 = (0.0463)(700) = 32.4 \text{ kPa}$$

Throttling Process

When a fluid flows through a restriction, such as an orifice, a partly closed valve, or a porous plug, without any appreciable change in kinetic energy, the primary result of the process is a pressure drop in the fluid. Such a *throttling process* produces no shaft work and results in negligible change in elevation. In the absence of heat transfer, Eq. (7.17) reduces to

$$\Delta H = 0$$

or

$$H_2 = H_1$$

The process therefore occurs at constant enthalpy.

Since the enthalpy of an ideal gas depends on temperature only, a throttling process does not change the temperature of an ideal gas. For most real gases at moderate conditions of temperature and pressure, a reduction in pressure at constant enthalpy results in a decrease in temperature. For example, if steam at 1,000 kPa and 300°C is throttled to 101.325 kPa (atmospheric pressure),

$$H_2 = H_1 = 3,052.1 \text{ kJ kg}^{-1}$$

Interpolation in the steam tables at this enthalpy and at a pressure of 101.325 kPa indicates a downstream temperature of 288.8°C. The temperature has decreased, but the effect is small. The following example illustrates the use of generalized correlations in calculations for a throttling process.

Example 7.9 Propane gas at 20 bar and 400 K is throttled in a steady-state flow process to 1 bar. Estimate the final temperature of the propane and its entropy change. Properties of propane can be found from suitable generalized correlations.

SOLUTION   Applying Eq. (6.68) to this constant-enthalpy process gives

$$\Delta H = \langle C_P^{ig} \rangle_H (T_2 - T_1) + H_2^R - H_1^R = 0$$

If propane in its final state at 1 bar is assumed an ideal gas, then $H_2^R = 0$, and the preceding equation gives

$$T_2 = \frac{H_1^R}{\langle C_P^{ig} \rangle_H} + T_1 \qquad (A)$$

For propane,

$$T_c = 369.8 \text{ K} \qquad P_c = 42.48 \text{ bar} \qquad \omega = 0.152$$

Thus for the initial state

$$T_{r_1} = \frac{400}{369.8} = 1.082 \qquad P_{r_1} = \frac{20}{42.48} = 0.471$$

At these conditions the generalized correlation based on second virial coefficients is satisfactory (see Fig. 3.15), and the computational procedure of Eqs. (3.50), (6.64), (3.51), and (6.65) is represented by (Sec. 6.6)

$$\texttt{HRB}(1.082,0.471,0.152) \equiv -0.452$$

whence

$$H_1^R = (8.314)(369.8)(-0.452) = -1,390 \text{ J mol}^{-1}$$

The only remaining quantity in Eq. (A) to be evaluated is $\langle C_P^{ig} \rangle_H$. Taking data for propane from Table C.1, we have

$$\frac{C_P^{ig}}{R} = 1.213 + 28.785 \times 10^{-3}T - 8.824 \times 10^{-6}T^2$$

For an initial calculation, we assume that $\langle C_P^{ig} \rangle_H$ is approximately the value of C_P^{ig} at the initial temperature of 400 K. This provides the value

$$\langle C_P^{ig} \rangle_H = 94.07 \text{ J mol}^{-1} \text{ K}^{-1}$$

Equation (A) now gives

$$T_2 = \frac{-1{,}390}{94.07} + 400 = 385.2 \text{ K}$$

Clearly, the temperature change is small, and we can reevaluate $\langle C_P^{ig}\rangle_H$ to an excellent approximation by calculating C_P^{ig} at the arithmetic mean temperature,

$$T_{am} = \frac{400 + 385.2}{2} = 392.6 \text{ K}$$

This gives

$$\langle C_P^{ig}\rangle_H = 92.73 \text{ J mol}^{-1} \text{ K}^{-1}$$

and recalculation of T_2 by Eq. (A) yields the final value:

$$T_2 = 385.0 \text{ K}$$

The entropy change of the propane is given by Eq. (6.69), which here becomes

$$\Delta S = \langle C_P^{ig}\rangle_S \ln \frac{T_2}{T_1} - R \ln \frac{P_2}{P_1} - S_1^R$$

Since the temperature change is so small, we can take

$$\langle C_P^{ig}\rangle_S = \langle C_P^{ig}\rangle_H = 92.73 \text{ J mol}^{-1} \text{ K}^{-1}$$

Calculation of S_1^R by Eq. (6.63) gives

$$S_1^R = -2.437 \text{ J mol}^{-1} \text{ K}^{-1}$$

Then

$$\Delta S = 92.73 \ln \frac{385.0}{400} - 8.314 \ln \frac{1}{20} + 2.437 = 23.80 \text{ J mol}^{-1} \text{ K}^{-1}$$

The positive value reflects the irreversibility of throttling processes.

When a wet vapor is throttled to a sufficiently low pressure, the liquid evaporates and the vapor becomes superheated. Thus if wet steam at 1,000 kPa ($t^{sat} = 179.88°C$) with a quality of 0.96 is throttled to 101.325 kPa,

$$H_2 = H_1 = (0.04)(762.6) + (0.96)(2{,}776.2) = 2{,}695.7 \text{ kJ kg}^{-1}$$

Steam with this enthalpy at 101.325 kPa has a temperature of 109.8°C, and is superheated ($t^{sat} = 100°C$). The considerable temperature drop here results from evaporation of liquid. If a saturated liquid is throttled to a lower pressure, some of the liquid vaporizes or *flashes*, producing a mixture of saturated liquid and saturated vapor at the lower pressure. Thus if saturated liquid water at 1,000 kPa ($t^{sat}=179.88°C$) is flashed to 101.325 kPa ($t^{sat}=100°C$),

$$H_2 = H_1 = 762.6 \text{ kJ kg}^{-1}$$

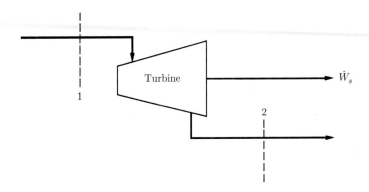

Figure 7.6: Steady-state flow through a turbine or expander.

At 101.325 kPa the quality of the resulting stream is found from Eq. (6.57):

$$762.6 = (1 - x)(419.1) + x(2{,}676.0)$$
$$= 419.1 + x(2{,}676.0 - 419.1)$$

Hence

$$x = 0.152$$

Thus 15.2 percent of the original liquid vaporized in the process. Again, the large temperature drop results from evaporation of liquid.

Throttling processes find frequent application in refrigeration (Chap. 9).

7.3 Turbines (Expanders)

The expansion of a gas in a nozzle to produce a high-velocity stream is a process that converts internal energy into kinetic energy. This kinetic energy can in turn be converted into shaft work when the stream impinges on blades attached to a rotating shaft. Thus a turbine (or expander) consists of alternate sets of nozzles and rotating blades through which vapor or gas flows in a steady-state expansion process whose overall effect is the efficient conversion of the internal energy of a high-pressure stream into shaft work. When steam provides the motive force as in a power plant, the device is called a turbine; when a high-pressure gas, such as ammonia or ethylene in a chemical or petrochemical plant, is the working fluid, the device is often called an expander. The process for either case is shown in Fig. 7.6.

Equations (7.16) and (7.17) are appropriate energy relations. However, the potential-energy term can be omitted, because there is little change in elevation. Moreover, in any properly designed turbine, heat transfer is negligible and the inlet and exit pipes are sized to make fluid velocities roughly equal. Equations (7.16) and (7.17) therefore reduce to

$$\dot{W}_s = \dot{m}\Delta H \tag{7.32}$$

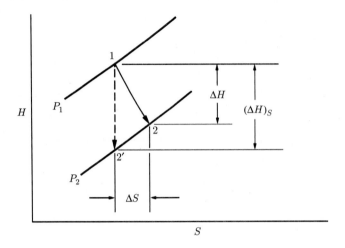

Figure 7.7: Adiabatic expansion process in a turbine or expander.

and
$$W_s = \Delta H \tag{7.33}$$

Normally, the inlet conditions T_1 and P_1 and the discharge pressure P_2 are known. Thus in Eq. (7.33) only H_1 is known, and we are left with both H_2 and W_s as unknowns. The energy equation alone does not allow any calculations to be made. However, if the fluid in the turbine undergoes an expansion process that is *reversible* as well as adiabatic, then the process is isentropic, and $S_2 = S_1$. This second equation allows us to determine the final state of the fluid and hence H_2. For this special case, we can evaluate W_s by Eq. (7.33), written as

$$W_s(\text{isentropic}) = (\Delta H)_S \tag{7.34}$$

The shaft work given by Eq. (7.34) is numerically the *maximum* that can be obtained from an adiabatic turbine with given inlet conditions and given discharge pressure. Actual turbines produce less work, because the actual expansion process is irreversible. We therefore define a turbine efficiency as

$$\eta \equiv \frac{W_s}{W_s(\text{isentropic})}$$

where W_s is the actual shaft work. By Eqs. (7.33) and (7.34)

$$\eta = \frac{\Delta H}{(\Delta H)_S} \tag{7.35}$$

Values of η for properly designed turbines or expanders are usually in the range of 0.7 to 0.8.

Figure 7.7 shows an HS diagram on which are compared an actual expansion process in a turbine and the reversible process for the same intake conditions and the same discharge pressure. The reversible path is a vertical line of constant entropy from point 1 at the intake pressure P_1 to point 2′ at the discharge pressure P_2. The line representing the actual irreversible process starts also from point 1, but is directed downward and to the right, in the direction of increasing entropy. Since the process is adiabatic, irreversibilities cause an increase in entropy of the fluid. The process terminates at point 2 on the isobar for P_2. The more irreversible the process, the further this point lies to the right on the P_2 isobar, and the lower the efficiency η of the process.

Example 7.10 A steam turbine with rated capacity of 56,400 kW operates with steam at inlet conditions of 8,600 kPa and 500°C, and discharges into a condenser at a pressure of 10 kPa. Assuming a turbine efficiency of 0.75, determine the state of the steam at discharge and the mass rate of flow of the steam.

SOLUTION   At the inlet conditions of 8,600 kPa and 500°C, the following values are given in the steam tables:

$$H_1 = 3,391.6 \text{ kJ kg}^{-1}$$

$$S_1 = 6.6858 \text{ kJ kg}^{-1} \text{ K}^{-1}$$

If the expansion to 10 kPa is isentropic, then

$$S_2' = S_1 = 6.6858$$

Steam with this entropy at 10 kPa is wet, and we apply Eq. (6.57), with $M = S$:

$$S = (1 - x^v)S^l + x^v S^v = S^l + x^v(S^v - S^l)$$

Then

$$6.6858 = 0.6493 + x_2'(8.1511 - 0.6493)$$

and

$$x_2' = 0.8047$$

This is the quality (fraction vapor) of the discharge stream at point 2′. The enthalpy H_2' is also given by Eq. (6.57), written

$$H = H^l + x^v(H^v - H^l)$$

Thus

$$H_2' = 191.8 + 0.8047(2,584.8 - 191.8) = 2,117.4 \text{ kJ kg}^{-1}$$

and

$$(\Delta H)_S = H_2' - H_1 = 2,117.4 - 3,391.6 = -1,274.2 \text{ kJ kg}^{-1}$$

By Eq. (7.35) we then have

$$\Delta H = \eta(\Delta H)_S = (0.75)(-1,274.2) = -955.6 \text{ kJ kg}^{-1}$$

whence

$$H_2 = H_1 + \Delta H = 3,391.6 - 955.6 = 2,436.0 \text{ kJ kg}^{-1}$$

Thus the steam in its actual final state is also wet, and its quality is found from the equation:

$$2{,}436.0 = 191.8 + x_2(2{,}584.8 - 191.8)$$

Solution gives

$$x_2 = 0.9378$$

Finally,

$$S_2 = 0.6493 + (0.9378)(8.1511 - 0.6493) = 7.6846 \text{ kJ kg}^{-1} \text{ K}^{-1}$$

This value may be compared with the initial value of $S_1 = 6.6858$.

The steam rate is found from Eq. (7.32). For $-\dot{W}_s = 56{,}400$ kW or $56{,}400$ kJ s^{-1}, we have

$$-56{,}400 = \dot{m}(2{,}436.0 - 3{,}391.6)$$

and

$$\dot{m} = 59.02 \text{ kg s}^{-1}$$

Example 7.10 was solved with the aid of the steam tables. When a comparable set of tables is not available for the motive fluid, the generalized correlations of Sec. 6.6 may be used in conjunction with Eqs. (6.68) and (6.69), as illustrated in the following example.

Example 7.11   A stream of ethylene gas at 300°C and 45 bar is expanded adiabatically in a turbine to 2 bar. Calculate the isentropic work produced. Determine the properties of ethylene by:

(*a*) Equations for an ideal gas.

(*b*) Appropriate generalized correlations.

SOLUTION   The enthalpy and entropy changes for the process are given by Eqs. (6.68) and (6.69):

$$\Delta H = \langle C_P^{ig} \rangle_H (T_2 - T_1) + H_2^R - H_1^R \qquad (6.68)$$

and

$$\Delta S = \langle C_P^{ig} \rangle_S \ln \frac{T_2}{T_1} - R \ln \frac{P_2}{P_1} + S_2^R - S_1^R \qquad (6.69)$$

As given values, we have $P_1 = 45$ bar, $P_2 = 2$ bar, and $T_1 = 300 + 273.15 = 573.15$ K.

(*a*) If ethylene is assumed an ideal gas, then all residual properties are zero, and the preceding equations reduce to

$$\Delta H = \langle C_P^{ig} \rangle_H (T_2 - T_1)$$

and

$$\Delta S = \langle C_P^{ig} \rangle_S \ln \frac{T_2}{T_1} - R \ln \frac{P_2}{P_1}$$

For an isentropic process, $\Delta S = 0$, and the last equation becomes

$$\frac{\langle C_P^{ig} \rangle_S}{R} \ln \frac{T_2}{T_1} = \ln \frac{P_2}{P_1} = \ln \frac{2}{45} = -3.1135$$

or

$$\ln T_2 = \frac{-3.1135}{\langle C_P^{ig} \rangle_S / R} + \ln 573.15$$

Then

$$T_2 = \exp\left(\frac{-3.1135}{\langle C_P^{ig} \rangle_S / R} + 6.3511\right) \qquad (A)$$

Equation (5.17) provides an expression for $\langle C_P^{ig} \rangle_S / R$, which for computational purposes is represented by

$$\text{MCPS(573.15,T2;1.424,14.394E-3,-4.392E-6,0.0)}$$

where the constants for ethylene come from Table C.1. Temperature T_2 is conveniently found by iteration. We assume an initial value for computation of $\langle C_P^{ig} \rangle_S / R$. Equation (A) then provides a new value of T_2 from which to recompute $\langle C_P^{ig} \rangle_S / R$, and the procedure continues to convergence on the final value:

$$T_2 = 370.8 \text{ K}$$

Then

$$W_s(\text{isentropic}) = (\Delta H)_S = \langle C_P^{ig} \rangle_H (T_2 - T_1)$$

The value of $\langle C_P^{ig} \rangle_H / R$ is given by Eq. (4.8); for computational purposes we have

$$\text{MCPH(573.15,370.8;1.424,14.394E-3,-4.392E-6,0.0)} \equiv 7.224$$

whence

$$W_s(\text{isentropic}) = (7.224)(8.314)(370.8 - 573.15) = -12{,}153 \text{ J mol}^{-1}$$

(b) For ethylene,

$$T_c = 282.3 \text{ K} \qquad P_c = 50.4 \text{ bar} \qquad \omega = 0.087$$

At the initial state,

$$T_{r_1} = \frac{573.15}{282.3} = 2.030 \qquad P_{r_1} = \frac{45}{50.4} = 0.893$$

According to Fig. 3.15, the generalized correlations based on second virial coefficients should be satisfactory. The computational procedures of Eqs. (3.50), (3.51), and (6.62) through (6.65) are represented by

$$\text{HRB(2.030,0.893,0.087)} \equiv -0.234$$

and

$$\text{SRB(2.030,0.893,0.087)} \equiv -0.097$$

Then

$$H_1^R = (-0.234)(8.314)(282.3) = -549 \text{ J mol}^{-1}$$

and

$$S_1^R = (-0.097)(8.314) = -0.806 \text{ J mol}^{-1} \text{ K}^{-1}$$

For the purpose of getting an initial estimate of S_2^R, we assume that $T_2 = 370.8$ K, the value determined in part (a). Then

$$T_{r_2} = \frac{370.8}{282.3} = 1.314 \qquad P_{r_2} = \frac{2}{50.4} = 0.040$$

whence

$$\text{SRB}(1.314,0.040,0.087) \equiv -0.0139$$

and

$$S_2^R = (-0.0139)(8.314) = -0.116 \text{ J mol}^{-1}\text{ K}^{-1}$$

If the expansion process is isentropic, Eq. (6.69) gives

$$0 = \langle C_P^{ig}\rangle_S \ln \frac{T_2}{573.15} - 8.314\ln\frac{2}{45} - 0.116 + 0.806$$

from which

$$\ln \frac{T_2}{573.15} = \frac{-26.576}{\langle C_P^{ig}\rangle_S}$$

or

$$T_2 = \exp\left(\frac{-26.576}{\langle C_P^{ig}\rangle_S} + 6.3511\right)$$

An iteration process exactly like that of part (a) yields the result

$$T_2 = 365.8 \text{ K}$$

For the recomputation of S_2, we now find

$$T_{r_2} = 1.296 \qquad P_{r_2} = 0.040$$

whence

$$\text{SRB}(1.296,0.040,0.087) \equiv -0.0144$$

and

$$S_2^R = (-0.0144)(8.314) = -0.120 \text{ J mol}^{-1}\text{ K}^{-1}$$

This result is so little changed from the initial value that another recalculation of T_2 is unnecessary. We therefore evaluate H_2^R at the reduced conditions already established,

$$\text{HRB}(1.296,0.040,0.087) \equiv -0.0262$$

and

$$H_2^R = (-0.0262)(8.314)(282.3) = -61 \text{ J mol}^{-1}$$

Equation (6.68) now gives

$$(\Delta H)_S = \langle C_P^{ig}\rangle_H(365.8 - 573.15) - 61 + 549$$

Evaluation of $\langle C_P^{ig}\rangle_H$ as in part (a) with $T_2 = 365.8$ K gives

$$\langle C_P^{ig}\rangle_H = 59.843 \text{ J mol}^{-1}\text{ K}^{-1}$$

whence

$$(\Delta H)_S = -11{,}920 \text{ J mol}^{-1}$$

and

$$W_s(\text{isentropic}) = (\Delta H)_S = -11{,}920 \text{ J mol}^{-1}$$

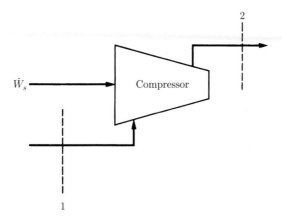

Figure 7.8: Steady-state compression process.

7.4 Compression Processes

Just as expansion processes result in pressure reductions in a flowing fluid, so compression processes bring about pressure increases. Compressors, pumps, fans, blowers, and vacuum pumps are all devices designed for this purpose. They are vital for the transport of fluids, for fluidization of particulate solids, for bringing fluids to the proper pressure for reaction or processing, etc. We are here concerned not with the design of such devices, but with specification of energy requirements for the steady-state compression of fluids from one pressure to a higher one.

Compressors

The compression of gases may be accomplished in equipment with rotating blades (like a turbine operating in reverse) or in cylinders with reciprocating pistons. Rotary equipment is used for high-volume flow where the discharge pressure is not too high. For high pressures, reciprocating compressors are required.

The energy equations are independent of the type of equipment; indeed, they are the same as for turbines or expanders, because here too potential- and kinetic-energy changes are presumed negligible. Thus Eqs. (7.32) through (7.34) apply to adiabatic compression, a process represented by Fig. 7.8.

In a compression process, the isentropic work, as given by Eq. (7.34), is the *minimum* shaft work required for compression of a gas from a given initial state to a given discharge pressure. Thus we define a compressor efficiency as

$$\eta \equiv \frac{W_s(\text{isentropic})}{W_s}$$

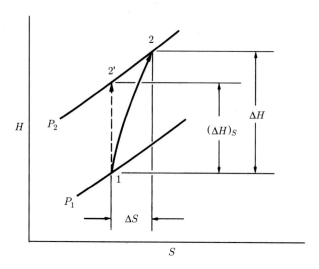

Figure 7.9: Adiabatic compression process.

In view of Eqs. (7.33) and (7.34), this is also given by

$$\eta \equiv \frac{(\Delta H)_S}{\Delta H} \tag{7.36}$$

Compressor efficiencies are usually in the range of 0.7 to 0.8. The compression process is shown on an HS diagram in Fig. 7.9. The vertical path rising from point 1 to point 2′ represents the isentropic compression process from P_1 to P_2. The actual compression process follows a path from point 1 upward and to the right in the direction of increasing entropy, terminating at point 2 on the isobar for P_2.

Example 7.12 Saturated-vapor steam at 100 kPa ($t^{\text{sat}} = 99.63°\text{C}$) is compressed adiabatically to 300 kPa. If the compressor efficiency is 0.75, what is the work required and what are the properties of the discharge stream?

SOLUTION   For saturated steam at 100 kPa,

$$S_1 = 7.3598 \text{ kJ kg}^{-1} \text{ K}^{-1}$$

$$H_1 = 2,675.4 \text{ kJ kg}^{-1}$$

For isentropic compression to 300 kPa,

$$S_2' = S_1 = 7.3598 \text{ kJ kg}^{-1} \text{ K}^{-1}$$

By interpolation in the tables for superheated steam at 300 kPa, we find that steam with this entropy has an enthalpy of

$$H_2' = 2,888.8 \text{ kJ kg}^{-1}$$

Thus

$$(\Delta H)_S = 2{,}888.8 - 2{,}675.4 = 213.4 \text{ kJ kg}^{-1}$$

By Eq. (7.36),

$$\Delta H = \frac{(\Delta H)_S}{\eta} = \frac{213.4}{0.75} = 284.5 \text{ kJ kg}^{-1}$$

whence

$$H_2 = H_1 + \Delta H = 2{,}675.4 + 284.5 = 2{,}959.9 \text{ kJ kg}^{-1}$$

Again by interpolation, we find that superheated steam with this enthalpy has the additional properties:

$$T_2 = 246.1°\text{C} \qquad S_2 = 7.5019 \text{ kJ kg}^{-1} \text{ K}^{-1}$$

Moreover, by Eq. (7.33), the work required is

$$W_s = \Delta H = 284.5 \text{ kJ kg}^{-1}$$

The direct application of Eqs. (7.32) through (7.34) presumes the availability of tables of data or an equivalent thermodynamic diagram for the fluid being compressed. Where such information is not available, the generalized correlations of Sec. 6.6 may be used in conjunction with Eqs. (6.68) and (6.69), exactly as illustrated in Example 7.11 for an expansion process.

The assumption of ideal gases leads to equations of relative simplicity. By Eq. (5.18) for an ideal gas

$$\Delta S = \langle C_P \rangle_S \ln \frac{T_2}{T_1} - R \ln \frac{P_2}{P_1} \qquad (5.18)$$

where for simplicity of notation the superscript "ig" has been omitted from the mean heat capacity. If the compression is isentropic, $\Delta S = 0$, and this equation becomes

$$T_2' = T_1 \left(\frac{P_2}{P_1} \right)^{R/\langle C_P' \rangle_S} \qquad (7.37)$$

where T_2' is the temperature that results when compression from T_1 and P_1 to P_2 is *isentropic* and where $\langle C_P' \rangle_S$ is the mean heat-capacity for the temperature range from T_1 to T_2'.

The enthalpy change for isentropic compression is given by Eq. (4.9), written as

$$(\Delta H)_S = \langle C_P' \rangle_H (T_2' - T_1)$$

In accord with Eq. (7.34), we then have

$$W_s(\text{isentropic}) = \langle C_P' \rangle_H (T_2' - T_1) \qquad (7.38)$$

This result may be combined with the compressor efficiency to give

$$W_s = \frac{W_s(\text{isentropic})}{\eta} \qquad (7.39)$$

The *actual* discharge temperature T_2 resulting from compression is also found from Eq. (4.9), now written

$$\Delta H = \langle C_P \rangle_H (T_2 - T_1)$$

whence

$$T_2 = T_1 + \frac{\Delta H}{\langle C_P \rangle_H} \tag{7.40}$$

where by Eq. (7.33) $\Delta H = W_s$. Here $\langle C_P \rangle_H$ is the mean heat-capacity for the temperature range from T_1 to T_2.

For the special case of an ideal gas with constant heat capacities,

$$\langle C_P' \rangle_H = \langle C_P \rangle_H = \langle C_P' \rangle_S = C_P$$

Equations (7.37) and (7.38) therefore become

$$T_2' = T_1 \left(\frac{P_2}{P_1} \right)^{R/C_P}$$

and

$$W_s(\text{isentropic}) = C_P (T_2' - T_1)$$

Combining these equations gives[6]

$$W_s(\text{isentropic}) = C_P T_1 \left[\left(\frac{P_2}{P_1} \right)^{R/C_P} - 1 \right] \tag{7.41}$$

For monatomic gases, such as argon and helium, $R/C_P = 2/5 = 0.4$. For diatomic gases, such as oxygen, nitrogen, and air at moderate temperatures, an approximate value is $R/C_P = 2/7 = 0.2857$. For gases of greater molecular complexity the ideal-gas heat capacity depends more strongly on temperature, and Eq. (7.41) is less likely to be suitable. One can easily show that the assumption of constant heat capacities also leads to the result

$$T_2 = T_1 + \frac{T_2' - T_1}{\eta} \tag{7.42}$$

---

[6]Since $R = C_P - C_V$ for an ideal gas, we can write

$$\frac{R}{C_P} = \frac{C_P - C_V}{C_P} = \frac{\gamma - 1}{\gamma}$$

An alternative form of Eq. (7.41) is therefore

$$W_s(\text{isentropic}) = \frac{\gamma R T_1}{\gamma - 1} \left[\left(\frac{P_2}{P_1} \right)^{(\gamma - 1)/\gamma} - 1 \right]$$

Although this form is the one most commonly encountered, Eq. (7.41) is simpler and more easily applied.

Example 7.13 If methane (assumed to be an ideal gas) is compressed adiabatically from 20°C and 140 kPa to 560 kPa, estimate the work requirement and the discharge temperature of the methane. The compressor efficiency is 0.75.

SOLUTION   Application of Eq. (7.37) requires evaluation of the exponent $R/\langle C_P' \rangle_S$. This can be accomplished with Eq. (5.17), which for the present computation is represented by

$$\text{MCPS}(293.15,T2;1.702,9.081\text{E-}3,-2.164\text{E-}6,0.0)$$

where the constants for methane come from Table C.1. We choose a value for T_2' somewhat higher than the initial temperature $T_1 = 293.15$ K. Evaluation of $\langle C_P' \rangle_S / R$ then provides a value for the exponent in Eq. (7.37). With $P_2/P_1 = 560/140 = 4.0$ and $T_1 = 293.15$ K, we then calculate T_2'. The procedure is repeated until no further significant change occurs in the value of T_2'. This process results in the values

$$T_2' = 397.37 \text{ K} \qquad \text{and} \qquad \frac{\langle C_P' \rangle_S}{R} = 4.5574$$

For the same T_1 and T_2', we evaluate $\langle C_P' \rangle_H / R$ by Eq. (4.9):

$$\text{MCPH}(293.15,397.37;1.702,9.081\text{E-}3,-2.164\text{E-}6,0.0) \equiv 4.5774$$

whence

$$\langle C_P' \rangle_H = (4.5774)(8.314) = 38.056 \text{ J mol}^{-1} \text{ K}^{-1}$$

Then by Eq. (7.38),

$$W_s(\text{isentropic}) = (38.056)(397.37 - 293.15) = 3{,}966.2 \text{ J mol}^{-1}$$

The actual work is found from Eq. (7.39) as

$$W_s = \frac{3{,}966.2}{0.75} = 5{,}288.3 \text{ J mol}^{-1}$$

Application of Eq. (7.40) for the calculation of T_2 gives

$$T_2 = 293.15 + \frac{5{,}288.3}{\langle C_P \rangle_H}$$

Since $\langle C_P \rangle_H$ depends on T_2, we again iterate. With T_2' as a starting value, this leads to the results:

$$T_2 = 428.65 \text{ K} \qquad \text{or} \qquad t_2 = 155.5°\text{C}$$

and

$$\langle C_P \rangle_H = 39.027 \text{ J mol}^{-1} \text{ K}^{-1}$$

Pumps

Liquids are usually moved by pumps, generally rotating equipment. The same equations apply to adiabatic pumps as to adiabatic compressors. Thus, Eqs. (7.32) through (7.34) and (7.36) are valid. However, application of Eq. (7.33) for the calculation of $W_s = \Delta H$ requires values of the enthalpy of compressed liquids, and these are seldom available. The fundamental property relation, Eq. (6.8), provides an alternative. For an isentropic process,

$$dH = V\,dP \qquad (\text{const } S)$$

Combining this with Eq. (7.34) gives

$$W_s(\text{isentropic}) = (\Delta H)_S = \int_{P_1}^{P_2} V\,dP$$

The usual assumption for liquids (at conditions well removed from the critical point) is that V is independent of P. Integration then gives

$$W_s(\text{isentropic}) = (\Delta H)_S = V(P_2 - P_1) \tag{7.43}$$

Also useful are the following equations from Chap. 6:

$$dH = C_P\,dT + V(1 - \beta T)dP \tag{6.27}$$

and

$$dS = C_P\frac{dT}{T} - \beta V\,dP \tag{6.28}$$

where the volume expansivity β is defined by Eq. (3.2). Since temperature changes in the pumped fluid are very small and since the properties of liquids are insensitive to pressure (again at conditions not close to the critical point), these equations are usually integrated on the assumption that C_P, V, and β are constant, usually at initial values. Thus, to a good approximation

$$\Delta H = C_P\Delta T + V(1 - \beta T)\Delta P \tag{7.44}$$

and

$$\Delta S = C_P \ln\frac{T_2}{T_1} - \beta V\Delta P \tag{7.45}$$

Example 7.14   Water at 45°C and 10 kPa enters an adiabatic pump and is discharged at a pressure of 8,600 kPa. Assume the pump efficiency to be 0.75. Calculate the work of the pump, the temperature change of the water, and the entropy change of the water.

SOLUTION   The following properties are available for saturated liquid water at 45°C (318.15 K):

$$V = 1,010 \text{ cm}^3 \text{ kg}^{-1}$$
$$\beta = 425 \times 10^{-6} \text{ K}^{-1}$$
$$C_P = 4.178 \text{ kJ kg}^{-1} \text{ K}^{-1}$$

By Eq. (7.43),

$$W_s(\text{isentropic}) = (\Delta H)_S = (1,010)(8,600 - 10) = 8.676 \times 10^6 \text{ kPa cm}^3 \text{ kg}^{-1}$$

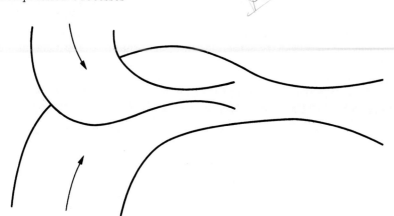

Figure 7.10: Single-stage ejector.

Since $1 \text{ kJ} = 10^6 \text{ kPa cm}^3$,

$$W_s(\text{isentropic}) = (\Delta H)_S = 8.676 \text{ kJ kg}^{-1}$$

By Eq. (7.36),

$$\Delta H = \frac{(\Delta H)_S}{\eta} = \frac{8.676}{0.75} = 11.57 \text{ kJ kg}^{-1}$$

Since $W_s = \Delta H$,

$$W_s = 11.57 \text{ kJ kg}^{-1}$$

The temperature change of the water during pumping is found from Eq. (7.44):

$$11.57 = 4.178 \, \Delta T + 1,010[1 - (425 \times 10^{-6})(318.15)]\frac{8,590}{10^6}$$

Solution for ΔT gives

$$\Delta T = 0.97 \text{ K} \qquad \text{or} \qquad 0.97°\text{C}$$

The entropy change of the water is given by Eq. (7.45):

$$\Delta S = 4.178 \ln \frac{319.12}{318.15} - (425 \times 10^{-6})(1,010)\frac{8,590}{10^6} = 0.0090 \text{ kJ kg}^{-1} \text{ K}^{-1}$$

Ejectors

Ejectors remove gases or vapors from an evacuated space and compress them for discharge at a higher pressure. Where the mixing of the gases or vapors with the driving fluid is allowable, ejectors are usually lower in first cost and maintenance costs than other types of vacuum punps. As illustrated in Fig. 7.10 an ejector consists of an inner converging-diverging nozzle through which the driving fluid (commonly steam) is fed, and an outer, larger nozzle through which both the extracted gases or vapors and the driving fluid pass. The momentum of the high-speed fluid leaving the driving nozzle is partly transferred to the extracted gases or vapors, and the mixture velocity is therefore less than that of the driving fluid leaving the smaller nozzle. It is nevertheless higher than the speed of sound, and the larger nozzle therefore acts as a converging/diverging *diffuser* in which the pressure rises and the velocity decreases, passing through the speed of sound at the throat. Although the usual energy equations for nozzles apply, the mixing process is complex, and as a result ejector design is largely empirical.

PROBLEMS

7.1. Two boilers, both operating at 200(psia), discharge equal amounts of steam into the same steam main. Steam from the first boiler is superheated at 420(°F) and steam from the second is wet with a quality of 96%. Assuming adiabatic mixing and negligable changes in potential and kinetic energies, what is the equilibrium condition after mixing and what is $S_{G,\text{total}}$ for each (lb$_m$) of discharge steam.

7.2. Two nonconducting tanks of negligible heat capacity and of equal volume initially contain equal quantities of the same ideal gas at the same T and P. Tank A discharges to the atmosphere through a small turbine in which the gas expands isentropically; tank B discharges to the atmosphere through a porous plug. Both devices operate until discharge ceases.

(*a*) When discharge ceases is the temperature in tank A less than, equal to, or greater than the temperature in tank B?

(*b*) When the pressures in both tanks have fallen to half the initial pressure, is the temperature of the gas discharging from the turbine less than, equal to, or greater than the temperature of the gas discharging from the porous plug?

(*c*) During the discharge process, is the temperature of the gas leaving the turbine less than, equal to, or greater than the temperature of the gas leaving tank A at the same instant?

(*d*) During the discharge process, is the temperature of the gas leaving the porous plug less than, equal to, or greater than the temperature of the gas leaving tank B at the same instant?

(*e*) When discharge ceases, is the mass of gas remaining in tank A less than, equal to, or greater than the mass of gas remaining in tank B?

7.3. A rigid tank of 80(ft)^3 capacity contains 4,180(lb$_m$) of saturated liquid water at 430(°F). This amount of liquid almost completely fills the tank, the small remaining volume being occupied by saturated-vapor steam. Since a bit more vapor space in the tank is wanted, a valve at the top of the tank is opened, and saturated-vapor steam is vented to the atmosphere until the temperature in the tank falls to 420(°F). Assuming no heat transfer to the contents of the tank, determine the mass of steam vented.

7.4. Liquid nitrogen is stored in 0.5-m^3 metal tanks that are thoroughly insulated. Consider the process of filling an evacuated tank, initially at 295 K. It is attached to a line containing liquid nitrogen at its normal boiling point of 77.3 K and at a pressure of several bars. At this condition, its enthalpy is -120.8 kJ kg^{-1}. When a valve in the line is opened, the nitrogen flowing into the tank at first evaporates in the process of cooling the tank. If the tank has a mass of 30 kg and the metal has a specific heat capacity of 0.43 kJ kg^{-1} K^{-1}, what mass of nitrogen must flow into the tank just to cool it to a temperature such that *liquid* nitrogen begins to accumulate in the tank? Assume that the nitrogen and the tank are always at the same temperature.

The properties of saturated nitrogen vapor at several temperatures are given as follows:

T/K	P/bar	$V^v/\text{m}^3\text{ kg}^{-1}$	$H^v/\text{kJ kg}^{-1}$
80	1.396	0.1640	78.9
85	2.287	0.1017	82.3
90	3.600	0.06628	85.0
95	5.398	0.04487	86.8
100	7.775	0.03126	87.7
105	10.83	0.02223	87.4
110	14.67	0.01598	85.6

7.5. A tank of 50-m^3 capacity contains steam at 4,500 kPa and 400°C. Steam is vented from the tank through a relief valve to the atmosphere until the pressure in the tank falls to 3,500 kPa. If the venting process is adiabatic, estimate the final temperature of the steam in the tank and the mass of steam vented.

7.6. A tank of 0.1-m^3 [3.53(ft)^3] volume contains air at 25°C [77(°F)] and 101.33 kPa [14.7(psia)]. The tank is connected to a compressed-air line which supplies air at the constant conditions of 45°C [113(°F)] and 1,500 kPa [217.6(psia)]. A valve in the line is cracked so that air flows slowly into the tank until the pressure equals the line pressure. If the process occurs slowly enough that the temperature in the tank remains at 25°C [77(°F)], how much heat is lost from the tank? Assume air an ideal gas for which $C_P = (7/2)R$ and $C_V = (5/2)R$.

7.7. A small adiabatic air compressor is used to pump air into a 20-m^3 [706.3(ft)^3] insulated tank. The tank initially contains air at 25°C [77(°F)] and 101.33 kPa [1(atm)], exactly the conditions at which air enters the compressor. The pumping process continues until the pressure in the tank reaches 1,000 kPa [9.87(atm)]. If the process is adiabatic and if compression is isentropic, what is the shaft work of the compressor? Assume air an ideal gas for which $C_P = (7/2)R$ and $C_V = (5/2)R$.

7.8. A tank of 4-m^3 capacity contains 1,500 kg of liquid water at 250°C in equilibrium with its vapor, which fills the rest of the tank. A quantity of 1,000 kg of water at 50°C is pumped into the tank. How much heat must be added during this process if the temperature in the tank is not to change?

7.9. Gas at constant T and P is contained in a supply line connected through a valve to a closed tank containing the same gas at a lower pressure. The valve is opened to allow flow of gas into the tank, and then is shut again.

(*a*) Develop a general equation relating n_1 and n_2, the moles (or mass) of gas in the tank at the beginning and end of the process, to the properties U_1 and U_2, the internal energy of the gas in the tank at the beginning and end of the process, and H', the enthalpy of the gas in the supply line, and to Q, the heat transferred to the material in the tank during the process.

(b) Reduce the general equation to its simplest form for the special case of an ideal gas with constant heat capacities.

(c) Further reduce the equation of (b) for the case of $n_1 = 0$.

(d) Further reduce the equation of (c) for the case in which, in addition, $Q = 0$.

(e) Treating nitrogen as an ideal gas for which $C_P = (7/2)R$, apply the appropriate equation to the case in which a steady supply of nitrogen at 25°C and 3 bar flows into an evacuated tank of 4-m^3 volume, and calculate the moles of nitrogen that flow into the tank to equalize the pressures for two cases:

  1. It is assumed that no heat flows from the gas to the tank or through the tank walls.

  2. The tank weighs 400 kg, is perfectly insulated, has an initial temperature of 25°C, has a specific heat of 0.46 kJ kg^{-1} K^{-1}, and is heated by the gas so as always to be at the temperature of the gas in the tank.

7.10. Develop equations which may be solved to give the final temperature of the gas remaining in a tank after the tank has been bled from an initial pressure P_1 to a final pressure P_2. Known quantities are the initial temperature, the tank volume, the heat capacity of the gas, the total heat capacity of the containing tank, P_1, and P_2. Assume the tank to be always at the temperature of the gas remaining in the tank, and the tank to be perfectly insulated.

7.11. A well-insulated tank of 50-m^3 volume initially contains 16,000 kg of water distributed between liquid and vapor phases at 25°C. Saturated steam at 1,500 kPa is admitted to the tank until the pressure reaches 800 kPa. What mass of steam is added?

7.12. An insulated evacuated tank of 1.75-m^3 volume is attached to a line containing steam at 400 kPa and 240°C. Steam flows into the tank until the pressure in the tank reaches 400 kPa. Assuming no heat flow from the steam to the tank, prepare graphs showing the mass of steam in the tank and its temperature as a function of pressure in the tank.

7.13. A 2-m^3 tank initially contains a mixture of saturated-vapor steam and saturated-liquid water at 3,000 kPa. Of the total mass, 10% is vapor. Saturated-liquid water is bled from the tank through a valve until the total mass in the tank is 40% of the initial total mass. If during the process the temperature of the contents of the tank is kept constant, how much heat is transferred?

7.14. A stream of water at 85°C, flowing at the rate of 5 kg s^{-1} is formed by mixing water at 24°C with saturated steam at 400 kPa. Assuming adiabatic operation, at what rates are the steam and water fed to the mixer?

7.15. In a desuperheater, liquid water at 3,100 kPa and 50°C is sprayed into a stream of superheated steam at 3,000 kPa and 375°C in an amount such that a single stream of saturated-vapor steam at 2,900 kPa flows from the desuperheater at the rate of 15 kg s^{-1}. Assuming adiabatic operation, what is the mass flow rate of the water? What is $\dot{S}_{G,\text{total}}$ for the process? What is the irreversible feature of the process?

7.16. Superheated steam at 700 kPa and 280°C flowing at the rate of 50 kg s^{-1} is mixed with liquid water at 40°C to produce steam at 700 kPa and 200°C. Assuming adiabatic operation, at what rate is water supplied to the mixer? What is $\dot{S}_{G,\text{total}}$ for the process? What is the irreversible feature of the process?

7.17. A stream of air at 12 bar and 900 K is mixed with another stream of air at 2 bar and 400 K with 2.5 times the mass flow rate. If this process were accomplished reversibly and adiabatically, what would be the temperature and pressure of the resulting air stream? Assume air an ideal gas for which $C_P = (7/2)R$.

7.18. Hot nitrogen gas at 750(°F) and atmospheric pressure flows into a waste-heat boiler at the rate of 40(lb$_m$)(s)$^{-1}$, and transfers heat to water boiling at 1(atm). The water feed to the boiler is saturated liquid at 1(atm), and it leaves the boiler as superheated steam at 1(atm) and 300(°F). If the nitrogen is cooled to 325(°F) and if heat is lost to the surroundings at a rate of 60(Btu) for each (lb$_m$) of steam generated, what is the steam-generation rate? If the surroundings are at 70(°F), what is $\dot{S}_{G,\text{total}}$ for the process? Assume nitrogen an ideal gas for which $C_P = (7/2)R$.

7.19. Hot nitrogen gas at 400°C and atmospheric pressure flows into a waste-heat boiler at the rate of 20 kg s^{-1}, and transfers heat to water boiling at 101.33 kPa. The water feed to the boiler is saturated liquid at 101.33 kPa, and it leaves the boiler as superheated steam at 101.33 kPa and 150°C. If the nitrogen is cooled to 170°C and if heat is lost to the surroundings at a rate of 80 kJ for each kilogram of steam generated, what is the steam-generation rate? If the surroundings are at 25°C, what is $\dot{S}_{G,\text{total}}$ for the process? Assume nitrogen an ideal gas for which $C_P = (7/2)R$.

7.20. Air expands adiabatically through a nozzle from a negligible initial velocity to a final velocity of 325 m s^{-1}. What is the temperature drop of the air, if air is assumed an ideal gas for which $C_P = (7/2)R$?

7.21. Steam enters a nozzle at 800 kPa and 280°C at negligible velocity and discharges at a pressure of 525 kPa. Assuming isentropic expansion of the steam in the nozzle, what is the exit velocity and what is the cross-sectional area at the nozzle exit for a flow rate of 0.75 kg s^{-1}?

7.22. Steam enters a converging nozzle at 800 kPa and 280°C with negligible velocity. If expansion is isentropic, what is the minimum pressure that can be reached in such a nozzle and what is the cross-sectional area at the nozzle throat at this pressure for a flow rate of 0.75 kg s^{-1}?

7.23. A gas enters a converging nozzle at pressure P_1 with negligible velocity, expands isentropically in the nozzle, and discharges into a chamber at pressure P_2. Sketch graphs showing the velocity at the throat and the mass flow rate as functions of the pressure ratio P_2/P_1.

7.24. For a converging/diverging nozzle with negligible entrance velocity in which expansion is isentropic, sketch graphs of mass flow rate $\dot{m}$, velocity u, and area ratio A/A_1 vs. the pressure ratio P/P_1. Here, A is the cross-sectional area of the nozzle at the point in the nozzle where the pressure is P, and subscript 1 denotes the nozzle entrance.

7.25. An ideal gas with constant heat capacities enters a converging/diverging nozzle with negligible velocity. If it expands isentropically within the nozzle, show that the throat velocity is given by

$$u_{\text{throat}}^2 = \frac{\gamma R T_1}{M}\left(\frac{2}{\gamma + 1}\right)$$

where T_1 is the temperature of the gas entering the nozzle, M is the molar mass, and R is the molar gas constant.

7.26. Steam expands isentropically in a converging/diverging nozzle from inlet conditions of 1,400 kPa, 325°C, and negligible velocity to a discharge pressure of 140 kPa. At the throat the cross-sectional area is 6 cm². Determine the mass flow rate of the steam and the state of the steam at the exit of the nozzle.

7.27. Steam expands adiabatically in a nozzle from inlet conditions of 130(psia), 420(°F), and a velocity of 230(ft)(s)$^{-1}$ to a discharge pressure of 35(psia) where its velocity is 2,000(ft)(s)$^{-1}$. What is the state of the steam at the nozzle exit, and what is $\dot{S}_{G,\text{total}}$ for the process?

7.28. Air discharges from an adiabatic nozzle at 15°C with a velocity of 580 m s^{-1}. What is the temperature at the entrance of the nozzle if the entrance velocity is negligible? Assume air an ideal gas for which $C_P = (7/2)R$.

7.29. A steam turbine operates adiabatically at a power level of 3,500 kW. Steam enters the turbine at 2,400 kPa and 500°C and exhausts from the turbine as saturated vapor at 20 kPa. What is the steam rate through the turbine, and what is the turbine efficiency?

7.30. A turbine operates adiabatically with superheated steam entering at T_1 and P_1 with a mass flow rate $\dot{m}$. The exhaust pressure is P_2 and the turbine efficiency is η. For one of the following sets of operating conditions, determine the power output of the turbine and the enthalpy and entropy of the exhaust steam.

(*a*) $T_1 = 450°C$, $P_1 = 8,000$ kPa, $\dot{m} = 80$ kg s^{-1}, $P_2 = 30$ kPa, $\eta = 0.80$.

(*b*) $T_1 = 550°C$, $P_1 = 9,000$ kPa, $\dot{m} = 90$ kg s^{-1}, $P_2 = 20$ kPa, $\eta = 0.77$.

(*c*) $T_1 = 600°C$, $P_1 = 8,600$ kPa, $\dot{m} = 70$ kg s^{-1}, $P_2 = 10$ kPa, $\eta = 0.82$.

(*d*) $T_1 = 400°C$, $P_1 = 7,000$ kPa, $\dot{m} = 65$ kg s^{-1}, $P_2 = 50$ kPa, $\eta = 0.75$.

(*e*) $T_1 = 200°C$, $P_1 = 1,400$ kPa, $\dot{m} = 50$ kg s^{-1}, $P_2 = 200$ kPa, $\eta = 0.75$.

(*f*) $T_1 = 900(°F)$, $P_1 = 1,100(\text{psia})$, $\dot{m} = 150(\text{lb}_m)(s)^{-1}$, $P_2 = 2(\text{psia})$, $\eta = 0.80$.

(*g*) $T_1 = 800(°F)$, $P_1 = 1,000(\text{psia})$, $\dot{m} = 100(\text{lb}_m)(s)^{-1}$, $P_2 = 4(\text{psia})$, $\eta = 0.75$.

7.31. The steam rate to a turbine for variable output is controlled by a throttle valve in the inlet line. Steam is supplied to the throttle valve at 1,700 kPa and 225°C. During a test run, the pressure at the turbine inlet is 1,000 kPa, the exhaust steam at 10 kPa has a quality of 0.95, the steam flow rate is 0.5 kg s^{-1}, and the power output of the turbine is 180 kW.

(*a*) What are the heat losses from the turbine?

(*b*) What would be the power output if the steam supplied to the throttle valve were expanded isentropically to the final pressure?

7.32. Saturated steam at 125 kPa is compressed adiabatically in a centrifugal compressor to 700 kPa at the rate of 2.5 kg s^{-1}. The compressor efficiency is 78 percent. What is the power requirement of the compressor and what are the enthalpy and entropy of the steam in its final state?

7.33. Carbon dioxide gas enters an adiabatic expander at 8 bar and 400°C and discharges at 1 bar. If the turbine efficiency is 0.75, what is the discharge temperature and what is the work output per mole of CO_2? Assume CO_2 an ideal gas at these conditions.

7.34. Nitrogen gas initially at 8.5 bar expands isentropically to 1 bar and 150°C. Assuming nitrogen an ideal gas, calculate the *initial* temperature and the work produced per mole of nitrogen.

7.35. Combustion products from a burner enter a gas turbine at 10 bar and 950°C and discharge at 1.5 bar. The turbine operates adiabatically with an efficiency of 77%. Assuming the combustion products to be an ideal-gas mixture with a heat capacity of 32 J mol^{-1} K^{-1}, what is the work output of the turbine per mole of gas, and what is the temperature of the gases discharging from the turbine?

7.36. An expander operates adiabatically with nitrogen entering at T_1 and P_1 with a molar flow rate $\dot{n}$. The exhaust pressure is P_2, and the expander efficiency is η. Estimate the power output of the expander and the temperature of the exhaust stream for one of the following sets of operating conditions.

(*a*) $T_1 = 480°C$, $P_1 = 6$ bar, $\dot{n} = 200$ mol s^{-1}, $P_2 = 1$ bar, $\eta = 0.80$.

(*b*) $T_1 = 400°C$, $P_1 = 5$ bar, $\dot{n} = 150$ mol s^{-1}, $P_2 = 1$ bar, $\eta = 0.75$.

(*c*) $T_1 = 500°C$, $P_1 = 7$ bar, $\dot{n} = 175$ mol s^{-1}, $P_2 = 1$ bar, $\eta = 0.78$.

(*d*) $T_1 = 450°C$, $P_1 = 8$ bar, $\dot{n} = 100$ mol s^{-1}, $P_2 = 2$ bar, $\eta = 0.85$.

(*e*) $T_1 = 900(°F)$, $P_1 = 95(psia)$, $\dot{n} = 0.5(lb\ mol)(s)^{-1}$, $P_2 = 15(psia)$, $\eta = 0.80$.

7.37. A compressor operates adiabatically with air entering at T_1 and P_1 with a molar flow rate $\dot{n}$. The discharge pressure is P_2 and the compressor efficiency is η. Estimate the power requirement of the compressor and the temperature of the discharge stream for one of the following sets of operating conditions.

(*a*) $T_1 = 25°C$, $P_1 = 101.33$ kPa, $\dot{n} = 100$ mol s^{-1}, $P_2 = 375$ kPa, $\eta = 0.75$.

(*b*) $T_1 = 80°C$, $P_1 = 375$ kPa, $\dot{n} = 100$ mol s^{-1}, $P_2 = 1,000$ kPa, $\eta = 0.70$.

(*c*) $T_1 = 30°C$, $P_1 = 100$ kPa, $\dot{n} = 150$ mol s^{-1}, $P_2 = 500$ kPa, $\eta = 0.80$.

(*d*) $T_1 = 100°C$, $P_1 = 500$ kPa, $\dot{n} = 50$ mol s^{-1}, $P_2 = 1,300$ kPa, $\eta = 0.75$.

(*e*) $T_1 = 80(°F)$, $P_1 = 14.7(psia)$, $\dot{n} = 0.5(lb\ mol)(s)^{-1}$, $P_2 = 55(psia)$, $\eta = 0.75$.

(*f*) $T_1 = 150(°F)$, $P_1 = 55(psia)$, $\dot{n} = 0.5(lb\ mol)(s)^{-1}$, $P_2 = 135(psia)$, $\eta = 0.70$.

7.38. Isobutane expands adiabatically in a turbine from 5,000 kPa and 250°C to 500 kPa at the rate of 0.7 kg mol s^{-1}. If the turbine efficiency is 0.80, what is the power output of the turbine and what is the temperature of the isobutane leaving the turbine?

7.39. Ammonia gas is compressed from 21°C and 200 kPa to 1,000 kPa in an adiabatic compressor with an efficiency of 0.82. Estimate the final temperature, the work required, and the entropy change of the ammonia.

7.40. Propylene is compressed adiabatically from 11.5 bar and 30°C to 18 bar at the rate of 1 kg mol s^{-1}. If the compressor efficiency is 0.8, what is the power requirement of the compressor and what is the discharge temperature of the propylene?

7.41. Methane is compressed adiabatically in a pipeline pumping station from 3,500 kPa and 35°C to 5,500 kPa at the rate of 1.5 kg mol s^{-1}. If the compressor efficiency is 0.78, what is the power requirement of the compressor and what is the discharge temperature of the methane?

7.42. A pump operates adiabatically with liquid water entering at T_1 and P_1 with a mass flow rate $\dot{m}$. The discharge pressure is P_2, and the pump efficiency is η. For one of the following sets of operating conditions, determine the power requirement of the pump and the temperature of the water discharged from the pump.

(a) $T_1 = 25°C$, $P_1 = 100$ kPa, $\dot{m} = 20$ kg s^{-1}, $P_2 = 2{,}000$ kPa, $\eta = 0.75$, $\beta = 257.2 \times 10^{-6}$ K^{-1}.

(b) $T_1 = 90°C$, $P_1 = 200$ kPa, $\dot{m} = 30$ kg s^{-1}, $P_2 = 5{,}000$ kPa, $\eta = 0.70$, $\beta = 696.2 \times 10^{-6}$ K^{-1}.

(c) $T_1 = 60°C$, $P_1 = 20$ kPa, $\dot{m} = 15$ kg s^{-1}, $P_2 = 5{,}000$ kPa, $\eta = 0.75$, $\beta = 523.1 \times 10^{-6}$ K^{-1}.

(d) $T_1 = 70(°F)$, $P_1 = 1(atm)$, $\dot{m} = 50(lb_m)(s)^{-1}$, $P_2 = 20(atm)$, $\eta = 0.70$, $\beta = 217.3 \times 10^{-6}$ K^{-1}.

(e) $T_1 = 200(°F)$, $P_1 = 15(psia)$, $\dot{m} = 80(lb_m)(s)^{-1}$, $P_2 = 1{,}500(psia)$, $\eta = 0.75$, $\beta = 714.3 \times 10^{-6}$ K^{-1}.

CHAPTER 8

PRODUCTION OF POWER FROM HEAT

Except for nuclear power, the sun is the source of all the mechanical energy used by man. The total rate at which energy reaches the earth from the sun is staggering, but the rate at which it falls on a square meter of surface is small. The difficulty is to concentrate the energy gathered over a large area so that its use in the production of work becomes practical; research in this area continues. While not widely used for large-scale power generation, solar radiation does find application for the direct heating of water, for generation of high temperatures for metallurgical applications (solar furnaces), and to evaporate water in the production of salt.

The kinetic energy associated with mass movement of air has been used to some extent for the production of work (windmills), especially in rural areas. Variations and uncertainties in wind speed, and the need for large-size equipment to produce significant quantities of work, are problems in this field.

Conceivably, the potential energy of tides could be exploited. Attempts in this direction on a large scale have been made in parts of the world where tides are particularly high. However, total power production from this source is unlikely to be significant in comparison with world demands for energy.

By far the most important sources of power are the chemical (molecular) energy of fuels and nuclear energy. Significant power generation also results from the conversion of the potential energy of water into work, a process that can in principle be accomplished with an efficiency of 100 percent. On the other hand, all present-day methods for the *large-scale* use of molecular or nuclear energy are based on the evolution of heat and subsequent conversion of part of the heat into useful work. Despite improvements in equipment design, the efficiency of conversion is relatively low (values greater than 35 percent are uncommon), a consequence of the second law. When it is possible to convert the energy in fuels into work without the intermediate generation of heat, conversion efficiency is considerably improved.

The usual device for the direct conversion of chemical energy into electrical energy is the electrolytic cell. Progress has been made in developing cells which operate on hydrogen and on carbonaceous fuels such as natural gas or coal. Such *fuel cells* are already in use to supply modest power requirements for special purposes. The efficiency of these cells ranges from 55 to 85 percent, about twice the value obtained by the conventional process of first converting the chemical energy into heat.

In a conventional power plant the molecular energy of fuel is released by a combustion process. The function of the work-producing device is to convert part of the heat of combustion into mechanical energy. In a nuclear power plant the fission or fusion process releases the energy of the nucleus of the atom as heat, and then this heat is partially converted into work. Thus, the thermodynamic analysis of heat engines, as presented in this chapter, applies equally well to conventional (fossil-fuel) and nuclear power plants.

The steam power plant is a large-scale heat engine in which the working fluid (H_2O) is in steady-state flow successively through a pump, a boiler, a turbine, and a condenser in a cyclic process. The working fluid is separated from the heat source, and heat is transferred across a physical boundary. In a coal-fired plant the combustion gases are separated from the steam by boiler-tube walls.

The *internal*-combustion engine is another form of heat engine, wherein high temperatures are attained by conversion of the chemical energy of a fuel directly into internal energy within the work-producing device. Examples of this type are the Otto engine and the gas turbine.[1]

To illustrate the calculation of thermal efficiencies, we analyze in this chapter several common heat-engine cycles.

8.1   The Steam Power Plant

The Carnot-engine cycle, described in Chap. 5, operates reversibly and consists of two isothermal steps connected by two adiabatic steps. In the isothermal step at higher temperature T_H, heat $|Q_H|$ is absorbed by the working fluid of the engine, and in the isothermal step at lower temperature T_C, heat $|Q_C|$ is discarded by the fluid. The work produced is $|W| = |Q_H| - |Q_C|$, and the thermal efficiency of the Carnot engine [Eq. (5.8)] is

$$\eta \equiv \frac{|W|}{|Q_H|} = 1 - \frac{T_C}{T_H}$$

Clearly, η increases as T_H increases and as T_C decreases. Although the efficiencies of practical heat engines are lowered by irreversibilities, it is still true that their efficiencies are increased when the average temperature at which heat is absorbed is increased and when the average temperature at which heat is rejected is decreased.

---

[1] Details of steam power plants and internal-combustion engines can be found in E. B. Woodruff, H. B. Lammers, and T. S. Lammers, *Steam Plant Operation*, 6th ed., McGraw-Hill, New York, 1992; and C. F. Taylor, *The Internal Combustion Engine in Theory and Practice: Thermodynamics, Fluid Flow, Performance*, MIT Press, Boston, 1984.

Figure 8.1 shows a simple steady-state steady-flow process in which steam generated in a boiler is expanded in an adiabatic turbine to produce work. The discharge stream from the turbine passes to a condenser from which it is pumped adiabatically back to the boiler. The power produced by the turbine is much greater than that required by the pump, and the net power output is equal to the difference between the rate of heat input in the boiler $|\dot{Q}_H|$ and the rate of heat rejection in the condenser $|\dot{Q}_C|$.

The property changes of the fluid as it flows through the individual pieces of equipment may be shown as lines on a TS diagram, as illustrated in Fig. 8.2. The sequence of lines represents a cycle. Indeed, the particular cycle shown is a *Carnot* cycle. In this idealization, step $1 \rightarrow 2$ is the isothermal absorption of heat at T_H, and is represented by a horizontal line on the TS diagram. This vaporization process occurs also at constant pressure and produces saturated-vapor steam from saturated-liquid water. Step $2 \rightarrow 3$ is a reversible, adiabatic expansion of saturated vapor to a pressure at which $T^{\text{sat}} = T_C$. This isentropic expansion process is represented by a vertical line on the TS diagram and produces a wet vapor. Step

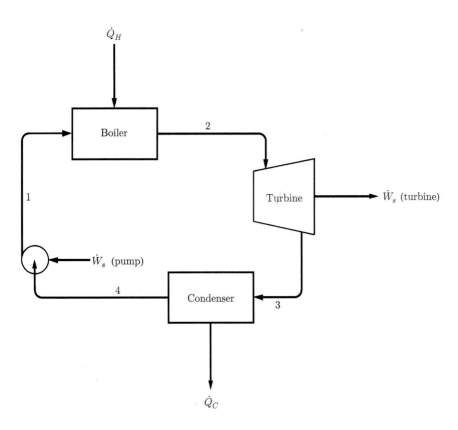

Figure 8.1: Simple steam power plant.

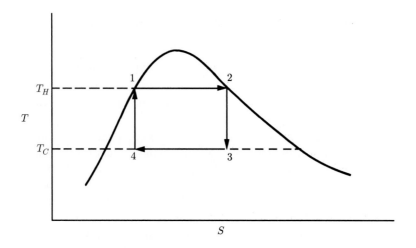

Figure 8.2: Carnot cycle on a TS diagram.

$3 \rightarrow 4$ is the isothermal rejection of heat at temperature T_C, and is represented by a horizontal line on the TS diagram. It is a condensation process, but is incomplete. Step $4 \rightarrow 1$ takes the cycle back to its origin, producing saturated-liquid water at point 1. It is an isentropic compression process represented by a vertical line on the TS diagram.

The thermal efficiency of this cycle is that of a Carnot engine, given by Eq. (5.8). As a reversible cycle, it could serve as a standard of comparison for actual steam power plants. However, severe practical difficulties attend the operation of equipment intended to carry out steps $2 \rightarrow 3$ and $4 \rightarrow 1$. Turbines that take in saturated steam produce an exhaust with high liquid content, which causes severe erosion problems.[2] Even more difficult is the design of a pump that takes in a mixture of liquid and vapor (point 4) and discharges a saturated liquid (point 1). For these reasons, an alternative model cycle is taken as the standard, at least for fossil-fuel-burning power plants. It is called the *Rankine cycle*, and differs from the cycle of Fig. 8.2 in two major respects. First, the heating step $1 \rightarrow 2$ is carried well beyond vaporization, so as to produce a superheated vapor, and second, the cooling step $3 \rightarrow 4$ brings about complete condensation, yielding saturated liquid to be pumped to the boiler. The Rankine cycle therefore consists of the four steps shown by Fig. 8.3, and described as follows:

$1 \rightarrow 2$ A constant-pressure heating process in a boiler. The step lies along an isobar (the pressure of the boiler), and consists of three sections: heating of subcooled liquid water to its saturation temperature, vaporization at

---

[2]Nevertheless, present-day nuclear power plants generate saturated steam and operate with turbines designed to eject liquid at various stages of expansion.

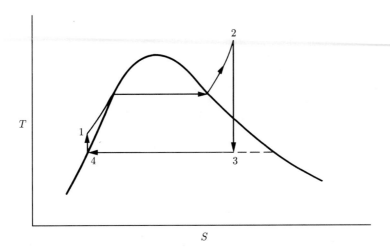

Figure 8.3: The Rankine cycle.

constant temperature and pressure, and superheating of the vapor to a temperature well above its saturation temperature.

$2 \rightarrow 3$ Reversible, adiabatic (isentropic) expansion of vapor in a turbine to the pressure of the condenser. The step normally crosses the saturation curve, producing a wet exhaust. However, the superheating accomplished in step $1 \rightarrow 2$ shifts the vertical line far enough to the right on Fig. 8.3 that the moisture content is not too large.

$3 \rightarrow 4$ A constant-pressure, constant-temperature process in a condenser to produce saturated liquid at point 4.

$4 \rightarrow 1$ Reversible, adiabatic (isentropic) pumping of the saturated liquid to the pressure of the boiler, producing subcooled liquid. The vertical line (whose length is exaggerated in Fig. 8.3) is very short, because the temperature rise associated with compression of a liquid is small.

Power plants can be built to operate on a cycle that departs from the Rankine cycle solely because of the irreversibilities of the work-producing and work-requiring steps. We show in Fig. 8.4 the effects of these irreversibilities on steps $2 \rightarrow 3$ and $4 \rightarrow 1$. The lines are no longer vertical, but tend in the direction of increasing entropy. The turbine exhaust is normally still wet, but as long as the moisture content is less than about 10 percent, erosion problems are not serious. Slight subcooling of the condensate in the condenser may occur, but the effect is inconsequential.

The boiler serves to transfer heat from a burning fuel to the cycle, and the condenser transfers heat from the cycle to the surroundings. Neglecting kinetic- and potential-energy changes reduces the energy relations, Eqs. (7.16) and (7.17),

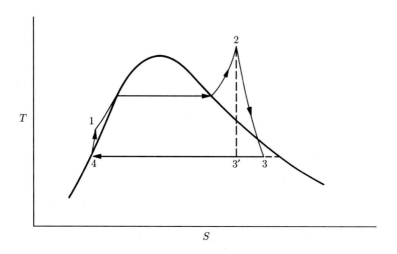

Figure 8.4: Simple practical power cycle.

in either case to

$$\dot{Q} = \dot{m}\Delta H \tag{8.1}$$

and

$$Q = \Delta H \tag{8.2}$$

Turbine and pump calculations are treated in detail in Chap. 7.

> **Example 8.1** Steam generated in a power plant at a pressure of 8,600 kPa and
> a temperature of 500°C is fed to a turbine. Exhaust from the turbine enters a
> condenser at 10 kPa, where it is condensed to saturated liquid, which is then pumped
> to the boiler.
>
> > (a) Determine the thermal efficiency of a Rankine cycle operating at these
> > conditions.
> >
> > (b) Determine the thermal efficiency of a practical cycle operating at these
> > conditions if the turbine efficiency and pump efficiency are both 0.75.
> >
> > (c) If the rating of the power cycle of part (b) is 80,000 kW, what is the steam
> > rate and what are the heat-transfer rates in the boiler and condenser?
>
> SOLUTION   (a) The turbine operates under the same conditions as the turbine of
> Example 7.10, where we found
>
> $$(\Delta H)_S = -1{,}274.2 \text{ kJ kg}^{-1}$$
>
> Thus,
>
> $$W_s(\text{isentropic}) = (\Delta H)_S = -1{,}274.2 \text{ kJ kg}^{-1}$$
>
> Moreover, we found the enthalpy at the end of isentropic expansion (H_2' in Exam-
> ple 7.10) to be
>
> $$H_3' = 2{,}117.4 \text{ kJ kg}^{-1}$$

The enthalpy of saturated liquid at 10 kPa (and $t^{sat} = 45.83°C$) is

$$H_4 = 191.8 \text{ kJ kg}^{-1}$$

Thus by Eq. (8.2) applied to the condenser,

$$Q(\text{condenser}) = H_4 - H'_3 = 191.8 - 2{,}117.4 = -1{,}925.6 \text{ kJ kg}^{-1}$$

where the minus sign signifies that the heat flows out of the system.

The pump operates under essentially the same conditions as the pump of Example 7.14, where we found

$$W_s(\text{isentropic}) = (\Delta H)_S = 8.7 \text{ kJ kg}^{-1}$$

Thus,

$$H_1 = H_4 + (\Delta H)_S = 191.8 + 8.7 = 200.5 \text{ kJ kg}^{-1}$$

The enthalpy of superheated steam at 8,600 kPa and 500°C is

$$H_2 = 3{,}391.6 \text{ kJ kg}^{-1}$$

By Eq. (8.2) applied to the boiler,

$$Q(\text{boiler}) = H_2 - H_1 = 3{,}391.6 - 200.5 = 3{,}191.1 \text{ kJ kg}^{-1}$$

The net work of the Rankine cycle is the sum of the turbine work and the pump work:

$$W_s(\text{Rankine}) = -1{,}274.2 + 8.7 = -1{,}265.5 \text{ kJ kg}^{-1}$$

This result is of course also given by

$$W_s(\text{Rankine}) = -Q(\text{boiler}) - Q(\text{condenser})$$
$$= -3{,}191.1 + 1{,}925.6 = -1{,}265.5 \text{ kJ kg}^{-1}$$

The thermal efficiency of the cycle is

$$\eta = \frac{|W_s(\text{Rankine})|}{Q(\text{boiler})} = \frac{1{,}265.5}{3{,}191.1} = 0.3966$$

(*b*) If the turbine efficiency is 0.75, then we also have from Example 7.10 that

$$W_s(\text{turbine}) = \Delta H = -955.6 \text{ kJ kg}^{-1}$$

and

$$H_3 = H_2 + \Delta H = 3{,}391.6 - 955.6 = 2{,}436.0 \text{ kJ kg}^{-1}$$

For the condenser,

$$Q(\text{condenser}) = H_4 - H_3 = 191.8 - 2{,}436.0 = -2{,}244.2 \text{ kJ kg}^{-1}$$

By Example 7.14 for the pump,

$$W_s(\text{pump}) = \Delta H = 11.6 \text{ kJ kg}^{-1}$$

The net work of the cycle is therefore

$$\dot{W}_s(\text{net}) = -955.6 + 11.6 = -944.0 \text{ kJ kg}^{-1}$$

and

$$H_1 = H_4 + \Delta H = 191.8 + 11.6 = 203.4 \text{ kJ kg}^{-1}$$

Then

$$Q(\text{boiler}) = H_2 - H_1 = 3{,}391.6 - 203.4 = 3{,}188.2 \text{ kJ kg}^{-1}$$

The thermal efficiency of the cycle is therefore

$$\eta = \frac{|W_s(\text{net})|}{Q(\text{boiler})} = \frac{944.0}{3{,}188.2} = 0.2961$$

which may be compared with the result of part (a).

(c) For a power rating of 80,000 kW, we have

$$\dot{W}_s(\text{net}) = \dot{m}W_s(\text{net})$$

or

$$\dot{m} = \frac{\dot{W}_s(\text{net})}{W_s(\text{net})} = \frac{-80{,}000 \text{ kJ s}^{-1}}{-944.0 \text{ kJ kg}^{-1}} = 84.75 \text{ kg s}^{-1}$$

Then by Eq. (8.1),

$$\dot{Q}(\text{boiler}) = (84.75)(3{,}188.2) = 270.2 \times 10^3 \text{ kJ s}^{-1}$$

and

$$\dot{Q}(\text{condenser}) = (84.75)(-2{,}244.2) = -190.2 \times 10^3 \text{ kJ s}^{-1}$$

Note that

$$\dot{Q}(\text{boiler}) + \dot{Q}(\text{condenser}) = -\dot{W}_s(\text{net})$$

The thermal efficiency of a steam power cycle is increased when the pressure and hence the vaporization temperature in the boiler is raised. It is also increased by increased superheating in the boiler. Thus, high boiler pressures and temperatures favor high efficiencies. However, these same conditions increase the capital investment in the plant, because they require heavier construction and more expensive materials of construction. Moreover, these costs increase ever more rapidly as more severe conditions are imposed. Thus, in practice power plants seldom operate at pressures much above 10,000 kPa and temperatures much above 600°C. The thermal efficiency of a power plant increases as the pressure and hence the temperature in the condenser is reduced. However, the condensation temperature must be higher than the temperature of the cooling medium, usually water, and this is controlled by local conditions of climate and geography. Power plants universally operate with the condenser pressure as low as practical.

Most modern power plants operate on a modification of the Rankine cycle that incorporates feedwater heaters. Water from the condenser, rather than being pumped directly back to the boiler, is first heated by steam extracted from the turbine. This is normally done in several stages, with steam taken from the turbine at several intermediate states of expansion. An arrangement with four feedwater

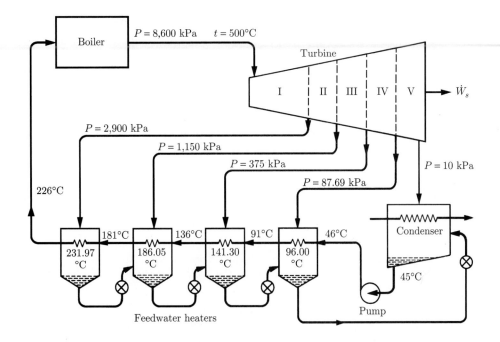

Figure 8.5: Steam power plant with feedwater heating.

heaters is shown in Fig. 8.5. The operating conditions indicated on this figure and described in the following paragraphs are typical, and are the basis for the illustrative calculations of Example 8.2.

The conditions of steam generation in the boiler are the same as in Example 8.1: 8,600 kPa and 500°C. The exhaust pressure of the turbine, 10 kPa, is also the same. The saturation temperature of the exhaust steam is therefore 45.83°C. Allowing for slight subcooling of the condensate, we fix the temperature of the liquid water from the condenser at 45°C. The feedwater pump, which operates under exactly the conditions of the pump in Example 7.14, causes a temperature rise of about 1°C, making the temperature of the feedwater entering the series of heaters equal to 46°C.

The saturation temperature of steam at the boiler pressure of 8,600 kPa is 300.06°C, and the temperature to which the feedwater can be raised in the heaters is certainly less. This temperature is a design variable, which is ultimately fixed by economic considerations. However, a value must be chosen before any thermodynamic calculations can be made. We have therefore arbitrarily specified a temperature of 226°C for the feedwater stream entering the boiler. We have also specified that each of the four feedwater heaters accomplishes the same temperature rise. Thus, the total temperature rise of $226 - 46 = 180$°C is divided into four 45°C increments. This establishes all intermediate feedwater temperatures at the values shown on Fig. 8.5.

The steam supplied to a given feedwater heater must be at a pressure high enough that its saturation temperature is higher than the temperature of the feedwater stream leaving the heater. We have here presumed a minimum temperature difference for heat transfer of no less than 5°C, and have chosen extraction steam pressures such that the T^{sat} values shown in each feedwater heater are at least 5°C greater than the exit temperature of the feedwater stream. The condensate from each feedwater heater is flashed through a throttle valve to the heater at the next lower pressure, and the collected condensate in the final heater of the series is flashed into the condenser. Thus, all condensate returns from the condenser to the boiler by way of the feedwater heaters.

The purpose of heating the feedwater in this manner is to raise the average temperature at which heat is added in the boiler. This raises the thermal efficiency of the plant, which is said to operate on a *regenerative cycle*.

Example 8.2   Determine the thermal efficiency of the power plant shown in Fig. 8.5, assuming turbine and pump efficiencies of 0.75. If its power rating is 80,000 kW, what is the steam rate from the boiler and the heat-transfer rates in the boiler and condenser?

SOLUTION   Initial calculations are made on the basis of 1 kg of steam entering the turbine from the boiler. The turbine is in effect divided into five sections, as indicated in Fig. 8.5. Because steam is extracted at the end of each section, the flow rate in the turbine decreases from one section to the next. The amounts of steam extracted from the first four sections are determined by energy balances.

For this, we need enthalpies of the compressed feedwater streams. The effect of pressure at constant temperature on a liquid is given by Eq. (7.44) written as

$$\Delta H = V(1 - \beta T)\Delta P \qquad (\text{const } T)$$

For saturated liquid water at 226°C (499.15 K), we find from the steam tables:

$$P^{\text{sat}} = 2{,}598.2 \text{ kPa}$$

$$H = 971.5 \text{ kJ kg}^{-1}$$

$$V = 1{,}201 \text{ cm}^3 \text{ kg}^{-1}$$

In addition, at this temperature

$$\beta = 1.582 \times 10^{-3} \text{ K}^{-1}$$

Thus, for a pressure change from the saturation pressure to 8,600 kPa,

$$\Delta H = 1{,}201[1 - (1.528 \times 10^{-3})(499.15)]\frac{(8{,}600 - 2{,}598.2)}{10^6} = 1.5 \text{ kJ kg}^{-1}$$

and

$$H = H(\text{sat. liq.}) + \Delta H = 971.5 + 1.5 = 973.0 \text{ kJ kg}^{-1}$$

Similar calculations yield the enthalpies of the feedwater at other temperatures. All pertinent values are given in the accompanying table.

$t/°C$	226	181	136	91	46
$H/\text{kJ kg}^{-1}$ for water at t and $P = 8{,}600$ kPa	973.0	771.3	577.4	387.5	200.0

Consider the first section of the turbine and the first feedwater heater, as shown by Fig. 8.6. The enthalpy and entropy of the steam entering the turbine are found from the tables for superheated steam. The assumption of isentropic expansion of steam in section I of the turbine to 2,900 kPa leads to the result

$$(\Delta H)_S = -320.5 \text{ kJ kg}^{-1}$$

If we assume that the turbine efficiency is independent of the pressure to which the steam expands, then Eq. (7.35) gives

$$\Delta H = \eta(\Delta H)_S = (0.75)(-320.5) = -240.4 \text{ kJ kg}^{-1}$$

By Eq.(7.33),
$$W_s(\text{I}) = \Delta H = -240.4 \text{ kJ}$$

In addition, the enthalpy of the steam discharged from this section of the turbine is

$$H = 3{,}391.6 - 240.4 = 3{,}151.2 \text{ kJ kg}^{-1}$$

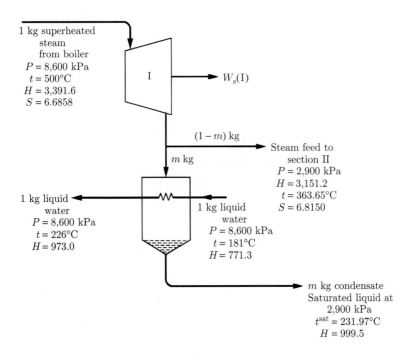

Figure 8.6: Section I of turbine and first feedwater heater. Enthalpy in kJ kg^{-1}; entropy in kJ kg^{-1} K^{-1}.

An energy balance on the feedwater heater requires application of Eq. (7.15). Neglecting kinetic- and potential-energy changes and noting that $\dot{Q} = -\dot{W}_s = 0$, we have

$$\Delta(\dot{m}H)_{fs} = 0$$

This equation gives mathematical expression to the requirement that the total enthalpy change for the process be zero. Thus on the basis of 1 kg of steam entering the turbine (see Fig. 8.6),

$$m(999.5 - 3,151.2) + (1)(973.0 - 771.3) = 0$$

whence

$$m = 0.09374 \text{ kg} \qquad \text{and} \qquad 1 - m = 0.90626 \text{ kg}$$

On the basis of 1 kg of steam entering the turbine, $1 - m$ is the mass of steam flowing into section II of the turbine.

Section II of the turbine and the second feedwater heater are shown in Fig. 8.7. In doing the same calculations as for section I, we assume that each kilogram of steam leaving section II expands from its state *at the turbine entrance* to the exit

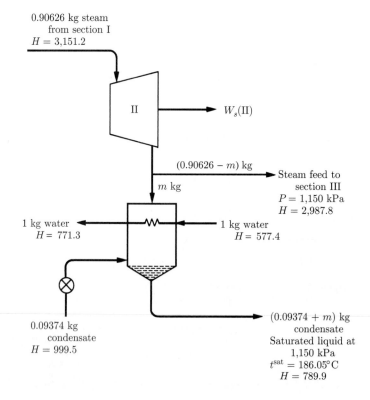

Figure 8.7: Section II of turbine and second feedwater heater. Enthalpy in kJ kg^{-1}; entropy in kJ kg^{-1} K^{-1}.

of section II with an efficiency of 0.75 compared with isentropic expansion. The enthalpy of the steam leaving section II found in this way is

$$H = 2{,}987.8 \text{ kJ kg}^{-1}$$

Then on the basis of 1 kg of steam entering the turbine,

$$W_s(\text{II}) = (2{,}987.8 - 3{,}151.2)(0.90626) = -148.08 \text{ kJ}$$

An energy balance on the feedwater heater (Fig. 8.7) gives

$$(0.09374 + m)(789.9) - (0.09374)(999.5) - m(2{,}987.8) + (1)(771.3 - 577.4) = 0$$

whence

$$m = 0.07971 \text{ kg}$$

Note that throttling the condensate stream does not change its enthalpy.

These results and those of similar calculations for the remaining sections of the turbine are listed in the following table:

	$H/\text{kJ kg}^{-1}$ at section exit	W_s/kJ for section	$t/°\text{C}$ at section exit	State	m/kg of steam extracted
Sec. I	3,151.2	−240.40	363.65	superheated vapor	0.09374
Sec. II	2,987.8	−148.08	272.48	superheated vapor	0.07928
Sec. III	2,827.4	−132.65	183.84	superheated vapor	0.06993
Sec. IV	2,651.3	−133.32	96.00	wet vapor $x = 0.9919$	0.06257
Sec. V	2,435.9	−149.59	45.83	wet vapor $x = 0.9378$	

From these results we have

$$\sum W_s = -804.0 \text{ kJ} \qquad \text{and} \qquad \sum m = 0.3055 \text{ kg}$$

Thus for every kilogram of steam entering the turbine, the work produced is 804.0 kJ, and 0.3055 kg of steam is extracted from the turbine for the feedwater heaters. The work required by the pump is exactly the work calculated for the pump in Example 7.14, that is, 11.6 kJ. The net work of the cycle is therefore

$$W_s(\text{net}) = -804.0 + 11.6 = -792.4 \text{ kJ}$$

on the basis of 1 kg of steam generated in the boiler. On the same basis, the heat added in the boiler is

$$Q(\text{boiler}) = \Delta H = 3{,}391.6 - 973.0 = 2{,}418.6 \text{ kJ}$$

The thermal efficiency of the cycle is therefore

$$\eta = \frac{|W_s(\text{net})|}{Q(\text{boiler})} = \frac{792.4}{2{,}418.6} = 0.3276$$

This is a significant improvement over the value of 0.2961 found in Example 8.1.

Since $\dot{W}_s(\text{net}) = -80{,}000$ kJ s^{-1},

$$\dot{m} = \frac{\dot{W}_s(\text{net})}{W_s(\text{net})} = \frac{-80{,}000}{-792.4} = 100.96 \text{ kg s}^{-1}$$

This is the steam rate to the turbine, and with it we can calculate the heat-transfer rate in the boiler:

$$\dot{Q}(\text{boiler}) = \dot{m}\Delta H = (100.96)(2{,}418.6) = 244.2 \times 10^3 \text{ kJ s}^{-1}$$

The heat-transfer rate to the cooling water in the condenser is

$$\dot{Q}(\text{condenser}) = -\dot{Q}(\text{boiler}) - \dot{W}_s(\text{net})$$

$$= -244.2 \times 10^3 - (-80.0 \times 10^3)$$

$$= -164.2 \times 10^3 \text{ kJ s}^{-1}$$

Although the steam generation rate is higher than was found in Example 8.1, the heat-transfer rates in the boiler and condenser are appreciably less, because their functions are partly taken over by the feedwater heaters.

8.2 Internal-Combustion Engines

In a steam power plant, the steam is an inert medium to which heat is transferred from a burning fuel or from a nuclear reactor. It is therefore characterized by large heat-transfer surfaces: (1) for the absorption of heat by the steam at a high temperature in the boiler, and (2) for the rejection of heat from the steam at a relatively low temperature in the condenser. The disadvantage is that when heat must be transferred through walls (as through the metal walls of boiler tubes) the ability of the walls to withstand high temperatures and pressures imposes a limit on the temperature of heat absorption. In an internal-combustion engine, on the other hand, a fuel is burned within the engine itself, and the combustion products serve as the working-medium, acting for example on a piston in a cylinder. High temperatures are internal, and do not involve heat-transfer surfaces.

Burning of fuel within the internal-combustion engine complicates thermodynamic analysis. Moreover, fuel and air flow steadily into an internal-combustion engine and combustion products flow steadily out of it; there is no working medium that undergoes a cyclic process, as does the steam in a steam power plant. However, for making simple analyses, one imagines cyclic engines with air as the working fluid that are equivalent in performance to actual internal-combustion engines. In addition, the combustion step is replaced by the addition to the air of an equivalent amount of heat. In each of the following sections, we first present a qualitative

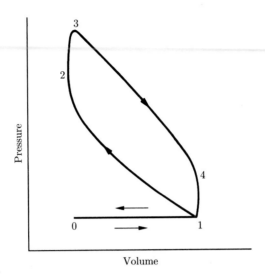

Figure 8.8: Otto internal-combustion-engine cycle.

description of an internal-combustion engine. Quantitative analysis is then made of an ideal cycle in which air, treated as an ideal gas with constant heat capacities, is the working medium.

8.3 The Otto Engine

The most common internal-combustion engine, because of its use in automobiles, is the Otto engine. Its cycle consists of four strokes, and starts with an intake stroke at essentially constant pressure, during which a piston moving outward draws a fuel/air mixture into a cylinder. This is represented by line $0 \rightarrow 1$ in Fig. 8.8. During the second stroke ($1 \rightarrow 2 \rightarrow 3$), all valves are closed, and the fuel/air mixture is compressed, approximately adiabatically, along line $1 \rightarrow 2$; the mixture is then ignited, and combustion occurs so rapidly that the volume remains nearly constant while the pressure rises along line $2 \rightarrow 3$. It is during the third stroke ($3 \rightarrow 4 \rightarrow 1$) that work is produced. The high-temperature, high-pressure products of combustion expand, approximately adiabatically, along line $3 \rightarrow 4$; the exhaust valve then opens and the pressure falls rapidly at nearly constant volume along line $4 \rightarrow 1$. During the fourth or exhaust stroke (line $1 \rightarrow 0$), the piston pushes the remaining combustion gases (except for the contents of the clearance volume) from the cylinder. The volume plotted in Fig. 8.8 is the total volume of gas contained in the engine between the piston and the cylinder head.

The effect of increasing the compression ratio, defined as the ratio of the volumes at the beginning and end of the compression stroke, is to increase the efficiency of the engine, i.e., to increase the work produced per unit quantity of fuel.

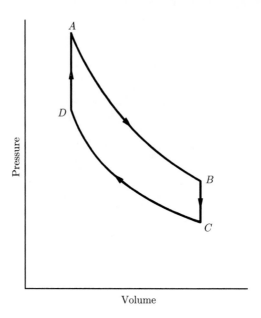

Figure 8.9: Air-standard Otto cycle.

We demonstrate this for an idealized cycle, called the air-standard cycle, shown in Fig. 8.9. It consists of two adiabatic and two constant-volume steps, which comprise a heat-engine cycle for which air is the working fluid. In step DA, sufficient heat is absorbed by the air at constant volume to raise its temperature and pressure to the values resulting from combustion in an actual Otto engine. Then the air is expanded adiabatically and reversibly (step AB), cooled at constant volume (step BC), and finally compressed adiabatically and reversibly to the initial state at D.

The thermal efficiency η of the air-standard cycle shown in Fig. 8.9 is simply

$$\eta = \frac{-W_s(\text{net})}{Q_{DA}} = \frac{Q_{DA} + Q_{BC}}{Q_{DA}} \tag{8.3}$$

For 1 mol of air with constant heat capacities,

$$Q_{DA} = C_V(T_A - T_D)$$

$$Q_{BC} = C_V(T_C - T_B)$$

Substituting these expressions in Eq. (8.3) gives

$$\eta = \frac{C_V(T_A - T_D) - C_V(T_B - T_C)}{C_V(T_A - T_D)}$$

or

$$\eta = 1 - \frac{T_B - T_C}{T_A - T_D} \tag{8.4}$$

The thermal efficiency is also related in a simple way to the compression ratio $r = V_C/V_D$. We replace each temperature in Eq. (8.4) by an appropriate group PV/R, in accord with the ideal-gas equation. Thus

$$T_B = \frac{P_B V_B}{R} = \frac{P_B V_C}{R}$$

$$T_C = \frac{P_C V_C}{R}$$

$$T_A = \frac{P_A V_A}{R} = \frac{P_A V_D}{R}$$

$$T_D = \frac{P_D V_D}{R}$$

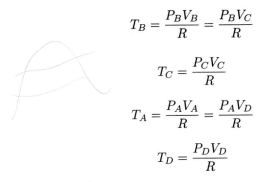

Substituting into Eq. (8.4) leads to

$$\eta = 1 - \frac{V_C}{V_D}\left(\frac{P_B - P_C}{P_A - P_D}\right) = 1 - r\left(\frac{P_B - P_C}{P_A - P_D}\right) \tag{8.5}$$

For the two adiabatic, reversible steps, we have $PV^\gamma = $ const. Hence

$$P_A V_D^\gamma = P_B V_C^\gamma \qquad (\text{since } V_D = V_A \text{ and } V_C = V_B)$$

$$P_C V_C^\gamma = P_D V_D^\gamma$$

These expressions are combined to eliminate the volumes:

$$\frac{P_B}{P_C} = \frac{P_A}{P_D}$$

Also

$$\frac{P_C}{P_D} = \left(\frac{V_D}{V_C}\right)^\gamma = \left(\frac{1}{r}\right)^\gamma$$

These equations transform Eq. (8.5) as follows:

$$\eta = 1 - r\frac{(P_B/P_C - 1)P_C}{(P_A/P_D - 1)P_D} = 1 - r\frac{P_C}{P_D}$$

or

$$\eta = 1 - r\left(\frac{1}{r}\right)^\gamma = 1 - \left(\frac{1}{r}\right)^{\gamma-1} \tag{8.6}$$

This equation shows that the thermal efficiency increases rapidly with the compression ratio r at low values of r, but more slowly at high compression ratios. This agrees with the results of actual tests on Otto engines.

8.4 The Diesel Engine

The Diesel engine differs from the Otto engine primarily in that the temperature at the end of compression is sufficiently high that combustion is initiated spontaneously. This higher temperature results because of a higher compression ratio that carries the compression step to a higher pressure. The fuel is not injected until the end of the compression step, and then is added slowly enough that the combustion process occurs at approximately constant pressure.

For the same compression ratio, the Otto engine has a higher efficiency than the Diesel engine. However, preignition limits the compression ratio attainable in the Otto engine. The Diesel engine therefore operates at higher compression ratios, and consequently at higher efficiencies.

Example 8.3 Sketch the air-standard Diesel cycle on a PV diagram, and derive an equation giving the thermal efficiency of this cycle in relation to the compression ratio r (ratio of volumes at the beginning and end of the compression step) and the expansion ratio r_e (ratio of volumes at the end and beginning of the expansion step).

SOLUTION   The air-standard Diesel cycle is the same as the air-standard Otto cycle, except that the heat-absorption step (corresponding to the combustion process in the actual engine) is at constant pressure, as indicated by line DA in Fig. 8.10.

On the basis of one mol of air, considered to be an ideal gas with constant heat capacities, the heat absorbed in the cycle is

$$Q_{DA} = C_P(T_A - T_D)$$

The heat rejected in step BC is

$$Q_{BC} = C_V(T_C - T_B)$$

By an energy balance, $-W_s = Q_{DA} + Q_{BC}$, and the thermal efficiency is given by

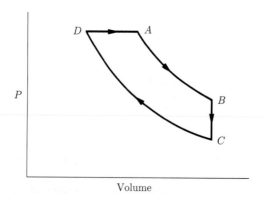

Figure 8.10: Air-standard Diesel cycle.

$$\eta = 1 - \frac{C_V}{C_P}\left(\frac{T_B - T_C}{T_A - T_D}\right) = 1 - \frac{1}{\gamma}\left(\frac{T_B - T_C}{T_A - T_D}\right) \qquad (A)$$

For reversible, adiabatic expansion (step AB) and reversible, adiabatic compression (step CD), Eq. (3.22) applies:

$$T_A V_A^{\gamma-1} = T_B V_B^{\gamma-1}$$

and

$$T_D V_D^{\gamma-1} = T_C V_C^{\gamma-1}$$

By definition, the compression ratio is $r = V_C/V_D$; in addition the expansion ratio is defined as $r_e = V_B/V_A$. Thus

$$T_B = T_A \left(\frac{1}{r_e}\right)^{\gamma-1} \qquad (B)$$

and

$$T_C = T_D \left(\frac{1}{r}\right)^{\gamma-1} \qquad (C)$$

Substituting Eqs. (B) and (C) into Eq. (A) gives

$$\eta = 1 - \frac{1}{\gamma}\left[\frac{T_A(1/r_e)^{\gamma-1} - T_D(1/r)^{\gamma-1}}{T_A - T_D}\right] \qquad (D)$$

Also $P_A = P_D$, and from the ideal-gas equation,

$$P_D V_D = R T_D \qquad \text{and} \qquad P_A V_A = R T_A$$

Moreover, $V_C = V_B$, and we may therefore write

$$\frac{T_D}{T_A} = \frac{V_D}{V_A} = \frac{V_D/V_C}{V_A/V_B} = \frac{r_e}{r}$$

This relation combines with Eq. (D) to give

$$\eta = 1 - \frac{1}{\gamma}\left[\frac{(1/r_e)^{\gamma-1} - (r_e/r)(1/r)^{\gamma-1}}{1 - r_e/r}\right]$$

or

$$\eta = 1 - \frac{1}{\gamma}\left[\frac{(1/r_e)^{\gamma} - (1/r)^{\gamma}}{1/r_e - 1/r}\right] \qquad (8.7)$$

8.5 The Gas-Turbine Power Plant

The Otto and Diesel engines exemplify direct use of the energy of high-temperature, high-pressure gases acting on a piston within a cylinder; no heat transfer with an external source is required. However, turbines are more efficient than reciprocating engines, and the advantages of internal combustion are combined with those of the turbine in the gas-turbine engine.

The gas turbine is driven by high-temperature gases from a combustion space, as indicated in Fig. 8.11. The entering air is compressed (supercharged) to a pressure of several bars before combustion. The centrifugal compressor operates on the

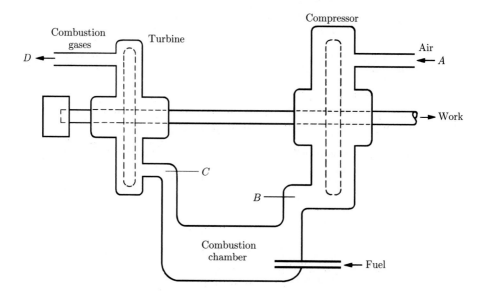

Figure 8.11: Gas-turbine power plant.

same shaft as the turbine, and part of the work of the turbine serves to drive the compressor. The unit shown in Fig. 8.11 is a complete power plant, as are Otto and Diesel engines. The gas turbine is just one part and performs the same function as the steam turbine in a steam power plant (Fig. 8.1).

The higher the temperature of the combustion gases entering the turbine, the higher the efficiency of the unit, i.e., the greater the work produced per unit of fuel burned. The limiting temperature is determined by the strength of the metal turbine blades, and is much lower than the theoretical flame temperature (Sec. 4.8) of the fuel. Sufficient excess air must be supplied to keep the combustion temperature at a safe level.

The idealization of the gas-turbine cycle (based on air, and called the Brayton cycle) is shown on a PV diagram in Fig. 8.12. The compression step AB is represented by an adiabatic, reversible (isentropic) path in which the pressure increases from P_A (atmospheric pressure) to P_B. The combustion process is replaced by the constant-pressure addition of an amount of heat Q_{BC}. Work is produced in the turbine as the result of isentropic expansion of the air to pressure P_D. Since the hot gases from the turbine are exhausted to the atmosphere, $P_D = P_A$. The thermal efficiency of the cycle is given by

$$\eta = \frac{-W_s(\text{net})}{Q_{BC}} = \frac{-W_{CD} - W_{AB}}{Q_{BC}} \tag{8.8}$$

where each energy quantity is based on 1 mol of air.

The work done as the air passes through the compressor is given by Eq. (7.33), and for air as an ideal gas with constant heat capacities,

$$W_{AB} = H_B - H_A = C_P(T_B - T_A)$$

Similarly, for the combustion and turbine processes,

$$Q_{BC} = C_P(T_C - T_B)$$

$$W_{CD} = C_P(T_D - T_C)$$

Substituting these equations into Eq. (8.8) and simplifying leads to

$$\eta = 1 - \frac{T_D - T_A}{T_C - T_B} \tag{8.9}$$

Since processes AB and CD are isentropic, the temperatures and pressures are related as follows [Eq. (3.23)]:

$$\frac{T_B}{T_A} = \left(\frac{P_B}{P_A}\right)^{(\gamma-1)/\gamma} \tag{8.10}$$

and

$$\frac{T_D}{T_C} = \left(\frac{P_D}{P_C}\right)^{(\gamma-1)/\gamma} = \left(\frac{P_A}{P_B}\right)^{(\gamma-1)/\gamma} \tag{8.11}$$

With these equations T_A and T_D may be eliminated to give

$$\eta = 1 - \left(\frac{P_A}{P_B}\right)^{(\gamma-1)/\gamma} \tag{8.12}$$

Example 8.4 A gas-turbine power plant with a compression ratio $P_A/P_B = 6$ operates with air entering the compressor at 25°C. If the maximum permissible temperature in the turbine is 760°C, determine:

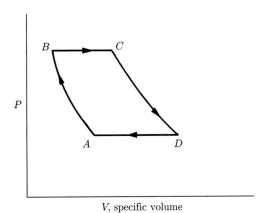

Figure 8.12: Ideal cycle for gas-turbine power plant.

(a) The efficiency η of the reversible ideal-gas cycle for these conditions if $\gamma = 1.4$.

(b) The thermal efficiency of the power plant for the given conditions if the compressor and turbine operate adiabatically but irreversibly with efficiencies $\eta_c = 0.83$ and $\eta_t = 0.86$.

SOLUTION   (a) Direct substitution in Eq. (8.12) gives the ideal-cycle efficiency:

$$\eta = 1 - (1/6)^{(1.4-1)/1.4} = 1 - 0.60 = 0.40$$

(b) Irreversibilities in the compressor and turbine greatly reduce the thermal efficiency of the power plant, because the net work is the difference between the work required by the compressor and the work produced by the turbine. The temperature of the air entering the compressor T_A and the temperature of the air entering the turbine, the specified maximum for T_C, are the same as for the ideal cycle. However, the temperature after irreversible compression in the compressor T_B is higher than the temperature after *isentropic* compression T_B', and the temperature after irreversible expansion in the turbine T_D is higher than the temperature after *isentropic* expansion T_D'.

The work required by the compressor is

$$W(\text{comp}) = C_P(T_B - T_A) \qquad (A)$$

Alternatively, this may be found from the isentropic work:

$$W(\text{comp}) = \frac{C_P(T_B' - T_A)}{\eta_c} \qquad (B)$$

Similarly, the work produced by the turbine is

$$W(\text{turb}) = C_P(T_D - T_C) = C_P \eta_t (T_D' - T_C) \qquad (C)$$

and the heat absorbed in place of combustion is

$$Q = C_P(T_C - T_B) \qquad (D)$$

These equations are combined to give the thermal efficiency of the power plant:

$$\eta = \frac{-W(\text{comp}) - W(\text{turb})}{Q} = \frac{-[(T_B' - T_A)/\eta_c] + \eta_t(T_C - T_D')}{T_C - T_B}$$

Combining Eqs. (A) and (B) and using the result to eliminate T_B from this equation gives after simplification

$$\eta = \frac{-(T_B'/T_A - 1) + \eta_t \eta_c (T_C/T_A - T_D'/T_A)}{\eta_c(T_C/T_A - 1) - (T_B'/T_A - 1)} \qquad (E)$$

The temperature ratio T_B'/T_A is related to the pressure ratio by Eq. (8.10). The ratio T_C/T_A depends on given conditions. In view of Eq. (8.11), the ratio T_D'/T_A can be written

$$\frac{T_D'}{T_A} = \frac{T_C T_D'}{T_A T_C} = \frac{T_C}{T_A}\left(\frac{P_A}{P_B}\right)^{(\gamma-1)/\gamma}$$

Substituting these expressions in Eq. (E) gives

$$\eta = \frac{\eta_t \eta_c (T_C/T_A)(1 - 1/\alpha) - (\alpha - 1)}{\eta_c(T_C/T_A - 1) - (\alpha - 1)} \tag{8.13}$$

where

$$\alpha = \left(\frac{P_B}{P_A}\right)^{(\gamma-1)/\gamma}$$

It can be shown from Eq. (8.13) that the thermal efficiency of the gas-turbine power plant increases as the temperature of the air entering the turbine (T_C) increases, and also as the compressor and turbine efficiencies η_c and η_t increase.

The given efficiency values are here

$$\eta_t = 0.86 \qquad \text{and} \qquad \eta_c = 0.83$$

Other given data provide

$$\frac{T_C}{T_A} = \frac{760 + 273.15}{25 + 273.15} = 3.47$$

and

$$\alpha = (6)^{(1.4-1)/1.4} = 1.67$$

Substituting these quantities in Eq. (8.13) gives

$$\eta = \frac{(0.86)(0.83)(3.47)(1 - 1/1.67) - (1.67 - 1)}{(0.83)(3.47 - 1) - (1.67 - 1)} = 0.235$$

This analysis shows that, even with a compressor and turbine of rather high efficiencies, the thermal efficiency (23.5 percent) is considerably reduced from the ideal-cycle value of 40 percent.

8.6 Jet Engines; Rocket Engines

In the power cycles so far considered the high-temperature, high-pressure gas expands in a turbine (steam power plant, gas turbine) or in the cylinders of an Otto or Diesel engine with reciprocating pistons. In either case, the power becomes available through a rotating shaft. Another device for expanding the hot gases is a nozzle. Here the power is available as kinetic energy in the jet of exhaust gases leaving the nozzle. The entire power plant, consisting of a compression device and a combustion chamber, as well as a nozzle, is known as a jet engine. Since the kinetic energy of the exhaust gases is directly available for propelling the engine and its attachments, jet engines are most commonly used to power aircraft. There are several types of jet-propulsion engines based on different ways of accomplishing the compression and expansion processes. Since the air striking the engine has kinetic energy (with respect to the engine), its pressure may be increased in a diffuser.

The turbojet engine illustrated in Fig. 8.13 takes advantage of a diffuser to reduce the work of compression. The axial-flow compressor completes the job of compression, and then the fuel is injected and burned in the combustion chamber.

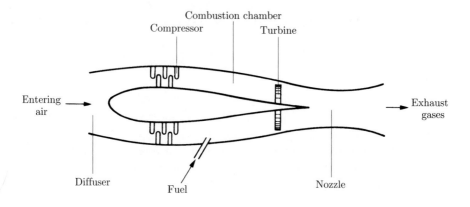

Figure 8.13: The turbojet power plant.

The hot combustion-product gases first pass through a turbine where the expansion provides just enough power to drive the compressor. The remainder of the expansion to the exhaust pressure is accomplished in the nozzle. Here, the velocity of the gases with respect to the engine is increased to a level above that of the entering air. This increase in velocity provides a thrust (force) on the engine in the forward direction. If the compression and expansion processes are adiabatic and reversible, the turbojet-engine cycle is identical to the ideal gas-turbine-power-plant cycle shown in Fig. 8.11. The only differences are that, physically, the compression and expansion steps are carried out in devices of different types.

A rocket engine differs from a jet engine in that the oxidizing agent is carried with the engine. Instead of depending on the surrounding air for burning the fuel, the rocket is self-contained. This means that the rocket can operate in a vacuum such as in outer space. In fact, the performance is better in a vacuum, because none of the thrust is required to overcome friction forces.

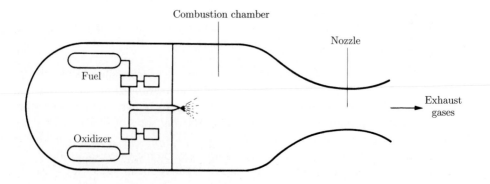

Figure 8.14: Liquid-fuel rocket engine.

In rockets burning liquid fuels the oxidizing agent (e.g., liquid oxygen) is pumped from tanks into the combustion chamber. Simultaneously, fuel (e.g., kerosene) is pumped into the chamber and burned. The combustion takes place at a constant high pressure and produces high-temperature product gases that are expanded in a nozzle, as indicated in Fig. 8.14.

In rockets burning solid fuels the fuel (organic polymers) and oxidizer (e.g., ammonium perchlorate) are contained together in a solid matrix and stored at the forward end of the combustion chamber.

In an ideal rocket, the combustion and expansion steps are the same as those for an ideal jet engine (Fig. 8.12). A solid-fuel rocket requires no compression work, and in a liquid-fuel rocket the compression energy is small, since the fuel and oxidizer are pumped as liquids.

PROBLEMS

8.1. The basic cycle for a steam power plant is shown by Fig. 8.1. Suppose that the turbine operates adiabatically with inlet steam at 6,800 kPa and 550°C and that the exhaust steam enters the condenser at 50°C with a quality of 0.96. Saturated liquid water leaves the condenser, and is pumped to the boiler. Neglecting pump work and kinetic- and potential-energy changes, determine the thermal efficiency of the cycle and the turbine efficiency.

8.2. A steam power plant operates on the cycle of Fig. 8.4. For one of the following sets of operating conditions, determine the steam rate, the heat-transfer rates in the boiler and condenser, and the thermal efficiency of the plant.

(*a*) $P_1 = P_2 = 10{,}000$ kPa; $T_2 = 600°C$; $P_3 = P_4 = 10$ kPa; $\eta(\text{turbine}) = 0.80$; $\eta(\text{pump}) = 0.75$; power rating $= 80{,}000$ kW.

(*b*) $P_1 = P_2 = 7{,}000$ kPa; $T_2 = 550°C$; $P_3 = P_4 = 20$ kPa; $\eta(\text{turbine}) = 0.75$; $\eta(\text{pump}) = 0.75$; power rating $= 100{,}000$ kW.

(*c*) $P_1 = P_2 = 8{,}500$ kPa; $T_2 = 600°C$; $P_3 = P_4 = 10$ kPa; $\eta(\text{turbine}) = 0.80$; $\eta(\text{pump}) = 0.80$; power rating $= 70{,}000$ kW.

(*d*) $P_1 = P_2 = 6{,}500$ kPa; $T_2 = 525°C$; $P_3 = P_4 = 101.33$ kPa; $\eta(\text{turbine}) = 0.78$; $\eta(\text{pump}) = 0.75$; power rating $= 50{,}000$ kW.

(*e*) $P_1 = P_2 = 950(\text{psia})$; $T_2 = 1{,}000(°F)$; $P_3 = P_4 = 14.7(\text{psia})$; $\eta(\text{turbine}) = 0.78$; $\eta(\text{pump}) = 0.75$; power rating $= 50{,}000$ kW.

(*f*) $P_1 = P_2 = 1{,}125(\text{psia})$; $T_2 = 1{,}100(°F)$; $P_3 = P_4 = 1(\text{psia})$; $\eta(\text{turbine}) = 0.80$; $\eta(\text{pump}) = 0.75$; power rating $= 80{,}000$ kW.

8.3. Steam enters the turbine of a power plant operating on the Rankine cycle (Fig. 8.3) at 3,300 kPa and exhausts at 50 kPa. To show the effect of superheating on the performance of the cycle, calculate the thermal efficiency of the cycle and the quality of the exhaust steam from the turbine for turbine-inlet steam temperatures of 450, 550, and 650°C.

8.4. Steam enters the turbine of a power plant operating on the Rankine cycle (Fig. 8.3) at 600°C and exhausts at 30 kPa. To show the effect of boiler pressure on the performance of the cycle, calculate the thermal efficiency of the cycle and the quality of the exhaust steam from the turbine for boiler pressures of of 5,000, 7,500, and 10,000 kPa.

8.5. A steam power plant employs two adiabatic turbines in series. Steam enters the first turbine at 650°C and 7,000 kPa and discharges from the second turbine at 20 kPa. The system is designed for equal power outputs from the two turbines, based on a turbine efficiency of 78% for *each* turbine. Determine the temperature and pressure of the steam in its intermediate state between the two turbines. What is the overall efficiency of the two turbines together with respect to isentropic expansion of the steam from the initial to the final state?

8.6. A steam power plant operating on a regenerative cycle, as illustrated in Fig. 8.5, includes just one feedwater heater. Steam enters the turbine at 4,500 kPa and 500°C and exhausts at 20 kPa. Steam for the feedwater heater is extracted from the turbine at 350 kPa, and in condensing raises the temperature of the feedwater to within 6°C of its condensation temperature at 350 kPa. If the turbine and pump efficiencies are both 0.78, what is the thermal efficiency of the cycle and what fraction of the steam entering the turbine is extracted for the feedwater heater?

8.7. A steam power plant operating on a regenerative cycle, as illustrated in Fig. 8.5, includes just one feedwater heater. Steam enters the turbine at 650(psia) and 900(°F) and exhausts at 1(psia). Steam for the feedwater heater is extracted from the turbine at 50(psia), and in condensing raises the temperature of the feedwater to within 11(°F) of its condensation temperature at 50(psia). If the turbine and pump efficiencies are both 0.78, what is the thermal efficiency of the cycle and what fraction of the steam entering the turbine is extracted for the feedwater heater?

8.8. A steam power plant operating on a regenerative cycle, as illustrated in Fig. 8.5, includes two feedwater heaters. Steam enters the turbine at 6,500 kPa and 600°C and exhausts at 20 kPa. Steam for the feedwater heaters is extracted from the turbine at pressures such that the feed water is heated to 190°C in two equal increments of temperature rise, with 5-°C approaches to the steam-condensation temperature in each feedwater heater. If the turbine and pump efficiencies are both 0.80, what is the thermal efficiency of the cycle and what fraction of the steam entering the turbine is extracted for each feedwater heater?

8.9. A power plant operating on heat recovered from the exhaust gases of internal combustion engines uses isobutane as the working medium in a modified Rankine cycle in which the upper pressure level is above the critical pressure of isobutane. Thus the isobutane does not undergo a change of phase as it absorbs heat prior to its entry into the turbine. Isobutane vapor is heated at 4,800 kPa to 260°C, and enters the turbine as a supercritical fluid at these conditions. Isentropic expansion in the turbine produces a superheated vapor at 450 kPa, which is cooled and condensed at constant pressure. The resulting saturated liquid enters the pump for return to the heater. If the power output of the modified Rankine cycle is 1,000 kW, what is the isobutane flow rate, the heat-transfer rates in the heater and condenser, and the thermal efficiency of the cycle?

The vapor pressure of isobutane is given by

$$\ln P^{\text{sat}}/\text{kPa} = 14.57100 - \frac{2{,}606.775}{t/^{\circ}\text{C} + 274.068}$$

8.10. A power plant operating on heat from a geothermal source uses isobutane as the working medium in a Rankine cycle (Fig. 8.3). Isobutane is heated at 3,400 kPa (a pressure just a little below its critical pressure) to a temperature of 140°C, at which conditions it enters the turbine. Isentropic expansion in the turbine produces superheated vapor at 450 kPa, which is cooled and condensed to saturated liquid and pumped to the heater/boiler. If the flow rate of isobutane is 75 kg s^{-1}, what is the power output of the Rankine cycle and what are the heat-transfer rates in the heater/boiler and cooler/condenser? What is the thermal efficiency of the cycle?

Repeat these calculations for a cycle in which the turbine and pump each have an efficiency of 80%.

The vapor pressure of isobutane is given in the preceding problem.

8.11. For comparison of Diesel- and Otto-engine cycles:

(a) Show that the thermal efficiency of the air-standard Diesel cycle can be expressed as

$$\eta = 1 - \left(\frac{1}{r}\right)^{\gamma-1} \frac{r_c^{\gamma} - 1}{\gamma(r_c - 1)}$$

where r is the compression ratio and r_c is the *cutoff ratio*, defined as $r_c = V_A/V_D$. (See Fig. 8.10.)

(b) Show that for the same compression ratio the thermal efficiency of the air-standard Otto engine is greater than the thermal efficiency of the air-standard Diesel cycle. (*Hint:* Show that the fraction which multiplies $(1/r)^{\gamma-1}$ in the above equation for η is greater than unity by expanding r_c^{γ} in a Taylor series with remainder taken to the first derivative.)

(c) If $\gamma = 1.4$, how does the thermal efficiency of an air-standard Otto cycle with a compression ratio of 8 compare with the thermal efficiency of an air-standard Diesel cycle with the same compression ratio and a cutoff ratio of 2? How is the comparison changed if the cutoff ratio is 3?

8.12. An air-standard Diesel cycle absorbs 1,500 J mol^{-1} of heat (step DA of Fig. 8.10, which simulates combustion). The pressure and temperature at the beginning of the compression step are 1 bar and 20°C, and the pressure at the end of the compression step is 4 bar. Assuming air to be an ideal gas for which $C_P = (7/2)R$ and $C_V = (5/2)R$, what are the compression ratio and the expansion ratio of the cycle?

8.13. Calculate the efficiency for an air-standard gas-turbine cycle (the Brayton cycle) operating with a pressure ratio of 3. Repeat for pressure ratios of 5, 7, and 9. Take $\gamma = 1.35$.

8.14. An air-standard gas-turbine cycle is modified by installation of a regenerative heat exchanger to transfer energy from the air leaving the turbine to the air leaving the compressor. In an optimum counter-current exchanger, the temperature of the air leaving the compressor is raised to that of point D in Fig. 8.12, and the temperature of the gas leaving the turbine is cooled to that of point B in Fig. 8.12. Show that the thermal efficiency of this cycle is given by

$$\eta = 1 - \frac{T_A}{T_C} \left(\frac{P_B}{P_A} \right)^{(\gamma-1)/\gamma}$$

8.15. Consider an air-standard cycle for the turbojet power plant shown in Fig. 8.13. The temperature and pressure of the air entering the compressor are 1 bar and 30°C. The pressure ratio in the compressor is 6.5, and the temperature at the turbine inlet is 1,100°C. If expansion in the nozzle is isentropic and if the nozzle exhausts at 1 bar, what is the pressure at the nozzle inlet (turbine exhaust) and what is the velocity of the air leaving the nozzle?

CHAPTER 9

REFRIGERATION AND LIQUEFACTION

Refrigeration is best known for its use in the air conditioning of buildings and in the treatment, transportation, and preservation of foods and beverages. It also finds large-scale industrial use, for example, in the manufacture of ice and the dehydration of gases. Applications in the petroleum industry include lubricating-oil purification, low-temperature reactions, and separation of volatile hydrocarbons. A closely related process is gas liquefaction, which has important commercial applications.

The purpose of this chapter is to present a thermodynamic analysis of refrigeration and liquefaction processes. However, the details of equipment design are left to specialized books.[1]

The word *refrigeration* implies the maintenance of a temperature below that of the surroundings. This requires continuous absorption of heat at a low temperature level, usually accomplished by evaporation of a liquid in a steady-state flow process. The vapor formed may be returned to its original liquid state for reevaporation in either of two ways. Most commonly, it is simply compressed and then condensed. Alternatively, it may be absorbed by a liquid of low volatility, from which it is subsequently evaporated at higher pressure. Before treating these practical refrigeration cycles, we consider the Carnot refrigerator, which provides a standard of comparison.

---

[1] *ASHRAE Handbook: Refrigeration*, 1994; *Fundamentals*, 1993; *HVAC Systems and Equipment*, 1992; *HVAC Applications*, 1991; American Society of Heating, Refrigerating, and Air-Conditioning Engineers, Inc., Atlanta; Shan K. Wang, *Handbook of Air Conditioning and Refrigeration*, McGraw-Hill, New York, 1993.

9.1   The Carnot Refrigerator

In a continuous refrigeration process, the heat absorbed at a low temperature is continuously rejected to the surroundings at a higher temperature. Basically, a refrigeration cycle is a reversed heat-engine cycle. Heat is transferred from a low temperature level to a higher one; according to the second law, this requires an external source of energy. The ideal refrigerator, like the ideal heat engine (Sec. 5.2), operates on a Carnot cycle, consisting in this case of two isothermal steps in which heat $|Q_C|$ is absorbed at the lower temperature T_C and heat $|Q_H|$ is rejected at the higher temperature T_H, and two adiabatic steps. The cycle requires the addition of net work W to the system. Since ΔU of the working fluid is zero for the cycle, the first law gives

$$W = |Q_H| - |Q_C| \tag{9.1}$$

The usual measure of the effectiveness of a refrigerator is called its *coefficient of performance*, defined as

$$\omega = \frac{\text{heat absorbed at the lower temperature}}{\text{net work}}$$

Thus

$$\omega \equiv \frac{|Q_C|}{W} \tag{9.2}$$

Division of Eq. (9.1) by $|Q_C|$ gives

$$\frac{W}{|Q_C|} = \frac{|Q_H|}{|Q_C|} - 1$$

But according to Eq. (5.7),

$$\frac{|Q_H|}{|Q_C|} = \frac{T_H}{T_C}$$

whence

$$\frac{W}{|Q_C|} = \frac{T_H}{T_C} - 1 = \frac{T_H - T_C}{T_C}$$

and Eq. (9.2) becomes

$$\omega = \frac{T_C}{T_H - T_C} \tag{9.3}$$

This equation applies only to a refrigerator operating on a Carnot cycle, and it gives the maximum possible value of ω for any refrigerator operating between given values of T_H and T_C. It shows clearly that the refrigeration effect per unit of work decreases as the temperature T_C of heat absorption decreases and as the temperature T_H of heat rejection increases. For refrigeration at a temperature level of 5°C and a surroundings temperature of 30°C, the value of ω for a Carnot refrigerator is

$$\omega = \frac{5 + 273.15}{(30 + 273.15) - (5 + 273.15)} = 11.13$$

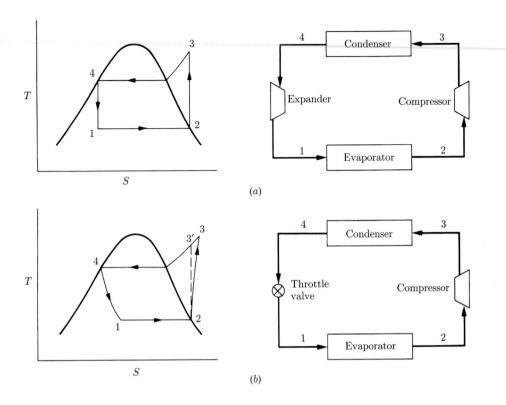

Figure 9.1: Vapor-compression refrigeration cycles.

9.2 The Vapor-Compression Cycle

A liquid evaporating at constant pressure provides a means for heat absorption at constant temperature. Likewise, condensation of the vapor, after compression to a higher pressure, provides for the rejection of heat at constant temperature. The liquid from the condenser is returned to its original state by an expansion process. This can be carried out in a turbine from which work is obtained. When compression and expansion are isentropic, this sequence of processes constitutes the cycle of Fig. 9.1a. It is equivalent to the Carnot cycle, except that superheated vapor from the compressor (point 3 in Fig. 9.1a) must be cooled to its saturation temperature before condensation begins.

On the basis of a unit mass of fluid, the heat absorbed in the evaporator is

$$|Q_C| = \Delta H = H_2 - H_1$$

This equation follows from Eq. (7.17) when the small changes in potential and kinetic energy are neglected. Likewise, the heat rejected in the condenser is

$$|Q_H| = H_3 - H_4$$

By Eq. (9.1),
$$W = (H_3 - H_4) - (H_2 - H_1)$$
and by Eq. (9.2), the coefficient of performance is

$$\omega = \frac{H_2 - H_1}{(H_3 - H_4) - (H_2 - H_1)} \qquad (9.4)$$

This process requires a turbine or expander that operates on a two-phase liquid/vapor mixture. Such a machine is impractical for small units. Therefore, the cycle of Fig. 9.1a is used only for large installations. More commonly, expansion is accomplished by throttling the liquid from the condenser through a partly opened valve. The pressure drop in this irreversible process results from fluid friction in the valve. In small units, such as household refrigerators and air conditioners, the simplicity and lower cost of the throttle valve outweigh the energy savings possible with a turbine. As shown in Sec. 7.2, the throttling process occurs at constant enthalpy.

The vapor-compression cycle that includes a valve through which expansion occurs is shown in Fig. 9.1b, where line 4 → 1 represents the constant-enthalpy throttling process. Line 2 → 3, representing an actual compression process, slopes in the direction of increasing entropy, reflecting the irreversibility inherent in the process. The dashed line 2 → 3′ is the path of isentropic compression (see Fig. 7.9). For this cycle, the coefficient of performance is simply

$$\omega = \frac{H_2 - H_1}{H_3 - H_2} \qquad (9.5)$$

Design of the evaporator, compressor, condenser, and auxiliary equipment requires knowledge of the rate of circulation of refrigerant $\dot{m}$. This is determined from the heat absorbed in the evaporator[2] by the equation

$$\dot{m} = \frac{|Q_C|}{H_2 - H_1} \qquad (9.6)$$

The vapor-compression cycle of Fig. 9.1b is shown on a PH diagram in Fig. 9.2. Such diagrams are more commonly used in the description of refrigeration processes than TS diagrams, because they show directly the required enthalpies. Although the evaporation and condensation processes are represented by constant-pressure paths, small pressure drops do occur because of fluid friction.

9.3 Comparison of Refrigeration Cycles

For given values of T_C and T_H, the highest possible value of ω is attained for Carnot-cycle refrigeration. The vapor-compression cycle with reversible compression and

---

[2]In the United States refrigeration equipment is commonly rated in *tons of refrigeration*; a ton of refrigeration is defined as heat absorption at the rate of 12,000(Btu) or 11,376 kJ per hour. This corresponds approximately to the rate of heat removal required to freeze 1(ton) of water, initially at 32(°F), per day.

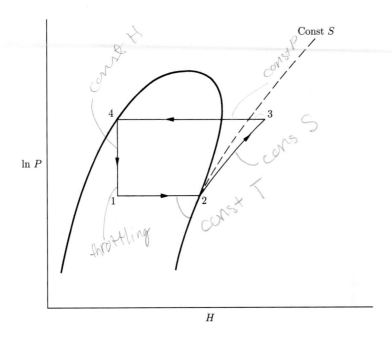

Figure 9.2: Vapor-compression refrigeration cycle on a PH diagram.

expansion approaches this upper limit. A vapor-compression cycle with expansion in a throttle valve has a somewhat lower value, and this is reduced further when compression is not isentropic. The following example provides an indication of typical values for coefficients of performance.

Example 9.1 A refrigerated space is maintained at $10(°F)$, and cooling water is available at $70(°F)$. The evaporator and condenser are of sufficient size that a $10(°F)$ minimum-temperature difference for heat transfer can be realized in each. The refrigeration capacity is $120{,}000(Btu)(hr)^{-1}$. The refrigerant is tetrafluoroethane(HFC-134a), for which data are given in Table 9.1 and Fig. 9.3.

(a) What is the value of ω for a Carnot refrigerator?

(b) Calculate ω and $\dot{m}$ for the vapor-compression cycle of Fig. 9.1a.

(c) Calculate ω and $\dot{m}$ for the vapor-compression cycle of Fig. 9.1b if the compressor efficiency is 0.80.

SOLUTION (a) By Eq. (9.3) for a Carnot refrigerator,

$$\omega = \frac{0 + 459.67}{(80 + 459.67) - (0 + 459.67)} = 5.75$$

(b) Since HFC-134a is the refrigerant, the enthalpies for states 1, 2, 3, and 4 of Fig. 9.1a are read from Table 9.1 and Fig. 9.3. From the entry in Table 9.1 at

Table 9.1: Thermodynamic properties of saturated tetrafluoroethane[†]

$t(°F)$	$P(psia)$	Volume $(ft)^3(lb_m)^{-1}$		Enthalpy $(Btu)(lb_m)^{-1}$		Entropy $(Btu)(lb_m)^{-1}(R)^{-1}$	
		V^l	V^v	H^l	H^v	S^l	S^v
−40	7.429	0.01132	5.782	0.000	97.050	0.00000	0.23125
−35	8.577	0.01139	5.053	1.489	97.804	0.00352	0.23032
−30	9.862	0.01145	4.432	2.984	98.556	0.00701	0.22945
−25	11.297	0.01152	3.901	4.484	99.306	0.01048	0.22863
−20	12.895	0.01158	3.445	,5.991	100.054	0.01392	0.22786
−15	14.667	0.01165	3.052	7.505	100.799	0.01733	0.22714
−10	16.626	0.01172	2.712	9.026	101.542	0.02073	0.22647
−5	18.787	0.01180	2.416	10.554	102.280	0.02409	0.22584
0	21.162	0.01187	2.159	12.090	103.015	0.02744	0.22525
5	23.767	0.01194	1.934	13.634	103.745	0.03077	0.22470
10	26.617	0.01202	1.736	15.187	104.471	0.03408	0.22418
15	29.726	0.01210	1.563	16.748	105.192	0.03737	0.22370
20	33.110	0.01218	1.410	18.318	105.907	0.04065	0.22325
25	36.785	0.01226	1.275	19.897	106.617	0.04391	0.22283
30	40.768	0.01235	1.155	21.486	107.320	0.04715	0.22244
35	45.075	0.01243	1.048	23.085	108.016	0.05018	0.22207
40	49.724	0.01252	0.953	24.694	108.705	0.05359	0.22172
45	54.732	0.01262	0.868	26.314	109.386	0.05679	0.22140
50	60.116	0.01271	0.792	27.944	110.058	0.05998	0.22110
55	65.895	0.01281	0.724	29.586	110.722	0.06316	0.22081
60	72.087	0.01291	0.663	31.239	111.376	0.06633	0.22054
65	78.712	0.01301	0.608	32.905	112.019	0.06949	0.22028
70	85.787	0.01312	0.558	34.583	112.652	0.07264	0.22003
75	93.333	0.01323	0.512	36.274	113.272	0.07578	0.21979
80	101.37	0.01335	0.472	37.978	113.880	0.07892	0.21957
85	109.92	0.01347	0.434	39.697	114.475	0.08205	0.21934
90	119.00	0.01359	0.400	41.430	115.055	0.08518	0.21912
95	128.63	0.01372	0.369	43.179	115.619	0.08830	0.21890
100	138.83	0.01386	0.341	44.943	116.166	0.09142	0.21868
105	149.63	0.01400	0.315	46.725	116.694	0.09454	0.21845
110	161.05	0.01415	0.292	48.524	117.203	0.09766	0.21822
115	173.11	0.01430	0.270	50.343	117.690	0.10078	0.21797
120	185.84	0.01447	0.250	52.181	118.153	0.10391	0.21772
125	199.25	0.01464	0.231	54.040	118.591	0.10704	0.21744
130	213.38	0.01482	0.214	55.923	119.000	0.11018	0.21715
135	228.25	0.01502	0.198	57.830	119.377	0.11333	0.21683
140	243.88	0.01522	0.184	59.764	119.720	0.11650	0.21648
150	277.57	0.01567	0.157	63.722	120.284	0.12288	0.21566
160	314.69	0.01620	0.134	67.823	120.650	0.12938	0.21463
170	355.51	0.01683	0.114	72.106	120.753	0.13603	0.21329

[†]Adapted by permission from *ASHRAE Handbook: Fundamentals*, p. 17.29, American Society of Heating, Refrigerating, and Air-Conditioning Engineers, Inc., Atlanta, 1993.

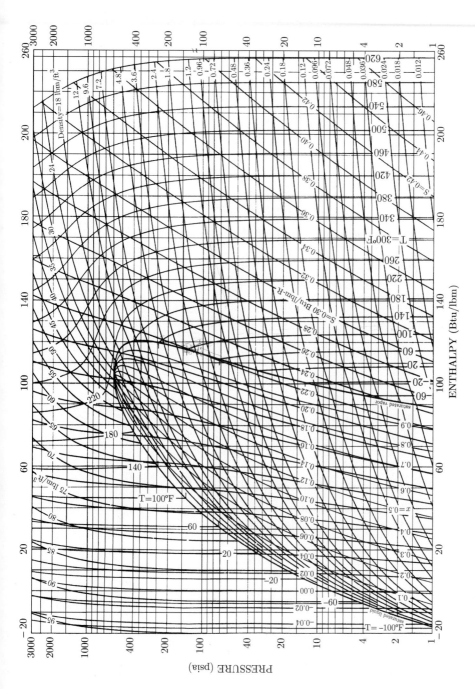

Figure 9.3: PH diagram for tetrafluoroethane(HFC-134a). (*Reproduced by permission. ASHRAE Handbook: Fundamentals*, p. 17.28, American Society of Heating, Refrigerating, and Air-Conditioning Engineers, Inc., Atlanta, 1993.)

$10 - 10 = 0(°F)$, we see that HFC-134a vaporizes in the evaporator at a pressure of $21.162(\text{psia})$. Its properties as a saturated vapor at these conditions are

$$H_2 = 103.015(\text{Btu})(\text{lb}_m)^{-1}$$
$$S_2 = 0.22525(\text{Btu})(\text{lb}_m)^{-1}(\text{R})^{-1}$$

From the entry at $70 + 10 = 80(°F)$ in Table 9.1, we find that HFC-134a condenses at $101.37(\text{psia})$; its properties as a saturated liquid at these conditions are

$$H_4 = 37.978(\text{Btu})(\text{lb}_m)^{-1}$$
$$S_4 = 0.07892(\text{Btu})(\text{lb}_m)^{-1}(\text{R})^{-1}$$

Since $S_3 = S_2 = 0.22525$, the enthalpy from Fig. 9.3 at this entropy and at a pressure of $101.37(\text{psia})$ is about

$$H_3 = 117(\text{Btu})(\text{lb}_m)^{-1}$$

State 1 is a two-phase mixture to which Eq. (6.57) applies. Written for the entropy, it is

$$S = (1 - x)S^l + xS^v$$

where x is the quality (mass fraction of the mixture that is vapor). Since $S_1 = S_4 = 0.07892$, this becomes

$$0.07892 = (1 - x)(0.02744) + x(0.22525)$$

Solution for x gives

$$x = 0.2602$$

Similarly,

$$H_1 = (1 - x)H^l + xH^v$$
$$= (0.7398)(12.090) + (0.2602)(103.015) = 35.75(\text{Btu})(\text{lb}_m)^{-1}$$

Evaluation of the coefficient of performance by Eq. (9.4) gives

$$\omega = \frac{103.015 - 35.75}{(117 - 37.978) - (103.015 - 35.75)} = 5.72$$

By Eq. (9.6), the HFC-134a circulaation rate is

$$\dot{m} = \frac{120,000}{103.015 - 35.75} = 1,784(\text{lb}_m)(\text{hr})^{-1}$$

(c) For the expansion step of the cycle shown in Fig. 9.1b,

$$H_1 = H_4 = 37.978(\text{Btu})(\text{lb}_m)^{-1}$$

For the compression step,

$$(\Delta H)_S = (H_3 - H_2)_S = 117 - 103.015 = 13.98(\text{Btu})(\text{lb}_m)^{-1}$$

By Eq. (7.36) for a compressor efficiency of 0.80,

$$\Delta H = H_3 - H_2 = \frac{(\Delta H)_S}{\eta} = \frac{13.98}{0.80} = 17.48(\text{Btu})(\text{lb}_m)^{-1}$$

The coefficient of performance is now found from Eq. (9.5):

$$\omega = \frac{H_2 - H_1}{H_3 - H_2} = \frac{103.015 - 37.978}{17.48} = 3.72$$

The HFC-134a circulation rate is

$$\dot{m} = \frac{120,000}{103.015 - 37.978} = 1,845(\text{lb}_\text{m})(\text{hr})^{-1}$$

Results are summarized as follows:

Cycle	ω	$\dot{m}(\text{lb}_\text{m})(\text{hr})^{-1}$
(a) Carnot	5.75	
(b) Fig. 9.1a	5.72	1,784
(c) Fig. 9.1b	3.72	1,845

9.4 The Choice of Refrigerant

As shown in Sec. 5.2, the efficiency of a Carnot heat engine is independent of the working medium of the engine. Similarly, the coefficient of performance of a Carnot refrigerator is independent of the refrigerant. However, the irreversibilities inherent in the vapor-compression cycle cause the coefficient of performance of practical refrigerators to depend to some extent on the refrigerant. Nevertheless, such characteristics as its toxicity, flammability, cost, corrosion properties, and vapor pressure in relation to temperature are of greater importance in the choice of refrigerant. So that air cannot leak into the refrigeration system, the vapor pressure of the refrigerant at the evaporator temperature should be greater than atmospheric pressure. On the other hand, the vapor pressure at the condenser temperature should not be unduly high, because of the initial cost and operating expense of high-pressure equipment. These two requirements limit the choice of refrigerant to relatively few fluids. The final selection then depends on the other characteristics mentioned.

Ammonia, methyl chloride, carbon dioxide, propane and other hydrocarbons can serve as refrigerants. Halogenated hydrocarbons came into common use as refrigerants in the 1930's. Most common were the fully halogenated chlorofluorocarbons, CCl_3F (trichlorofluoromethane or CFC-11)[3] and CCl_2F_2 (dichlorodifluoromethane or CFC-12). These are very stable molecules that persist in the atmosphere for hundreds of years, causing severe ozone depletion. Their production has now mostly ended. Replacements are certain hydrochlorofluorocarbons, less than fully halogenated hydrocarbons which cause relatively little ozone depletion, and hydrofluorocarbons, which contain no chlorine and cause no ozone depletion. Examples are $CHCl_2CF_3$ (dichlorotrifluoroethane or HCFC-123), CF_3CH_2F

---

[3]The abbreviated designation is nomenclature of the American Society of Heating, Refrigerating, and Air-Conditioning Engineers.

(tetrafluoroethane or HFC-134a), and CHF_2CF_3 (pentafluoroethane or HFC-125). A pressure/enthalpy diagram for tetrafluoroethane (HFC-134a) is shown in Fig. 9.3; Table 9.1 provides saturation data for the same refrigerant. Tables and diagrams for a variety of other refrigerants are readily available.[4]

Limits placed on the operating pressures of the evaporator and condenser of a refrigeration system also limit the temperature difference $T_H - T_C$ over which a simple vapor-compression cycle can operate. With T_H fixed by the temperature of the surroundings, a lower limit is placed on the temperature level of refrigeration. This can be overcome by the operation of two or more refrigeration cycles employing different refrigerants in a *cascade*. A two-stage cascade is shown in Fig. 9.4.

---

[4] *ASHRAE Handbook: Fundamentals*, Chapt. 17, 1989; R. H. Perry and D. Green, *Perry's Chemical Engineers' Handbook*, 7th ed., sec. 2, 1996. Extensive data for ammonia are given by L. Haar and J. S. Gallagher, *J. Phys. Chem. Ref. Data*, vol. 7, pp. 635–792, 1978.

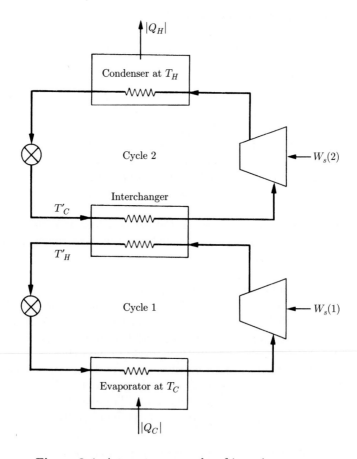

Figure 9.4: A two-stage cascade refrigeration system.

Here, the two cycles operate so that the heat absorbed in the interchanger by the refrigerant of the higher-temperature cycle (cycle 2) serves to condense the refrigerant in the lower-temperature cycle (cycle 1). The two refrigerants are so chosen that at the required temperature levels each cycle operates at reasonable pressures. For example, assume the following operating temperatures (Fig. 9.4):

$$T_H = 86(°F)$$
$$T_C' = 0(°F)$$
$$T_H' = 10(°F)$$
$$T_C = -50(°F)$$

If tetrafluoroethane(HFC-134a) is the refrigerant in cycle 2, then the intake and discharge pressures for the compressor are about 21(psia) and 112(psia), and the pressure ratio is about 5.3. If propylene is the refrigerent in cycle 1, these pressures are about 16 and 58(psia), and the pressure ratio is about 3.6. These are all reasonable values. On the other hand, for a single cycle operating between -50 and $86(°F)$ with HFC-134a as refrigerant, the intake pressure to the condenser is about 5.6(psia), well below atmospheric pressure. Moreover, for a discharge pressure of about 112(psia) the pressure ratio is 20, too high a value for a single-stage compressor.

9.5 Absorption Refrigeration

In vapor-compression refrigeration the work of compression is usually supplied by an electric motor. But the source of the electric energy for the motor is probably a heat engine (central power plant) used to drive a generator. Thus the work for refrigeration comes ultimately from heat at a high temperature level. This suggests the direct use of heat as the energy source for refrigeration. The absorption-refrigeration machine is based on this idea.

The work required by a Carnot refrigerator absorbing heat at temperature T_C and rejecting heat at the temperature of the surroundings, here designated T_S, follows from Eqs. (9.2) and (9.3):

$$W = \frac{T_S - T_C}{T_C}|Q_C|$$

where $|Q_C|$ is the heat absorbed. If a source of heat is available at a temperature above that of the surroundings, say at T_H, then work can be obtained from a Carnot engine operating between this temperature and the surroundings temperature T_S. The heat required $|Q_H|$ for the production of work $|W|$ is found from Eq. (5.8):

$$\eta = \frac{|W|}{|Q_H|} = 1 - \frac{T_S}{T_H}$$

whence

$$|Q_H| = |W|\frac{T_H}{T_H - T_S}$$

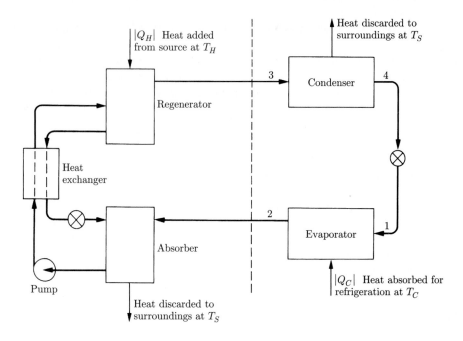

Figure 9.5: Schematic diagram of an absorption-refrigeration unit.

Substitution for $|W|$ gives

$$|Q_H| = |Q_C| \frac{T_H}{T_H - T_S} \frac{T_S - T_C}{T_C} \tag{9.7}$$

The value of $|Q_H|/|Q_C|$ given by this equation is of course a minimum, because Carnot cycles cannot be achieved in practice.

A schematic diagram for a typical absorption refrigerator is shown in Fig. 9.5. The essential difference between a vapor-compression and an absorption refrigerator is in the different means employed for compression. The section of the absorption unit to the right of the dashed line in Fig. 9.5 is the same as in a vapor-compression refrigerator, but the section to the left accomplishes compression by what amounts to a heat engine. Refrigerant as vapor from the evaporator is absorbed in a relatively nonvolatile liquid solvent at the pressure of the evaporator and at relatively low temperature. The heat given off in the process is discarded to the surroundings at T_S. This is the lower temperature level of the heat engine. The liquid solution from the absorber, which contains a relatively high concentration of refrigerant, passes to a pump, which raises the pressure of the liquid to that of the condenser. Heat from the higher temperature source at T_H is transferred to the compressed liquid solution, raising its temperature and evaporating the refrigerant from the solvent. Vapor passes from the regenerator to the condenser, and solvent, which now contains a relatively low concentration of refrigerant, returns to the absorber.

The heat exchanger conserves energy and also adjusts stream temperatures toward proper values. Low-pressure steam is the usual source of heat for the regenerator.

The most commonly used absorption-refrigeration system operates with water as the refrigerant and a lithium bromide solution as the absorbent. This system is obviously limited to refrigeration temperatures above the freezing point of water. It is treated in detail by Perry and Green.[5] For lower temperatures the usual system operates with ammonia as refrigerant and water as the solvent.

As an example, one might have refrigeration at a temperature level of $-10°C$ ($T_C = 263.15$ K) and a heat source of condensing steam at atmospheric pressure ($T_H = 373.15$ K). For a surroundings temperature of $30°C$ ($T_S = 303.15$ K), the minimum possible value of $|Q_H|/|Q_C|$ is found from Eq. (9.7):

$$\frac{|Q_H|}{|Q_C|} = \left(\frac{373.15}{373.15 - 303.15}\right)\left(\frac{303.15 - 263.15}{263.15}\right) = 0.81$$

For an actual absorption refrigerator, the value would be on the order of three times this result.

9.6 The Heat Pump

The heat pump, a reversed heat engine, is a device for heating houses and commercial buildings during the winter and cooling them during the summer. In the winter it operates so as to absorb heat from the surroundings and reject heat into the building. Refrigerant is evaporated in coils placed underground or in the outside air, and the vapor is compressed for condensation by air or water, used to heat the building, at temperatures above the required heating level. The operating cost of the installation is the cost of electric power to run the compressor. If the unit has a coefficient of performance, $|Q_C|/W = 4$, the heat available to heat the house $|Q_H|$ is equal to five times the energy input to the compressor. Any economic advantage of the heat pump as a heating device depends on the cost of electricity in comparison with the cost of fuels such as oil and natural gas.

The heat pump also serves for air conditioning during the summer. The flow of refrigerant is simply reversed, and heat is absorbed from the building and rejected through underground coils or to the outside air.

Example 9.2 A house has a winter heating requirement of 30 kJ s^{-1} and a summer cooling requirement of 60 kJ s^{-1}. Consider a heat-pump installation to maintain the house temperature at $20°C$ in winter and $25°C$ in summer. This requires circulation of the refrigerant through interior exchanger coils at $30°C$ in winter and $5°C$ in summer. Underground coils provide the heat source in winter and the heat sink in summer. For a year-round ground temperature of $15°C$, the heat-transfer characteristics of the coils necessitate refrigerant temperatures of $10°C$ in winter and $25°C$ in summer. What are the minimum power requirements for winter heating and summer cooling?

---

[5]R. H. Perry and D. Green, *op. cit.*, sec. 11.

SOLUTION The minimum power requirements are provided by a Carnot heat pump. For winter heating, the house coils are at the higher-temperature level T_H, and we know that $|Q_H| = 30$ kJ s^{-1}. Application of Eq. (5.7) gives

$$|Q_C| = |Q_H| \frac{T_C}{T_H} = 30 \left(\frac{10 + 273.15}{30 + 273.15} \right) = 28.02 \text{ kJ s}^{-1}$$

This is the heat absorbed in the ground coils. By Eq. (9.1) we now have

$$W = |Q_H| - |Q_C| = 30 - 28.02 = 1.98 \text{ kJ s}^{-1}$$

Thus the power requirement is 1.98 kW.

For summer cooling, $|Q_C| = 60$ kJ s^{-1}, and the house coils are at the lower-temperature level T_C. Combining Eqs. (9.2) and (9.3) and solving for W, we get

$$W = |Q_C| \frac{T_H - T_C}{T_C}$$

whence

$$W = 60 \left(\frac{25 - 5}{5 + 273.15} \right) = 4.31 \text{ kJ s}^{-1}$$

The power requirement here is therefore 4.31 kW.

9.7 Liquefaction Processes

Liquefied gases are in common use for a variety of purposes. For example, liquid propane in cylinders serves as a domestic fuel, liquid oxygen is carried in rockets, natural gas is liquefied for ocean transport, and liquid nitrogen is used for low-temperature refrigeration. In addition, gas mixtures (e.g., air) are liquefied for separation into their component species by fractionation.

Liquefaction results when a gas is cooled to a temperature in the two-phase region. This may be accomplished in several ways:

1. By heat exchange at constant pressure.

2. By expansion in a turbine from which work is obtained.

3. By a throttling process.

The first method requires a heat sink at a temperature lower than that to which the gas is cooled, and is most commonly used to precool a gas prior to its liquefaction by the other two methods. An external refrigerator is required for a gas temperature below that of the surroundings.

The three methods are illustrated in Fig. 9.6. The constant-pressure process (1) approaches the two-phase region (and liquefaction) most closely for a given drop in temperature. The throttling process (3) does not result in liquefaction unless the initial state is at a high enough pressure and low enough temperature for the constant-enthalpy process to cut into the two-phase region. This does not occur when the initial state is at A. If the initial state is at A', where the temperature is the same but the pressure is higher than at A, then isenthalpic expansion by

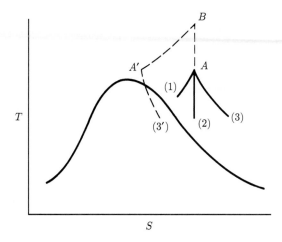

Figure 9.6: Cooling processes on a TS diagram.

process (3′) does result in the formation of liquid. The change of state from A to A' is most easily accomplished by compression of the gas to the final pressure at B, followed by constant-pressure cooling to A'. Liquefaction by isentropic expansion along process (2) may be accomplished from lower pressures (for given temperature) than by throttling. For example, continuation of process (2) from initial state A ultimately results in liquefaction.

The throttling process (3) is the one commonly employed in small-scale commercial liquefaction plants. The temperature of the gas must of course decrease during expansion. This is indeed what happens with most gases at usual conditions of temperature and pressure. The exceptions are hydrogen and helium, which increase in temperature upon throttling unless the initial temperature is below about 100 K for hydrogen and 20 K for helium. Liquefaction of these gases by throttling requires initial reduction of the temperature to lower values by method 1 or 2.

As already mentioned, the temperature must be low enough and the pressure high enough prior to throttling that the constant-enthalpy path cuts into the two-phase region. For example, reference to a TS diagram for air[6] shows that at a pressure of 100(atm) the temperature must be less than 305(R) for any liquefaction to occur along a path of constant enthalpy. In other words, if air is compressed to 100(atm) and cooled to below 305(R), it can be partly liquefied by throttling. The most economical way to cool the air is by countercurrent heat exchange with the unliquefied portion of the air from the expansion process.

This simplest kind of liquefaction system, known as the Linde process, is shown in Fig. 9.7. After compression, the gas is precooled to ambient temperature.

---

[6]R. H. Perry and D. Green, *op. cit.*, sec. 2.

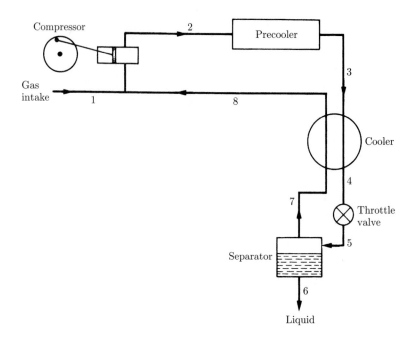

Figure 9.7: Linde liquefaction process.

It may even be further cooled by refrigeration. The lower the temperature of the gas entering the throttle valve, the greater the fraction of gas that is liquefied. For example, evaporating a refrigerant in the precooler at $-40(°F)$ gives a lower temperature into the valve than if water at $70(°F)$ is the cooling medium.

Under steady-state conditions, an energy balance [Eq. (7.15)] around the separator, valve, and cooler gives $\Delta(\dot{m}H)_{\text{fs}} = 0$, or

$$H_6 z + H_8(1 - z) = H_3 \tag{9.8}$$

where the enthalpies are for a unit mass of fluid at the positions indicated in Fig. 9.7. Knowledge of the enthalpies allows solution of Eq. (9.8) for z, the fraction of the gas that is liquefied.

The flow diagram for the Claude process, shown by Fig. 9.8, is the same as for the Linde process, except that an expansion engine or turbine replaces the throttle valve. The energy balance here becomes

$$H_6 z + H_8(1 - z) - W_s = H_3 \tag{9.9}$$

where W_s is the work of the expansion engine on the basis of a unit mass of fluid entering the cooler at point 3. If the engine operates adiabatically, the work is given by Eq. (7.33), which here becomes

$$W_s = (H_5 - H_4) \tag{9.10}$$

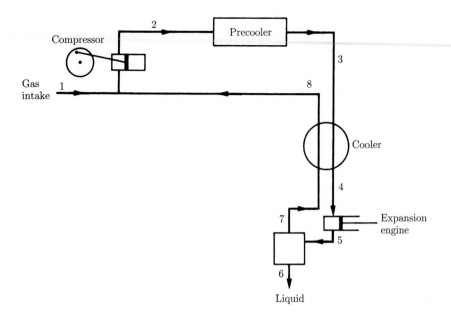

Figure 9.8: Claude liquefaction process.

Equations (9.8) through (9.10) suppose that no heat leaks into the apparatus from the surroundings. This can never be exactly true, and heat leakage may be significant when temperatures are very low, even with well-insulated equipment.

Example 9.3 Natural gas, assumed here to be pure methane, is liquefied in a simple Linde process (Fig. 9.7). Compression is to 60 bar and precooling is to 300 K. The separator is maintained at a pressure of 1 bar, and unliquefied gas at this pressure leaves the cooler at 295 K. What fraction of the gas is liquefied in the process, and what is the temperature of the high-pressure gas entering the throttle valve?

SOLUTION   Data for methane are given by Perry and Green.[7] From the table of properties for superheated methane,

$$H_3 = 1140.0 \text{ kJ kg}^{-1} \qquad \text{(at 300 K and 60 bar)}$$

$$H_8 = 1188.9 \text{ kJ kg}^{-1} \qquad \text{(at 295 K and 1 bar)}$$

By interpolation in the table of properties for saturated liquid and vapor, we find for a pressure of 1 bar that

$$T^{\text{sat}} = 111.45 \text{ K}$$
$$H_6 = 285.4 \text{ kJ kg}^{-1} \qquad \text{(saturated liquid)}$$
$$H_7 = 796.9 \text{ kJ kg}^{-1} \qquad \text{(saturated vapor)}$$

[7]R. H. Perry and D. Green, *op. cit.*, sec. 2.

Solution of Eq. (9.8) for z gives

$$z = \frac{H_8 - H_3}{H_8 - H_6} = \frac{1,188.9 - 1,140.0}{1,188.9 - 285.4} = 0.0541$$

Thus 5.41% of the gas entering the throttle valve emerges as liquid.

The temperature of the gas at point 4 is found from its enthalpy, which is calculated by an energy balance around the cooler:

$$(1)(H_4 - H_3) + (1 - z)(H_8 - H_7) = 0$$

Solution for H_4 and substitution of known values yields

$$H_4 = 1,140.0 - (0.9459)(1,188.9 - 796.9) = 769.2 \text{ kJ kg}^{-1}$$

Interpolation in the tables for superheated methane at 60 bar gives the temperature of the gas entering the throttle valve as 206.5 K.

PROBLEMS

9.1. A Carnot engine is coupled to a Carnot refrigerator so that all of the work produced by the engine is used by the refrigerator in extraction of heat from a heat reservoir at 0°C at the rate of 35 kJ s^{-1}. The source of energy for the Carnot engine is a heat reservoir at 250°C. If both devices discard heat to the surroundings at 25°C, how much heat does the engine absorb from its heat-source reservoir?

If the actual coefficient of performance of the refrigerator is $\omega = 0.6\,\omega_{\text{Carnot}}$ and if the thermal efficiency of the engine is $\eta = 0.6\,\eta_{\text{Carnot}}$, how much heat does the engine absorb from its heat-source reservoir?

9.2. A refrigeration system requires 1.5 kW of power for a refrigeration rate of 4 kJ s^{-1}.

(a) What is the coefficient of performance?

(b) How much heat is rejected in the condenser?

(c) If heat rejection is at 40°C, what is the lowest temperature the system can possibly maintain?

9.3. A conventional vapor-compression refrigeration system operates on the cycle of Fig. 9.1b. The refrigerant is tetrafluoroethane (Table 9.1, Fig. 9.3). For one of the following sets of operating conditions, determine the circulation rate of the refrigerant, the heat-transfer rate in the condenser, the power requirement, the coefficient of performance of the cycle, and the coefficient of performance of a Carnot refrigeration cycle operating between the same temperature levels.

(a) Evaporation $t = 30(°\text{F})$; condensation $t = 80(°\text{F})$; η(compressor) $= 0.79$; refrigeration rate $= 600(\text{Btu})(\text{s})^{-1}$.

(b) Evaporation $t = 20(°\text{F})$; condensation $t = 80(°\text{F})$; η(compressor) $= 0.78$; refrigeration rate $= 500(\text{Btu})(\text{s})^{-1}$.

(c) Evaporation $t = 10(°F)$; condensation $t = 80(°F)$; η(compressor) $= 0.77$; refrigeration rate $= 400(\text{Btu})(\text{s})^{-1}$.

(d) Evaporation $t = 0(°F)$; condensation $t = 80(°F)$; η(compressor) $= 0.76$; refrigeration rate $= 300(\text{Btu})(\text{s})^{-1}$.

(e) Evaporation $t = -10(°F)$; condensation $t = 80(°F)$; η(compressor) $= 0.75$; refrigeration rate $= 200(\text{Btu})(\text{s})^{-1}$.

9.4 A conventional vapor-compression refrigeration system operates on the cycle of Fig. 9.1*b*. The refrigerant is water. Given that the evaporation $t = 4°C$, the condensation $t = 34°C$, η(compressor) $= 0.76$, and the refrigeration rate $= 1,200 \text{ kJ s}^{-1}$, determine the circulation rate of the refrigerant, the heat-transfer rate in the condenser, the power requirement, the coefficient of performance of the cycle, and the coefficient of performance of a Carnot refrigeration cycle operating between the same temperature levels.

9.5. A refrigerator with tetrafluoroethane (Table 9.1, Fig. 9.3) as refrigerant operates with an evaporation temperature of $-15(°F)$ and a condensation temperature of $80(°F)$. Saturated liquid refrigerant from the condenser flows through an expansion valve into the evaporator, from which it emerges as saturated vapor.

(a) For a cooling rate of $5(\text{Btu})(\text{s})^{-1}$, what is the circulation rate of the refrigerant?

(b) By how much would the circulation rate be reduced if the throttle valve were replaced by a turbine in which the refrigerant expands isentropically?

(c) Suppose the cycle of (a) is modified by the inclusion of a countercurrent heat exchanger between the condenser and the throttle valve in which heat is transferred to vapor returning from the evaporator. If liquid from the condenser enters the exchanger at $80(°F)$ and if vapor from the evaporator enters the exchanger at $-15(°F)$ and leaves at $70(°F)$, what is the circulation rate of the refrigerant?

(d) For each of (a), (b), and (c), determine the coefficient of performance for isentropic compression of the vapor.

9.6. A vapor-compression refrigeration system is conventional except that a countercurrent heat exchanger is installed to subcool the liquid from the condenser by heat exchange with the vapor stream from the evaporator. The minimum temperature difference for heat transfer is $10(°F)$. Tetrafluoroethane is the refrigerant (Table 9.1, Fig. 9.3), evaporating at $20(°F)$ and condensing at $80(°F)$. The heat load on the evaporator is $2,000(\text{Btu})(\text{s})^{-1}$. If the compressor efficiency is 75%, what is the power requirement?

How does this result compare with the power required by the compressor if the system operates without the heat exchanger? How do the refrigerant circulation rates compare for the two cases?

9.7. Consider the vapor-compression refrigeration cycle of Fig. 9.1*b* with tetrafluoroethane as refrigerant (Table 9.1, Fig. 9.3). If the evaporation temperature is $10(°F)$, show the effect of condensation temperature on the coefficient of performance by making calculations for condensation temperatures of 60, 80, and $100(°F)$.

(a) Assume isentropic compression of the vapor.

(b) Assume a compressor efficiency of 75%.

9.8. A heat pump is used to heat a house in the winter and to cool it in the summer. During the winter, the outside air serves as a low-temperature heat source; during the summer, it acts as a high-temperature heat sink. The heat-transfer rate through the walls and roof of the house is 0.75 kJ s^{-1} for each °C of temperature difference between the inside and outside of the house, summer and winter. The heat-pump motor is rated at 1.5 kW. Determine the minimum outside temperature for which the house can be maintained at 20°C during the winter and the maximum outside temperature for which the house can be maintained at 25°C during the summer.

9.9. Dry methane is supplied by a compressor and precooling system to the cooler of a Linde liquefaction system (Fig. 9.7) at 180 bar and 300 K. The low-pressure methane leaves the cooler at a temperature 6°C lower than the temperature of the incoming high-pressure stream. The separator operates at 1 bar, and the product is saturated liquid at this pressure. What is the maximum fraction of the methane entering the cooler that can be liquefied. Thermodynamic properties of methane are given by R. H. Perry and D. Green, *Perry's Chemical Engineers' Handbook*, 7th ed., Sec. 2, McGraw-Hill, New York, 1996.

9.10. Rework the preceding problem for methane entering at 200 bar, and precooled to 240 K by external refrigeration.

9.11. An advertisement is noted in a rural newspaper for a dairy-barn unit that combines a milk cooler with a water heater. Milk must, of course, be refrigerated, and hot water is required for washing purposes. The usual barn is equipped with a conventional air-cooled electric refrigerator and an electric-resistance water heater. The new unit is said to provide both the necessary refrigeration and the required hot water at a cost for electricity about the same as the cost of running just the refrigerator in the usual installation. To assess this claim, compare two refrigeration units: The advertized unit takes 50,000(Btu)(h)$^{-1}$ from a milk cooler at 30(°F), and discards heat through a condenser at 150(°F) to raise the temperature of water from 56 to 146(°F). The conventional unit takes the same amount of heat from the same milk cooler at 30(°F) and discards heat through an air-cooled condenser at 120(°F); in addition, the same amount of water is heated electrically from 56 to 146(°F). Estimate the *total* electric power requirements for the two cases, assuming that the actual work in both is 50% greater than required by Carnot refrigerators operating between the given temperatures.

CHAPTER 10

SOLUTION THERMODYNAMICS: THEORY

In Chap. 6 we treated the thermodynamic properties of constant-composition fluids. However, many applications of chemical-engineering thermodynamics are to systems wherein multicomponent gases or liquids undergo composition changes as the result of mixing or separation processes, the transfer of species from one phase to another, or chemical reaction. The properties of such systems depend on composition as well as on temperature and pressure.

Our first task in this chapter is therefore to develop a fundamental property relation for homogeneous solutions of variable composition. We then introduce a new class of thermodynamic properties known as *partial properties*. The mathematical definition of these quantities endows them with all the characteristics of properties of the individual species as they exist in solution. For example, in a liquid solution of ethanol and water we speak of the partial molar volume of ethanol and the partial molar volume of water in the solution, and their values are, in general, different from the molar volumes of pure ethanol and pure water at the same temperature and pressure.

A partial property of fundamental importance, because of its application in phase and chemical-reaction equilibria, is the chemical potential. Useful also are the property relations valid for ideal-gas mixtures, which lead to development of the concepts of fugacity and the ideal solution.

Finally, we present a general treatment of a class of solution properties known as excess properties. Of particular interest for vapor/liquid equilibrium is the excess Gibbs energy and a related property, the activity coefficient.

315

10.1   Fundamental Property Relation

Equation (6.6) expresses the basic relation connecting the Gibbs energy to the temperature and pressure in any closed system:

$$d(nG) = (nV)dP - (nS)dT \qquad\qquad (6.6)$$

We apply this equation to the case of a single-phase fluid that does not undergo chemical reaction. This closed system is then of constant composition, and we can write immediately that

$$\left[\frac{\partial(nG)}{\partial P}\right]_{T,n} = nV \qquad \text{and} \qquad \left[\frac{\partial(nG)}{\partial T}\right]_{P,n} = -nS$$

where the subscript n indicates that the numbers of moles of *all* chemical species are held constant.

We are now prepared to treat the more general case of a single-phase, *open* system that can interchange matter with its surroundings. The total Gibbs energy nG is still a function of T and P. Since material may be taken from or added to the system, nG is now also a function of the numbers of moles of the chemical species present. Thus

$$nG = g(P, T, n_1, n_2, \ldots, n_i, \ldots)$$

where the n_i are mole numbers of the species. The total differential of nG is

$$d(nG) = \left[\frac{\partial(nG)}{\partial P}\right]_{T,n} dP + \left[\frac{\partial(nG)}{\partial T}\right]_{P,n} dT + \sum_i \left[\frac{\partial(nG)}{\partial n_i}\right]_{P,T,n_j} dn_i$$

where the summation is over all species present, and subscript n_j indicates that all mole numbers except the ith are held constant. As shown above, we can replace the first two partial derivatives by (nV) and $-(nS)$:

$$d(nG) = (nV)dP - (nS)dT + \sum_i \left[\frac{\partial(nG)}{\partial n_i}\right]_{P,T,n_j} dn_i$$

The derivative of nG with respect to the number of moles of species i has a special significance, and is given its own symbol and name. Thus, we define the *chemical potential* of species i in the mixture as

$$\mu_i \equiv \left[\frac{\partial(nG)}{\partial n_i}\right]_{P,T,n_j} \qquad\qquad (10.1)$$

Expressed in terms of μ_i, the general equation for $d(nG)$ is

$$\boxed{d(nG) = (nV)dP - (nS)dT + \sum_i \mu_i\, dn_i} \qquad\qquad (10.2)$$

Equation (10.2) is the fundamental property relation for single-phase fluid systems of constant or variable mass and constant or variable composition. It is the foundation equation upon which the structure of solution thermodynamics is built. It may be written for the special case of one mole of solution, in which case $n = 1$ and the n_i are replaced by mole fractions x_i:

$$dG = V\,dP - S\,dT + \sum_i \mu_i\,dx_i \qquad (10.3)$$

Therefore

$$G = G(T, P, x_1, x_2, \ldots, x_i, \ldots)$$

an expression which displays the functional relationship of the molar Gibbs energy to its *canonical* variables, T, P, and $\{x_i\}$. Equation (6.10) is a special case of Eq. (10.3), applicable to a constant-composition solution. Although the n_i of Eq. (10.2) are all independent variables, this is not true of the x_i in Eq. (10.3), because their sum must be unity: $\sum_i x_i = 1$. This precludes certain mathematical operations which depend upon independence of the variables. Nevertheless, Eq. (10.3) does imply that

$$S = -\left(\frac{\partial G}{\partial T}\right)_{P,x} \qquad (10.4)$$

and

$$V = \left(\frac{\partial G}{\partial P}\right)_{T,x} \qquad (10.5)$$

Other solution properties are found from definitions; the enthalpy, for example, from $H = G + TS$. Thus we see again that whenever the Gibbs energy is expressed as a function of its canonical variables, it plays the role of a generating function, providing the means for calculation of all other thermodynamic properties by simple mathematical operations (differentiation and elementary algebra).

10.2 Chemical Potential as a Criterion for Phase Equilibria

Consider a closed system consisting of two phases in equilibrium. Within this *closed* system, each of the individual phases is an *open* system, free to transfer mass to the other. Equation (10.2) may therefore be written for each phase:

$$d(nG)^\alpha = (nV)^\alpha dP - (nS)^\alpha dT + \sum_i \mu_i^\alpha\,dn_i^\alpha$$

$$d(nG)^\beta = (nV)^\beta dP - (nS)^\beta dT + \sum_i \mu_i^\beta\,dn_i^\beta$$

where superscripts α and β identify the phases. In writing these expressions, we have supposed that at equilibrium T and P are uniform throughout the entire system. The change in the total Gibbs energy of the two-phase system is the sum

of these equations. When each total-system property is expressed by an equation
of the form

$$nM = (nM)^\alpha + (nM)^\beta$$

this sum is given by

$$d(nG) = (nV)dP - (nS)dT + \sum_i \mu_i^\alpha dn_i^\alpha + \sum_i \mu_i^\beta dn_i^\beta$$

Since the two-phase system is closed, Eq. (6.6) must also be valid. Comparison of
the two equations shows that at equilibrium

$$\sum_i \mu_i^\alpha dn_i^\alpha + \sum_i \mu_i^\beta dn_i^\beta = 0$$

The changes dn_i^α and dn_i^β result from mass transfer between the phases, and mass
conservation requires that

$$dn_i^\alpha = -dn_i^\beta$$

Therefore

$$\sum_i (\mu_i^\alpha - \mu_i^\beta)dn_i^\alpha = 0$$

Since the dn_i^α are independent and arbitrary, the only way the left-hand side of
this equation can in general be zero is for each term in parentheses separately to
be zero. Hence

$$\mu_i^\alpha = \mu_i^\beta \qquad (i = 1, 2, \ldots, N)$$

where N is the number of species present in the system. Although not given here,
a similar but more comprehensive derivation shows (as we have supposed) that T
and P must also be the same for the two phases at equilibrium.

By successively considering pairs of phases, we may readily generalize to more
than two phases the equality of chemical potentials; the result for π phases is

$$\boxed{\mu_i^\alpha = \mu_i^\beta = \cdots = \mu_i^\pi} \qquad (i = 1, 2, \ldots, N) \qquad (10.6)$$

Thus multiple phases at the same T and P are in equilibrium when the chemical
potential of each species is the same in all phases.

The application of Eq. (10.6) to specific phase-equilibrium problems (Chaps.
11–14) requires use of *models* of solution behavior, which provide expressions for G
or for the μ_i as functions of temperature, pressure, and composition. The simplest
of these, the ideal-gas mixture and the ideal solution, are treated in Secs. 10.4 and
10.8 of this chapter.

10.3 Partial Properties

The definition of the chemical potential by Eq. (10.1) as the mole-number derivative of nG suggests that such derivatives may be of particular use in solution thermodynamics. Thus we write

$$\bar{M}_i \equiv \left[\frac{\partial(nM)}{\partial n_i} \right]_{P,T,n_j} \tag{10.7}$$

This equation defines the *partial molar property* $\bar{M}_i$ of species i in solution, where $\bar{M}_i$ may represent the partial molar internal energy $\bar{U}_i$, the partial molar enthalpy $\bar{H}_i$, the partial molar entropy $\bar{S}_i$, the partial molar Gibbs energy $\bar{G}_i$, etc. It is a *response function*, representing the change of total property nM due to addition at constant T and P of a differential amount of species i to a finite amount of solution.

Comparison of Eq. (10.1) with Eq. (10.7) written for the Gibbs energy shows that the chemical potential and the partial molar Gibbs energy are identical; i.e.,

$$\mu_i \equiv \bar{G}_i \tag{10.8}$$

Example 10.1 By definition, the partial molar volume is

$$\bar{V}_i \equiv \left[\frac{\partial(nV)}{\partial n_i} \right]_{P,T,n_j} \tag{A}$$

What physical interpretation can be given to this equation?

SOLUTION Consider an open beaker containing an equimolar mixture of alcohol and water. The mixture occupies a total volume nV at room temperature T and atmospheric pressure P. Now add to this solution a drop of pure water, also at T and P, containing Δn_w moles, and mix it thoroughly into the solution, allowing sufficient time for heat exchange so that the contents of the beaker return to the initial temperature. What is the volume change of the solution in the beaker? One might suppose that the volume increases by an amount equal to the volume of the water added, i.e., by $V_w \Delta n_w$, where V_w is the molar volume of pure water at T and P. If this were true, we would have

$$\Delta(nV) = V_w \Delta n_w$$

However, we find by experiment that the actual value of $\Delta(nV)$ is somewhat less than that given by this equation. Evidently, the *effective* molar volume of the added water in solution is less than the molar volume of pure water at the same T and P. Designating the effective molar volume in solution by $\tilde{V}_w$, we can write

$$\Delta(nV) = \tilde{V}_w \Delta n_w \tag{B}$$

or

$$\tilde{V}_w = \frac{\Delta(nV)}{\Delta n_w} \tag{C}$$

If this effective molar volume is to represent the property of species i in the original equimolar solution, it must be based on data for a solution of this composition. However, in the process described a finite drop of water is added to the equimolar solution, causing a small but finite change in composition. We may, however, consider the limiting case for which $\Delta n_w \to 0$. Then Eq. (C) becomes

$$\widetilde{V}_w = \lim_{\Delta n_w \to 0} \frac{\Delta(nV)}{\Delta n_w} = \frac{d(nV)}{dn_w}$$

Since T, P, and n_a (the number of moles of alcohol) are constant, this equation is more appropriately written

$$\widetilde{V}_w = \left[\frac{\partial(nV)}{\partial n_w}\right]_{P,T,n_a}$$

Comparison with Eq. (A) shows that the effective molar volume $\widetilde{V}_w$ is the partial molar volume $\bar{V}_w$ of the water in solution, that is, the rate of change of the total solution volume with n_w at constant T, P, and n_a. Equation (B), written for the addition of dn_w moles of water to the solution, then becomes

$$d(nV) = \bar{V}_w \, dn_w \tag{D}$$

When $\bar{V}_w$ is considered the molar property of water as it exists in solution, the total volume change $d(nV)$ is merely this molar property multiplied by the number of moles of water added.

If dn_w moles of water is added to a volume of *pure* water, then we have every reason to expect the volume change of the system to be given by

$$d(nV) = V_w \, dn_w \tag{E}$$

where V_w is the molar volume of pure water at T and P. Comparison of Eqs. (D) and (E) indicates that $\bar{V}_w = V_w$ when the "solution" is pure water.

The definition of a partial molar property, Eq. (10.7), provides the means for calculation of partial properties from solution-property data. Implicit in this definition is another, equally important, equation that allows the reverse, that is, the calculation of solution properties from knowledge of the partial properties. The derivation of this equation starts with the observation that the thermodynamic properties of a homogeneous phase are functions of temperature, pressure, and the numbers of moles of the individual species which comprise the phase. For thermodynamic property M we may therefore write[1]

$$nM = \mathcal{M}(T, P, n_1, n_2, \ldots, n_i, \ldots)$$

The total differential of nM is

$$d(nM) = \left[\frac{\partial(nM)}{\partial P}\right]_{T,n} dP + \left[\frac{\partial(nM)}{\partial T}\right]_{P,n} dT + \sum_i \left[\frac{\partial(nM)}{\partial n_i}\right]_{P,T,n_j} dn_i$$

---

[1]Mere functionality does not make a set of variables into *canonical* variables. These are the canonical variables only for G.

where subscript n indicates that *all* mole numbers are held constant, and subscript n_j that all mole numbers *except* n_i are held constant. Because the first two partial derivatives on the right are evaluated at constant n and because the partial derivative of the last term is given by Eq. (10.7), this equation may be written more simply as

$$d(nM) = n\left(\frac{\partial M}{\partial P}\right)_{T,x} dP + n\left(\frac{\partial M}{\partial T}\right)_{P,x} dT + \sum_i \bar{M}_i\, dn_i \qquad (10.9)$$

where subscript x denotes differentiation at constant composition.

Since $n_i = x_i n$,

$$dn_i = x_i\, dn + n\, dx_i$$

Replacing dn_i by this expression and replacing $d(nM)$ by the identity

$$d(nM) \equiv n\, dM + M\, dn$$

we write Eq. (10.9) as

$$n\, dM + M\, dn = n\left(\frac{\partial M}{\partial P}\right)_{T,x} dP + n\left(\frac{\partial M}{\partial T}\right)_{P,x} dT + \sum_i \bar{M}_i(x_i\, dn + n\, dx_i)$$

When the terms containing n are collected and separated from those containing dn, this equation becomes

$$\left[dM - \left(\frac{\partial M}{\partial P}\right)_{T,x} dP - \left(\frac{\partial M}{\partial T}\right)_{P,x} dT - \sum_i \bar{M}_i\, dx_i\right]n + \left[M - \sum_i x_i\bar{M}_i\right]dn = 0$$

In application, one is free to choose a system of any size, as represented by n, and to choose any variation in its size, as represented by dn. Thus n and dn are independent and arbitrary. The only way that the left-hand side of this equation can then, in general, be zero is for *each* term in brackets to be zero. We therefore have

$$dM = \left(\frac{\partial M}{\partial P}\right)_{T,x} dP + \left(\frac{\partial M}{\partial T}\right)_{P,x} dT + \sum_i \bar{M}_i dx_i \qquad (10.10)$$

and

$$\boxed{M = \sum_i x_i \bar{M}_i} \qquad (10.11)$$

Multiplication of Eq. (10.11) by n yields the alternative expression

$$\boxed{nM = \sum_i n_i \bar{M}_i} \qquad (10.12)$$

Equation (10.10) is in fact just a special case of Eq. (10.9), obtained by setting $n = 1$, which also makes $n_i = x_i$. Equations (10.11) and (10.12) on the other hand are new and vital. Known as *summability relations*, they allow calculation of mixture properties from partial properties, playing a role opposite to that of Eq. (10.7), which provides for the calculation of partial properties from mixture properties.

One further important equation follows directly from Eqs. (10.10) and (10.11). Since Eq. (10.11) is a general expression for M, differentiation yields a general expression for dM:

$$dM = \sum_i x_i \, d\bar{M}_i + \sum_i \bar{M}_i \, dx_i$$

Comparison of this equation with Eq. (10.10), another general equation for dM, yields the *Gibbs/Duhem*[2] *equation*:

$$\boxed{\left(\frac{\partial M}{\partial P}\right)_{T,x} dP + \left(\frac{\partial M}{\partial T}\right)_{P,x} dT - \sum_i x_i \, d\bar{M}_i = 0} \qquad (10.13)$$

This equation must be satisfied for all changes in P, T, and the $\bar{M}_i$ caused by changes of state in a homogeneous phase. For the important special case of changes at constant T and P, it simplifies to

$$\boxed{\sum_i x_i \, d\bar{M}_i = 0} \qquad (\text{const } T, P) \qquad (10.14)$$

Equation (10.11) implies that a molar solution property is given as a sum of its parts and that $\bar{M}_i$ is the molar property of species i as it exists in solution. This is a proper interpretation provided one understands that the defining equation for $\bar{M}_i$, Eq. (10.7), is an apportioning formula which *arbitrarily* assigns to each species i a share of the mixture property, subject to the constraint of Eq. (10.11).[3]

The constituents of a solution are in fact intimately intermixed, and owing to molecular interactions cannot have private properties of their own. Nevertheless, partial molar properties, as defined by Eq. (10.7), have all the characteristics of properties of the individual species as they exist in solution. Thus for all practical purposes they may be *assigned* as property values to the individual species.

The properties of solutions as represented by the symbol M may be on a unit-mass basis as well as on a mole basis. The equations relating solution properties are the same in form on either basis; one merely replaces the various n's, representing moles, by m's, representing mass, and speaks of partial *specific* properties rather than of partial *molar* properties. In order to accommodate either, we generally speak simply of partial properties.

Since we are concerned here primarily with the properties of solutions, we represent molar (or unit-mass) properties of the solution by the plain symbol M.

----

[2]Pierre-Maurice-Marie Duhem (1861–1916), French physicist.

[3]Other apportioning equations, which make different allocations of the mixture property, are possible and are equally valid.

Partial properties are denoted by an overbar, with a subscript to identify the species; the symbol is therefore $\bar{M}_i$. In addition, we need a symbol for the properties of the individual species as they exist in the *pure state at the T and P of the solution.* These molar (or unit-mass) properties are identified by only a subscript, and the symbol is M_i. In summary, the three kinds of properties used in solution thermodynamics are distinguished by the following symbolism:

Solution properties	M,	for example: $U,\ H,\ S,\ G$
Partial properties	$\bar{M}_i$,	for example: $\bar{U}_i,\ \bar{H}_i,\ \bar{S}_i,\ \bar{G}_i$
Pure-species properties	M_i,	for example: $U_i,\ H_i,\ S_i,\ G_i$

Equations for partial properties can always be derived from an equation for the solution property as a function of composition by direct application of Eq. (10.7). For binary systems, however, an alternative procedure may be more convenient. Written for a binary solution, the summability relation, Eq. (10.11), becomes

$$M = x_1 \bar{M}_1 + x_2 \bar{M}_2 \tag{A}$$

whence

$$dM = x_1\, d\bar{M}_1 + \bar{M}_1\, dx_1 + x_2\, d\bar{M}_2 + \bar{M}_2\, dx_2 \tag{B}$$

When M is known as a function of x_1 at constant T and P, the appropriate form of the Gibbs/Duhem equation is Eq. (10.14), expressed here as

$$x_1\, d\bar{M}_1 + x_2\, d\bar{M}_2 = 0 \tag{C}$$

Since $x_1 + x_2 = 1$, it follows that $dx_1 = -dx_2$. Eliminating dx_2 in favor of dx_1 in Eq. (B) and combining the result with Eq. (C) gives

$$dM = \bar{M}_1\, dx_1 - \bar{M}_2\, dx_1$$

or

$$\frac{dM}{dx_1} = \bar{M}_1 - \bar{M}_2 \tag{D}$$

Eliminating $\bar{M}_2$ from Eqs. (A) and (D), and solving for $\bar{M}_1$, we get

$$\boxed{\bar{M}_1 = M + x_2 \frac{dM}{dx_1}} \tag{10.15}$$

Similarly, elimination of $\bar{M}_1$ and solution for $\bar{M}_2$ gives

$$\boxed{\bar{M}_2 = M - x_1 \frac{dM}{dx_1}} \tag{10.16}$$

Thus for binary systems, the partial properties are readily calculated directly from an expression for the solution property as a function of composition at constant T

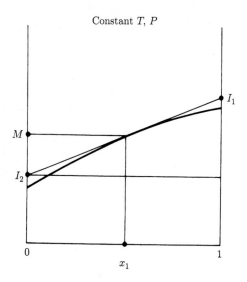

Figure 10.1: Graphical construction of Example 10.2.

and P. The corresponding equations for multicomponent systems are much more complex, and are given in detail by Van Ness and Abbott.[4]

Example 10.2   Describe a graphical interpretation of Eqs. (10.15) and (10.16).

SOLUTION   Figure 10.1 shows a representative plot of M vs. x_1 for a binary system. Values of the derivative dM/dx_1 are given by the slopes of lines drawn tangent to the curve of M vs. x_1. One such line drawn tangent at a particular value of x_1 is shown in Fig. 10.1. Its intercepts with the boundaries of the figure at $x_1 = 1$ and $x_1 = 0$ are labeled I_1 and I_2. As is evident from the figure, two equivalent expressions can be written for the slope of this line:

$$\frac{dM}{dx_1} = \frac{M - I_2}{x_1 - 0} \quad \text{and} \quad \frac{dM}{dx_1} = \frac{I_1 - I_2}{1 - 0} = I_1 - I_2$$

Solving the first equation for I_2 and the second for I_1 (with elimination of I_2) gives

$$I_2 = M - x_1 \frac{dM}{dx_1} \quad \text{and} \quad I_1 = M + (1 - x_1) \frac{dM}{dx_1}$$

Comparison of these expressions with Eqs. (10.15) and (10.16) shows that

$$I_1 = \bar{M}_1 \quad \text{and} \quad I_2 = \bar{M}_2$$

Thus the tangent intercepts give directly the values of the two partial properties. These intercepts of course shift as the point of tangency moves along the curve, and the limiting values are indicated by the constructions shown in Fig. 10.2. The tangent drawn at $x_1 = 0$ (pure species 2) gives $\bar{M}_2 = M_2$, consistent with the

[4]H. C. Van Ness and M. M. Abbott, *Classical Thermodynamics of Nonelectrolyte Solutions: With Applications to Phase Equilibria*, pp. 46–54, McGraw-Hill, New York, 1982.

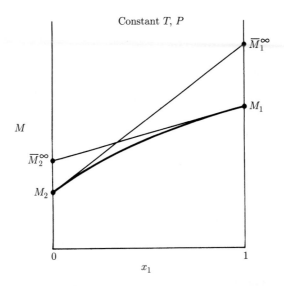

Figure 10.2: Infinite-dilution values of partial properties.

conclusion reached in Example 10.1 regarding the partial property of a pure species. The opposite intercept gives $\bar{M}_1 = \bar{M}_1^\infty$, the partial property of species 1 when it is present at *infinite dilution* $(x_1 = 0)$. Similar comments apply to the tangent drawn at $x_1 = 1$ (pure species 1). In this case $\bar{M}_1 = M_1$ and $\bar{M}_2 = \bar{M}_2^\infty$, since it is species 2 that is present at infinite dilution $(x_1 = 1, x_2 = 0)$.

Example 10.3   The need arises in a laboratory for 2,000 cm^3 of an antifreeze consisting of a 30-mole-percent solution of methanol in water. What volumes of pure methanol and of pure water at 25°C must be mixed to form the 2,000 cm^3 of antifreeze, also at 25°C? Partial molar volumes for methanol and water in a 30-mole-percent methanol solution at 25°C are:

$$\text{Methanol(1):} \quad \bar{V}_1 = 38.632 \text{ cm}^3 \text{ mol}^{-1}$$
$$\text{Water(2):} \quad \bar{V}_2 = 17.765 \text{ cm}^3 \text{ mol}^{-1}$$

For the pure species at 25°C:

$$\text{Methanol(1):} \quad V_1 = 40.727 \text{ cm}^3 \text{ mol}^{-1}$$
$$\text{Water(2):} \quad V_2 = 18.068 \text{ cm}^3 \text{ mol}^{-1}$$

SOLUTION   Equation (10.11) written for the volume of a binary solution is

$$V = x_1\bar{V}_1 + x_2\bar{V}_2$$

All quantities on the right are known, and we calculate the molar volume of the antifreeze solution:

$$V = (0.3)(38.632) + (0.7)(17.765) = 24.025 \text{ cm}^3 \text{ mol}^{-1}$$

The required total volume of solution is

$$V^t = nV = 2,000 \text{ cm}^3$$

Thus the total number of moles required is

$$n = \frac{V^t}{V} = \frac{2,000}{24.025} = 83.246 \text{ mol}$$

Of this, 30 percent is methanol, and 70 percent is water:

$$n_1 = (0.3)(83.246) = 24.974 \text{ mol}$$
$$n_2 = (0.7)(83.246) = 58.272 \text{ mol}$$

The volume of each pure species is $V_i^t = n_i V_i$; thus

$$V_1^t = (24.974)(40.727) = 1,017 \text{ cm}^3$$
$$V_2^t = (58.272)(18.068) = 1,053 \text{ cm}^3$$

Values of $\bar{V}_1$, $\bar{V}_2$, and V for the binary solution methanol(1)/water(2) at 25°C are plotted in Fig. 10.3 as functions of x_1. The line drawn tangent to the V-vs.-x_1 curve at $x_1 = 0.3$ illustrates the procedure by which values of $\bar{V}_1$ and $\bar{V}_2$ are obtained. We note that the curve for $\bar{V}_1$ becomes horizontal $(d\bar{V}_1/dx_1 = 0)$ at $x_1 = 1$ and the curve for $\bar{V}_2$ becomes horizontal at $x_1 = 0$ or $x_2 = 1$. This is a requirement of Eq. (10.14), the Gibbs/Duhem equation, which here becomes

$$x_1 \, d\bar{V}_1 + x_2 \, d\bar{V}_2 = 0$$

Division of this equation by dx_1 and rearrangement gives

$$\frac{d\bar{V}_1}{dx_1} = -\frac{x_2}{x_1} \frac{d\bar{V}_2}{dx_1}$$

This result shows that the slopes $d\bar{V}_1/dx_1$ and $d\bar{V}_2/dx_1$ must be of opposite sign. When $x_1 = 1$, $x_2 = 0$ and $d\bar{V}_1/dx_1 = 0$, provided $d\bar{V}_2/dx_1$ remains finite. When $x_1 = 0$, $x_2 = 1$ and $d\bar{V}_2/dx_1 = 0$. The curves for $\bar{V}_1$ and $\bar{V}_2$ in Fig. 10.3 appear to be horizontal at *both* ends; this is a peculiarity of the system considered.

For an *ideal solution*, treated in detail in Sec. 10.8, the partial molar volumes of the species in solution are equal to the molar volumes of the pure species at the same T and P. In this event, the summability relation becomes

$$V^{id} = x_1 V_1 + x_2 V_2$$

implying a linear relation between V^{id} and x_1:

$$V^{id} = (V_1 - V_2)x_1 + V_2$$

When the methanol/water system is assumed an ideal solution, the resulting V-vs.-x_1 relation is represented by the straight dashed line shown in Fig. 10.3 connecting the pure-species volumes (V_1 at $x_1 = 1$ and V_2 at $x_1 = 0$). For the specific problem posed here, the use of V_1 and V_2 in place of the partial properties yields

$$V_1^t = 983 \qquad V_2^t = 1,017 \text{ cm}^3$$

Both values are about 3.4 percent low.

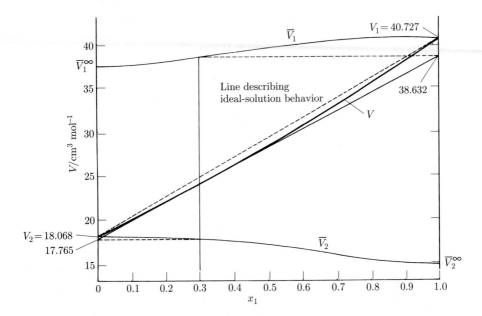

Figure 10.3: Molar volumes for methanol(1)/water(2) at 25°C and 1(atm).

Example 10.4   The enthalpy of a binary liquid system of species 1 and 2 at fixed T and P is represented by the equation

$$H = 400x_1 + 600x_2 + x_1x_2(40x_1 + 20x_2)$$

where H is in J mol^{-1}. Determine expressions for $\bar{H}_1$ and $\bar{H}_2$ as functions of x_1, numerical values for the pure-species enthalpies H_1 and H_2, and numerical values for the partial enthalpies at infinite dilution $\bar{H}_1^\infty$ and $\bar{H}_2^\infty$.

SOLUTION   Replacing x_2 by $1 - x_1$ in the given equation for H yields

$$H = 600 - 180x_1 - 20x_1^3 \tag{A}$$

whence

$$\frac{dH}{dx_1} = -180 - 60x_1^2$$

By Eq. (10.15),

$$\bar{H}_1 = H + x_2\frac{dH}{dx_1}$$

Substitution for H and dH/dx_1 gives

$$\bar{H}_1 = 600 - 180x_1 - 20x_1^3 - 180x_2 - 60x_1^2x_2$$

Replacing x_2 by $1 - x_1$ and simplifying, we get

$$\bar{H}_1 = 420 - 60x_1^2 + 40x_1^3 \tag{B}$$

Similarly, by Eq. (10.16)

$$\bar{H}_2 = H - x_1 \frac{dH}{dx_1}$$

whence

$$\bar{H}_2 = 600 - 180x_1 - 20x_1^3 + 180x_1 + 60x_1^3$$

or

$$\bar{H}_2 = 600 + 40x_1^3 \tag{C}$$

We could equally well have started with the given equation for H. With dH/dx_1 a *total* derivative, x_2 is not a constant. Since $x_2 = 1 - x_1$, $dx_2/dx_1 = -1$. Differentiation of the given equation for H therefore gives

$$\frac{dH}{dx_1} = 400 - 600 + x_1x_2(40 - 20) + (40x_1 + 20x_2)(-x_1 + x_2)$$

When x_2 is replaced by $1 - x_1$, this reduces to the expression previously obtained.

A numerical value for H_1 results when we substitute $x_1 = 1$ in either Eq. (A) or (B). Both equations yield $H_1 = 400$ J mol^{-1}. Similarly H_2 is found from either Eq. (A) or (C) when $x_1 = 0$. The result is $H_2 = 600$ J mol^{-1}. The infinite-dilution values $\bar{H}_1^\infty$ and $\bar{H}_2^\infty$ are found from Eqs. (B) and (C) when $x_1 = 0$ in Eq. (B) and $x_1 = 1$ in Eq. (C). The results are

$$\bar{H}_1^\infty = 420 \qquad \text{and} \qquad \bar{H}_2^\infty = 640 \text{ J mol}^{-1}$$

We show now how partial properties are related to one another. Since by Eq. (10.8) $\mu_i = \bar{G}_i$, we may write Eq. (10.2) as

$$d(nG) = (nV)dP - (nS)dT + \sum_i \bar{G}_i \, dn_i \tag{10.17}$$

Application of the criterion of exactness, Eq. (6.12), to this equation yields the Maxwell relation,

$$\left(\frac{\partial V}{\partial T}\right)_{P,n} = -\left(\frac{\partial S}{\partial P}\right)_{T,n} \tag{6.16}$$

plus the two additional equations

$$\left(\frac{\partial \bar{G}_i}{\partial T}\right)_{P,n} = -\left[\frac{\partial (nS)}{\partial n_i}\right]_{P,T,n_j}$$

and

$$\left(\frac{\partial \bar{G}_i}{\partial P}\right)_{T,n} = \left[\frac{\partial (nV)}{\partial n_i}\right]_{P,T,n_j}$$

where subscript n indicates constancy of all n_i, and therefore of composition, and subscript n_j indicates that all mole numbers except the ith are held constant. In view of Eq. (10.7), these last two equations are most simply written as

$$\left(\frac{\partial \bar{G}_i}{\partial T}\right)_{P,x} = -\bar{S}_i \tag{10.18}$$

and

$$\left(\frac{\partial \bar{G}_i}{\partial P}\right)_{T,x} = \bar{V}_i \tag{10.19}$$

These equations allow calculation of the effect of temperature and pressure on the partial Gibbs energy (or chemical potential). They are the partial-property analogs of Eqs. (10.4) and (10.5). Indeed, for every equation providing a *linear* relation among the thermodynamic properties of a *constant-composition* solution there exists a corresponding equation connecting the corresponding partial properties of each species in the solution. We demonstrate this by example.

Consider the equation that defines enthalpy,

$$H = U + PV \tag{2.5}$$

For n moles,

$$nH = nU + P(nV)$$

Differentiation with respect to n_i at constant T, P, and n_j yields

$$\left[\frac{\partial(nH)}{\partial n_i}\right]_{P,T,n_j} = \left[\frac{\partial(nU)}{\partial n_i}\right]_{P,T,n_j} + P\left[\frac{\partial(nV)}{\partial n_i}\right]_{P,T,n_j}$$

By Eq. (10.7) this becomes

$$\bar{H}_i = \bar{U}_i + P\bar{V}_i$$

which is the partial-property analog of Eq. (2.5).

In a constant-composition solution, $\bar{G}_i$ is a function of P and T. We may therefore write

$$d\bar{G}_i = \left(\frac{\partial \bar{G}_i}{\partial P}\right)_{T,x} dP + \left(\frac{\partial \bar{G}_i}{\partial T}\right)_{P,x} dT$$

As a result of Eqs. (10.18) and (10.19) this becomes

$$d\bar{G}_i = \bar{V}_i\, dP - \bar{S}_i\, dT$$

which may be compared with Eq. (6.10). These examples illustrate the parallelism that exists between equations for a constant-composition solution and the corresponding equations for the partial properties of the species in solution. We can therefore write simply by analogy many equations that relate partial properties.

10.4 Ideal-Gas Mixtures

If n moles of an ideal-gas mixture occupy a total volume V^t at temperature T, the pressure is

$$P = \frac{nRT}{V^t}$$

If the n_i moles of species i in this mixture occupy the same total volume alone at the same temperature, the pressure is

$$p_i = \frac{n_i RT}{V^t}$$

Dividing the latter equation by the former gives

$$\frac{p_i}{P} = \frac{n_i}{n} = x_i$$

or

$$p_i = x_i P \qquad (i = 1, 2, \ldots, N) \tag{10.20}$$

where x_i is the mole fraction of species i in the ideal-gas mixture, and p_i is known as the *partial pressure* of species i. The sum of the partial pressures as given by Eq. (10.20) equals the total pressure.

An ideal gas (Sec. 3.3) is a model gas comprised of imaginary molecules of zero volume that do not interact. Each chemical species in an ideal-gas mixture therefore has its own private properties, uninfluenced by the presence of other species. This is the basis for the following statement of *Gibbs's theorem*:

> *A partial molar property (other than the volume) of a constituent species in an ideal-gas mixture is equal to the corresponding molar property of the species as a pure ideal gas at the mixture temperature but at a pressure equal to its partial pressure in the mixture.*

This is expressed mathematically for generic partial property $\bar{M}_i^{ig}$ by the equation

$$\bar{M}_i^{ig}(T, P) = M_i^{ig}(T, p_i) \tag{10.21}$$

where superscript ig denotes an ideal-gas property, and $\bar{M}_i^{ig} \neq \bar{V}_i^{ig}$.

Since the enthalpy of an ideal gas is independent of pressure

$$H_i^{ig}(T, p_i) = H_i^{ig}(T, P)$$

whence

$$\bar{H}_i^{ig}(T, P) = H_i^{ig}(T, P)$$

or more simply,

$$\bar{H}_i^{ig} = H_i^{ig} \tag{10.22}$$

where H_i^{ig} is the pure-species value at the *mixture T and P*. Application of the summability relation, Eq. (10.11), yields

$$\boxed{H^{ig} = \sum_i x_i H_i^{ig}} \tag{10.23}$$

Analogous equations apply for U^{ig} and other properties that are *independent of pressure*. [See Eq. (4.6) for C_P^{ig}.]

When Eq. (10.23) is written

$$H^{ig} - \sum_i x_i H_i^{ig} = 0$$

the difference on the left is the enthalpy change associated with a process in which appropriate amounts of the pure species at T and P are mixed to form one mole of mixture at the same T and P. For ideal gases, this *enthalpy change of mixing* (Sec. 11.3) is zero.

The entropy of an ideal gas does depend on pressure, and by Eq. (6.23),

$$dS_i^{ig} = -R\, d\ln P \qquad (\text{const } T)$$

Integration from p_i to P gives

$$S_i^{ig}(T,P) - S_i^{ig}(T,p_i) = -R\ln\frac{P}{p_i} = -R\ln\frac{P}{x_i P} = R\ln x_i$$

whence

$$S_i^{ig}(T,p_i) = S_i^{ig}(T,P) - R\ln x_i$$

Substituting this result into Eq. (10.21) written for the entropy gives

$$\bar{S}_i^{ig}(T,P) = S_i^{ig}(T,P) - R\ln x_i$$

or simply,

$$\bar{S}_i^{ig} = S_i^{ig} - R\ln x_i \tag{10.24}$$

where S_i^{ig} is the pure-species value at the mixture T and P. By the summability relation,

$$\boxed{S^{ig} = \sum_i x_i S_i^{ig} - R\sum_i x_i \ln x_i} \tag{10.25}$$

When this equation is rearranged as

$$S^{ig} - \sum_i x_i S_i^{ig} = R\sum_i x_i \ln\frac{1}{x_i}$$

we have on the left the *entropy change of mixing* for ideal gases. Since $1/x_i > 1$, this quantity is always positive, in agreement with the second law. The mixing process is inherently irreversible, and for ideal gases mixing at constant T and P is not accompanied by heat transfer [Eq. (10.23)].

For the Gibbs energy of an ideal-gas mixture, $G^{ig} = H^{ig} - TS^{ig}$; the parallel relation for partial properties is

$$\bar{G}_i^{ig} = \bar{H}_i^{ig} - T\bar{S}_i^{ig}$$

In combination with Eqs. (10.22) and (10.24) this becomes

$$\bar{G}_i^{ig} = H_i^{ig} - TS_i^{ig} + RT\ln x_i$$

or

$$\boxed{\mu_i^{ig} \equiv \bar{G}_i^{ig} = G_i^{ig} + RT\ln x_i} \tag{10.26}$$

An alternative expression for the chemical potential results by elimination of G_i^{ig} from this equation. As a result of Eq. (6.10) we may write for pure species i

$$dG_i^{ig} = V_i^{ig} dP \qquad \text{(const } T\text{)}$$

or

$$dG_i^{ig} = \frac{RT}{P} dP = RT \, d\ln P \qquad \text{(const } T\text{)}$$

Integration gives

$$G_i^{ig} = \Gamma_i(T) + RT \ln P \tag{10.27}$$

where $\Gamma_i(T)$, the integration constant at constant T, is a function of temperature only.[5] Equation (10.26) may therefore be written

$$\boxed{\mu_i^{ig} = \Gamma_i(T) + RT \ln x_i P} \tag{10.28}$$

Application of the summability relation, Eq. (10.11), produces an expression for the Gibbs energy of an ideal-gas mixture:

$$\boxed{G^{ig} = \sum_i x_i \Gamma_i(T) + RT \sum_i x_i \ln x_i P} \tag{10.29}$$

These equations, remarkable in their simplicity, represent a complete description of ideal-gas behavior.

10.5   Fugacity and Fugacity Coefficient for a Pure Species

As evident from Eq. (10.6), the chemical potential μ_i is fundamental to the formulation of criteria for phase equilibria. This is true as well for chemical-reaction equilibria. However, the chemical potential exhibits certain unfortunate characteristics which discourage its use in the solution of practical problems. The Gibbs energy, and hence μ_i, is defined in relation to the internal energy and entropy, both primitive quantities for which absolute values are unknown. As a result, we have no unequivocal absolute values for the chemical potential. Moreover, Eq. (10.28) shows that for an ideal-gas mixture μ_i approaches negative infinity when either P or x_i approaches zero. This observation is not limited to ideal gases, but is true for any gas . While these characteristics do not preclude the use of chemical potentials, the application of equilibrium criteria is facilitated by introduction of the *fugacity*,

---

[5]A dimensional ambiguity is evident with Eq. (10.27) and with analogous equations to follow in that P has units, whereas $\ln P$ must be dimensionless. This difficulty is more apparent than real, because the Gibbs energy is always expressed on a relative scale, absolute values being unknown. Thus in application only *differences* in Gibbs energy appear, leading to *ratios* of quantities with units of pressure in the argument of the logarithm. The only requirement is that consistency of pressure units be maintained.

a quantity that takes the place of μ_i but which does not exhibit its less desirable characteristics.

The origin of the fugacity concept resides in Eq. (10.27), an equation valid only for pure species i in the ideal-gas state. For a real fluid, we write an analogous equation:

$$G_i \equiv \Gamma_i(T) + RT \ln f_i \qquad (10.30)$$

in which pressure P is replaced by a new property f_i, which has units of pressure. This equation serves as a partial definition of f_i, which is called the *fugacity*[6] of pure species i.

Subtraction of Eq. (10.27) from Eq. (10.30), both written for the same temperature and pressure, gives

$$G_i - G_i^{ig} = RT \ln \frac{f_i}{P}$$

According to the definition of Eq. (6.39), $G_i - G_i^{ig}$ is the *residual Gibbs energy*, G_i^R. The dimensionless ratio f_i/P is a new property called the *fugacity coefficient* and given the symbol ϕ_i. Thus,

$$\boxed{G_i^R = RT \ln \phi_i} \qquad (10.31)$$

where

$$\boxed{\phi_i \equiv \frac{f_i}{P}} \qquad (10.32)$$

We now complete the definition of fugacity by setting the ideal-gas-state fugacity of pure species i equal to its pressure:

$$f_i^{ig} = P \qquad (10.33)$$

Thus for the special case of an ideal-gas, $G_i^R = 0$, $\phi_i = 1$, and Eq. (10.27) is recovered from Eq. (10.30).

The identification of $\ln \phi_i$ with G_i^R/RT by Eq. (10.31) allows Eq. (6.44) to be rewritten as

$$\boxed{\ln \phi_i = \int_0^P (Z_i - 1) \frac{dP}{P}} \qquad (\text{const } T) \qquad (10.34)$$

Fugacity coefficients (and therefore fugacities) for pure species are evaluated by this equation from PVT data or from an equation of state. For example, when the compressibility factor is given by Eq. (3.31), we have

$$Z_i - 1 = \frac{B_{ii}P}{RT}$$

---

[6]Introduced by Gilbert Newton Lewis (1875–1946), American physical chemist, who also developed the concepts of the partial property and the ideal solution.

where the second virial coefficient B_{ii} is a function of temperature only for a pure species. Substitution into Eq. (10.34) gives

$$\ln \phi_i = \frac{B_{ii}}{RT} \int_0^P dP \qquad (\text{const } T)$$

whence

$$\ln \phi_i = \frac{B_{ii}P}{RT} \tag{10.35}$$

Equation (10.30), which defines the fugacity of pure species i, may be written for species i as a saturated vapor

$$G_i^v = \Gamma_i(T) + RT \ln f_i^v \tag{10.36}$$

and for species i as a saturated liquid at the same temperature

$$G_i^l = \Gamma_i(T) + RT \ln f_i^l \tag{10.37}$$

By difference

$$G_i^v - G_i^l = RT \ln \frac{f_i^v}{f_i^l}$$

an equation applicable to the change of state from saturated liquid to saturated vapor, both at temperature T and at the vapor pressure P_i^{sat}. According to Eq. (6.51), $G_i^v - G_i^l = 0$; therefore

$$f_i^v = f_i^l = f_i^{\text{sat}} \tag{10.38}$$

where f_i^{sat} indicates the value for either saturated liquid or saturated vapor. The corresponding fugacity coefficient is

$$\phi_i^{\text{sat}} = \frac{f_i^{\text{sat}}}{P_i^{\text{sat}}} \tag{10.39}$$

whence

$$\phi_i^v = \phi_i^l = \phi_i^{\text{sat}} \tag{10.40}$$

Since coexisting phases of saturated liquid and saturated vapor are in equilibrium, the equality of fugacities as expressed by Eqs. (10.38) and (10.40) is a criterion of vapor/liquid equilibrium for pure species.

Because of the equality of fugacities of saturated liquid and vapor, the calculation of fugacity for species i as a compressed liquid is done in two steps. First, one calculates the fugacity coefficient of saturated vapor $\phi_i^v = \phi_i^{\text{sat}}$ by an integrated form of Eq. (10.34), evaluated for $P = P_i^{\text{sat}}$. Then by Eqs. (10.38) and (10.39)

$$f_i^l = f_i^{\text{sat}} = \phi_i^{\text{sat}} P_i^{\text{sat}}$$

The second step is the evaluation of the change in fugacity of the liquid with an increase in pressure above P_i^{sat}. The required equation follows directly from Eq. (10.30) in combination with Eq. (6.10). For the isothermal change of state from saturated liquid to compressed liquid at pressure P, Eq. (6.10) may be integrated to give

$$G_i - G_i^{\text{sat}} = \int_{P_i^{\text{sat}}}^{P} V_i \, dP$$

Next, Eq. (10.30) is written twice: for G_i and for G_i^{sat}. Subtraction provides a second expression for $G_i - G_i^{\text{sat}}$:

$$G_i - G_i^{\text{sat}} = RT \ln \frac{f_i}{f_i^{\text{sat}}}$$

Equating the two expressions for $G_i - G_i^{\text{sat}}$ yields

$$\ln \frac{f_i}{f_i^{\text{sat}}} = \frac{1}{RT} \int_{P_i^{\text{sat}}}^{P} V_i \, dP$$

Since V_i, the liquid-phase molar volume, is a very weak function of P at temperatures well below T_c, an excellent approximation is often obtained when evaluation of the integral is based on the assumption that V_i is constant at the value for saturated liquid, V_i^l:

$$\ln \frac{f_i}{f_i^{\text{sat}}} = \frac{V_i^l (P - P_i^{\text{sat}})}{RT}$$

Substituting $f_i^{\text{sat}} = \phi_i^{\text{sat}} P_i^{\text{sat}}$ and solving for f_i gives

$$f_i = \phi_i^{\text{sat}} P_i^{\text{sat}} \exp \frac{V_i^l (P - P_i^{\text{sat}})}{RT} \tag{10.41}$$

The exponential is known as the Poynting[7] factor.

Example 10.5   For H_2O at a temperature of 300°C and for pressures up to 10,000 kPa (100 bar) plot values of f_i and ϕ_i calculated from data in the steam tables vs. P.

SOLUTION   Equation (10.30) may be written for a low-pressure reference state (designated by *) at temperature T:

$$G_i^* = \Gamma_i(T) + RT \ln f_i^*$$

Subtracting this equation from Eq. (10.30) itself, representing the state at pressure P but at the same temperature T, gives after rearrangement

$$\ln \frac{f_i}{f_i^*} = \frac{1}{RT} (G_i - G_i^*)$$

Since by definition $G_i = H_i - TS_i$ and $G_i^* = H_i^* - TS_i^*$ this becomes

$$\ln \frac{f_i}{f_i^*} = \frac{1}{R} \left[\frac{H_i - H_i^*}{T} - (S_i - S_i^*) \right]$$

[7] John Henry Poynting (1852–1914), British physicist.

If the reference-state pressure P^* is low enough that the fluid closely approximates an ideal gas, then $f_i^* = P^*$, and

$$\ln \frac{f_i}{P^*} = \frac{1}{R} \left[\frac{H_i - H_i^*}{T} - (S_i - S_i^*) \right] \qquad (A)$$

The lowest pressure for which data at 300°C are given in the steam tables is 1 kPa, and we assume that steam at these conditions is for practical purposes an ideal gas. Data for this state provide the following reference values:

$$P^* = 1 \text{ kPa}$$
$$H_i^* = 3,076.8 \text{ J g}^{-1}$$
$$S_i^* = 10.3450 \text{ J g}^{-1} \text{ K}^{-1}$$

Equation (A) may now be applied to states of superheated steam at 300°C for various values of P from 1 kPa to the saturation pressure of 8,592.7 kPa. For example, at $P = 4,000$ kPa and 300°C

$$H_i = 2,962.0 \text{ J g}^{-1}$$
$$S_i = 6.3642 \text{ J g}^{-1} \text{ K}^{-1}$$

These values must be multiplied by the molar mass of water (18.015) to put them on a molar basis for substitution into Eq. (A):

$$\ln \frac{f_i}{P^*} = \frac{18.015}{8.314} \left[\frac{2,962.0 - 3,076.8}{573.15} - (6.3642 - 10.3450) \right] = 8.1917$$

and $f_i/P^* = 3,611.0$. Since $P^* = 1$ kPa, $f_i = 3,611.0$ kPa. Thus the fugacity coefficient at 4,000 kPa is given by

$$\phi_i = \frac{f_i}{P} = \frac{3,611.0}{4,000} = 0.9028$$

Similar calculations at other pressures lead to the values plotted in Fig. 10.4 at pressures up to the saturation pressure of 8,592.7 kPa, where $f_i^{\text{sat}} = 6,738.9$ kPa and $\phi_i^{\text{sat}} = 0.7843$. According to Eqs. (10.38) and (10.40), the saturation values are unchanged by condensation.

Values of f_i and ϕ_i for liquid water at higher pressures are found by application of Eq. (10.41). Taking V_i^l equal to the molar volume of saturated liquid water at 300°C, we have

$$V_i^l = (1.403)(18.015) = 25.28 \text{ cm}^3 \text{ mol}^{-1}$$

For a pressure of 10,000 kPa, Eq. (10.41) then gives

$$f_i = 6,738.9 \exp \frac{25.28(10,000 - 8,592.7)}{(8,314)(573.15)} = 6,789.4 \text{ kPa}$$

The fugacity coefficient for liquid water at these conditions is then

$$\phi_i = f_i/P = 6,789.4/10,000 = 0.6789$$

Such calculations allow completion of Fig. 10.4, where the solid lines show how f_i and ϕ_i vary with pressure.

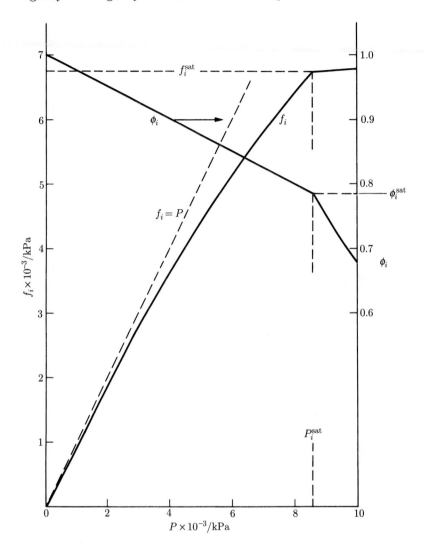

Figure 10.4: Fugacity and fugacity coefficient of steam at 300°C.

The curve for f_i deviates increasingly with increasing pressure from ideal-gas behavior, which is shown by the dashed line, $f_i = P$. At P_i^{sat} there is a sharp break, and the curve then rises very slowly with increasing pressure. Thus the fugacity of liquid water at 300°C is a weak function of pressure. This behavior is characteristic of liquids at temperatures well below the critical temperature. The fugacity coefficient ϕ_i decreases steadily from its zero-pressure value of unity as the pressure rises. Its rapid decrease in the liquid region is a consequence of the near constancy of the fugacity itself.

10.6   Fugacity and Fugacity Coefficient for Species in Solution

The definition of the fugacity of a species in solution is parallel to the definition of the pure-species fugacity. We simply write for species i in a mixture of real gases or in a solution of liquids an equation analogous to the ideal-gas expression, Eq. (10.28):

$$\mu_i \equiv \Gamma_i(T) + RT \ln \hat{f}_i \qquad (10.42)$$

where $\hat{f}_i$ is the fugacity of species i in solution, replacing the product $x_i P$. Since it is not a partial property, we identify it by a circumflex rather than an overbar.

A direct application of this definition indicates its potential utility. In Sec. 10.2 we found that the chemical potential provides a criterion for phase equilibrium according to the equation

$$\mu_i^\alpha = \mu_i^\beta = \cdots = \mu_i^\pi \qquad (i = 1, 2, ..., N) \qquad (10.6)$$

Since all phases in equilibrium are at the same temperature, an alternative and equally general criterion follows immediately from Eq. (10.42):

$$\boxed{\hat{f}_i^\alpha = \hat{f}_i^\beta = \cdots = \hat{f}_i^\pi} \qquad (i = 1, 2, ..., N) \qquad (10.43)$$

Thus multiple phases at the same T and P are in equilibrium when the fugacity of each constituent species is the same in all phases. This criterion of equilibrium is the one usually applied by chemical engineers in the solution of phase-equilibrium problems.

For the specific case of multicomponent vapor/liquid equilibrium, Eq. (10.43) becomes

$$\hat{f}_i^v = \hat{f}_i^l \qquad (i = 1, 2, \ldots, N) \qquad (10.44)$$

Equation (10.38) results as a special case when this relation is applied to the vapor/liquid equilibrium of *pure* species i.

The definition of the residual Gibbs energy as given by Eq. (6.36) along with Eq. (10.7), the definition of a partial property, provides a defining equation for the partial residual Gibbs energy. Thus, upon multiplication by n, Eq. (6.36) becomes

$$nG^R = nG - nG^{ig}$$

This equation applies to n moles of mixture. Differentiation with respect to n_i at constant T, P, and the n_j gives

$$\left[\frac{\partial(nG^R)}{\partial n_i} \right]_{P,T,n_j} = \left[\frac{\partial(nG)}{\partial n_i} \right]_{P,T,n_j} - \left[\frac{\partial(nG^{ig})}{\partial n_i} \right]_{P,T,n_j}$$

Reference to Eq. (10.7) shows that each term has the form of a partial molar property. Thus,

$$\boxed{\bar{G}_i^R = \bar{G}_i - \bar{G}_i^{ig}} \qquad (10.45)$$

an equation which defines the *partial residual Gibbs energy*, $\bar{G}_i^R$.

Subtracting Eq. (10.28) from Eq. (10.42), both written for the same temperature and pressure, yields

$$\mu_i - \mu_i^{ig} = RT \ln \frac{\hat{f}_i}{x_i P}$$

This result combined with Eq. (10.45) and the identity $\mu_i \equiv \bar{G}_i$ gives

$$\boxed{\bar{G}_i^R = RT \ln \hat{\phi}_i} \tag{10.46}$$

where by definition

$$\boxed{\hat{\phi}_i \equiv \frac{\hat{f}_i}{x_i P}} \tag{10.47}$$

The dimensionless ratio $\hat{\phi}_i$ is called the *fugacity coefficient of species i in solution*.

Equation (10.46) is the analog of Eq. (10.31), which relates ϕ_i to G_i^R. For an ideal gas, $\bar{G}_i^R$ is necessarily zero; therefore $\hat{\phi}_i^{ig} = 1$, and

$$\hat{f}_i^{ig} = x_i P \tag{10.48}$$

Thus the fugacity of species i in an ideal-gas mixture is equal to its partial pressure. The definition of a residual property is given by Eq. (6.39),

$$M^R \equiv M - M^{ig} \tag{6.39}$$

where M is the molar (or unit-mass) value of a thermodynamic property of a fluid and M^{ig} is the value that the property would have if the fluid were an ideal gas of the same composition at the same T and P. From this we have immediately [see the development of Eq. (10.45)]

$$\bar{M}_i^R = \bar{M}_i - \bar{M}_i^{ig} \tag{10.49}$$

These equations are the basis for extension of the fundamental property relation, given by Eq. (10.2), to residual properties.

We first develop an alternative form of Eq. (10.2), just as was done in Sec. 6.1, where the fundamental property relation was restricted to phases of constant composition. We make use of the same mathematical identity:

$$d\left(\frac{nG}{RT}\right) \equiv \frac{1}{RT}d(nG) - \frac{nG}{RT^2}dT$$

Substitution for $d(nG)$ by Eq. (10.2) and for G by Eq. (6.3) gives, after algebraic reduction,

$$\boxed{d\left(\frac{nG}{RT}\right) = \frac{nV}{RT}dP - \frac{nH}{RT^2}dT + \sum_i \frac{\bar{G}_i}{RT}dn_i} \tag{10.50}$$

We note with respect to this equation that all terms have the units of moles; moreover, in contrast to Eq. (10.2), the enthalpy rather than the entropy appears on the right-hand side. Equation (10.50) is a general relation expressing G/RT as a function of *all* of its canonical variables, T, P, and the mole numbers. It reduces to Eq. (6.29) for the special case of 1 mole of a constant-composition phase. Equations (6.30) and (6.31) follow from either equation, and equations for the other thermodynamic properties then come from appropriate defining equations. Knowledge of G/RT as a function of its canonical variables allows evaluation of all other thermodynamic properties, and therefore implicitly contains complete property information. However, we cannot directly exploit this characteristic, and in practice we deal with related properties, such as the residual Gibbs energy.

Since Eq. (10.50) is general, it may be written for the special case of an ideal gas:

$$d\left(\frac{nG^{ig}}{RT}\right) = \frac{nV^{ig}}{RT}dP - \frac{nH^{ig}}{RT^2}dT + \sum_i \frac{\bar{G}_i^{ig}}{RT}dn_i$$

In view of Eqs. (6.39) and (10.49), this equation may be subtracted from Eq. (10.50) to give

$$d\left(\frac{nG^R}{RT}\right) = \frac{nV^R}{RT}dP - \frac{nH^R}{RT^2}dT + \sum_i \frac{\bar{G}_i^R}{RT}dn_i \qquad (10.51)$$

Equation (10.51) is the *fundamental residual-property relation*. Its derivation from Eq. (10.2) parallels the derivation in Chap. 6 that led from Eq. (6.10) to Eq. (6.40). Indeed, Eqs. (6.10) and (6.40) are special cases of Eqs. (10.2) and (10.51), valid for one mole of a constant-composition fluid. An alternative form of Eq. (10.51) follows by introduction of the fugacity coefficient as given by Eq. (10.46):

$$d\left(\frac{nG^R}{RT}\right) = \frac{nV^R}{RT}dP - \frac{nH^R}{RT^2}dT + \sum_i \ln\hat{\phi}_i \, dn_i \qquad (10.52)$$

Equations so general as Eqs. (10.51) and (10.52) are useful for practical application only in their restricted forms. Division of Eqs. (10.51) and (10.52) by dP and restriction to constant T and composition leads to

$$\frac{V^R}{RT} = \left[\frac{\partial(G^R/RT)}{\partial P}\right]_{T,x} \qquad (10.53)$$

Similarly, division by dT and restriction to constant P and composition gives

$$\frac{H^R}{RT} = -T\left[\frac{\partial(G^R/RT)}{\partial T}\right]_{P,x} \qquad (10.54)$$

These equations are restatements of Eqs. (6.41) and (6.42) wherein the restriction of the derivatives to constant composition is shown explicitly. They lead to Eqs. (6.44),

(6.45), and (6.46) for the calculation of residual properties from volumetric data. Moreover Eq. (10.53) leads to Eq. (10.34), from which fugacity coefficients are calculated from volumetric data. It is through the residual properties that this kind of experimental information enters into the practical application of thermodynamics.

In addition, from Eq. (10.52) we have

$$\ln \hat{\phi}_i = \left[\frac{\partial(nG^R/RT)}{\partial n_i} \right]_{P,T,n_j} \tag{10.55}$$

This equation demonstrates that $\ln \hat{\phi}_i$ is a partial property with respect to G^R/RT. The partial-property analogs of Eqs. (10.53) and (10.54) are therefore

$$\left(\frac{\partial \ln \hat{\phi}_i}{\partial P} \right)_{T,x} = \frac{\bar{V}_i^R}{RT} \tag{10.56}$$

and

$$\left(\frac{\partial \ln \hat{\phi}_i}{\partial T} \right)_{P,x} = -\frac{\bar{H}_i^R}{RT^2} \tag{10.57}$$

Since the $\ln \hat{\phi}_i$ values are partial properties with respect to G^R/RT, the summability relation, Eq. (10.11), here takes the form

$$\boxed{\frac{G^R}{RT} = \sum_i x_i \ln \hat{\phi}_i} \tag{10.58}$$

Moreover, the $\ln \hat{\phi}_i$ values must conform to the Gibbs/Duhem equation, and as a special case of Eq. (10.14) we have

$$\boxed{\sum_i x_i \, d\ln \hat{\phi}_i = 0} \qquad (\text{const } T, P) \tag{10.59}$$

Example 10.6 Develop a general equation for calculation of $\ln \hat{\phi}_i$ values from compressibility-factor data.

SOLUTION For n moles of a constant-composition mixture, Eq. (6.44) becomes

$$\frac{nG^R}{RT} = \int_0^P (nZ - n) \frac{dP}{P}$$

Differentiation with respect to n_i at constant T, P, and n_j in accord with Eq. (10.55) yields

$$\ln \hat{\phi}_i = \int_0^P \left[\frac{\partial(nZ - n)}{\partial n_i} \right]_{P,T,n_j} \frac{dP}{P}$$

Since $\partial(nZ)/\partial n_i = \bar{Z}_i$ and $\partial n/\partial n_i = 1$, this reduces to

$$\ln \hat{\phi}_i = \int_0^P (\bar{Z}_i - 1) \frac{dP}{P} \tag{10.60}$$

where integration is at constant temperature and composition. This equation is the partial-property analog of Eq. (10.34). It allows the calculation of $\hat{\phi}_i$ values from PVT data.

10.7 Generalized Correlations for the Fugacity Coefficient

The generalized methods developed in Sec. 3.6 for the compressibility factor Z and in Sec. 6.6 for the residual enthalpy and entropy of pure gases are applied here to the fugacity coefficient. Equation (10.34) is put into generalized form by substitution of the relations,

$$P = P_c P_r \qquad dP = P_c\, dP_r$$

Hence

$$\ln \phi_i = \int_0^{P_r} (Z_i - 1) \frac{dP_r}{P_r} \tag{10.61}$$

where integration is at constant T_r. Substitution for Z_i by Eq. (3.46) yields

$$\ln \phi = \int_0^{P_r} (Z^0 - 1) \frac{dP_r}{P_r} + \omega \int_0^{P_r} Z^1 \frac{dP_r}{P_r}$$

where for simplicity we have dropped subscript i. This equation may be written in alternative form as

$$\ln \phi = \ln \phi^0 + \omega \ln \phi^1 \tag{10.62}$$

where

$$\ln \phi^0 \equiv \int_0^{P_r} (Z^0 - 1) \frac{dP_r}{P_r}$$

and

$$\ln \phi^1 \equiv \int_0^{P_r} Z^1 \frac{dP_r}{P_r}$$

The integrals in these equations may be evaluated numerically or graphically for various values of T_r and P_r from the data for Z^0 and Z^1 given in Tables E.1 through E.4. Another method, and the one adopted by Lee and Kesler to extend their correlation to fugacity coefficients, is based on an equation of state.

Since Eq. (10.62) may also be written

$$\phi = (\phi^0)(\phi^1)^\omega \tag{10.63}$$

we have the option of presenting correlations for ϕ^0 and ϕ^1 rather than for their logarithms. This is the choice made here, and Tables E.13 through E.16 present values for these quantities as derived from the Lee/Kesler correlation as functions of T_r and P_r, thus providing a three-parameter generalized correlation for fugacity coefficients. Tables E.13 and E.15 for ϕ^0 can be used alone as a two-parameter correlation which does not incorporate the refinement introduced by the acentric factor.

Example 10.7 Estimate from Eq. (10.63) a value for the fugacity of 1-butene vapor at 200°C and 70 bar.

SOLUTION These are the same conditions given in Example 6.6, where we found

$$T_r = 1.127 \qquad P_r = 1.731 \qquad \omega = 0.191$$

By interpolation in Tables E.15 and E.16 at these conditions,

$$\phi^0 = 0.627 \qquad \text{and} \qquad \phi^1 = 1.096$$

Equation (10.63) then gives

$$\phi = (0.627)(1.096)^{0.191} = 0.638$$

and

$$f = \phi P = (0.638)(70) = 44.7 \text{ bar}$$

A useful generalized correlation for $\ln \phi$ results when the simplest form of the virial equation is valid. Equations (3.47) and (3.48) combine to give

$$Z - 1 = \frac{P_r}{T_r}(B^0 + \omega B^1)$$

Substitution in Eq. (10.61) and integration yield

$$\ln \phi = \frac{P_r}{T_r}(B^0 + \omega B^1)$$

or

$$\phi = \exp\left[\frac{P_r}{T_r}(B^0 + \omega B^1)\right] \tag{10.64}$$

This equation, used in conjunction with Eqs. (3.50) and (3.51), provides reliable values of ϕ for any nonpolar or slightly polar gas when applied at conditions where Z is approximately linear in pressure. Figure 3.15 again serves as a guide to its suitability.

Just as we gave names in Sec. 6.6 to the functions used in evaluation of H^R and S^R by the generalized virial-coefficient correlation, we also name the function used to evaluate ϕ. Equation (10.64) in combination with Eqs. (3.50) and (3.51) provides this function, which for computational purposes is named

PHIB(TR,PR,OMEGA)

It denotes numerical values of ϕ. Representative computer programs for its evaluation are given in App. D. From Example 6.6, Step (b), we have for instance

PHIB(0.650,0.0316,0.191) ≡ 0.956

The generalized correlation just described is for *pure* gases only. In the remainder of this section we show how the virial equation may be generalized to allow calculation of fugacity coefficients $\hat{\phi}_i$ of species in gas *mixtures*.

The virial equation is written for a gas mixture exactly as it is for a pure species:

$$Z = 1 + \frac{BP}{RT} \qquad (3.31)$$

The mixture second virial coefficient B is a function of temperature and composition. Its *exact* composition dependence is given by statistical mechanics, and this makes the virial equation preeminent among equations of state where it is applicable, i.e., to gases at low to moderate pressures. The equation giving this composition dependence is

$$B = \sum_i \sum_j y_i y_j B_{ij} \qquad (10.65)$$

where y represents mole fractions in a gas mixture. The indices i and j identify species, and both run over all species present in the mixture. The virial coefficient B_{ij} characterizes a bimolecular interaction between molecule i and molecule j, and therefore $B_{ij} = B_{ji}$. The summations account for all possible bimolecular interactions.

For a binary mixture $i = 1, 2$ and $j = 1, 2$; the expansion of Eq. (10.65) then gives

$$B = y_1 y_1 B_{11} + y_1 y_2 B_{12} + y_2 y_1 B_{21} + y_2 y_2 B_{22}$$

or

$$B = y_1^2 B_{11} + 2 y_1 y_2 B_{12} + y_2^2 B_{22} \qquad (10.66)$$

Two types of virial coefficients have appeared: B_{11} and B_{22}, for which the successive subscripts are the same, and B_{12}, for which the two subscripts are different. The first type represents the virial coefficient of a pure species; the second is a mixture property, known as a *cross coefficient*. Both are functions of temperature only.

Equation (10.66) allows us to find expressions for $\ln \hat{\phi}_1$ and $\ln \hat{\phi}_2$ for a binary gas mixture that obeys Eq. (3.31), the simplest form of the virial equation. Written for n moles of gas mixture, Eq. (3.31) becomes

$$nZ = n + \frac{nBP}{RT}$$

Differentiation with respect to n_1 gives

$$\bar{Z}_1 \equiv \left[\frac{\partial(nZ)}{\partial n_1} \right]_{P,T,n_2} = 1 + \frac{P}{RT} \left[\frac{\partial(nB)}{\partial n_1} \right]_{T,n_2}$$

Substitution for $\bar{Z}_1$ in Eq. (10.60) yields

$$\ln \hat{\phi}_1 = \frac{1}{RT} \int_0^P \left[\frac{\partial(nB)}{\partial n_1} \right]_{T,n_2} dP = \frac{P}{RT} \left[\frac{\partial(nB)}{\partial n_1} \right]_{T,n_2}$$

where the integration is elementary, because B is not a function of pressure. All that remains is evaluation of the derivative.

The second virial coefficient as given by Eq. (10.66) may be written

$$B = y_1(1-y_2)B_{11} + 2y_1y_2B_{12} + y_2(1-y_1)B_{22}$$
$$= y_1 B_{11} - y_1y_2B_{11} + 2y_1y_2B_{12} + y_2B_{22} - y_1y_2B_{22}$$

or

$$B = y_1 B_{11} + y_2 B_{22} + y_1 y_2 \delta_{12}$$

where

$$\delta_{12} \equiv 2B_{12} - B_{11} - B_{22}$$

Since $y_i = n_i/n$,

$$nB = n_1 B_{11} + n_2 B_{22} + \frac{n_1 n_2}{n}\delta_{12}$$

Differentiation gives

$$\left[\frac{\partial(nB)}{\partial n_1}\right]_{T,n_2} = B_{11} + \left(\frac{1}{n} - \frac{n_1}{n^2}\right)n_2\delta_{12}$$
$$= B_{11} + (1-y_1)y_2\delta_{12} = B_{11} + y_2^2\delta_{12}$$

Therefore

$$\ln\hat\phi_1 = \frac{P}{RT}(B_{11} + y_2^2\delta_{12}) \tag{10.67}$$

and similarly,

$$\ln\hat\phi_2 = \frac{P}{RT}(B_{22} + y_1^2\delta_{12}) \tag{10.68}$$

Equations (10.67) and (10.68) are readily extended for application to multicomponent mixtures; the general equation is[8]

$$\ln\hat\phi_k = \frac{P}{RT}\left[B_{kk} + \frac{1}{2}\sum_i\sum_j y_i y_j(2\delta_{ik} - \delta_{ij})\right] \tag{10.69}$$

where the dummy indices i and j run over all species, and

$$\delta_{ik} \equiv 2B_{ik} - B_{ii} - B_{kk}$$
$$\delta_{ij} \equiv 2B_{ij} - B_{ii} - B_{jj}$$

with $\delta_{ii} = 0$, $\delta_{kk} = 0$, etc., and $\delta_{ki} = \delta_{ik}$, etc.

Values of the pure-species virial coefficients B_{kk}, B_{ii}, etc., can be determined from the generalized correlation represented by Eqs. (3.48), (3.50), and (3.51). The

---

[8]H. C. Van Ness and M. M. Abbott, *Classical Thermodynamics of Nonelectrolyte Solutions: With Applications to Phase Equilibria*, pp. 135–140, McGraw-Hill, New York, 1982.

cross coefficients B_{ik}, B_{ij}, etc., are found from an extension of the same correlation. For this purpose, Eq. (3.48) can be rewritten in the more general form[9]

$$B_{ij} = \frac{RT_{cij}}{P_{cij}}(B^0 + \omega_{ij}B^1) \tag{10.70}$$

where B^0 and B^1 are the same functions of T_r as given by Eqs. (3.50) and (3.51). The combining rules proposed by Prausnitz for calculation of ω_{ij}, T_{cij}, and P_{cij} are

$$\omega_{ij} = \frac{\omega_i + \omega_j}{2} \tag{10.71}$$

$$T_{cij} = (T_{ci}T_{cj})^{1/2}(1 - k_{ij}) \tag{10.72}$$

and

$$P_{cij} = \frac{Z_{cij}RT_{cij}}{V_{cij}} \tag{10.73}$$

where

$$Z_{cij} = \frac{Z_{ci} + Z_{cj}}{2} \tag{10.74}$$

and

$$V_{cij} = \left(\frac{V_{ci}^{1/3} + V_{cj}^{1/3}}{2}\right)^3 \tag{10.75}$$

In Eq. (10.72), k_{ij} is an empirical interaction parameter specific to an i-j molecular pair. When $i = j$ and for chemically similar species, $k_{ij} = 0$. Otherwise, it is a small positive number evaluated from minimal PVT data or in the absence of data set equal to zero.

When $i = j$, all equations reduce to the appropriate values for a pure species. When $i \neq j$, these equations define a set of interaction parameters having no physical significance. Reduced temperature is given for each ij pair by $T_{rij} \equiv T/T_{cij}$.

For a mixture, values of B_{ij} from Eq. (10.70) substituted into Eq. (10.65) yield the mixture second virial coefficient B, and substituted into Eq. (10.69) [Eqs. (10.67) and (10.68) for a binary] they yield values of $\ln \hat{\phi}_i$.

The primary virtue of the generalized correlation for second virial coefficients presented here is simplicity; more accurate, but more complex, correlations appear in the literature.[10]

[9] J. M. Prausnitz, R. N. Lichtenthaler, and E. G. de Azevedo, *Molecular Thermodynamics of Fluid-Phase Equilibria*, 2nd ed., pp. 132 and 162, Prentice-Hall, Englewood Cliffs, NJ, 1986.

[10] C. Tsonopoulos, *AIChE J.*, vol. 20, pp. 263–272, 1974, vol. 21, pp. 827–829, 1975, vol. 24, pp. 1112–1115, 1978.; C. Tsonopoulos, *Adv. in Chemistry Series 182*, pp. 143–162, 1979; J. G. Hayden and J. P. O'Connell, *Ind. Eng. Chem. Proc. Des. Dev.*, vol. 14, pp. 209–216, 1975; D. W. McCann and R. P. Danner, *Ibid.*, vol. 23, pp. 529–533, 1984; J. A. Abusleme and J. H. Vera, *AIChE J.*, vol. 35, pp. 481–489, 1989.

Example 10.8 Estimate $\hat{\phi}_1$ and $\hat{\phi}_2$ by Eqs. (10.67) and (10.68) for an equimolar mixture of methyl ethyl ketone(1)/toluene(2) at 50°C and 25 kPa. Set all $k_{ij} = 0$.

SOLUTION The required data are as follows:

ij	$T_{cij}/$K	$P_{cij}/$bar	$V_{cij}/$cm^3 mol^{-1}	Z_{cij}	ω_{ij}
11	535.5	41.5	267.	0.249	0.323
22	591.8	41.1	316.	0.264	0.262
12	563.0	41.3	291.	0.256	0.293

where values in the last row have been calculated by Eqs. (10.71) through (10.75). The values of T_{rij}, together with B^0, B^1, and B_{ij} calculated for each ij pair by Eqs. (3.50), (3.51), and (10.70), are as follows:

ij	T_{rij}	B^0	B^1	$B_{ij}/$cm^3 mol^{-1}
11	0.603	−0.865	−1.300	−1,387.
22	0.546	−1.028	−2.045	−1,860.
12	0.574	−0.943	−1.632	−1,611.

Calculating δ_{12} according to its definition, we get

$$\delta_{12} = 2B_{12} - B_{11} - B_{22} = (2)(-1{,}611) + 1{,}387 + 1{,}860 = 25 \text{ cm}^3 \text{ mol}^{-1}$$

Equations (10.67) and (10.68) then yield

$$\ln \hat{\phi}_1 = \frac{P}{RT}(B_{11} + y_2^2 \delta_{12}) = \frac{25}{(8{,}314)(323.15)}[-1{,}387 + (0.5)^2(25)] = -0.0128$$

$$\ln \hat{\phi}_2 = \frac{P}{RT}(B_{22} + y_1^2 \delta_{12}) = \frac{25}{(8{,}314)(323.15)}[-1{,}860 + (0.5)^2(25)] = -0.0172$$

whence

$$\hat{\phi}_1 = 0.987 \qquad \text{and} \qquad \hat{\phi}_2 = 0.983$$

These results are representative of values obtained for vapor phases at typical conditions of low-pressure vapor/liquid equilibrium.

10.8 The Ideal Solution

The ideal gas is a useful model of the behavior of gases, and serves as a standard to which real-gas behavior can be compared. This is formalized by the introduction of residual properties. Another useful model is the *ideal solution*, which serves as a standard to which real-solution behavior can be compared. We will see in the following section how this is formalized by introduction of *excess properties*.

Equation (10.26) characterizes the behavior of a constituent species in an ideal-gas mixture:

$$\bar{G}_i^{ig} = G_i^{ig} + RT \ln x_i \tag{10.26}$$

This equation takes on a new dimension if we replace G_i^{ig}, the Gibbs energy of pure species i in the ideal-gas state, by G_i, the Gibbs energy of pure species i as it actually exists at the mixture T and P and in the same physical state (*real* gas, liquid, or solid) as the mixture. We can then apply it to species in real solutions, indeed to liquids and solids as well as to gases. We therefore *define* an ideal solution as one for which

$$\bar{G}_i^{id} = G_i + RT \ln x_i \qquad (10.76)$$

where superscript *id* denotes an ideal-solution property.

All other thermodynamic properties for an ideal solution follow from this equation. Thus when Eq. (10.76) is differentiated with respect to temperature at constant pressure and composition and then combined with Eq. (10.18) written for an ideal solution, we get

$$\bar{S}_i^{id} = -\left(\frac{\partial \bar{G}_i^{id}}{\partial T}\right)_{P,x} = -\left(\frac{\partial G_i}{\partial T}\right)_P - R \ln x_i$$

By Eq. (10.4), $(\partial G_i / \partial T)_P$ is simply $-S_i$, and this becomes

$$\bar{S}_i^{id} = S_i - R \ln x_i \qquad (10.77)$$

Similarly, as a result of Eq. (10.19),

$$\bar{V}_i^{id} = \left(\frac{\partial \bar{G}_i^{id}}{\partial P}\right)_{T,x} = \left(\frac{\partial G_i}{\partial P}\right)_T$$

and by Eq. (10.5)

$$\bar{V}_i^{id} = V_i \qquad (10.78)$$

Since $\bar{H}_i^{id} = \bar{G}_i^{id} + T\bar{S}_i^{id}$, substitutions by Eqs. (10.76) and (10.77) yield

$$\bar{H}_i^{id} = G_i + RT \ln x_i + TS_i - RT \ln x_i$$

or

$$\bar{H}_i^{id} = H_i \qquad (10.79)$$

The summability relation, Eq. (10.11), applied to the special case of an ideal solution is written

$$M^{id} = \sum_i x_i \bar{M}_i^{id}$$

Application to Eqs. (10.76) through (10.79) yields

$$G^{id} = \sum_i x_i G_i + RT \sum_i x_i \ln x_i \qquad (10.80)$$

$$S^{id} = \sum_i x_i S_i - R \sum_i x_i \ln x_i \qquad (10.81)$$

$$V^{id} = \sum_i x_i V_i \qquad (10.82)$$

$$H^{id} = \sum_i x_i H_i \qquad (10.83)$$

A simple equation for the fugacity of a species in an ideal solution follows from Eq. (10.76). Written for the special case of species i in an ideal solution, Equation (10.42) becomes

$$\mu_i^{id} = \bar{G}_i^{id} = \Gamma_i(T) + RT \ln \hat{f}_i^{id}$$

When this equation and Eq. (10.30) are combined with Eq. (10.76), $\Gamma_i(T)$ is eliminated, and the resulting expression reduces to

$$\hat{f}_i^{id} = x_i f_i \qquad (10.84)$$

This equation, known as the *Lewis/Randall rule*, applies to each species in an ideal solution at all conditions of temperature, pressure, and composition. It shows that the fugacity of each species in an ideal solution is proportional to its mole fraction; the proportionality constant is the fugacity of *pure* species i in the same physical state as the solution and at the same T and P. Division of both sides of Eq. (10.84) by Px_i and substitution of $\hat{\phi}_i^{id}$ for $\hat{f}_i^{id}/x_i P$ [Eq. (10.47)] and of ϕ_i for f_i/P [Eq. (10.32)] gives an alternative form:

$$\hat{\phi}_i^{id} = \phi_i \qquad (10.85)$$

Thus the fugacity coefficient of species i in an ideal solution is equal to the fugacity coefficient of *pure* species i in the same physical state as the solution and at the same T and P.

Ideal-solution behavior is often approximated by solutions comprised of molecules not too different in size and of the same chemical nature. Thus, a mixture of isomers, such as *ortho-*, *meta-*, and *para-*xylene, conforms very closely to ideal-solution behavior. So do mixtures of adjacent members of a homologous series, as for example, *n*-hexane/*n*-heptane, ethanol/propanol, and benzene/toluene. Other examples are acetone/acetonitrile and acetonitrile/nitromethane.

10.9 Excess Properties

The residual Gibbs energy and the fugacity coefficient are directly related to experimental PVT data by Eqs. (6.44), (10.34) and (10.60). Where such data can be adequately correlated by equations of state, thermodynamic-property information is advantageously provided by residual properties. Indeed, if convenient treatment of all fluids by means of equations of state were possible, the thermodynamic-property relations already presented would suffice. However, *liquid* solutions are often more easily dealt with through properties that measure their deviations, not from ideal-gas behavior, but from ideal-solution behavior. Thus the mathematical formalism of *excess* properties is analogous to that of the residual properties.

If M represents the molar (or unit-mass) value of any extensive thermodynamic property (e.g., V, U, H, S, G, etc.), then an excess property M^E is defined as the difference between the actual property value of a solution and the value it would have as an ideal solution at the same temperature, pressure, and composition. Thus,

$$M^E \equiv M - M^{id} \qquad (10.86)$$

This definition is analogous to the definition of a residual property as given by Eq. (6.39). However, excess properties have no meaning for pure species, whereas residual properties exist for pure species as well as for mixtures. In addition, we have analogous to Eq. (10.49) the partial-property relation,

$$\bar{M}_i^E = \bar{M}_i - \bar{M}_i^{id} \qquad (10.87)$$

where $\bar{M}_i^E$ is a partial excess property. The fundamental excess-property relation is derived in exactly the same way as the fundamental residual-property relation and leads to analogous results. Equation (10.50), written for the special case of an ideal solution, is subtracted from Eq. (10.50) itself, yielding

$$\boxed{d\left(\frac{nG^E}{RT}\right) = \frac{nV^E}{RT}dP - \frac{nH^E}{RT^2}dT + \sum_i \frac{\bar{G}_i^E}{RT}dn_i} \qquad (10.88)$$

This is the *fundamental excess-property relation*, analogous to Eq. (10.51), the fundamental residual-property relation.

The excess Gibbs energy is of particular interest. Equation (10.42) may be written

$$\bar{G}_i = \Gamma_i(T) + RT \ln \hat{f}_i$$

In accord with Eq. (10.84) for an ideal solution, this becomes

$$\bar{G}_i^{id} = \Gamma_i(T) + RT \ln x_i f_i$$

By difference

$$\bar{G}_i - \bar{G}_i^{id} = RT \ln \frac{\hat{f}_i}{x_i f_i}$$

The difference on the left is the partial excess Gibbs energy $\bar{G}_i^E$; the dimensionless ratio $\hat{f}_i/x_i f_i$ appearing on the right is called the *activity coefficient of species i in solution*, and is given the symbol γ_i. Thus, *by definition*,

$$\gamma_i \equiv \frac{\hat{f}_i}{x_i f_i} \tag{10.89}$$

and

$$\bar{G}_i^E = RT \ln \gamma_i \tag{10.90}$$

Comparison with Eq. (10.46) shows that Eq. (10.90) relates γ_i to $\bar{G}_i^E$ exactly as Eq. (10.46) relates $\hat{\phi}_i$ to $\bar{G}_i^R$. For an ideal solution, $\bar{G}_i^E = 0$, and therefore $\gamma_i = 1$.

An alternative form of Eq. (10.88) follows by introduction of the activity coefficient through Eq. (10.90):

$$d\left(\frac{nG^E}{RT}\right) = \frac{nV^E}{RT} dP - \frac{nH^E}{RT^2} dT + \sum_i \ln \gamma_i \, dn_i \tag{10.91}$$

Again, the generality of these equations precludes their direct practical application. Rather, we make use of restricted forms, which are written by inspection:

$$\frac{V^E}{RT} = \left[\frac{\partial(G^E/RT)}{\partial P}\right]_{T,x} \tag{10.92}$$

$$\frac{H^E}{RT} = -T \left[\frac{\partial(G^E/RT)}{\partial T}\right]_{P,x} \tag{10.93}$$

and

$$\ln \gamma_i = \left[\frac{\partial(nG^E/RT)}{\partial n_i}\right]_{P,T,n_j} \tag{10.94}$$

The last relation demonstrates that $\ln \gamma_i$ is a partial property with respect to G^E/RT. The partial-property analogues of Eqs. (10.92) and (10.93) are:

$$\left(\frac{\partial \ln \gamma_i}{\partial P}\right)_{T,x} = \frac{\bar{V}_i^E}{RT} \tag{10.95}$$

and

$$\left(\frac{\partial \ln \gamma_i}{\partial T}\right)_{P,x} = -\frac{\bar{H}_i^E}{RT^2} \tag{10.96}$$

Since $\ln \gamma_i$ is a partial property with respect to G^E/RT, we can write the following forms of the summability and Gibbs/Duhem equations:

$$\frac{G^E}{RT} = \sum_i x_i \ln \gamma_i \tag{10.97}$$

and

$$\boxed{\sum_i x_i d\ln\gamma_i = 0} \qquad \text{(const } T, P) \qquad\qquad (10.98)$$

These equations are analogs of Eqs. (10.53) through (10.59). Whereas the fundamental *residual*-property relation derives its usefulness from its direct relation to experimental PVT data and equations of state, the *excess*-property formulation is useful because V^E, H^E, and γ_i are all experimentally accessible. Activity coefficients are found from vapor/liquid equilibrium data, and V^E and H^E values come from mixing experiments, topics treated in the following chapter.

Equations (10.92) and (10.93) allow direct calculation of the effects of pressure and temperature on the excess Gibbs energy. For example, an equimolar mixture of benzene and cyclohexane at 25°C and 1 bar has an excess volume of about 0.65 cm^3 mol^{-1}and an excess enthalpy of about 800 J mol^{-1}. Thus at these conditions,

$$\left[\frac{\partial(G^E/RT)}{\partial P}\right]_{T,x} = \frac{0.65}{(83.14)(298.15)} = 2.62\times10^{-5}\ \text{bar}^{-1}$$

and

$$\left[\frac{\partial(G^E/RT)}{\partial T}\right]_{P,x} = \frac{-800}{(8.314)(298.15)^2} = -1.08\times10^{-3}\ \text{K}^{-1}$$

The most striking observation about these results is that it takes a pressure change of more than 40 bar to have an effect on the excess Gibbs energy equivalent to that of a temperature change of 1 K. This is the reason that for liquids at low pressures the effect of pressure on the excess Gibbs energy (and therefore on the activity coefficients) is usually neglected.

Just as the fundamental property relation of Eq. (10.50) provides complete property information from a canonical equation of state expressing G/RT as a function of T, P, and composition, so the fundamental *residual*-property relation, Eq. (10.51) or (10.52), provides complete *residual*-property information from a PVT equation of state, from PVT data, or from generalized PVT correlations. However, for complete *property* information, one needs in addition to PVT data the ideal-gas-state heat capacities of the species that comprise the system.

Given an equation for G^E/RT as a function of its canonical variables, T, P, and composition, the fundamental *excess*-property relation, Eq. (10.88) or (10.91), provides complete *excess*-property information. However, this formulation represents less-complete property information than does the residual-property formulation, because it tells us nothing about the properties of the pure constituent chemical species.

10.10 Hydrogen Bonding and Charge-Transfer Complexing

The intermolecular potential is dominated at small separations by repulsions, and at large separations by attractions varying approximately as r^{-6} (Sec. 3.8). These

Table 10.1: Pauling electronegativity $\mathcal{X}_P$ for some nonmetalic elements.

Element	$\mathcal{X}_P$	Element	$\mathcal{X}_P$
F	4.0	I	2.5
O	3.5	C	2.5
N	3.0	S	2.5
Cl	3.0	Se	2.4
Br	2.8	H	2.1

interactions are called "physical," because their origins are explained on the presumption that interacting species preserve their identities. For some systems another class of interactions, called "quasichemical," operates primarily at intermediate separations, i.e., at $r \approx r_0$ (see Fig. 3.17). As the name suggests, quasichemical forces are manifested as strong attractive interactions, in which participating species combine to form new chemical entities. We describe below two important classes of quasichemical interactions: hydrogen bonding and charge-transfer complexing.

Essential to a discussion of hydrogen bonding is the concept of *electronegativity*. According to valence-bond theory, the atoms which combine to form a molecule share electrons. If the bonded atoms are identical (e.g., the Cl atoms in a Cl_2 molecule), the bonding electrons are shared equally between the atoms. However, if the atoms are different (e.g., the H and Cl atoms of HCl), the shared electrons are generally attracted more strongly by one of the atoms (Cl in the case of HCl), and this atom is said to be *more electronegative* than the other. Thus, electronegativity is a measure of the relative ability of an atom in a molecule to attact electrons to itself.

The notion of electronegativity was introduced in 1932 by Pauling;[11] he was the first of several to propose a quantitative scale for its expression. Based largely on thermochemical data, Pauling's electronegativity $\mathcal{X}_P$ assumes values between about 0.7 and 4.0 for those elements known to participate in compound formation. Metallic elements have values less than about 2.0; nonmetals, values greater than about 2.0. Table 10.1 shows Pauling electronegativities for ten nonmetallic elements. Of these, fluorine is the most electronegative ($\mathcal{X}_P = 4.0$), and hydrogen, the least ($\mathcal{X}_P = 2.1$).

An intermolecular hydrogen bond forms between a hydrogen-donor molecule (conventionally represented as A-H) and an electron-rich acceptor site (conventionally denoted by the letter B). Entity A is an atom (possibly attached to other atoms), which is more electronegative than hydrogen. Hydrogen-acceptor site B may be an atom more electronegative than hydrogen; the site may also be a double or triple bond, or it may be an aromatic hydrocarbon ring. The hydrogen-bonded complex is conventionally represented as A-H $\cdots$ B, where the three dots denote the hydrogen bond.

---

[11]Linus Pauling (1901–1994), American chemist and (twice) Nobel laureate.

Examples of strong hydrogen donors include hydrogen fluoride (HF), water (HOH), hydrogen peroxide (HOOH), alcohols (ROH), carboxylic acids (RCOOH), ammonia (H_2NH), primary amines (RNH_2), and secondary amines (R_2NH). In each of these molecules, one or more hydrogen atoms is attached to an atom of a highly electronegative element (F, O, or N; see Table 10.1). The halogen acids HCl, HBr, and HI are also hydrogen donors, as are a few species containing the C-H bond. However, the difference in electronegativity between carbon and hydrogen is not large (Table 10.1), and the ability of the H in C-H to function as a donor hydrogen seems possible only when the carbon atom is itself attached to highly electronegative atoms or electron-rich sites. Verified examples of C-H hydrogen donors thus include chloroform (Cl_3CH), dichloromethane (Cl_2CH_2), and hydrogen cyanide (NCH).

The highly electronegative elements F, O, and N serve as atomic hydrogen-acceptor sites. Hence HF, HOH, HOOH, ROH, RCOOH, H_2NH, R_2NH, and NCH are hydrogen acceptors. But so are aldehydes (ROCH), ketones (ROCR), ethers (ROR), esters (ROCOR), and tertiary amines (R_3N), species which have no active hydrogens.

The phenomenon of hydrogen bonding is easily rationalized. The H in donor species A-H is electron deficient because of the higher electronegativity of A. Hence the H is attracted to the electron-rich acceptor site B. Unfortunately, such a simple electrostatic picture is unable to account quantitatively for some important features of the hydrogen bond.[12] As a result, an algebraic contribution cannot generally be ascribed to the intermolecular potential function $\mathcal{U}(r)$ for hydrogen-bonding interactions. Nevertheless, we can by example indicate the kinds of intermolecular pairs for which hydrogen-bonding interactions are important. Convenience here suggests division of hydrogen-bonding interactions into two classes: *association* and *solvation*.

Association is an attractive interaction between molecules *of the same kind*. In the context of hydrogen bonding, an associating species must have both an active hydrogen and a hydrogen-acceptor site. Examples include water (the O is an acceptor site), ammonia (with N the acceptor site), alcohols, primary and secondary amines, and carboxylic acids. Hydrogen bonding by association is often reflected dramatically in the properties (e.g., boiling points, heats of vaporization, and viscosities) of the pure species.

Solvation is an attractive interaction between *unlike* molecular species. With respect to hydrogen bonding, solvation occurs between a species that is a hydrogen donor and another species that is a hydrogen acceptor. In "pure" solvation, neither species associates; an example is the acetone/chloroform system, in which chloroform is (only) a hydrogen donor and acetone (only) a hydrogen acceptor. However, solvation may occur between two associators (e.g., ethanol and water), between an associator and a hydrogen donor (e.g., ethanol and chloroform), and between an associator and a hydrogen acceptor (e.g., ethanol and acetone).

---

[12]See, e.g., J. E. Huheey, *Inorganic Chemistry*, 3rd ed., pp. 268–272, Harper & Row, New York, 1983.

Table 10.2: Hydrogen-bonding interactions among pairs of species. ($\mathcal{D}$ = nonassociating H-donor; $\mathcal{A}$ = nonassociating H-acceptor; $\mathcal{AD}$ = associating species)

	$\mathcal{D}(1)$	$\mathcal{A}(1)$	$\mathcal{AD}(1)$
$\mathcal{D}(2)$	No H-bonding	$\mathcal{D}(2) \cdots \mathcal{A}(1)$	$\mathcal{D}(2) \cdots \mathcal{AD}(1)$ $\mathcal{AD}(1) \cdots \mathcal{AD}(1)$
$\mathcal{A}(2)$	$\mathcal{D}(1) \cdots \mathcal{A}(2)$	No H-bonding	$\mathcal{AD}(1) \cdots \mathcal{A}(2)$ $\mathcal{AD}(1) \cdots \mathcal{AD}(1)$
$\mathcal{AD}(2)$	$\mathcal{D}(1) \cdots \mathcal{AD}(2)$ $\mathcal{AD}(2) \cdots \mathcal{AD}(2)$	$\mathcal{AD}(2) \cdots \mathcal{A}(1)$ $\mathcal{AD}(2) \cdots \mathcal{AD}(2)$	$\mathcal{AD}(1) \cdots \mathcal{AD}(2)$ $\mathcal{AD}(2) \cdots \mathcal{AD}(1)$ $\mathcal{AD}(1) \cdots \mathcal{AD}(1)$ $\mathcal{AD}(2) \cdots \mathcal{AD}(2)$

Table 10.2 suggests the types of hydrogen-bonding interactions that can occur between molecules of various kinds. Here, $\mathcal{D}$ denotes a nonassociating hydrogen-donor species (e.g., Cl_3CH or Br_3CH), $\mathcal{A}$ is a nonassociating hydrogen acceptor species [e.g., $(CH_3)_2CO$ or $(C_2H_5)_2O$], and $\mathcal{AD}$ is a species that can associate by hydrogen bonding [e.g., CH_3OH or $(C_2H_5)_2NH$]. Mixtures containing two different associating species offer the richest variety of opportunities for hydrogen bonding. For example, in a binary mixture of ammonia(1) and water(2), hydrogen-bonded dimers may be formed in four ways, two by solvation and two by association:

$$
\begin{array}{ll}
\text{H} \quad\quad \text{H} & \quad\quad\quad \text{H} \\
\text{N-H} \cdots \text{O} & \text{HO-H} \cdots \text{NH} \\
\text{H} \quad\quad \text{H} & \quad\quad\quad \text{H}
\end{array}
$$

$$
\begin{array}{ll}
\text{H} \quad\quad \text{H} & \quad\quad\quad \text{H} \\
\text{N-H} \cdots \text{NH} & \text{HO-H} \cdots \text{O} \\
\text{H} \quad\quad \text{H} & \quad\quad\quad \text{H}
\end{array}
$$

Compelling experimental evidence exists for quasichemical interactions between certain non-hydrogen-donor polar compounds (e.g., pyridine, ketones, and aldehydes) and aromatic hydrocarbons (e.g., benzene). In these cases, the polar compounds have no active hydrogens, and hence the interaction cannot be hydrogen bonding. Nevertheless, a complex appears to be formed. Mulliken[13] gave the name *electron donor-acceptor complex* to these and other such entities; they are more commonly called *charge-transfer complexes*. Proper explanation of charge-transfer complexing requires use of concepts from molecular orbital theory, which we cannot

[13]R. S. Mulliken and W. B. Person, *Molecular Complexes: A Lecture and Reprint Volume*, Wiley-Interscience, New York, 1969.

develop here. Additionally, because of the apparent near-ubiquity of charge-transfer phenomena, and of the widely varying strengths of the interactions, it is often difficult to *predict* when they will make significant contributions to intermolecular forces. Mulliken and Person (*loc. cit.*) offer guidance, but most engineers would view the invocation of charge-transfer complexing as a helpful explanatory, rather than a predictive, exercise. Its role is clearest for the kinds of systems mentioned at the beginning of this paragraph.

10.11   Behavior of Excess Properties of Liquid Mixtures

Peculiarities of liquid-mixture behavior are most dramatically revealed in the excess properties. Those of primary interest are G^E (or G^E/RT) together with H^E and S^E, which are related to the temperature derivative of G^E. As suggested in Sec. 10.9, liquid properties at normal temperatures are not strongly influenced by pressure.

The excess Gibbs energy comes from experiment through reduction of vapor/liquid equilibrium data, and H^E is determined by mixing experiments (see Chapt. 11). The excess entropy is not measured directly, but is found from the general equation

$$G^E = H^E - TS^E \qquad (10.99)$$

which follows from Eqs. (6.3) and (10.86), the definition of G and the definition of an excess property.

Figure 10.5 illustrates the composition dependence of G^E, H^E, and TS^E for six binary liquid mixtures at 50°C and approximately atmospheric pressure. For consistency with Eq. (10.99), we plot the product TS^E rather than S^E itself. Although the systems exhibit a diversity of behavior, we may note the following common features:

1. All excess properties become zero as either species approaches purity.

2. Although G^E vs. x_1 is approximately parabolic in shape, both H^E and TS^E exhibit individualistic composition dependencies.

3. When an excess property M^E has a single sign (as does G^E in all six cases), the extreme value of M^E (maximum or minimum) often occurs near the equimolar composition.

Feature 1 is a consequence of the definition, Eq. (10.86); as any x_i approaches unity, M and M^{id} both approach M_i, the corresponding property of pure i. Features 2 and 3 are generalizations based on observation, and admit exceptions (note, e.g., the behavior of H^E for the ethanol/water system).

As suggested by Fig. 10.5, the principal excess properties (G^E, H^E, and S^E) can exhibit a variety of combinations of signs. The signs and relative magnitudes of these quantities are useful for qualitative engineering purposes and for elucidating

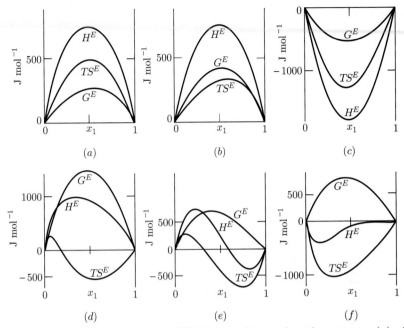

Figure 10.5: Excess properties at 50°C for six binary liquid systems: (*a*) chloroform(1)/*n*-heptane(2); (*b*) acetone(1)/methanol(2); (*c*) acetone(1)/chloroform(2); (*d*) ethanol(1)/*n*-heptane(2); (*e*) ethanol(1)/chloroform(2); (*f*) ethanol(1)/water(2).

the molecular phenomena which are the basis for observed solution behavior. Abbott et al.[14] have organized $G^E/H^E/S^E$ data for about 400 binary liquid mixtures in a visual scheme which permits identification of patterns, trends, and norms of behavior with respect to *mixture type*. In what follows, we describe the basis for the scheme, and present a few important generalizations which follow from examination of representative data.

Excess properties for liquid mixtures depend primarily on temperature and composition, and therefore comparison of data for different mixtures is best done at *fixed* T and x. Since many M^E data are available at near-ambient temperatures, T is chosen as 298.15 K (25°C). As noted above, extreme values for M^E often occur near equimolar composition; we therefore fix $x_1 = x_2 = 0.5$.

Division of Eq. (10.99) by RT puts it into dimensionless form:

$$\frac{G^E}{RT} = \frac{H^E}{RT} - \frac{S^E}{R} \tag{10.100}$$

The six possible combinations of sign for the three excess properties are enumerated in Table 10.3. Each combination defines a region on the G^E/RT vs. H^E/RT *diagram* shown in skeleton form as Fig. 10.6.

[14]M. M. Abbott, J. P. O'Connell, and Twenty Rensselaer Students, *Chem. Eng. Educ.*, vol. 28, pp. 18–23 and 77, 1994.

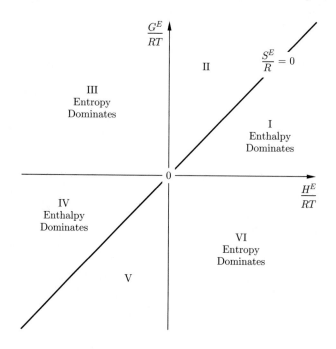

Figure 10.6: Diagram of G^E/RT vs. H^E/RT in skeleton form.

Table 10.3: Definition of regions on the G^E/RT vs. H^E/RT diagram.

Region	Sign G^E	Sign H^E	Sign S^E
I	+	+	+
II	+	+	−
III	+	−	−
IV	−	−	−
V	−	−	+
VI	−	+	+

Along the diagonal line on Fig. 10.6, $S^E = 0$. In regions to the right of the diagonal (V, VI, and I), S^E is positive; for regions to the left of the diagonal (II, III, and IV), S^E is negative. Lines of constant non-zero S^E are parallel to the diagonal.

It is convenient in modeling and rationalizing the behavior of G^E to focus on *enthalpic* (energetic) and *entropic* contributions, a separation suggested by Eqs. (10.99) and (10.100). According to these equations, G^E can be positive *or* negative if H^E and S^E have the same sign. If H^E and S^E are positive and if G^E is also positive, then $H^E > TS^E$ and "enthalpy dominates"; If H^E and S^E are positive and G^E is negative, then $TS^E > H^E$ and "entropy dominates". Similar

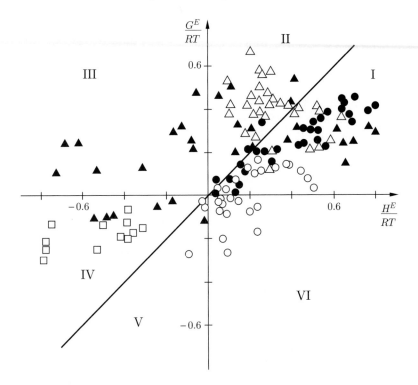

Figure 10.7: Equimolar excess properties for 135 binary mixtures at 298.15 K. Legend: ○ NP/NP mixtures; ● NA/NP mixtures; △ AS/NP mixtures; ▲ AS/NA and AS/AS mixtures; □ solvating NA/NA mixtures.

reasoning applies when both H^E and S^E are negative, leading to the identification of Regions I and IV on the G^E/RT vs. H^E/RT diagram as regions of *enthalpy domination*, and of Regions III and VI as regions of *entropy domination*. The notions of enthalpy and entropy domination can be helpful for explaining the molecular origins of observed mixture-property behavior.

Abbott et al. classify binary organic and aqueous/organic mixtures by a simple scheme based on hydrogen-bonding concepts (see Sec. 10.10). A pure species is categorized as nonpolar ("NP": e.g., benzene, carbon tetrachloride, or n-heptane); polar but non-associating ("NA": e.g., acetone, chloroform, or diethylether); or polar and associating ("AS": e.g., acetic acid, ethanol, or water). With these categories for pure species, there are then six binary mixture types: NP/NP (e.g., benzene/n-heptane); NA/NP (e.g., chloroform/n-heptane, Fig. 10.5a); AS/NP (e.g., ethanol/n-heptane, Fig. 10.5d); NA/NA (e.g., acetone/chloroform, Fig. 10.5c); AS/NA (e.g., ethanol/chloroform, Fig. 10.5e); and AS/AS (e.g., ethanol/water, Fig. 10.5f).

Figure 10.7 is a G^E/RT vs. H^E/RT plot of data for 135 different binary mixtures at 298 K, with systems distinguished according to the NP/NA/AS classification scheme. For clarity, we mainly omit data for which the three principal excess properties are very small. The figure appears chaotic at first glance, but on inspection important patterns emerge. We list here a few; statistics, where given, apply to the full data set analyzed by Abbott et al.

1. About 85% of all mixtures exhibit positive G^E *or* positive H^E (Regions I, II, III, and VI); about 70% have positive G^E *and* positive H^E (Regions I and II). Thus positive G^E and positive H^E are the "norms".

2. About 60% of all mixtures fall in Regions I and IV, with only about 15% in Regions III and VI. Thus the enthalpy is more likely to dominate solution behavior than is the entropy.

3. NP/NP mixtures (the open circles on Fig. 10.7) tend to concentrate in Regions I and VI: for such mixtures, H^E and S^E are normally positive. When G^E is positive (enthalpy domination), G^E/RT rarely exceeds about 0.2. If G^E is negative (entropy domination), G^E/RT is rarely less than -0.2.

4. NA/NP mixtures (the filled circles) usually fall in Region I, with occasional significant excursions into Region II. Thus, G^E and H^E are positive, as is (normally) S^E. Both G^E and H^E can be large.

5. AS/NP mixtures (the open triangles) invariably occupy Region I or II, with Region II behavior (negative S^E) favored when the polar species is a very strong associator, such as an alcohol or a carboxylic acid. In the latter case, G^E can be extremely large owing to the reinforcing effects of positive H^E and negative S^E [see Eq. (10.99)].

6. Mixtures containing two polar species exhibit a diversity of behaviors. Perhaps the easiest class to categorize involves pure solvation, in which one species is a non-associating hydrogen donor and the other a non-associating hydrogen acceptor. Here, unless one of the species has extremely high effective polarity (e.g., acetonitrile), Region IV behavior obtains: G^E, H^E, and S^E are all negative (enthalpy dominates). Examples are represented by the open squares on Fig. 10.7. For AS/NA and AS/AS mixtures (the filled triangles on Fig. 10.7), a variety of hydrogen-bonding possibilities is available (see Table 10.2), and it is impossible to make easy generalizations. We note however that these are the mixture types with significant representation in Region III; here, both H^E and S^E are negative, but entropy dominates.

PROBLEMS

10.1. What is the change in entropy when 0.7 m^3 of CO_2 and 0.3 m^3 of N_2, each at 1 bar and 25°C blend to form a homogeneous gas mixture at the same conditions? Assume ideal gases.

10.2. A vessel, divided into two parts by a partition, contains 4 mol of nitrogen gas at 75°C and 30 bar on one side and 2.5 mol of argon gas at 130°C and 20 bar on the other. If the partition is removed and the gases mix adiabatically and completely, what is the change in entropy? Assume nitrogen an ideal gas with $C_V = (5/2)R$ and argon an ideal gas with $C_V = (3/2)R$.

10.3. A stream of nitrogen flowing at the rate of 2 kg s^{-1} and and a stream of hydrogen flowing at the rate of 0.5 kg s^{-1} mix adiabatically in a steady-flow process. If the gases are assumed ideal, what is the rate of entropy increase as a result of the process?

10.4. Show that the "partial molar mass" of a species in solution is equal to its molar mass (molecular weight).

10.5. Show that a partial *specific* property of a species in solution is obtained by division of the partial *molar* property by the molar mass (molecular weight) of the species.

10.6. If the molar density of a binary mixture is given by the empirical expression

$$\rho = a_0 + a_1 x_1 + a_2 x_1^2$$

find the corresponding expressions for $\bar{V}_1$ and $\bar{V}_2$.

10.7. From the following compressibility-factor data for CO_2 at 150°C prepare plots of the fugacity and fugacity coefficient of CO_2 vs. P for pressures up to 500 bar. Compare the resulting curves with those found from the generalized correlation represented by Eq. (10.64).

P/bar	Z	P/bar	Z
10	0.985	100	0.869
20	0.970	200	0.765
40	0.942	300	0.762
60	0.913	400	0.824
80	0.885	500	0.910

10.8. For SO_2 at 600 K and 300 bar, determine good estimates of the fugacity and of G^R/RT.

10.9. Estimate the fugacity of isobutylene as a gas

(*a*) At 280°C and 20 bar.

(*b*) At 280°C and 100 bar.

10.10. Estimate the fugacity of cyclopentane at 110°C and 275 bar. At 110°C the vapor pressure of cyclopentane is 5.267 bar.

10.11. Estimate the fugacity of liquid 1-butene at 120°C and 34 bar. At 120°C the vapor pressure of *n*-butane is 25.83 bar.

10.12. From data in the steam tables, determine a good estimate for f/f^{sat} for liquid water at 150°C and 150 bar, where f^{sat} is the fugacity of saturated liquid at 150°C.

10.13. Steam at 9,000 kPa and 400°C undergoes an isothermal change of state to a pressure of 300 kPa. Determine the ratio of the fugacity in the final state to that in the initial state.

10.14. Steam at 1,000(psia) and 800(°F) undergoes an isothermal change of state to a pressure of 50(psia). Determine the ratio of the fugacity in the final state to that in the initial state.

10.15. Estimate the fugacity of liquid n-pentane at its normal-boiling-point temperature and 200 bar.

10.16. Estimate the fugacity of liquid isobutylene at its normal-boiling-point temperature and 300 bar.

10.17. Estimate the fugacity of liquid 1-butene at its normal-boiling-point temperature and 150 bar.

10.18. Prepare plots of f vs. P and of ϕ vs. P for chloroform at $200°C$ for the pressure range from 0 to 40 bar. At $200°C$ the vapor pressure of chloroform is 22.27 bar. Assume that Eq. (10.64) is valid for the vapor phase and that the molar volume of saturated liquid is given by Eq. (3.52).

10.19. Prepare plots of f vs. P and of ϕ vs. P for isobutane at $40°C$ for the pressure range from 0 to 10 bar. At $40°C$ the vapor pressure of isobutane is 5.28 bar. Assume that Eq. (10.64) is valid for the vapor phase and that the molar volume of saturated liquid is given by Eq. (3.52).

10.20. Humidity, relating to the quantity of moisture in atmospheric air, is accurately given by equations derived from the ideal-gas law.

(a) The *absolute humidity* $\mathcal{H}$ is defined as the mass of water vapor in a unit mass of dry air. Show that it is given by

$$\mathcal{H} = \frac{M_{H_2O}}{M_{air}} \frac{p_{H_2O}}{P - p_{H_2O}}$$

where M represents a molar mass and p_{H_2O} is the partial pressure of the water vapor, i.e., $p_{H_2O} = y_{H_2O}P$.

(b) The *saturation humidity* $\mathcal{H}^{sat}$ is defined as the value of $\mathcal{H}$ when air is in equilibrium with a large body of pure water. Show that it is given by

$$\mathcal{H}^{sat} = \frac{M_{H_2O}}{M_{air}} \frac{P_{H_2O}^{sat}}{P - P_{H_2O}^{sat}}$$

where $P_{H_2O}^{sat}$ is the vapor pressure of water at the ambient temperature.

(c) The *percentage humidity* is defined as the ratio of $\mathcal{H}$ to its saturation value, expressed as a percentage. On the other hand, the *relative humidity* is defined as the ratio of the partial pressure of water vapor in air to its vapor pressure, expressed as a percentage. What is the relation between these two quantities?

10.21. For a particular binary liquid solution at constant T and P, the molar enthalpies of mixtures are represented by the equation

$$H = x_1(a_1 + b_1x_1) + x_2(a_2 + b_2x_2)$$

where the a_i and b_i are constants. Since the equation has the form of Eq. (10.11), it might be that

$$\bar{H}_1 = a_1 + b_1x_1 \qquad \text{and} \qquad \bar{H}_2 = a_2 + b_2x_2$$

Derive expressions for $\bar{H}_1$ and $\bar{H}_2$ to show whether this is true.

10.22. Assume that the following equation is valid for a binary liquid solution at constant T and P:

$$\frac{G^E}{RT} = x_1 \ln\left(\frac{1}{x_1 + x_2 A}\right) + x_2 \ln\left(\frac{1}{x_2 + x_1 B}\right)$$

where A and B are constants for given T and P. Since the equation has the form of Eq. (10.97), it might be that

$$\gamma_1 = \frac{1}{x_1 + x_2 A} \qquad \text{and} \qquad \gamma_2 = \frac{1}{x_2 + x_1 B}$$

Derive expressions for γ_1 and γ_2 to show whether this is true.

10.23. The following expressions have been reported for the activity coefficients of species 1 and 2 in a binary liquid mixture at given T and P:

$$\ln \gamma_1 = x_2^2(0.273 + 0.096\, x_1)$$
$$\ln \gamma_2 = x_1^2(0.273 - 0.096\, x_2)$$

(*a*) Determine the implied expression for G^E/RT.

(*b*) *Generate* expressions for $\ln \gamma_1$ and $\ln \gamma_2$ from the result of (*a*).

(*c*) Compare the results of (*b*) with the reported expressions for $\ln \gamma_1$ and $\ln \gamma_2$. Discuss any discrepancy. Can the reported expressions possibly be correct?

10.24. For the system ethylene(1)/propylene(2) as a gas, estimate $\hat{f}_1$, $\hat{f}_2$, $\hat{\phi}_1$, and $\hat{\phi}_2$ at $t = 150°C$, $P = 30$ bar, and $y_1 = 0.35$:

(*a*) Through application of Eqs. (10.67) and (10.68).

(*b*) Assuming that the mixture is an ideal solution.

10.25. For the system methane(1)/ethane(2)/propane(3) as a gas, estimate $\hat{f}_1$, $\hat{f}_2$, $\hat{f}_3$, $\hat{\phi}_1$, $\hat{\phi}_2$, and $\hat{\phi}_3$ at $t = 100°C$, $P = 35$ bar, $y_1 = 0.21$, and $y_2 = 0.43$:

(*a*) Through application of Eq. (10.69).

(*b*) Assuming that the mixture is an ideal solution.

10.26. With reference to Example 10.4,

(*a*) Apply Eq. (10.7) to Eq. (A) to verify Eqs. (B) and (C).

(*b*) Show that Eqs. (B) and (C), when combined in accord with Eq. (10.11), regenerate Eq. (A).

(*c*) Show that Eqs. (B) and (C) satisfy Eq. (10.14), the Gibbs/Duhem equation.

(*d*) Show that at constant T and P

$$(d\bar{H}_1/dx_1)_{x_1=1} = (d\bar{H}_2/dx_1)_{x_1=0} = 0$$

(*e*) Plot values of H, $\bar{H}_1$, and $\bar{H}_2$, calculated by Eqs. (A), (B), and (C), vs. x_1. Label points H_1, H_2, $\bar{H}_1^\infty$, and $\bar{H}_2^\infty$, and show their values.

10.27. The molar volume ($cm^3\ mol^{-1}$) of a binary liquid mixture at T and P is given by

$$V = 120x_1 + 70x_2 + (15x_1 + 8x_2)x_1 x_2$$

For the given T and P,

(a) Find expressions for the partial molar volumes of species 1 and 2.

(b) Show that when these expressions are combined in accord with Eq. (10.11) the given equation for V is recovered.

(c) Show that these expressions satisfy Eq. (10.14), the Gibbs/Duhem equation.

(d) Show that $(d\bar{V}_1/dx_1)_{x_1=1} = (d\bar{V}_2/dx_1)_{x_1=0} = 0$

(e) Plot values of V, $\bar{V}_1$, and $\bar{V}_2$ calculated by the given equation for V and by the equations developed in (a) vs. x_1. Label points V_1, V_2, $\bar{V}_1^\infty$, and $\bar{V}_2^\infty$, and show their values.

10.28. The excess Gibbs energy of a binary liquid mixture at T and P is given by

$$G^E/RT = (-2.6x_1 - 1.8x_2)x_1x_2$$

For the given T and P,

(a) Find expressions for $\ln\gamma_1$ and $\ln\gamma_2$.

(b) Show that when these expressions are combined in accord with Eq. (10.97) the given equation for G^E/RT is recovered.

(c) Show that these expressions satisfy Eq. (10.98), the Gibbs/Duhem equation.

(d) Show that $(d\ln\gamma_1/dx_1)_{x_1=1} = (d\ln\gamma_2/dx_1)_{x_1=0} = 0$.

(e) Plot values of G^E/RT, $\ln\gamma_1$, and $\ln\gamma_2$ calculated by the given equation for G^E/RT and by the equations developed in (a) vs. x_1. Label points $\ln\gamma_1^\infty$, and $\ln\gamma_2^\infty$, and show their values.

10.29. For a ternary solution at constant T and P, the composition dependence of molar property M is given by

$$M = x_1M_1 + x_2M_2 + x_3M_3 + x_1x_2x_3C$$

where M_1, M_2, and M_3 are the values of M for pure species 1, 2, and 3, and C is a parameter independent of composition. Determine an expression for $\bar{M}_2$ by application of Eq. (10.7).

10.30. If for a binary solution one starts with an expression for M (or M^R or M^E) as a function of x_1 and applies Eqs. (10.15) and (10.16) to find $\bar{M}_1$ and $\bar{M}_2$ (or $\bar{M}_1^R$ and $\bar{M}_2^R$ or $\bar{M}_1^E$ and $\bar{M}_2^E$) and then combines these expressions by Eq. (10.11), the initial expression for M is regenerated. On the other hand, if one starts with expressions for $\bar{M}_1$ and $\bar{M}_2$, combines them in accord with Eq. (10.11), and then applies Eqs. (10.15) and (10.16), the initial expressions for $\bar{M}_1$ and $\bar{M}_2$ are regenerated if and only if the initial expressions for these quantities meet a specific condition. What is the condition?

10.31. A *pure-component pressure* p_i for species i in a gas mixture may be defined as the pressure that species i would exert if it alone occupied the mixture volume. Thus

$$p_i \equiv \frac{y_iZ_iRT}{V}$$

where y_i is the mole fraction of species i in the gas mixture and V is the molar volume of the gas mixture. Note that p_i as defined here is not a partial pressure y_iP, except for an ideal gas. Dalton's "law" of additive pressures states that the total pressure exerted by a gas mixture is equal to the sum of the pure-component pressures of its consitituent species: $P = \sum_i p_i$.

(a) Show that Dalton's "law" implies that $Z = \sum_i y_i Z_i$, where Z_i is the compressibility factor of pure species i evaluated at the mixture temperature but at its pure-component pressure.

(b) Show that Dalton's "law" implies that $\hat{\phi}_i = \phi_i$, where ϕ_i is the fugacity coefficient of pure species i evaluated at the mixture temperature but at its pure-component pressure.

(c) How is the assumption of Dalton's "law" the same as and how is it different from the assumption of an ideal solution?

(d) Rework Example 10.8 on the assumption that Dalton's "law" applies.

(e) Rework Example 10.8 on the assumption that the solution is an ideal solution.

10.32. The similar forms of the definitions of M^E and M^R suggest that excess properties and residual properties should be related in a simple way.

(a) Show that

$$M^E = M^R - \sum_i x_i M_i^R$$

(b) From this result show that

$$\bar{M}_i^E = \bar{M}_i^R - M_i^R$$

(c) Specialize this to the Gibbs energy, and show that the result leads to

$$\gamma_i = \frac{\hat{\phi}_i}{\phi_i}$$

10.33. Naive numerology suggests that there should be $2^3 = 8$ possible combinations of sign for G^E, H^E, and S^E. Table 10.3 shows only *six*. Why?

10.34. What *signs* would you expect to observe for G^E, H^E, and S^E for equimolar liquid solutions of the following pairs of species at 298 K? Explain your answers.

(a) Acetone/cyclohexane

(b) Acetone/dichloromethane

(c) Aniline/cyclohexane

(d) Benzene/carbon disulfide

(e) Benzene/n-hexane

(f) Chloroform/1,4-dioxane

(g) Chloroform/n-hexane

(h) Ethanol/n-nonane

CHAPTER 11

SOLUTION THERMODYNAMICS: APPLICATIONS

All of the fundamental equations and necessary definitions of solution thermodynamics are given in the preceding chapter. In this chapter we examine what can be learned from experiment. First, we consider measurements of vapor/liquid equilibrium (VLE) data. This material finds primary use in the next chapter. Second, we treat mixing experiments, which provide data for property changes of mixing. In particular, practical applications of the enthalpy change of mixing, called the heat of mixing, are presented in detail in Sec. 11.4.

11.1 Liquid-Phase Properties from VLE Data

Fugacity

Figure 11.1 shows a vessel in which a vapor mixture and a liquid solution coexist in vapor/liquid equilibrium. The temperature T and pressure P are uniform throughout the vessel, and can be measured with appropriate instruments. Samples of the vapor and liquid phases may be withdrawn for analysis, and this provides experimental values for the mole fractions in the vapor $\{y_i\}$ and the mole fractions in the liquid $\{x_i\}$. For species i in the vapor mixture, Eq. (10.47) is written

$$\hat{f}_i^v = y_i \hat{\phi}_i P$$

For vapor/liquid equilibrium Eq. (10.44) requires that $\hat{f}_i^l = \hat{f}_i^v$ for each species. Therefore

$$\hat{f}_i^l = y_i \hat{\phi}_i P$$

We could calculate values of $\hat{\phi}_i$ by Eq. (10.60), but for low-pressure VLE (up to at least 1 bar) vapor phases usually approximate ideal gases, for which $\hat{\phi}_i = 1$. This

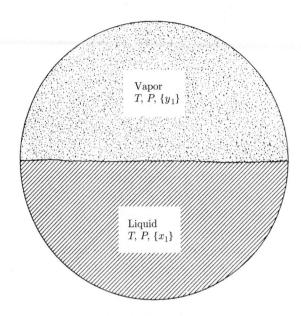

Figure 11.1: Schematic representation of VLE.

assumption introduces little error, and reduces the preceding equation to

$$\hat{f}_i^l = y_i P \qquad P < 1 \ bar$$

Thus, the fugacity of species i in the liquid phase is given to a good approximation by the partial pressure of species i in the vapor phase. In the limit where $x_i = y_i = 1$, the total pressure equals the vapor pressure of pure species i, and $\hat{f}_i^l = f_i^l = P_i^{\text{sat}}$.

In Table 11.1 the first three columns contain a set of experimental $P\text{-}x_1\text{-}y_1$ data for the methyl ethyl ketone(1)/toluene(2) system at 50°C.[1] Values of the liquid-phase fugacities are found from

$$\hat{f}_1 = y_1 P \qquad \text{and} \qquad \hat{f}_2 = y_2 P$$

where superscript l has for simplicity been dropped. These values are shown in columns 4 and 5 of Table 11.1, and are plotted in Fig. 11.2 as the solid lines. The straight dashed lines represent Eq. (10.84), the Lewis/Randall rule, which expresses the composition dependence of the constituent fugacities in an ideal solution:

$$\hat{f}_i^{id} = x_i f_i \qquad (10.84)$$

Figure 11.2, derived from a specific set of data, illustrates the general characteristics of the $\hat{f}_1$ and $\hat{f}_2$ vs. x_1 relationships for a binary liquid solution at constant

---

[1]M. Diaz Peña, A. Crespo Colin, and A. Compostizo, *J. Chem. Thermodyn.*, vol. 10, pp. 337–341, 1978.

Table 11.1: VLE Data for methyl ethyl ketone(1)/toluene(2) at 50°C

P/kPa	x_1	y_1	$\hat{f}_1 = y_1 P$	$\hat{f}_2 = y_2 P$	γ_1	γ_2
$12.30(P_2^{\text{sat}})$	0.0000	0.0000	0.000	12.300		1.000
15.51	0.0895	0.2716	4.212	11.298	1.304	1.009
18.61	0.1981	0.4565	8.496	10.114	1.188	1.026
21.63	0.3193	0.5934	12.835	8.795	1.114	1.050
24.01	0.4232	0.6815	16.363	7.697	1.071	1.078
25.92	0.5119	0.7440	19.284	6.636	1.044	1.105
27.96	0.6096	0.8050	22.508	5.542	1.023	1.135
30.12	0.7135	0.8639	26.021	4.099	1.010	1.163
31.75	0.7934	0.9048	28.727	3.023	1.003	1.189
34.15	0.9102	0.9590	32.750	1.400	0.997	1.268
$36.09(P_1^{\text{sat}})$	1.0000	1.0000	36.090	0.000	1.000	

T. Although the equilibrium P varies with composition, its influence on $\hat{f}_1$ and $\hat{f}_2$ is negligible, and a plot at constant T *and* P would look the same. Thus in Fig. 11.3 we show a schematic diagram of the $\hat{f}_i$ vs. x_i relation for species i $(i = 1, 2)$ in a binary solution at constant T and P.

Activity Coefficient

The lower dashed line in Fig. 11.3 again represents the Lewis/Randall rule, characteristic of ideal-solution behavior. It provides the simplest possible model for the composition dependence of $\hat{f}_i$, representing a standard to which actual behavior may be compared. Indeed, the activity coefficient formalizes this comparison:

$$\gamma_i \equiv \frac{\hat{f}_i}{x_i f_i} = \frac{\hat{f}_i}{\hat{f}_i^{id}}$$

Thus the activity coefficient of a species in solution is simply the ratio of its actual fugacity to the value given by the Lewis/Randall rule at the same T, P, and composition. For calculational purposes we substitute for both $\hat{f}_i$ and $\hat{f}_i^{id}$ to get

$$\gamma_i = \frac{y_i P}{x_i f_i} = \frac{y_i P}{x_i P_i^{\text{sat}}} \qquad (i = 1, 2, \dots, N) \tag{11.1}$$

This simple equation is adequate to our present purpose, allowing easy calculation of activity coefficients from experimental low-pressure VLE data. Values found by this equation are given in the last two columns of Table 11.1.

   We note in Fig. 11.3 that the solid line representing the actual composition dependence of $\hat{f}_i$ becomes tangent to the Lewis/Randall line at $x_i = 1$. This is a consequence of the Gibbs/Duhem equation, as will be shown presently. We also

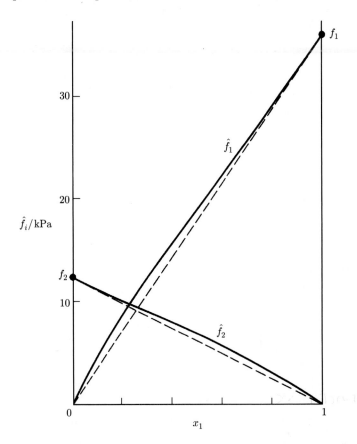

Figure 11.2: Fugacities for methyl ethyl ketone(1)/toluene(2) at 50°C. The dashed lines represent the Lewis/Randall rule.

note that in the other limit $\hat{f}_i$ becomes zero at $x_i = 0$. Thus, the ratio $\hat{f}_i/x_i$ is indeterminate in this limit, and application of l'Hôpital's rule yields

$$\lim_{x_i \to 0} \frac{\hat{f}_i}{x_i} = \left(\frac{d\hat{f}_i}{dx_i}\right)_{x_i=0} \equiv k_i \tag{11.2}$$

This equation defines *Henry's constant* k_i, the limiting slope of the $\hat{f}_i$-vs.-x_i curve at $x_i = 0$. As shown on Fig. 11.3, this is the slope of a line drawn tangent to the curve at $x_i = 0$. Thus *Henry's law* $\hat{f}_i = x_i k_i$ applies in the limit as $x_i \to 0$, and must also be of approximate validity for small values of x_i.

Henry's law is related to the Lewis/Randall rule through the Gibbs/Duhem equation. Writing Eq. (10.14) for a binary solution, replacing M by G, and noting that $d\mu_i = d\bar{G}_i$ gives

$$x_1 \, d\mu_1 + x_2 \, d\mu_2 = 0 \qquad (\text{const } T, P)$$

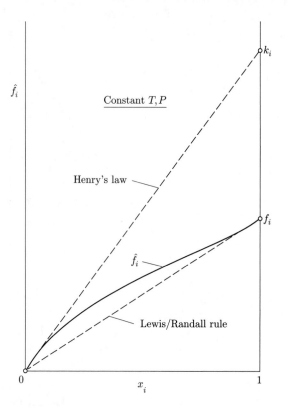

Figure 11.3: Composition dependence of fugacity for species i in a binary solution.

From Eq. (10.42) by differentiation at constant T,

$$d\mu_i = RT \, d\ln \hat{f}_i$$

whence

$$x_1 \, d\ln \hat{f}_1 + x_2 \, d\ln \hat{f}_2 = 0$$

Upon division by dx_1 this becomes

$$\boxed{x_1 \frac{d\ln \hat{f}_1}{dx_1} + x_2 \frac{d\ln \hat{f}_2}{dx_1} = 0 \qquad (\text{const } T, P)} \qquad (11.3)$$

and substitution of $-dx_2$ for dx_1 in the second term produces

$$x_1 \frac{d\ln \hat{f}_1}{dx_1} = x_2 \frac{d\ln \hat{f}_2}{dx_2}$$

Equivalently,

$$\frac{d\hat{f}_1/dx_1}{\hat{f}_1/x_1} = \frac{d\hat{f}_2/dx_2}{\hat{f}_2/x_2}$$

In the limit as $x_1 \rightarrow 1$ and $x_2 \rightarrow 0$,

$$\lim_{x_1 \rightarrow 1} \frac{d\hat{f}_1/dx_1}{\hat{f}_1/x_1} = \lim_{x_2 \rightarrow 0} \frac{d\hat{f}_2/dx_2}{\hat{f}_2/x_2}$$

Since $\hat{f}_1 = f_1$ when $x_1 = 1$, this may be rewritten as

$$\frac{1}{f_1} \left(\frac{d\hat{f}_1}{dx_1} \right)_{x_1=1} = \frac{(d\hat{f}_2/dx_2)_{x_2=0}}{\lim_{x_2 \rightarrow 0} (\hat{f}_2/x_2)}$$

According to Eq. (11.2), the numerator and denominator on the right-hand side of this equation are equal; therefore it reduces to

$$\left(\frac{d\hat{f}_1}{dx_1} \right)_{x_1=1} = f_1 \qquad (11.4)$$

This equation is the exact expression of the Lewis/Randall rule as it applies to real solutions. It shows that Eq. (10.84) becomes valid in the limit as $x_1 \rightarrow 1$, and that it is approximately correct for values of x_1 near unity. From this derivation we conclude that when Henry's law is valid for one species in a binary solution, the Lewis/Randall rule is valid for the other species.

We see by Eq. (10.89), $\gamma_i = \hat{f}_i/x_i f_i$, that $\gamma_i = 1$ whenever the Lewis/Randall rule is valid; this result therefore obtains for $x_i = 1$. Moreover, differentiation at constant T and P of the defining equation for γ_i yields

$$\frac{d\gamma_i}{dx_i} = \frac{1}{f_i} \left(\frac{1}{x_i} \frac{d\hat{f}_i}{dx_i} - \frac{\hat{f}_i}{x_i^2} \right)$$

In the limit as $x_i \rightarrow 1$, $\hat{f}_i \rightarrow f_i$, and this becomes

$$\left(\frac{d\gamma_i}{dx_i} \right)_{x_i=1} = \frac{1}{f_i} \left[\left(\frac{d\hat{f}_i}{dx_i} \right)_{x_i=1} - f_i \right]$$

However, Eq. (11.4) shows that the terms enclosed in the square brackets cancel; therefore

$$\left(\frac{d\gamma_i}{dx_i} \right)_{x_i=1} = 0$$

Thus another consequence of the Gibbs/Duhem equation is that each γ_i-vs.-x_i curve approaches unity with zero slope at $x_i = 1$. This is seen in Fig. 11.4 where the γ_1 and γ_2 values of Table 11.1 are plotted vs. x_1.

Figure 11.3 is drawn for a species that shows positive deviations from ideality in the sense of the Lewis/Randall rule. Negative deviations from ideality are less common, but are also observed. In this case the $\hat{f}_i$-vs.-x_i curve lies below the

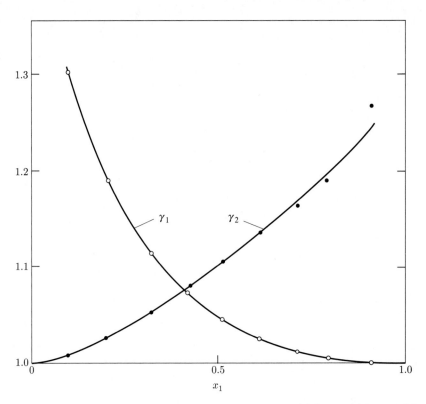

Figure 11.4: Activity coefficients for methyl ethyl ketone(1)/toluene(2) at 50°C as a function of composition.

Lewis/Randall line. In Fig. 11.5 we show the composition dependence of the fugacity of acetone in two different binary solutions at 50°C. When the second species is methanol, acetone shows positive deviations from ideality. When the second species is chloroform, acetone shows negative deviations from ideality. The fugacity of pure acetone f_{acetone} is of course the same regardless of the identity of the second species. However, Henry's constants, represented by the slopes of the two dotted lines, are very different for the two cases.

Excess Gibbs Energy

In Table 11.2 the first three columns repeat the P-x_1-y_1 data of Table 11.1 for the system methyl ethyl ketone(1)/toluene(2). These data points are also shown as circles on Fig. 11.6. In columns 4 and 5 we list values of $\ln \gamma_1$ and $\ln \gamma_2$, which are shown by the open squares and triangles on Fig. 11.7. They are combined in accord with Eq. (10.97), written for a binary system:

$$\boxed{\frac{G^E}{RT} = x_1 \ln \gamma_1 + x_2 \ln \gamma_2}$$

$$(11.5)$$

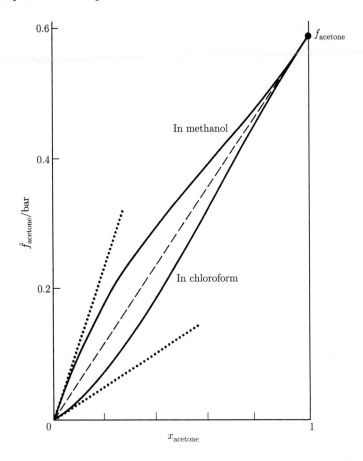

Figure 11.5: Composition dependence of the fugacity of acetone in two binary liquid solutions at 50°C.

The values of G^E/RT so calculated are divided by $x_1 x_2$ to provide in addition values of $G^E/x_1 x_2 RT$; the two sets of numbers are listed in columns 6 and 7 of Table 11.2 and appear as solid circles on Fig. 11.7.

The four thermodynamic functions for which we have experimental values, $\ln \gamma_1$, $\ln \gamma_2$, G^E/RT, and $G^E/x_1 x_2 RT$, are properties of the liquid phase. Figure 11.7 shows how each varies with composition for a particular binary system at a specified temperature. This figure is characteristic of systems for which

$$\gamma_i \geq 1 \qquad \text{and} \qquad \ln \gamma_i \geq 0 \qquad (i = 1, 2)$$

In such a case the liquid phase shows *positive deviations* from ideal-solution behavior. This is seen also in Fig. 11.6, where the P-x_1 data points all lie above the dashed straight line which represents ideal-solution behavior.[2]

---

[2] A linear P-x_1 relation is a consequence of Raoult's law (Sec. 12.5).

Table 11.2: VLE Data for methyl ethyl ketone(1)/toluene(2) at 50°C

P/kPa	x_1	y_1	$\ln\gamma_1$	$\ln\gamma_2$	G^E/RT	G^E/x_1x_2RT
$12.30(P_2^{\text{sat}})$	0.0000	0.0000		0.000	0.000	
15.51	0.0895	0.2716	0.266	0.009	0.032	0.389
18.61	0.1981	0.4565	0.172	0.025	0.054	0.342
21.63	0.3193	0.5934	0.108	0.049	0.068	0.312
24.01	0.4232	0.6815	0.069	0.075	0.072	0.297
25.92	0.5119	0.7440	0.043	0.100	0.071	0.283
27.96	0.6096	0.8050	0.023	0.127	0.063	0.267
30.12	0.7135	0.8639	0.010	0.151	0.051	0.248
31.75	0.7934	0.9048	0.003	0.173	0.038	0.234
34.15	0.9102	0.9590	−0.003	0.237	0.019	0.227
$36.09(P_1^{\text{sat}})$	1.0000	1.0000	0.000		0.000	

Since the activity coefficient of a species in solution becomes unity as the species becomes pure, each $\ln\gamma_i$ $(i = 1, 2)$ tends to zero as $x_i \to 1$. This is evident in Fig. 11.7. At the other limit, where $x_i \to 0$ and species i becomes infinitely dilute, $\ln\gamma_i$ is seen to approach some finite limit, which we denote by $\ln\gamma_i^\infty$.

In the limit as $x_1 \to 0$, the dimensionless excess Gibbs energy G^E/RT as given by Eq. (11.5) becomes

$$\lim_{x_1 \to 0} \frac{G^E}{RT} = (0)\ln\gamma_1^\infty + (1)(0) = 0$$

The same result is obtained for $x_2 \to 0$ $(x_1 \to 1)$. The value of G^E/RT (and G^E) therefore goes to zero at both $x_1 = 0$ and $x_1 = 1$.

The quantity G^E/x_1x_2RT becomes indeterminate both at $x_1 = 0$ and $x_1 = 1$, because G^E is zero in both limits, as is the product x_1x_2. Thus for $x_1 \to 0$, we have by l'Hôpital's rule

$$\lim_{x_1 \to 0} \frac{G^E}{x_1x_2RT} = \lim_{x_1 \to 0} \frac{G^E/RT}{x_1} = \lim_{x_1 \to 0} \frac{d(G^E/RT)}{dx_1} \tag{A}$$

The derivative of the final member is found by differentiation of Eq. (11.5) with respect to x_1:

$$\frac{d(G^E/RT)}{dx_1} = x_1\frac{d\ln\gamma_1}{dx_1} + \ln\gamma_1 + x_2\frac{d\ln\gamma_2}{dx_1} - \ln\gamma_2 \tag{B}$$

The minus sign preceding the last term comes from $dx_2/dx_1 = -1$, a consequence of the equation $x_1 + x_2 = 1$. Equation (10.98), the Gibbs/Duhem equation, may

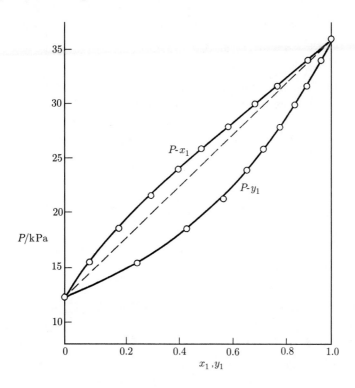

Figure 11.6: *Pxy* data at 50°C for methyl ethyl ketone(1)/toluene(2).

be written for a binary system and divided by dx_1 to give

$$x_1 \frac{d \ln \gamma_1}{dx_1} + x_2 \frac{d \ln \gamma_2}{dx_1} = 0 \qquad (\text{const } T, P) \qquad (11.6)$$

Although the data set treated here is at constant T, the pressure varies, and Eq. (11.6) strictly does not apply. However, as shown in Sec. 10.9, the activity coefficients for liquid phases at low pressure are very nearly independent of P, and negligible error is introduced by use of this equation. We therefore combine Eq. (11.6) with Eq. (B):

$$\frac{d(G^E/RT)}{dx_1} = \ln \frac{\gamma_1}{\gamma_2}$$

In the limiit as $x_1 \to 0$ $(x_2 \to 1)$, this becomes

$$\lim_{x_1 \to 0} \frac{d(G^E/RT)}{dx_1} = \lim_{x_1 \to 0} \ln \frac{\gamma_1}{\gamma_2} = \ln \gamma_1^\infty$$

and by Eq. (A),

$$\lim_{x_1 \to 0} \frac{G^E}{x_1 x_2 RT} = \ln \gamma_1^\infty$$

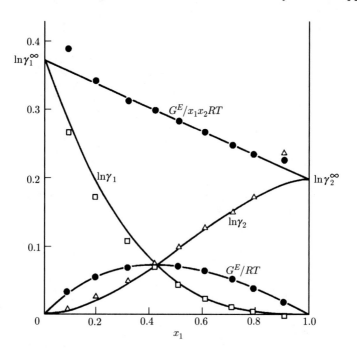

Figure 11.7: Liquid-phase properties for methyl ethyl ketone(1)/toluene(2) at 50°C from VLE data.

Similarly, as $x_1 \to 1$ $(x_2 \to 0)$,

$$\lim_{x_1 \to 1} \frac{G^E}{x_1 x_2 RT} = \ln \gamma_2^\infty$$

Thus the limiting values of $G^E/x_1 x_2 RT$ are equal to the infinite-dilution limits of $\ln \gamma_1$ and $\ln \gamma_2$. This result is illustrated in Fig. 11.7.

Equation (11.6), the Gibbs/Duhem equation, has further influence on the nature of Fig. 11.7. Rewritten as

$$\frac{d \ln \gamma_1}{dx_1} = -\frac{x_2}{x_1} \frac{d \ln \gamma_2}{dx_1}$$

it shows the direct relation required between the slopes of curves drawn through the data points for $\ln \gamma_1$ and $\ln \gamma_2$. Qualitatively, we observe that at every composition the slope of the $\ln \gamma_1$ curve is of opposite sign to the slope of the $\ln \gamma_2$ curve. Furthermore, when $x_2 \to 0$ (and $x_1 \to 1$), the slope of the $\ln \gamma_1$ curve is zero. Similarly, when $x_1 \to 0$, the slope of the $\ln \gamma_2$ curve is zero. Thus, each $\ln \gamma_i$ $(i = 1, 2)$ curve becomes horizontal at $x_i = 1$.

Data Reduction

Of the sets of points shown in Fig. 11.7, those for G^E/x_1x_2RT most closely conform to a simple mathematical relation. Thus we draw a straight line as a reasonable approximation to this set of points, and we give mathematical expression to this linear relation by the equation

$$\frac{G^E}{x_1x_2RT} = A_{21}x_1 + A_{12}x_2 \tag{11.7a}$$

where A_{21} and A_{12} are constants in any particular application. Alternatively,

$$\frac{G^E}{RT} = (A_{21}x_1 + A_{12}x_2)x_1x_2 \tag{11.7b}$$

Expressions for $\ln \gamma_1$ and $\ln \gamma_2$ are derived from Eq. (11.7b) by application of Eq. (10.94). Since this requires differentiation of nG^E/RT with respect to a mole number, we multiply Eq. (11.7b) by n and convert all mole fractions to mole numbers. Thus on the right-hand side x_1 is replaced by $n_1/(n_1 + n_2)$, and x_2, by $n_2/(n_1 + n_2)$. Since $n \equiv n_1 + n_2$, this gives

$$\frac{nG^E}{RT} = (A_{21}n_1 + A_{12}n_2)\frac{n_1n_2}{(n_1 + n_2)^2}$$

Differentiating with respect to n_1 in accord with Eq. (10.94) yields

$$\ln \gamma_1 = \left[\frac{\partial(nG^E/RT)}{\partial n_1}\right]_{P,T,n_2}$$

$$= n_2\left[(A_{21}n_1 + A_{12}n_2)\left(\frac{1}{(n_1 + n_2)^2} - \frac{2n_1}{(n_1 + n_2)^3}\right) + \frac{n_1A_{21}}{(n_1 + n_2)^2}\right]$$

Reconversion of the n_i's to x_i's gives

$$\ln \gamma_1 = x_2\left[(A_{21}x_1 + A_{12}x_2)(1 - 2x_1) + A_{21}x_1\right]$$

Further reduction, noting that $x_2 = 1 - x_1$, leads to

$$\ln \gamma_1 = x_2^2[A_{12} + 2(A_{21} - A_{12})x_1] \tag{11.8a}$$

Similarly, differentiation of Eq. (11.7b) with respect to n_2 yields

$$\ln \gamma_2 = x_1^2[A_{21} + 2(A_{12} - A_{21})x_2] \tag{11.8b}$$

These are the *Margules*[3] *equations*, and they represent a commonly used empirical model of solution behavior. For the limiting conditions of infinite dilution, they

---

[3]Max Margules (1856–1920), Austrian meteorologist and physicist.

show that when $x_1 = 0$, $\ln \gamma_1^\infty = A_{12}$, and when $x_2 = 0$, $\ln \gamma_2^\infty = A_{21}$. For the methyl ethyl ketone/toluene system considered here, the curves of Fig. 11.7 for G^E/RT, $\ln \gamma_1$, and $\ln \gamma_2$ represent Eqs. (11.7b), (11.8a), and (11.8b) with $A_{12} = 0.372$ and $A_{21} = 0.198$. These are values of the intercepts at $x_1 = 0$ and $x_1 = 1$ of the straight line drawn to represent the $G^E/x_1 x_2 RT$ data points.

What we have accomplished is the *reduction* of a set of VLE data to a simple mathematical equation for the dimensionless excess Gibbs energy,

$$\frac{G^E}{RT} = (0.198 x_1 + 0.372 x_2) x_1 x_2$$

which concisely stores the information of the data set. Indeed, with the Margules equations for $\ln \gamma_1$ and $\ln \gamma_2$, we can easily construct a correlation of the original P-x_1-y_1 data set.

Equation (11.1) may be rearranged and written for species 1 and 2 of a binary system as

$$y_1 P = x_1 \gamma_1 P_1^{\text{sat}} \qquad \text{and} \qquad y_2 P = x_2 \gamma_2 P_2^{\text{sat}}$$

Addition gives

$$P = x_1 \gamma_1 P_1^{\text{sat}} + x_2 \gamma_2 P_2^{\text{sat}} \tag{11.9}$$

whence

$$y_1 = \frac{x_1 \gamma_1 P_2^{\text{sat}}}{x_1 \gamma_1 P_1^{\text{sat}} + x_2 \gamma_2 P_2^{\text{sat}}} \tag{11.10}$$

Finding values of γ_1 and γ_2 from Eqs. (11.8) with A_{12} and A_{21} as determined for the methyl ethyl ketone(1)/toluene(2) system and taking P_1^{sat} and P_2^{sat} as the experimental values, we calculate P and y_1 by Eqs. (11.9) and (11.10) at various values of x_1. The results are shown by the solid lines of Fig. 11.6, which represent the calculated P-x_1 and P-y_1 relations. They clearly provide an adequate correlation of the experimental data points.

A second set of P-x_1-y_1 data, for chloroform(1)/1,4-dioxane(2) at 50°C,[4] is given in Table 11.3, along with values of pertinent thermodynamic functions. Figures 11.8 and 11.9 display as points all of the experimentally determined values. This system shows *negative deviations from solution ideality*; since γ_1 and γ_2 are less than unity, values of $\ln \gamma_1$, $\ln \gamma_2$, G^E/RT, and $G^E/x_1 x_2 RT$ are negative. Moreover, the P-x_1 data points in Fig. 11.8 all lie below the dashed line representing the ideal-solution relation. Again the data points for $G^E/x_1 x_2 RT$ are reasonably well correlated by Eq. (11.7a), and the Margules equations [Eqs. (11.8)] again apply, here with $A_{12} = -0.72$ and $A_{21} = -1.27$. Values of G^E/RT, $\ln \gamma_1$, $\ln \gamma_2$, P, and y_1 calculated by Eqs. (11.7b), (11.8a), (11.8b), (11.9), and (11.10) provide the curves shown for these quantities in Figs. 11.8 and 11.9. Again, the experimental P-x_1-y_1 data are adequately correlated.

---

[4]M. L. McGlashan and R. P. Rastogi, *Trans. Faraday Soc.*, vol. 54, p. 496, 1958.

Table 11.3: VLE Data for chloroform(1)/1,4-dioxane(2) at 50°C

P/kPa	x_1	y_1	$\ln\gamma_1$	$\ln\gamma_2$	G^E/RT	G^E/x_1x_2RT
$15.79(P_2^{\text{sat}})$	0.0000	0.0000		0.000	0.000	
17.51	0.0932	0.1794	−0.722	0.004	−0.064	−0.758
18.15	0.1248	0.2383	−0.694	−0.000	−0.086	−0.790
19.30	0.1757	0.3302	−0.648	−0.007	−0.120	−0.825
19.89	0.2000	0.3691	−0.636	−0.007	−0.133	−0.828
21.37	0.2626	0.4628	−0.611	−0.014	−0.171	−0.882
24.95	0.3615	0.6184	−0.486	−0.057	−0.212	−0.919
29.82	0.4750	0.7552	−0.380	−0.127	−0.248	−0.992
34.80	0.5555	0.8378	−0.279	−0.218	−0.252	−1.019
42.10	0.6718	0.9137	−0.192	−0.355	−0.245	−1.113
60.38	0.8780	0.9860	−0.023	−0.824	−0.120	−1.124
65.39	0.9398	0.9945	−0.002	−0.972	−0.061	−1.074
$69.36(P_1^{\text{sat}})$	1.0000	1.0000	0.000		0.000	

Although the correlations provided by the Margules equations for the two sets of VLE data presented here are satisfactory, they are not perfect. The two possible reasons are, first, that the Margules equations are not precisely suited to the data set; second, that the P-x_1-y_1 data themselves are systematically in error such that they do not conform to the requirements of the Gibbs/Duhem equation.

We have presumed in applying the Margules equations that the deviations of the experimental points for G^E/x_1x_2RT from the straight lines drawn to represent them result from random error in the data. Indeed, the straight lines do provide excellent correlations of all but a few data points. Only toward edges of a diagram are there significant deviations, and these have been discounted, because the error bounds widen rapidly as the edges of a diagram are approached. In the limits as $x_1 \to 0$ and $x_1 \to 1$, G^E/x_1x_2RT becomes indeterminate; experimentally this means that the values are subject to unlimited error and are not measurable. However, we cannot rule out the possibility that the correlation would be improved were the G^E/x_1x_2RT points represented by an appropriate *curve*. Finding the correlation that best represents the data is a trial procedure.

The Gibbs/Duhem equation imposes a constraint on activity coefficients that may not be satisfied by experimental values containing systematic error. If this is the case, the experimental values of $\ln\gamma_1$ and $\ln\gamma_2$ used for calculation of G^E/RT by Eq. (11.5), which does not depend on the Gibbs/Duhem equation, will not agree with values of $\ln\gamma_1$ and $\ln\gamma_2$ later calculated by equations derived from Eq. (10.94), which do implicitly contain the Gibbs/Duhem equation. In this event, no correlating equation exists that precisely represents the original P-x_1-y_1 data. Such data are said to be *inconsistent* with the Gibbs/Duhem equation, and are necessarily incorrect.

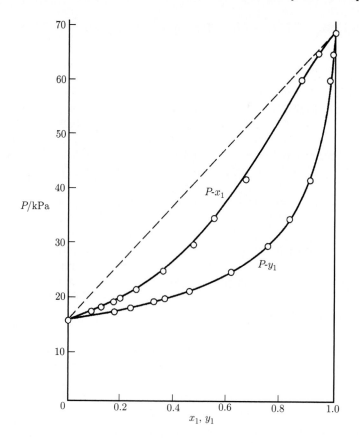

Figure 11.8: Pxy data at $50°$C for chloroform(1)/1,4-dioxane(2).

Our purpose now is to develop a simple test for the *consistency* with respect to the Gibbs/Duhem equation of a P-x_1-y_1 data set. Applied to a binary liquid phase at constant temperature and pressure, Eq. (10.91) becomes

$$d\left(\frac{nG^E}{RT}\right) = \ln\gamma_1\,dn_1 + \ln\gamma_2\,dn_2$$

This equation may be applied to a liquid phase at constant T but variable P, because liquid-phase properties are insensitive to pressure. If $n = 1$, $dn_1 = dx_1$ and $dn_2 = dx_2 = -dx_1$. The preceding equation may then be written

$$\frac{d(G^E/RT)}{dx_1} = \ln\frac{\gamma_1}{\gamma_2} \qquad (A)$$

Equation (A) is applied to *derived* property values, i.e., those given by a correlation, such as represented by the Margules equations, Eqs. (11.7) and (11.8). For the

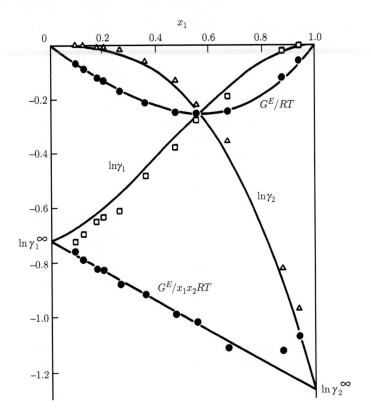

Figure 11.9: Liquid-phase properties from VLE data at $50°C/$ for chloroform(1)/1,4-dioxane(2).

corresponding *experimental* values, we write Eq. (11.5) as

$$\left(\frac{G^E}{RT}\right)^* = x_1 \ln \gamma_1^* + x_2 \ln \gamma_2^*$$

where the asterisk denotes experimental values. Both γ_1^* and γ_2^* are calculated from P-x_1-y_1 data by Eq. (11.1). Differentiation gives

$$\frac{d(G^E/RT)^*}{dx_1} = x_1 \frac{d\ln \gamma_1^*}{dx_1} + \ln \gamma_1^* + x_2 \frac{d\ln \gamma_2^*}{dx_1} - \ln \gamma_2^*$$

or

$$\frac{d(G^E/RT)^*}{dx_1} = \ln \frac{\gamma_1^*}{\gamma_2^*} + x_1 \frac{d\ln \gamma_1^*}{dx_1} + x_2 \frac{d\ln \gamma_2^*}{dx_1} \qquad (B)$$

Equation (B) is subtracted from Eq. (A) to yield

$$\frac{d(G^E/RT)}{dx_1} - \frac{d(G^E/RT)^*}{dx_1} = \ln \frac{\gamma_1}{\gamma_2} - \ln \frac{\gamma_1^*}{\gamma_2^*} - \left(x_1 \frac{d\ln \gamma_1^*}{dx_1} + x_2 \frac{d\ln \gamma_2^*}{dx_1} \right)$$

Table 11.4: VLE Data for diethyl ketone(1)/n-hexane(2) at 65°C

P/kPa	x_1	y_1	$\ln\gamma_1^*$	$\ln\gamma_2^*$	$\left(\dfrac{G^E}{x_1 x_2 RT}\right)^*$
$90.15(P_2^{\text{sat}})$	0.000	0.000		0.000	
91.78	0.063	0.049	0.901	0.033	1.481
88.01	0.248	0.131	0.472	0.121	1.114
81.67	0.372	0.182	0.321	0.166	0.955
78.89	0.443	0.215	0.278	0.210	0.972
76.82	0.508	0.248	0.257	0.264	1.043
73.39	0.561	0.268	0.190	0.306	0.977
66.45	0.640	0.316	0.123	0.337	0.869
62.95	0.702	0.368	0.129	0.393	0.993
57.70	0.763	0.412	0.072	0.462	0.909
50.16	0.834	0.490	0.016	0.536	0.740
45.70	0.874	0.570	0.027	0.548	0.844
$29.00(P_1^{\text{sat}})$	1.000	1.000	0.000		

The differences between like terms represent *residuals* between derived and experimental values. When we represent these residuals by δ, this equation takes the form

$$\frac{d\,\delta(G^E/RT)}{dx_1} = \delta\ln\frac{\gamma_1}{\gamma_2} - \left(x_1\frac{d\ln\gamma_1^*}{dx_1} + x_2\frac{d\ln\gamma_2^*}{dx_1}\right)$$

If a data set is reduced so as to make the residuals in G^E/RT scatter about zero, then the derivative $d\,\delta(G^E/RT)/dx_1$ is effectively zero, and the preceding equation becomes

$$\boxed{\;\delta\ln\frac{\gamma_1}{\gamma_2} = x_1\frac{d\ln\gamma_1^*}{dx_1} + x_2\frac{d\ln\gamma_2^*}{dx_1}\;}\qquad(11.11)$$

The right-hand side of this equation is exactly the quantity that Eq. (11.6), the Gibbs/Duhem equation, requires to be zero for consistent data. The residual on the left is therefore a direct measure of deviations from the Gibbs/Duhem equation. The extent to which values of this residual fail to scatter about zero measures the departure of the data from consistency with respect to this equation.[5]

Example 11.1   Reduce the VLE data set for diethyl ketone(1)/n-hexane(2) at 65°C reported by Maripuri and Ratcliff,[6] and given in the first three columns of Table 11.4.

---

[5]This test and other aspects of VLE data reduction are fully treated by H. C. Van Ness, *J. Chem. Thermodyn.*, vol. 27, pp. 113–134, 1995; *Pure & Appl. Chem.*, vol. 67, pp. 859–872, 1995.

[6]V. C. Maripuri and G. A. Ratcliff, *J. Appl. Chem. Biotechnol.*, vol. 22, pp. 899–903, 1972.

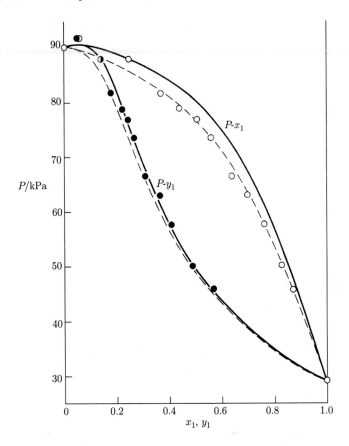

Figure 11.10: Correlations of VLE data for diethyl ketone(1)/n-hexane(2) at 65°C.

SOLUTION   The measured values of P, x_1, and y_1 appear in the first three columns of Table 11.4. The remaining columns present the *experimental* values, $\ln\gamma_1^*$, $\ln\gamma_2^*$, and $(G^E/x_1x_2RT)^*$, calculated from the data by Eqs. (11.1) and (11.5). All values are shown as points on Figs. 11.10 and 11.11. Our objective here is to find an equation for G^E/RT which provides a suitable correlation of the data. The data points of Fig. 11.11 for $(G^E/x_1x_2RT)^*$ show scatter, but are adequate to define a straight line, drawn here by eye and represented by the equation

$$\frac{G^E}{x_1x_2RT} = 0.70x_1 + 1.35x_2$$

This is Eq. (11.7a) with $A_{21} = 0.70$ and $A_{12} = 1.35$. Values of $\ln\gamma_1$ and $\ln\gamma_2$ at the given values of x_1, *derived* from this equation, are calculated by Eqs. (11.8), and *derived* values of P and y_1 at the same values of x_1 come from Eqs. (11.9) and (11.10). These results are plotted as the solid lines of Figs. 11.10 and 11.11. They clearly do not represent a good correlation of the data.

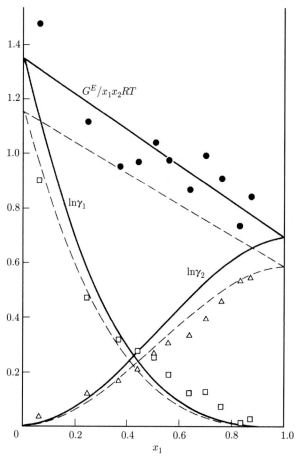

Figure 11.11: Liquid-phase properties from VLE data for diethyl ketone(1)/n-hexane(2) at 65°C.

The difficulty is that the data are not *consistent* with the Gibbs/Duhem equation. That is, the set of *experimental* values, $\ln \gamma_1^*$ and $\ln \gamma_2^*$, shown in Table 11.4 is not in accord with Eq. (11.6). However, the values of $\ln \gamma_1$ and $\ln \gamma_2$ *derived from the correlation* necessarily obey this equation; the two sets of values therefore cannot possibly agree, and the resulting correlation cannot provide a precise representation of the complete set of P-x_1-y_1 data.

Application of the test for consistency represented by Eq. (11.11) requires calculation of the residuals $\delta(G^E/RT)$ and $\delta \ln(\gamma_1/\gamma_2)$, values of which are plotted vs. x_1 in Fig. 11.12. The residuals $\delta(G^E/RT)$ distribute themselves about zero,[7] as

---

[7]The simple procedure used here to find a correlation for G^E/RT would no doubt be improved by a regression procedure that determines the values of A_{21} and A_{12} that minimize the sum of squares of the residuals $\delta(G^E/RT)$.

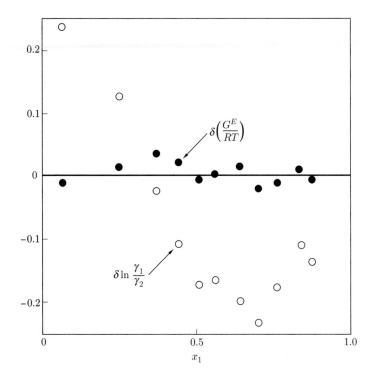

Figure 11.12: Consistency test of data for diethyl ketone(1)/n-hexane(2) at 65°C.

is required by the test, but the residuals $\delta \ln(\gamma_1/\gamma_2)$, which show the extent to which the data fail to satisfy the Gibbs/Duhem equation, clearly do not. Average absolute values of this residual less than 0.03 indicate data of a high degree of consistency; average absolute values of less than 0.10 are probably acceptable. The data set considered here shows an average absolute deviation of about 0.15, and must therefore contain significant error. Although one cannot be certain where the error lies, the values of y_1 are usually most suspect.

The method just described produces a correlation that is unnecessarily divergent from the experimental values. An alternative is to process just the P-x_1 data; this is possible because the P-x_1-y_1 data set includes more information than necessary. The procedure requires a computer, but in principle is simple enough. Assuming that the Margules equation is appropriate to the data, one merely searches for values of the parameters A_{12} and A_{21} that yield pressures by Eq. (11.9) that are as close as possible to the measured values. The method is applicable regardless of the correlating equation assumed, and is known as *Barker's method*.[8] Applied to the

[8] J. A. Barker, *Austral. J. Chem.*, vol. 6, pp. 207–210, 1953.

present data set, it yields the parameters

$$A_{21} = 0.596 \qquad \text{and} \qquad A_{12} = 1.153$$

Use of these parameters in Eqs. (11.7a), (11.8), (11.9), and (11.10) produces the results described by the dashed lines of Figs. 11.10 and 11.11. The correlation cannot be precise, but it clearly provides a better overall representation of the experimental P-x_1-y_1 data.

11.2 Models for the Excess Gibbs Energy

In general G^E/RT is a function of T, P, and composition, but for liquids at low to moderate pressures it is a very weak function of P. Therefore the pressure dependence of activity coefficients is usually neglected. Thus we have for data *at constant T*

$$\frac{G^E}{RT} = g(x_1, x_2, \ldots, x_N) \qquad (\text{const } T)$$

The Margules equation, Eq. (11.7), is an example of this functionality.

Other equations are also in common use for the correlation of activity coefficients. For binary systems the function often most conveniently represented by an equation is G^E/x_1x_2RT, and one procedure is to express this function as a power series in x_1:

$$\frac{G^E}{x_1x_2RT} = a + bx_1 + cx_1^2 + \cdots \qquad (\text{const } T)$$

Since $x_2 = 1 - x_1$ for a binary system of species 1 and 2, x_1 can be taken as the single independent variable. An equivalent power series with certain advantages is known as the Redlich/Kister expansion:[9]

$$\frac{G^E}{x_1x_2RT} = B + C(x_1 - x_2) + D(x_1 - x_2)^2 + \cdots$$

In application, different truncations of this series are appropriate. For each particular expression representing G^E/x_1x_2RT, specific expressions for $\ln\gamma_1$ and $\ln\gamma_2$ result from application of Eq. (10.94). Thus, when $B = C = D = \cdots = 0$, $G^E/RT = 0$, $\ln\gamma_1 = 0$, and $\ln\gamma_2 = 0$. In this event $\gamma_1 = \gamma_2 = 1$, and the solution is ideal.

If $C = D = \cdots = 0$, then

$$\frac{G^E}{x_1x_2RT} = B$$

where B is a constant for a given temperature. The corresponding equations for $\ln\gamma_1$ and $\ln\gamma_2$ are

$$\ln\gamma_1 = Bx_2^2 \qquad (11.12a)$$

[9]O. Redlich, A. T. Kister, and C. E. Turnquist, *Chem. Eng. Progr. Symp. Ser. No. 2*, vol. 48, pp. 49–61, 1952.

and

$$\ln \gamma_2 = Bx_1^2 \tag{11.12b}$$

The symmetrical nature of these relations is evident. The infinite-dilution values of the activity coefficients are given by $\ln \gamma_1^\infty = \ln \gamma_2^\infty = B$.

If $D = \cdots = 0$, then

$$\frac{G^E}{x_1 x_2 RT} = B + C(x_1 - x_2) = B + C(2x_1 - 1)$$

and in this case $G^E/x_1 x_2 RT$ is linear in x_1, and we recover the Margules equation by the substitutions, $B + C = A_{21}$ and $B - C = A_{12}$.

Another well-known equation is obtained when we write the reciprocal expression $x_1 x_2 RT/G^E$ as a linear function of x_1:

$$\frac{x_1 x_2}{G^E/RT} = B' + C'(x_1 - x_2) = B' + C'(2x_1 - 1)$$

This may also be written

$$\frac{x_1 x_2}{G^E/RT} = B'(x_1 + x_2) + C'(x_1 - x_2) = (B' + C')x_1 + (B' - C')x_2$$

We now let $B' + C' = 1/A'_{21}$ and $B' - C' = 1/A'_{12}$. Then

$$\frac{x_1 x_2}{G^E/RT} = \frac{x_1}{A'_{21}} + \frac{x_2}{A'_{12}} = \frac{A'_{12}x_1 + A'_{21}x_2}{A'_{12}A'_{21}}$$

or

$$\frac{G^E}{x_1 x_2 RT} = \frac{A'_{12}A'_{21}}{A'_{12}x_1 + A'_{21}x_2} \tag{11.13}$$

The activity coefficients implied by this equation are given by

$$\ln \gamma_1 = A'_{12}\left(1 + \frac{A'_{12}x_1}{A'_{21}x_2}\right)^{-2} \tag{11.14}$$

$$\ln \gamma_2 = A'_{21}\left(1 + \frac{A'_{21}x_2}{A'_{12}x_1}\right)^{-2} \tag{11.15}$$

These are known as the van Laar[10] equations. When $x_1 = 0$, $\ln \gamma_1^\infty = A'_{12}$; when $x_2 = 0$, $\ln \gamma_2^\infty = A'_{21}$.

The Redlich/Kister expansion, the Margules equations, and the van Laar equations are all special cases of a very general treatment based on rational functions, i.e., on equations for G^E given by ratios of polynomials. These are presented

---

[10] Johannes Jacobus van Laar (1860–1938), Dutch physical chemist.

in detail by Van Ness and Abbott.[11] They provide great flexibility in the fitting of VLE data for binary systems. However, they have scant theoretical foundation, and as a result there is no rational basis for their extension to multicomponent systems. Moreover, they do not incorporate an explicit temperature dependence for the parameters, though this can be supplied on an *ad hoc* basis.

Modern theoretical developments in the molecular thermodynamics of liquid-solution behavior are based on the concept of *local composition*. Within a liquid solution, local compositions, different from the overall mixture composition, are presumed to account for the short-range order and non-random molecular orientations that result from differences in molecular size and intermolecular forces. The concept was introduced by G. M. Wilson in 1964 with the publication of a model of solution behavior since known as the Wilson equation.[12] The success of this equation in the correlation of VLE data prompted the development of alternative local-composition models, most notably the NRTL (**N**on-**R**andom-**T**wo-**L**iquid) equation of Renon and Prausnitz[13] and the UNIQUAC (**UNI**versal **QUA**si-**C**hemical) equation of Abrams and Prausnitz.[14] A further significant development, based on the UNIQUAC equation, is the UNIFAC method,[15] in which activity coefficients are calculated from contributions of the various groups making up the molecules of a solution.

The Wilson equation, like the Margules and van Laar equations, contains just two parameters for a binary system (Λ_{12} and Λ_{21}), and is written

$$\frac{G^E}{RT} = -x_1 \ln(x_1 + x_2\Lambda_{12}) - x_2 \ln(x_2 + x_1\Lambda_{21}) \qquad (11.16)$$

$$\ln\gamma_1 = -\ln(x_1 + x_2\Lambda_{12}) + x_2\left(\frac{\Lambda_{12}}{x_1 + x_2\Lambda_{12}} - \frac{\Lambda_{21}}{x_2 + x_1\Lambda_{21}}\right) \qquad (11.17)$$

$$\ln\gamma_2 = -\ln(x_2 + x_1\Lambda_{21}) - x_1\left(\frac{\Lambda_{12}}{x_1 + x_2\Lambda_{12}} - \frac{\Lambda_{21}}{x_2 + x_1\Lambda_{21}}\right) \qquad (11.18)$$

For infinite dilution, these equations become

$$\ln\gamma_1^{\infty} = -\ln\Lambda_{12} + 1 - \Lambda_{21}$$

[11]H. C. Van Ness and M. M. Abbott, *Classical Thermodynamics of Nonelectrolyte Solutions: With Applications to Phase Equilibria*, Sec. 5-7, McGraw-Hill, New York, 1982.

[12]G. M. Wilson, *J. Am. Chem. Soc.*, vol. 86, pp. 127–130, 1964.

[13]H. Renon and J. M. Prausnitz, *AIChE J.*, vol. 14, p. 135–144, 1968.

[14]D. S. Abrams and J. M. Prausnitz, *AIChE J.*, vol. 21, p. 116–128, 1975.

[15]**UNIQUAC** **F**unctional-group **A**ctivity **C**oefficients; proposed by Aa. Fredenslund, R. L. Jones, and J. M. Prausnitz, *AIChE J.*, vol. 21, p. 1086–1099, 1975; given detailed treatment in the monograph: Aa. Fredenslund, J. Gmehling, and P. Rasmussen, *Vapor-Liquid Equilibrium using UNIFAC*, Elsevier, Amsterdam, 1977.

and

$$\ln \gamma_2^\infty = -\ln \Lambda_{21} + 1 - \Lambda_{12}$$

We note that Λ_{12} and Λ_{21} must always be positive numbers.

The NRTL equation contains three parameters for a binary system and is written

$$\frac{G^E}{x_1 x_2 RT} = \frac{G_{21}\tau_{21}}{x_1 + x_2 G_{21}} + \frac{G_{12}\tau_{12}}{x_2 + x_1 G_{12}} \tag{11.19}$$

$$\ln \gamma_1 = x_2^2 \left[\tau_{21} \left(\frac{G_{21}}{x_1 + x_2 G_{21}} \right)^2 + \frac{G_{12}\tau_{12}}{(x_2 + x_1 G_{12})^2} \right] \tag{11.20}$$

$$\ln \gamma_2 = x_1^2 \left[\tau_{12} \left(\frac{G_{12}}{x_2 + x_1 G_{12}} \right)^2 + \frac{G_{21}\tau_{21}}{(x_1 + x_2 G_{21})^2} \right] \tag{11.21}$$

Here

$$G_{12} = \exp(-\alpha\tau_{12}) \qquad G_{21} = \exp(-\alpha\tau_{21})$$

and

$$\tau_{12} = \frac{b_{12}}{RT} \qquad \tau_{21} = \frac{b_{21}}{RT}$$

where α, b_{12}, and b_{21}, parameters specific to a particular pair of species, are independent of composition and temperature. The infinite-dilution values of the activity coefficients are given by the equations

$$\ln \gamma_1^\infty = \tau_{21} + \tau_{12} \exp(-\alpha\tau_{12})$$

$$\ln \gamma_2^\infty = \tau_{12} + \tau_{21} \exp(-\alpha\tau_{21})$$

The UNIQUAC equation and the UNIFAC method are models of greater complexity and are treated in App. G.

The local-composition models have limited flexibility in the fitting of data, but they are adequate for most engineering purposes. Moreover, they are implicitly generalizable to multicomponent systems without the introduction of any parameters beyond those required to describe the constitutent binary systems. For example, the Wilson equation for multicomponent systems is written

$$\frac{G^E}{RT} = -\sum_i x_i \ln \sum_j x_j \Lambda_{ij} \tag{11.22}$$

and

$$\ln \gamma_i = 1 - \ln \sum_j x_j \Lambda_{ij} - \sum_k \frac{x_k \Lambda_{ki}}{\sum_j x_j \Lambda_{kj}} \tag{11.23}$$

where $\Lambda_{ij} = 1$ for $i = j$, etc. All indices in these equations refer to the same species, and all summations are over *all* species. For each ij pair there are two parameters, because $\Lambda_{ij} \neq \Lambda_{ji}$. For example, in a ternary system the three possible ij pairs are associated with the parameters $\Lambda_{12}, \Lambda_{21}; \Lambda_{13}, \Lambda_{31};$ and $\Lambda_{23}, \Lambda_{32}$.

The temperature dependence of the parameters is given by

$$\Lambda_{ij} = \frac{V_j}{V_i} \exp \frac{-a_{ij}}{RT} \qquad (i \neq j) \tag{11.24}$$

where V_j and V_i are the molar volumes at temperature T of pure liquids j and i, and a_{ij} is a constant independent of composition and temperature. Thus the Wilson equation, like all other local-composition models, has built into it an *approximate* temperature dependence for the parameters. Moreover, all parameters are found from data for binary (in contrast to multicomponent) systems. This makes parameter determination for the local-composition models a task of manageable proportions.

11.3   Property Changes of Mixing

Equations (10.80) through (10.83) are expressions for the properties of *ideal solutions*. Each may be combined with the defining equation for an excess property, Eq. (10.86), to yield

$$G^E = G - \sum_i x_i G_i - RT \sum_i x_i \ln x_i \tag{11.25}$$

$$S^E = S - \sum_i x_i S_i + R \sum_i x_i \ln x_i \tag{11.26}$$

$$V^E = V - \sum_i x_i V_i \tag{11.27}$$

$$H^E = H - \sum_i x_i H_i \tag{11.28}$$

In each of these equations there appears to the right of the equals sign a difference that is expressed in general as $M - \sum_i x_i M_i$. We call this quantity a *property change of mixing* and give it the symbol ΔM. Thus by definition,

$$\Delta M \equiv M - \sum_i x_i M_i \tag{11.29}$$

where M is a molar (or unit-mass) property of a solution and the M_i are molar (or unit-mass) properties of the pure species, all at the same T and P. Equations (11.25) through (11.28) are now rewritten

$$G^E = \Delta G - RT \sum_i x_i \ln x_i \tag{11.30}$$

$$S^E = \Delta S + R \sum_i x_i \ln x_i \tag{11.31}$$

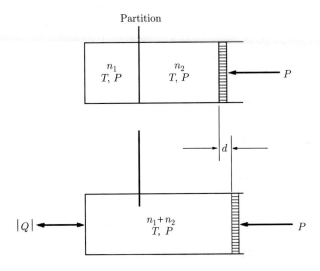

Figure 11.13: Schematic diagram of experimental mixing process.

$$V^E = \Delta V \tag{11.32}$$

$$H^E = \Delta H \tag{11.33}$$

where ΔG, ΔS, ΔV, and ΔH are the Gibbs energy change of mixing, the entropy change of mixing, the volume change of mixing, and the enthalpy change of mixing. For an ideal solution, each excess property is zero, and for this special case Eqs. (11.30) through (11.33) become

$$\Delta G^{id} = RT \sum_i x_i \ln x_i \tag{11.34}$$

$$\Delta S^{id} = -R \sum_i x_i \ln x_i \tag{11.35}$$

$$\Delta V^{id} = 0 \tag{11.36}$$

$$\Delta H^{id} = 0 \tag{11.37}$$

These equations are just restatements of Eqs. (10.80) through (10.83).

Equations (11.30) through (11.33) show that excess properties and property changes of mixing are readily calculated one from the other. Although historically the property changes of mixing were introduced first, because of their direct relation to experiment, it is the excess properties that more readily fit into the theoretical framework of solution thermodynamics. The property changes of mixing of major interest, because of their direct measurability, are ΔV and ΔH, and these two properties are identical to the corresponding excess properties.

An experimental mixing process for a binary system is represented schematically in Fig. 11.13. The two pure species, both at T and P, are initially separated by a partition, withdrawal of which allows mixing. As mixing occurs, expansion or contraction of the system is accompanied by movement of the piston so that the pressure is constant. In addition, heat is added or extracted to maintain a constant temperature. When mixing is complete, the total volume change of the system (as measured by piston displacement d) is

$$\Delta V^t = (n_1 + n_2)V - n_1 V_1 - n_2 V_2$$

Since the process occurs at constant pressure, the total heat transfer Q is equal to the total enthalpy change of the system:

$$Q = \Delta H^t = (n_1 + n_2)H - n_1 H_1 - n_2 H_2$$

Division of these equations by $n_1 + n_2$ gives

$$\Delta V \equiv V - x_1 V_1 - x_2 V_2 = \frac{\Delta V^t}{n_1 + n_2}$$

and

$$\Delta H \equiv H - x_1 H_1 - x_2 H_2 = \frac{Q}{n_1 + n_2}$$

Thus the *volume change of mixing* ΔV and the *enthalpy change of mixing* ΔH are found from the measured quantities ΔV^t and Q. Because of its association with Q, ΔH is usually called the *heat of mixing*.

Figure 11.14 shows experimental heats of mixing ΔH (or excess enthalpies H^E) for the ethanol/water system as a function of composition for several temperatures between 30 and 110°C. This figure illustrates much of the variety of behavior found for $H^E = \Delta H$ and $V^E = \Delta V$ data for binary liquid systems. Such data are also often represented by equations similar to those used for G^E data, in particular by the Redlich/Kister expansion.

Example 11.2   The excess enthalpy (heat of mixing) for a liquid mixture of species 1 and 2 at fixed T and P is represented by the equation

$$H^E = x_1 x_2 (40x_1 + 20x_2)$$

where H^E is in J mol^{-1}. Determine expressions for $\bar{H}_1^E$ and $\bar{H}_2^E$ as functions of x_1.

SOLUTION   The partial properties are found by application of Eqs. (10.15) and (10.16) with $M = H^E$. Thus

$$\bar{H}_1^E = H^E + (1 - x_1)\frac{dH^E}{dx_1} \qquad (A)$$

and

$$\bar{H}_2^E = H^E - x_1\frac{dH^E}{dx_1} \qquad (B)$$

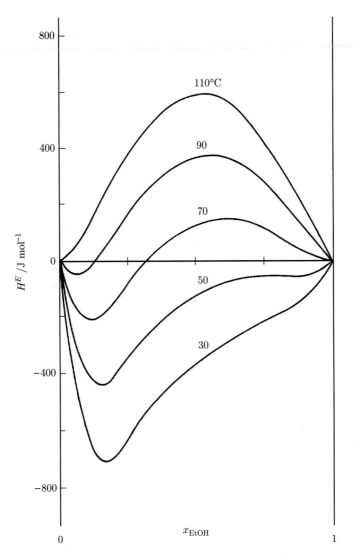

Figure 11.14: Excess enthalpies for ethanol/water.

Elimination of x_2 in favor of x_1 in the given equation for H^E yields

$$H^E = 20x_1 - 20x_1^3 \tag{C}$$

whence

$$\frac{dH^E}{dx_1} = 20 - 60x_1^2 \tag{D}$$

Substitution of Eqs. (C) and (D) into Eq. (A) leads to

$$\bar{H}_1^E = 20 - 60x_1^2 + 40x_1^3$$

Similarly, by Eqs. (B), (C), and (D),

$$\bar{H}_2^E = 40x_1^3$$

These equations contain much the same information as the equations of Example 10.4. Since the last term of the equation for H in the statement of Example 10.4 is identical to the expression given here for H^E, we may write

$$H = 400x_1 + 600x_2 + H^E$$

Clearly, $H_1 = 400$ J mol^{-1} and $H_2 = 600$ J mol^{-1}. The partial properties of Example 10.4 are related to $\bar{H}_1^E$ and $\bar{H}_2^E$ by the equations

$$\bar{H}_1 = \bar{H}_1^E + H_1^{id} = \bar{H}_1^E + H_1 = \bar{H}_1^E + 400$$

and

$$\bar{H}_2 = \bar{H}_2^E + H_2^{id} = \bar{H}_2^E + H_2 = \bar{H}_2^E + 600$$

These two equations follow from combination of Eq. (10.79) with Eq. (10.87).

Excess volumes (volume changes of mixing) for the methanol(1)/water(2) system at 25°C can be calculated from the volumetric data of Fig. 10.3. Equation (10.87) specializes to

$$\bar{V}_i^E = \bar{V}_i - \bar{V}_i^{id}$$

According to Eq. (10.78), $\bar{V}_i^{id} = V_i$. Therefore

$$\bar{V}_1^E = \bar{V}_1 - V_1 \qquad \text{and} \qquad \bar{V}_2^E = \bar{V}_2 - V_2$$

Equation (10.11) written for the excess volume of a binary system becomes

$$V^E = x_1\bar{V}_1^E + x_2\bar{V}_2^E$$

The results are shown in Fig. 11.15. The values on the figure for $x_1 = 0.3$ come from Example 10.3. Thus

$$\bar{V}_1^E = 38.632 - 40.727 = -2.095 \text{ cm}^3 \text{ mol}^{-1}$$

$$\bar{V}_2^E = 17.765 - 18.068 = -0.303 \text{ cm}^3 \text{ mol}^{-1}$$

and

$$V^E = (0.3)(-2.095) + (0.7)(-0.303) = -0.841 \text{ cm}^3 \text{ mol}^{-1}$$

The tangent line drawn at $x_1 = 0.3$ illustrates the determination of partial excess volumes by the method of tangent intercepts. Whereas the values of V in Fig. 10.3 range from 18.068 to 40.727 cm^3 mol^{-1}, the values of $V^E = \Delta V$ go from zero at $x_1 = 0$ and at $x_1 = 1$ to a value of about -1 cm^3 mol^{-1} at a mole fraction of about 0.5. The curves showing $\bar{V}_1^E$ and $\bar{V}_2^E$ are nearly symmetrical for the methanol/water system, but this is by no means so for all systems.

Figure 11.16 illustrates the composition dependence of ΔG, ΔH, and $T\Delta S$ for six binary liquid systems at 50°C and approximately atmospheric pressure. The related quantities G^E, H^E, and TS^E were shown for the same systems in Fig. 10.5. As with the excess properties, property changes of mixing exhibit diverse behavior, but again all systems have certain common features:

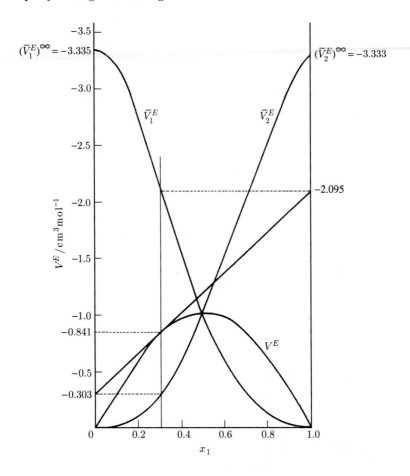

Figure 11.15: Excess volumes for methanol(1)/water(2) at 25°C.

1. Each ΔM is zero for a pure species.

2. The Gibbs energy change of mixing ΔG is always negative.

3. The entropy change of mixing ΔS is positive.

Feature 1 follows from Eq. (11.29). Feature 2 is a consequence of the requirement that the Gibbs energy be a minimum for equilibrium states at specified T and P (Sec. 14.1). Feature 3 reflects the fact that negative entropy changes of mixing are *unusual*; it is *not* a consequence of the second law of thermodynamics, which merely forbids negative entropy changes of mixing for systems *isolated* from their surroundings. Defined for conditions of constant T and P, ΔS is observed to be negative for certain special classes of mixtures, none of which is represented in Fig. 11.16.

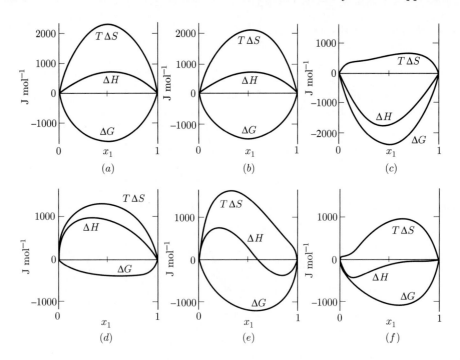

Figure 11.16: Property changes of mixing at $50°$C for six binary liquid systems: (a) chloroform(1)/n-heptane(2); (b) acetone(1)/methanol(2); (c) acetone(1)/chloroform(2); (d) ethanol(1)/n-heptane(2); (e) ethanol(1)/chloroform(2); (f) ethanol(1)/water(2).

Example 11.3 The property changes of mixing and excess properties are interrelated. Show how Figs. 10.5 and 11.16 are generated from correlated data for $\Delta H(x)$ and $G^E(x)$.

SOLUTION For Fig. 11.16, $\Delta H(x)$ comes from curve-fits of calorimetric data, and for Fig. 10.5, $G^E(x)$ is found from reduction of isothermal VLE data by Barker's method (see Example 11.1). By Eq. (11.33),

$$H^E = \Delta H$$

and, by Eq. (10.99),

$$S^E = \frac{H^E - G^E}{T}$$

These equations allow completion of Fig. 10.5. The property changes of mixing ΔG and ΔS follow from G^E and S^E by application of Eqs. (11.30) and (11.31):

$$\Delta G = G^E + RT \sum_i x_i \ln x_i$$

$$\Delta S = S^E - R \sum_i x_i \ln x_i$$

These permit completion of Fig. 11.16.

11.4 Heat Effects of Mixing Processes

The heat of mixing, defined in accord with Eq. (11.29), is

$$\Delta H = H - \sum_i x_i H_i \qquad (11.38)$$

It gives the enthalpy change when pure species are mixed at constant T and P to form one mole (or a unit mass) of solution. Data are most commonly available for binary systems, for which Eq. (11.38) solved for H becomes

$$H = x_1 H_1 + x_2 H_2 + \Delta H \qquad (11.39)$$

This equation provides for the calculation of the enthalpies of binary mixtures from enthalpy data for pure species 1 and 2 and from the heats of mixing. Treatment is here restricted to binary systems.

Data for heats of mixing are usually available for a very limited number of temperatures. If the heat capacities of the pure species and of the mixture are known, heats of mixing are calculated for other temperatures by a method analogous to the calculation of standard heats of reaction at elevated temperatures from the value at 25°C.

Heats of mixing are similar in many respects to heats of reaction. When a chemical reaction occurs, the energy of the products is different from the energy of the reactants at the same T and P because of the chemical rearrangement of the constituent atoms. When a mixture is formed, a similar energy change occurs because interactions between the force fields of like and unlike molecules are different. These energy changes are generally much smaller than those associated with chemical bonds; thus heats of mixing are generally much smaller than heats of reaction.

When solids or gases are dissolved in liquids, the heat effect is called a *heat of solution*, and is based on the dissolution of *1 mole of solute*. If we take species 1 as the solute, then x_1 is the moles of solute per mole of solution. Since ΔH is the heat effect per mole of solution, $\Delta H / x_1$ is the heat effect per mole of solute. Thus

$$\widetilde{\Delta H} = \frac{\Delta H}{x_1}$$

where $\widetilde{\Delta H}$ is the heat of solution on the basis of a mole of *solute*.

Solution processes are conveniently represented by *physical-change* equations analogous to chemical-reaction equations. Thus if 1 mole of LiCl is dissolved in 12 moles of H_2O, the process is represented as

$$\text{LiCl}(s) + 12\text{H}_2\text{O}(l) \rightarrow \text{LiCl}(12\text{H}_2\text{O})$$

The designation $\text{LiCl}(12\text{H}_2\text{O})$ means that the product is a solution of 1 mole of LiCl in 12 moles of H_2O. The enthalpy change accompanying this process at 25°C and 1 bar is $\widetilde{\Delta H} = -33{,}614$ J. That is, a solution of 1 mole of LiCl in 12 moles of H_2O has an enthalpy 33,614 J less than that of 1 mole of pure LiCl(s) and 12 moles of pure $H_2O(l)$. Equations for physical changes such as this are readily combined with equations for chemical reactions. This is illustrated in the following example.

Example 11.4   Calculate the heat of formation of LiCl in 12 moles of H_2O at 25°C.

SOLUTION   The process implied by the problem statement results in the formation from its constituent elements of 1 mole of LiCl *in solution* in 12 moles of H_2O. The equation representing this process is obtained as follows:

$$Li + \tfrac{1}{2}Cl_2 \rightarrow LiCl(s) \qquad \Delta H^\circ_{298} = -408{,}610 \text{ J}$$

$$LiCl(s) + 12H_2O(l) \rightarrow LiCl(12H_2O) \quad \widetilde{\Delta H}_{298} = -33{,}614 \text{ J}$$

$$\overline{Li + \tfrac{1}{2}Cl_2 + 12H_2O(l) \rightarrow LiCl(12H_2O) \quad \Delta H^\circ_{298} = -442{,}224 \text{ J}}$$

The first reaction describes a chemical change resulting in the formation of LiCl(s) from its elements, and the enthalpy change accompanying this reaction is the standard heat of formation of LiCl(s) at 25°C. The second reaction represents the physical change resulting in the solution of 1 mole of LiCl(s) in 12 moles of $H_2O(l)$. The enthalpy change accompanying this reaction is a heat of solution. The enthalpy change of $-442{,}224$ J for the overall process is known as the heat of formation of LiCl *in* 12 moles of H_2O. This figure does *not* include the heat of formation of the H_2O.

Often heats of solution are not reported directly and must be calculated from heats of formation by the reverse of the calculation just illustrated. The data given by the Bureau of Standards[16] for the heats of formation of 1 mole of LiCl are:

LiCl(s)	$-408{,}610$ J
LiCl·H_2O(s)	$-712{,}580$ J
LiCl·$2H_2O$(s)	$-1{,}012{,}650$ J
LiCl·$3H_2O$(s)	$-1{,}311{,}300$ J
LiCl in 3 moles H_2O	$-429{,}366$ J
LiCl in 5 moles H_2O	$-436{,}805$ J
LiCl in 8 moles H_2O	$-440{,}529$ J
LiCl in 10 moles H_2O	$-441{,}579$ J
LiCl in 12 moles H_2O	$-442{,}224$ J
LiCl in 15 moles H_2O	$-442{,}835$ J

From these data heats of solution are readily calculated. Take the case of the solution of 1 mole of LiCl in 5 moles of H_2O. The reaction representing this process is obtained as follows:

$$Li + \tfrac{1}{2}Cl_2 + 5H_2O(l) \rightarrow LiCl(5H_2O) \quad \Delta H^\circ_{298} = -436{,}805 \text{ J}$$

$$LiCl(s) \rightarrow Li + \tfrac{1}{2}Cl_2 \qquad \Delta H^\circ_{298} = 408{,}610 \text{ J}$$

$$\overline{LiCl(s) + 5H_2O(l) \rightarrow LiCl(5H_2O) \quad \widetilde{\Delta H}_{298} = -28{,}195 \text{ J}}$$

---

[16] "The NBS Tables of Chemical Thermodynamic Properties," *J. Phys. Chem. Ref. Data*, vol. 11, suppl. 2, pp. 2-291 and 2-292, 1982.

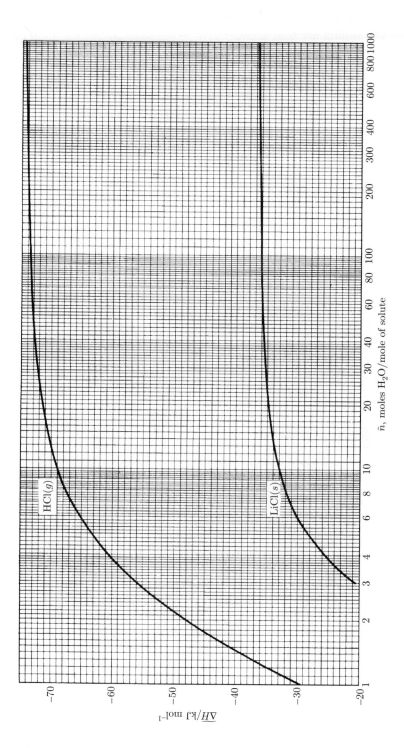

Figure 11.17: Heats of solution at 25°C. (Based on data from "The NBS Tables of Chemical Thermodynamic Properties," *J. Phys. Chem. Ref. Data*, vol. 11, suppl. 2, 1982.)

This calculation can be carried out for each quantity of H_2O for which data are given. The results are then conveniently represented graphically by a plot of $\widetilde{\Delta H}$, the heat of solution per mole of solute, vs. $\tilde{n}$, the moles of solvent per mole of solute. The composition variable, $\tilde{n} \equiv n_2/n_1$, is related to x_1:

$$\tilde{n} = \frac{x_2(n_1 + n_2)}{x_1(n_1 + n_2)} = \frac{1 - x_1}{x_1}$$

whence

$$x_1 = \frac{1}{1 + \tilde{n}}$$

We therefore have the following relations between ΔH, the heat of mixing based on 1 mole of solution, and $\widetilde{\Delta H}$, the heat of solution based on 1 mole of solute:

$$\widetilde{\Delta H} = \frac{\Delta H}{x_1} = \Delta H(1 + \tilde{n})$$

or

$$\Delta H = \frac{\widetilde{\Delta H}}{1 + \tilde{n}}$$

Figure 11.17 shows plots of $\widetilde{\Delta H}$ vs. $\tilde{n}$ for $LiCl(s)$ and $HCl(g)$ dissolved in water at 25°C. Data in this form are readily applied to the solution of practical problems.

Example 11.5 A single-effect evaporator operating at atmospheric pressure concentrates a 15% (by weight) LiCl solution to 40%. The feed enters the evaporator at the rate of 2 kg s^{-1} at 25°C. The normal boiling point of a 40% LiCl solution is about 132°C, and its specific heat is estimated as 2.72 kJ kg^{-1} °C^{-1}. What is the heat-transfer rate in the evaporator?

SOLUTION   The 2 kg of 15% LiCl solution entering the evaporator each second consists of 0.30 kg LiCl and 1.70 kg H_2O. A material balance shows that 1.25 kg of H_2O is evaporated and that 0.75 kg of 40% LiCl solution is produced. The process is indicated schematically in Fig. 11.18.

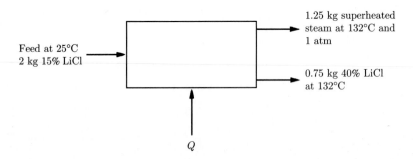

Figure 11.18: Process of Example 11.5.

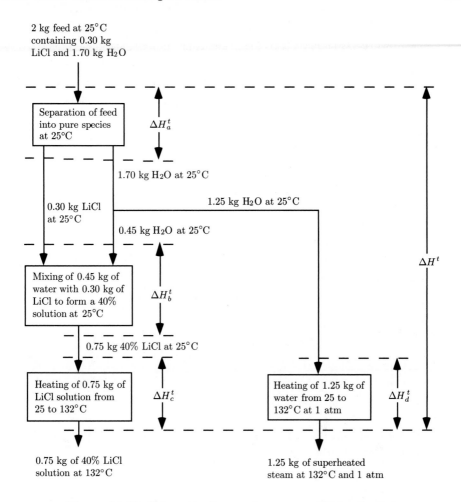

2 kg feed at 25°C
containing 0.30 kg
LiCl and 1.70 kg H_2O

Separation of feed into pure species at 25°C

ΔH_a^t

1.70 kg H_2O at 25°C

0.30 kg LiCl at 25°C

1.25 kg H_2O at 25°C

0.45 kg H_2O at 25°C

Mixing of 0.45 kg of water with 0.30 kg of LiCl to form a 40% solution at 25°C

ΔH_b^t

ΔH^t

0.75 kg 40% LiCl at 25°C

Heating of 0.75 kg of LiCl solution from 25 to 132°C

ΔH_c^t

Heating of 1.25 kg of water from 25 to 132°C at 1 atm

ΔH_d^t

0.75 kg of 40% LiCl solution at 132°C

1.25 kg of superheated steam at 132°C and 1 atm

Figure 11.19: Schematic diagram for process of Example 11.5.

The energy balance for this flow process gives $\Delta H^t = Q$, where ΔH^t is the total enthalpy of the product streams minus the total enthalpy of the feed stream. Thus the problem reduces to finding ΔH^t from the available data. Since enthalpy is a state function, the path used for the calculation of ΔH^t is immaterial and may be selected as convenience dictates and without reference to the actual path followed in the evaporator. The data available are heats of solution of LiCl in H_2O at 25°C (Fig. 11.17), and the calculational path, shown in Fig. 11.19, allows their direct use.

The enthalpy changes for the individual steps shown in Fig. 11.19 must add up to the total enthalpy change:

$$\Delta H^t = \Delta H_a^t + \Delta H_b^t + \Delta H_c^t + \Delta H_d^t$$

The individual enthalpy changes are determined as follows.

ΔH_a^t: This step involves the separation of 2 kg of a 15% LiCl solution into its pure constituents at 25°C. This is an "unmixing" process, and the heat effect is the same as for the corresponding mixing process, but is of opposite sign. For 2 kg of 15% LiCl solution, the moles of material entering are

$$\frac{(0.3)(1{,}000)}{42.39} = 7.077 \text{ mol LiCl}$$

and

$$\frac{(1.70)(1{,}000)}{18.015} = 94.366 \text{ mol H}_2\text{O}$$

Thus the solution contains 13.33 moles of H_2O per mole of LiCl. From Fig. 11.17 the heat of solution per mole of LiCl for $\tilde{n} = 13.33$ is $-33{,}800$ J. For the "unmixing" of 2 kg of solution,

$$\Delta H_a^t = (+33{,}800)(7.077) = 239{,}250 \text{ J}$$

ΔH_b^t: This step results in the mixing of 0.45 kg of water with 0.30 kg of LiCl to form a 40% solution at 25°C. This solution is made up of

$$0.30 \text{ kg} \quad \text{or} \quad 7.077 \text{ mol LiCl}$$

and

$$0.45 \text{ kg} \quad \text{or} \quad 24.979 \text{ mol H}_2\text{O}$$

Thus the final solution contains 3.53 moles of H_2O per mole of LiCl. From Fig. 11.17 the heat of solution per mole of LiCl at this value of $\tilde{n}$ is $-23{,}260$ J. Therefore

$$\Delta H_b^t = (-23{,}260)(7.077) = -164{,}630 \text{ J}$$

ΔH_c^t: For this step 0.75 kg of 40% LiCl solution is heated from 25 to 132°C. Since $\Delta H_c^t = mC_P\Delta T$,

$$\Delta H_c^t = (0.75)(2.72)(132 - 25) = 218.28 \text{ kJ}$$

or

$$\Delta H_c^t = 218{,}280 \text{ J}$$

ΔH_d^t: In this step liquid water is vaporized and heated to 132°C. The enthalpy change is obtained from the steam tables:

$$\Delta H_d^t = (1.25)(2{,}740.3 - 104.8) = 3{,}294.4 \text{ kJ}$$

or

$$\Delta H_d^t = 3{,}294{,}400 \text{ J}$$

Adding the individual enthalpy changes gives

$$\begin{aligned} \Delta H &= \Delta H_a^t + \Delta H_b^t + \Delta H_c^t + \Delta H_d^t \\ &= 239{,}250 - 164{,}630 + 218{,}280 + 3{,}294{,}400 \\ &= 3{,}587{,}300 \text{ J} \end{aligned}$$

The required heat-transfer rate is therefore 3,587.3 kJ s^{-1}.

The most convenient method for representation of enthalpy data for binary solutions is by *enthalpy/concentration (Hx) diagrams*. These diagrams are graphs of the enthalpy plotted as a function of composition (mole fraction or mass fraction of one species) with temperature as parameter. The pressure is a constant and is usually 1 atmosphere. Figure 11.20 shows a partial diagram for the H_2SO_4/H_2O system.

The enthalpy values are based on a mole or a unit mass of *solution*, and Eq. (11.39) is directly applicable. Values of H for the solution depend not only on the heats of mixing, but also on the enthalpies H_1 and H_2 of the pure species. Once these are known for a given T and P, H is fixed for all solutions at the same T and P, because ΔH has a unique and measurable value at each composition. Since absolute enthalpies are unknown, arbitrary zero points are chosen for the enthalpies of the pure species. Thus, the *basis* of an enthalpy/concentration diagram is $H_1 = 0$ for some specified state of species 1 and $H_2 = 0$ for some specified state of species 2. The same temperature need not be selected for these states for both species. In the case of the H_2SO_4/H_2O diagram shown in Fig. 11.20, $H_1 = 0$ for pure liquid H_2O at the triple point [$\simeq32(°F)$], and $H_2 = 0$ for pure liquid H_2SO_4 at 25°C [77(°F)]. In this case the 32(°F) isotherm terminates at $H = 0$ at the end of the diagram representing pure liquid H_2O, and the 77(°F) isotherm terminates at $H = 0$ at the other end of the diagram representing pure liquid H_2SO_4.

The advantage of taking $H = 0$ for pure liquid water at its triple point is that this is the base of the steam tables. Enthalpy values from the steam tables can then be used in conjunction with values taken from the enthalpy/concentration diagram. Were some other base used for the diagram, one would have to apply a correction to the steam-table values to put them on the same basis as the diagram.

For an *ideal solution*, isotherms on an enthalpy/concentration diagram are straight lines connecting the enthalpy of pure species 2 at $x_1 = 0$ with the enthalpy of pure species 1 at $x_1 = 1$. This follows immediately from Eq. (10.83),

$$H^{id} = x_1 H_1 + (1 - x_1)H_2 = x_1(H_1 - H_2) + H_2$$

and is illustrated for a single isotherm in Fig. 11.21 by the dashed line. The solid curve shows how the isotherm might appear for a real solution. Also shown is a tangent line from which partial enthalpies may be determined in accord with Eqs. (10.15) and (10.16). Comparison of Eq. (10.83) with Eq. (11.39) shows that $\Delta H = H - H^{id}$; that is, ΔH is the vertical distance between the curve and the dashed line of Fig. 11.21. Here, the actual isotherm lies below the ideal-solution isotherm, and ΔH is everywhere negative. This means that heat is evolved whenever the pure species at the given temperature are mixed to form a solution at the same temperature. Such a system is said to be *exothermic*. The H_2SO_4/H_2O system is an example. An *endothermic* system is one for which the heats of solution are positive; in this case heat is absorbed to keep the temperature constant. An example is the methanol/benzene system.

One feature of an enthalpy/concentration diagram which makes it particularly useful is the ease with which problems involving adiabatic mixing may be solved. This results from the fact that adiabatic mixing may be represented by a straight

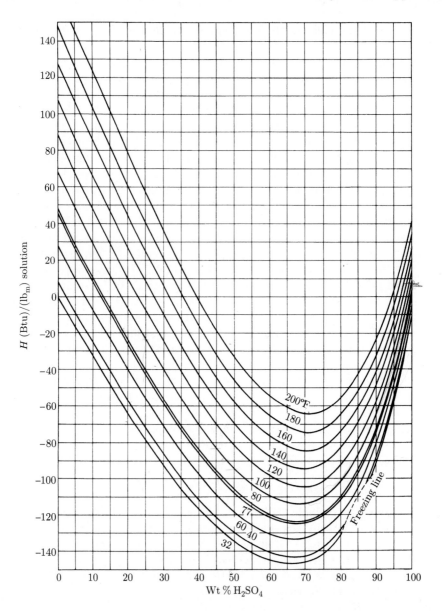

Figure 11.20: Hx diagram for H_2SO_4/H_2O. (Redrawn from the data of W. D. Ross, *Chem. Eng. Prog.*, vol. 48, pp. 314 and 315, 1952. By premission.)

line on the Hx diagram. More precisely, the point on an Hx diagram which represents a solution formed by adiabatic mixing of two other solutions must lie on the straight line connecting the points representing the two initial solutions. This is shown as follows.

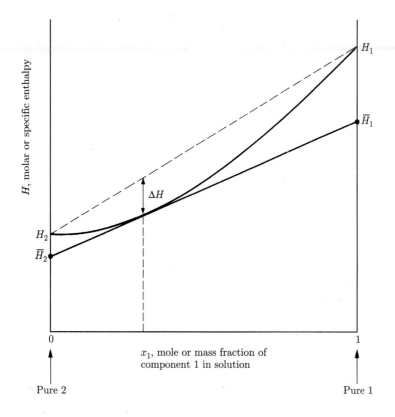

Figure 11.21: Graphical constructions on an Hx diagram.

Let the superscripts a and b denote two initial binary solutions, consisting of n^a and n^b moles respectively. Let superscript c denote the final solution obtained by simple mixing of solutions a and b in an adiabatic process. This process may be batch mixing at constant pressure or a steady-flow process involving no shaft work or change in potential or kinetic energy. In either case,

$$\Delta H^t = Q = 0$$

We may therefore write for the overall change in state

$$(n^a + n^b)H^c = n^a H^a + n^b H^b$$

In addition, a material balance for species 1 gives

$$(n^a + n^b)x_1^c = n^a x_1^a + n^b x_1^b$$

These two equations may be written

$$n^a(H^c - H^a) = -n^b(H^c - H^b)$$

and

$$n^a(x_1^c - x_1^a) = -n^b(x_1^c - x_1^b)$$

Division of the first equation by the second gives

$$\frac{H^c - H^a}{x_1^c - x_1^a} = \frac{H^c - H^b}{x_1^c - x_1^b} \tag{A}$$

We now show that the three points c, a, and b represented by (H^c, x_1^c), (H^a, x_1^a), and (H^b, x_1^b) lie along a straight line on an Hx diagram. The general equation for a straight line in these coordinates is

$$H = mx_1 + k \tag{B}$$

Assuming that this line passes through points a and b, we can write

$$H^a = mx_1^a + k \tag{C}$$

and

$$H^b = mx_1^b + k \tag{D}$$

Subtraction first of Eq. (C) and then Eq. (D) from Eq. (B) gives

$$H - H^a = m(x_1 - x_1^a)$$

and

$$H - H^b = m(x_1 - x_1^b)$$

Dividing the first of these by the second, we obtain

$$\frac{H - H^a}{H - H^b} = \frac{x_1 - x_1^a}{x_1 - x_1^b}$$

or

$$\frac{H - H^a}{x_1 - x_1^a} = \frac{H - H^b}{x_1 - x_1^b}$$

Any point with the coordinates (H, x_1) which satisfies this equation lies on the straight line connecting points a and b. Equation (A) clearly shows that the point (H^c, x_1^c) satisfies this requirement.

The use of enthalpy/concentration diagrams is illustrated in the following examples for the NaOH/H$_2$O system; an Hx diagram is shown in Fig. 11.22.

Example 11.6   A single-effect evaporator concentrates 10,000(lb$_m$)(hr)$^{-1}$ of a 10% (by weight) aqueous solution of NaOH to 50%. The feed enters at 70($^\circ$F). The evaporator operates at an absolute pressure of 3(in Hg), and under these conditions the boiling point of a 50% solution of NaOH is 190($^\circ$F). What is the heat-transfer rate in the evaporator?

SOLUTION   On the basis of 10,000(lb$_m$) of 10% NaOH fed to the evaporator, a material balance shows that the product stream consists of 8,000(lb$_m$) of superheated steam at 3(in Hg) and 190($^\circ$F), and 2,000(lb$_m$) of 50% NaOH at 190($^\circ$F). The

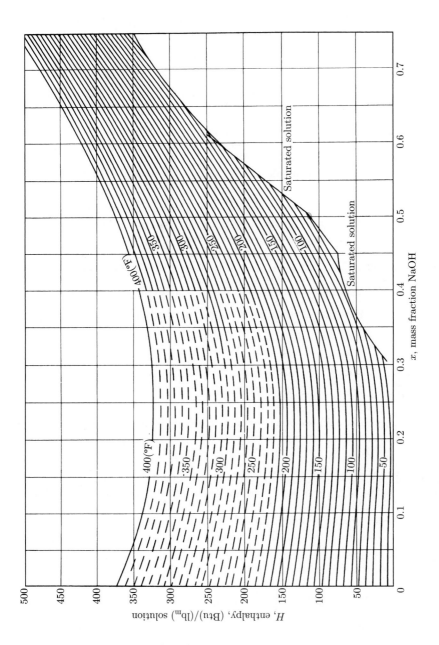

Figure 11.22: Hx diagram for NaOH/H$_2$O. (Reproduced by permission. W. L. McCabe, *Trans. AIChE.*, vol. 31, pp. 129–164, 1935; R. H. Wilson and W. L. McCabe, *Ind. Eng. Chem.*, vol. 34, pp. 558–566, 1942.)

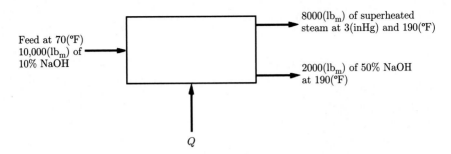

Figure 11.23: Schematic diagram for process of Example 11.6.

process is indicated schematically in Fig. 11.23. The energy balance for this flow process is

$$\Delta H^t = Q$$

In this case ΔH^t is easily determined from enthalpy values taken from the Hx diagram of Fig. 11.22 and from the steam tables:

Enthalpy of superheated steam at 3(in Hg) and 190(°F) = $1{,}146(\text{Btu})(\text{lb}_m)^{-1}$

Enthalpy of 10% NaOH solution at 70(°F) = $34(\text{Btu})(\text{lb}_m)^{-1}$

Enthalpy of 50% NaOH solution at 190(°F) = $215(\text{Btu})(\text{lb}_m)^{-1}$

Thus

$$Q = \Delta H^t = (8{,}000)(1{,}146) + (2{,}000)(215) - (10{,}000)(34)$$
$$= 9{,}260{,}000(\text{Btu})(\text{hr})^{-1}$$

A comparison of this example with Example 11.5 shows the simplification introduced by use of an enthalpy/concentration diagram.

Example 11.7   A 10% aqueous NaOH solution at 70(°F) is mixed with a 70% aqueous NaOH solution at 200(°F) to form a solution containing 40% NaOH.

(a) If the mixing is done adiabatically, what is the final temperature of the solution?

(b) If the final temperature is brought to 70(°F), how much heat must be removed during the process?

SOLUTION   (a) A straight line drawn on Fig. 11.22 connecting the points representing the two initial solutions must contain the point representing the final solution obtained by adiabatic mixing. The particular solution represented by a point on this line at a concentration of 40% NaOH has an enthalpy of $192(\text{Btu})(\text{lb}_m)^{-1}$. Moreover, the isotherm for 220(°F) passes through this point. Thus the final temperature, obtained graphically, is 220(°F).

(b) The overall process cannot be represented by a single straight line on Fig. 11.22. However, we may select any convenient path for calculating ΔH of the process and hence Q, since the energy balance gives $Q = \Delta H$. Thus the process may be considered as occurring in two steps: adiabatic mixing, followed by simple cooling of the resulting solution to the final temperature. The first step is considered in part (a). It results in a solution at 220(°F) with an enthalpy of 192(Btu)(lb$_m$)$^{-1}$. When this solution is cooled to 70(°F), the resulting enthalpy from Fig. 11.22 is 70(Btu)(lb$_m$)$^{-1}$. Therefore

$$Q = \Delta H = 70 - 192 = -122(\text{Btu})(\text{lb}_m)^{-1}$$

Thus 122(Btu) is *evolved* for each pound *mass* of solution formed.

Example 11.8 Determine the enthalpy of solid NaOH at 68(°F) on the basis used for the NaOH/H$_2$O enthalpy/concentration diagram of Fig. 11.22.

SOLUTION   The isotherms on an Hx diagram for a system such as NaOH/H$_2$O terminate at points where the limit of solubility of the solid in water is reached. Thus the isotherms in Fig. 11.22 do not extend to a mass fraction representing pure NaOH. How, then, is the basis of the diagram with respect to NaOH selected? In the case of the water the basis is $H_{\text{H}_2\text{O}} = 0$ for liquid water at 32(°F), consistent with the base of the steam tables. For NaOH the basis is $\bar{H}_{\text{NaOH}} = 0$ for NaOH in an infinitely dilute solution at 68(°F).

This means that the partial specific enthalpy of NaOH at infinite dilution (i.e., at $x_{\text{NaOH}} \to 0$) is arbitrarily set equal to zero at 68(°F). The graphical interpretation is that the diagram is constructed in such a way that a tangent drawn to the 68(°F) isotherm at $x_{\text{NaOH}} = 0$ intersects the $x_{\text{NaOH}} = 1$ ordinate (not shown) at an enthalpy of zero. The selection of $\bar{H}_{\text{NaOH}}^{\infty}$ as zero at 68(°F) automatically fixes the values of the enthalpy of NaOH in all other states.

In particular, the enthalpy of solid NaOH at 68(°F) can be calculated for the basis selected. If 1(lb$_m$) of solid NaOH at 68(°F) is dissolved in an infinite amount of water at 68(°F), and if the temperature is held constant by extraction of the heat of solution, the result is an infinitely dilute solution at 68(°F). Since the water is pure in both the initial and final states, its enthalpy does not change. The heat of solution at 68(°F) is therefore

$$\widetilde{\Delta H}_{\text{NaOH}}^{\infty} = \bar{H}_{\text{NaOH}}^{\infty} - H_{\text{NaOH}}$$

However, $\bar{H}_{\text{NaOH}}^{\infty} = 0$ at 68(°F). Therefore

$$\widetilde{\Delta H}_{\text{NaOH}}^{\infty} = -H_{\text{NaOH}} \qquad [68(°\text{F})]$$

The enthalpy of solid NaOH at 68(°F), H_{NaOH}, is therefore equal to the negative of the heat of solution of NaOH in an infinite amount of water at 68(°F). A literature value[17] for this heat of solution at 25°C on the basis of 1 mole of NaOH is

$$\widetilde{\Delta H}_{\text{NaOH}}^{\infty} = -10,637(\text{cal}) \qquad [25°\text{C}]$$

---

[17]M. W. Chase, Jr., et al., "JANAF Thermochemical Tables," 3d ed., *J. Phys. Chem. Ref. Data*, vol. 14, suppl. 1, p. 1243, 1985.

If the difference in temperature between $25°C$ [$77(°F)$] and $68(°F)$ is neglected, the enthalpy of solid NaOH at $68(°F)$ is

$$H_{NaOH} = -\widetilde{\Delta H}_{NaOH}^{\infty} = \frac{-(-10{,}637)(1.8)}{40.00} = 478.7(Btu)(lb_m)^{-1}$$

This figure represents the enthalpy of solid NaOH at $68(°F)$ on the same basis as was selected for the NaOH/H_2O enthalpy/concentration diagram of Fig. 11.22.

Example 11.9 Solid NaOH at $70(°F)$ is mixed with H_2O at $70(°F)$ to produce a solution containing 45% NaOH at $70(°F)$. How much heat must be transferred per pound *mass* of solution formed?

SOLUTION   On the basis of $1(lb_m)$ of 45% NaOH solution, $0.45(lb_m)$ of solid NaOH must be dissolved in $0.55(lb_m)$ of H_2O. The energy balance is $\Delta H = Q$. The enthalpy of H_2O at $70(°F)$ may be taken from the steam tables, or it may be read from Fig. 11.22 at $x_1 = 0$. In either case, $H_{H_2O} = 38(Btu)(lb_m)^{-1}$. The enthalpy of 45% NaOH at $70(°F)$ is read from Fig. 11.22 as $H = 93(Btu)(lb_m)^{-1}$. We assume that the enthalpy of solid NaOH at $70(°F)$ is essentially the same as the value calculated in the preceding example for $68(°F)$: $H_{NaOH} = 478.7(Btu)(lb_m)^{-1}$. Therefore

$$Q = \Delta H = (1)(93) - (0.55)(38) - (0.45)(478.7) = -143(Btu)$$

Thus, $143(Btu)$ is *evolved* for each pound *mass* of solution formed.

11.5   Molecular Basis for Mixture Behavior

The relations between excess properties and property changes of mixing (Sec. 11.3) facilitate discussion of the molecular phenomena which give rise to observed excess-property behavior. An essential connection is provided by Eq. (11.33), which asserts the identity of H^E and ΔH. Thus we may focus on the *mixing process* (and hence on ΔH) for explaining the behavior of H^E.

The sign and magnitude of ΔH roughly reflect differences in the strengths of intermolecular attractions between pairs of unlike species on the one hand, and pairs of like species on the other. In the standard mixing process (Sec. 11.3) interactions between like species are disrupted, and interactions between unlike species are promoted. If the unlike attractions are *weaker* than the average of those between species of the same kind, then in the mixing process more energy is required to break like attractions than is made available by formation of unlike attractions. In this case $\Delta H(= H^E)$ is positive, i.e., the mixing process is endothermic. If the unlike attractions are *stronger*, then ΔH is negative, and the mixing process is exothermic.

In Secs. 3.8 and 10.10 we identified intermolecular attractive interactions of four kinds: dispersion, induction, direct electrostatic, and quasichemical. A summary list of important points follows:

1. Of the four attractive interactions, the dispersion force is always present. It dominates when interacting molecules are non-polar or slightly polar. (See Table 3.2 and the accompanying discussion.)

2. The induction force requires that at least one of the interacting species be polar. It is often the weakest of the "physical" intermolecular attractive forces (Table 3.2).

3. For neutral molecules, the simplest and normally the strongest direct electrostatic force is that operating between two permanent dipoles. This force can dominate "physical" attractive interactions if the molecules have high effective polarity, i.e., if they are small and have large permanent dipoles.

4. Quasichemical forces, when present, can be the strongest of the four attractive interactions. However, their existence requires special chemical make-up of the interacting molecules. Hydrogen bonding is the most important interaction of this type, although charge-transfer complexing can play a major role in some kinds of systems.

With these notions in mind, we offer some rationalizations of the observed signs and magnitudes of $H^E (= \Delta H)$ for binary liquid mixtures of the kinds discussed in Sec. 10.11.

H^E of NP/NP Mixtures

Here, dispersion forces are usually the only significant attractive intermolecular forces. Thus H^E reflects energetic effects associated with disruption of dispersion interactions between like species, and simultaneous promotion of dispersion interaction between unlike species. Molecular theory[18] suggests that dispersion forces between unlike species are weaker than the average of dispersion forces between the like species. (This is the molecular basis of the "like prefers like" rule of elementary chemistry.) Here, then, one expects H^E to be positive. This is what is usually observed for NP/NP mixtures (Fig. 10.7).

H^E of NA/NP Mixtures

For this class of mixtures, interactions between molecules of like species are *different in kind* for the two species. In particular, two molecules of the polar species experience a direct-electrostatic interaction and a (usually weak) induction interaction, in addition to the usual dispersion interaction; here, the attractive forces are stronger than would be observed for a non-polar species of similar size and geometry. Interaction between unlike species, on the other hand, involves only the dispersion and (weak) induction forces. One therefore expects H^E to be positive, only more so than for otherwise similar NP/NP mixtures. Experiment bears this out, on average (Fig. 10.7).

H^E of AS/NP Mixtures

As for NP/NP and NA/NP mixtures, one expects positive H^E; this is what Fig. 10.7

---

[18]See, e.g., J. M. Prausnitz, R. N. Lichtenthaler, and E. G. de Azevedo, *Molecular Thermodynamics of Fluid-Phase Equilibria*, 2nd ed., Sec. 4.4, Prentice Hall, Englewood Cliffs, NJ, 1986.

shows. However, H^E is often observed to be only modest in magnitude, frequently less than H^E for otherwise similar NA/NP mixtures. The reason for this is the unusual strength of the like interactions for the associating polar species. Here, hydrogen-bonded complexes for the polar species can persist in solution up to rather high dilution, thus mitigating the otherwise very large positive values of H^E expected from simple disruption/promotion arguments.

H^E of Solvating NA/NA Mixtures

These mixtures are the major occupants of Region IV in Fig. 10.7. Since neither species associates by hydrogen bonding, attractions between like species result from dispersion, induction, and dipole/dipole interactions. The same kinds of interaction obtain for unlike molecules, but in addition there is superposed a strong attraction owing to the formation of a hydrogen-bonded solvation complex. The net effect is a negative value for H^E; the system is exothermic.

H^E of AS/NA and AS/AS Mixtures

All four types of attractive interaction occur between unlike species, and for at least one of the pure species. Thus the sign and magnitude of H^E reflect a balance between competing effects of dipole/dipole interactions, association, and solvation. Qualitative prediction of enthalpic behavior is difficult, except by analogy. Figure 10.7 suggests the diversity of behavior observed for such mixtures.

The excess entropy is related to ΔS through Eqs. (11.31) and (11.35). Thus

$$S^E = \Delta S - \Delta S^{id} \tag{11.40}$$

where

$$\Delta S^{id} = -R \sum_i x_i \ln x_i \tag{11.35}$$

An ideal solution is one comprising molecules of identical size and shape, and for which intermolecular forces are the same for all molecular pairs, whether like or unlike. For such a hypothetical solution, the entropy change of mixing, given by Eq. (11.35), is always positive.

In a real mixture, molecules of different species have different sizes and/or shapes, and the intrinsic strengths of molecular interactions are different for different molecular pairs. As a result, ΔS for a real mixture may be greater or less than ΔS^{id}, and by Eq. (11.40), S^E may be positive or negative. For rationalizing the behavior of S^E, we find it convenient to consider separately *size/shape* effects on the one hand and *structural* effects on the other. (The word "structure" refers to the order brought about at the molecular level by intermolecular forces.)

Pure size/shape effects result in a ΔS greater than ΔS^{id}, and hence provide a positive contribution to S^E. Prausnitz et al.[19] discuss the relative roles of size

---

[19] J. M. Prausnitz, R. N. Lichtenthaler, and E. G. de Azevedo, *op. cit.*, Sec. 7.4.

and shape, and give references to the relevant literature. If size effects alone are considered, an approximate upper bound to this contribution to S^E is given by the *Flory-Huggins equation*

$$S^E = -R\sum_i x_i \ln \frac{\Phi_i}{x_i}$$

where the Φ_i are apparent volume fractions,

$$\Phi_i \equiv \frac{x_i V_i}{\sum_j x_j V_j}$$

and the V_i are molar volumes of the pure species.

Structural contributions to ΔS (hence to S^E) reflect primarily the relative strengths of competing intermolecular attractions. Consider the mixing of a nonassociating polar species (e.g., acetone) with a nonpolar species (e.g., n-hexane). Energetically, the net result of the mixing process is determined primarily by the energy associated with disruption of dipole/dipole interactions, as discussed earlier with respect to H^E for NA/NP mixtures. With respect to entropy, this is a *structure-breaking* process wherein molecular aggregates promoted by a strong dipole/dipole interaction are *broken up* by mixing.

Consider instead the mixing of two nonassociating polar species, one a hydrogen donor and the other a hydrogen acceptor (e.g., chloroform/acetone, Fig. 10.5c). Energetically, the net result of the mixing process is determined primarily by the energy associated with formation of a solvation complex, as discussed earlier regarding H^E for solvating NA/NA mixtures. This is a *structure-making* process, wherein molecular aggregates promoted by a strong quasichemical interaction are *formed* on mixing.

Structure-breaking implies a positive contribution to S^E ($\Delta S > \Delta S^{id}$), and structure-making a negative contribution to S^E ($\Delta S < \Delta S^{id}$). When used in conjunction with size/shape arguments, these simple notions help to explain observed signs for S^E. By way of example, we consider again binary liquid mixtures of the kinds discussed in Sec. 10.11.

S^E of NP/NP Mixtures

In the absence of significant size/shape effects, S^E is usually positive, owing to the relative weakness of unlike vs. like intermolecular attractions. Thus structure breaking on mixing is a stronger effect than structure making. However, the enthalpy contribution to G^E often dominates, and Region I behavior obtains. For mixtures of species of significantly different size (e.g., n-hexane/n-hexadecane), positive size/shape contributions can reinforce structural effects, producing values of S^E large enough for entropy to dominate; G^E is then negative and Region VI behavior is observed.

S^E of NA/NP Mixtures

As already noted, the mixing process here primarily involves structure-breaking (positive contributions to S^E). Size/shape effects can have an augmenting influence to produce substantial positive S^E. However, H^E often is also large, and enthalpy usually dominates (Region I).

S^E of AS/NP Mixtures

Mixing nominally promotes a structure-*breaking* disruption of hydrogen-bonded complexes of the associating species. However, the persistence of these complexes in solution up to rather high dilution can greatly reduce this positive contribution to S^E, leading to negative values of S^E over much of the composition range. This effect is observed for mixtures of strong associators (e.g., alcohols and carboxylic acids) with hydrocarbons. An example is the ethanol/n-heptane system of Fig. 10.5d, which shows Region II behavior.

S^E of Solvating NA/NA Mixtures

As noted in earlier discussion, this is predominately a structure-*making* situation, and S^E is negative. However, H^E is also negative (and large), whence enthalpy usually dominates, making G^E negative (Region IV).

S^E of AS/NA and AS/AS Mixtures

The complexities discussed with respect to H^E also apply to S^E; structure-breaking and structure-making effects compete to provide a variety of sign combinations and a range of magnitudes for S^E.

PROBLEMS

11.1. Equations analogous to Eqs. (10.15) and (10.16) apply for excess properties. Since $\ln \gamma_i$ is a partial property with respect to G^E/RT, these analogous equations can be written for $\ln \gamma_1$ and $\ln \gamma_2$ in a binary system.

   (*a*) Write these equations, and apply them to Eq. (11.13) to show that Eqs. (11.14) and (11.15) are indeed obtained.

   (*b*) The alternative procedure is to apply Eq. (10.94). Proceding in the manner that led to Eqs. (11.8), show that Eqs. (11.14) and (11.15) are again reproduced.

11.2. The following is a set of VLE data for the system methanol(1)/water(2) at 333.15 K (extracted from K. Kurihara et al., *J. Chem. Eng. Data*, vol. 40, pp. 679–684, 1995):

P/kPa	x_1	y_1	P/kPa	x_1	y_1
19.953	0.0000	0.0000	60.614	0.5282	0.8085
39.223	0.1686	0.5714	63.998	0.6044	0.8383
42.984	0.2167	0.6268	67.924	0.6804	0.8733
48.852	0.3039	0.6943	70.229	0.7255	0.8922
52.784	0.3681	0.7345	72.832	0.7776	0.9141
56.652	0.4461	0.7742	84.562	1.0000	1.0000

(a) Basing calculations on Eq. (11.1), find parameter values for the Margules equation that provide the best fit of G^E/RT to the data, and prepare a Pxy diagram that compares the experimental points with curves determined from the correlation.

(b) Repeat (a) for the van Laar equation.

(c) Repeat (a) for the Wilson equation.

(d) Using Barker's method, find parameter values for the Margules equation that provide the best fit of the P-x_1 data. Prepare a diagram showing the residuals δP and δy_1 plotted vs. x_1.

(e) Repeat (d) for the van Laar equation.

(f) Repeat (d) for the Wilson equation.

11.3. The following is a set of VLE data for the system acetone(1)/methanol(2) at 55°C (extracted from D. C. Freshwater and K. A. Pike, *J. Chem. Eng. Data*, vol. 12, pp. 179–183, 1967):

P/kPa	x_1	y_1	P/kPa	x_1	y_1
68.728	0.0000	0.0000	97.646	0.5052	0.5844
72.278	0.0287	0.0647	98.462	0.5432	0.6174
75.279	0.0570	0.1295	99.811	0.6332	0.6772
77.524	0.0858	0.1848	99.950	0.6605	0.6926
78.951	0.1046	0.2190	100.278	0.6945	0.7124
82.528	0.1452	0.2694	100.467	0.7327	0.7383
86.762	0.2173	0.3633	100.999	0.7752	0.7729
90.088	0.2787	0.4184	101.059	0.7922	0.7876
93.206	0.3579	0.4779	99.877	0.9080	0.8959
95.017	0.4050	0.5135	99.799	0.9448	0.9336
96.365	0.4480	0.5512	96.885	1.0000	1.0000

(a) Basing calculations on Eq. (11.1), find parameter values for the Margules equation that provide the best fit of G^E/RT to the data, and prepare a Pxy diagram that compares the experimental points with curves determined from the correlation.

(b) Repeat (a) for the van Laar equation.

(c) Repeat (a) for the Wilson equation.

(d) Using Barker's method, find parameter values for the Margules equation that provide the best fit of the P-x_1 data. Prepare a diagram showing the residuals δP and δy_1 plotted vs. x_1.

(e) Repeat (d) for the van Laar equation.

(f) Repeat (d) for the Wilson equation.

11.4. The following is a set of activity-coefficient data for a binary liquid system as determined from VLE data:

x_1	γ_1	γ_2	x_1	γ_1	γ_2
0.0523	1.202	1.002	0.5637	1.120	1.102
0.1299	1.307	1.004	0.6469	1.076	1.170
0.2233	1.295	1.006	0.7832	1.032	1.298
0.2764	1.228	1.024	0.8576	1.016	1.393
0.3482	1.234	1.022	0.9388	1.001	1.600
0.4187	1.180	1.049	0.9813	1.003	1.404
0.5001	1.129	1.092			

Inspection of these experimental values suggests that they are *noisy*, but the question is whether they are *consistent*, and therefore possibly on average correct.

(a) Find experimental values for G^E/RT and plot them along with the experimental values of $\ln\gamma_1$ and $\ln\gamma_2$ on a single graph.

(b) Develop a valid correlation for the composition dependence of G^E/RT and show lines on the graph of part (a) that represent this correlation for all three of the quantities plotted there.

(c) Apply the consistency test described in Example 11.1 to these data, and draw a conclusion with respect to this test.

11.5. VLE data for methyl *tert*-butyl ether(1)/dichloromethane(2) at 308.15 K (extracted from F. A. Mato, C. Berro, and A. Péneloux, *J. Chem. Eng. Data*, vol. 36, pp. 259–262, 1991) are as follows:

P/kPa	x_1	y_1	P/kPa	x_1	y_1
85.265	0.0000	0.0000	59.651	0.5036	0.3686
83.402	0.0330	0.0141	56.833	0.5749	0.4564
82.202	0.0579	0.0253	53.689	0.6736	0.5882
80.481	0.0924	0.0416	51.620	0.7676	0.7176
76.719	0.1665	0.0804	50.455	0.8476	0.8238
72.422	0.2482	0.1314	49.926	0.9093	0.9002
68.005	0.3322	0.1975	49.720	0.9529	0.9502
65.096	0.3880	0.2457	49.624	1.0000	1.0000

The data are well correlated by the three-parameter Margules equation [an extension of Eq. (11.7)]:

$$\frac{G^E}{RT} = (A_{21}x_1 + A_{12}x_2 - Cx_1x_2)x_1x_2$$

Implied by this equation are the expressions:

$$\ln\gamma_1 = x_2^2[A_{12} + 2(A_{21} - A_{12} - C)x_1 + 3Cx_1^2]$$

$$\ln\gamma_2 = x_1^2[A_{21} + 2(A_{12} - A_{21} - C)x_2 + 3Cx_2^2]$$

(a) Basing calculations on Eq. (11.1), find the values of parameters A_{12}, A_{21}, and C that provide the best fit of G^E/RT to the data.

(b) Prepare a plot of $\ln\gamma_1$, $\ln\gamma_2$, and G^E/x_1x_2RT vs. x_1 showing both the correlation and experimental values.

(c) Prepare a Pxy diagram (see Fig. 11.10) that compares the experimental data with the the correlation determined in (a).

(d) Prepare a consistency-test diagram like Fig. 11.12.

(e) Using Barker's method, find the values of parameters A_{12}, A_{21}, and C that provide the best fit of the P-x_1 data. Prepare a diagram showing the residuals δP and δy_1 plotted vs. x_1.

11.6. Following are VLE data for the system acetonitrile(1)/benzene(2) at $45°C$ (extracted from I. Brown and F. Smith, *Austral. J. Chem.*, vol. 8, p. 62, 1955):

P/kPa	x_1	y_1	P/kPa	x_1	y_1
29.819	0.0000	0.0000	36.978	0.5458	0.5098
31.957	0.0455	0.1056	36.778	0.5946	0.5375
33.553	0.0940	0.1818	35.792	0.7206	0.6157
35.285	0.1829	0.2783	34.372	0.8145	0.6913
36.457	0.2909	0.3607	32.331	0.8972	0.7869
36.996	0.3980	0.4274	30.038	0.9573	0.8916
37.068	0.5069	0.4885	27.778	1.0000	1.0000

The data are well correlated by the three-parameter Margules equation (see Problem 11.5).

(a) Basing calculations on Eq. (11.1), find the values of parameters A_{12}, A_{21}, and C that provide the best fit of G^E/RT to the data.

(b) Prepare a plot of $\ln \gamma_1$, $\ln \gamma_2$, and $G^E/x_1 x_2 RT$ vs. x_1 showing both the correlation and experimental values.

(c) Prepare a Pxy diagram (see Fig. 11.10) that compares the experimental data with the the correlation determined in (a).

(d) Prepare a consistency-test diagram like Fig. 11.12.

(e) Using Barker's method, find the values of parameters A_{12}, A_{21}, and C that provide the best fit of the P-x_1 data. Prepare a diagram showing the residuals δP and δy_1 plotted vs. x_1.

11.7. If Eq. (11.1) is valid for isothermal VLE in a binary system, show that

$$\left(\frac{dP}{dx_1}\right)_{x_1=0} \geq -P_2^{\text{sat}} \qquad \left(\frac{dP}{dx_1}\right)_{x_1=1} \leq P_1^{\text{sat}}$$

11.8. At $25°C$ and atmospheric pressure the excess volumes of binary liquid mixtures of species 1 and 2 are given by the equation

$$V^E = x_1 x_2 (45x_1 + 25x_2)$$

where V is in $cm^3 \; mol^{-1}$. At these conditions, $V_1 = 110$ and $V_2 = 90 \; cm^3 \; mol^{-1}$. Determine the partial molar volumes $\bar{V}_1$ and $\bar{V}_2$ in a mixture containing 40 mole-% of species 1 at the given conditions.

11.9. Excess volumes ($cm^3 \; mol^{-1}$) for the system ethanol(1)/methyl butyl ether(2) at $25°C$ are given by the equation

$$V^E = x_1 x_2 [-1.026 + 0.220(x_1 - x_2)]$$

Given that $V_1 = 58.63$ and $V_2 = 118.46$ cm^3 mol^{-1}, what volume of mixture is formed when 750 cm^3 of pure species 1 is mixed with 1,500 cm^3 of species 2 at 25°C? What would be the volume if an ideal solution were formed?

11.10. If LiCl·2H$_2$O(s) and H$_2$O(l) are mixed isothermally at 25°C to form a solution containing 10 moles of water for each mole of LiCl, what is the heat effect per mole of solution?

11.11. If a liquid solution of HCl in water, containing 1 mol of HCl and 4.5 mol of H$_2$O, absorbs an additional 1 mol of HCl(g) at the constant temperature of 25°C, what is the heat effect?

11.12. What is the heat effect when 20 kg of LiCl(s) is added to 125 kg of an aqueous solution containing 10-wt-% LiCl in an isothermal process at 25°C?

11.13. A mass of 12 kg s^{-1} of Cu(NO$_3$)$_2$·6H$_2$O along with 15 kg s^{-1} of water, both at 25°C, are fed to a tank where mixing takes place. The resulting solution passes through a heat exchanger which adjusts its temperature to 25°C. What is the rate of heat transfer in the exchanger?

- For Cu(NO$_3$)$_2$, $\Delta H^\circ_{f_{298}} = -302.9$ kJ
- For Cu(NO$_3$)$_2$·6H$_2$O, $\Delta H^\circ_{f_{298}} = -2{,}110.8$ kJ
- The heat of solution of 1 mol of Cu(NO$_3$)$_2$ in water at 25°C is -47.84 kJ, independent of $\tilde{n}$ for values of interest here.

11.14. A liquid solution of LiCl in water at 25°C contains 1 mol of LiCl and 7 mol of water. If 1 mol of LiCl·3H$_2$O(s) is dissolved isothermally in this solution, what is the heat effect?

11.15. It is required to produce an aqueous LiCl solution by mixing LiCl·2H$_2$O(s) with water. The mixing occurs *both* adiabatically and without change in temperature at 25°C. Determine the mole fraction of LiCl in the final solution.

11.16. Data from the Bureau of Standards (*J. Phys. Chem. Ref. Data*, vol. 11, suppl. 2, 1982) include the following heats of formation for 1 mol of CaCl$_2$ in water at 25°C:

CaCl$_2$ in 10 mol H$_2$O	-862.74 kJ
CaCl$_2$ in 15 mol H$_2$O	-867.85 kJ
CaCl$_2$ in 20 mol H$_2$O	-870.06 kJ
CaCl$_2$ in 25 mol H$_2$O	-871.07 kJ
CaCl$_2$ in 50 mol H$_2$O	-872.91 kJ
CaCl$_2$ in 100 mol H$_2$O	-873.82 kJ
CaCl$_2$ in 300 mol H$_2$O	-874.79 kJ
CaCl$_2$ in 500 mol H$_2$O	-875.13 kJ
CaCl$_2$ in 1,000 mol H$_2$O	-875.54 kJ

From these data prepare a plot of $\widetilde{\Delta H}$, the heat of solution at 25°C of CaCl$_2$ in water, vs. $\tilde{n}$, the mole ratio of water to CaCl$_2$.

11.17. A liquid solution contains 1 mol of CaCl$_2$ and 25 mol of water. Using data from the preceding problem, determine the heat effect when an additional 1 mol of CaCl$_2$ is dissolved isothermally in this solution.

11.18. Solid CaCl$_2$·6H$_2$O and liquid water at 25°C are mixed *adiabatically* in a continuous process to form a brine of 15-wt-% CaCl$_2$. Using data from Problem 11.16, determine the temperature of the brine solution formed. The specific heat of a 15-wt-% aqueous CaCl$_2$ solution at 25°C is 3.28 kJ kg^{-1} °C^{-1}.

11.19. Consider a plot of $\widetilde{\Delta H}$, the heat of solution based on one mole of solute (species 1), vs. $\tilde{n}$, the moles of solvent per mole of solute, at constant T and P. Figure 11.17 is an example of such a plot, except that the plot considered here has a linear rather than logarithmic scale along the abscissa. Let a tangent drawn to the $\widetilde{\Delta H}$ vs. $\tilde{n}$ curve intercept the ordinate at point I.

(a) Prove that the slope of the tangent at a particular point is equal to the partial excess enthalpy of the solvent in a solution with the composition represented by $\tilde{n}$; i.e., prove that

$$\frac{d\widetilde{\Delta H}}{d\tilde{n}} = \bar{H}_2^E$$

(b) Prove that the intercept I equals the partial excess enthalpy of the solute in the same solution; i.e., prove that

$$I = \bar{H}_1^E$$

11.20. Suppose that ΔH for a particular solute(1)/solvent(2) system is represented by the equation

$$\Delta H = x_1 x_2 (A_{21} x_1 + A_{12} x_2) \tag{A}$$

Relate the behavior of a plot of $\widetilde{\Delta H}$ vs. $\tilde{n}$ to the features of this equation. Specifically, rewrite Eq. (A) in the form $\widetilde{\Delta H}(\tilde{n})$, and then show that

(a) $\displaystyle\lim_{\tilde{n}\to 0} \widetilde{\Delta H} = 0$.

(b) $\displaystyle\lim_{\tilde{n}\to\infty} \widetilde{\Delta H} = A_{12}$

(c) $\displaystyle\lim_{\tilde{n}\to 0} d\widetilde{\Delta H}/d\tilde{n} = A_{21}$

11.21. If the heat of mixing at temperature t_0 is ΔH_0 and if the heat of mixing of the same solution at temperature t is ΔH, show that the two heats of mixing are related by

$$\Delta H = \Delta H_0 + \int_{t_0}^{t} \Delta C_P dt$$

where ΔC_P is the heat-capacity change of mixing, defined by Eq. (11.29).

11.22. What is the heat effect when 150(lb$_m$) of H_2SO_4 is mixed with 350(lb$_m$) of an aqueous solution containing 25-wt-% H_2SO_4 in an isothermal process at 100(°F)?

11.23. For a 50-wt-% aqueous solution of H_2SO_4 at 140(°F), what is the excess enthalpy H^E in (Btu)(lb$_m$)$^{-1}$?

11.24. A mass of 400(lb$_m$) of 35-wt-% aqueous NaOH solution at 130(°F) is mixed with 175(lb$_m$) of 10-wt-% solution at 200(°F).

(a) What is the heat effect if the final temperature is 80(°F)?

(b) If the mixing is adiabatic, what is the final temperature?

11.25. A single-effect evaporator concentrates a 20-wt-% aqueous solution of H_2SO_4 to 70%. The feed rate is 25(lb$_m$)(s)$^{-1}$, and the feed temperature is 80(°F). The evaporator is maintained at an absolute pressure of 1.5(psia), at which pressure the boiling point of 70-% H_2SO_4 is 217(°F). What is the heat-transfer rate in the evaporator?

11.26. What temperature results when sufficient NaOH(s) at 68($°$F) is dissolved adiabatically in a 10-wt-% aqueous NaOH solution, originally at 80($°$F), to bring the concentration up to 35%?

11.27. What is the heat effect when sufficient $SO_3(l)$ at 25$°$C is reacted with H_2O at 25$°$C to give a 50-wt-% H_2SO_4 solution at 60$°$C?

11.28. A mass of 140(lb_m) of 15-wt-% solution of H_2SO_4 in water at 160($°$F) is mixed at atmospheric pressure with 230(lb_m) of 80-wt-% H_2SO_4 at 100($°$F). During the process heat in the amount of 20,000(Btu) is transferred from the system. Determine the temperature of the product solution.

11.29. An insulated tank, open to the atmosphere, contains 1,500(lb_m) of 40-wt-% sulfuric acid at 60($°$F). It is heated to 180($°$F) by injection of live saturated steam at 1(atm), which fully condenses in the process. How much steam is required, and what is the final concentration of H_2SO_4 in the tank?

11.30. Saturated steam at 40(psia) is throttled to 1(atm) and mixed adiabatically with (and condensed by) 45-wt-% sulfuric acid at 80($°$F) in a flow process that raises the temperature of the acid to 160($°$F). How much steam is required for each pound *mass* of entering acid, and what is the concentration of the hot acid?

11.31. A batch of 40-wt-% NaOH solution in water at atmospheric pressure and 80($°$F) is heated in an insulated tank by injection of live steam drawn through a valve from a line containing saturated steam at 35(psia). The process is stopped when the NaOH solution reaches a concentration of 38 wt-%. At what temperature does this occur?

11.32. For a 35-wt-% aqueous solution of H_2SO_4 at 100($°$F), what is the heat of mixing ΔH in $(Btu)(lb_m)^{-1}$?

11.33. If pure liquid H_2SO_4 at 80($°$F) is added adiabatically to pure liquid water at 80($°$F) to form a 40-wt-% solution, what is the final temperature of the solution?

11.34. A liquid solution containing 2(lb mol) H_2SO_4 and 15(lb mol) H_2O at 100($°$F) absorbs 1(lb mol) of $SO_3(g)$, also at 100($°$F), forming a more concentrated sulfuric acid solution. If the process occurs isothermally, determine the heat transferred.

11.35. Determine the heat of mixing ΔH of sulfuric acid in water and the partial specific enthalpies of H_2SO_4 and H_2O for a solution containing 65-wt-% H_2SO_4 at 77($°$F).

11.36. It is proposed to cool a stream of 75-wt-% sulfuric acid solution at 140($°$F) by diluting it with chilled water at 40($°$F). Determine the amount of water that must be added to 1(lb_m) of 75-% acid before cooling below 140($°$F) actually occurs.

11.37. The following liquids, all at atmospheric pressure and 120($°$F), are mixed: 25(lb_m) of pure water, 40(lb_m) of pure sulfuric acid, and 75(lb_m) of 25-wt-% sulfuric acid.

(*a*) How much heat is liberated if mixing is isothermal at 120($°$F)?

(*b*) The mixing process is carried out in two steps: First, the pure sulfuric acid and the 25-% solution are mixed, and the total heat of part (*a*) is extracted; second, the pure water is added adiabatically. What is the temperature of the intermediate solution formed in the first step?

11.38. A large quantity of very dilute aqueous NaOH solution is neutralized by addition of the stoichiometric amount of a 10-mole-% aqueous HCl solution. Estimate the heat effect per mole of NaOH neutralized if the tank is maintained at 25$°$C and 1(atm) and the neutralization reaction goes to completion. Data:

- For NaCl, $\lim\limits_{\tilde{n}\to\infty} \widetilde{\Delta H} = 3.88$ kJ mol^{-1}

- For NaOH, $\lim\limits_{\tilde{n}\to\infty} \widetilde{\Delta H} = -44.50$ kJ mol^{-1}

11.39. A large quantity of very dilute aqueous HCl solution is neutralized by addition of the stoichiometric amount of a 10-mole-% aqueous NaOH solution. Estimate the heat effect per mole of HCl neutralized if the tank is maintained at 25°C and 1(atm) and the neutralization reaction goes to completion.

- For NaCl, $\lim\limits_{\tilde{n}\to\infty} \widetilde{\Delta H} = 3.88$ kJ mol^{-1}

11.40. The heat of mixing (or heat of solution) is *negative* for the systems represented on Figs. 11.17, 11.20, and 11.22. Offer molecular explanations of *why* this is so.

11.41. Listed below are excess-enthalpy data at 25°C for two series of equimolar binary liquid mixtures. Explain why the mixture containing benzene is the "outlier" in each series.

Series	Mixture	H^E/J mol^{-1}
A	CH$_2$Cl$_2$/benzene	−18
	/cyclohexane	1,188
	/n-hexane	1,311
B	acetone/benzene	144
	/cyclohexane	1,574
	/n-hexane	1,555

CHAPTER 12

VLE AT LOW TO MODERATE PRESSURES

A number of industrially important processes, such as distillation, absorption, and extraction, bring two phases into contact. When the phases are not in equilibrium, mass transfer occurs between the phases. The rate of transfer of each species depends on the departure of the system from equilibrium. Quantitative treatment of mass-transfer rates requires knowledge of the equilibrium states (T, P, and compositions) of the system.

In most industrial processes coexisting phases are vapor and liquid, although liquid/liquid, vapor/solid, and liquid/solid systems are also encountered. In this chapter we present a general qualitative discussion of vapor/liquid phase behavior (Sec. 12.3) and describe the calculation of temperatures, pressures, and phase compositions for systems in vapor/liquid equilibrium at low to moderate pressures (Sec. 12.4).[1] Comprehensive expositions are given of dewpoint, bubblepoint, and P,T-flash calculations.

12.1   The Nature of Equilibrium

Equilibrium is a static condition in which no changes occur in the macroscopic properties of a system with time. This implies a balance of all potentials that may cause change. In engineering practice, the assumption of equilibrium is justified when it leads to results of satisfactory accuracy. For example, in the reboiler for a distillation column, equilibrium between vapor and liquid phases is commonly assumed. For finite vaporization rates this is an approximation, but it does not introduce significant error into engineering calculations.

---

[1]For VLE at high pressures, see Chap. 13.

If a system containing fixed amounts of chemical species and consisting of liquid and vapor phases in intimate contact is completely isolated, then in time there is no further tendency for any change to occur within the system. The temperature, pressure, and phase compositions reach final values which thereafter remain fixed. The system is in equilibrium. Nevertheless, at the microscopic level, conditions are not static. The molecules comprising one phase at a given instant are not the same molecules as those in that phase at a later time. Molecules with sufficiently high velocities that are near the interphase boundary overcome surface forces and pass into the other phase. However, the average rate of passage of molecules is the same in both directions, and there is no net transfer of material between the phases.

12.2 The Phase Rule. Duhem's Theorem

The phase rule for nonreacting systems, presented without proof in Sec. 2.8, results from application of a rule of algebra. The number of phase-rule variables which must be arbitrarily specified in order to fix the intensive state of a system at equilibrium, called the degrees of freedom F, is the difference between the total number of phase-rule variables and the number of independent equations that can be written connecting these variables.

The *intensive* state of a PVT system containing N chemical species and π phases in equilibrium is characterized by the temperature T, the pressure P, and $N - 1$ mole fractions[2] for each phase. These are the phase-rule variables, and their number is $2 + (N - 1)(\pi)$. The masses of the phases are not phase-rule variables, because they have no influence on the intensive state of the system.

The phase-equilibrium equations that may be written connecting the phase-rule variables are given by Eqs. (10.6) or Eqs. (10.43):

$$\mu_i^\alpha = \mu_i^\beta = \cdots = \mu_i^\pi \qquad (i = 1, 2, ..., N) \qquad (10.6)$$

$$\hat{f}_i^\alpha = \hat{f}_i^\beta = \cdots = \hat{f}_i^\pi \qquad (i = 1, 2, ..., N) \qquad (10.43)$$

Equations (10.6) and (10.43) contain $(\pi - 1)(N)$ independent phase-equilibrium equations. They are equations connecting the phase-rule variables, because the chemical potentials and fugacities are functions of temperature, pressure, and composition. The difference between the number of phase-rule variables and the number of equations connecting them is the degrees of freedom:

$$F = 2 + (N - 1)(\pi) - (\pi - 1)(N)$$

This reduces to Eq. (2.11):

$$\boxed{F = 2 - \pi + N} \qquad (2.11)$$

---

[2] Only $N - 1$ mole fractions are required, because $\sum_i x_i = 1$.

Applications of the phase rule were discussed in Sec. 2.8.

Duhem's theorem is another rule, similar to the phase rule, but less celebrated. It applies to closed systems for which the extensive state as well as the intensive state of the system is fixed. The state of such a system is said to be *completely determined*, and is characterized not only by the $2 + (N - 1)\pi$ intensive phase-rule variables but also by the π extensive variables represented by the masses (or mole numbers) of the phases. Thus the total number of variables is

$$2 + (N - 1)\pi + \pi = 2 + N\pi$$

If the system is closed and formed from specified amounts of the chemical species present, then we can write a material-balance equation for each of the N chemical species. These in addition to the $(\pi - 1)N$ phase-equilibrium equations provide a total number of independent equations equal to

$$(\pi - 1)N + N = \pi N$$

The difference between the number of variables and the number of equations is therefore

$$2 + N\pi - \pi N = 2$$

On the basis of this result, Duhem's theorem is stated as follows:

> *For any closed system formed initially from given masses of prescribed chemical species, the equilibrium state is completely determined when any two independent variables are fixed.*

The two independent variables subject to specification may in general be either intensive or extensive. However, the number of *independent intensive* variables is given by the phase rule. Thus when $F = 1$, at least one of the two variables must be extensive, and when $F = 0$, both must be extensive.

12.3   VLE: Qualitative Behavior

Vapor/liquid equilibrium (VLE) refers to systems in which a single liquid phase is in equilibrium with its vapor. In this qualitative discussion, we limit consideration to systems comprised of two chemical species, because systems of greater complexity cannot be adequately represented graphically.

When $N = 2$, the phase rule becomes $F = 4 - \pi$. Since there must be at least one phase ($\pi = 1$), the maximum number of phase-rule variables which must be specified to fix the intensive state of the system is *three*: namely, P, T, and one mole (or mass) fraction. All equilibrium states of the system can therefore be represented in three-dimensional P-T-composition space. Within this space, the states of *pairs* of phases coexisting at equilibrium ($F = 4 - 2 = 2$) define surfaces. A schematic three-dimensional diagram illustrating these surfaces for VLE is shown in Fig. 12.1.

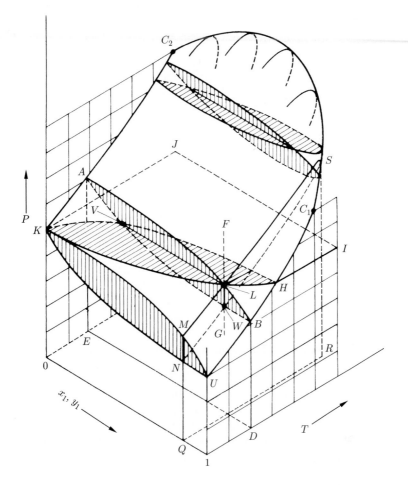

Figure 12.1: $PTxy$ diagram for vapor/liquid equilibrium.

This figure shows schematically the P-T-composition surfaces which represent equilibrium states of saturated vapor and saturated liquid for a binary system. The under surface represents saturated-vapor states; it is the P-T-y_1 surface. The upper surface represents saturated-liquid states; it is the P-T-x_1 surface. These surfaces intersect along the lines $UBHC_1$ and KAC_2, which represent the vapor pressure-vs.-T curves for pure species 1 and 2. Moreover, the under and upper surfaces form a continuous rounded surface across the top of the diagram between C_1 and C_2, the critical points of pure species 1 and 2; the critical points of the various mixtures of the two species lie along a line on the rounded edge of the surface between C_1 and C_2. This critical locus is defined by the points at which vapor and liquid phases in equilibrium become identical. Further discussion of the critical region is given later.

The region lying above the upper surface of Fig. 12.1 is the subcooled-liquid

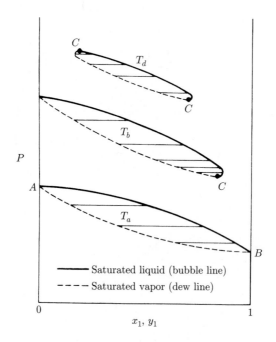

Figure 12.2: Pxy diagram for three temperatures.

region; that below the under surface is the superheated-vapor region. The interior space between the two surfaces is the region of coexistence of both liquid and vapor phases. If one starts with a liquid at F and reduces the pressure at constant temperature and composition along vertical line FG, the first bubble of vapor appears at point L, which lies on the upper surface. Thus, L is a *bubblepoint*, and the upper surface is the bubblepoint surface. The state of the vapor bubble in equilibrium with the liquid at L must be represented by a point on the under surface at the temperature and pressure of L. This point is indicated by the letter V. Line VL is an example of a *tie line*, which connects points representing phases in equilibrium.

As the pressure is further reduced along line FG, more and more liquid vaporizes until at W the process is complete. Thus W lies on the under surface and represents a state of saturated vapor having the mixture composition. Since W is the point at which the last drops of liquid (dew) disappear, it is a *dewpoint*, and the lower surface is the dewpoint surface. Continued reduction of pressure merely leads into the superheated vapor region.

Because of the complexity of Fig. 12.1, the detailed characteristics of binary VLE are usually depicted by two-dimensional graphs that display what is seen on various planes that cut the three-dimensional diagram. The three principal planes, each perpendicular to one of the coordinate axes, are illustrated in Fig. 12.1. Thus a vertical plane perpendicular to the temperature axis is outlined as $ALBDEA$. The lines on this plane represent a P-x_1-y_1 phase diagram at constant T, of which

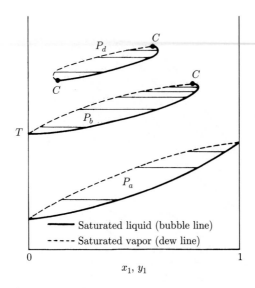

Figure 12.3: Txy diagram for three pressures.

we have already seen examples in Figs. 11.6, 11.8, and 11.10. If the lines from several such planes are projected on a single parallel plane, a diagram like Fig. 12.2 is obtained. It shows P-x_1-y_1 plots for three different temperatures. The one for T_a represents the section of Fig. 12.1 indicated by $ALBDEA$. The horizontal lines are tie lines connecting the compositions of phases in equilibrium. The temperature T_b lies between the two pure-species critical temperatures identified by C_1 and C_2 in Fig. 12.1, and temperature T_d is above both critical temperatures. The curves for these two temperatures therefore do not extend all the way across the diagram. However, the first passes through one mixture critical point, and the second through two such points. All three of these critical points are denoted by the letter C. Each is a tangent point at which a horizontal line touches the curve. This is so because all tie lines connecting phases in equilibrium are horizontal, and the tie line connecting *identical* phases (the definition of a critical point) must therefore be the last such line to cut the diagram.

A horizontal plane passed through Fig. 12.1 perpendicular to the P axis is identified by $HIJKLH$. Viewed from the top, the lines on this plane represent a T-x_1-y_1 diagram. When lines for several pressures are projected on a parallel plane, the resulting diagram appears as in Fig. 12.3. This figure is analogous to Fig. 12.2, except that it represents values for three constant pressures, P_a, P_b, and P_d.

It is also possible to plot the vapor mole fraction y_1 vs. the liquid mole fraction x_1 for either the constant-temperature conditions of Fig. 12.2 or the constant-pressure conditions of Fig. 12.3. Examples of such x_1-y_1 diagrams are shown later.

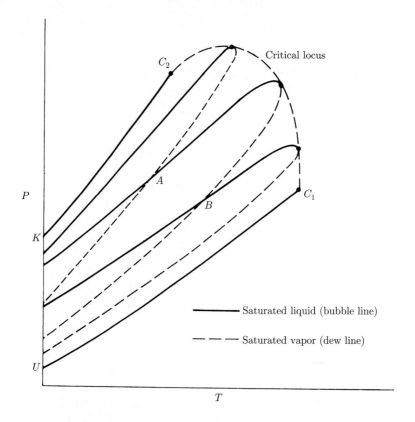

Figure 12.4: *PT* diagram for several compositions.

The third plane identified in Fig. 12.1 is the vertical one perpendicular to the composition axis and indicated by $MNQRSLM$. When projected on a parallel plane, the lines from several such planes present a diagram such as that shown by Fig. 12.4. This is the *P-T* diagram; lines UC_1 and KC_2 are vapor-pressure curves for the pure species, identified by the same letters as in Fig. 12.1. Each interior loop represents the *P-T* behavior of saturated liquid and of saturated vapor for a *mixture of fixed composition*; the different loops are for different compositions. Clearly, the *P-T* relation for saturated liquid is different from that for saturated vapor of the same composition. This is in contrast with the behavior of a pure species, for which the bubble and dew lines coincide. At points A and B in Fig. 12.4 saturated-liquid and saturated-vapor lines intersect. At such points a saturated liquid of one composition and a saturated vapor of another composition have the same T and P, and the two phases are therefore in equilibrium. The tie lines connecting the coinciding points at A and at B are perpendicular to the *P-T* plane, as illustrated by the tie line VL in Fig. 12.1.

The critical point of a binary mixture occurs where the nose of a loop in Fig. 12.4 is tangent to the envelope curve. Put another way, the envelope curve

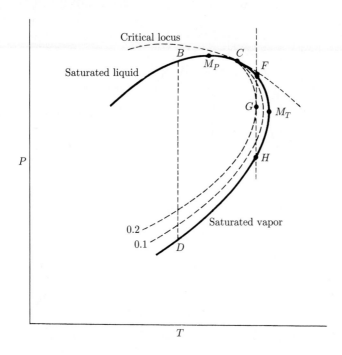

Figure 12.5: Portion of a PT diagam in the critical region.

is the critical locus. One can verify this by considering two closely adjacent loops and noting what happens to the point of intersection as their separation becomes infinitesimal. Figure 12.4 illustrates that the location of the critical point on the nose of the loop varies from one composition to another. For a pure species the critical point is the highest temperature and highest pressure at which vapor and liquid phases can coexist, but for a mixture it is in general neither. Therefore under certain conditions a condensation process occurs as the result of a *reduction* in pressure.

Consider the enlarged nose section of a single P-T loop shown in Fig.12.5. The critical point is at C. The points of maximum pressure and maximum temperature are identified as M_P and M_T. The dashed curves of Fig. 12.5 indicate the fraction of the overall system that is liquid in a two-phase mixture of liquid and vapor. To the left of the critical point C a reduction in pressure along a line such as BD is accompanied by vaporization of liquid from bubblepoint to dewpoint, as would be expected. However, if the original condition corresponds to point F, a state of saturated *vapor*, liquefaction occurs upon reduction of the pressure and reaches a maximum at G, after which vaporization takes place until the dewpoint is reached at H. This phenomenon is called *retrograde condensation*. It is of considerable importance in the operation of certain deep natural-gas wells where the pressure and temperature in the underground formation are approximately the conditions represented by point F. If one then maintains the pressure at the wellhead at

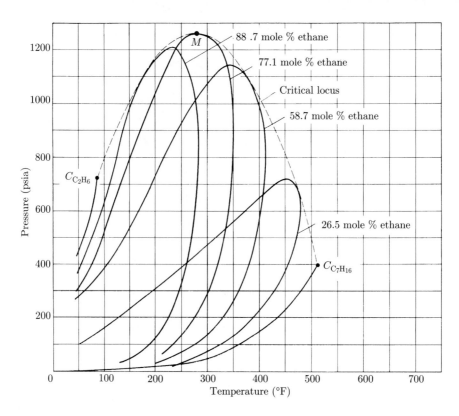

Figure 12.6: *PT* diagram for ethane/*n*-heptane. (Reproduced by permission from F. H. Barr-David, *AIChE J.*, vol. 2, pp. 426–427, 1956.)

a value near that of point *G*, considerable liquefaction of the product stream is accomplished along with partial separation of the heavier species of the mixture. Within the underground formation itself, the pressure tends to drop as the gas supply is depleted. If not prevented, this leads to the formation of a liquid phase and a consequent reduction in the production of the well. Repressuring is therefore a common practice; i.e., lean gas (gas from which the heavier species have been removed) is returned to the underground reservoir to maintain an elevated pressure.

A *P-T* diagram for the ethane(1)/heptane(2) system is shown in Fig. 12.6, and a y_1-x_1 diagram for several pressures for the same system appears in Fig. 12.7. According to convention, one plots as y_1 and x_1 the mole fractions of the more volatile species in the mixture. The maximum and minimum concentrations of the more volatile species obtainable by distillation at a given pressure are indicated by the points of intersection of the appropriate y_1-x_1 curve with the diagonal, for at these points the vapor and liquid have the same composition. They are in fact mixture critical points, unless $y_1 = x_1 = 0$ or $y_1 = x_1 = 1$. Point *A* in Fig. 12.7 represents the composition of the vapor and liquid phases at the maximum pressure at which the phases can coexist in the ethane/heptane system. The composition

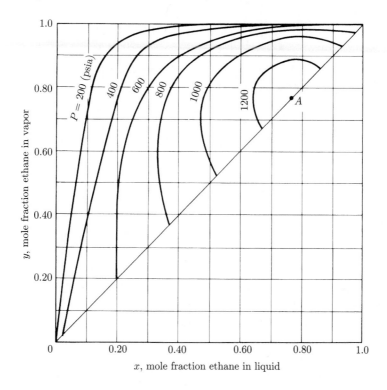

Figure 12.7: yx diagram for ethane/n-heptane. (Reproduced by permission from F. H. Barr-David, *AIChE J.*, vol. 2, p. 426–427, 1956.)

is about 77 mole-percent ethane and the pressure is about 1,263(psia). The corresponding point on Fig. 12.6 is labeled M. Barr-David[3] has prepared a complete set of consistent phase diagrams for this system.

The P-T diagram of Fig. 12.6 is typical for mixtures of nonpolar substances such as hydrocarbons. An example of a diagram for a highly nonideal system, methanol(1)/benzene(2), is shown in Fig. 12.8. The nature of the curves in this figure suggests how difficult it can be to predict phase behavior, particularly for species so dissimilar as methanol and benzene.

Although VLE in the critical region is of considerable importance in the petroleum and natural-gas industries, most chemical processing is accomplished at much lower pressures. The systems of interest rarely conform to ideal-solution behavior, and phase behavior is conveniently classified according to the sign and magnitude of deviations from ideal-solution behavior. In addition to the examples shown in Chap. 11, we present here data for four systems that represent the common types of behavior.

Data for tetrahydrofuran(1)/carbon tetrachloride(2) at 30°C are shown in

---

[3]F. H. Barr-David, *AIChE J.*, vol. 2, p. 426, 1956.

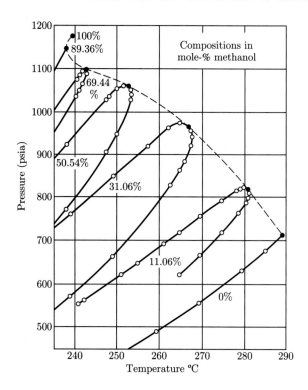

Figure 12.8: *PT* diagram for methanol/benzene. (Reprinted from *Chem. Eng. Sci.*, vol. 19, J. M. Skaates and W. B. Kay, "The phase relations of binary systems that form azeotropes," pp. 431–444, copyright 1964, with permission from Elsevier Science Ltd., Pergamon Imprint, The Boulevard, Langford Lane, Kidlington OX5 1GB, UK)

Fig. 12.9a. Here, the P-x_1 or bubblepoint curve on a P-x_1-y_1 diagram lies below a linear P-x_1 relation, and the system therefore exhibits negative deviations from ideal-solution behavior. When the deviations become sufficiently large relative to the difference between the two pure-species vapor pressures, the Px curve exhibits a minimum, as illustrated in Fig. 12.9b for the chloroform(1)/tetrahydrofuran(2) system at 30°C. This figure shows that the P-y_1 curve also has a minimum at the same point. Thus at this point where $x_1 = y_1$ the dewpoint and bubblepoint curves are tangent to the same horizontal line. A boiling liquid of this composition produces a vapor of exactly the same composition, and the liquid therefore does not change in composition as it evaporates. No separation of such a constant-boiling solution is possible by distillation. The term *azeotrope* is used to describe this state.

The data for furan(1)/carbon tetrachloride(2) at 30°C shown by Fig. 12.9c provide an example of a system for which the P-x_1 curve exhibits small positive deviations from linearity. Ethanol(1)/toluene(2) is a system for which the positive deviations are sufficiently large to lead to a *maximum* in the P-x_1 curve, as shown for 65°C by Fig. 12.9d. Just as a minimum on the P-x_1 curve represents an azeotrope, so does a maximum. Thus there are minimum-pressure and maximum-pressure azeotropes. In either case the vapor and liquid phases at the azeotropic state are of identical composition.

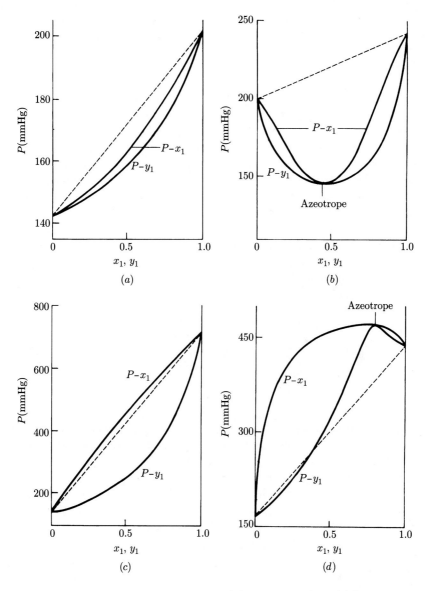

Figure 12.9: *Pxy* diagrams at constant *T*. (*a*) Tetrahydrofuran(1)/carbon tetrachloride(2) at 30°C; (*b*) chloroform(1)/tetrahydrofuran(2) at 30°C(*c*) furan(1)/carbon tetrachloride(2) at 30°C; (*d*) ethanol(1)/toluene(2) at 65°C. Dashed lines: *Px* relation for ideal liquid solutions.

At the molecular level, appreciable negative deviations from solution ideality in the liquid phase reflect stronger intermolecular attractions between unlike than between like pairs of molecules (Sec. 11.5). Conversely, appreciable positive deviations result for solutions in which intermolecular forces between like molecules are stronger than between unlike. In this latter case the forces between like molecules may be so strong as to prevent complete miscibility, and the system then forms two separate liquid phases over a range of compositions. Systems of limited miscibility are treated in Chap. 14.

Since distillation processes are carried out more nearly at constant pressure than at constant temperature, t-x_1-y_1 diagrams of data at constant P are of practical interest. The four such diagrams corresponding to those of Fig. 12.9 are shown for atmospheric pressure in Fig. 12.10. Note that the dewpoint (t-y_1) curves lie above the bubblepoint (t-x_1) curves. Moreover, the minimum-pressure azeotrope of Fig. 12.9b appears as a maximum-temperature (or maximum-boiling) azeotrope on Fig. 12.10b. There is an analogous correspondence between Figs. 12.9d and 12.10d. The y_1-x_1 diagrams at constant P for the same four systems are shown in Fig. 12.11. The point at which a curve crosses the diagonal line of the diagram represents an azeotrope, for at such a point $y_1 = x_1$.

12.4   The Gamma/Phi Formulation of VLE

For species i in a vapor mixture, Eq. (10.47) is written

$$\hat{f}_i^v = y_i \hat{\phi}_i P$$

and for species i in the liquid solution, Eq. (10.89) becomes

$$\hat{f}_i^l = x_i \gamma_i f_i$$

According to Eq. (10.44) these two expressions must be equal, whence

$$y_i \hat{\phi}_i P = x_i \gamma_i f_i \qquad (i = 1, 2, \ldots, N)$$

Superscripts v and l are not used here because of a presumption that $\hat{\phi}_i$ refers to the vapor phase and that γ_i and f_i are liquid-phase properties. Substituting for f_i by Eq. (10.41) gives

$$\boxed{y_i \Phi_i P = x_i \gamma_i P_i^{\text{sat}} \qquad (i = 1, 2, \ldots, N)} \qquad (12.1)$$

where

$$\Phi_i \equiv \frac{\hat{\phi}_i}{\phi_i^{\text{sat}}} \exp\left[-\frac{V_i^l (P - P_i^{\text{sat}})}{RT} \right]$$

Since the Poynting factor (represented by the exponential) at low to moderate pressures differs from unity by only a few parts per thousand, its omission introduces negligible error, and this equation simplifies to

$$\Phi_i = \frac{\hat{\phi}_i}{\phi_i^{\text{sat}}} \qquad (12.2)$$

Systematic application of Eqs. (12.1) and (12.2) depends on the availability of correlations of data from which values may be obtained for the P_i^{sat}, Φ_i, and γ_i.

The vapor pressures of the pure species are usually calculated from equations that give the P_i^{sat} as functions of temperature. Most commonly used is the

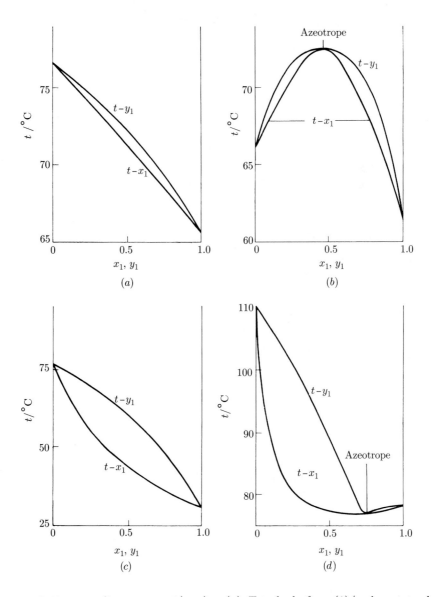

Figure 12.10: *txy* diagrams at 1(atm). (*a*) Tetrahydrofuran(1)/carbon tetrachloride(2); (*b*) chloroform(1)/tetrahydrofuran(2); (*c*) furan(1)/carbon tetrachloride(2); (*d*) ethanol(1)/toluene(2).

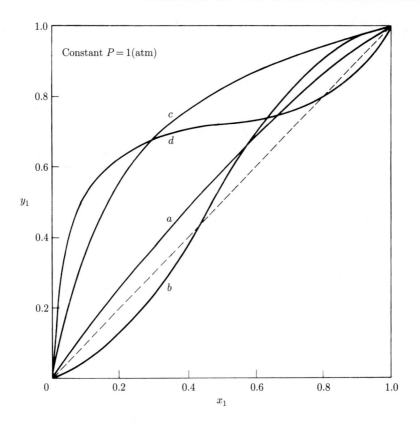

Figure 12.11: Equilibrium yx diagrams at 1(atm). (a) Tetrahydrofuran(1)/carbon tetrachloride(2); (b) chloroform(1)/tetrahydrofuran(2); (c) furan(1)/carbon tetrachloride(2); (d) ethanol(1)/toluene(2).

Antoine equation, Eq. (6.55), which we rewrite for species i as

$$\ln P_i^{\text{sat}} = A_i - \frac{B_i}{T + C_i} \tag{12.3}$$

Restriction to relatively low pressures allows calculation of the fugacity coefficients in Eq. (12.2) from the simplest form of the virial equation of state, the two-term expansion in P [Eq. (3.31)]. In this case the expression for $\hat{\phi}_i$, the fugacity coefficient for species i in solution, follows from Eq. (10.69):

$$\hat{\phi}_i = \exp \frac{P}{RT} \left[B_{ii} + \frac{1}{2} \sum_j \sum_k y_j y_k (2\delta_{ji} - \delta_{jk}) \right] \tag{12.4}$$

where

$$\delta_{ji} \equiv 2B_{ji} - B_{jj} - B_{ii}$$
$$\delta_{jk} \equiv 2B_{jk} - B_{jj} - B_{kk}$$

with $\delta_{ii} = 0$, $\delta_{jj} = 0$, etc., and $\delta_{ij} = \delta_{ji}$, etc. Values of the virial coefficients come from a generalized correlation, such as the one represented by Eqs. (10.70) through (10.75).

The fugacity coefficient for pure i as a saturated vapor ϕ_i^{sat} is obtained from Eq. (12.4) with all δ_{ji} and δ_{jk} set equal to zero:

$$\phi_i^{\text{sat}} = \exp \frac{B_{ii} P_i^{\text{sat}}}{RT} \qquad (12.5)$$

This result also follows from Eq. (10.35).

Combination of Eqs. (12.2), (12.4), and (12.5) gives

$$\Phi_i = \exp \frac{B_{ii}(P - P_i^{\text{sat}}) + \frac{1}{2} P \sum_j \sum_k y_j y_k (2\delta_{ji} - \delta_{jk})}{RT} \qquad (12.6)$$

For a binary system comprised of species 1 and 2, this becomes

$$\Phi_1 = \exp \frac{B_{11}(P - P_1^{\text{sat}}) + P y_2^2 \delta_{12}}{RT} \qquad (12.7)$$

and

$$\Phi_2 = \exp \frac{B_{22}(P - P_2^{\text{sat}}) + P y_1^2 \delta_{12}}{RT} \qquad (12.8)$$

Activity coefficients [γ_i in Eq. (12.1)] are evaluated from models for G^E as discussed in Sec. 11.2. Thus for data *at constant* T we presume the availability of a correlation giving

$$\frac{G^E}{RT} = g(x_1, x_2, \ldots, x_N) \qquad (\text{const } T)$$

12.5 Dewpoint and Bubblepoint Calculations

Although VLE problems with other combinations of variables are possible, those of engineering interest are usually dewpoint or bubblepoint calculations; there are four classes:

> BUBL P: Calculate $\{y_i\}$ and P, given $\{x_i\}$ and T
> DEW P: Calculate $\{x_i\}$ and P, given $\{y_i\}$ and T
> BUBL T: Calculate $\{y_i\}$ and T, given $\{x_i\}$ and P
> DEW T: Calculate $\{x_i\}$ and T, given $\{y_i\}$ and P

Thus, one specifies either T or P *and* either the liquid-phase or the vapor-phase composition, fixing $1 + (N - 1)$ or N phase-rule variables, exactly the number of degrees of freedom F for vapor/liquid equilibrium. All of these calculations require iterative schemes because of the complex functionality implicit in Eqs. (12.1) and

(12.2). In particular, we have the following functional relationships for low-pressure VLE:

$$\Phi_i = \Phi(T, P, y_1, y_2, \ldots, y_{N-1})$$
$$\gamma_i = \gamma(T, x_1, x_2, \ldots, x_{N-1})$$
$$P_i^{\text{sat}} = f(T)$$

For example, when solving for $\{y_i\}$ and P, we do not have values necessary for calculation of the Φ_i, and when solving for $\{x_i\}$ and T, we can evaluate neither the P_i^{sat} nor the γ_i. Simple iterative procedures, described in the following paragraphs, allow efficient solution of each of the four types of problem.

In all cases Eq. (12.1) provides the basis of calculation. This equation, valid for each species i in a multicomponent system, may be written either as

$$y_i = \frac{x_i \gamma_i P_i^{\text{sat}}}{\Phi_i P} \tag{12.9}$$

or as

$$x_i = \frac{y_i \Phi_i P}{\gamma_i P_i^{\text{sat}}} \tag{12.10}$$

Since $\sum_i y_i = 1$ and $\sum_i x_i = 1$, we also have

$$1 = \sum_i \frac{x_i \gamma_i P_i^{\text{sat}}}{\Phi_i P}$$

or

$$P = \sum_i \frac{x_i \gamma_i P_i^{\text{sat}}}{\Phi_i} \tag{12.11}$$

and

$$1 = \sum_i \frac{y_i \Phi_i P}{\gamma_i P_i^{\text{sat}}}$$

or

$$P = \frac{1}{\sum_i y_i \Phi_i / \gamma_i P_i^{\text{sat}}} \tag{12.12}$$

BUBL P. The iteration scheme for this simple and direct bubblepoint calculation is shown in Fig. 12.12. With reference to a computer program for carrying it out, one reads and stores the given values of T and $\{x_i\}$, along with all constants required in evaluation of the P_i^{sat}, γ_i, and Φ_i. Since $\{y_i\}$ is not given, we cannot yet determine values for the Φ_i, and each is set equal to unity. Values for $\{P_i^{\text{sat}}\}$ are found from the Antoine equation [Eq. (12.3)] and values of $\{\gamma_i\}$ come from an activity-coefficient correlation. Equations (12.11) and (12.9) are now solved for P and $\{y_i\}$. Values of Φ_i from Eq. (12.6) allow recalculation of P by Eq. (12.11). Iteration leads to final values for P and $\{y_i\}$.

DEW P. The calculational scheme here is shown in Fig. 12.13. We read and store T and $\{y_i\}$, along with appropriate constants. Since we can calculate neither

BUBL P

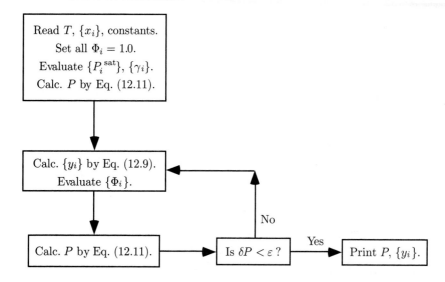

Figure 12.12: Block diagram for the calculation BUBL P.

the Φ_i nor the γ_i, all values of each are set equal to unity. Values of $\{P_i^{\text{sat}}\}$ are found from the Antoine equation, and Eqs. (12.12) and (12.10) are then solved for P and $\{x_i\}$. Evaluation of $\{\gamma_i\}$ now allows recalculation of P by Eq. (12.12). With this rather good estimate of P, we evaluate $\{\Phi_i\}$ and enter an inner iteration loop that converges on values for $\{x_i\}$ and $\{\gamma_i\}$. Subsequent recalculation of P by Eq. (12.12) leads to the outer iteration loop that establishes the final value of P. Since the x_i calculated within the inner loop are not constrained to sum to unity, each value is divided by $\sum_i x_i$:

$$x_i = \frac{x_i}{\displaystyle\sum_i x_i}$$

This yields a set of *normalized* x_i values, which do sum to unity. Actually, the inner loop can be omitted; it is included simply to make the calculational procedure more efficient.

   In the BUBL P and DEW P calculations, the temperature is known initially, and this allows immediate calculation of the key quantities P_i^{sat}. This is not the case for the two remaining procedures, BUBL T and DEW T, where the temperature is to be found. Although the individual vapor pressures P_i^{sat} are strong functions of temperature, vapor-pressure *ratios* are weak functions of T, and calculations are greatly facilitated by introduction of these ratios. We therefore multiply the right-hand sides of Eqs. (12.11) and (12.12) by P_j^{sat} (outside the summation) and divide by P_j^{sat} (inside the summation). Solution for the P_j^{sat} outside the summation then

DEW P

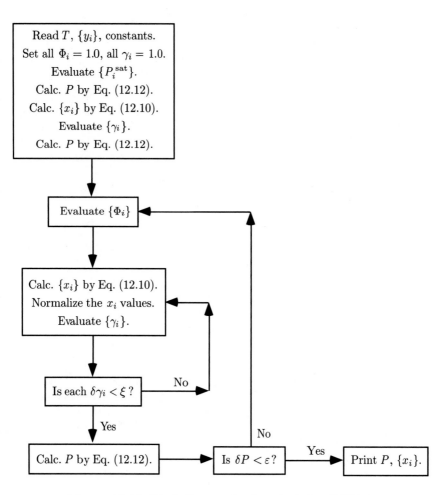

Figure 12.13: Block diagram for the calculation DEW P.

gives

$$P_j^{\text{sat}} = \frac{P}{\sum_i (x_i \gamma_i / \Phi_i)(P_i^{\text{sat}} / P_j^{\text{sat}})} \tag{12.13}$$

and

$$P_j^{\text{sat}} = P \sum_i \frac{y_i \Phi_i}{\gamma_i} \left(\frac{P_j^{\text{sat}}}{P_i^{\text{sat}}} \right) \tag{12.14}$$

In these equations the summations are over all species including j, which is an *arbitrarily selected species* from the set $\{i\}$. The temperature corresponding to the

BUBL T

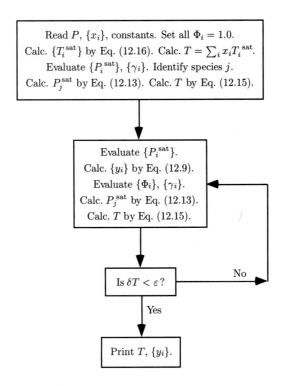

Figure 12.14: Block diagram for the calculation BUBL T.

vapor pressure P_j^{sat} is found from an appropriate equation giving vapor pressure as a function of T. Equation (12.3) solved for T is

$$T = \frac{B_j}{A_j - \ln P_j^{\text{sat}}} - C_j \tag{12.15}$$

where A_j, B_j, and C_j are the Antoine constants for species j. For purposes of finding an initial temperature to start an iteration procedure, we need values of the saturation *temperatures* of the pure species T_i^{sat} at pressure P. These are also given by the Antoine equation, written as

$$T_i^{\text{sat}} = \frac{B_i}{A_i - \ln P} - C_i \tag{12.16}$$

BUBL T. Figure 12.14 shows the iterative scheme for this bubblepoint calculation. The given values of P and $\{x_i\}$ along with appropriate constants are read and stored. In the absence of T and the y_i values, all Φ_i are set equal to unity.

Iteration is controlled by T, and for an initial estimate we set

$$T = \sum_i x_i T_i^{\text{sat}} \qquad (12.17)$$

where the T_i^{sat} are found from Eq. (12.16). With this initial value of T, we find values for $\{P_i^{\text{sat}}\}$ from the Antoine equations and values of $\{\gamma_i\}$ from the activity-coefficient correlation. Species j is identified, P_j^{sat} is calculated by Eq. (12.13), and a new value of T is found from Eq. (12.15). The P_i^{sat} are immediately reevaluated, and the y_i are calculated by Eq. (12.9). Values can now be found for both $\{\Phi_i\}$ and $\{\gamma_i\}$, allowing a revised value of P_j^{sat} to be calculated by Eq. (12.13) and a better estimate of T to be found from Eq. (12.15). Iteration then leads to final values of T and $\{y_i\}$.

DEW T. The scheme for this dewpoint calculation is shown in Fig. 12.15. Since we know neither the x_i values nor the temperature, all values of both Φ_i and γ_i are set equal to unity. Iteration is again controlled by T, and here we find an initial value by

$$T = \sum_i y_i T_i^{\text{sat}} \qquad (12.18)$$

With this value of T, we determine $\{P_i^{\text{sat}}\}$ from the Antoine equations. All quantities on the right-hand side of Eq. (12.14) are now fixed; we identify species j and solve for P_j^{sat}, from which we get a new value for T by Eq. (12.15). We immediately reevaluate $\{P_i^{\text{sat}}\}$, which together with $\{\Phi_i\}$ permits calculation of the x_i by Eq. (12.10). This allows recalculation of P_j^{sat} by Eq. (12.14) and of T by Eq. (12.15). With this rather good estimate of T, we again evaluate $\{P_i^{\text{sat}}\}$ and $\{\Phi_i\}$, and enter an inner iteration loop that converges on values of $\{x_i\}$ and $\{\gamma_i\}$. Subsequent recalculation of P_j^{sat} and T then leads to the outer iteration loop that produces a final value of T. As in the DEW P procedure, the x_i calculated within the inner loop are not constrained to sum to unity, and each value is divided by $\sum_i x_i$:

$$x_i = \frac{x_i}{\displaystyle\sum_i x_i}$$

This set of normalized x_i values does sum to unity. Again, the inner loop is included simply to make the calculational procedure more efficient.

Raoult's Law

When Eq. (12.1) is applied to vapor/liquid equilibrium for which the ideal-gas model applies to the vapor phase and the ideal-solution model applies to the liquid phase, a very simple expression is obtained for VLE. For ideal gases, fugacity coefficients $\hat{\phi}_i$ and ϕ_i^{sat} are unity, and Eq. (12.2) becomes $\Phi_i = 1$. For ideal solutions, the activity coefficients γ_i are also unity. Equation (12.1) therefore reduces to

$$\boxed{y_i P = x_i P_i^{\text{sat}} \qquad (i = 1, 2, \ldots, N)} \qquad (12.19)$$

DEW T

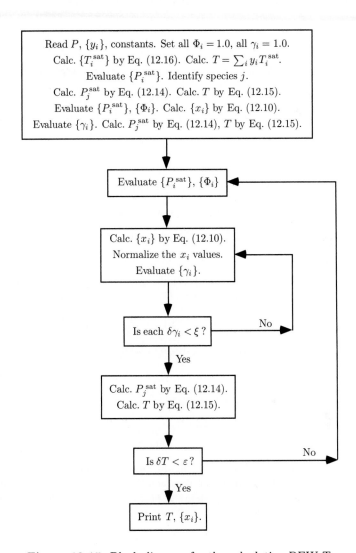

Figure 12.15: Block diagram for the calculation DEW T.

an equation which expresses *Raoult's law*.[4] It is the simplest possible equation for VLE, and as such fails to provide a realistic representation of real behavior for most systems. Nevertheless, it is useful for displaying VLE calculations in their least complex form, as is evident in the following example.

[4] François Marie Raoult (1830–1901), French chemist.

Example 12.1   Binary system acetonitrile(1)/nitromethane(2) conforms closely to Raoult's law. Vapor pressures for the pure species are given by the following Antoine equations:

$$\ln P_1^{\text{sat}}/\text{kPa} = 14.2724 - \frac{2,945.47}{t/°\text{C} + 224.00}$$

$$\ln P_2^{\text{sat}}/\text{kPa} = 14.2043 - \frac{2,972.64}{t/°\text{C} + 209.00}$$

(a) Prepare a graph showing P vs. x_1 and P vs. y_1 for a temperature of 75°C.

(b) Prepare a graph showing t vs. x_1 and t vs. y_1 for a pressure of 70 kPa.

SOLUTION   (a) BUBL P calculations are required, and Eq. (12.11) in this case reduces to

$$P = x_1 P_1^{\text{sat}} + x_2 P_2^{\text{sat}}$$

When $1 - x_1$ is substituted for x_2, this becomes

$$P = P_2^{\text{sat}} + (P_1^{\text{sat}} - P_2^{\text{sat}})x_1 \tag{A}$$

Thus a plot of P vs. x_1 is a straight line connecting P_2^{sat} at $x_1 = 0$ with P_1^{sat} at $x_1 = 1$. At 75°C, vapor pressures calculated from the given equations are

$$P_1^{\text{sat}} = 83.21 \qquad \text{and} \qquad P_2^{\text{sat}} = 41.98 \text{ kPa}$$

We can, of course, calculate P for a single value of x_1. For example, when $x_1 = 0.6$,

$$P = 41.98 + (83.21 - 41.98)(0.6) = 66.72 \text{ kPa}$$

The corresponding value of y_1 is then found from Eq. (12.19):

$$y_1 = \frac{x_1 P_1^{\text{sat}}}{P} = \frac{(0.6)(83.21)}{66.72} = 0.7483$$

These results mean that at 75°C a liquid mixture of 60 mole-percent acetonitrile and 40 mole-percent nitromethane is in equilibrium with a vapor containing 74.83 mole-percent acetonitrile at a pressure of 66.72 kPa. The results of calculations for 75°C at a number of values of x_1 are tabulated as follows:

x_1	y_1	P/kPa
0.0	0.0	41.98
0.2	0.3313	50.23
0.4	0.5692	58.47
0.6	0.7483	66.72
0.8	0.8880	74.96
1.0	1.0	83.21

These same results are shown by the P-x_1-y_1 diagram of Fig. 12.16.

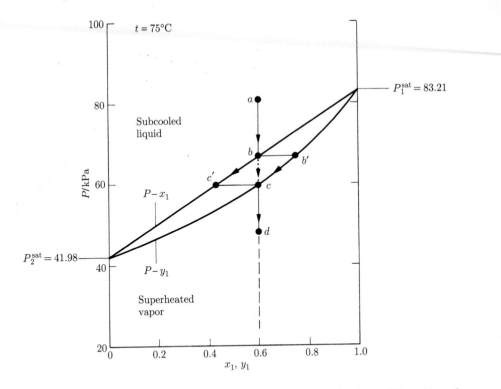

Figure 12.16: *Pxy* diagram for acetonitrile(1)/nitromethane(2) at 75°C as given by Raoult's law.

This figure is a phase diagram on which the line labeled $P\text{-}x_1$ represents states of saturated liquid; the subcooled-liquid region lies above this line. The curve labeled $P\text{-}y_1$ represents states of saturated vapor; the superheated-vapor region lies below the $P\text{-}y_1$ curve. Points lying between the saturated-liquid and saturated-vapor lines are in the two-phase region, where saturated liquid and saturated vapor coexist in equilibrium. The $P\text{-}x_1$ and $P\text{-}y_1$ lines meet at the edges of the diagram, where saturated liquid and saturated vapor of the pure species coexist at the vapor pressures P_1^{sat} and P_2^{sat}.

We can illustrate the nature of phase behavior in this binary (two constituent species) system by following the course of a constant-temperature process on the $P\text{-}x_1\text{-}y_1$ diagram. Imagine a subcooled liquid mixture of 60 mole-percent acetonitrile and 40 mole-percent nitromethane existing in a piston/cylinder arrangement at 75°C. Its state is represented by point a in Fig. 12.16. The pressure is reduced slowly enough to maintain the system at equilibrium at 75°C. Since the system is closed, the overall composition remains constant during the process, and the states of the system *as a whole* fall on the vertical line descending from point a. When the pressure decreases to the state represented by point b, the system is saturated liquid on the verge of vaporizing. A minute further decrease in pressure is accompanied by the appearance of a bubble of vapor, represented by point b'. The two points b and b' together represent the equilibrium state at $x_1 = 0.6$, $P = 66.72$ kPa, and

$y_1 = 0.7483$ for which calculations were illustrated. Point b is a bubblepoint, and the P-x_1 line is the locus of bubblepoints.

As the pressure is further reduced, the amount of vapor increases and the amount of liquid decreases, with the states of the two phases following paths $b'c$ and bc', respectively. The dotted line from b to c represents the *overall* states of the two-phase system. Finally, as point c is approached, the liquid phase, represented by point c', has almost disappeared, with only minute drops (dew) remaining. Point c is therefore a dewpoint, and the P-y_1 line is the locus of dewpoints. Once the dew has evaporated, only saturated vapor at point c remains, and further pressure reduction leads to superheated vapor at point d.

The composition of the vapor at point c is $y_1 = 0.6$, but the composition of the liquid at point c' and the pressure must either be read from the graph or calculated. This is a DEW P calculation, and Eq. (12.12) yields

$$P = \frac{1}{y_1/P_1^{\text{sat}} + y_2/P_2^{\text{sat}}}$$

For $y_1 = 0.6$ and $t = 75°\text{C}$,

$$P = \frac{1}{0.6/83.21 + 0.4/41.98} = 59.74 \text{ kPa}$$

By Eq. (12.19),

$$x_1 = \frac{y_1 P}{P_1^{\text{sat}}} = \frac{(0.6)(59.74)}{83.21} = 0.4308$$

This is the liquid-phase composition at point c'.

(b) When pressure P is fixed, the temperature varies along with x_1 and y_1. For a given pressure, the temperature range is bounded by the saturation temperatures t_1^{sat} and t_2^{sat}, the temperatures at which the pure species exert vapor pressures equal to P. For the present system, these temperatures are calculated from the Antoine equations with $P_j^{\text{sat}} = P = 70$ kPa:

$$t_1^{\text{sat}} = 69.84 \qquad \text{and} \qquad t_2^{\text{sat}} = 89.58°\text{C}$$

For the purpose of preparing a t-x_1-y_1 diagram, the simplest procedure is to select values of t between t_1^{sat} and t_2^{sat}, calculate P_1^{sat} and P_2^{sat} for these temperatures, and evaluate x_1 by Eq. (A):

$$x_1 = \frac{P - P_2^{\text{sat}}}{P_1^{\text{sat}} - P_2^{\text{sat}}}$$

For example, at 78°C,

$$P_1^{\text{sat}} = 91.76 \qquad \text{and} \qquad P_2^{\text{sat}} = 46.84 \text{ kPa}$$

whence

$$x_1 = \frac{70 - 46.84}{91.76 - 46.84} = 0.5156$$

By Eq. (12.19),

$$y_1 = \frac{x_1 P_1^{\text{sat}}}{P} = \frac{(0.5156)(91.76)}{70} = 0.6759$$

x_1	y_1	$t°C$
0.0	0.0	$89.58(t_2^{\text{sat}})$
0.1424	0.2401	86
0.3184	0.4742	82
0.5156	0.6759	78
0.7378	0.8484	74
1.0	1.0	$69.84(t_1^{\text{sat}})$

The results of this and similar calculations for $P = 70$ kPa are given in the accompanying table. Figure 12.17 is the t-x_1-y_1 diagram showing these results.

On this phase diagram, drawn for a constant pressure of 70 kPa, the t-y_1 curve represents states of saturated vapor, with states of superheated vapor lying above it. The t-x_1 curve represents states of saturated liquid, with states of subcooled liquid lying below it. The two-phase region lies between these curves.

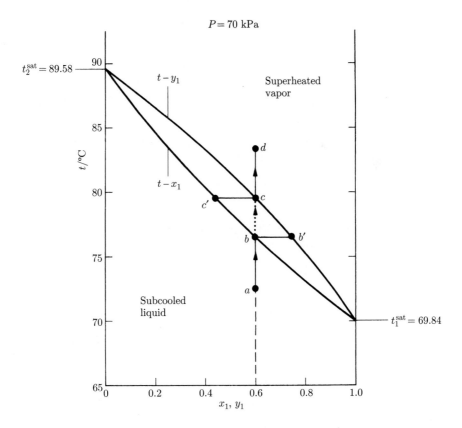

Figure 12.17: txy diagram for acetonitrile(1)/nitromethane(2) at 70 kPa as given by Raoult's law.

With reference to Fig. 12.17, we describe the course of a constant-pressure heating process leading from a state of subcooled liquid at point a to a state of superheated vapor at point d. The path shown on the figure is for a constant composition of 60 mole-percent acetonitrile. The temperature of the liquid increases as the result of heating from point a to point b, where the first bubble of vapor appears. Thus point b is a bubblepoint, and the t-x_1 curve is the locus of bubblepoints.

We know $x_1 = 0.6$ and $P = 70$ kPa; t is therefore determined by a BUBL T calculation, which requires iteration. Equation (12.13) is here written

$$P_2^{\text{sat}} = \frac{P}{x_1 \alpha + x_2} \qquad (B)$$

where $\alpha \equiv P_1^{\text{sat}}/P_2^{\text{sat}}$. Subtracting $\ln P_2^{\text{sat}}$ from $\ln P_1^{\text{sat}}$ as given by the Antoine equations, we get

$$\ln \alpha = 0.0681 - \frac{2{,}945.47}{t + 224.00} + \frac{2{,}972.64}{t + 209.00} \qquad (C)$$

Iteration is as follows:

1. Choosing a value of α for some intermediate t, calculate P_2^{sat} by Eq. (B).

2. Calculate t from the Antoine equation for species 2:

$$t = \frac{2{,}972.64}{14.2043 - \ln P_2^{\text{sat}}} - 209.00$$

3. Determine a new value of α by Eq. (C) and a new value of P_2^{sat} by Eq. (B).

4. Return to step 2, and iterate to convergence.

The result is $t = 76.42°\text{C}$, the temperature of points b and b'. At this temperature, $P_1^{\text{sat}} = 87.17$ kPa, and by Eq. (12.19) we find the composition of point b':

$$y_1 = \frac{x_1 P_1^{\text{sat}}}{P} = \frac{(0.6)(87.17)}{70} = 0.7472$$

Vaporization of a mixture at constant pressure, unlike vaporization of a pure species, does not in general occur at constant temperature. As the heating process continues beyond point b, the temperature rises, the amount of vapor increases, and the amount of liquid decreases. During this process, the vapor- and liquid-phase compositions change as indicated by paths $b'c$ and bc', until the dewpoint is reached at point c, where the last droplets of liquid disappear. The t-y_1 curve is the locus of dewpoints.

The vapor composition at point c is $y = 0.6$; since the pressure is also known ($P = 70$ kPa), we may carry out a DEW T calculation. Equation (12.14) here becomes

$$P_1^{\text{sat}} = P(y_1 + y_2 \alpha)$$

The iteration procedure is as before, except that it is based on P_1^{sat} rather than P_2^{sat}, and

$$t = \frac{2{,}945.47}{14.2724 - \ln P_1^{\text{sat}}} - 224.00$$

The result here is $t = 79.58°C$, the temperature of points c and c'. With $P_1^{sat} = 96.53$ kPa, we find by Eq. (12.19) that the composition at point c' is

$$x_1 = \frac{y_1 P}{P_1^{sat}} = \frac{(0.6)(70)}{96.53} = 0.4351$$

Thus the temperature rises from 76.42 to 79.58°C during the vaporization step from point b to point c. Continued heating simply superheats the vapor to point d.

Modified Raoult's Law

For low to moderate pressures a much more realistic VLE equation results when Eq. (12.1) is simplified just by the assumption of the ideal-gas model for the vapor phase, thus making the $\Phi_i = 1.0$ by Eq. (12.2). When this assumption is made, Eq. (12.1) reduces to

$$\boxed{y_i P = x_i \gamma_i P_i^{sat} \qquad (i = 1, 2, \ldots, N)} \qquad (12.20)$$

This equation provides a *modified Raoult's law*, and it was used for data reduction as Eq. (11.1) in Sec. 11.1. Bubblepoint and dewpoint calculations made with it are, of course, somewhat simpler than those shown by Figs. 12.12 through 12.15. Indeed, the BUBL P calculation yields final results in a single step, without iteration.

Example 12.2 For the system 2-propanol(1)/water(2), the following parameter values are recommended for the Wilson equation:

$$a_{12} = 437.98 \qquad a_{21} = 1{,}238.00 \text{ cal mol}^{-1}$$

$$V_1 = 76.92 \qquad V_2 = 18.07 \text{ cm}^3 \text{ mol}^{-1}$$

In addition, we have the following Antoine equations:

$$\ln P_1^{sat} = 16.6780 - \frac{3{,}640.20}{T - 53.54}$$

$$\ln P_2^{sat} = 16.2887 - \frac{3{,}816.44}{T - 46.13}$$

where T is in kelvins and the vapor pressures are in kPa. Assuming the validity of Eq. (12.20), calculate:

(a) P and $\{y_i\}$, for $T = 353.15$ K and $x_1 = 0.25$.

(b) P and $\{x_i\}$, for $T = 353.15$ K and $y_1 = 0.60$.

(c) T and $\{y_i\}$, for $P = 101.33$ kPa and $x_1 = 0.85$.

(d) T and $\{x_i\}$, for $P = 101.33$ kPa and $y_1 = 0.40$.

(e) P^{az}, the azeotropic pressure, and $x_1^{az} = y_1^{az}$, the azeotropic composition, for $T = 353.15$ K.

SOLUTION Since we have assumed the validity of Eq. (12.20), $\Phi_i = 1.0$ throughout this problem. This, together with the fact that we are considering a binary system, makes the solution simple enough that the steps can be explained as though carried out by hand calculations.

(a) A BUBL P calculation. For $T = 353.15$ K, the Antoine equations yield the following vapor pressures:

$$P_1^{\text{sat}} = 92.59 \qquad P_2^{\text{sat}} = 47.38 \text{ kPa}$$

Activity coefficients as given by the Wilson equation are calculated by Eqs. (11.17) and (11.18). First, we find values of Λ_{12} and Λ_{21} by Eq. (11.24). Thus

$$\Lambda_{12} = \frac{V_2}{V_1} \exp \frac{-a_{12}}{RT} = \frac{18.07}{76.92} \exp \frac{-437.98}{(1.987)(353.15)} = 0.1258$$

and

$$\Lambda_{21} = \frac{V_1}{V_2} \exp \frac{-a_{21}}{RT} = \frac{76.92}{18.07} \exp \frac{-1,238.00}{(1.987)(353.15)} = 0.7292$$

Substituting known values into Eqs. (11.17) and (11.18) gives:

$$\ln \gamma_1 = -\ln(0.25 + 0.75 \times 0.1258)$$

$$+ 0.75 \left(\frac{0.1258}{0.25 + 0.75 \times 0.1258} - \frac{0.7292}{0.75 + 0.25 \times 0.7292} \right)$$

or

$$\ln \gamma_1 = 1.0661 + 0.75(-0.4168) = 0.7535$$

and

$$\ln \gamma_2 = -\ln(0.75 + 0.25 \times 0.7292) - 0.25(-0.4168) = 0.1743$$

whence

$$\gamma_1 = 2.1244 \qquad \gamma_2 = 1.1904$$

By Eq. (12.11) with all $\Phi_i = 1.0$,

$$P = (0.25)(2.1244)(92.59) + (0.75)(1.1904)(47.38) = 91.47 \text{ kPa}$$

From Eq. (12.20), written as $y_i = x_i \gamma_i P_i^{\text{sat}}/P$, we get

$$y_1 = 0.538 \qquad y_2 = 0.462$$

(b) A DEW P calculation. With T unchanged from part (a), the values of P_1^{sat}, P_2^{sat}, Λ_{12}, and Λ_{21} are the same as already calculated. However, here the liquid-phase composition is unknown. To initiate calculations, we take the liquid phase to be an ideal solution, and set all $\gamma_i = 1.0$. Equation (12.12) then becomes

$$P = \frac{1}{y_1/P_1^{\text{sat}} + y_2/P_2^{\text{sat}}}$$

From this we find $P = 67.01$ kPa. Equation (12.20), written $x_1 = y_1 P/P_1^{\text{sat}}$ now gives

$$x_1 = \frac{(0.6)(67.01)}{92.59} = 0.434$$

whence $x_2 = 1 - x_1 = 0.566$. The resulting values of γ_1 and and γ_2, calculated by Eqs. (11.17) and (11.18) are

$$\gamma_1 = 1.4277 \qquad \gamma_2 = 1.4558$$

We next use these initial values of the activity coefficients to recompute P, writing Eq. (12.12) as

$$P = \frac{1}{y_1/\gamma_1 P_1^{\text{sat}} + y_2/\gamma_2 P_2^{\text{sat}}}$$

The result is $P = 96.73$ kPa. Recalculation of x_1 by Eq. (12.10) gives

$$x_1 = \frac{y_1 P}{\gamma_1 P_1^{\text{sat}}} = \frac{(0.60)(96.73)}{(1.4277)(92.59)} = 0.439$$

and $x_2 = 0.561$. Equations (11.17) and (11.18) now provide new values of the activity coefficients:

$$\gamma_1 = 1.4167 \qquad \gamma_2 = 1.4646$$

Iteration within the inner loop of Fig. (12.13) leads to the values

$$x_1 = 0.449 \qquad \gamma_1 = 1.3957 \qquad \gamma_2 = 1.4821$$

and by Eq. (12.12) $P = 96.72$ kPa. Since the Φ_i are fixed at unity, no further iteration is required, and we have the final values

$$P = 96.72 \text{ kPa} \qquad x_1 = 0.449 \qquad x_2 = 0.551$$

(c) A BUBL T calculation. Application of Eq. (12.16) with the given Antoine constants and $P = 101.33$ kPa leads to the values

$$T_1^{\text{sat}} = 355.39 \qquad T_2^{\text{sat}} = 373.15 \text{ K}$$

An initial value for T then follows from Eq. (12.17):

$$T = (0.85)(355.39) + (0.15)(373.15) = 358.05 \text{ K}$$

The P_i^{sat} values at this temperature are evaluated by the given Antoine equations

$$P_1^{\text{sat}} = 112.60 \qquad P_2^{\text{sat}} = 57.60 \text{ kPa}$$

The activity coefficients at this temperature come from the Wilson equations with Λ_{12} and Λ_{21} provided by Eq. (11.24):

$$\Lambda_{12} = 0.1269 \qquad \Lambda_{21} = 0.7471$$

Equations (11.17) and (11.18) then yield

$$\gamma_1 = 1.0197 \qquad \gamma_2 = 2.5265$$

Substitution of values into Eq. (12.13), with $i = 1$ and each $\Phi_i = 1$ gives

$$P_1^{\text{sat}} = \frac{101.33}{(0.85)(1.0197) + (0.15)(2.5265)(57.60/112.60)} = 95.54 \text{ kPa}$$

The temperature, recalculated by Eq. (12.15), is $T = 353.92$ K. The sequence of calculations is now repeated for this temperature, yielding

$$P_2^{\text{sat}} = 48.73 \text{ kPa} \qquad \Lambda_{12} = 0.1260 \qquad \Lambda_{21} = 0.7321$$

$$\gamma_1 = 1.0203 \qquad \gamma_2 = 2.5624$$

$$P_1^{\text{sat}} = 95.24 \text{ kPa} \qquad T = 353.85 \text{ K}$$

The change in T is small, and additional iteration leads to no significant further change in values. We therefore calculate y_1 by Eq. (12.20):

$$y_1 = \frac{x_1 \gamma_1 P_1^{\text{sat}}}{P} = \frac{(0.85)(1.0203)(95.24)}{(101.33)} = 0.815$$

Thus for final results we have

$$T = 353.85 \text{ K} \qquad y_1 = 0.815 \qquad y_2 = 0.185$$

(d) A DEW T calculation. Since $P = 101.33$ kPa, the saturation temperatures are the same as those of part (c), but the initial T is given by Eq. (12.18):

$$T = (0.40)(355.39) + (0.60)(373.15) = 366.05 \text{ K}$$

The P_i^{sat} values at this temperature found from the Antoine equations are

$$P_1^{\text{sat}} = 152.89 \qquad P_2^{\text{sat}} = 78.19 \text{ kPa}$$

For $j = 1$ and with $\gamma_i = \Phi_i = 1.0$, we find an initial value of P_1^{sat} by Eq. (12.14):

$$P_1^{\text{sat}} = 101.33 \left[0.40 + 0.60 \left(\frac{152.89}{78.19} \right) \right] = 159.41 \text{ kPa}$$

Writing Eq. (12.15) for species 1 gives the new estimate, $T = 367.17$ K. At this temperature $P_2^{\text{sat}} = 81.54$ kPa, and Λ_{12} and Λ_{21} by Eq. (11.24) are

$$\Lambda_{12} = 0.1289 \qquad \Lambda_{21} = 0.7801$$

Application of the Wilson equation for evaluation of activity coefficients requires knowledge of the liquid-phase composition. We therefore calculate x_1 by Eq. (12.20):

$$x_1 = \frac{y_1 P}{\gamma_1 P_1^{\text{sat}}} = \frac{(0.40)(101.33)}{(1)(159.41)} = 0.254$$

and $x_2 = 0.746$. Equations (11.17) and (11.18) then give

$$\gamma_1 = 2.0276 \qquad \gamma_2 = 1.1902$$

We now recalculate P_1^{sat} by Eq. (12.14):

$$P_1^{\text{sat}} = 101.33 \left[\frac{0.40}{2.0276} + \frac{0.60}{1.1902} \left(\frac{159.41}{81.54} \right) \right] = 119.86 \text{ kPa}$$

The temperature, reevaluated by Eq. (12.15), is $T = 359.65$ K. At this temperature,

$$P_2^{\text{sat}} = 61.31 \text{ kPa} \qquad \Lambda_{12} = 0.1273 \qquad \Lambda_{21} = 0.7529$$

These values remain fixed while the iterations of the inner loop of Fig. 12.15 are carried out. Calculation of x_1 by Eq. (12.20) gives

$$x_1 = \frac{(0.40)(101.33)}{(2.0276)(119.86)} = 0.167$$

and $x_2 = 0.833$. By Eqs. (11.17) and (11.18),

$$\gamma_1 = 2.8103 \qquad \gamma_2 = 1.0999$$

Equation (12.20) yields new values of x_1 and x_2, which are then normalized, and γ_1 and γ_2 are again calculated by Eqs. (11.17) and (11.18). The process is repeated until the γ_1 and γ_2 values do not change appreciably in successive iterations. The results of this procedure are

$$x_1 = 0.0658 \qquad \gamma_1 = 5.1369 \qquad \gamma_2 = 1.0203$$

Leaving the inner loop, we calculate P_1^{sat} by Eq. (12.14):

$$P_1^{\text{sat}} = 101.33 \left[\frac{0.40}{5.1369} + \frac{0.60}{1.0203} \left(\frac{119.86}{61.31} \right) \right] = 124.38 \text{ kPa}$$

Equation (12.15), written for species 1, yields $T = 360.61$ K. At this temperature,

$$P_2^{\text{sat}} = 63.62 \text{ kPa} \qquad \Lambda_{12} = 0.1275 \qquad \Lambda_{21} = 0.7563$$

We now return to the inner loop, and iteration for x_1, γ_1, and γ_2 leads to the values

$$x_1 = 0.0639 \qquad \gamma_1 = 5.0999 \qquad \gamma_2 = 1.0205$$

A return to the outer loop produces no significant change in these results. Thus we find

$$T = 360.61 \text{ K} \qquad x_1 = 0.0639 \qquad x_2 = 0.9361$$

  (*e*) First we determine whether or not an azeotrope exists at the given temperature. This calculation is facilitated by the definition of a quantity called the *relative volatility*

$$\boxed{\alpha_{12} \equiv \frac{y_1/x_1}{y_2/x_2}} \tag{12.21}$$

This quantity becomes unity at an azeotrope. By Eq. (11.20),

$$\frac{y_i}{x_i} = \frac{\gamma_i P_i^{\text{sat}}}{P}$$

Therefore

$$\alpha_{12} = \frac{\gamma_1 P_1^{\text{sat}}}{\gamma_2 P_2^{\text{sat}}} \tag{12.22}$$

When $x_1 = 0$, $\gamma_2 = 1$ and $\gamma_1 = \gamma_1^\infty$, and when $x_1 = 1$, $\gamma_1 = 1$ and $\gamma_2 = \gamma_2^\infty$. Therefore in these limits α_{12} is given by

$$(\alpha_{12})_{x_1=0} = \frac{\gamma_1^\infty P_1^{\text{sat}}}{P_2^{\text{sat}}} \qquad \text{and} \qquad (\alpha_{12})_{x_1=1} = \frac{P_1^{\text{sat}}}{\gamma_2^\infty P_2^{\text{sat}}}$$

These values are readily calculated from the given information. If one of them is less than 1 and the other is greater than 1, then an azeotrope exists, because α_{12} is a continuous function of x_1 and must then pass through the value of 1.0 at some intermediate composition.

Values of P_1^{sat} and P_2^{sat} and values of Λ_{12} and Λ_{21} for the Wilson equation are given in part (a) for the temperature of interest here. Expressions for the infinite-dilution activity coefficients appear following Eqs. (11.17) and (11.18). Thus

$$\ln\gamma_1^\infty = -\ln\Lambda_{12} + 1 - \Lambda_{21} = -\ln 0.1258 + 1 - 0.7292 = 2.3439$$

$$\ln\gamma_2^\infty = -\ln\Lambda_{21} + 1 - \Lambda_{12} = -\ln 0.7292 + 1 - 0.1258 = 1.1900$$

and

$$\gamma_1^\infty = 10.422 \qquad \gamma_2^\infty = 3.287$$

The limiting values of α_{12} are

$$(\alpha_{12})_{x_1=0} = \frac{(10.422)(92.59)}{47.38} = 20.37$$

and

$$(\alpha_{12})_{x_1=1} = \frac{92.59}{(3.287)(47.38)} = 0.595$$

From these results, we conclude that an azeotrope does indeed exist.

For $\alpha_{12} = 1$, Eq. (12.22) becomes

$$\frac{\gamma_1^{\text{az}}}{\gamma_2^{\text{az}}} = \frac{P_2^{\text{sat}}}{P_1^{\text{sat}}} = \frac{47.38}{92.59} = 0.5117$$

The difference between Eqs. (11.18) and (11.17), the Wilson equations for γ_2 and γ_1, gives the general expression

$$\ln\frac{\gamma_1}{\gamma_2} = \ln\frac{x_2 + x_1\Lambda_{21}}{x_1 + x_2\Lambda_{12}} + \frac{\Lambda_{12}}{x_1 + x_2\Lambda_{12}} - \frac{\Lambda_{21}}{x_2 + x_1\Lambda_{21}}$$

Thus the azeotropic composition is the value of x_1 (with $x_2 = 1 - x_1$) for which this equation is satisfied when

$$\ln\frac{\gamma_1}{\gamma_2} = \ln 0.5117 = -0.6700$$

and

$$\Lambda_{12} = 0.1258 \qquad \Lambda_{21} = 0.7292$$

Solution by trial gives $x_1^{\text{az}} = 0.7173$. For this value of x_1, we find from Eq. (11.17) that $\gamma_1^{\text{az}} = 1.0787$. With $x_1^{\text{az}} = y_1^{\text{az}}$, Eq. (12.20) becomes

$$P^{\text{az}} = \gamma_1^{\text{az}} P_1^{\text{sat}} = (1.0787)(92.59)$$

Thus

$$P^{\text{az}} = 99.83 \text{ kPa} \qquad x_1^{\text{az}} = y_1^{\text{az}} = 0.7173$$

Dewpoint and bubblepoint calculations are readily made with software packages such as Mathcad$^{\circledR}$ and Maple$^{\circledR}$, in which iteration is an integral part of an equation-solving routine. Mathcad programs for solution of Example 12.2 are given in App. D.2.

Calculations for multicomponent systems made without simplifying assumptions are readily carried out in like manner by computer. Table 12.1 shows results for the system n-hexane(1)/ethanol(2)/methylcyclopentane(3)/benzene(4) of a complete BUBL T calculation. The given pressure P is 1(atm), and the given liquid-phase mole fractions x_i are listed in the second column of Table 12.1. Parameters for the Antoine equations[5] [T in kelvins, P in (atm)], supplied as input data, are

$$
\begin{array}{lll}
A_1 = 9.2033 & B_1 = 2697.55 & C_1 = -48.78 \\
A_2 = 12.2786 & B_2 = 3803.98 & C_2 = -41.68 \\
A_3 = 9.1690 & B_3 = 2731.00 & C_3 = -47.11 \\
A_4 = 9.2675 & B_4 = 2788.51 & C_4 = -52.36
\end{array}
$$

In addition, the following virial coefficients[6] (in $cm^3 \ mol^{-1}$) are provided:

$$
\begin{array}{llll}
B_{11} = -1360.1 & B_{12} = -657.0 & B_{13} = -1274.2 & B_{14} = -1218.8 \\
B_{22} = -1174.7 & B_{23} = -621.8 & B_{24} = -589.7 & \\
B_{33} = -1191.9 & B_{34} = -1137.9 & & \\
B_{44} = -1086.9 & & &
\end{array}
$$

Finally, input information includes parameters for the UNIFAC method (App. G). The calculated values of T and the vapor-phase mole fractions y_i compare favorably with experimental values.[7] Also listed in Table 12.1 are final computed values of P_i^{sat}, Φ_i, and γ_i.

The BUBL T calculations for which results are given in Table 12.1 are for a pressure of 1(atm), a pressure for which vapor phases are often assumed to be ideal gases and for which Φ_i is unity for each species. In fact, these values here lie between 0.98 and 1.00. This illustrates the fact that at pressures of 1 bar and less, the assumption of ideal gases usually introduces little error. The additional assumption of liquid-phase ideality ($\gamma_i = 1$), on the other hand, is justified only infrequently. We note that γ_i for ethanol in Table 12.1 is greater than 8.

Values of parameters for the Margules, van Laar, Wilson, NRTL, and UNIQUAC equations are given for many binary pairs by Gmehling et al.[8] in a summary collection of the world's published VLE data for low to moderate pressures.

---

[5]R. C. Reid, J. M. Prausnitz, and T. K. Sherwood, *The Properties of Gases and Liquids*, 3d ed., App. A, McGraw-Hill, New York, 1977.

[6]From the correlation of J. G. Hayden and J. P. O'Connell, *Ind. Eng. Chem. Proc. Des. Dev.*, vol. 14, pp. 209–216, 1975.

[7]J. E. Sinor and J. H. Weber, *J. Chem. Eng. Data*, vol. 5, pp. 243–247, 1960.

[8]J. Gmehling, U. Onken, and W. Arlt, *Vapor-Liquid Equilibrium Data Collection*, Chemistry Data Series, vol. I, Parts 1–8, DECHEMA, Frankfurt/Main, 1977–1990.

Table 12.1: Results of BUBL T Calculations at 1(atm) for the System n-Hexane/Ethanol/Methylcyclopentane(MCP)/Benzene

Species k	x_i	y_i(calc)	y_i(exp)	P_i^{sat}/(atm)	Φ_i	γ_i
n-Hexane(1)	0.162	0.139	0.140	0.797	0.993	1.073
Ethanol(2)	0.068	0.279	0.274	0.498	0.999	8.241
MCP(3)	0.656	0.500	0.503	0.725	0.990	1.042
Benzene(4)	0.114	0.082	0.083	0.547	0.983	1.289

T(calc) = 334.82 K T(exp) = 334.85 K   Iterations = 4

These values are based on reduction of experimental data through application of Eq. (12.20). On the other hand, data reduction for determination of parameters in the UNIFAC method (App. G) does not include the ideal-gas assumption, and is carried out with Eq. (12.1).

12.6   Flash Calculations

An important application of VLE is the *flash calculation*. The name originates from the fact that a liquid at a pressure equal to or greater than its bubblepoint pressure "flashes" or partially evaporates when the pressure is reduced below the bubblepoint pressure, producing a two-phase system of vapor and liquid in equilibrium. We consider here only the P, T-flash, which refers to any calculation of the quantities and compositions of the vapor and liquid phases making up a two-phase system in equilibrium at known T, P, and *overall* composition. This poses a problem known to be determinate on the basis of Duhem's theorem, because two independent variables (T and P) are specified for a system of fixed overall composition, that is, a system formed from given masses of non-reacting chemical species.

Consider a system containing one mole of nonreacting chemical species with an *overall* composition represented by the set of mole fractions $\{z_i\}$. Let $\mathcal{L}$ be the moles of liquid, with mole fractions $\{x_i\}$, and let $\mathcal{V}$ be the moles of vapor, with mole fractions $\{y_i\}$. The material-balance equations are

$$\mathcal{L} + \mathcal{V} = 1$$

$$z_i = x_i \mathcal{L} + y_i \mathcal{V} \qquad (i = 1, 2, \ldots, N)$$

Choosing to eliminate $\mathcal{L}$ from these equations, we get

$$z_i = x_i(1 - \mathcal{V}) + y_i \mathcal{V} \qquad (i = 1, 2, \ldots, N) \tag{12.23}$$

A convenient measure of the tendency of a given chemical species to partition itself preferentially between liquid and vapor phases is the equilibrium ratio K_i, defined as

$$\boxed{K_i \equiv \frac{y_i}{x_i}} \tag{12.24}$$

This quantity is usually called simply a *K-value*. Although it adds nothing to our thermodynamic knowledge of VLE, it does serve as a measure of the "lightness" of a constituent species, that is, of its tendency to concentrate in the vapor phase. When K_i is greater than unity, species i concentrates in the vapor phase; when less, it concentrates in the liquid phase, and is considered a "heavy" constituent. Moreover, the use of K-values makes for computational convenience, allowing elimination of one set of mole fractions $\{y_i\}$ or $\{x_i\}$ in favor of the other.

Thus in Eq. (12.23) we substitute $x_i = y_i/K_i$, and solve for y_i:

$$y_i = \frac{z_i K_i}{1 + \mathcal{V}(K_i - 1)} \qquad (i = 1, 2, \ldots, N) \qquad (12.25)$$

Since $x_i = y_i/K_i$, an alternative equation is

$$x_i = \frac{z_i}{1 + \mathcal{V}(K_i - 1)} \qquad (i = 1, 2, \ldots, N) \qquad (12.26)$$

Since both sets of mole fractions must sum to unity, $\sum_i x_i = \sum_i y_i = 1$. Thus, if we sum Eq. (12.25) over all species and subtract unity from this sum, the difference F_y must be zero; that is,

$$F_y = \sum_i \frac{z_i K_i}{1 + \mathcal{V}(K_i - 1)} - 1 = 0 \qquad (12.27)$$

Similar treatment of Eq. (12.26) yields the difference F_x, which must also be zero:

$$F_x = \sum_i \frac{z_i}{1 + \mathcal{V}(K_i - 1)} - 1 = 0 \qquad (12.28)$$

Solution to a P, T-flash problem is accomplished when a value of $\mathcal{V}$ is found that makes *either* the function F_y or F_x equal to zero. However, a more convenient function for use in a *general* solution procedure[9] is the difference $F_y - F_x = F$:

$$F = \sum_i \frac{z_i(K_i - 1)}{1 + \mathcal{V}(K_i - 1)} = 0 \qquad (12.29)$$

The advantage of this function is apparent from its derivative:

$$\frac{dF}{d\mathcal{V}} = -\sum_i \frac{z_i(K_i - 1)^2}{[1 + \mathcal{V}(K_i - 1)]^2} \qquad (12.30)$$

Since $dF/d\mathcal{V}$ is always negative, the F vs. $\mathcal{V}$ relation is monotonic, and this makes Newton's method (App. H), a rapidly converging iteration procedure, well suited to solution for $\mathcal{V}$. Equation (H.1) for the nth iteration here becomes

$$F + \left(\frac{dF}{d\mathcal{V}}\right)\Delta\mathcal{V} = 0 \qquad (12.31)$$

---

[9]H. H. Rachford, Jr., and J. D. Rice, *J. Petrol. Technol.*, vol. 4(10), sec. 1, p. 19 and sec. 2, p. 3, October, 1952.

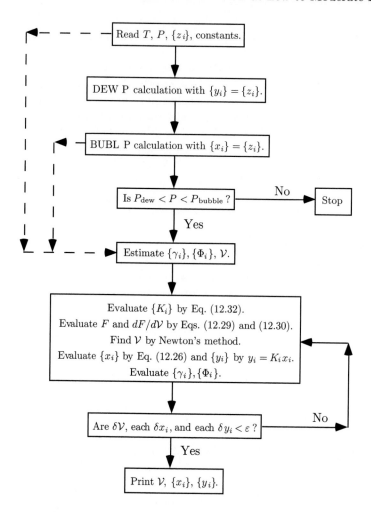

Figure 12.18: Block diagram for a P, T-flash calculation.

where $\Delta \mathcal{V} \equiv \mathcal{V}_{n+1} - \mathcal{V}_n$, and F and $(dF/d\mathcal{V})$ are found by Eqs. (12.29) and (12.30). In these equations the K-values come from Eq. (12.1) written

$$K_i = \frac{y_i}{x_i} = \frac{\gamma_i P_i^{\text{sat}}}{\Phi_i P} \qquad (i = 1, 2, \dots, N) \qquad (12.32)$$

where Φ_i is given by Eq. (12.2). The K-values contain all of the thermodynamic information, and are related in a complex way to T, P, $\{y_i\}$, and $\{x_i\}$. Since we are solving for $\{y_i\}$ and $\{x_i\}$, the P,T-flash calculation inevitably requires iteration.

A general solution scheme is shown by the block diagram of Fig. 12.18. The given information is read and stored. Since we do not know in advance whether the system of stated composition at the stated T and P is in fact a mixture of saturated liquid and saturated vapor and not entirely liquid or entirely vapor, we do preliminary calculations to establish the nature of the system. At the given T and overall composition, the system exists as a superheated vapor if its pressure is less than the dewpoint pressure P_{dew}. On the other hand, it exists as a subcooled liquid if its pressure is greater than the bubblepoint pressure P_{bubl}. Only for pressures between P_{dew} and P_{bubl} is the system an equilibrium mixture of vapor and liquid. We therefore determine P_{dew} by a DEW P calculation (Fig. 12.13) at the given T and for $\{y_i\} = \{z_i\}$ and P_{bubl} by a BUBL P calculation (Fig. 12.12) at the given T and for $\{x_i\} = \{z_i\}$. The P, T-flash calculation is performed only if the given pressure P lies between P_{dew} and P_{bubl}. If this is the case, then we make use of the results of the preliminary DEW P and BUBL P calculations to provide initial estimates of $\{\gamma_i\}$, $\{\hat{\phi}_i\}$, and $\mathcal{V}$. For the dewpoint, $\mathcal{V} = 1$, and we have calculated values of P_{dew}, $\gamma_{i,\text{dew}}$, and $\hat{\phi}_{i,\text{dew}}$; for the bubblepoint, $\mathcal{V} = 0$, and we have calculated values of P_{bubl}, $\gamma_{i,\text{bubl}}$, and $\hat{\phi}_{i,\text{bubl}}$. The simplest procedure is to interpolate between dewpoint and bubblepoint values in relation to the location of P between P_{dew} and P_{bubl}:

$$\frac{\gamma_i - \gamma_{i,\text{dew}}}{\gamma_{i,\text{bubl}} - \gamma_{i,\text{dew}}} = \frac{\hat{\phi}_i - \hat{\phi}_{i,\text{dew}}}{\hat{\phi}_{i,\text{bubl}} - \hat{\phi}_{i,\text{dew}}} = \frac{P - P_{\text{dew}}}{P_{\text{bubl}} - P_{\text{dew}}}$$

and

$$\frac{\mathcal{V} - 1}{0 - 1} = \frac{P - P_{\text{dew}}}{P_{\text{bubl}} - P_{\text{dew}}} \qquad \text{or} \qquad \mathcal{V} = \frac{P_{\text{bubl}} - P}{P_{\text{bubl}} - P_{\text{dew}}}$$

With these initial values of the γ_i and $\hat{\phi}_i$, initial values of the K_i can be calculated by Eq. (12.32). The P_i^{sat} and ϕ_i^{sat} values are already available from the preliminary DEW P and BUBL P calculations. Equations (12.29) and (12.30) now provide starting values of F and $dF/d\mathcal{V}$ for Newton's method as represented by Eq. (12.31). Repeated application of this equation leads to the value of $\mathcal{V}$ for which Eq. (12.29) is satisfied for the present estimates of the K_i. The remaining calculations serve to provide new estimates of the γ_i and Φ_i from which to reevaluate the K_i. This sequence of steps (an outer iteration) is repeated until there is no significant change in results from one iteration to the next. After the first outer iteration, the values of $\mathcal{V}$ and $(dF/d\mathcal{V})$ used to start Newton's method (an inner iteration) are simply the most recently calculated values. Once a value of $\mathcal{V}$ is established, values of x_i are calculated by Eq. (12.26) and values of y_i are given by $y_i = K_i x_i$. The nature of these calculations is well illustrated in the following example, where Raoult's law provides the VLE relationships.

Example 12.3 The system acetone(1)/acetonitrile(2)/nitromethane(3) at $80°C$ and 110 kPa has the overall composition, $z_1 = 0.45$, $z_2 = 0.35$, $z_3 = 0.20$. Assuming that Raoult's law is appropriate to this system, determine $\mathcal{L}$, $\mathcal{V}$, $\{x_i\}$, and $\{y_i\}$.

SOLUTION   The vapor pressures of the pure species at $80°C$ are

$$P_1^{\text{sat}} = 195.75 \qquad P_2^{\text{sat}} = 97.84 \qquad P_3^{\text{sat}} = 50.32 \text{ kPa}$$

First, we do a BUBL P calculation with $\{z_i\} = \{x_i\}$ to determine P_{bubl}. When Raoult's law applies, Eq. (12.11) becomes

$$P_{\text{bubl}} = x_1 P_1^{\text{sat}} + x_2 P_2^{\text{sat}} + x_3 P_3^{\text{sat}}$$

Numerically,

$$P_{\text{bubl}} = (0.45)(195.75) + (0.35)(97.84) + (0.20)(50.32) = 132.40 \text{ kPa}$$

Second, we do a DEW P calculation with $\{z_i\} = \{y_i\}$ to determine P_{dew}. For Raoult's law Eq. (12.12) is written

$$P_{\text{dew}} = \frac{1}{y_1/P_1^{\text{sat}} + y_2/P_2^{\text{sat}} + y_3/P_3^{\text{sat}}}$$

Substituting numerical values, we get

$$P_{\text{dew}} = 101.52 \text{ kPa}$$

Since the given pressure lies between P_{bubl} and P_{dew}, the system is in the two-phase region, and we proceed to the flash calculation.

Since Raoult's law applies, Eq. (12.32) here becomes $K_i = P_i^{\text{sat}}/P$, whence

$$K_1 = \frac{195.75}{110} = 1.7795$$

Similarly,

$$K_2 = 0.8895 \qquad \text{and} \qquad K_3 = 0.4575$$

Substitution of known values into Eq. (12.27) gives

$$\frac{(0.45)(1.7795)}{1 + 0.7795\mathcal{V}} + \frac{(0.35)(0.8895)}{1 - 0.1105\mathcal{V}} + \frac{(0.20)(0.4575)}{1 - 0.5425\mathcal{V}} = 1 \qquad (A)$$

Solution for $\mathcal{V}$ by trial yields

$$\mathcal{V} = 0.7364 \text{ mol}$$

whence

$$\mathcal{L} = 1 - \mathcal{V} = 0.2636 \text{ mol}$$

It is clear from Eq. (12.25) that each term on the left side of Eq. (A) is a value of y_i. We therefore find that $y_1 = 0.5087$, $y_2 = 0.3389$, and $y_3 = 0.1524$. Then from Eq. (12.24),

$$x_1 = \frac{y_1}{K_1} = \frac{0.5087}{1.7795} = 0.2859$$

Similarly,

$$x_2 = 0.3810 \qquad \text{and} \qquad x_3 = 0.3331$$

Obviously, $\sum_i y_i = \sum_i x_i = 1$. The procedure of this example is valid regardless of the number of species present.

Flash calculations for multicomponent systems made without the simplifying assumptions inherent in Raoult's law are readily carried out by computer as outlined in Fig. 12.18. Table 12.2 shows the results of a complete P, T-flash calculation for the system n-hexane(1)/ethanol(2)/methylcyclopentane(3)/benzene(4). This is the same system for which results of a BUBL T calculation were presented (Table 12.1),

Table 12.2: Results of a P, T-Flash Calculation at 1(atm) and 334.15 K for n-Hexane/Ethanol/Methylcyclopentane(MCP)/Benzene

Species(i)	z_i	x_i	y_i	K_i
n-Hexane(1)	0.250	0.160	0.270	1.694
Ethanol(2)	0.400	0.569	0.362	0.636
MCP(3)	0.200	0.129	0.216	1.668
Benzene(4)	0.150	0.142	0.152	1.070
$P = 1$(atm)	$T = 334.15$ K	$\mathcal{V} = 0.8166$		

and the same correlations and parameter values are used here. The given P and T are here 1(atm) and 334.15 K. The given overall mole fractions for the system $\{z_i\}$ are listed in the table along with the calculated values of the liquid-phase and vapor-phase mole fractions and the K-values. The molar fraction of the system that is vapor is here found to be $\mathcal{V} = 0.8166$.

12.7 Solute (1) / Solvent (2) Systems

The gamma/phi approach to VLE calculations, as presented in Sec. 12.4, presumes knowledge of the vapor pressure of each species at the temperature of interest. We consider here binary systems for which species 1, designated the solute, is either unstable at the system temperature or is *supercritical*, that is, the system temperature exceeds its critical temperature. Therefore its vapor pressure cannot be measured, and its fugacity f_1 as a pure liquid at the system temperature cannot be calculated by Eq. (10.41).

Although Eqs. (12.1) and (12.2) can be applied to species 2, designated the solvent, they are not applicable to the solute, and an alternative approach is required. We show in Fig. 12.19 a typical plot of the liquid-phase fugacity of the solute $\hat{f}_1$ vs. its mole fraction x_1 at constant temperature. This figure differs from Fig. 11.3 in that the curve representing $\hat{f}_1$ does not extend all the way to $x_1 = 1$. Thus the location of f_1, the liquid-phase fugacity of pure species 1, is not established, and the line representing the Lewis/Randall rule cannot be drawn. The tangent line at the origin, representing Henry's law (Sec. 11.1), provides alternative information. We recall that the slope of the tangent line is Henry's constant, defined by Eq. (11.2). Thus

$$k_1 \equiv \lim_{x_1 \to 0} \frac{\hat{f}_1}{x_1} \tag{12.33}$$

Henry's constant is a strong function of temperature, but only weakly dependent on pressure. Nevertheless, we note that the definition of k_1 is for temperature T at the VLE pressure for $x_1 \to 0$, namely, the vapor pressure of the pure solvent P_2^{sat}.

The activity coefficient of the solute at infinite dilution is

$$\lim_{x_1 \to 0} \gamma_1 = \lim_{x_1 \to 0} \frac{\hat{f}_1}{x_1 f_1} = \frac{1}{f_1} \lim_{x_1 \to 0} \frac{\hat{f}_1}{x_1}$$

In view of Eq. (12.33), this becomes $\gamma_1^\infty = k_1/f_1$, or

$$f_1 = \frac{k_1}{\gamma_1^\infty} \tag{12.34}$$

where γ_1^∞ represents the infinite-dilution value of the activity coefficient of the solute. Since both k_1 and γ_1^∞ are evaluated at P_2^{sat}, this pressure also applies to f_1. However, the effect of P on a liquid-phase fugacity, given by a Poynting factor, is very small, and for practical purposes may usually be neglected. Since for VLE $\hat{f}_1^l = \hat{f}_1^v = \hat{f}_1$, the activity coefficient of the solute is given by

$$\gamma_1 \equiv \frac{\hat{f}_1}{x_1 f_1} = \frac{y_1 P \hat{\phi}_1}{x_1 f_1}$$

Thus

$$\gamma_1 = \frac{y_1 P \hat{\phi}_1 \gamma_1^\infty}{x_1 k_1} \tag{12.35}$$

For the solvent, this equation takes the place of Eqs. (12.1) and (12.2).

In application to a BUBL P calculation for a binary system, this equation is written

$$y_1 = \frac{x_1 (\gamma_1/\gamma_1^\infty) k_1}{\hat{\phi}_1 P} \tag{12.36}$$

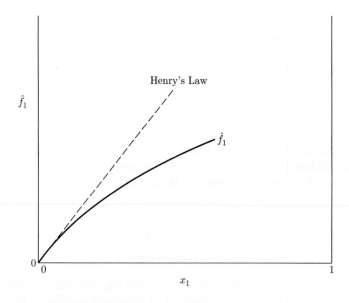

Figure 12.19: Plot of solute fugacity $\hat{f}_1$ vs. x_1.

For the solvent, species 2, Eq.(12.9) becomes

$$y_2 = \frac{x_2 \gamma_2 P_2^{\text{sat}}}{\Phi_2 P} \tag{12.37}$$

Since $y_1 + y_2 = 1$,

$$P = \frac{x_1 \left(\gamma_1/\gamma_1^\infty\right) k_1}{\hat{\phi}_1} + \frac{x_2 \gamma_2 P_2^{\text{sat}}}{\Phi_2} \tag{12.38}$$

Note that the same correlation that provides for the evaluation of γ_1 also allows evaluation of γ_1^∞.

As a simple example, consider a system for which

$$\frac{G^E}{RT} = B x_1 x_2 \qquad \hat{\phi}_1 = 1.0 \qquad \Phi_2 = 1.0$$

Then by Eqs. (11.12)

$$\gamma_1 = \exp\left(B x_2^2\right) \qquad \gamma_2 = \exp\left(B x_1^2\right)$$

and

$$\gamma_1^\infty = \exp(B) \qquad \text{and} \qquad \left(\gamma_1/\gamma_1^\infty\right) = \exp\left[B\left(x_2^2 - 1\right)\right]$$

Equation (12.38) here becomes

$$P = x_1 k_1 \exp\left[B\left(x_2^2 - 1\right)\right] + x_2 P_2^{\text{sat}} \exp\left[B x_1^2\right]$$

and by Eq. (12.36) we have

$$y_1 = \frac{x_1 k_1 \exp\left[B\left(x_2^2 - 1\right)\right]}{P}$$

For comparison, if the vapor pressure of species 1 were known, the resulting formulation would be

$$P = x_1 P_1^{\text{sat}} \exp\left[B x_2^2\right] + x_2 P_2^{\text{sat}} \exp\left[B x_1^2\right]$$

$$y_1 = \frac{x_1 P_1^{\text{sat}} \exp\left[B x_2^2\right]}{P}$$

The only difference in the input data for the two formulations is that Henry's constant for species 1 is required in the former case whereas the vapor pressure of species 1 appears in the latter.

There remains the problem of finding Henry's constant from the available VLE data. For equilibrium

$$\hat{f}_1 = \hat{f}_1^l = \hat{f}_1^v = y_1 P \hat{\phi}_1$$

Division by x_1 gives

$$\frac{\hat{f}_1}{x_1} = P \hat{\phi}_1 \frac{y_1}{x_1}$$

Henry's constant is defined as the limit as $x_1 \to 0$ of the ratio on the left; therefore

$$k_1 = P_2^{\text{sat}} \hat{\phi}_1^{\infty} \lim_{x_1 \to 0} \frac{y_1}{x_1}$$

The limiting value of y_1/x_1 can be found by plotting y_1/x_1 vs. x_1 and extrapolating to zero.

PROBLEMS

Solutions to some of the problems of this chapter require vapor pressures as a function of temperature for species which constitute systems in VLE. The table below lists parameter values for the Antoine equation,

$$\ln P^{\text{sat}}/\text{kPa} = A - \frac{B}{t/^{\circ}\text{C} + C}$$

	A	B	C
Acetone	14.3916	2,795.82	230.00
Acetonitrile	14.7258	3,271.24	241.85
Benzene	13.8594	2,773.78	220.07
Chlorobenzene	13.9926	3,295.12	217.55
1-Chlorobutane	13.9600	2,826.26	224.10
1,4-Dioxane	14.1177	2,966.88	210.00
Ethanol	16.6758	3,674.49	226.45
Ethylbenzene	14.0045	3,279.47	213.20
n-Heptane	13.8587	2,991.32	216.64
Methanol	16.5938	3,644.30	239.76
Methyl acetate	14.4015	2,739.17	223.12
n-Pentane	13.8183	2,477.07	233.21
1-Propanol	16.0692	3,448.66	204.09
Toluene	14.0098	3,103.01	219.79
Water	16.2620	3,799.89	226.35

12.1. Assuming Raoult's law to be valid, prepare a Pxy diagram for a temperature of 90°C and a txy diagram for a pressure of 90 kPa for one of the following systems:

(a) Benzene(1)/ethylbenzene(2).

(b) 1-Chlorobutane(1)/chlorobenzene(2).

12.2. Assuming Raoult's law to apply to the system n-pentane(1)/n-heptane(2)

(a) What are the values of x_1 and y_1 at $t = 55^{\circ}$C and $P = \frac{1}{2}(P_1^{\text{sat}} + P_2^{\text{sat}})$? For these conditions plot the fraction of the system that is vapor $\mathcal{V}$ vs. overall composition z_1.

(b) For $t = 55^{\circ}$C and $z_1 = 0.5$, plot P, x_1, and y_1 vs. $\mathcal{V}$.

12.3. Work Problem 12.2 for one of the following:

(a) $t = 65°C$; (b) $t = 75°C$; (c) $t = 85°C$; (d) $t = 95°C$.

12.4. A liquid mixture of cyclohexanone(1)/phenol(2) for which $x_1 = 0.6$ is in equilibrium with its vapor at 144°C. Determine the equilibrium pressure P and vapor composition y_1 from the following information:

- $G^E/RT = A(T)x_1x_2$.
- At 144°C, $P_1^{sat} = 75.20$ and $P_2^{sat} = 31.66$ kPa.
- The system forms an azeotrope at 144°C for which $x_1^{az} = y_1^{az} = 0.294$.

12.5. A binary system of species 1 and 2 consists of vapor and liquid phases in equilibrium at temperature T. The *overall* mole fraction of species 1 in the system is $z_1 = 0.65$. At temperature T,

- $P_1^{sat} = 32.27$ kPa.
- $P_2^{sat} = 73.14$ kPa.
- $G^E/RT = 0.67\, x_1x_2$.

Assuming the validity of Eq. (12.20),

(a) Over what range of pressures can this system exist as two phases at the given T and z_1?

(b) For a liquid-phase mole fraction $x_1 = 0.75$, what is the pressure P and what molar fraction V of the system is vapor?

(c) Show whether or not the system exhibits an azeotrope.

12.6. For the system ethyl ethanoate(1)/n-heptane(2) at 343.15 K,

- $P_1^{sat} = 79.8$ kPa.
- $P_2^{sat} = 40.5$ kPa.
- $G^E/RT = 0.95\, x_1x_2$.

Assuming the validity of Eq. (12.20),

(a) Make a BUBL P calculation for $T = 343.15$ K, $x_1 = 0.05$.

(b) Make a DEW P calculation for $T = 343.15$ K, $y_1 = 0.05$.

(c) What is the azeotrope composition and pressure at $T = 343.15$ K?

(d) Rework (a), but assume Henry's law to hold for species 1, and assume the Lewis/Randall rule for species 2.

12.7. A binary system of species 1 and 2 consists of vapor and liquid phases in equilibrium at temperature T, for which

- $P_1^{sat} = 1.24$ bar.
- $P_2^{sat} = 0.89$ bar.
- $G^E/RT = 1.8\, x_1x_2$.

Assuming the validity of Eq. (12.20),

(a) For what range of values of the *overall* mole fraction z_1 can this two-phase system exist with a *liquid* mole fraction $x_1 = 0.65$?

(b) What is the pressure P and vapor mole fraction y_1 within this range?

(c) What are the pressure and composition of the azeotrope at temperature T?

12.8. The excess Gibbs energy for the system chloroform(1)/ethanol(2) at 55°C is well represented by the Margules equation, written:

$$G^E/RT = (1.42\,x_1 + 0.59\,x_2)x_1 x_2$$

The vapor pressures of chloroform and ethanol at 55°C are

- $P_1^{\text{sat}} = 82.37$ kPa.
- $P_2^{\text{sat}} = 37.31$ kPa.

(a) Assuming the validity of Eq. (12.20), make BUBL P calculations at 55°C for liquid-phase mole fractions of 0.25, 0.50, and 0.75.

(b) For comparison, repeat the calculations using Eqs. (12.1) and (12.2) with virial coeffieients:

- $B_{11} = -963$ cm^3 mol^{-1}.
- $B_{22} = -1,523$ cm^3 mol^{-1}.
- $B_{12} = 52$ cm^3 mol^{-1}.

12.9. For the acetone(1)/methanol(2) system a vapor mixture for which $z_1 = 0.25$ and $z_2 = 0.75$ is cooled to temperature T in the two-phase region and flows into a separation chamber at a pressure of 1 bar. If the composition of the liquid product is to be $x_1 = 0.175$, what is the required value of T, and what is the value of y_1? For liquid mixtures of this system to a good approximation

$$G^E/RT = 0.64x_1 x_2$$

12.10 The following is a rule of thumb: For a binary system in VLE at low pressure, the equilibrium vapor-phase mole fraction y_1 corresponding to an equimolar liquid mixture is approximately

$$y_1 = \frac{P_1^{\text{sat}}}{P_1^{\text{sat}} + P_2^{\text{sat}}}$$

where P_i^{sat} is a pure-species vapor pressure. Clearly, this equation is valid if Raoult's law applies. Prove that it is also valid for VLE described by Eq. (12.20) when $G^E/RT = Ax_1 x_2$.

Problems 12.10–12.21 following require parameter values for the Wilson or NRTL equation for liquid-phase activity coefficients. The following table gives parameter values for both equations. Parameters a_{12}, a_{21}, b_{12}, and b_{21} have units of cal mol^{-1}, and V_1 and V_2 have units of cm^3 mol^{-1}. Values are those recommended by Gmehling et al. *Vapor-Liquid Equilibrium Data Collection*, Chemistry Data Series, vol. I, parts 1a, 1b, 2c and 2e, DECHEMA, Frankfurt/Main, 1981–1988.

System	V_1 V_2	Wilson equation		NRTL equation		
		a_{12}	a_{21}	b_{12}	b_{21}	α
Acetone(1)	74.05	291.27	1,448.01	631.05	1,197.41	0.5343
Water(2)	18.07					
Methanol(1)	40.73	107.38	469.55	−253.88	845.21	0.2994
Water(2)	18.07					
1-Propanol(1)	75.14	775.48	1,351.90	500.40	1,636.57	0.5081
Water(2)	18.07					
Water(1)	18.07	1,696.98	−219.39	715.96	548.90	0.2920
1,4-Dioxane	85.71					
Methanol(1)	40.73	504.31	196.75	343.70	314.59	0.2981
Acetonitrile(2)	66.30					
Acetone(1)	74.05	−161.88	583.11	184.70	222.64	0.3084
Methanol(2)	40.73					
Methyl acetate(1)	79.84	−31.19	813.18	381.46	346.54	0.2965
Methanol(2)	40.73					
Methanol(1)	40.73	1,734.42	183.04	730.09	1,175.41	0.4743
Benzene(2)	89.41					
Ethanol(1)	58.68	1,556.45	210.52	713.57	1,147.86	0.5292
Toluene(2)	106.85					

12.11. For one of the binary systems listed in the preceding table, based on the Wilson equation make the following calculations.

(a) BUBL P: $t = 60°C$, $x_1 = 0.3$.

(b) DEW P: $t = 60°C$, $y_1 = 0.3$.

(c) P, T-flash: $t = 60°C$, $P = \frac{1}{2}(P_{\text{bubble}} + P_{\text{dew}})$, $z_1 = 0.3$.

(d) If an azeotrope exists at $t = 60°C$, find P^{az} and $x_1^{\text{az}} = y_1^{\text{az}}$.

12.12. Work the preceding problem for the NRTL equation.

12.13. For one of the binary systems listed in the preceding table, based on the Wilson equation make the following calculations.

(a) BUBL T: $P = 101.33$ kPa, $x_1 = 0.3$.

(b) DEW T: $P = 101.33$ kPa, $y_1 = 0.3$.

(c) P, T-flash: $P = 101.33$ kPa, $T = \frac{1}{2}(T_{\text{bubble}} + T_{\text{dew}})$, $z_1 = 0.3$.

(d) If an azeotrope exists at $P = 101.33$ kPa, find T^{az} and $x_1^{\text{az}} = y_1^{\text{az}}$.

12.14. Work the preceding problem for the NRTL equation.

12.15. For one of the binary systems listed in the preceding table, based on the Wilson equation prepare a Pxy diagram for $t = 60°C$.

12.16. For one of the binary systems listed in the preceding table, based on the Wilson equation prepare a txy diagram for $P = 101.33$ kPa.

12.17. For one of the binary systems listed in the preceding table, based on the NRTL equation prepare a Pxy diagram for $t = 60°C$.

12.18. For one of the binary systems listed in the preceding table, based on the NRTL equation prepare a txy diagram for $P = 101.33$ kPa.

12.19. For the acetone(1)/methanol(2)/water(3) system, based on the Wilson equation make the following calculations.

(a) BUBL P: $t = 65°C$, $x_1 = 0.3$, $x_2 = 0.4$.

(b) DEW P: $t = 65°C$, $y_1 = 0.3$, $y_2 = 0.4$.

(c) P,T-flash: $t = 65°C$, $P = \frac{1}{2}(P_{\text{bubble}} + P_{\text{dew}})$, $z_1 = 0.3$, $z_2 = 0.4$.

12.20. Work the preceding problem for the NRTL equation.

12.21. For the acetone(1)/methanol(2)/water(3) system, based on the Wilson equation make the following calculations.

(a) BUBL T: $P = 101.33$ kPa, $x_1 = 0.3$, $x_2 = 0.4$.

(b) DEW T: $P = 101.33$ kPa, $y_1 = 0.3$, $y_2 = 0.4$.

(c) P,T-flash: $P = 101.33$ kPa, $T = \frac{1}{2}(T_{\text{bubble}} + T_{\text{dew}})$, $z_1 = 0.3$, $z_2 = 0.2$.

12.22. Work the preceding problem for the NRTL equation.

12.23. For a binary system the excess Gibbs energy of the liquid phase is given by an equation of the form $G^E/RT = Ax_1x_2$, where A is a function of temperature only. Assuming the validity of Eq. (12.20), show that

(a) The relative volatility of species 1 to species 2 at infinite dilution of species 1 is given by

$$\alpha_{12}(x_1 = 0) = \frac{P_1^{\text{sat}}}{P_2^{\text{sat}}}(\exp A)$$

(b) Henry's constant for species 1 is given by

$$k_1 = P_1^{\text{sat}}(\exp A)$$

(c) At every temperature for which an azeotrope exists the azeotropic composition x_1^{az} and azeotropic pressure P^{az} are related by

$$\frac{1}{x_1^{\text{az}}} = 1 + \left[\frac{\ln(P^{\text{az}}/P_1^{\text{sat}})}{\ln(P^{\text{az}}/P_2^{\text{sat}})}\right]^{1/2}$$

12.24. The excess Gibbs energy for binary systems consisting of liquids not too dissimilar in chemical nature is represented to a reasonable approximation by the equation

$$G^E/RT = Ax_1x_2$$

where A is a function of temperature only. For such systems, it is often observed that the ratio of the vapor pressures of the pure species is nearly constant over a considerable temperature range. Let this ratio be r, and determine the range of values of A, expressed as a function of r, for which no azeotrope can exist. Assume the vapor phase an ideal gas.

12.25. A concentrated binary solution containing mostly species 2 (but $x_2 \neq 1$) is in equilibrium with a vapor phase containing both species 1 and 2. The pressure of this two-phase system is 1 bar; the temperature is 25°C. Starting with Eq. (10.44), determine from the following data good estimates of x_1 and y_1.

$$k_1 = 200 \text{ bar} \qquad P_2^{\text{sat}} = 0.10 \text{ bar}$$

State and justify all assumptions.

12.26. A system formed of methane(1) and a light oil(2) at 200 K and 30 bar consists of a vapor phase containing 95 mole-% methane and a liquid phase containing oil and dissolved methane. The fugacity of the methane is given by Henry's law, and at the temperature of interest Henry's constant is $k_1 = 200$ bar. Stating any assumptions, estimate the equilibrium mole fraction of methane in the liquid phase. The second virial coefficient of pure methane at 200 K is $-105 \text{ cm}^3 \text{ mol}^{-1}$.

12.27. Assume that the last three data points (including the value of P_1^{sat}) of Table 11.1 can not be measured. Nevertheless, a correlation based on the remaining data points is required. Assuming the validity of Eq. (12.20), Eq. (12.38) may be written

$$P = x_1(\gamma_1/\gamma_1^{\infty})k_1 + x_2\gamma_2 P_2^{\text{sat}}$$

Data reduction may be based on Barker's method, i.e., minimizing the sum of squares of the residuals between the experimental values of P and the values predicted by this equation (see Example 11.1). Assume that the activity coefficients can be adequately represented by the Margules equation.

(a) Show that $\ln(\gamma_1/\gamma_1^{\infty}) = x_2^2[A_{12} + 2(A_{21} - A_{12})x_1] - A_{12}$.

(b) Find a value for Henry's constant k_1.

(c) Determine values for parameters A_{12} and A_{21} by Barker's method.

(d) Find values for δy_1 for the data points.

How could the regression be done so as to minimize the sum of squares of the residuals in G^E/RT, thus including the y_1 values in the data-reduction process?

12.28. Assume that the first three data points (including the value of P_2^{sat}) of Table 11.1 can not be measured. Nevertheless, a correlation based on the remaining data points is required. Assuming the validity of Eq. (12.20), Eq. (12.38) may be written

$$P = x_1\gamma_1 P_1^{\text{sat}} + x_2(\gamma_2/\gamma_2^{\infty})k_2$$

Data reduction may be based on Barker's method, i.e., minimizing the sum of squares of the residuals between the experimental values of P and the values predicted by this equation (see Example 11.1). Assume that the activity coefficients can be adequately represented by the Margules equation.

(a) Show that $\ln(\gamma_2/\gamma_2^{\infty}) = x_1^2[A_{21} + 2(A_{12} - A_{21})x_2] - A_{21}$.

(b) Find a value for Henry's constant k_2.

(c) Determine values for parameters A_{12} and A_{21} by Barker's method.

(d) Find values for δy_1 for the data points.

How could the regression be done so as to minimize the sum of squares of the residuals in G^E/RT, thus including the y_1 values in the data-reduction process?

12.29. Work Problem 12.27 with the data set of Table 11.3.

12.30. Work Problem 12.28 with the data set of Table 11.3.

CHAPTER 13

THERMODYNAMIC PROPERTIES AND VLE FROM EQUATIONS OF STATE

As discussed in Chap. 3, equations of state provide concise descriptions of the PVT behavior for pure fluids. The only equation of state that we have used extensively is the two-term virial equation,

$$Z = 1 + \frac{BP}{RT} \tag{3.31}$$

suitable for gases at pressures up to several bars. In reduced form for pure gases, this equation leads to generalized correlations for Z [Eqs. (3.47) and (3.48)], H^R [Eq. (6.62)], S^R [Eq. (6.63)], and ϕ [Eq. (10.64)]. When extended to gas mixtures, it yields a general expression for $\hat{\phi}_i$ [Eq. (12.4)], which is useful for low-pressure VLE calculations.

In this chapter we first present a general treatment of the calculation of thermodynamic properties of fluids and fluid mixtures from equations of state. Then the use of an equation of state for VLE calculations is described. The gamma/phi approach to VLE (Chap. 12) finds use primarily where pressures are no more than a few bars. An equation of state, applied to both the liquid and vapor phases, provides an alternative approach, valid to high pressures.

13.1 Properties of Fluids from the Virial Equations of State

Equations of state written for fluid mixtures are exactly the same as the equations of state presented for pure fluids in Secs. 3.4 and 3.5. The additional information needed for application to mixtures is the composition dependence of the parameters. For the virial equations, which apply only to gases, this dependence is given by exact equations arising out of statistical mechanics. The expression for B, the second virial coefficient, is

$$B = \sum_i \sum_j y_i y_j B_{ij} \qquad (10.65)$$

As indicated in Sec. 10.7, generalized methods are available for evaluation of the B_{ij}. For a binary mixture, Eq. (10.65) reduces to

$$B = y_1^2 B_{11} + 2y_1 y_2 B_{12} + y_2^2 B_{22} \qquad (10.66)$$

The third virial coefficient C is expressed as

$$C = \sum_i \sum_j \sum_k y_i y_j y_k C_{ijk} \qquad (13.1)$$

where C's with the same subscripts, regardless of order, are equal. For a binary mixture, Eq. (13.1) becomes

$$C = y_1^3 C_{111} + 3y_1^2 y_2 C_{112} + 3y_1 y_2^2 C_{122} + y_2^3 C_{222} \qquad (13.2)$$

Here C_{111} and C_{222} are the third virial coefficients for pure species 1 and 2, whereas C_{112} and C_{122} are cross-coefficients. Published generalized correlations for third virial coefficients[1] are based on very limited experimental data. Consistent with the mixing rules of Eqs. (10.65) and (13.1), the temperature derivatives of B and C are given exactly by

$$\frac{dB}{dT} = \sum_i \sum_j y_i y_j \frac{dB_{ij}}{dT} \qquad (13.3)$$

and

$$\frac{dC}{dT} = \sum_i \sum_j \sum_k y_i y_j y_k \frac{dC_{ijk}}{dT} \qquad (13.4)$$

As explained in Secs. 10.6 and 10.7, residual properties are readily calculated from equations of state. By Eq. (6.44), applicable to constant-composition fluids,

$$\frac{G^R}{RT} = \int_0^P (Z-1)\frac{dP}{P} \qquad (\text{const } T, x) \qquad (13.5)$$

When the compressibility factor is given by the two-term virial equation,

$$Z - 1 = \frac{BP}{RT} \qquad (3.31)$$

Equation (13.5) then yields

$$\frac{G^R}{RT} = \frac{BP}{RT} \qquad (13.6)$$

[1]R. deSantis and B. Grande, *AIChE J.*, vol. 25, pp. 931–938, 1979; H. Orbey and J. H. Vera, *ibid.*, vol. 29, pp. 107–113, 1983.

By Eq. (10.54),

$$\frac{H^R}{RT} = -T\left[\frac{\partial(G^R/RT)}{\partial T}\right]_{P,x} = -T\left(\frac{P}{R}\right)\left(\frac{1}{T}\frac{dB}{dT} - \frac{B}{T^2}\right)$$

or

$$\frac{H^R}{RT} = \frac{P}{R}\left(\frac{B}{T} - \frac{dB}{dT}\right) \qquad (13.7)$$

Substitution of Eqs. (13.6) and (13.7) into Eq. (6.43) gives

$$\frac{S^R}{R} = -\frac{P}{R}\frac{dB}{dT} \qquad (13.8)$$

The evaluation of residual enthalpies and residual entropies by Eqs. (13.7) and (13.8) is straightforward for given values of T, P, and composition, provided one has sufficient data to evaluate B and dB/dT by Eqs. (10.65) and (13.3). The range of applicability of these equations is the same as for Eq. (3.31), as discussed in Sec. 3.4.

The required values of B_{ij} in Eq. (10.65) can be determined from the generalized correlation for second virial coefficients according to the equation,

$$B_{ij} = \frac{RT_{cij}}{P_{cij}}(B^0 + \omega_{ij}B^1) \qquad (10.70)$$

where B^1 and B^1 are given by Eqs. (3.50) and (3.51), and ω_{ij}, T_{cij}, and P_{cij} come from the combining rules of Eqs. (10.71) through (10.75). An equation for dB_{ij}/dT, from which to determine values required in Eq. (13.3), results from differentiation of Eq. (10.70):

$$\frac{dB_{ij}}{dT} = \frac{RT_{cij}}{P_{cij}}\left(\frac{dB^0}{dT} + \omega_{ij}\frac{dB^1}{dT}\right)$$

or

$$\frac{dB_{ij}}{dT} = \frac{R}{P_{cij}}\left(\frac{dB^0}{dT_{rij}} + \omega_{ij}\frac{dB^1}{dT_{rij}}\right) \qquad (13.9)$$

where $T_{rij} = T/T_{cij}$. The derivatives dB^0/dT_{rij} and dB^1/dT_{rij} are given as functions of reduced temperature by Eqs. (6.64) and (6.65).

Example 13.1 Estimate V, H^R, and S^R for an equimolar mixture of methyl ethyl ketone(1) and toluene(2) at 50°C and 25 kPa.

SOLUTION   The required data are given with Example 10.8, along with calculated values of the B_{ij}. In addition, values of dB_{ij}/dT are needed. We assume in Eq. (10.72) that all $k_{ij} = 0$. Values of T_{rij}, together with dB^0/dT_{rij}, dB^1/dT_{rij}, and dB_{ij}/dT calculated for each ij pair by Eqs. (6.64), (6.65), and (13.9), are as follows:

ij	T_{rij}	dB^0/dT_{rij}	dB^1/dT_{rij}	dB_{ij}/dT $\text{cm}^3 \text{ mol}^{-1} \text{ K}^{-1}$
11	0.603	2.515	10.020	11.643
22	0.546	3.255	16.793	15.315
12	0.574	2.858	12.948	13.391

With values of B_{ij} calculated in Example 10.8 and values of dB_{ij}/dT calculated here, Eqs. (10.66) and (13.3) yield

$$B = (0.5)^2(-1{,}387) + (2)(0.5)(0.5)(-1{,}611) + (0.5)^2(-1{,}860)$$
$$= -1{,}617 \text{ cm}^3 \text{ mol}^{-1}$$

$$dB/dT = (0.5)^2(11.643) + (2)(0.5)(0.5)(13.391) + (0.5)^2(15.315)$$
$$= 13.435 \text{ cm}^3 \text{ mol}^{-1} \text{ K}^{-1}$$

Substitution of these values in Eqs. (3.31), (13.7), and (13.8) gives for $T = 323.15$ K and $P = 25$ kPa:

$$Z = 1 + \frac{BP}{RT} = 1 + \frac{(-1{,}617)(25)}{(8{,}314)(323.15)} = 0.9850$$

$$\frac{H^R}{RT} = \frac{P}{R}\left(\frac{B}{T} - \frac{dB}{dT}\right) = \frac{25}{8{,}314}\left(\frac{-1{,}617}{323.15} - 13.435\right) = -0.05545$$

$$\frac{S^R}{R} = -\frac{P}{R}\frac{dB}{dT} = \frac{-25}{8{,}314}(13.435) = -0.04040$$

From these values we find

$$V = \frac{ZRT}{P} = \frac{(0.9850)(8{,}314)(323.15)}{25} = 105{,}850 \text{ cm}^3 \text{ mol}^{-1}$$

$$H^R = (-0.05545)(8.314)(323.15) = -149.0 \text{ J mol}^{-1}$$

$$S^R = (-0.04040)(8.314) = -0.3359 \text{ J mol}^{-1} \text{ K}^{-1}$$

Since Eq. (3.31) expresses Z as a function of P and T, the mathematical operations of Eqs. (13.5) and (10.54) are readily carried out. However, when the equation of state expresses Z as a function of V and T, as is most often the case, Eqs. (13.5) and (10.54) are inappropriate, and must be transformed so that V rather than P is the independent variable. This is accomplished by application of the general equation $PV = ZRT$. Differentiation at constant T gives

$$P \, dV + V \, dP = RT \, dZ \qquad (\text{const } T)$$

which may be rewritten as

$$\frac{dP}{P} = \frac{dZ}{Z} - \frac{dV}{V} \qquad (\text{const } T)$$

Substitution into Eq. (13.5) leads to

$$\frac{G^R}{RT} = Z - 1 - \ln Z - \int_{\infty}^{V} (Z-1)\frac{dV}{V} \qquad (13.10)$$

The corresponding equation for H^R follows from Eq. (6.40), which in view of Eq. (6.38) may be written

$$d\left(\frac{G^R}{RT}\right) = (Z-1)\frac{dP}{P} - \frac{H^R}{RT^2}dT$$

Upon division by dT, restriction to constant V, and rearrangement, this becomes

$$\frac{H^R}{RT^2} = \frac{Z-1}{P}\left(\frac{\partial P}{\partial T}\right)_V - \left[\frac{\partial(G^R/RT)}{\partial T}\right]_V$$

Differentiation of $P = ZRT/V$ provides the first derivative on the right, and differentiation of Eq. (13.10) provides the second. Substitution leads to

$$\frac{H^R}{RT} = Z - 1 + T\int_{\infty}^{V}\left(\frac{\partial Z}{\partial T}\right)_V \frac{dV}{V} \qquad (13.11)$$

The residual entropy is found from Eq. (6.43).

When Z is given by the three-term virial equation,

$$Z - 1 = \frac{B}{V} + \frac{C}{V^2} \qquad (3.33)$$

Eqs. (13.10) and (13.11) become

$$\frac{G^R}{RT} = \frac{2B}{V} + \frac{(3/2)C}{V^2} - \ln Z \qquad (13.12)$$

and

$$\frac{H^R}{RT} = T\left[\left(\frac{B}{T} - \frac{dB}{dT}\right)\frac{1}{V} + \left(\frac{C}{T} - \frac{1}{2}\frac{dC}{dT}\right)\frac{1}{V^2}\right] \qquad (13.13)$$

Application of these equations, useful for gases up to moderate pressures, requires values of all B_{ij}, C_{ijk}, and their temperature derivatives for substitution into Eqs. (10.65), (13.1), (13.3), and (13.4).

13.2 Properties of Fluids from Cubic Equations of State

As discussed in Sec. 3.5 and illustrated by Fig. 3.12, equations of state that are cubic in molar volume are capable of describing the behavior of both liquid and vapor phases of pure fluids.

The application of such equations to mixtures requires that the equation-of-state parameters be expressed as functions of composition. No exact theory like that for the virial equations prescribes this composition dependence, which instead is often imposed by empirical *mixing rules*. For the Redlich/Kwong equation,

$$P = \frac{RT}{V-b} - \frac{a}{T^{1/2}V(V+b)} \tag{3.35}$$

the mixing rules that have found frequent use are:

$$a = \sum_i \sum_j y_i y_j a_{ij} \tag{13.14}$$

with $a_{ij} = a_{ji}$, and

$$b = \sum_i y_i b_i \tag{13.15}$$

The a_{ij} are of two types: pure-species parameters (like subscripts) and interaction parameters (unlike subscripts). The b_i are parameters for the pure species.

One procedure for evaluation of parameters is a generalization of Eqs. (3.40) and (3.41):

$$a_{ij} = \frac{0.42748R^2 T_{cij}^{2.5}}{P_{cij}} \tag{13.16}$$

and

$$b_i = \frac{0.08664 RT_{ci}}{P_{ci}} \tag{13.17}$$

where Eqs. (10.72) through (10.75) provide for the calculation of the T_{cij} and P_{cij}.

Multiplication of the Redlich/Kwong equation [Eq. (3.35)] by V/RT leads to its expression in alternative form:

$$Z = \frac{1}{1-h} - \frac{a}{bRT^{1.5}}\left(\frac{h}{1+h}\right) \tag{13.18}$$

Equivalently,

$$Z - 1 = \frac{h}{1-h} - \frac{a}{bRT^{1.5}}\left(\frac{h}{1+h}\right) \tag{13.19}$$

where

$$h \equiv \frac{bP}{ZRT} \tag{13.20}$$

Equations (13.10) and (13.11) in combination with Eq. (13.19) lead to

$$\frac{G^R}{RT} = Z - 1 - \ln(1-h)Z - \left(\frac{a}{bRT^{1.5}}\right)\ln(1+h) \tag{13.21}$$

and

$$\frac{H^R}{RT} = Z - 1 - \left(\frac{3a}{2bRT^{1.5}}\right)\ln(1+h) \tag{13.22}$$

Once a and b for the mixture are determined by Eqs. (13.14) through (13.17), then for given T and P we find Z, G^R/RT, and H^R/RT by Eqs. (13.18) through (13.22) and S^R/R by Eq. (6.43). The procedure requires initial solution of Eqs. (13.18) and (13.20), usually by an iterative scheme, as described in connection with Eq. (3.44) for a gas or vapor phase.

13.3   Fluid Properties from Correlations of the Pitzer Type

Generalized correlations of the Pitzer type provide an alternative to the use of a cubic equation of state for the calculation of thermodynamic properties. However, no theoretical basis exists for general extension of these correlations to mixtures. Nevertheless, Z, as given by

$$Z = Z^0 + \omega Z^1 \qquad (3.46)$$

depends on T_r, P_r, and ω, and approximate results for mixtures can often be obtained with critical parameters for the mixture and a simple linear mixing rule for the acentric factor. Since values for the actual critical properties T_c and P_c for mixtures are rarely known, use is made of the pseudoparameters T_{pc} and P_{pc}, determined again by a simple linear mixing rule. Thus, by definition,

$$T_{pc} = \sum_i y_i T_{c_i} \qquad (13.23)$$

$$P_{pc} = \sum_i y_i P_{c_i} \qquad (13.24)$$

$$\omega = \sum_i y_i \omega_i \qquad (13.25)$$

The pseudoreduced temperature and pseudoreduced pressure, which replace T_r and P_r, are determined by

$$T_{pr} = \frac{T}{T_{pc}} \qquad (13.26)$$

$$P_{pr} = \frac{P}{P_{pc}} \qquad (13.27)$$

Thus, for a mixture at given T_{pr} and P_{pr} we may determine a value of Z by Eq. (3.46) and Tables E.1–E.4, of H^R/RT_{pc} by Eq. (6.60) and Tables E.5–E.8, and of S^R/R by Eq. (6.61) and Tables E.9–E.12.

Example 13.2   Estimate V, H^R, and S^R for an equimolar mixture of carbon dioxide(1) and propane(2) at 450 K and 140 bar by:

    (a) The Redlich/Kwong equation.

    (b) The Lee/Kesler correlations.

SOLUTION   (a) The required data are given as follows:

ij	T_{cij}/K	P_{cij}/bar	$V_{cij}/\mathrm{cm^3\,mol^{-1}}$	Z_{cij}	ω_{ij}
11	304.2	73.83	94.0	0.274	0.224
22	369.8	42.48	200.0	0.276	0.152
12	335.4	54.62	140.4	0.275	0.188

where the values in the last row are calculated by Eqs. (10.71) through (10.75) with $k_{12} = 0$. Substitution of appropriate values into Eqs. (13.16) and (13.17) gives:

ij	a_{ij}/bar cm^6 K$^{1/2}$ mol^{-2}	b_i/cm^3 mol^{-1}
11	64.595×10^6	29.68
22	182.923×10^6	62.71
12	111.453×10^6	

Parameters a and b for the mixture are given by Eqs. (13.14) and (13.15):

$$a = y_1^2 a_{11} + 2 y_1 y_2 a_{12} + y_2^2 a_{22}$$

$$= (0.5)^2 (64.595 \times 10^6) + (2)(0.5)(0.5)(111.453 \times 10^6)$$

$$+ (0.5)^2 (182.923 \times 10^6)$$

$$a = 117.61 \times 10^6 \text{ bar cm}^6 \text{ K}^{1/2} \text{ mol}^{-2}$$

$$b = y_1 b_1 + y_2 b_2 = (0.5)(29.68) + (0.5)(62.71)$$

$$b = 46.195 \text{ cm}^3 \text{ mol}^{-1}$$

The dimensionless quantity $a/bRT^{1.5}$ is evaluated as

$$\frac{a}{bRT^{1.5}} = \frac{117.61 \times 10^6}{(46.195)(83.14)(450)^{1.5}} = 3.2079$$

Similarly,

$$\frac{bP}{RT} = \frac{(46.195)(140)}{(83.41)(450)} = 0.17286$$

Therefore Eq. (13.18) becomes

$$Z = \frac{1}{1-h} - 3.2079 \left(\frac{h}{1+h} \right)$$

and Eq. (13.20) gives

$$h = \frac{0.17286}{Z}$$

Solution for Z and h yields

$$Z = 0.6918 \qquad \text{and} \qquad h = 0.2499$$

The molar volume is therefore

$$V = \frac{ZRT}{P} = \frac{(0.6918)(83.14)(450)}{140} = 184.9 \text{ cm}^3 \text{ mol}^{-1}$$

By Eq. (13.21),

$$\frac{G^R}{RT} = 0.6918 - 1 - \ln[(0.7501)(0.6918)] - 3.2079 \ln 1.2499 = -0.3678$$

By Eq. (13.22),

$$\frac{H^R}{RT} = 0.6918 - 1 - (1.5)(3.2079) \ln 1.2499 = -1.382$$

whence
$$H^R = (-1.382)(8.314)(450) = -5{,}170 \text{ J mol}^{-1}$$

By Eq. (6.43)
$$\frac{S^R}{R} = \frac{H^R}{RT} - \frac{G^R}{RT} = -1.382 + 0.368 = -1.014$$

whence
$$S^R = (-1.014)(8.314) = -8.43 \text{ J mol}^{-1} \text{ K}^{-1}$$

(b) The pseudocritical constants are found by Eqs. (13.23) and (13.24):
$$T_{pc} = y_1 T_{c11} + y_2 T_{c22} = (0.5)(304.2) + (0.5)(369.8) = 337.0 \text{ K}$$

and
$$P_{pc} = y_1 P_{c11} + y_2 P_{c22} = (0.5)(73.83) + (0.5)(42.48) = 58.15 \text{ bar}$$

Therefore
$$T_{pr} = \frac{450}{337.0} = 1.335 \qquad \text{and} \qquad P_{pr} = \frac{140}{58.15} = 2.41$$

Values of Z^0 and Z^1 from Tables E.3 and E.4 at these reduced conditions are:
$$Z^0 = 0.697 \qquad \text{and} \qquad Z^1 = 0.205$$

With ω given by
$$\omega = y_1 \omega_1 + y_2 \omega_2 = (0.5)(0.224) + (0.5)(0.152) = 0.188$$

we apply Eq. (3.46):
$$Z = Z^0 + \omega Z^1 = 0.697 + (0.188)(0.205) = 0.736$$

from which
$$V = \frac{ZRT}{P} = \frac{(0.736)(83.14)(450)}{140} = 196.7 \text{ cm}^3 \text{ mol}^{-1}$$

Similarly, from Tables E.7 and E.8,
$$\left(\frac{H^R}{RT_{pc}}\right)^0 = -1.730 \qquad \left(\frac{H^R}{RT_{pc}}\right)^1 = -0.169$$

Substitution into Eq. (6.60) gives
$$\frac{H^R}{RT_{pc}} = -1.730 + (0.188)(-0.169) = -1.762$$

whence
$$H^R = (8.314)(337.0)(-1.762) = -4{,}937 \text{ J mol}^{-1}$$

By Tables E.11 and E.12 and Eq. (6.61),
$$\frac{S^R}{R} = -0.967 + (0.188)(-0.330) = -1.029$$

whence
$$S^R = (8.314)(-1.029) = -8.56 \text{ J mol}^{-1} \text{ K}^{-1}$$

13.4 VLE from Cubic Equations of State

In Sec. 10.6 we showed that phases at the same T and P are in equilibrium when the fugacity of each species is the same in all phases. For VLE, this requirement is written

$$\hat{f}_i^v = \hat{f}_i^l \qquad (i = 1, 2, \ldots, N) \tag{10.44}$$

An alternative form of Eq. (10.44) results from introduction of the fugacity coefficient, as defined by Eq. (10.47):

$$y_i P \hat{\phi}_i^v = x_i P \hat{\phi}_i^l$$

or

$$\boxed{y_i \hat{\phi}_i^v = x_i \hat{\phi}_i^l \qquad (i = 1, 2, \ldots, N)} \tag{13.28}$$

For the special case of pure species i, this becomes

$$\phi_i^v = \phi_i^l \tag{13.29}$$

a relation already expressed by Eq. (10.40).

Vapor Pressures for a Pure Species

We consider first application of Eq. (13.29) to an equation of state for the calculation of the saturation or equilibrium vapor pressure of pure species i at given temperature T.

As discussed in Sec. 3.5 with respect to cubic equations of state for pure species, a subcritical isotherm on a PV diagram exhibits a smooth transition from the liquid to the vapor region, shown by the curve labeled $T_2 < T_c$ on Fig. 3.12. We tacitly assumed in that discussion independent knowledge of the equilibrium vapor pressure at this temperature. In fact, this value is implicit in the equation of state. We reproduce in Fig. 13.1 the subcritical isotherm of Fig. 3.12, without any indication of the location of the equilibrium pressure P_i^{sat}. However, it clearly must lie between the pressures P' and P'' shown on the figure.

The equilibrium criterion expressed by Eq. (13.29) may be written

$$\ln \phi_i^l - \ln \phi_i^v = 0 \tag{13.30}$$

The fugacity coefficient of a pure liquid or vapor is a function of its temperature and pressure. For a *saturated* liquid or vapor, the equilibrium pressure is P_i^{sat}. Therefore Eq. (13.30) implicitly expresses the functional relation,

$$g(T, P_i^{\text{sat}}) = 0$$

or

$$P_i^{\text{sat}} = f(T)$$

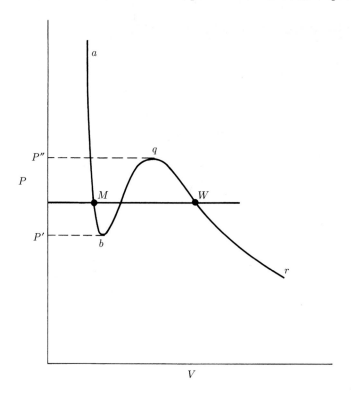

Figure 13.1: Isotherm for $T < T_c$ on PV diagram for a pure fluid.

If the isotherm of Fig. 13.1 is represented by a cubic equation of state, then its roots for a specific P between P' and P'' include both a liquid-like volume on branch ab of the isotherm and a vapor-like volume on branch qr, represented for example by points M and W. Moreover, an equation of state for pure species i implies an expression for $\ln \phi_i$. For example, Eq. (13.21), written for pure i and combined with Eq. (10.31), yields

$$\ln \phi_i = Z_i - 1 - \ln(1 - h_i)Z_i - \left(\frac{a_i}{b_i RT^{1.5}} \right) \ln(1 + h_i)$$

where

$$h_i \equiv \frac{b_i P}{Z_i RT}$$

Values for $\ln \phi_i^l$ and $\ln \phi_i^v$ may therefore be calculated corresponding to points M and W. If these values satisfy Eq. (13.30), then $P = P_i^{\text{sat}}$ and points M and W represent the saturated-liquid and saturated-vapor states at temperature T. If Eq. (13.30) is not satisfied, the value of P must be found for which it is satisfied, either by trial or by a suitable iteration scheme.

VLE from Equations of State

The application of Eq. (13.28) to the determination of *mixture* VLE is in principle the same as the calculation of pure-species VLE, but is much more difficult. Since $\hat{\phi}_i^v$ is a function of T, P, and $\{y_i\}$, and $\hat{\phi}_i^l$ is a function of T, P, and $\{x_i\}$, Eq. (13.28) represents N complex relations among the $2N$ variables T, P, $(N-1)$ y_i's and $(N-1)$ x_i's. Thus, specification of N of these variables, usually either T or P *and* either the vapor- or liquid-phase compositions, allows solution for the remaining N variables. These are BUBL P, DEW P, BUBL T, and DEW T calculations (Sec. 12.5).

Values of $\hat{\phi}_i^l$ and $\hat{\phi}_i^v$ are readily found from equations of state. Equation (10.60) is a general equation for calculation of $\ln\hat{\phi}_i$ from equations that express Z as a function of T and P. However, cubic and higher-order equations of state all lead to expressions for Z as functions of T and V. The derivation of an equation in which the independent variable is V rather than P starts with Eq. (10.52), which may be written:

$$d\left(\frac{nG^R}{RT}\right) = \frac{n(Z-1)}{P}dP - \frac{nH^R}{RT^2}dT + \sum_i \ln\hat{\phi}_i\,dn_i$$

Division by dn_i and restriction to constant T, nV, and n_j $(j\neq i)$ leads to

$$\ln\hat{\phi}_i = \left[\frac{\partial(nG^R/RT)}{\partial n_i}\right]_{T,nV,n_j} - \frac{n(Z-1)}{P}\left(\frac{\partial P}{\partial n_i}\right)_{T,nV,n_j}$$

But $P = (nZ)RT/nV$, and therefore

$$\left(\frac{\partial P}{\partial n_i}\right)_{T,nV,n_j} = \frac{P}{nZ}\left[\frac{\partial(nZ)}{\partial n_i}\right]_{T,nV,n_j}$$

Combination of the last two equations gives

$$\ln\hat{\phi}_i = \left[\frac{\partial(nG^R/RT)}{\partial n_i}\right]_{T,nV,n_j} - \left(\frac{Z-1}{Z}\right)\left[\frac{\partial(nZ)}{\partial n_i}\right]_{T,nV,n_j}$$

When the first derivative on the right-hand side is found from Eq. (13.10), this equation reduces to

$$\ln\hat{\phi}_i = -\int_\infty^V \left\{\left[\frac{\partial(nZ)}{\partial n_i}\right]_{T,nV,n_j} - 1\right\}\frac{dV}{V} - \ln Z \qquad (13.31)$$

where the derivative and $\ln Z$ are evaluated by an equation of state.

The original Redlich/Kwong equation of state as given by Eq. (3.35) is rarely satisfactory for VLE calculations. Two widely used cubic equations, developed specifically for this purpose, are the Soave/Redlich/Kwong (SRK) equation[2] and

---

[2]G. Soave, *Chem. Eng. Sci.*, vol. 27, pp. 1197–1203, 1972.

the Peng/Robinson (PR) equation.[3] Both are special cases of the following expression, written for pure species i:

$$Z_i = \frac{PV_i}{RT} = \frac{V_i}{V_i - b_i} - \frac{a_i(T)V_i}{RT(V_i + \varepsilon b_i)(V_i + \sigma b_i)} \tag{13.32}$$

where

$$a_i(T) = \frac{\Omega_a \alpha(T_{ri}; \omega_i)R^2 T_{ci}^2}{P_{ci}} \tag{13.33}$$

$$b_i = \frac{\Omega_b R T_{ci}}{P_{ci}} \tag{13.34}$$

and ε, σ, Ω_a, and Ω_b are equation-specific constants. For the Soave/Redlich/Kwong equation:

$$\alpha(T_{ri}; \omega_i) = \left[1 + (0.480 + 1.574\,\omega_i - 0.176\,\omega_i^2)\left(1 - T_{ri}^{1/2}\right)\right]^2 \tag{13.35}$$

For the Peng/Robinson equation:

$$\alpha(T_{ri}; \omega_i) = \left[1 + (0.37464 + 1.54226\,\omega_i - 0.26992\,\omega_i^2)\left(1 - T_{ri}^{1/2}\right)\right]^2 \tag{13.36}$$

Written for a mixture, Eq. (13.32) becomes

$$Z = \frac{V}{V - b} - \frac{a(T)V}{RT(V + \varepsilon b)(V + \sigma b)} \tag{13.37}$$

where a and b are mixture values, related to the a_i and b_i by *mixing rules*. Equation (13.31) applied to Eq. (13.37) leads to

$$\ln \hat{\phi}_i = \frac{\bar{b}_i}{b}(Z - 1) - \ln \frac{(V - b)Z}{V} + \frac{a/bRT}{\varepsilon - \sigma}\left(1 + \frac{\bar{a}_i}{a} - \frac{\bar{b}_i}{b}\right)\ln \frac{V + \sigma b}{V + \varepsilon b} \tag{13.38}$$

where $\bar{a}_i$ and $\bar{b}_i$ are "partial parameters" for species i, defined by

$$\bar{a}_i = \left[\frac{\partial(na)}{\partial n_i}\right]_{T,n_j} \tag{13.39}$$

and

$$\bar{b}_i = \left[\frac{\partial(nb)}{\partial n_i}\right]_{T,n_j} \tag{13.40}$$

These equations do not depend on the particular mixing rules adopted for the composition dependence of a and b. Solution of Eq. (13.38) for fugacity coefficient

[3]D.-Y. Peng and D. B. Robinson, *Ind. Eng. Chem. Fundam.*, vol. 15, pp. 59–64, 1976.

$\hat{\phi}_i$ at given T and P requires prior solution of Eq. (13.37) for V, from which is found $Z = PV/RT$. For pure species i, Eq. (13.38) reduces to

$$\ln \phi_i = Z_i - 1 - \ln \frac{(V_i - b_i)Z_i}{V_i} + \frac{a_i/b_i RT}{\varepsilon - \sigma} \ln \frac{V_i + \sigma b_i}{V_i + \varepsilon b_i} \qquad (13.41)$$

Because of inadequacies in empirical mixing rules, such as those given by Eqs. (13.14) and (13.15), the equation-of-state approach to VLE was long limited to systems exhibiting modest and well-behaved deviations from ideal-solution behavior in the liquid phase, e.g., to systems containing hydrocarbons and cryogenic fluids. However, the introduction by Wong and Sandler[4] of a new class of theoretically based mixing rules for cubic equations of state has greatly expanded their useful application.

The first of the Wong/Sandler mixing rules relates the difference in mixture quantities b and a/RT to the corresponding differences (identified by subscripts) for the pure species:

$$b - \frac{a}{RT} = \sum_p \sum_q x_p x_q E_{pq} \qquad (13.42)$$

where

$$E_{pq} \equiv \frac{1}{2} \left(b_p - \frac{a_p}{RT} + b_q - \frac{a_q}{RT} \right) (1 - k_{pq}) \qquad (13.43)$$

Binary interaction parameters k_{pq} are determined for the pq pairs ($p \neq q$) from experimental data. Note that $k_{pq} = k_{qp}$ and $k_{pp} = k_{qq} = 0$. Since the quantity on the left-hand side of Eq. (13.42) represents the second virial coefficient as predicted by Eq. (13.37), the basis for Eq. (13.42) lies in Eq. (10.65), which expresses the quadratic dependence of the mixture second virial coefficient on mole fraction.

The second Wong/Sandler mixing rule relates ratios of a/RT to b:

$$\frac{a}{bRT} = 1 - D \qquad (13.44)$$

where

$$D \equiv 1 + \frac{G^E}{cRT} - \sum_p x_p \frac{a_p}{b_p RT} \qquad (13.45)$$

The quantity G^E/RT is given by an appropriate correlation for the excess Gibbs energy of the liquid phase, and is evaluated at the mixture composition, regardless of whether the mixture is liquid or vapor. The constant c is specific to the equation of state.

Elimination of a from Eq. (13.42) by Eq. (13.44) provides an expression for b:

$$b = \frac{1}{D} \sum_p \sum_q x_p x_q E_{pq} \qquad (13.46)$$

---

[4]D. S. H. Wong and S. I. Sandler, *AIChE J.*, vol. 38, pp. 671–680, 1992; *Ind. Eng. Chem. Res.*, vol. 31, pp. 2033–2039, 1992.

Since E_{pq} as given by Eq. (13.43) is a function of T, b also depends on T. However, in the limit for pure species i, Eq. (13.46) yields b_i, which is independent of T. The net result is that b for mixtures is a very weak function of T. Mixture parameter a follows from Eq. (13.44):

$$a = bRT(1 - D) \tag{13.47}$$

Equations (13.39) and (13.40) may now be applied for the evaluation of partial parameters $\bar{a}_i$ and $\bar{b}_i$:

$$\bar{b}_i = \frac{1}{D}\left[2\sum_j x_j E_{ij} - b\left(1 + \frac{\ln\gamma_i}{c} - \frac{a_i}{b_i RT}\right)\right] \tag{13.48}$$

and

$$\bar{a}_i = bRT\left(\frac{a_i}{b_i RT} - \frac{\ln\gamma_i}{c}\right) + a\left(\frac{\bar{b}_i}{b} - 1\right) \tag{13.49}$$

Fugacity coefficients for pure species i are given by Eq. (13.41), which may be applied separately to the liquid phase and to the vapor phase to yield the pure-species values ϕ_i^l and ϕ_i^v. For pure-species VLE [Eq. (13.29)], these two quantities are equal. Given parameters a_i and b_i, the pressure P in Eq. (13.32) that makes these two values equal is P_i^{sat}, the equilibrium vapor pressure of pure species i as predicted by the equation of state.

The correlations for $\alpha(T_{ri}; \omega_i)$ given by Eqs. (13.35) and (13.36) are designed to provide values of a_i that yield pure-species vapor pressures which on average are in reasonable agreement with experiment. However, reliable correlations for P_i^{sat} as a function of temperature are available for many pure species. Thus when P_i^{sat} is known for a particular temperature, a_i should be evaluated so that the equation of state correctly predicts this known value. The procedure is to write Eq. (13.41) for each of the phases, combining the two equations in accord with Eq. (13.29), written

$$\ln\phi_i^l = \ln\phi_i^v$$

The resulting expression may be solved for a_i:

$$a_i = \frac{b_i RT(\varepsilon - \sigma)\left(\ln\dfrac{V_i^l - b_i}{V_i^v - b_i} + Z_i^v - Z_i^l\right)}{\ln\dfrac{(V_i^l + \sigma b_i)(V_i^v + \varepsilon b_i)}{(V_i^l + \varepsilon b_i)(V_i^v + \sigma b_i)}} \tag{13.50}$$

where $Z_i^v = P_i^{\text{sat}} V_i^v / RT$ and $Z_i^l = P_i^{\text{sat}} V_i^l / RT$. Values of V_i^v and V_i^l come from solution of Eq. (13.32) for each phase with $P = P_i^{\text{sat}}$ at temperature T. Since a value of a_i is *required* for these calculations, an iterative procedure is implemented with an initial value for a_i from the appropriate correlation for $\alpha(T_{ri}; \omega_i)$.

The binary interaction parameters k_{pq} are evaluated from liquid-phase G^E correlations for the pq binaries. The most satisfactory procedure is to apply at infinite dilution the relation between a liquid-phase activity coefficient and its underlying fugacity coefficients,[5] $\gamma_i^\infty = \hat{\phi}_i^\infty/\phi_i$. Rearrangement of the logarithmic form yields

$$\ln \hat{\phi}_i^\infty = \ln \gamma_i^\infty + \ln \phi_i \tag{13.51}$$

where $\ln \gamma_i^\infty$ comes from the G^E correlation and $\ln \phi_i$ is given by Eq. (13.41) written for the liquid phase. Equation (13.51) supplies a value for $\ln \hat{\phi}_i^\infty$ which when used with Eq. (13.38) ultimately leads to values for k_{pq}, as shown in what follows.

For a binary system comprised of species p and q, Eqs. (13.38), (13.49), and (13.51) may be written for species p at infinite dilution. The three resulting equations are then combined to yield

$$\frac{\bar{b}_p^\infty}{b_q} = \frac{\ln \gamma_p^\infty + \ln \phi_p - M_p}{Z_q - 1} \tag{13.52}$$

where

$$M_p \equiv -\ln \frac{(V_q - b_q)Z_q}{V_q} + \frac{1}{\varepsilon - \sigma}\left(\frac{a_p}{b_p RT} - \frac{\ln \gamma_p^\infty}{c}\right)\ln \frac{V_q + \sigma b_q}{V_q + \varepsilon b_q} \tag{13.53}$$

By Eq. (13.48) written for species p at infinite dilution in a pq binary,

$$\frac{\bar{b}_p^\infty}{b_q} = \frac{\dfrac{2E_{pq}}{b_q} - 1 - \dfrac{\ln \gamma_p^\infty}{c} + \dfrac{a_p}{b_p RT}}{1 - \dfrac{a_q}{b_q RT}} \tag{13.54}$$

Equations (13.52) and (13.54) are now equated, and E_{pq} is eliminated by Eq. (13.43). We replace k_{pq} by k_p, the infinite-dilution value at $x_p \to 0$, and solve for k_p:

$$k_p = 1 - \frac{\left(b_q - \dfrac{a_q}{RT}\right)\left(\dfrac{\ln \gamma_p^\infty + \ln \phi_p - M_p}{Z_q - 1}\right) + b_q\left(1 + \dfrac{\ln \gamma_p^\infty}{c} - \dfrac{a_p}{b_p RT}\right)}{b_p - \dfrac{a_p}{RT} + b_q - \dfrac{a_q}{RT}} \tag{13.55}$$

where $\ln \phi_p$ comes from Eq. (13.41). All values in Eq. (13.55) are for the liquid phase at $P = P_q^{\text{sat}}$. The analogous equation for k_q, the infinite-dilution value of k_{pq} at $x_q \to 0$, is written

$$k_q = 1 - \frac{\left(b_p - \dfrac{a_p}{RT}\right)\left(\dfrac{\ln \gamma_q^\infty + \ln \phi_q - M_q}{Z_p - 1}\right) + b_p\left(1 + \dfrac{\ln \gamma_q^\infty}{c} - \dfrac{a_q}{b_q RT}\right)}{b_p - \dfrac{a_p}{RT} + b_q - \dfrac{a_q}{RT}} \tag{13.56}$$

where M_q is given by an equation analogous to Eq. (13.53) but with subscripts reversed. All values in Eq. (13.56) are for the liquid phase at $P = P_p^{\text{sat}}$.

---

[5]Dividing Eq. (10.47) by Eq. (10.32) and comparing the result with Eq. (10.89) gives $\gamma_i = \hat{\phi}_i/\phi_i$.

One advantage of this procedure is that k_p and k_q are found directly from the pure-species parameters a_p, a_q, b_p, and b_q. In addition, the required values of $\ln \gamma_p^\infty$ and $\ln \gamma_q^\infty$ can be found from experimental data for the pq binary system, independent of the correlating expression used for G^E.

A second advantage is that the procedure, applied for infinite dilution of each species, yields two values of k_{pq} from which a composition-dependent function can be generated. A simple linear relation proves satisfactory:

$$k_{pq} = k_p x_q + k_q x_p \tag{13.57}$$

The two values k_p and k_q are usually not very different, and k_{pq} is not strongly composition dependent. Nevertheless, the quadratic dependence of $b - (a/RT)$ on composition indicated by Eq. (13.42) is not exactly preserved. Since this quantity is not a *true* second virial coefficient, only a value predicted by a cubic equation of state, a strict quadratic dependence is not required. Moreover, the composition-dependent k_{pq} leads to better results than does use of a constant value.

The equation-specific constants for the SRK and PR equations are given by the following table:

	SRK Equation	PR Equation
ε	0	-0.414214
σ	1	2.414214
Ω_a	0.42748	0.457235
Ω_b	0.08664	0.077796
c	0.69315	0.62323

Given the means to calculate values of $\hat{\phi}_i$, computation schemes are readily devised for solution of VLE problems. For this purpose Eq. (13.28) is often rewritten as

$$y_i = K_i x_i \tag{13.58}$$

where K_i, the K-value, is given by

$$K_i = \frac{\hat{\phi}_i^l}{\hat{\phi}_i^v} \tag{13.59}$$

Since $\sum_i y_i = 1$, we can write as a result of Eq. (13.58) that

$$\sum_i K_i x_i = 1 \tag{13.60}$$

Thus for bubblepoint calculations, where the x_i are known, the problem is to find the set of K-values that satisfies Eq. (13.60).

Alternatively, Eq. (13.58) may be written $x_i = y_i/K_i$. Since $\sum_i x_i = 1$, it follows that

$$\sum_i \frac{y_i}{K_i} = 1 \tag{13.61}$$

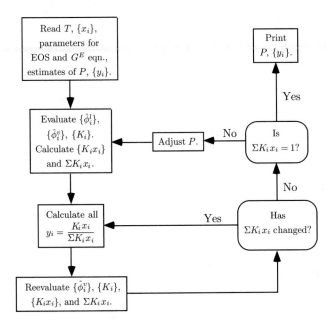

Figure 13.2: Block diagram for BUBL P calculation.

Thus for dewpoint calculations, where the y_i are known, the problem is to find the set of K-values that satisfies Eq. (13.61).

Considered here is the BUBL P calculation, for which a block diagram of a computer program is shown by Fig. 13.2. A choice must be made of an equation of state. We consider here only the Soave/Redlich/Kwong and Peng/Robinson equations, as represented by Eq. (13.32) and the information following it. These two equations usually give comparable results. A choice must also be made of a 2-parameter correlating expression to represent the liquid-phase composition dependence of G^E for each pq binary. The Wilson, NRTL, and UNIQUAC equations (Sec. 11.2) are of general applicability; for binary systems the Margules and van Laar equations may also be used. The equation selected depends on evidence of its suitability to the particular system treated. Reasonable estimates of the parameters in the equation must also be known at the temperature of interest. These parameters are directly related to infinite-dilution values of the activity coefficients for each pq binary.

As indicated in Fig. 13.2, the input information includes the known values of T and $\{x_i\}$ and parameters for the equation of state and the G^E-expression. In addition, we provide estimates of P and $\{y_i\}$, the quantities to be evaluated. These require some preliminary calculations:

1. For the chosen equation of state (with appropriate values of Ω_a, Ω_b, ε, σ, and c), for each species find values of b_i and preliminary values of a_i from Eqs. (13.33) through (13.36).

2. If the vapor pressure P_i^{sat} for species i at temperature T is known, determine a new value for a_i by Eqs. (13.50) and (13.32).

3. Evaluate k_p and k_q by Eqs. (13.55) and (13.56) for each pq binary.

4. Although pressure P is to be determined, an estimate is required to permit *any* VLE calculations. A reasonable initial value is the sum of the pure-species vapor pressures, each weighted by its known liquid-phase mole fraction.

5. The vapor-phase composition is also to be determined, but it must be estimated in order to initiate calculations. Assuming both the liquid and vapor phases to be ideal solutions, Eqs. (10.85) and (13.28) combine to give

$$y_i = x_i \frac{\phi_i^l}{\phi_i^v}$$

Evaluation of ϕ_i^l and ϕ_i^v for the pure species by Eq. (13.41) then provides estimates for y_i. Since these are not constrained to sum to unity, they should be normalized to yield an initial vapor-phase composition.

The iterative process indicated in Fig. 13.2 can now be started. The calculation of $\{\hat{\phi}_i^l\}$ by Eq. (13.38) requires prior evaluation at the known *liquid*-phase composition of D by Eq. (13.45), of b and a by Eqs. (13.46) and (13.47), of $\{\bar{b}_i\}$ and $\{\bar{a}_i\}$ by Eqs. (13.48) and (13.49). These calculations are repeated for $\{\hat{\phi}_i^v\}$ by application of the same equations at the *vapor*-phase composition. In these calculations the mixture volume V is determined from the equation of state, Eq. (13.37), applied to the appropriate phase at a given composition, temperature T, and pressure P. Values for $\{K_i\}$ now come from Eq. (13.59). These allow calculation of $\{K_i x_i\}$; according to Eq. (13.60) this set should be identical to $\{y_i\}$. However, the constraint $\sum_i y_i = 1$ has not yet been imposed, and it is likely that $\sum_i K_i x_i \neq 1$. We therefore normalize the values of y_i:

$$y_i = \frac{K_i x_i}{\sum\limits_i K_i x_i}$$

and this insures that the set of y_i's used in subsequent calculations does sum to unity.

This new set of y_i's is used to reevaluate $\{\hat{\phi}_i^v\}$, $\{K_i\}$, $\{K_i x_i\}$, and hence $\sum_i K_i x_i$. If the value of $\sum_i K_i x_i$ has changed, we again calculate the y_i and repeat the sequence of calculations. Iteration leads to a stable value of $\sum_i K_i x_i$, and we then ask whether $\sum_i K_i x_i$ is unity. If not, then the value of P is adjusted according to some rational scheme. When $\sum_i K_i x_i > 1$, P is too low; when $\sum_i K_i x_i < 1$, P is too high. The entire iterative procedure is then repeated with a new pressure P. The last calculated values of y_i are used as the initial estimate of $\{y_i\}$.

A vast store of liquid-phase excess-property data for binary systems at temperatures near 30°C and somewhat higher is available in the literature. Effective use of these data to extend G^E correlations to higher temperatures is critical to the procedure considered here. The key relations are Eq. (10.93),

$$d\left(\frac{G^E}{RT}\right) = -\frac{H^E}{RT^2}dT \qquad (\text{const } P, x)$$

and the excess-property analog of Eq. (2.24),

$$dH^E = C_P^E \, dT \qquad (\text{const } P, x)$$

Integration of the first of these equations from T_0 to T gives

$$\frac{G^E}{RT} = \left(\frac{G^E}{RT}\right)_{T_0} - \int_{T_0}^{T} \frac{H^E}{RT^2}dT \tag{13.62}$$

Similarly, the second equation may be integrated from T_1 to T:

$$H^E = H_1^E + \int_{T_1}^{T} C_P^E \, dT \tag{13.63}$$

In addition, we may write

$$dC_P^E = \left(\frac{\partial C_P^E}{\partial T}\right)_{P,x} dT$$

Integration from T_2 to T yields

$$C_P^E = C_{P_2}^E + \int_{T_2}^{T} \left(\frac{\partial C_P^E}{\partial T}\right)_{P,x} dT$$

Combining this equation with Eqs. (13.62) and (13.63) leads to

$$\frac{G^E}{RT} = \left(\frac{G^E}{RT}\right)_{T_0} - \left(\frac{H^E}{RT}\right)_{T_1}\left(\frac{T}{T_0} - 1\right)\frac{T_1}{T}$$

$$- \frac{C_{P_2}^E}{R}\left[\ln\frac{T}{T_0} - \left(\frac{T}{T_0} - 1\right)\frac{T_1}{T}\right] - I \tag{13.64}$$

where

$$I \equiv \int_{T_0}^{T} \frac{1}{RT^2} \int_{T_1}^{T} \int_{T_2}^{T} \left(\frac{\partial C_P^E}{\partial T}\right)_{P,x} dT \, dT \, dT$$

This general equation makes use of excess Gibbs-energy data at temperature T_0, excess enthalpy (heat-of-mixing) data at T_1, and excess heat-capacity data at T_2.

Evaluation of the integral I requires information with respect to the temperature dependence of C_P^E. Because of the relative paucity of excess-heat-capacity data, the usual assumption is that this property is constant, independent of T.

In this event, integral I is zero, and the closer T_0 and T_1 are to T, the less the influence of this assumption. When no information is available with respect to C_P^E, and excess enthalpy data are available at only a single temperature, the excess heat capacity must be assumed zero. In this case only the first two terms on the right-hand side of Eq. (13.64) are retained, and it more rapidly becomes imprecise as T increases.

Our primary interest in Eq. (13.64) is its application to binary systems at infinite dilution of one of the constituent species. For this purpose, we divide Eq. (13.64) by the product $x_1 x_2$. For C_P^E independent of T (and thus with $I = 0$), it then becomes

$$\frac{G^E}{x_1 x_2 RT} = \left(\frac{G^E}{x_1 x_2 RT}\right)_{T_0} - \left(\frac{H^E}{x_1 x_2 RT}\right)_{T_1} \left(\frac{T}{T_0} - 1\right) \frac{T_1}{T}$$

$$- \frac{C_P^E}{x_1 x_2 R} \left[\ln \frac{T}{T_0} - \left(\frac{T}{T_0} - 1\right) \frac{T_1}{T}\right]$$

As shown in Sec. 11.1,

$$\left(\frac{G^E}{x_1 x_2 RT}\right)_{x_i = 0} \equiv \ln \gamma_i^\infty$$

The preceding equation applied at infinite dilution of species i may therefore be written

$$\ln \gamma_i^\infty = (\ln \gamma_i^\infty)_{T_0} - \left(\frac{H^E}{x_1 x_2 RT}\right)_{T_1, x_i = 0} \left(\frac{T}{T_0} - 1\right) \frac{T_1}{T}$$

$$- \left(\frac{C_P^E}{x_1 x_2 R}\right)_{x_i = 0} \left[\ln \frac{T}{T_0} - \left(\frac{T}{T_0} - 1\right) \frac{T_1}{T}\right] \qquad (13.65)$$

The ethanol(1)/water(2) binary system serves as a specific illustration, with the Peng/Robinson equation serving as the equation of state. At a base temperature T_0 of 363.15 K (90°C), the VLE data of Pemberton and Mash[6] provide infinite-dilution values of the activity coefficients:

$$(\ln \gamma_1^\infty)_{T_0} = 1.7720 \qquad \text{and} \qquad (\ln \gamma_2^\infty)_{T_0} = 0.9042$$

Correlation of the excess enthalpy data of J. A. Larkin[7] at $T_1 = 383.15$ yields the values:

$$\left(\frac{H^E}{x_1 x_2 RT}\right)_{T_1, x_1 = 0} = -0.0598 \qquad \text{and} \qquad \left(\frac{H^E}{x_1 x_2 RT}\right)_{T_1, x_2 = 0} = 0.6735$$

---

[6] R. C. Pemberton and C. J. Mash, *Int. DATA Series, Ser. B*, vol. 1, p. 66, 1978.

[7] As reported in *Heats of Mixing Data Collection*, Chemistry Data Series, vol. III, part 1, pp. 457–459, DECHEMA, Frankfurt/Main, 1984.

Correlations of the excess enthalpy for the temperature range from 50 to 110°C lead to infinite-dilution values of C_P^E/x_1x_2R, which are nearly constant and equal to

$$\left(\frac{C_P^E}{x_1x_2R}\right)_{x_1=0} = 13.8 \quad \text{and} \quad \left(\frac{C_P^E}{x_1x_2R}\right)_{x_2=0} = 7.2$$

These data allow direct application of Eq. (13.65) to estimate $\ln\gamma_1^\infty$ and $\ln\gamma_2^\infty$ for $T > 363.15$ K. The van Laar equations [Eqs. (11.13), (11.14), and (11.15)] are well suited to this system, and the parameters for this equation are given as

$$A'_{12} = \ln\gamma_1^\infty \quad \text{and} \quad A'_{21} = \ln\gamma_2^\infty$$

The available data allow prediction of VLE at 363.15 K and at the higher temperatures, 473.15 and 523.15 K, for which measured VLE data are given by Barr-David and Dodge.[8] Pure-species vapor pressures at 363.15 K are the measured values reported with the data set of Pemberton and Mash. The data of Barr-David and Dodge do not include these values, but for water at 473.15 and 523.15 K and for ethanol at 473.15 K they can be calculated from reliable correlations. At 523.15 K ethanol is supercritical, and parameter a_i is determined from an alternative generalized correlation for $\alpha(T_{ri};\omega_i)$.[9] Furthermore, extrapolation of excess-property data into the supercritical region cannot be expected to produce reasonable values. Better results are obtained when the parameters, A_{12}, A_{21}, k_1, and k_2, are assumed to remain constant at the values used for the lower temperature where both species are subcritical. Calculated parameters and root-mean-square (RMS) deviations between computed and experimental values for P and y_1 are shown in Table 13.1.

Table 13.1: VLE Results for Ethanol(1)/Water(2).

T/K	A'_{12}	A'_{21}	k_1	k_2	RMS % δP	RMS δy_1
363.15	1.7720	0.9042	0.2317	0.2936	0.38	
473.15	1.5204	0.6001	0.2158	0.1799	2.17	0.007
523.15	1.5204	0.6001	0.2158	0.1799	1.36	0.008

The small value of RMS % δP shown for $T = 363.15$ K (90°C) indicates both the suitability of the van Laar equation for G^E and the capability of the equation-of-state method to reproduce the data. The RMS results at 473.15 K (200°C) indicate the quality of predictions based only on vapor-pressure data for the pure species and on mixture data at 90°C. An extrapolation based on the same data to 523.15 K (250°C) produces comparable results. Extrapolations to still higher temperatures can be expected to become progressively less accurate. The quality of prediction for 200 and 250°C is indicated by the Pxy diagram of Fig. 13.3.

[8]F. H. Barr-David and B. F. Dodge, *J. Chem. Eng. Data*, vol. 4, pp. 107–121, 1959.

[9]R. Stryjek and J. H. Vera, *Canadian J. Chem. Eng.*, vol. 64, pp. 323–333, 1986.

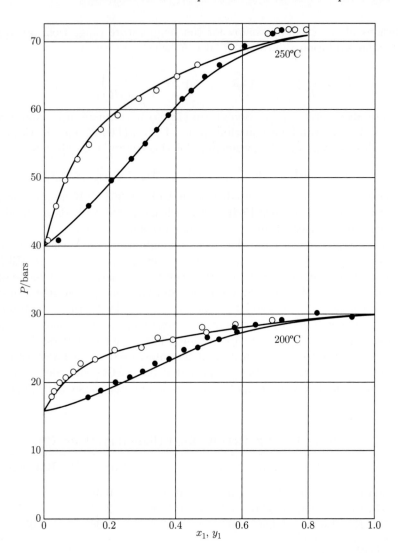

Figure 13.3: Pxy diagram for ethanol(1)/water(2). The lines represent predicted values; the points are experimental values.

VLE from K-value Correlations

Because of the complex functionality of the K-values, VLE calculations in general require iterative procedures suited only to computer solution (Sec. 12.6). However, in the case of mixtures of light hydrocarbons, in which the molecular force fields are relatively uncomplicated, we may assume as a reasonable approximation that both the liquid and the vapor phases are ideal solutions. Equation (10.85) shows that $\hat{\phi}_i^{id} = \phi_i$, and the assumption of ideal solutions therefore reduces Eq. (13.28) to

$$K_i = \frac{\phi_i^l(T, P)}{\phi_i^v(T, P)} = \frac{f_i^l(T, P)}{P\phi_i^v(T, P)}$$

The fugacity $f_i^l(T, P)$ is given by Eq. (10.41), which here becomes

$$f_i^l(T, P) = P_i^{\text{sat}}\phi_i^{\text{sat}}(T, P_i^{\text{sat}}) \exp \frac{V_i^l(P - P_i^{\text{sat}})}{RT}$$

where V_i^l is the molar volume of pure species i as a saturated liquid. Thus the K-value is given by

$$K_i = \frac{P_i^{\text{sat}}\phi_i^{\text{sat}}(T, P_i^{\text{sat}})}{P\phi_i^v(T, P)} \exp \frac{V_i^l(P - P_i^{\text{sat}})}{RT} \tag{13.66}$$

The great attraction of of Eq. (13.66) is that it contains just properties of the *pure* species and therefore expresses K-values as functions of T and P, independent of the compositions of the liquid and vapor phases. Moreover, ϕ_i^{sat} and ϕ_i^v can be evaluated from equations of state for the pure species or from generalized correlations. This allows K-values for light hydrocarbons to be calculated and correlated as functions of T and P. However, the method is limited for any species to subcritical temperatures, because the vapor-pressure curve terminates at the critical point.

In Figs. 13.4 and 13.5, we present nomographs for the K-values of light hydrocarbons as functions of T and P, prepared by DePriester[10] on the basis of earlier equation-of-state calculations. They allow for an *average* effect of composition, and are suitable for approximate calculations.

Example 13.3 For a mixture of 10 mole-% methane, 20 mole-% ethane, and 70 mole-% propane at 50($°$F), determine:

(a) The dewpoint pressure.

(b) The bubblepoint pressure.

The K-values are given by Fig. 13.4.

SOLUTION (a) When the system is at its dewpoint, only a minute amount of liquid is present, and the given mole fractions are values of y_i. Since the temperature is specified, the K-values depend on the choice of P, and by trial we find the value for which Eq. (13.61) is satisfied. Results for several values of P are given in the following table:

---

[10]C. L. DePriester, *Chem. Eng. Progr. Symp. Ser. No. 7*, vol. 49, pp. 1–43, 1953. They have been published in modified form for direct use with SI units ($°$C and kPa) by D. B. Dadyburjor, *Chem. Eng. Progr.*, vol. 74(4), pp. 85–86, April, 1978.

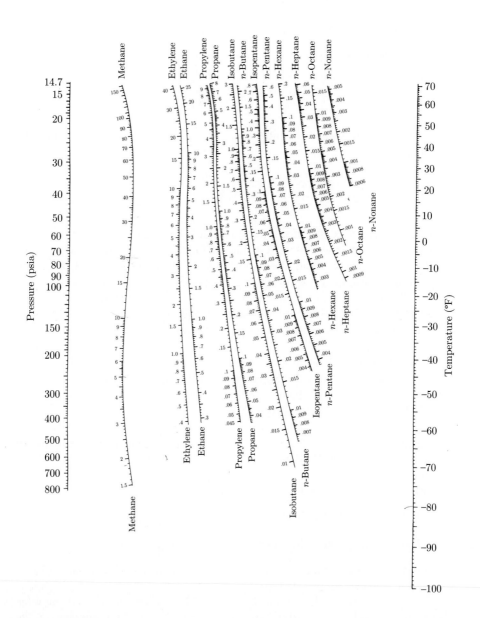

Figure 13.4: K-values for systems of light hydrocarbons. Low-temperature range. (Reproduced by permission from C. L. DePriester, *Chem. Eng. Progr. Symp. Ser. No. 7*, vol. 49, p. 41, 1953.)

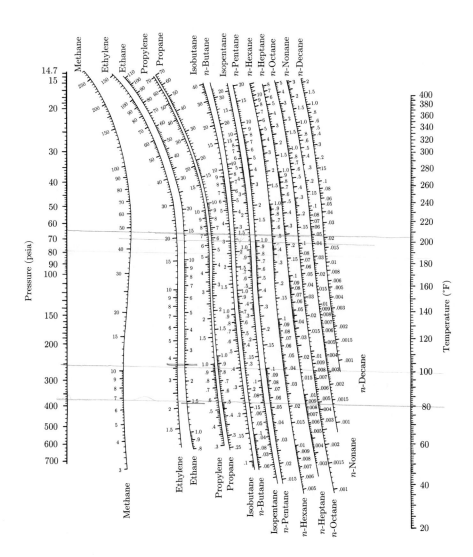

Figure 13.5: K-values for systems of light hydrocarbons. High-temperature range. (Reproduced by permission from C. L. DePriester, *Chem. Eng. Progr. Symp. Ser. No. 7*, vol. 49, p. 42, 1953.)

| | | $P = 100$(psia) | | $P = 150$(psia) | | $P = 126$(psia) | |
Species	y_i	K_i	y_i/K_i	K_i	y_i/K_i	K_i	y_i/K_i
Methane	0.10	20.0	0.005	13.2	0.008	16.0	0.006
Ethane	0.20	3.25	0.062	2.25	0.089	2.65	0.075
Propane	0.70	0.92	0.761	0.65	1.077	0.762	0.919
		$\Sigma(y_i/K_i) = 0.828$		$\Sigma(y_i/K_i) = 1.174$		$\Sigma(y_i/K_i) = 1.000$	

From the results given in the last two columns we see that Eq. (13.61) is satisfied when $P = 126$(psia). This is the dewpoint pressure, and the composition of the minute amount of liquid is given by the values of $x_i = y_i/K_i$ listed in the last column of the table.

(b) When the system is almost completely condensed, it is at its bubblepoint, and the given mole fractions become values of x_i. In this case we find by trial the value of P for which the K_i values satisfy Eq. (13.60). Results for several values of P are given in the following table:

| | | $P = 380$(psia) | | $P = 400$(psia) | | $P = 385$(psia) | |
Species	x_i	K_i	$K_i x_i$	K_i	$K_i x_i$	K_i	$K_i x_i$
Methane	0.10	5.60	0.560	5.25	0.525	5.49	0.549
Ethane	0.20	1.11	0.222	1.07	0.214	1.10	0.220
Propane	0.70	0.335	0.235	0.32	0.224	0.33	0.231
		$\Sigma K_i x_i = 1.017$		$\Sigma K_i x_i = 0.963$		$\Sigma K_i x_i = 1.000$	

We see that Eq. (13.60) is satisfied when $P = 385$(psia). This is the bubblepoint pressure. The composition of the bubble of vapor is given by $y_i = K_i x_i$, as shown in the last column.

Flash calculations can also be made for light hydrocarbons with the data of Figs. 13.4 and 13.5. The procedure here is exactly as described in Example 12.3, where Raoult's law applied. We recall that the problem is to calculate for a system of given *overall* composition $\{z_i\}$ at given T and P the fraction of the system that is vapor $\mathcal{V}$ and the compositions of the vapor phase $\{y_i\}$ and the liquid phase $\{x_i\}$. The equation to be satisfied is Eq. (12.27), here written

$$\sum_i \frac{z_i K_i}{1 + \mathcal{V}(K_i - 1)} = 1 \tag{13.67}$$

Since T and P are specified, the K_i for light hydrocarbons as given by Figs. 13.4 and 13.5 are known, and $\mathcal{V}$, the only unknown in Eq. (13.67), is found by trial.

Example 13.4   For the system described in Example 13.3, what fraction of the system is vapor when the pressure is 200(psia) and what are the compositions of the equilibrium vapor and liquid phases?

SOLUTION The given pressure lies between the dewpoint and bubblepoint pressures established for this system in Example 13.3. The system therefore consists of two phases. The procedure is to find by trial that value of $\mathcal{V}$ for which Eq. (13.67) is satisfied. We note that there is always a trivial solution for $\mathcal{V} = 1$. The results of several trials are shown in the following table. The columns headed y_i give values of the terms in the sum of Eq. (13.67), because each such term is in fact a y_i value.

Species	z_i	K_i	y_i for $\mathcal{V} = 0.35$	y_i for $\mathcal{V} = 0.25$	y_i for $\mathcal{V} = 0.273$	$x_i = y_i/K_i$ for $\mathcal{V} = 0.273$
Methane	0.10	10.0	0.241	0.308	0.289	0.029
Ethane	0.20	1.76	0.278	0.296	0.292	0.166
Propane	0.70	0.52	0.438	0.414	0.419	0.805
			$\Sigma y_i = 0.957$	$\Sigma y_i = 1.018$	$\Sigma y_i = 1.000$	$\Sigma x_i = 1.000$

Thus Eq. (13.67) is satisfied when $\mathcal{V} = 0.273$. The phase compositions are given in the last two columns of the table.

PROBLEMS

13.1. Estimate Z, H^R, and S^R at 75°C and 2 bar for an equimolar vapor mixture of propane(1) and n-pentane(2). Second virial coefficients, in cm^3 mol^{-1}:

$t/°C$	B_{11}	B_{22}	B_{12}
50	−331	−980	−558
75	−276	−809	−466
100	−235	−684	−399

13.2. Use the data of Problem 13.1 to determine $\hat{\phi}_1$ and $\hat{\phi}_2$ as functions of composition for binary vapor mixtures of propane(1) and n-pentane(2) at 75°C and 2 bar. Plot the results on a single graph. Discuss the features of this plot.

13.3. For a binary gas mixture described by Eqs. (3.31) and (10.66), prove that

$$G^E = \delta_{12} P y_1 y_2$$

$$S^E = -\frac{d\delta_{12}}{dT} P y_1 y_2$$

$$H^E = \left(\delta_{12} - T\frac{d\delta_{12}}{dT}\right) P y_1 y_2$$

$$C_P^E = -T\frac{d^2\delta_{12}}{dT^2} P y_1 y_2$$

Hint: See Problem 10.32, Part (a).

13.4. The quantity $\delta_{ij} \equiv 2B_{ij} - B_{ii} - B_{jj}$ plays a role in the thermodynamics of gas mixtures at low pressures: see, e.g., Eq. (12.4) for $\hat{\phi}_i$. This quantity can be positive or negative, depending on the chemical nature of species i and j. For what kinds of molecular pairs would one expect δ_{ij} to be

(a) positive; (b) negative; (c) essentially zero?

13.5. Equations (13.6), (13.7), and (13.8) provide expressions for G^R, H^R, and S^R for a gas described by Eq. (3.31), the two-term virial equation of state. Use these results in conjunction with the trends exhibited on Figs. 3.11 and 3.19 to rationalize that G^R, H^R, and S^R are generally *negative* for a gas at normal T and P. Similarly, rationalize that the residual heat capacity C_P^R is generally *positive*. (*Note*: No numbers are needed or wanted here. Base reasoning solely on the geometric features of B vs. T at normal reduced temperatures.)

13.6. Determine expressions for G^R, H^R, and S^R implied by the van der Waals equation of state, Eq. (3.34).

13.7. Determine expressions for G^R, H^R, and S^R implied by the Dieterici equation:

$$P = \frac{RT}{V - b} \exp\left(-\frac{a}{VRT}\right)$$

Here, parameters a and b are functions of composition only.

13.8. Determine expressions for G^R, H^R, and S^R implied by the three-term virial equation in volume, Eq. (3.33).

13.9. Find expressions for $\hat{\phi}_1$ and $\hat{\phi}_2$ for a binary gas mixture described by Eq. (3.33). Mixing rules for B and C are given by Eqs. (10.66) and (13.2).

13.10. An equimolar mixture of methane and propane is discharged from a compressor at 5,500 kPa and 90°C at the rate of 1.4 kg s^{-1}. If the velocity in the discharge line is not to exceed 30 m s^{-1}, what is the minimum diameter of the discharge line?

13.11. Estimate V, H^R, S^R, and G^R for one of the following binary vapor mixtures:

(a) Acetone(1)/1,3-butadiene(2) with mole fractions $y_1 = 0.28$ and $y_2 = 0.72$ at $t = 60°C$ and $P = 170$ kPa.

(b) Acetonitrile(1)/diethyl ether(2) with mole fractions $y_1 = 0.37$ and $y_2 = 0.63$ at $t = 50°C$ and $P = 120$ kPa.

(c) Methyl chloride(1)/ethyl chloride(2) with mole fractions $y_1 = 0.45$ and $y_2 = 0.55$ at $t = 25°C$ and $P = 100$ kPa.

(d) Nitrogen(1)/ammonia(2) with mole fractions $y_1 = 0.83$ and $y_2 = 0.17$ at $t = 20°C$ and $P = 300$ kPa.

(e) Sulfur dioxide(1)/ethylene(2) with mole fractions $y_1 = 0.32$ and $y_2 = 0.68$ at $t = 25°C$ and $P = 420$ kPa.

13.12. For the binary vapor mixture nitrogen(1)/isobutane(2), calculate V, H^R, S^R, and G^R with $y_1 = 0.35$, $y_2 = 0.65$, $t = 150°C$, and $P = 60$ bar by the following methods:

(a) Assume the mixture an ideal solution with properties of the pure species given by the Lee/Kesler correlations.

(*b*) Apply the Lee/Kesler correlations directly to the mixture.

(*c*) Use the Redlich/Kwong equation of state with Eqs. (13.16) and (13.17). In Eq. (10.72), set $k_{12} = 0.11$.

13.13. For the binary vapor mixture hydrogen sulfide(1)/ethane(2), calculate V, H^R, S^R, and G^R with $y_1 = 0.20$, $y_2 = 0.80$, $t = 140°C$, and $P = 80$ bar by the following methods:

(*a*) Assume the mixture an ideal solution with properties of the pure species given by the Lee/Kesler correlations.

(*b*) Apply the Lee/Kesler correlations directly to the mixture.

(*c*) Use the Redlich/Kwong equation of state with Eqs. (13.16) and (13.17). In Eq. (10.72), set $k_{12} = 0.06$.

13.14. Using the parameter values calculated in part (*c*) of Problem 13.12, estimate $\hat{\phi}_1$ and $\hat{\phi}_2$ for the nitrogen(1)/isobutane(2) mixture of Problem 13.12.

13.15. Using the parameter values calculated in part (*c*) of Problem 13.13, estimate $\hat{\phi}_1$ and $\hat{\phi}_2$ for the hydrogen sulfide(1)/ethane(2) mixture of Problem 13.13.

13.16. If a system exhibits VLE, at least one of the K-values must be greater than 1.0 and at least one must be less than 1.0. Offer a proof of this observation.

13.17. Flash calculations are simpler for binary systems than for the general multicomponent case, because the equilibrium compositions for a binary are independent of the overall composition. Show that, for a binary system in VLE,

$$x_1 = \frac{1 - K_2}{K_1 - K_2} \qquad y_1 = \frac{K_1(1 - K_2)}{K_1 - K_2}$$

$$\mathcal{V} = \frac{z_1(K_1 - K_2) - (1 - K_2)}{(K_1 - 1)(1 - K_2)}$$

13.18. Assuming the validity of the DePriester charts, make the following VLE calculations for the methane(1)/ethylene(2)/ethane(3) system:

(*a*) BUBL P, given $x_1 = 0.10$, $x_2 = 0.50$, and $t = -60(°F)$.

(*b*) DEW P, given $y_1 = 0.50$, $y_2 = 0.25$, and $t = -60(°F)$.

(*c*) BUBL T, given $x_1 = 0.12$, $x_2 = 0.40$, and $P = 250(psia)$.

(*d*) DEW T, given $y_1 = 0.43$, $y_2 = 0.36$, and $P = 250(psia)$.

13.19. Assuming the validity of the DePriester charts, make the following VLE calculations for the ethane(1)/propane(2)/isobutane(3)/isopentane(4) system:

(*a*) BUBL P, given $x_1 = 0.10$, $x_2 = 0.20$, $x_3 = 0.30$, and $t = 60°C$.

(*b*) DEW P, given $y_1 = 0.48$, $y_2 = 0.25$, $y_3 = 0.15$, and $t = 60°C$.

(*c*) BUBL T, given $x_1 = 0.14$, $x_2 = 0.13$, $x_3 = 0.25$, and $P = 15$ bar.

(*d*) DEW T, given $y_1 = 0.42$, $y_2 = 0.30$, $y_3 = 0.15$, and $P = 15$ bar.

13.20. The stream from a gas well consists of 50-mole-% methane, 10-mole-% ethane, 20-mole-% propane, and 20-mole-% n-butane. This stream is fed into a partial condenser maintained at a pressure of 250(psia), where its temperature is brought to 80(°F). Determine the molar fraction of the gas that condenses and the compositions of the liquid and vapor phases leaving the condenser.

13.21. An equimolar mixture of n-butane and n-hexane at pressure P is brought to a temperature of $95°C$, where it exists as a vapor/liquid mixture in equilibrium. If the mole fraction of n-hexane in the liquid phase is 0.75, what is pressure P (in bar), what is the molar fraction of the system that is liquid, and what is the composition of the vapor phase?

13.22. A mixture containing 25-mole-% n-pentane, 45-mole-% n-hexane, and 30-mole-% n-heptane is brought to a condition of $200(°F)$ and $2(atm)$. What molar fraction of the system is liquid, and what are the phase compositions?

13.23. A mixture containing 15-mole-% ethane, 35-mole-% propane, and 50-mole-% n-butane is brought to a condition of $40°C$ at pressure P. If the molar fraction of liquid in the system is 0.40, what is pressure P (in bar) and what are the compositions of the liquid and vapor phases?

13.24. A mixture containing 1-mole-% ethane, 5-mole-% propane, 44-mole-% n-butane, and 50-mole-% isobutane is brought to a condition of $70(°F)$ at pressure P. If the molar fraction of the system that is vapor is 0.2, what is pressure P, and what are the compositions of the vapor and liquid phases?

13.25. A mixture of 30-mole-% methane, 10-mole-% ethane, 30-mole-% propane, and 30-mole-% n-butane is brought to a condition of $-15°C$ at pressure P, where it exists as a vapor/liquid mixture in equilibrium. If the mole fraction of the methane in the vapor phase is 0.80, what is pressure P (in bar)?

13.26. The top tray of a distillation column and the condenser are at a pressure of $20(psia)$. The liquid on the top tray is an equimolar mixture of n-butane and n-pentane. The vapor from the top tray, assumed to be in equilibrium with the liquid, goes to the condenser where 50 mole percent of the vapor is condensed. What is the temperature on the top tray? What are the temperature and composition of the vapor leaving the condenser?

13.27. n-Butane is separated from an equimolar methane/n-butane gas mixture by compression of the gas to pressure P at $40°C$. If 40% of the feed on a mole basis is condensed, what is pressure P (in bar) and what are the compositions of the resulting vapor and liquid phases?

CHAPTER 14

TOPICS IN PHASE EQUILIBRIA

14.1 Equilibrium and Stability

Consider a closed system containing an arbitrary number of species and comprised of an arbitrary number of phases in which the temperature and pressure are uniform (though not necessarily constant). The system is assumed to be initially in a nonequilibrium state with respect to mass transfer between phases and chemical reaction. Any changes which occur in the system are necessarily irreversible, and they take the system ever closer to an equilibrium state. We may imagine that the system is placed in surroundings such that the system and surroundings are always in thermal and mechanical equilibrium. Heat exchange and expansion work are then accomplished reversibly. Under these circumstances the entropy change of the surroundings is given by

$$dS_{\text{surr}} = \frac{dQ_{\text{surr}}}{T_{\text{surr}}} = \frac{-dQ}{T}$$

The final term applies to the system, for which the heat transfer dQ has a sign opposite to that of dQ_{surr}, and the temperature of the system T replaces T_{surr}, because both must have the same value for reversible heat transfer. The second law requires that

$$dS^t + dS_{\text{surr}} \geq 0$$

where S^t is the total entropy of the system. Combination of these expressions yields, upon rearrangement:

$$dQ \leq T \, dS^t \tag{14.1}$$

Application of the first law provides

$$dU^t = dQ + dW = dQ - P \, dV^t$$

or

$$dQ = dU^t + P\,dV^t$$

Combining this equation with Eq. (14.1) gives

$$dU^t + P\,dV^t \leq T\,dS^t$$

or

$$\boxed{dU^t + P\,dV^t - T\,dS^t \leq 0} \tag{14.2}$$

Since this relation involves properties only, it must be satisfied for changes in state of *any* closed system of uniform T and P, without restriction to the conditions of mechanical and thermal reversibility assumed in its derivation. The inequality applies to every incremental change of the system between nonequilibrium states, and it dictates the direction of change that leads toward equilibrium. The equality holds for changes between equilibrium states (reversible processes). Thus Eq. (6.1) is just a special case of Eq. (14.2).

Equation (14.2) is so general that application to practical problems is difficult; restricted versions are much more useful. For example, by inspection we see that

$$(dU^t)_{S^t,V^t} \leq 0$$

where the subscripts specify properties held constant. Similarly, for processes that occur at constant U^t and V^t,

$$(dS^t)_{U^t,V^t} \geq 0$$

An *isolated* system is necessarily constrained to constant internal energy and volume, and for such a system it follows directly from the second law that the last equation is valid.

If a process is restricted to occur at constant T and P, then Eq. (14.2) may be written:

$$dU^t_{T,P} + d(PV^t)_{T,P} - d(TS^t)_{T,P} \leq 0$$

or

$$d(U^t + PV^t - TS^t)_{T,P} \leq 0$$

From the definition of the Gibbs energy [Eq. (6.3)],

$$G^t = H^t - TS^t = U^t + PV^t - TS^t$$

Therefore

$$\boxed{(dG^t)_{T,P} \leq 0} \tag{14.3}$$

Of the possible specializations of Eq. (14.2), this is the most useful, because T and P, which are easily measured, are more convenient as constants than are other pairs of variables, such as U^t and V^t.

Equation (14.3) indicates that all irrevesible processes occurring at constant T and P proceed in such a direction as to cause a decrease in the Gibbs energy of the system. Therefore:

> The equilibrium state of a closed system is that state for which the total Gibbs energy is a minimum with respect to all possible changes at the given T and P.

This criterion of equilibrium provides a general method for determination of equilibrium states. One writes an expression for G^t as a function of the numbers of moles (mole numbers) of the species in the several phases, and then finds the set of values for the mole numbers that minimizes G^t, subject to the constraints of mass conservation. This procedure can be applied to problems of phase, chemical-reaction, or combined phase and chemical-reaction equilibrim; it is most useful for complex equilibrium problems, and is illustrated for chemical-reaction equilibrium in Sec. 15.9.

At the equilibrium state differential variations can occur in the system at constant T and P without producing any change in G^t. This is the meaning of the equality in Eq. (14.3). Thus another form of this criterion of equilibrium is

$$\boxed{(dG^t)_{T,P} = 0} \tag{14.4}$$

To apply this equation, one develops an expression for dG^t as a function of the mole numbers of the species in the various phases, and sets it equal to zero. The resulting equation along with those representing the conservation of mass provide working equations for the solution of equilibrium problems. Equation (14.4) leads directly to Eq. (10.6) for phase equilibrium and it is applied to chemical-reaction equilibrium in Sec. 15.3.

Equation (14.3) provides a criterion that must be satisfied by any single phase that is *stable* with respect to the alternative that it split into two phases. It requires that the Gibbs energy of an equilibrium state be the minimum value with respect to all possible changes at the given T and P. Thus, e.g., when mixing of two liquids occurs at constant T and P, the total Gibbs energy must decrease, because the mixed state must be the one of lower Gibbs energy with respect to the unmixed state. We can write:

$$G^t \equiv nG < \sum_i n_i G_i$$

from which

$$G < \sum_i x_i G_i$$

or

$$G - \sum_i x_i G_i < 0 \qquad \text{(const } T, P)$$

According to the definition of Eq. (11.29), the quantity on the left is the Gibbs energy change of mixing. Therefore

$$\Delta G < 0$$

Thus, as noted in Sec. 11.3, the Gibbs energy change of mixing must always be negative, and a plot of ΔG vs. x_1 for a binary system must appear as shown by

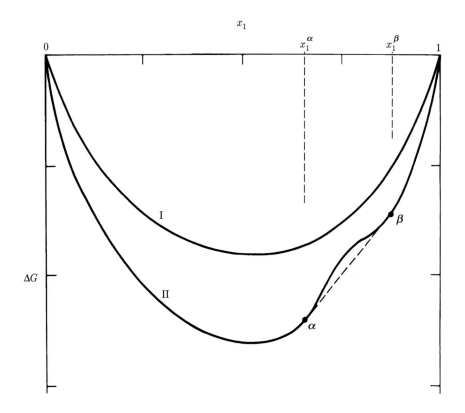

Figure 14.1: Gibbs energy change of mixing. Curve I, complete miscibility; curve II, two phases exist between a and b.

one of the curves of Fig. 14.1. With respect to curve II, however, there is a further consideration. If, when mixing occurs, a system can achieve a lower value of the Gibbs energy by forming *two* phases than by forming a single phase, then the system splits into two phases. This is in fact the situation represented between points α and β on curve II of Fig. 14.1, because the straight dashed line connecting points α and β represents the ΔG that would obtain for the range of states consisting of two phases of compositions x_1^{α} and x_1^{β} in various proportions. Thus the solid curve shown between points α and β cannot represent a stable phase with respect to phase splitting. The equilibrium states between α and β consist of two phases.

These considerations lead to the following criterion of stability for a single-phase binary system. At constant temperature and pressure, ΔG and its first and second derivatives must be continuous functions of x_1, and the second derivative must everywhere satisfy the inequality

$$\frac{d^2 \, \Delta G}{dx_1^2} > 0 \qquad (\text{const } T, P)$$

Since T is constant, we may equally well write

$$\boxed{\frac{d^2(\Delta G/RT)}{dx_1^2} > 0 \qquad (\text{const } T, P)} \qquad (14.5)$$

This requirement has a number of consequences. Equation (11.30), rearranged and written for a binary system, becomes

$$\frac{\Delta G}{RT} = x_1 \ln x_1 + x_2 \ln x_2 + \frac{G^E}{RT}$$

from which

$$\frac{d(\Delta G/RT)}{dx_1} = \ln x_1 - \ln x_2 + \frac{d(G^E/RT)}{dx_1}$$

and

$$\frac{d^2(\Delta G/RT)}{dx_1^2} = \frac{1}{x_1 x_2} + \frac{d^2(G^E/RT)}{dx_1^2}$$

Hence, equivalent to Eq. (14.5), stability requires that

$$\frac{d^2(G^E/RT)}{dx_1^2} > -\frac{1}{x_1 x_2} \qquad (\text{const } T, P) \qquad (14.6)$$

Further, we note by Eq. (11.5) that, for a binary mixture,

$$\frac{G^E}{RT} = x_1 \ln \gamma_1 + x_2 \ln \gamma_2$$

whence

$$\frac{d(G^E/RT)}{dx_1} = \ln \gamma_1 - \ln \gamma_2 + x_1 \frac{d \ln \gamma_1}{dx_1} + x_2 \frac{d \ln \gamma_2}{dx_1}$$

Invoking Eq. (11.6), the activity-coefficient form of Gibbs/Duhem equation, reduces this to

$$\frac{d(G^E/RT)}{dx_1} = \ln \gamma_1 - \ln \gamma_2$$

A second differentiation gives

$$\frac{d^2(G^E/RT)}{dx_1^2} = \frac{d \ln \gamma_1}{dx_1} - \frac{d \ln \gamma_2}{dx_1}$$

and a second application of the Gibbs/Duhem equation produces

$$\frac{d^2(G^E/RT)}{dx_1^2} = \frac{1}{x_2} \frac{d \ln \gamma_1}{dx_1}$$

This equation in combination with Eq. (14.6) yields

$$\frac{d \ln \gamma_1}{dx_1} > -\frac{1}{x_1} \qquad (\text{const } T, P)$$

which is yet another condition for stability. It is equivalent to Eq. (14.5), from which it ultimately derives. Other stability criteria follow directly, e.g.,

$$\frac{d\hat{f}_1}{dx_1} > 0 \qquad (\text{const } T, P)$$

and

$$\frac{d\mu_1}{dx_1} > 0 \qquad (\text{const } T, P)$$

The last three stability conditions can equally well be written for species 2; thus for *either* species in a binary mixture we have

$$\frac{d\ln\gamma_i}{dx_i} > -\frac{1}{x_i} \qquad (\text{const } T, P) \tag{14.7}$$

$$\frac{d\hat{f}_i}{dx_i} > 0 \qquad (\text{const } T, P) \tag{14.8}$$

$$\frac{d\mu_i}{dx_i} > 0 \qquad (\text{const } T, P) \tag{14.9}$$

Example 14.1   The stability criteria apply to a *particular* phase. However, there is nothing to preclude their application to problems in phase equilibria, where the phase of interest (e.g., a liquid mixture) is in equilibrium with another phase (e.g., a vapor mixture). Consider binary isothermal vapor/liquid equilibria at pressures low enough that the vapor phase may be considered an ideal-gas mixture. What are the implications of liquid-phase stability to the features of isothermal Pxy diagrams such as those in Fig. 12.9?

SOLUTION   Focus initially on the *liquid* phase. By Eq. (14.8) applied to species 1,

$$\frac{d\hat{f}_1}{dx_1} = \hat{f}_1 \frac{d\ln\hat{f}_1}{dx_1} > 0$$

whence, since $\hat{f}_1$ cannot be negative,

$$\frac{d\ln\hat{f}_1}{dx_1} > 0$$

Similarly, applying Eq. (14.8) to species 2 and noting that $dx_2 = -dx_1$, we find that

$$\frac{d\ln\hat{f}_2}{dx_1} < 0$$

Combination of the last two inequalities gives

$$\frac{d\ln\hat{f}_1}{dx_1} - \frac{d\ln\hat{f}_2}{dx_1} > 0 \qquad (\text{const } T, P) \tag{A}$$

which is the basis for the first part of this analysis. Since $\hat{f}_i^v = y_i P$ for an ideal-gas mixture and since $\hat{f}_i^l = \hat{f}_i^v$ for VLE, the left side of Eq. (A) may be written

$$\frac{d \ln \hat{f}_1}{dx_1} - \frac{d \ln \hat{f}_2}{dx_1} = \frac{d \ln y_1 P}{dx_1} - \frac{d \ln y_2 P}{dx_1} = \frac{d \ln y_1}{dx_1} - \frac{d \ln y_2}{dx_1}$$

$$= \frac{1}{y_1} \frac{dy_1}{dx_1} - \frac{1}{y_2} \frac{dy_2}{dx_1} = \frac{1}{y_1} \frac{dy_1}{dx_1} + \frac{1}{y_2} \frac{dy_1}{dx_1} = \frac{1}{y_1 y_2} \frac{dy_1}{dx_1}$$

Thus Eq. (A) yields

$$\frac{dy_1}{dx_1} > 0 \qquad (B)$$

which is an essential feature of binary VLE. We note that, although P is not constant for isothermal VLE, Eq. (A) is still approximately valid, because its application is to the *liquid* phase, for which properties are relatively insensitive to pressure.

The second part of this analysis draws on the fugacity form of the Gibbs/Duhem equation, Eq. (11.3), applied again to the *liquid* phase:

$$x_1 \frac{d \ln \hat{f}}{dx_1} + x_2 \frac{d \ln \hat{f}_2}{dx_1} = 0 \qquad (\text{const } T, P) \qquad (11.3)$$

Again we note that $\hat{f}_i = y_i P$ for low-pressure VLE. Hence

$$x_1 \frac{d \ln y_1 P}{dx_1} + x_2 \frac{d \ln y_2 P}{dx_1} = 0$$

from which we find by manipulations similar to those used to develop Eq. (B) that

$$\frac{1}{P} \frac{dP}{dx_1} = \frac{(y_1 - x_1)}{y_1 y_2} \frac{dy_1}{dx_1} \qquad (C)$$

Since by Eq. (B) $dy_1/dx_1 > 0$, Eq. (C) asserts that the sign of dP/dx_1 is the same as the sign of the quantity $y_1 - x_1$.

The last part of this analysis is based on simple mathematics, according to which, at constant T,

$$\frac{dP}{dy_1} = \frac{dP/dx_1}{dy_1/dx_1} \qquad (D)$$

But by Eq. (B), $dy_1/dx_1 > 0$. Thus dP/dy_1 has the same sign as dP/dx_1.

In summary, the stability requirement implies the following for VLE in binary systems at constant temperature:

$$\boxed{\frac{dy_1}{dx_1} > 0 \qquad \frac{dP}{dx_1}, \ \frac{dP}{dy_1}, \text{ and } (y_1 - x_1) \text{ have the same sign}}$$

At an azeotrope, where $y_1 = x_1$,

$$\frac{dP}{dx_1} = 0 \qquad \text{and} \qquad \frac{dP}{dy_1} = 0$$

Although derived for conditions of low pressure, these results are of general validity, as illustrated by the VLE data shown in Fig. 12.9.

14.2   Liquid/Liquid Equilibrium (LLE)

Many pairs of chemical species, were they to mix to form a single liquid phase in a certain composition range, would not satisfy the stability criterion of Eq. (14.5). Such systems therefore split in this composition range into two liquid phases of different compositions. If the phases are at thermodynamic equilibrium, the phenomenon is an example of *liquid/liquid equilibrium* (LLE), which is important for industrial operations such as solvent extraction.

The equilibrium criteria for LLE are the same as for VLE, namely, uniformity of T, P, and of the fugacity $\hat{f}_i$ for each chemical species throughout both phases. For LLE in a system of N species at uniform T and P, we denote the liquid phases by superscripts α and β, and write the equilibrium criteria as:

$$\hat{f}_i^{\alpha} = \hat{f}_i^{\beta} \qquad (i = 1, 2, \ldots, N)$$

With the introduction of activity coefficients, this becomes

$$x_i^{\alpha} \gamma_i^{\alpha} f_i^{\alpha} = x_i^{\beta} \gamma_i^{\beta} f_i^{\beta}$$

If each pure species can exist as liquid at the temperature of the system, $f_i^{\alpha} = f_i^{\beta} = f_i$, and the last equation becomes

$$\boxed{x_i^{\alpha} \gamma_i^{\alpha} = x_i^{\beta} \gamma_i^{\beta} \qquad (i = 1, 2, \ldots, N)} \qquad (14.10)$$

In Eq. (14.10), the activity coefficients γ_i^{α} and γ_i^{β} derive from the *same function* G^E/RT; thus they are functionally identical, distinguished mathematically only by the mole fractions to which they apply. For a liquid/liquid system containing N chemical species

$$\gamma_i^{\alpha} = \gamma_i(x_1^{\alpha}, x_2^{\alpha}, \ldots, x_{N-1}^{\alpha}, T, P) \qquad (14.11a)$$

$$\gamma_i^{\beta} = \gamma_i(x_1^{\beta}, x_2^{\beta}, \ldots, x_{N-1}^{\beta}, T, P) \qquad (14.11b)$$

According to Eqs. (14.10) and (14.11), we can write N equilibrium equations in $2N$ intensive variables (T, P, and $N-1$ independent mole fractions for each phase). Solution of the equilibrium equations for LLE therefore requires prior specification of numerical values for N of the intensive variables. This is in accord with the phase rule, Eq. (2.11), for which $F = 2 - \pi + N = 2 - 2 + N = N$. The same result is obtained for VLE with no special constraints on the equilibrium state.

In the general description of LLE, any number of species may be considered, and pressure may be a significant variable. We treat here a simpler (but important) special case, that of *binary* LLE either at constant pressure or at reduced temperatures low enough that the effect of pressure on the activity coefficients may be ignored. With but one independent mole fraction per phase, Eq. (14.10) gives

$$x_1^{\alpha} \gamma_1^{\alpha} = x_1^{\beta} \gamma_1^{\beta} \qquad (14.12a)$$

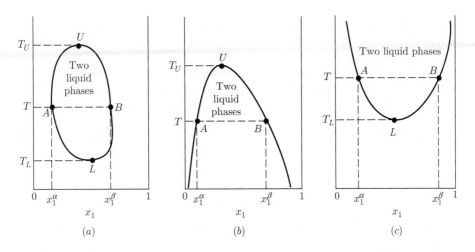

Figure 14.2: Three types of constant-pressure liquid/liquid solubility diagram.

and

$$(1 - x_1^\alpha)\gamma_2^\alpha = (1 - x_1^\beta)\gamma_2^\beta \tag{14.12b}$$

where

$$\gamma_i^\alpha = \gamma_i(x_1^\alpha, T) \tag{14.13a}$$

$$\gamma_i^\beta = \gamma_i(x_1^\beta, T) \tag{14.13b}$$

Here we have two equations and three variables (x_1^α, x_1^β, and T); fixing one of the variables allows solution of Eqs. (14.12) for the remaining two. Since $\ln \gamma_i$, rather than γ_i, is a more natural thermodynamic function, application of Eqs. (14.12) often proceeds from the rearrangements

$$\ln \frac{\gamma_1^\alpha}{\gamma_1^\beta} = \ln \frac{x_1^\beta}{x_1^\alpha} \tag{14.14a}$$

and

$$\ln \frac{\gamma_2^\alpha}{\gamma_2^\beta} = \ln \frac{1 - x_1^\beta}{1 - x_1^\alpha} \tag{14.14b}$$

For conditions of constant pressure, or when pressure effects are negligible, binary LLE is conveniently displayed on a *solubility diagram*, a plot of T vs. x_1. Figure 14.2 shows binary solubility diagrams of three types. The first diagram (Fig. 14.2a) shows curves (*binodal curves*) that define an "island." They represent the compositions of coexisting phases: curve UAL, those of the α phase (rich in species 2), and curve UBL, those of the β phase (rich in species 1). Equilibrium compositions x_1^α and x_1^β at a particular T are defined by the intersections of a horizontal *tie line* with the binodal curves. Temperature T_L is a lower *consolute temperature*, or lower *critical solution temperature* (LCST); temperature T_U is an upper *consolute temperature*, or upper *critical solution temperature* (UCST). At temperatures between T_L and T_U, LLE is possible; for $T < T_L$ and $T > T_U$, a single liquid phase is obtained for the full range of compositions. The consolute points are analogous to the liquid/gas critical point of a pure fluid; they are limiting states of two-phase equilibrium for which all properties of the two equilibrium phases are identical.

Actually, the behavior shown on Fig. 14.2a is infrequently observed; the LLE binodal curves are often interrupted by curves for yet another phase transition. When the binodal curves intersect the freezing curve, only a UCST can exist (Fig. 14.2b); when they intersect the VLE bubblepoint curve, only an LCST can exist (Fig. 14.2c); when they intersect both, no consolute point exists, and a fourth type of behavior is observed.[1]

Thus it is apparent that real systems exhibit a diversity of LLE behavior. The thermodynamic basis for calculation or correlation of LLE is an expression for G^E/RT, from which activity coefficients are derived. The suitability of a particular expression is determined by its ability to accommodate the various features illustrated by Fig. 14.2. This is a *severe* test, because, unlike their role in low-pressure VLE where they represent corrections to Raoult's law, the activity coefficients here are the *only* thermodynamic contribution to an LLE calculation.

Example 14.2 A limiting case of binary LLE is that for which the α phase is very dilute in species 1 and the β phase is very dilute in species 2. In this event, to a good approximation,

$$\gamma_1^\alpha \simeq \gamma_1^\infty \qquad \gamma_2^\alpha \simeq 1 \qquad \gamma_1^\beta \simeq 1 \qquad \gamma_2^\beta \simeq \gamma_2^\infty$$

Substitution into the equilibrium equations, Eqs. (14.12), gives

$$x_1^\alpha \gamma_1^\infty \simeq x_1^\beta \qquad 1 - x_1^\alpha \simeq (1 - x_1^\beta)\gamma_2^\infty$$

and solution for the mole fractions yields the approximate expressions

$$x_1^\alpha = \frac{\gamma_2^\infty - 1}{\gamma_1^\infty \gamma_2^\infty - 1} \qquad\qquad (A)$$

$$x_1^\beta = \frac{\gamma_1^\infty(\gamma_2^\infty - 1)}{\gamma_1^\infty \gamma_2^\infty - 1} \qquad\qquad (B)$$

Alternatively, we may solve for the infinite-dilution activity coefficients, obtaining

$$\gamma_1^\infty = \frac{x_1^\beta}{x_1^\alpha} \qquad\qquad (C)$$

---

[1] A comprehensive treatment of LLE is given by J. M. Sørensen, T. Magnussen, P. Rasmussen, and Aa. Fredenslund, *Fluid Phase Equilibria*, vol. 2, pp. 297–309, 1979; vol. 3, pp. 47–82, 1979; vol. 4, pp. 151–163, 1980. For a compilation of data see W. Arlt, M. E. A. Macedo, P. Rasmussen, and J. M. Sørensen, *Liquid-Liquid Equilibrium Data Collection*, Chemistry Data Series, vol. V, Parts 1–4, DECHEMA, Frankfurt/Main, 1979–1987.

$$\gamma_2^\infty = \frac{1 - x_1^\alpha}{1 - x_1^\beta} \tag{D}$$

Equations (A) and (B) provide order-of-magnitude estimates of equilibrium compositions from two-parameter expressions for G^E/RT, where the γ_i^∞ are usually related to the parameters in a simple way. Equations (C) and (D) serve the opposite function; they provide simple explicit expressions for the γ_i^∞ in terms of measurable equilibrium compositions. Equations (C) and (D) show that positive deviations from ideal-solution behavior promote LLE, for

$$\gamma_1^\infty \simeq \frac{1}{x_1^\alpha} > 1 \qquad \text{and} \qquad \gamma_2^\infty \simeq \frac{1}{x_2^\beta} > 1$$

The extreme example of binary LLE is that of *complete immiscibility* of the two species. When $x_1^\alpha = x_2^\beta = 0$, γ_1^α and γ_2^α are uniity, and Eqs. (14.12) therefore require that

$$\gamma_1^\alpha = \gamma_2^\beta = \infty$$

Strictly speaking, probably no two liquids are completely immiscible. However, actual solubilities may be so small (e.g., for some hydrocarbon/water systems) that the idealizations $x_1^\alpha = x_2^\beta = 0$ provide suitable approximations for practical calculations (Example 14.7).

Example 14.3 The simplest expression for G^E/RT capable of predicting LLE is the one-parameter equation

$$\frac{G^E}{RT} = Ax_1x_2 \tag{A}$$

for which

$$\ln \gamma_1 = Ax_2^2 = A(1 - x_1)^2 \qquad \text{and} \qquad \ln \gamma_2 = Ax_1^2$$

Specializing these two expressions to the α and β phases and combining them with Eqs. (14.14) gives

$$A\left[(1 - x_1^\alpha)^2 - (1 - x_1^\beta)^2\right] = \ln \frac{x_1^\beta}{x_1^\alpha} \tag{B}$$

and

$$A\left[(x_1^\alpha)^2 - (x_1^\beta)^2\right] = \ln \frac{1 - x_1^\beta}{1 - x_1^\alpha} \tag{C}$$

Given a value of parameter A, one finds equilibrium compositions x_1^α and x_1^β as the solution to Eqs. (B) and (C).

Solubility curves implied by Eq. (A) are symmetrical about $x_1 = 0.5$, for substitution of the relation

$$x_1^\beta = 1 - x_1^\alpha \tag{D}$$

into Eqs. (B) and (C) reduces them both to the *same* equation:

$$A(1 - 2x_1) = \ln \frac{1 - x_1}{x_1} \tag{E}$$

When $A > 2$, this equation has three real roots: $x_1 = 1/2$, $x_1 = r$, and $x_1 = 1 - r$, where $0 < r < 1/2$. The latter two roots are the *equilibrium* compositions (x_1^α and x_1^β), whereas the first root is a trivial solution. For $A < 2$ only the trivial solution exists; the value $A = 2$ corresponds to a consolute point, where the three roots converge to the value $1/2$. The following table shows values of A as calculated from Eq. (E) for various values of x_1^α ($= 1 - x_1^\beta$). We note particularly the sensitivity of x_1^α to small increases in A from its limiting value of 2.

A	x_1^α	A	x_1^α
2.0	0.5	2.4780	0.15
2.0067	0.45	2.7465	0.1
2.0273	0.4	3.2716	0.05
2.0635	0.35	4.6889	0.01
2.1182	0.3	5.3468	0.005
2.1972	0.25	6.9206	0.001
2.3105	0.2	7.6080	0.0005

The actual *shape* of a solubility curve is determined by the temperature dependence of G^E/RT. To illustrate this, we assume the following T dependence of parameter A in Eq. (A):

$$A = \frac{a}{T} + b - c \ln T \tag{F}$$

where a, b, and c are constants. By Eq. (10.93), this implies that the excess enthalpy H^E is linear in T, and that the excess heat capacity C_P^E is independent of T:

$$H^E = R(a + cT)x_1 x_2 \tag{G}$$

$$C_P^E = \left(\frac{\partial H^E}{\partial T}\right)_{P,x} = Rc x_1 x_2 \tag{H}$$

The excess enthalpy and the temperature dependence of A are directly related. From Eq. (F),

$$\frac{dA}{dT} = -\frac{1}{T^2}(a + cT)$$

Combination of this equation with Eq. (G) yields

$$\frac{dA}{dT} = -\frac{H^E}{x_1 x_2 RT^2}$$

Thus dA/dT is negative for an endothermic system (positive H^E) and positive for an exothermic system (negative H^E). A negative value of dA/dT at a consolute point implies a UCST, because A decreases to 2.0 as T *increases*. Conversely, a positive value implies an LCST, because A decreases to 2.0 as T *decreases*. Hence a system described by Eqs. (A) and (F) exhibits a UCST if endothermic at the consolute point and an LCST if exothermic at the consolute point. Equation (F) written for a consolute point ($A = 2$) becomes

$$T \ln T = \frac{a}{c} - \left(\frac{2 - b}{c}\right) T \tag{I}$$

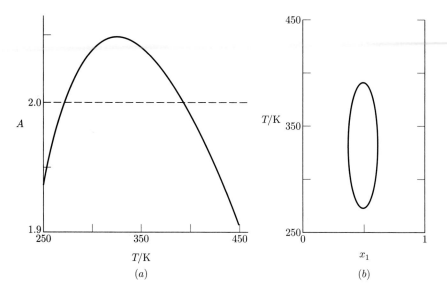

Figure 14.3: (*a*) A vs. T; (*b*) Solubility diagram for a binary system described by $G^E/RT = Ax_1x_2$ with $A = -975/T + 22.4 - 3\ln T$. ($H^E$ changes sign.)

Depending on the values of of a, b, and c, this equation has zero, one, or two temperature roots.

By way of example, we treat hypothetical binary systems described by Eqs. (A) and (F) and for which LLE obtains in the temperature range 250 to 450 K. Setting $c = 3.0$ makes the excess heat capacity positive, independent of T, for which by Eq. (H) the maximum value (at $x_1 = x_2 = 0.5$) is 6.24 J mol^{-1} K^{-1}. Consider first the case for which

$$A = \frac{-975}{T} + 22.4 - 3\ln T$$

Here, Eq. (I) has two roots, corresponding to an LCST and a UCST:

$$T_L = 272.9 \qquad \text{and} \qquad T_U = 391.2 \text{ K}$$

Values of A are plotted vs. T in Fig. 14.3a and the solubility curve [from Eq. (E)] is shown by Fig. 14.3b. This case—that of a closed solubility loop—is of the type shown by Fig. 14.2a. It requires that H^E *change sign* in the temperature interval for which LLE obtains.

As a second case, let

$$A = \frac{-540}{T} + 21.1 - 3\ln T$$

Here, Eq. (I) has only *one* root in the temperature range 250 to 450 K. It is a UCST, $T_U = 346.0$ K, because Eq. (G) gives a positive H^E at this temperature. Values of A and the corresponding solubility curve are given by Fig. 14.4.

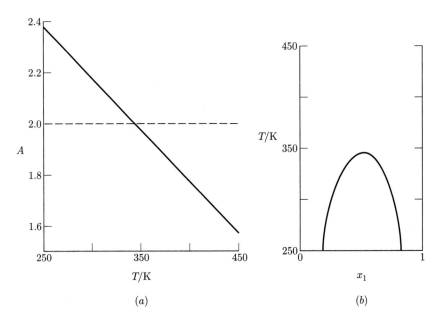

Figure 14.4: (*a*) A vs. T; (*b*) Solubility diagram for a binary system described by $G^E/RT = Ax_1x_2$ with $A = -540/T + 21.1 - 3\ln T$. ($H^E$ is positive.)

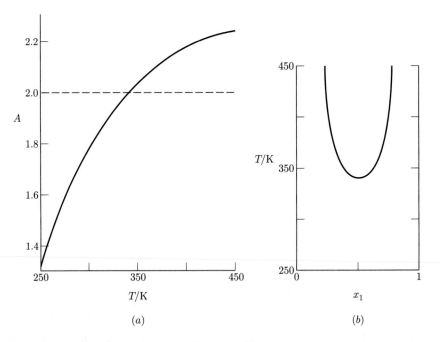

Figure 14.5: (*a*) A vs. T; (*b*) Solubility diagram for a binary system described by $G^E/RT = Ax_1x_2$ with $A = -1,500/T + 23.9 - 3\ln T$. ($H^E$ is negative.)

Finally, let

$$A = \frac{-1,500}{T} + 23.9 - 3\ln T$$

This case is similar to the second, there being only one T (339.7 K) that solves Eq. (I) for the temperature range considered. However, this is an LCST, because H^E is now negative. Values of A and the solubility curve are shown in Fig. 14.5.

Example 14.3 demonstrates in a "brute-force" way that LLE cannot be predicted by the expression $G^E/RT = Ax_1x_2$ for values of $A < 2$. If the goal is merely to determine under what conditions LLE can occur, but not to find the compositions of the coexisting phases, then one may instead invoke the stability criteria of Sec. 14.1, and determine under what conditions they are satisfied. We illustrate the procedure by example.

Example 14.4 The result that LLE is predicted by the expression $G^E/RT = Ax_1x_2$ only for $A \geq 2$ is readily obtained by stability analysis. For example, to apply inequality (14.6) we find

$$\frac{d^2(G^E/RT)}{dx_1^2} = \frac{d^2(Ax_1x_2)}{dx_1^2} = -2A$$

Stability therefore requires that

$$2A < \frac{1}{x_1x_2}$$

The minimum value of the right-hand side of this inequality is 4, obtained for $x_1 = x_2 = 1/2$; thus $A < 2$ yields stability of single-phase mixtures over the entire composition range. Conversely, if $A > 2$, then binary mixtures described by $G^E/RT = Ax_1x_2$ form two liquid phases over some part of the composition range.

Example 14.5 Some expressions for G^E/RT are incapable of representing LLE. An example is the Wilson equation, Eq. (11.16):

$$\frac{G^E}{RT} = -x_1\ln(x_1 + x_2\Lambda_{12}) - x_2\ln(x_2 + x_1\Lambda_{21}) \qquad (11.16)$$

Show that the stability criteria are satisfied for all values of Λ_{12}, Λ_{21}, and x_1.

SOLUTION We work here with inequality (14.7), written for species 1 as

$$\frac{d\ln(x_1\gamma_1)}{dx_1} > 0$$

For the Wilson equation, $\ln\gamma_1$ is given by Eq. (11.17). Addition of $\ln x_1$ to both sides of that equation yields

$$\ln(x_1\gamma_1) = -\ln\left(1 + \frac{x_2}{x_1}\Lambda_{12}\right) + x_2\left(\frac{\Lambda_{12}}{x_1 + x_2\Lambda_{12}} - \frac{\Lambda_{21}}{x_2 + x_1\Lambda_{21}}\right)$$

from which we obtain

$$\frac{d\ln(x_1\gamma_1)}{dx_1} = \frac{x_2\Lambda_{12}^2}{x_1(x_1 + x_2\Lambda_{12})^2} + \frac{\Lambda_{21}^2}{(x_2 + x_1\Lambda_{21})^2}$$

All quantities on the right-hand side of this equation are positive, and thus

$$\frac{d\ln(x_1\gamma_1)}{dx_1} > 0$$

for all x_1 and for all nonzero Λ_{12} and Λ_{21}.[2] Thus inequality (14.7) is always satisfied, and LLE cannot be represented by the Wilson equation.

14.3   Vapor/Liquid/Liquid Equilibrium (VLLE)

We noted in Sec. 14.2 that the binodal curves representing LLE can intersect the VLE bubblepoint curve. When this happens, it gives rise to the phenomenon of vapor/liquid/liquid equilibrium (VLLE). A binary system consisting of two liquid phases and one vapor phase in equilibrium has (according to the phase rule) but one degree of freedom. For a given pressure, the temperature and the compositions of all three phases are therefore fixed. On a temperature/composition diagram the points representing the states of the three phases in equilibrium fall on a horizontal line at T^*. In Fig. 14.6, points C and D represent the two liquid phases, and point E represents the vapor phase. If more of either species is added to a system whose overall composition lies between points C and D, and if the three-phase equilibrium pressure is maintained, the phase rule requires that the temperature and the compositions of the phases be unchanged. However, the relative amounts of the phases adjust themselves to reflect the change in overall composition of the system.

At temperatures above T^* in Fig. 14.6, the system may be a single liquid phase, two phases (liquid and vapor), or a single vapor phase, depending on the overall composition. In region α the system is a single liquid rich in species 2; in region β it is a single liquid, rich in species 1. In region α–V, liquid and vapor are in equilibrium. The states of the individual phases fall on lines AC and AE. In region β–V, liquid and vapor phases, described by lines BD and BE, also exist at equilibrium. Finally, in the region designated V, the system is a single vapor phase. Below the three-phase temperature T^*, the system is entirely liquid, with features described in Sec. 14.2; this is the region of LLE.

When a vapor is cooled at constant pressure, it follows a path represented on Fig. 14.6 by a vertical line. Several such lines are shown. If one starts at point k, the vapor first reaches its dewpoint at line BE and then its bubblepoint at line BD, where condensation into single liquid phase β is complete. This is the same process that takes place when the species are completely miscible. If one starts at point n, no condensation of the vapor occurs until temperature T^* is reached. Then condensation occurs entirely at this temperature, producing the two liquid phases represented by points C and D. If one starts at an intermediate point m, the process is a combination of the two just described. After the dewpoint is reached, the vapor, tracing a path along line BE, is in equilibrium with a liquid tracing

---

[2]Both Λ_{12} and Λ_{21} are positive *definite*, because $\Lambda_{12} = \Lambda_{21} = 0$ yields infinite values for γ_1^∞ and γ_2^∞.

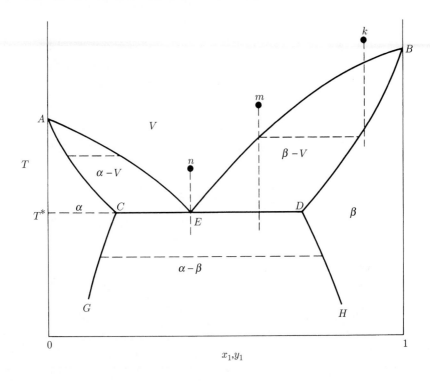

Figure 14.6: Txy diagram at constant P for a binary system exhibiting VLLE.

a path along Line BD. However, at temperature T^* the vapor phase is at point E. All remaining condensation therefore occurs at this temperature, producing the two liquids of points C and D.

Figure 14.6 is drawn for a single constant pressure; equilibrium phase compositions, and hence the locations of the lines, change with pressure, but the general nature of the diagram is the same over a range of pressures. For most systems the species become more soluble in one another as the temperature increases, as indicated by lines CG and DH of Fig. 14.6. If this diagram is drawn for successively higher pressures, the corresponding three-phase equilibrium temperatures increase, and lines CG and DH extend further and further until they meet at the liquid/liquid consolute point M, as shown by Fig. 14.7.

As the pressure increases, line CD becomes shorter and shorter (as indicated in Fig. 14.7 by lines $C'D'$ and $C''D''$, until at point M it diminishes to a differential length. For still higher pressures (P_4) the temperature is above the critical-solution temperature, and there is but a single liquid phase. The diagram then represents two-phase VLE, and it has the form of Fig. 12.10d, exhibiting a minimum-boiling azeotrope.

For an intermediate range of pressures, the vapor phase in equilibrium with the two liquid phases has a composition that does not lie between the compositions of the two liquids. This is illustrated in Fig. 14.7 by the curves for P_3, which

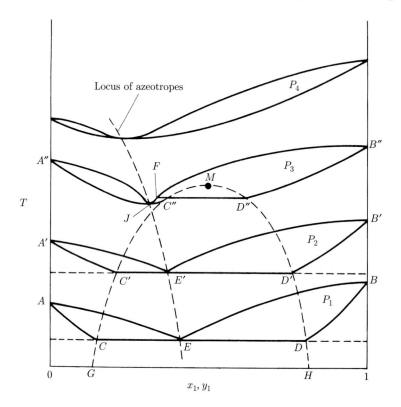

Figure 14.7: Txy diagram for several pressures.

terminate at A'' and B''. The vapor in equilibrium with the two liquids at C'' and D'' is at point F. In addition the system exhibits an azeotrope, as indicated at point J.

Not all systems behave as described in the preceding paragraphs. Sometimes the upper critical-solution temperature is never attained, because a vapor/liquid critical temperature is reached first. In other cases the liquid solubilities decrease with an increase in temperature. In this event a lower critical-solution temperature exists, unless solid phases appear first. There are also systems which exhibit both upper and lower critical-solution temperatures.[3]

Figure 14.8 is the phase diagram drawn at *constant* T corresponding to the constant-P diagram of Fig. 14.6. On it we identify the three-phase-equilibrium pressure as P^*, the three-phase-equilibrium vapor composition as y_1^*, and the compositions of the two liquid phases that contribute to the vapor/liquid/liquid equilibrium state as x_1^α and x_1^β. The phase boundaries separating the three liquid-phase

[3]For a comprehensive discussion of binary fluid-phase behavior, see J. S. Rowlinson and F. L. Swinton, *Liquids and Liquid Mixtures*, 3d ed., Butterworth Scientific, London, 1982.

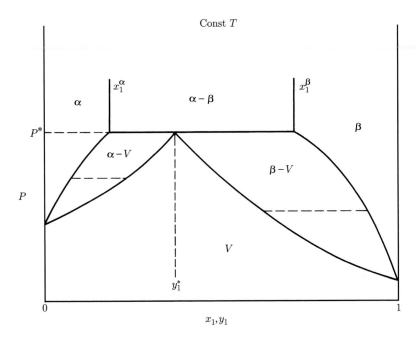

Figure 14.8: Pxy diagram at constant T for two partially miscible liquids.

regions are nearly vertical, because pressure has only a weak influence on liquid solubilities.

The compositions of the vapor and liquid phases in equilibrium for partially miscible systems are calculated in the same way as for miscible systems. In the regions where a single liquid is in equilibrium with its vapor, the general nature of Fig. 14.8 is not different in any essential way from that of Fig. 12.9d. Since limited miscibility implies highly nonideal behavior, any general assumption of liquid-phase ideality is excluded. Even a combination of Henry's law, valid for a species at infinite dilution, and Raoult's law, valid for a species as it approaches purity, is not very useful, because each approximates actual behavior for only a very small composition range. Thus G^E is large, and its composition dependence is often not adequately represented by simple equations. Nevertheless, the NRTL and UNIQUAC equations and the UNIFAC method (App. G) provide suitable correlations for activity coefficients.

Example 14.6 Careful equilibrium measurements for the diethyl ether(1)/water(2) system at 35°C have been reported.[4] Discuss the correlation and behavior of the phase-equilibrium data for this system.

SOLUTION The Pxy behavior of this system is shown by Fig. 14.9, where the very

[4]M. A. Villamañán, A. J. Allawi, and H. C. Van Ness, *J. Chem. Eng. Data*, vol. 29, pp. 431–435, 1984.

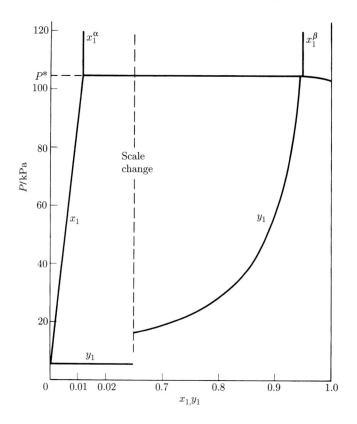

Figure 14.9: Pxy diagram at 35°C for diethyl ether(1)/water(2).

rapid rise in pressure with increasing liquid-phase ether concentration in the dilute-ether region is apparent. The three-phase pressure $P^* = 104.6$ kPa is reached at an ether mole fraction of only 0.0117. Here, y_1 also increases very rapidly to its three-phase value of $y_1^* = 0.946$. In the dilute-water region, on the other hand, rates of change are quite small, as shown to an expanded scale in Fig. 14.10.

The curves in Figs. 14.9 and 14.10 provide an excellent correlation of the VLE data. They result from BUBL P calculations carried out as indicated in Fig. 12.12. The excess Gibbs energy and activity coefficients are here expressed as functions of liquid-phase composition by a 4-parameter modified Margules equation [see Eqs. (11.7) and (11.8)]:

$$\frac{G^E}{RT} = A_{21}x_1 + A_{12}x_2 - Q$$

$$\ln\gamma_1 = x_2^2\left[A_{12} + 2(A_{21} - A_{12})x_1 - Q - x_1\frac{dQ}{dx_1}\right]$$

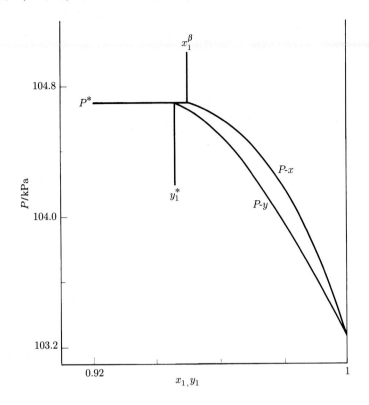

Figure 14.10: Pxy diagram for diethyl ether(1)/water(2), ether-rich region.

$$\ln \gamma_2 = x_1^2 \left[A_{21} + 2(A_{12} - A_{21})x_2 - Q + x_2 \frac{dQ}{dx_1} \right]$$

where

$$Q = \frac{\alpha_{12} x_1 \alpha_{21} x_2}{\alpha_{12} x_1 + \alpha_{21} x_2}$$

$$\frac{dQ}{dx_1} = \frac{\alpha_{12} \alpha_{21} (\alpha_{21} x_2^2 - \alpha_{12} x_1^2)}{(\alpha_{12} x_1 + \alpha_{21} x_2)^2}$$

and

$$A_{21} = 3.35629 \qquad A_{12} = 4.62424$$

$$\alpha_{12} = 3.78608 \qquad \alpha_{21} = 1.81775$$

The BUBL P calculations also require values of Φ_1 and Φ_2, which come from Eqs. (12.7) and (12.8) with virial coefficients:

$$B_{11} = -996 \qquad B_{22} = -1245 \qquad B_{12} = -567 \text{ cm}^3 \text{ mol}^{-1}$$

In addition, the vapor pressures of the pure species at $35°C$ are

$$P_1^{\text{sat}} = 103.264 \qquad P_2^{\text{sat}} = 5.633 \text{ kPa}$$

The high degree of nonideality of the liquid phase is indicated by the values of the activity coefficients of the dilute species, which range for diethyl ether from $\gamma_1 = 81.8$ at $x_1^\alpha = 0.0117$ to $\gamma_1^\infty = 101.9$ at $x_1 = 0$ and for water from $\gamma_2 = 19.8$ at $x_1^\beta = 0.9500$ to $\gamma_2^\infty = 28.7$ at $x_1 = 1$.

Thermodynamic insight into the phenomenon of low-pressure VLLE is provided by the modified Raoult's-law expression, Eq. (12.20). For temperature T and the three-phase-equilibrium pressure P^*, Eq. (12.20) has a double application:

$$x_i^\alpha \gamma_i^\alpha P_i^{\text{sat}} = y_i^* P^* \qquad \text{and} \qquad x_i^\beta \gamma_i^\beta P_i^{\text{sat}} = y_i^* P^*$$

Implicit in these equations is the LLE requirement of Eq. (14.10). Thus for a binary system we have four equations:

$$x_1^\alpha \gamma_1^\alpha P_1^{\text{sat}} = y_1^* P^* \tag{A}$$

$$x_1^\beta \gamma_1^\beta P_1^{\text{sat}} = y_1^* P^* \tag{B}$$

$$x_2^\alpha \gamma_2^\alpha P_2^{\text{sat}} = y_2^* P^* \tag{C}$$

$$x_2^\beta \gamma_2^\beta P_2^{\text{sat}} = y_2^* P^* \tag{D}$$

All of these equations are correct, but two of them are preferred over the others. Consider the expressions for $y_1^* P^*$:

$$x_1^\alpha \gamma_1^\alpha P_1^{\text{sat}} = x_1^\beta \gamma_1^\beta P_1^{\text{sat}} = y_1^* P^*$$

For the case of two species that approach complete immiscibility (Example 14.2),

$$x_1^\alpha \to 0 \qquad \gamma_1^\alpha \to \gamma_1^\infty \qquad x_1^\beta \to 1 \qquad \gamma_1^\beta \to 1$$

Thus

$$(0)(\gamma_1^\infty) P_1^{\text{sat}} = P_1^{\text{sat}} = y_1^* P^*$$

This equation implies that $\gamma_1^\infty \to \infty$; a similar derivation shows that $\gamma_2^\infty \to \infty$. Thus Eqs. (B) and (C), which include neither γ_1^α nor γ_2^β, are chosen as the more useful expressions. They may be added to give the three-phase pressure,

$$P^* = x_1^\beta \gamma_1^\beta P_1^{\text{sat}} + x_2^\alpha \gamma_2^\alpha P_2^{\text{sat}} \tag{14.15}$$

In addition, the three-phase vapor composition is given by Eq. (B):

$$y_1^* = \frac{x_1^\beta \gamma_1^\beta P_1^{\text{sat}}}{P^*} \tag{14.16}$$

For the diethyl ether(1)/water(2) system at 35°C (Example 14.6), the correlation for G^E/RT provides the values

$$\gamma_1^\beta = 1.0095 \qquad \gamma_2^\alpha = 1.0013$$

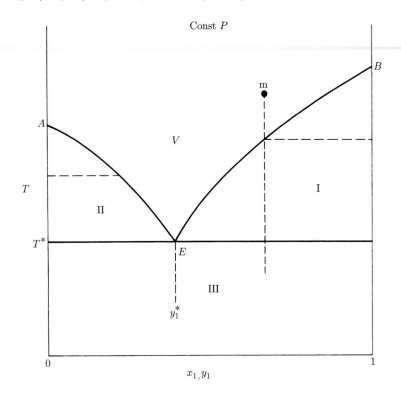

Figure 14.11: Txy diagram for a binary system of immiscible liquids.

These allow calculation of P^* and y_1^* by Eqs. (14.15) and (14.16):

$$P^* = (0.9500)(1.0095)(103.264) + (0.9883)(1.0013)(5.633) = 104.6 \text{ kPa}$$

and

$$y_1^* = \frac{(0.9500)(1.0095)(103.264)}{104.6} = 0.946$$

Although no two liquids are totally immiscible, this condition is so closely approached in some instances that the assumption of complete immiscibility does not lead to appreciable error. The phase characteristics of an immiscible system are illustrated by the temperature/composition diagram of Fig. 14.11. This diagram is a special case of Fig. 14.6 wherein phase α is pure species 2 and phase β is pure species 1. Thus lines ACG and BDH of Fig. 14.6 have in Fig. 14.11 become vertical lines at $x_1 = 0$ and $x_1 = 1$.

In region I, vapor phases with compositions represented by line BE are in equilibrium with pure liquid 1. Similarly, in region II, vapor phases whose compositions lie along line AE are in equilibrium with pure liquid 2. Liquid/liquid equilibrium exists in Region III, where the two phases are pure liquid 1 and pure

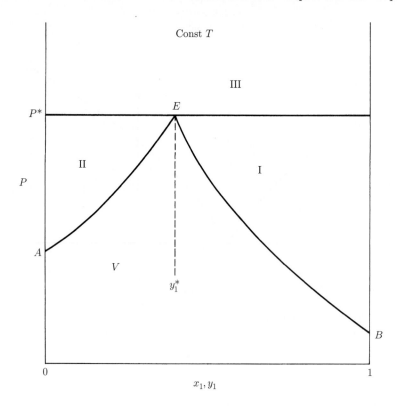

Figure 14.12: Pxy diagram for a binary system of immiscible liquids.

liquid 2. If one cools a vapor mixture starting at point m, the constant-composition path is represented by the vertical line shown in the figure. At the dewpoint, where this line crosses line BE, pure liquid 1 begins to condense. Further reduction in temperature toward T^* causes continued condensation of pure liquid 1; the vapor-phase composition progresses along line BE until it reaches point E. Here, the remaining vapor condenses at temperature T^*, producing two liquid phases, one of pure species 1 and the other of pure species 2. A similar process, carried out to the left of point E, is the same, except that pure liquid 2 condenses initially. The constant-temperature phase diagram for an immiscible system is represented by Fig. 14.12.

Numerical calculations for immiscible systems are particularly simple, because of the following equalities:

$$x_2^\alpha = 1 \qquad \gamma_2^\alpha = 1 \qquad x_1^\beta = 1 \qquad \gamma_1^\beta = 1$$

The three-phase-equilibrium pressure P^* as given by Eq. (14.15) is therefore

$$P^* = P_1^{\text{sat}} + P_2^{\text{sat}} \tag{A}$$

from which, by Eq. (14.16),

$$y_1^* = \frac{P_1^{\text{sat}}}{P_1^{\text{sat}} + P_2^{\text{sat}}} \qquad (B)$$

For region I where vapor is in equilibrium with pure liquid 1, Eq. (12.20) becomes

$$y_1(\text{I})P = P_1^{\text{sat}}$$

or

$$y_1(\text{I}) = \frac{P_1^{\text{sat}}}{P} \qquad (C)$$

Similarly, for region II where vapor is in equilibrium with pure liquid 2,

$$y_2(\text{II})P = [1 - y_1(\text{II})]P = P_2^{\text{sat}}$$

or

$$y_1(\text{II}) = 1 - \frac{P_2^{\text{sat}}}{P} \qquad (D)$$

Example 14.7 Prepare a table of temperature/composition data for the benzene(1)/water(2) system at a pressure of 101.33 kPa (1 atm) from the following vapor-pressure data:

$t/°C$	$P_1^{\text{sat}}/\text{kPa}$	$P_2^{\text{sat}}/\text{kPa}$	$P_1^{\text{sat}} + P_2^{\text{sat}}/\text{kPa}$
60	52.22	19.92	72.14
70	73.47	31.16	104.63
75	86.40	38.55	124.95
80	101.05	47.36	148.41
80.1	101.33	47.56	148.89
90	136.14	70.11	206.25
100.0	180.04	101.33	281.37

SOLUTION We assume that benzene and water are completely immiscible as liquids. Then the three-phase equilibrium temperature t^* is estimated from Eq. (A), here written

$$P(t^*) = P_1^{\text{sat}} + P_2^{\text{sat}} = 101.33 \text{ kPa}$$

The last column of the preceding table shows that t^* lies between 60 and 70°C. By interpolation, we find that $t^* = 69.0°C$, and at this temperature we find, again by interpolation, that $P_1^{\text{sat}}(t^*) = 71.31$ kPa. Thus by Eq. (B),

$$y_1^* = \frac{71.31}{101.33} = 0.704$$

For the two regions of vapor/liquid equilibrium, Eqs. (C) and (D) become

$$y_1(\text{I}) = \frac{P_1^{\text{sat}}}{P} = \frac{P_1^{\text{sat}}}{101.33}$$

and

$$y_1(\text{II}) = 1 - \frac{P_2^{\text{sat}}}{P} = 1 - \frac{P_2^{\text{sat}}}{101.33}$$

Application of these equations for a number of temperatures gives the results summarized in the table that follows.

$t/°C$	$y_1(\text{II})$	$y_1(\text{I})$
100.0	0.000	$\cdots$
90	0.308	$\cdots$
80.1	0.531	1.000
80	0.533	0.997
75	0.620	0.853
70	0.693	0.725
69.0	0.704	0.704

14.4   Solid/Liquid Equilibrium (SLE)

Phase behavior involving the solid and liquid states is the basis for separation processes (e.g., crystallization) in chemical and materials engineering. Indeed, a wide variety of binary phase behavior is observed for systems exhibiting solid/solid, solid/liquid, and solid/solid/liquid equilibria. We develop here a rigorous formulation of solid/liquid equilibrium (SLE), and present as applications analyses of two limiting classes of behavior. Comprehensive treatments can be found elsewhere.[5]

The basis for representing SLE is

$$\hat{f}_i^l = \hat{f}_i^s \qquad (\text{all } i)$$

where uniformity of T and P is understood. As with LLE, we eliminate each $\hat{f}_i$ in favor of an activity coefficient. Thus

$$x_i \gamma_i^l f_i^l = z_i \gamma_i^s f_i^s \qquad (\text{all } i)$$

where x_i and z_i are, respectively, the mole fractions of species i in the liquid and solid solutions. Equivalently,

$$x_i \gamma_i^l = z_i \gamma_i^s \psi_i \qquad (\text{all } i) \tag{14.17}$$

where

$$\psi_i \equiv f_i^s / f_i^l \tag{14.18}$$

The right-hand side of this equation, defining ψ_i as the ratio of fugacities at the T and P of the system, may be written in expanded form as

$$\frac{f_i^s(T, P)}{f_i^l(T, P)} = \frac{f_i^s(T, P)}{f_i^s(T_{m_i}, P)} \cdot \frac{f_i^s(T_{m_i}, P)}{f_i^l(T_{m_i}, P)} \cdot \frac{f_i^l(T_{m_i}, P)}{f_i^l(T, P)}$$

---

[5]See, e.g., R. T. DeHoff, *Thermodynamics in Materials Science*, Chaps. 9 and 10, McGraw-Hill, New York, 1993. A data compilation is given by H. Knapp, M. Teller, and R. Langhorst, *Solid-Liquid Equilibrium Data Collection*, Chemistry Data Series, vol. VIII, DECHEMA, Frankfurt/Main, 1987.

where T_{m_i} is the melting temperature ("freezing point") of pure species i, i.e., the temperature at which pure-species SLE obtains. Thus the second ratio on the right-hand side is *unity* because $f_i^l = f_i^s$ at the melting point of pure species i. Hence

$$\psi_i = \frac{f_i^s(T, P)}{f_i^s(T_{m_i}, P)} \cdot \frac{f_i^l(T_{m_i}, P)}{f_i^l(T, P)} \tag{14.19}$$

According to Eq. (14.19), evaluation of ψ_i requires expressions for the effect of temperature on fugacity. Here, we recall that, by Eq. (10.31),

$$\ln \phi_i = \frac{G_i^R}{RT}$$

whence, since $\phi_i = f_i/P$,

$$\ln f_i = \frac{G_i^R}{RT} + \ln P$$

Thus

$$\left(\frac{\partial \ln f_i}{\partial T}\right)_P = \left[\frac{\partial (G_i^R/RT)}{\partial T}\right]_P = -\frac{H_i^R}{RT^2}$$

where the second equality comes from Eq. (10.54). Integration of this equation for a *phase* from T_{m_i} to T gives

$$\frac{f_i(T, P)}{f_i(T_{m_i}, P)} = \exp \int_{T_{m_i}}^{T} -\frac{H_i^R}{RT^2} dT \tag{14.20}$$

Applying Eq. (14.20) separately to the solid and liquid phases, substituting the expressions into Eq. (14.19), and noting that

$$-(H_i^{R,s} - H_i^{R,l}) = -[(H_i^s - H_i^{ig}) - (H_i^l - H_i^{ig})] = H_i^l - H_i^s$$

we obtain the exact expression

$$\psi_i = \exp \int_{T_{m_i}}^{T} \frac{H_i^l - H_i^s}{RT^2} dT \tag{14.21}$$

To evaluate the integeral, we note for each phase that

$$H_i(T) = H_i(T_{m_i}) + \int_{T_{m_i}}^{T} C_{P_i} dT$$

and

$$C_{P_i}(T) = C_{P_i}(T_{m_i}) + \int_{T_{m_i}}^{T} \left(\frac{\partial C_{P_i}}{\partial T}\right)_P dT$$

Hence, for a *phase*,

$$H_i(T) = H_i(T_{m_i}) + C_{P_i}(T_{m_i})[T - T_{m_i}] + \int_{T_{m_i}}^{T} \int_{T_{m_i}}^{T} \left(\frac{\partial C_{P_i}}{\partial T}\right)_P dT\, dT \tag{14.22}$$

Applying Eq. (14.22) separately to the solid and liquid phases and performing the integration required by Eq. (14.21) yields

$$\int_{T_{m_i}}^{T} \frac{H_i^l - H_i^s}{RT^2} dT = \frac{\Delta H_i^{sl}}{RT_{m_i}} \left(\frac{T - T_{m_i}}{T} \right)$$

$$+ \frac{\Delta C_{P_i}^{sl}}{R} \left[\ln \frac{T}{T_{m_i}} - \left(\frac{T - T_{m_i}}{T} \right) \right] + I \qquad (14.23)$$

where integral I is defined by

$$I \equiv \int_{T_{m_i}}^{T} \frac{1}{RT^2} \int_{T_{m_i}}^{T} \int_{T_{m_i}}^{T} \left[\frac{\partial (C_{P_i}^l - C_{P_i}^s)}{\partial T} \right]_P dT \, dT \, dT$$

In Eq. (14.23), ΔH_i^{sl} is the enthalpy change of melting ("heat of fusion") and $\Delta C_{P_i}^{sl}$ is the heat-capacity change of melting. Both quantities are evaluated at the melting temperature T_{m_i}.

Equations (14.17), (14.21), and (14.23) provide a formal basis for solution of problems in solid/liquid equilibria. For purposes of development, pressure has been carried through as a thermodynamic variable. However, its effect is rarely important for engineering applications, and subsequently we ignore it. The full rigor of Eq. (14.23) is rarely maintained. The triple integral represented by I is a second-order contribution, and is normally neglected. The heat-capacity change of melting can be significant, but is not always available; moreover, inclusion of the term involving $\Delta C_{P_i}^{sl}$ adds little to a qualitative understanding of SLE. Hence in what follows we assume that

$$\psi_i = \exp \frac{\Delta H_i^{sl}}{RT_{m_i}} \left(\frac{T - T_{m_i}}{T} \right) \qquad (14.24)$$

With ψ_i given by Eq. (14.24), all that is required for formulating an SLE problem is a set of statements about the temperature and composition dependence of the activity coefficients γ_i^l and γ_i^s. In the general case, this requires algebraic expressions for $G^E(T, \text{composition})$ for both liquid and solid solutions. Consider two limiting special cases:

I.   Assume ideal-solution behavior for both phases, i.e., let $\gamma_i^l = 1$ and $\gamma_i^s = 1$ for all T and compositions.

II. Assume ideal-solution behavior for the liquid phase ($\gamma_i^l = 1$), and complete immiscibility for all species in the solid state (i.e., set $z_i \gamma_i^s = 1$).

These two cases, restricted to binary systems, are considered in the following.

Case I

The two equilibrium equations which follow from Eq. (14.17) are

$$x_1 = z_1 \psi_1 \tag{14.25a}$$

$$x_2 = z_2 \psi_2 \tag{14.25b}$$

where ψ_1 and ψ_2 are given by Eq. (14.24) with $i = 1$ and $i = 2$. Since $x_2 = 1 - x_1$ and $z_2 = 1 - z_1$, Eqs. (14.25) can be solved to give x_1 and z_1 as explicit functions of the ψ_i's and thus of T:

$$x_1 = \frac{\psi_1(1 - \psi_2)}{\psi_1 - \psi_2} \tag{14.26}$$

$$z_1 = \frac{1 - \psi_2}{\psi_1 - \psi_2} \tag{14.27}$$

with

$$\psi_1 = \exp \frac{\Delta H_1^{sl}}{R T_{m_1}} \left(\frac{T - T_{m_1}}{T} \right) \tag{14.28a}$$

$$\psi_2 = \exp \frac{\Delta H_2^{sl}}{R T_{m_2}} \left(\frac{T - T_{m_2}}{T} \right) \tag{14.28b}$$

Inspection of these results verifies that $x_i = z_i = 1$ for $T = T_{m_i}$. Moreover, analysis shows that both x_i and z_i vary monotonically with T. Hence systems described by Eqs. (14.25) exhibit lens-shaped SLE diagrams, as shown on Fig. 14.13, where the upper line is the freezing curve and the lower line is the melting curve. The liquid-solution region lies above the freezing curve, and the solid-solution region lies below the melting curve. Examples of systems exhibiting diagrams of this type range from nitrogen/carbon monoxide at low temperature to copper/nickel at high temperature. Comparison of this figure with Fig. (12.17) suggests that Case I-SLE behavior is analogous to Raoult's-law behavior for VLE. Comparison of the assumptions leading to Eqs. (14.25) and (12.19) confirms the analogy. As with Raoult's law, Eq. (14.25) rarely describes the behavior of actual systems. However, it is an important limiting case, and serves as a standard against which observed SLE can be compared.

Case II

The two equilibrium equations resulting from Eq. (14.17) are here

$$x_1 = \psi_1 \tag{14.29}$$

$$x_2 = \psi_2 \tag{14.30}$$

where ψ_1 and ψ_2 are given as functions solely of temperature by Eqs. (14.28). Thus x_1 and x_2 are also solely functions of temperature, and Eqs. (14.29) and (14.30) can apply simultaneously only for the particular temperature where $\psi_1 +$

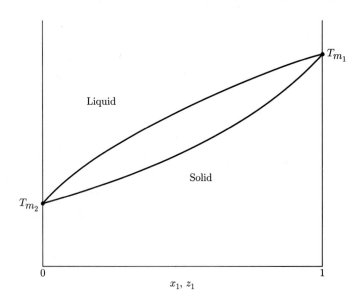

Figure 14.13: Txz diagram for Case I (ideal liquid and solid solutions).

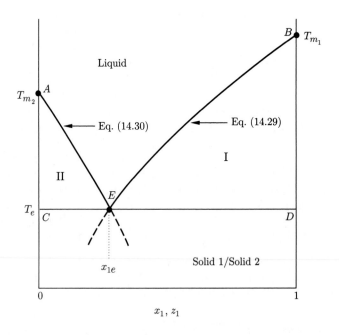

Figure 14.14: Txz diagram for Case II (ideal liquid solution; immiscible solids).

$\psi_2 = 1$ and hence $x_1 + x_2 = 1$. This is the *eutectic temperature* T_e. Thus, three distinct equilibrium situations exist: one where Eq. (14.29) alone applies, one where Eq. (14.30) alone applies, and the special case where they apply together at T_e.

1. Equation (14.29) alone applies. By this equation and Eq. (14.28a),

$$x_1 = \exp \frac{\Delta H_1^{sl}}{RT_{m_1}} \left(\frac{T - T_{m_1}}{T} \right) \tag{14.31}$$

   This equation has validity only from $T = T_{m_1}$, where $x_1 = 1$ to $T = T_e$, where $x_1 = x_{1e}$, the *eutectic composition*. (Note that $x_1 = 0$ only for $T = 0$.) Equation (14.31) therefore applies where a liquid solution is in equilibrium with pure species 1 as a solid phase. This is represented by Region I on Fig. 14.14, where liquid solutions with compositions x_1 given by line BE are in equilibrium with pure solid 1.

2. Equation (14.30) alone applies. By this equation and Eq. (14.28b), with $x_2 = 1 - x_1$, we obtain

$$x_1 = 1 - \exp \frac{\Delta H_2^{sl}}{RT_{m_2}} \left(\frac{T - T_{m_2}}{T} \right) \tag{14.32}$$

   This equation has validity only from $T = T_{m_2}$, where $x_1 = 0$ to $T = T_e$, where $x_1 = x_{1e}$, the eutectic composition. Equation (14.32) therefore applies where a liquid solution is in equilibrium with pure species 2 as a solid phase. This is represented by Region II on Fig. 14.14, where liquid solutions with compositions x_1 given by line AE are in equilibrium with pure solid 2.

3. Equations (14.29) and (14.30) apply simultaneously, and are set equal since they must both give the eutectic composition x_{1e}. The resulting expression

$$\exp \frac{\Delta H_1^{sl}}{RT_{m_1}} \left(\frac{T - T_{m_1}}{T} \right) = 1 - \exp \frac{\Delta H_2^{sl}}{RT_{m_2}} \left(\frac{T - T_{m_2}}{T} \right) \tag{14.33}$$

   is satisfied for the single temperature $T = T_e$. Substitution of T_e into either Eq. (14.31) or (14.32) yields the eutectic composition. Coordinates T_e and x_{1e} define a *eutectic state*, a special state of three-phase equilibrium, lying along line CED on Fig. 14.14, for which liquid of composition x_{1e} coexists with pure solid 1 and pure solid 2. This is a state of solid/solid/liquid equilibrium. At temperatures below T_e the two pure immiscible solids coexist.

Figure 14.14, the phase diagram for Case II, is an exact analog of Fig. 14.11 for immiscible liquids, because the assumptions upon which its generating equations are based are analogs of the corresponding VLLE assumptions.

14.5 Solid/Vapor Equilibrium (SVE)

At temperatures below its triple point, a pure solid can vaporize. Solid/vapor equilibrium for a pure species is represented on a PT diagram by the *sublimation curve* (see Fig. 3.1); here, as for VLE, the equilibrium pressure for a particular temperature is called the (solid/vapor) saturation pressure P^{sat}.

We consider in this section the equilibrium of a pure solid (species 1) with a binary vapor *mixture* containing species 1 and a second species (species 2), assumed insoluble in the solid phase. Since it is usually the major constituent of the vapor phase, species 2 is conventionally called the *solvent* species. Hence species 1 is the *solute* species, and its mole fraction y_1 in the vapor phase is its *solubility* in the solvent. Our goal is to develop a procedure for computing y_1 as a function of T and P for vapor solvents.

Only one phase-equilibrium equation can be written for this system, because species 2, by assumption, does not distribute between the two phases. The solid is *pure* species 1. Thus

$$f_1^s = \hat{f}_1^v$$

Equation (10.41) for a pure liquid is, with minor change of notation, appropriate here:

$$f_1^s = \phi_1^{\text{sat}} P_1^{\text{sat}} \exp \frac{V_1^s(P - P_1^{\text{sat}})}{RT}$$

where P_1^{sat} is the solid/vapor saturation pressure at temperature T and V_1^s is the molar volume of the solid. For the vapor phase, we write, by Eq. (10.47),

$$\hat{f}_1^v = y_1 \hat{\phi}_1 P$$

Combining the three preceding equations and solving for y_1 gives

$$y_1 = \frac{P_1^{\text{sat}}}{P} F_1 \tag{14.34}$$

where

$$F_1 \equiv \frac{\phi_1^{\text{sat}}}{\hat{\phi}_1} \exp \frac{V_1^s(P - P_1^{\text{sat}})}{RT} \tag{14.35}$$

Function F_1 reflects vapor-phase nonidealities through ϕ^{sat} and $\hat{\phi}_1$ and the effect of pressure on the fugacity of the solid through the exponential Poynting factor. For sufficiently low pressures, both effects are negligible, in which case $F_1 \approx 1$ and $y_1 \approx P_1^{\text{sat}}/P$. At moderate and high pressures, vapor-phase nonidealities become important, and for very high pressures even the Poynting factor cannot be ignored. Since F_1 is generally observed to be greater than unity, it is sometimes called an "enhancement factor," because according to Eq. (14.34) it leads to a solid solubility *greater* than would obtain in the absence of these pressure-induced effects.

Estimation of Solid Solubility at High Pressure

Solubilities at temperatures and pressures above the critical values of the solvent have important applications for supercritical separation processes. Examples are extraction of caffeine from coffee beans and separation of asphaltenes from heavy petroleum fractions. For a typical solid/vapor (SVE) problem, the solid/vapor saturation pressure P_1^{sat} is very small, and the saturated vapor is for practical purposes an ideal gas. Hence ϕ_1^{sat} for pure solute vapor at this pressure is close to unity. Moreover, except for very low values of the system pressure P, the solid solubility y_1 is small, and $\hat{\phi}_1$ can be approximated by $\hat{\phi}_1^\infty$, the vapor-phase fugacity coefficient of the solute at infinite dilution. Finally, since P_1^{sat} is very small, the pressure difference $P - P_1^{\text{sat}}$ in the Poynting factor is nearly equal to P at any pressure where this factor is important. With these usually reasonable approximations, Eq. (14.35) reduces to

$$F_1 = \frac{1}{\hat{\phi}_1^\infty} \exp \frac{PV_1^s}{RT} \tag{14.36}$$

an expression suitable for engineering applications. In this equation, P_1^{sat} and V_1^s are pure-species properties, found in a handbook or estimated from a suitable correlation. Quantity $\hat{\phi}_1^\infty$, on the other hand, must be computed from a *PVT* equation of state—one suitable for vapor mixtures at high pressures.

Cubic equations of state are usually satisfactory for this kind of calculation. Two widely used examples, introduced in Sec. 13.4, are the Soave/Redlich/Kwong (SRK) and the Peng/Robinson (PR) equations. The expression for $\hat{\phi}_i$ developed there is applicable here, but with the specialized Wong/Sandler mixing rules replaced by those of Eqs. (13.14) and (13.15), rewritten here as

$$a(T) = \sum_p \sum_q y_p y_q a_{pq}(T) \tag{14.37}$$

and

$$b = \sum_p y_p b_p \tag{14.38}$$

In Eq. (14.38), b_p is a parameter for pure species p, determined from Eq. (13.34). In Eq. (14.37), a_{pq} is estimated from pure-species parameters a_p and a_q by the empirical *combining rule*

$$a_{pq} = (1 - l_{pq})(a_p a_q)^{1/2} \tag{14.39}$$

Pure-species parameters are found from Eqs. (13.33) and (13.35) or (13.36). The binary interaction parameter l_{pq} must be found for each pq pair $(p \neq q)$ from experimental data. By convention, $l_{pq} = l_{qp}$ and $l_{pp} = l_{qq} = 0$.

From Eqs. (14.37) and (14.38) we obtain the following expressions for partial parameters $\bar{a}_i$ and $\bar{b}_i$, as defined by Eqs. (13.39) and (13.40):

$$\bar{a}_i = -a + 2\sum_p y_p a_{pi} \tag{14.40}$$

$$\bar{b}_i = b_i \tag{14.41}$$

Substitution of these expressions into Eq. (13.38) yields a prescription for $\hat{\phi}_i$ appropriate for both the SRK and PR equations of state:

$$\ln \hat{\phi}_i = \frac{b_i}{b}(Z - 1) - \ln \frac{(V - b)Z}{V} + \frac{a/bRT}{\varepsilon - \sigma}\left(\frac{2\sum_p y_p a_{pi}}{a} - \frac{b_i}{b}\right)\ln \frac{V + \sigma b}{V + \varepsilon b} \tag{14.42}$$

Note that all unsubscripted quantities are for the *mixture*, evaluated at the T, P, and composition of the mixture.

For species 1 at infinite dilution in a binary system, the "mixture" is pure species 2. In this event, Eqs. (14.42) and (14.39) yield an expression for $\hat{\phi}_1^\infty$:

$$\ln \hat{\phi}_1^\infty = \frac{b_1}{b_2}(Z_2 - 1) - \ln \frac{(V_2 - b_2)Z_2}{V_2}$$

$$+ \frac{a_2/b_2 RT}{\varepsilon - \sigma}\left[2(1 - l_{12})\left(\frac{a_1}{a_2}\right)^{1/2} - \frac{b_1}{b_2}\right]\ln \frac{V_2 + \sigma b_2}{V_2 + \varepsilon b_2} \qquad (14.43)$$

Equation (14.43) is used in conjunction with the equation of state, Eq. (13.32), which provides values of Z_2 and V_2 corresponding to a particular T and P.

As an example, we consider the calculation of the solubility of naphthalene(1) in carbon dioxide(2) at 35°C and pressures up to 300 bar. Strictly, this is not solid/*vapor* equilibrium, because the critical temperature of CO_2 is 31.1°C. However, the development of this section remains valid.

We take as a basis Eq. (14.36), with $\hat{\phi}_1^\infty$ determined from the SRK equation of state. For solid naphthalene at 35°C,

$$P_1^{\text{sat}} = 2.9 \times 10^{-4} \text{ bar} \qquad \text{and} \qquad V_1^s = 125 \text{ cm}^3 \text{ mol}^{-1}$$

Equations (14.43) and (13.32) reduce to the SRK expressions on assignment of the values $\varepsilon = 0$ and $\sigma = 1$. Evaluation of parameters a_1, a_2, b_1, and b_2 requires values for T_c, P_c, and ω, which are found in App. B. Thus Eqs. (13.33), (13.34), and (13.35) give

$$a_1 = 7.299 \times 10^7 \text{ bar cm}^6 \text{ mol}^{-2} \qquad b_1 = 133.1 \text{ cm}^3 \text{ mol}^{-1}$$

$$a_2 = 3.664 \times 10^6 \text{ bar cm}^6 \text{ mol}^{-2} \qquad b_2 = 29.68 \text{ cm}^3 \text{ mol}^{-1}$$

and Eqs. (14.43) and (13.32) become

$$\ln \hat{\phi}_1^\infty = 4.485(Z_2 - 1) - \ln\left(\frac{V_2 - 29.68}{V_2}\right)Z_2$$

$$+ [21.61 - 43.01(1 - l_{12})]\ln\left(\frac{V_2 + 29.68}{V_2}\right) \qquad (A)$$

and

$$Z_2 = \frac{V_2}{V_2 - 29.68} - \frac{143.0}{V_2 + 29.68} \qquad (B)$$

where

$$V_2 = 25{,}620\frac{Z_2}{P} \qquad (C)$$

In Eqs. (A), (B), and (C), V_2 has units of cm^3 mol^{-1} and P has units of bar. To

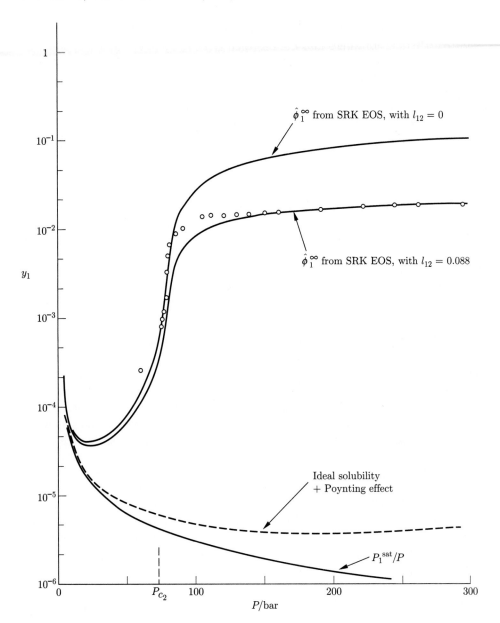

Figure 14.15: Solubility of naphthalene(1) in carbon dioxide(2) at 35°C. Circles are data. Curves are computed from Eqs. (14.34) and (14.36) under various assumptions.

find $\hat{\phi}_1^\infty$ for a given pressure P, one first solves Eqs. (B) and (C) for V_2 and Z_2. Substitution of these values into Eq. (A) then gives $\hat{\phi}_1^\infty$. For example, suppose that $P = 200$ bar. Then, from Eqs. (B) and (C), we find $V_2 = 56.71$ cm^3 mol^{-1} and $Z_2 = 0.4428$; hence, from Eq. (A), if $l_{12} = 0$, we get $\hat{\phi}_1^\infty = 4.767 \times 10^{-5}$. This small value leads by Eq. (14.36) to a very large enhancement factor F_1.

Tsekhanskaya et al.[6] report solubility data for naphthalene in carbon dioxide at 35°C and high pressures, given as circles on Fig. 14.15. The sharp increase in solubility as the pressure approaches the critical value (73.83 bar for CO_2) is typical of supercritical systems. Shown for comparison are the results of calculations based on Eqs. (14.34) and (14.36), under various assumptions. The lowest curve shows the "ideal solubility" P_1^{sat}/P, for which the enhancement factor F_1 is unity. The dashed curve incorporates the Poynting effect, which is significant at the higher pressures. The topmost curve includes the Poynting effect as well as $\hat{\phi}_1^\infty$, estimated from Eq. (14.43) with SRK constants and with $l_{12} = 0$; this purely predictive result captures the general trends of the data, but over-estimates the solubility at the higher pressures. *Correlation* of the data requires a non-zero value for the interaction parameter; with $l_{12} = 0.088$, we obtain the semi-quantitative representation shown on Fig. 14.15 as the second curve from the top.

14.6   Equilibrium Adsorption of Gases on Solids

The process by which certain porous solids bind large numbers of molecules to their surfaces is known as adsorption. Not only does it serve as a separation process, but it is also a vital part of catalytic-reaction processes. As a separation process, adsorption is used most often for removal of low-concentration impurities and pollutants from fluid streams. It is also the basis for chromatography. In surface-catalyzed reactions, the initial step is adsorption of reactant species; the final step is the reverse process, desorption of product species. Since most industrially important reactions are catalytic, adsorption plays a fundamental role in reaction engineering.

The nature of the adsorbing surface is the determining factor in adsorption. The molecular characterization of solid surfaces is not yet fully developed; however, current knowledge allows a helpful description. To be useful as an *adsorbent*, a solid must present a large surface area per unit mass (up to 1,500 m^2 per gram). This can only be achieved with porous solids such as activated carbon, silica gels, aluminas, and zeolites, which contain many cavities or pores with diameters as small as a fraction of a nanometer. Surfaces of such solids are necessarily irregular at the molecular level, and they contain *sites* of particular attraction for adsorbing molecules. If the sites are close together, the adsorbed molecules may interact with one another; if they are sufficiently dispersed, the adsorbed molecules may interact only with the sites. Depending upon the strength of the forces binding them to the sites, these *adsorbate* molecules may be mobile or fixed in position. The

[6]Y. V. Tsekhanskaya, M. B. Iomtev, and E. V. Mushkina, *Russian J. Phys. Chem.*, vol. 38, pp. 1173–1176, 1964.

relatively weak electrostatic, induction, and dispersion forces discussed in Sec. 3.8 favor mobility and result in *physical adsorption.* On the other hand, much stronger quasichemical forces can act to fix molecules to the surface, promoting *chemisorption.* Although adsorption may be classified in several ways, the usual distinction is between physical adsorption and chemisorption. Based on the strength of the binding forces, this division is observed experimentally in the magnitudes of the heat of adsorption.

In the adsorption of gases, the number of molecules attracted to a solid surface depends on conditions in the gas phase. For very low pressures, relatively few molecules are adsorbed, and only a fraction of the solid surface is covered. As the gas pressure increases at a given temperature, surface coverage increases. When all sites become occupied, the adsorbed molecules are said to form a *monolayer.* Further increase in pressure promotes *multilayer* adsorption. It is also possible for multilayer adsorption to occur on one part of a porous surface when vacant sites still remain on another part.

The complexities of solid surfaces and our inability to characterize exactly their interactions with adsorbed molecules limits our understanding of the adsorption process. It does not, however, prevent development of an exact thermodynamic description of adsorption equilibrium, applicable alike to physical adsorption and chemisorption and equally to monolayer and multilayer adsorption. The thermodynamic framework is independent of any *particular* theoretical or empirical description of material behavior. However, in application such a description is essential, and meaningful results require appropriate models of behavior.

The thermodynamic treatment of gas/adsorbate equilibrium is in many respects analogous to that of vapor/liquid equilibrium as presented in Sec. 12.4. However, the definition of a system to which the equations of thermodynamics apply presents a problem. The force field of the solid adsorbent influences properties in the adjacent gas phase, but its effect decreases rapidly with distance. Thus the properties of the gas change rapidly in the immediate neighborhood of the solid surface, but they do not change abruptly. A region of change exists which contains gradients in the properties of the gas, but the distance into the gas phase that the solid makes its influence felt cannot be precisely established.

This problem is circumvented by a construct devised by J. W. Gibbs. Imagine that the gas-phase properties extend unchanged up to the solid surface. Differences between the actual and the unchanged properties can then be attributed to a mathematical surface, treated as a two-dimensional phase with its own thermodynamic properties. This provides not only a precisely defined surface phase to account for the singularities of the interfacial region, but it also extracts them from the three-dimensional gas phase so that it too may be treated precisely. The solid, despite the influence of its force field, is presumed inert and not otherwise to participate in the gas/adsorbate equilibrium. Thus for purposes of thermodynamic analysis the adsorbate is treated as a two-dimensional phase, inherently an *open* system because it is in equilibrium with the gas phase.

The fundamental property relation for an open PVT system is given by

Eq. (10.2):

$$d(nG) = (nV)dP - (nS)dT + \sum_i \mu_i dn_i$$

An analogous equation may be written for a two-dimensional phase. The only difference is that pressure and molar volume are not in this case appropriate variables. Pressure is replaced by the *spreading pressure* Π, and the molar volume by the *molar area a*:

$$d(nG) = (na)d\Pi - (nS)dT + \sum_i \mu_i dn_i \qquad (14.44)$$

This equation is written on the basis of a unit mass, usually a gram or a kilogram, of solid adsorbent. Thus n is the *specific* amount adsorbed, i.e., the number of moles of adsorbate *per unit mass of adsorbent*. Moreover, area A is defined as the specific surface area, i.e., the area *per unit mass of adsorbent*, a quantity characteristic of a particular adsorbent. The molar area, $a \equiv A/n$, is the surface area per mole of adsorbate.

The spreading pressure is the two-dimensional analog of pressure, having units of force per unit length, akin to surface tension. It can be pictured as the force in the plane of the surface that must be exerted perpendicular to each unit length of edge to keep the surface from spreading, i.e., to keep it in mechanical equilibrium. It is not subject to direct experimental measurement, and must be calculated, significantly complicating the treatment of adsorbed-phase equilibrium.

Since the spreading pressure adds an extra variable, the number of degrees of freedom for gas/adsorbate equilibrium is given by an altered version of the phase rule. For gas/adsorbate equilibrium, $\pi = 2$; therefore

$$F = N - \pi + 3 = N - 2 + 3 = N + 1$$

Thus for adsorption of a pure species,

$$F = 1 + 1 = 2$$

and two phase-rule variables, e.g., T and P or T and n, must be fixed independently to establish an equilibrium state. Note that the inert solid phase is counted neither as a phase nor as a species.

We recall the summability relation for the Gibbs energy, which follows from Eqs. (10.8) and (10.12):

$$nG = \sum_i n_i \mu_i$$

Differentiation gives

$$d(nG) = \sum_i \mu_i dn_i + \sum_i n_i d\mu_i$$

Comparison with Eq. (14.44) shows that

$$(nS)dT - (na)d\Pi + \sum_i n_i d\mu_i = 0$$

or

$$S \, dT - a \, d\Pi + \sum_i x_i d\mu_i = 0$$

This is the Gibbs/Duhem equation for the adsorbate. Restricting it to constant temperature produces the *Gibbs adsorption isotherm*:

$$- a \, d\Pi + \sum_i x_i d\mu_i = 0 \qquad (\text{const } T) \qquad (14.45)$$

The condition of equilibrium between adsorbate and gas presumes the same temperature for the two phases and requires

$$\mu_i = \mu_i^g$$

where μ_i^g represents the gas-phase chemical potential. For a change in equilibrium conditions,

$$d\mu_i = d\mu_i^g$$

If the gas phase is an *ideal gas* (the usual assumption), then differentiation of Eq. (10.28) at constant temperature yields

$$d\mu_i^g = RT d \ln y_i P$$

Combining the last two equations with the Gibbs adsorption isotherm gives

$$- \frac{a}{RT} d\Pi + d \ln P + \sum_i x_i d \ln y_i = 0 \qquad (\text{const } T) \qquad (14.46)$$

where x_i and y_i represent adsorbate and gas-phase mole fractions respectively.

Pure-Gas Adsorption

Basic to the experimental study of pure-gas adsorption are measurements at constant temperature of n, the moles of gas adsorbed, as a function of P, the pressure in the gas phase. Each set of data represents an *adsorption isotherm* for the pure gas on a particular solid adsorbent. Available data are summarized by Valenzuela and Myers.[7] The correlation of such data requires an analytical relation between n and P, and such a relation should be consistent with Eq. (14.46).

Written for a pure chemical species, this equation becomes

$$\frac{a}{RT} d\Pi = d \ln P \qquad (\text{const } T) \qquad (14.47)$$

The compressibility-factor analog for an adsorbate is defined by the equation

$$z \equiv \frac{\Pi a}{RT} \qquad (14.48)$$

[7]D. P. Valenzuela and A. L. Myers, *Adsorption Equilibrium Data Handbook*, Prentice Hall, Englewood Cliffs, NJ, 1989.

Differentiation at constant T yields

$$dz = \frac{\Pi}{RT}da + \frac{a}{RT}d\Pi$$

Replacing the last term by Eq. (14.47) and eliminating Π/RT in favor of z/a in accord with Eq. (14.48), we rewrite this equation as

$$-d\ln P = z\frac{da}{a} - dz$$

Substituting $a = A/n$ and $da = -A\,dn/n^2$ gives

$$-d\ln P = -z\frac{dn}{n} - dz$$

Adding dn/n to both sides of this equation and rearranging,

$$d\ln \frac{n}{P} = (1-z)\frac{dn}{n} - dz$$

Integration from $P = 0$ (where $n = 0$ and $z = 1$) to $P = P$ and $n = n$ yields

$$\ln \frac{n}{P} - \ln \lim_{P \to 0} \frac{n}{P} = \int_0^n (1-z)\frac{dn}{n} + 1 - z$$

The limiting value of n/P as $n \to 0$ and $P \to 0$ must be found by extrapolation of experimental data. Applying l'Hôpital's rule to this limit gives

$$\lim_{P \to 0} \frac{n}{P} = \lim_{P \to 0} \frac{dn}{dP} \equiv k$$

Thus k is defined as the limiting slope of an isotherm as $P \to 0$, and is known as Henry's constant for adsorption. It is a function of temperature only for a given adsorbent and adsorbate, and is characteristic of the specific interaction between a particular adsorbent and a particular adsorbate.

The preceding equation may therefore be written

$$\ln \frac{n}{kP} = \int_0^n (1-z)\frac{dn}{n} + 1 - z$$

or

$$n = kP \exp\left[\int_0^n (1-z)\frac{dn}{n} + 1 - z\right] \tag{14.49}$$

This general relation between n, the moles adsorbed, and P, the gas-phase pressure, includes z, the adsorbate compressibility factor, which may be represented by an equation of state for the adsorbate. The simplest such equation is the ideal-gas analog, $z = 1$, and in this case Eq. (14.49) yields

$$n = kP$$

which is Henry's law for adsorption.

An equation of state known as the ideal-lattice-gas equation[8] has been developed specifically for an adsorbate:

$$z = -\frac{m}{n} \ln\left(1 - \frac{n}{m}\right)$$

where m is a constant. This equation is based on the presumptions that the surface of the adsorbate is a two-dimensional lattice of energetically equivalent sites, each of which may bind an adsorbate molecule, and that the bound molecules do not interact with each other. The validity of this model is therefore limited to no more than monolayer coverage. Substitution of this equation into Eq. (14.49) and integration leads to the *Langmuir isotherm*:[9]

$$n = \left(\frac{m - n}{m}\right) kP$$

Solution for n yields

$$n = \frac{mP}{\dfrac{m}{k} + P} \tag{14.50}$$

Alternatively

$$n = \frac{kbP}{b + P} \tag{14.51}$$

where $b \equiv m/k$, and k is Henry's constant. Note that when $P \to 0$, n/P properly approaches k. At the other extreme, where $P \to \infty$, n approaches m, the *saturation* value of the specific amount absorbed, representing full monolayer coverage.

Based on the same assumptions as for the ideal-lattice-gas equation, Langmuir in 1918 derived Eq. (14.50) by noting that at equilibrium the rate of adsorption of gas molecules must be equal to the rate of desorption of adsorbed molecules.[10] For monolayer adsorption, the number of sites may be divided into the fraction occupied θ and the fraction vacant $1 - \theta$. By definition,

$$\theta \equiv \frac{n}{m} \qquad \text{and} \qquad 1 - \theta = \frac{m - n}{m}$$

where m is the value of n for full monolayer coverage. For the assumed conditions, the rate of adsorption is proportional to the rate at which molecules strike the surface, which in turn is proportional to the pressure, and also proportional to the fraction $1 - \theta$ of surface sites not occupied by adsorbed molecules. The rate of

---

[8]See, e.g., T. L. Hill, *An Introduction to Statistical Mechanics*, Sec. 7-1, Addison-Wesley, Reading, MA, 1960.

[9]Irving Langmuir (1881–1957), the 2nd American to receive the Nobel Prize in chemistry, awarded for his contributions in the field of surface chemistry.

[10]I. Langmuir, *J. Am. Chem. Soc.*, vol. 40, p. 1361, 1918.

desorption is proportional to the occupied fraction θ of sites. Equating the two rates gives

$$\kappa P \frac{m-n}{m} = \kappa' \frac{n}{m}$$

where κ and κ' are proportionality (rate) constants. Solving for n and rearranging gives

$$n = \frac{\kappa m P}{\kappa P + \kappa'} = \frac{m P}{\dfrac{1}{K} + P}$$

where $K \equiv \kappa/\kappa'$, the ratio of the forward and reverse adsorption rate constants, is the conventional adsorption equilibrium constant. The second equality in this equation is equivalent to Eq. (14.50), and indicates that the adsorption equilibrium constant is equal to Henry's constant divided by m, i.e., $K = k/m$.

Since the asumptions upon which it is based are fulfilled at low surface coverage, the Langmuir isotherm is always valid as $\theta \to 0$ and as $n \to 0$. As surface coverage increases, these assumptions become increasingly unrealistic. Nevertheless, the Langmuir isotherm may provide an approximate overall fit to n vs. P data; however, it does not lead to reasonable values for m.

Substituting $a = A/n$ in Eq. (14.47), we get

$$\frac{A\,d\Pi}{RT} = n\,d\ln P$$

Integration at constant temperature from $P = 0$ (where $\Pi = 0$) to $P = P$ and $\Pi = \Pi$ yields

$$\frac{\Pi A}{RT} = \int_0^P \frac{n}{P}\,dP \tag{14.52}$$

This equation provides the *only* means for evaluation of spreading pressure. The integration may be carried out numerically or graphically with experimental data, or the data may be fit to an equation for an isotherm. For example, if the integrand n/P is given by Eq. (14.51), the Langmuir isotherm, then

$$\frac{\Pi A}{RT} = kb \ln \frac{P+b}{b} \tag{14.53}$$

an equation valid for $n \to 0$.

No equation of state is known that leads to an adsorption isotherm which in general fits experimental data over the entire range of n from zero to full monolayer coverage. Isotherms that find practical use are often 3-parameter empirical extensions of the Langmuir isotherm. An example is the Toth equation:[11]

$$n = \frac{mP}{(b+P^t)^{1/t}} \tag{14.54}$$

---

[11] Valenzuela and Myers, *op. cit.*

which reduces to the Langmuir equation for $t = 1$. When the integrand of Eq. (14.52) is expressed by the Toth equation and most other 3-parameter equations, its integration requires numerical methods. Moreover, the empirical element of such equations often introduces a singularity that makes them behave improperly in the limit as $P \to 0$. Thus for the Toth equation ($t < 1$) the second derivative d^2n/dP^2 approaches $-\infty$ in this limit, making values of Henry's constant as calculated by this equation too large. Nevertheless, the Toth equation finds frequent practical use as an adsorption isotherm. However, it is not always suitable, and a number of other adsorption isotherms are in use, as discussed by Suzuki.[12] Among them, the Freundlich equation,

$$\theta = \frac{n}{m} = \alpha P^{1/\beta} \qquad (\beta > 1) \tag{14.55}$$

is a 2-parameter (α and β) isotherm that often successfully correlates experimental data for low and intermediate values of θ.

> **Example 14.8**   Our purpose here is to illustrate numerically the concepts developed for pure-gas adsorption. Nakahara et al.[13] report data for ethylene adsorbed on a carbon molecular sieve ($A = 650$ m^2 g^{-1}) at 50°C. The data, shown as filled circles on Fig. 14.16, consist of pairs of values (n, P), where n is moles of adsorbate per kg of adsorbent and P is the equilibrium gas pressure in kPa. Trends shown by the data are typical for physical adsorption on a heterogeneous adsorbent at low-to-moderate surface coverage. The solid line on Fig. 14.16 represents a curve-fit to the data by Eq. (14.54), the Toth equation, with parameter values as reported by Valenzuela and Myers (*loc. cit.*).
>
> $$m = 4.7087$$
> $$b = 2.1941$$
> $$t = 0.3984$$
>
> These imply an apparent value of Henry's constant:
>
> $$k(\text{Toth}) = \lim_{P \to 0} \frac{n}{P} = \frac{m}{b^{1/t}} = 0.6551 \text{ mol kg}^{-1} \text{ kPa}^{-1}$$
>
> Although the overall quality of the fit is excellent, the value of Henry's constant is too large, as we will show.
>
> Extraction of Henry's constant from an adsorption isotherm is facilitated when n/P (rather than n) is considered the dependent variable and n (rather than P) the independent variable. The data plotted in this form are shown by Fig. 14.17. On this plot, Henry's constant is the extrapolated intercept:
>
> $$k = \lim_{P \to 0} \frac{n}{P} = \lim_{n \to 0} \frac{n}{P}$$

[12]M. Suzuki, *Adsorption Engineering*, pp. 35–51, Elsevier, Amsterdam, 1990.

[13]T. Nakahara, M. Hirata, and H. Mori, *J. Chem. Eng. Data*, vol. 27, pp. 317–320, 1982.

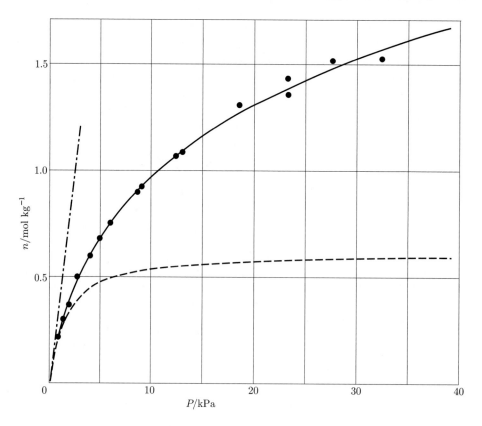

Figure 14.16: Adsorption isotherm for ethylene on a carbon molecular sieve at $50°$C. Legend: ● experimental data; ——— Toth equation; — — — Langmuir equation for $n \to 0$; — • — • — Henry's law.

where the second equality follows from the first because $n \to 0$ as $P \to 0$. Evaluation of the intercept (and hence of k) is done in this case by fitting all of the n/P data by a cubic polynomial in n:

$$\frac{n}{P} = C_0 + C_1 n + C_2 n^2 + C_3 n^3$$

The evaluated parameters are

$$
\begin{aligned}
C_0 &= 0.4016 \\
C_1 &= -0.6471 \\
C_2 &= 0.4567 \\
C_3 &= -0.1200
\end{aligned}
$$

whence

$$k = C_0 = 0.4016 \text{ mol kg}^{-1} \text{ kPa}^{-1}$$

Representation of n/P by the cubic polynomial appears as the solid curve on Fig. 14.17, and the extrapolated intercept ($C_0 = k = 0.4016$) is indicated by an

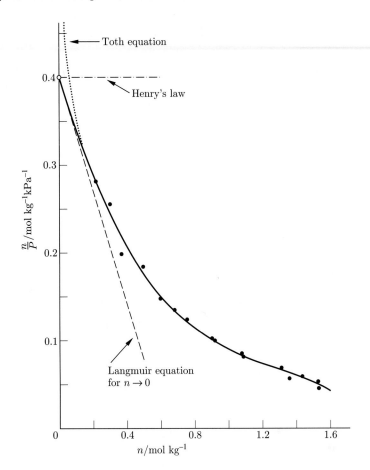

Figure 14.17: Plot of n/P vs. n for ethylene on a carbon molecular sieve at $50°C$. Legend: ● experimental data; ——— cubic polynomial fit of n/P vs. n; — — — Langmuir equation for $n \to 0$; — • — • — Henry's law; ● ● ● ● ● Toth equation for small n.

open circle. For comparison, the dotted line is the low-n portion of the n/P curve given by the Toth equation. Here it is apparent that the extrapolated intercept k(Toth), off-scale on this figure, is unreasonably high. The Toth equation cannot provide an accurate representation of adsorption behavior at very low values of n or P.

The Langmuir equation, on the other hand, is always suitable for sufficiently small n or P. Rearrangement of Eq. (14.51) gives

$$\frac{n}{P} = k - \frac{1}{b}n$$

which shows that the Langmuir equation implies a linear variation of n/P with n. Hence the limiting tangent to the "true" isotherm on a plot of n/P vs. n represents the Langmuir approximation to the isotherm for small n, and is shown by the dashed lines on Figs. 14.16 and 14.17. It is given by the equation

$$\frac{n}{P} = 0.4016 - 0.6471n$$

or, equivalently, by

$$n = \frac{0.6206P}{1.5454 + P}$$

Figures 14.16 and 14.17 show that Henry's law (represented by the dot-dash lines) and the limiting form of the Langmuir equation provide, respectively, in this example upper and lower bounds for the actual isotherm. The Langmuir isotherm when fit to *all* the experimental data yields a curve in Fig. 14.16 that fits the data reasonably well, but not so well as the 3-parameter Toth expression.

Neither the spreading pressure nor the adsorbate equation of state is required for an empirical correlation of single-species adsorption data. However, a set of (n, P) data *implies* an equation of state for the adsorbed phase, and hence a relationship between the spreading pressure Π and the moles adsorbed. By Eq. (14.52),

$$\Pi = \frac{RT}{A} \int_0^P \frac{n}{P}\, dP = \frac{RT}{A} \int_0^n \frac{n}{P} \frac{dP}{dn}\, dn$$

Equation (14.48) may be written

$$z = \frac{\Pi A}{nRT}$$

whence

$$z = \frac{1}{n} \int_0^P \frac{n}{P}\, dP = \frac{1}{n} \int_0^n \frac{n}{P} \frac{dP}{dn}\, dn$$

Finding numerical values for z and Π therefore depends on evaluation of the integral

$$I \equiv \int_0^P \frac{n}{P}\, dP = \int_0^n \frac{n}{P} \frac{dP}{dn}\, dn$$

Choice of the form depends on whether P or n is the independent variable. The Toth equation gives the integrand n/P as a function of P, and therefore

$$I(\text{Toth}) = \int_0^P \frac{m\, dP}{(b + P^t)^{1/t}}$$

The cubic polynomial gives n/P as a function of n; whence

$$I(\text{cubic}) = \int_0^n \left(\frac{C_0 - C_2 n^2 - 2C_3 n^3}{C_0 + C_1 n + C_2 n^2 + C_3 n^3} \right) dn$$

These two expressions permit numerical determination of $z(n)$ and $\Pi(n)$ as a result of correlations presented in this example. Thus, for $n = 1$ mol kg^{-1} and $A = 650$ m^2 g^{-1}, we find by both the Toth and cubic-polynomial equations that $z = 1.69$. From this result,

$$\Pi = \frac{nRT}{A} z = \frac{1 \text{ mol kg}^{-1} \times 83.14 \text{ cm}^3 \text{ bar mol}^{-1} \text{ K}^{-1} \times 323.15 \text{ K}}{650{,}000 \text{ m}^2 \text{ kg}^{-1}}$$

$$\times 1.69 \times 10^{-6} \text{ m}^3 \text{ cm}^{-3} \times 10^5 \text{ N m}^{-2} \text{ bar}^{-1}$$

$$= 6.99 \times 10^{-3} \text{ N m}^{-1} = 6.99 \text{ mN m}^{-1} = 6.99 \text{ dyn cm}^{-1}$$

The adsorption capacity of an adsorbent depends directly on its specific surface area A, but determination of these large values is not a trivial matter. The means is provided by the adsorption process itself. The basic idea is to measure the quantity of a gas adsorbed at full monolayer coverage and to multiply the number of molecules adsorbed by the area occupied by a single molecule. Two difficulties attend this procedure. First is the problem of detecting of the point of full monolayer coverage. Second, one finds that with different gases as adsorbates different area values result. The latter problem is circumvented by the adoption of nitrogen as a standard adsorbate. The procedure is to make measurements of the (physical) adsorption of N_2 at its normal boiling point ($-195.8°C$) for pressures up to its vapor pressure of 1(atm). The result is a curve the first part of which is like that in Fig. 14.16. When monolayer coverage is nearly complete, multilayer adsorption begins, and the curve changes direction, with n increasing ever more rapidly with pressure. Finally, as the pressure approaches 1(atm), the vapor pressure of the N_2 adsorbate, the curve becomes nearly vertical because of condensation in the pores of the adsorbent. The problem is to identify the point on the curve that represents full monolayer coverage. The usual procedure is to fit the Brunauer/Emmett/Teller (BET) equation, a 2-parameter extension of the Langmuir isotherm to multilayer adsorption, to the n vs. P data. From this, one can determine a value for m.[14] Once m is known, multiplication by Avogadro's number and by the area occupied by one adsorbed N_2 molecule (16.2 $Å^2$) yields the surface area. The method has its uncertainties, particularly for molecular sieves where the pores may contain unadsorbed molecules. Nevertheless, it is a useful and widely used tool for characterizing and comparing adsorption capacities.

Heat of Adsorption

The Clapeyron equation, derived in Sec. 6.3 for the latent heat of phase transition of pure chemical species, is also applicable to pure-gas adsorption equilibrium. Here, however, the two-phase equilibrium pressure depends not only on temperature, but on surface coverage or the amount adsorbed. Thus the analogous equation for adsorption is written

$$\left(\frac{\partial P}{\partial T}\right)_n = \frac{\Delta H^{av}}{T\Delta V^{av}} \tag{14.56}$$

where subscript n signifies that the derivative is taken at constant amount adsorbed. Superscript av denotes a property change of *desorption*, i.e., the difference between the vapor-phase and the adsorbed-phase property. The quantity $\Delta H^{av} \equiv H^v - H^a$ is defined as the *isosteric heat of adsorption*, and is usually a positive quantity.[15] The heat of adsorption is a useful indication of the strength of the forces binding

---

[14]J. M. Smith, *Chemical Kinetics*, 3d ed., sec. 8-1, McGraw-Hill, New York, 1981.

[15]Other heats of adsorption, defined differently, are also in use. However, the isosteric heat is the most common, and is the one needed for energy balances on adsorption columns.

adsorbed molecules to the surface of the adsorbent, and its magnitude can therefore often be used to distinguish between physical adsorption and chemisorption.

The dependence of heats of adsorption on surface coverage has its basis in the energetic heterogeneity of most solid surfaces. The first sites on a surface to be occupied are those which attract adsorbate molecules most strongly and with the greatest release of energy. Thus the heat of adsorption decreases with surface coverage. Once all sites are occupied and multilayer adsorption begins, the dominant forces become those between adsorbate molecules, and for subcritical species the decreasing heat of adsorption approaches the heat of vaporization.

Assumed in the derivation of the Langmuir isotherm is the energetic equivalence of all adsorption sites, implying that the heat of adsorption is independent of surface coverage. This explains in part the inability of the Langmuir isotherm to provide a close fit to most experimental data over a wide range of surface coverage. The Freundlich isotherm, Eq. (14.55), implies a logarithmic decrease in the heat of adsorption with surface coverage.

As in the development of the Clausius/Clapeyron equation (Example 6.4), if for low pressures one assumes that the gas phase is ideal and that the adsorbate is of negligible volume compared with the gas-phase volume, Eq. (14.56) becomes

$$\left(\frac{\partial \ln P}{\partial T}\right)_n = \frac{\Delta H^{av}}{RT^2} \tag{14.57}$$

Application of this equation requires the measurement of isotherms, such as the one at 50°C in Fig. 14.16, at several temperatures. Cross plotting yields sets of P vs. T relations at constant n, from which values for the partial derivative of Eq. (14.57) can be obtained. For chemisorption, ΔH^{av} values usually range from 60 to 170 kJ mol^{-1}. For physical adsorption, they are smaller. For example, measured values at very low coverage for the physical adsorption of nitrogen and n-butane on 5A zeolite are 18.0 and 43.1 kJ mol^{-1}, respectively.[16]

Mixed-Gas Adsorption

Mixed-gas adsorption is treated in a similar way to the gamma/phi formulation of VLE (Sec. 12.4). Denoting a gas-phase property by superscript g, we rewrite Eqs. (10.30) and (10.42), defining equations for fugacity, as

$$G_i^g = \Gamma_i^g(T) + RT \ln f_i^g \tag{14.58}$$

and

$$\mu_i^g = \Gamma_i^g(T) + RT \ln \hat{f}_i^g \tag{14.59}$$

We note as a result of Eqs. (10.33) and (10.48) that

$$\lim_{P \to 0} \frac{f_i^g}{P} = 1 \qquad \text{and} \qquad \lim_{P \to 0} \frac{\hat{f}_i^g}{y_i P} = 1$$

---

[16]N. Hashimoto and J. M. Smith, *Ind. Eng. Chem. Fund.*, vol. 12, p. 353, 1973.

For the adsorbate analogous equations are

$$G_i = \Gamma_i(T) + RT \ln f_i \tag{14.60}$$

and

$$\mu_i = \Gamma_i(T) + RT \ln \hat{f}_i \tag{14.61}$$

with

$$\lim_{\Pi \to 0} \frac{f_i}{\Pi} = 1 \qquad \text{and} \qquad \lim_{\Pi \to 0} \frac{\hat{f}_i}{x_i \Pi} = 1$$

The Gibbs energies as given by Eqs. (14.58) and (14.60) may be equated for pure-gas/adsorbate equilibrium:

$$\Gamma_i^g(T) + RT \ln f_i^g = \Gamma_i(T) + RT \ln f_i$$

Rearrangement gives

$$\frac{f_i}{f_i^g} = \exp\left(\frac{\Gamma_i^g(T) - \Gamma_i(T)}{RT}\right) \equiv F_i(T) \tag{14.62}$$

We evaluate $F_i(T)$ from the limiting value of f_i/f_i^g as both P and Π approach zero; thus

$$\lim_{\substack{P \to 0 \\ \Pi \to 0}} \frac{f_i}{f_i^g} = \lim_{\substack{P \to 0 \\ \Pi \to 0}} \frac{\Pi}{P} = \lim_{\substack{n_i \to 0 \\ P \to 0}} \frac{n_i}{P} \lim_{\substack{\Pi \to 0 \\ n_i \to 0}} \frac{\Pi}{n_i}$$

The first limit of the last member is Henry's constant k_i; the second limit is evaluated from Eq. (14.48), written $\Pi/n_i = z_i RT/A$; thus

$$\lim_{\substack{\Pi \to 0 \\ n_i \to 0}} \frac{\Pi}{n_i} = \frac{RT}{A}$$

In combination with Eq. (14.62) these equations give

$$F_i(T) = \frac{k_i RT}{A} \tag{14.63}$$

and

$$f_i = \frac{k_i RT}{A} f_i^g \tag{14.64}$$

Similarly, we may equate Eqs. (14.59) and (14.61):

$$\Gamma_i^g(T) + RT \ln \hat{f}_i^g = \Gamma_i(T) + RT \ln \hat{f}_i$$

from which

$$\frac{\hat{f}_i}{\hat{f}_i^g} = \exp\left(\frac{\Gamma_i^g(T) - \Gamma_i(T)}{RT}\right) \equiv F_i(T)$$

Then by Eq. (14.63),

$$\hat{f}_i = \frac{k_i RT}{A} \hat{f}_i^g \tag{14.65}$$

These equations show that equality of fugacities is not a proper criterion for gas/adsorbate equilibrium. This is also evident from the fact that the units of gas-phase fugacities are those of pressure, while the units of adsorbate fugacities are those of spreading pressure. In most applications the fugacities appear as ratios, and the factor $k_i RT/A$ cancels. Nevertheless it is instructive to note that equality of chemical potentials, not fugacities, is the fundamental criterion of phase equilibrium.

An activity coefficient for the constituent species of a mixed-gas adsorbate is defined by the equation

$$\gamma_i \equiv \frac{\hat{f}_i}{x_i f_i^\circ}$$

where $\hat{f}_i$ and f_i° are evaluated at the same T and *spreading pressure* Π. The degree sign (°) denotes values for the equilibrium adsorption of *pure i* at the spreading pressure of the *mixture*. Substitution for the fugacities by Eqs. (14.64) and (14.65) gives

$$\gamma_i = \frac{\hat{f}_i^g(P)}{x_i f_i^g(P_i^\circ)}$$

The fugacities are evaluated at the pressures indicated in parentheses, where P is the equilibrium mixed-gas pressure and P_i° is the equilibrium pure-gas pressure that produces the same spreading pressure. If the gas-phase fugacities are eliminated in favor of fugacity coefficients [Eqs. (10.32) and (10.47)], then

$$\gamma_i = \frac{y_i \hat{\phi}_i P}{x_i \phi_i P_i^\circ}$$

or

$$y_i \hat{\phi}_i P = x_i \phi_i P_i^\circ \gamma_i \tag{14.66}$$

The usual assumption is that the gas phase is ideal, in which case the fugacity coefficients are unity, and

$$y_i P = x_i P_i^\circ \gamma_i \tag{14.67}$$

These equations provide the means for calculation of activity coefficients from mixed-gas adsorption data. Alternatively, if γ_i values can be predicted, they allow calculation of adsorbate composition. In particular, if the mixed-gas adsorbate forms an ideal solution, then $\gamma_i = 1$, and the resulting equation is the adsorption analog of Raoult's law:

$$y_i P = x_i P_i^\circ \tag{14.68}$$

This equation is always valid as $P \to 0$ and within the pressure range for which Henry's law is a suitable approximation.

Equation (14.52) is applicable not only for pure-gas adsorption but also for adsorption of a constant-composition gas mixture. Applied in the range where Henry's law is valid, it yields

$$\frac{\Pi A}{RT} = kP \tag{14.69}$$

where k is the mixed-gas Henry's constant. For adsorption of pure species i at the same spreading pressure, this becomes

$$\frac{\Pi A}{RT} = k_i P_i^{\circ}$$

Combining these two equations with Eq. (14.68) gives

$$y_i k_i = x_i k$$

Summing over all i, we find

$$k = \sum_i y_i k_i \tag{14.70}$$

Eliminating k between these two equations yields

$$x_i = \frac{y_i k_i}{\sum\limits_i y_i k_i} \tag{14.71}$$

This simple equation, requiring only data for pure-gas adsorption, provides adsorbate compositions in the limit as $P \to 0$.

For an ideal adsorbed solution, in analogy with Eq. (10.82) for volumes,

$$a = \sum_i x_i a_i^{\circ}$$

where a is the molar area for the mixed-gas adsorbate and a_i° is the molar area of the pure-gas adsorbate at the same temperature and spreading pressure. Since $a = A/n$ and $a_i^{\circ} = A/n_i^{\circ}$, this equation may be written

$$\frac{1}{n} = \sum_i \frac{x_i}{n_i^{\circ}}$$

or

$$n = \frac{1}{\sum\limits_i (x_i/n_i^{\circ})} \tag{14.72}$$

where n is the specific amount of mixed-gas adsorbate and n_i° is the specific amount of pure-i adsorbate at the same spreading pressure. The amount of species i in the *mixed-gas* adsorbate is of course $n_i = x_i n$.

The prediction of mixed-gas adsorption equilibria by *ideal-adsorbed-solution theory*[17] is based on Eqs. (14.68) and (14.72). We give here a brief outline of the

---

[17] A. L. Myers and J. M. Prausnitz, *AIChE J.*, vol. 11, pp. 121–127, 1965; D. P. Valenzuela and A. L. Myers, *op. cit.*

procedure. Since there are $N + 1$ degrees of freedom, both T and P, as well as the gas-phase composition, must be specified. Solution is for the adsorbate composition and the specific amount absorbed. Adsorption isotherms for *each pure species* must be known over the pressure range from zero to the value that produces the spreading pressure of the mixed-gas adsorbate. For purposes of illustration we assume Eq. (14.51), the Langmuir isotherm, to apply for each pure species, writing it:

$$n_i^\circ = \frac{k_i b_i P_i^\circ}{b_i + P_i^\circ} \qquad (A)$$

The inverse of Eq. (14.53) provides an expression for P_i°, which yields values of P_i° corresponding to the spreading pressure of the mixed-gas adsorbate:

$$P_i^\circ = b_i \left(\exp \frac{\psi}{k_i b_i} - 1 \right) \qquad (B)$$

where

$$\psi \equiv \frac{\Pi A}{RT}$$

The following steps then constitute a solution procedure:

1. An initial estimate of ψ is found from the Henry's-law equations. Combining the definition of ψ with Eqs. (14.69) and (14.70) yields

$$\psi = P \sum_i y_i k_i$$

2. With this estimate of ψ, calculate P_i° for each species i by Eq. (B) and n_i° for each species i by Eq. (A).

3. One can show that the error in ψ is approximated by

$$\delta\psi = \frac{P \sum\limits_i \dfrac{y_i}{P_i^\circ} - 1}{P \sum\limits_i \dfrac{y_i}{P_i^\circ n_i^\circ}}$$

Moreover, the approximation becomes increasingly exact as the error decreases. If $\delta\psi$ is smaller than some preset tolerence (say $\delta\psi < \psi \times 10^{-7}$), the calculation goes to the final step; if not, a new value, $\psi = \psi + \delta\psi$, is determined, and the calculation returns to the preceding step.

4. Calculate x_i for each species i by Eq. (14.68):

$$x_i = \frac{y_i P}{P_i^\circ}$$

Calculate the specific amount absorbed by Eq. (14.72).

Use of the Langmuir isotherm has made this computational scheme appear quite simple, because direct solution for P_i° (step 2) is possible. However, most equations for the adsorption isotherm are less tractable, and this calculation must be done numerically. This significantly increases the computational task, but does not alter the general procedure.

Predictions of adsorption equilibria by ideal-adsorbed-solution theory are usually satisfactory when the specific amount adsorbed is less than a third of the saturation value for monolayer coverage. At higher adsorbed amounts, appreciable negative deviations from ideality are promoted by differences in size of the adsorbate molecules and by adsorbent heterogeneity. One must then have recourse to Eq. (14.67). The difficulty is in obtaining values of the activity coefficients, which are strong functions of both spreading pressure and temperature. This is in contrast to activity coefficients for liquid phases, which for most applications are insensitive to pressure. This topic is treated by Talu et al.[18]

14.7 VLE by Molecular Simulation

In Secs. 3.8, 4.2, 5.9, and 6.1 we describe how macroscopic thermodynamic properties, such as internal energy, entropy, and the Helmholtz energy can be calculated from properties of individual molecules and their assemblies. This approach has also been applied directly to VLE, primarily to pure species and to binary and simple ternary mixtures. Descriptions of assumptions, methods, and calculational procedures are given by Panagiotopoulos[19] and in an extended review by Gubbins.[20] Their extensive bibliographies are a guide to relevant literature.

The method requires suitable intermolecular potential energy functions $\mathcal{U}(r)$ and solution of the equations of statistical mechanics for the assemblies of molecules. As mentioned in Sec. 3.8, potential energy functions are as yet primarily empirical. Except for the simplest molecules, $\mathcal{U}(r)$ cannot be predicted by *ab initio*[21] calculations, because of still-inadequate computer speed. Therefore, semi-empirical functions based on quantum-mechanical theory and experimental data are employed.

Two procedures are used for the very large number of computer calculations required to treat molecular assemblies. The first, proposed by Panagiotopoulos,[22] is called the direct Gibbs-ensemble Monte Carlo method. One considers two separate phases, each represented by a finite number of molecules contained in a simulation box. Both are at the same specified temperature but are of different initial densities

---

[18] O. Talu, J. Li, and A. L. Myers, *Adsorption*, vol. 1, pp. 103–112, 1995.

[19] A. Z. Panagiotopoulos, *Molecular Simulation*, vol. 9, pp. 1–23, 1992.

[20] K. E. Gubbins, "Applications of Molecular Theory to Phase Equilibrium Predictions" in *Models for Thermodynamic and Phase Equilibrium Calculations*, S. I. Sandler, ed., pp. 507–600, Marcel Dekker, Inc., New York, 1994.

[21] Meaning: from the beginning , i.e., from first principles.

[22] A. Z. Panagiotopoulos, *Molecular Physics*, vol. 61, pp. 813–826, 1987.

and compositions. The idea is to implement a sequence of perturbations which gradually bring these two phases into equilibrium. This ultimately insures internal equilibrium for each phase separately, equality of pressures between the two phases, and equality of the chemical potentials for each species in the two phases. The perturbations ("moves") designed to bring about these conditions of equilibrium are therefore:

1. Random displacement of molecules within each box. These are the usual moves of Monte Carlo simulation, insuring internal equilibrium and generating the ensemble upon which the partition function is based, thus leading to a set of thermodynamic properties for the molecules of each box.

2. Random equal and opposite volume changes in the two boxes. These moves alter the pressures in the boxes and ultimately bring about their equality. As the pressures change, the ongoing Monte Carlo simulations in the boxes generate evolving thermodynamic properties.

3. Random transfer of molecules between the two boxes. These moves alter compositions and chemical potentials μ_i of the species in the boxes, ultimately bringing about equality of the chemical potentials for each species in the two boxes. These moves also contribute to the evolution of the thermodynamic properties of the molecules in the boxes.

Moves may be accepted or rejected on the basis of certain probabilities that insure progression on average to states of lower Gibbs energy for the two boxes considered together. Phase equilibrium obtains at the state of minimum total Gibbs energy.

A limitation of the Gibbs-ensemble Monte Carlo method is that the successful transfer of molecules between phases (item 3) becomes difficult (improbable) for dense fluids, leading to excessive calculation time. In this event, a second, more indirect, procedure can be employed. The idea here is to calculate the chemical potentials of the species in each box for a range of states. Equilibrium compositions are then identified as those states having the same temperature, pressure, and chemical potential for each species in the two boxes. Much more computation is needed by this procedure than for the direct procedure, except where the molecular-transfer step becomes difficult. For high-density fluids and for solids, the indirect procedure is preferred.

Molecular simulation is not a routine method for VLE calculations nor does it substitute for experimental data. At present, it is most useful for extrapolating to conditions not accessible by other means.

The Gibbs ensemble procedure has also been employed to estimate adsorption isotherms for simple systems. The approach is illustrated[23] by calculations for a straight cylindrical pore where both fluid/fluid and fluid/adsorbent molecular interactions can be represented by the Lennard-Jones potential-energy function [Eq. (3.55)]. Simulation calculations have also been made for isotherms of methane

---

[23] A. Z. Panagiotopoulos, *Molecular Physics*, vol. 62, pp. 701–719, 1987

and ethane adsorbed on a model carbonaceous slit pore.[24] Isosteric heats of adsorption have also been calculated.[25]

PROBLEMS

14.1. A binary liquid system exhibits LLE at $25°C$. Determine from each of the following sets of miscibility data estimates for parameters A_{12} and A_{21} in the Margules equation at $25°C$:

(a) $x_1^\alpha = 0.10$, $x_1^\beta = 0.90$.

(b) $x_1^\alpha = 0.20$, $x_1^\beta = 0.90$.

(c) $x_1^\alpha = 0.10$, $x_1^\beta = 0.80$.

14.2. Work Problem 14.1 for the van Laar equation.

14.3. Consider a binary vapor-phase mixture described by Eqs. (3.31) and (10.66). Under what (highly unlikely) conditions would one expect the mixture to split into two immiscible vapor phases?

14.4. Pure liquid species 2 and 3 are for practical purposes immiscible in one another. Liquid species 1 is soluble in both liquid 2 and liquid 3. One mole each of liquids 1, 2, and 3 are shaken together to form an equilibrium mixture of two liquid phases: an α-phase containing species 1 and 2, and a β-phase containing species 1 and 3. What are the mole fractions of species 1 in the α and β phases, if at the temperature of the experiment, the excess Gibbs energies of the phases are given by

$$\frac{(G^E)^\alpha}{RT} = 0.4\, x_1^\alpha x_2^\alpha \qquad \text{and} \qquad \frac{(G^E)^\beta}{RT} = 0.8\, x_1^\beta x_3^\beta$$

14.5. Figures 14.3, 14.4, and 14.5 are based on Eqs. (A) and (F) of Example 14.3 with C_P^E assumed *positive* and given by $C_P^E/R = 3\,x_1 x_2$. Graph the corresponding figures for the following cases, in which C_P^E is assumed *negative*.

(a) $A = \dfrac{975}{T} - 18.4 + 3\ln T$

(b) $A = \dfrac{540}{T} - 17.1 + 3\ln T$

(c) $A = \dfrac{1{,}500}{T} - 19.9 + 3\ln T$

14.6. It has been suggested that a value for G^E of at least $0.5\,RT$ is required for liquid/liquid phase splitting in a binary system. Offer some justification for this statement.

[24]R. F. Cracknell and David Nicholson, *Adsorption*, vol. 1, p. 7, 1995.

[25]R. F. Cracknell, D. Nicholson, and N. Quirke, *Molecular Simulation*, vol. 13, p. 161, 1994.

14.7. It is demonstrated in Example 14.5 that the Wilson equation for G^E is incapable of representing LLE. Show that the simple modification of Wilson's equation given by

$$G^E/RT = -C[x_1 \ln(x_1 + x_2\Lambda_{12}) + x_2 \ln(x_2 + x_1\Lambda_{21})]$$

can represent LLE. Here, C is a constant.

14.8. In Example 14.2 a plausibility argument was developed from the LLE *equilibrium* equations to demonstrate that positive deviations from ideal-solution behavior are conducive to liquid/liquid phase splitting.

(*a*) Use one of the binary stability criteria to reach this same conclusion.

(*b*) Is it possible *in principle* for a system exhibiting negative deviations from ideality to form two liquid phases?

14.9. Vapor sulfur hexafluoride SF_6 at pressures of about 1,600 kPa is used as a dielectric in large primary circuit breakers for electric transmission systems. As liquids, SF_6 and H_2O are essentially immiscible, and it is therefore necessary to specify a low enough moisture content in the vapor SF_6 so that if condensation occurs in cold weather a liquid-water phase will not form first in the system. For a preliminary determination, assume the vapor phase an ideal gas and prepare the phase diagram (see Fig. 14.11) for $H_2O(1)/SF_6(2)$ at 1,600 kPa in the composition range up to 1,000 parts per million of water (mole basis). The following approximate equations for vapor pressure are adequate:

- $\ln P_1^{\text{sat}}/\text{kPa} = 19.1478 - \dfrac{5{,}363.70}{T/\text{K}}$

- $\ln P_2^{\text{sat}}/\text{kPa} = 14.6511 - \dfrac{2{,}048.97}{T/\text{K}}$

14.10. Starting with Eq. (14.7), detrive the stability criteria of Eqs. (14.8) and (14.9).

14.11. Toluene(1) and water(2) are essentially immiscible as liquids. Determine the dew-point temperatures and the compositions of the first drops of liquid formed when vapor mixtures of these species with mole fractions $z_1 = 0.2$ and $z_1 = 0.7$ are cooled at the constant pressure of 101.33 kPa. What is the bubble-point temperature and the composition of the last drop of vapor in each case? See the Problems section of Chapt. 12 for vapor pressure equations.

14.12. n-Heptane(1) and water(2) are essentially immiscible as liquids. A vapor mixture containing 65-mole-% water at 100°C and 101.33 kPa is cooled slowly at constant pressure until condensation is complete. Construct a plot for the process showing temperature vs. the equilibrium mole fraction of heptane in the residual vapor. See the Problems section of Chapt. 12 for vapor pressure equations.

14.13. Consider a binary system of species 1 and 2 in which the liquid phase exhibits partial miscibility. In the regions of miscibility, the excess Gibbs energy at a particular temperature is expressed by the equation,

$$G^E/RT = 2.25\, x_1 x_2$$

In addition, the vapor pressures of the pure species are

$$P_1^{\text{sat}} = 75 \text{ kPa} \qquad \text{and} \qquad P_2^{\text{sat}} = 110 \text{ kPa}$$

Making the usual assumptions for low-pressure VLE, prepare a Pxy diagram for this system at the given temperature.

14.14. The system water(1)/n-pentane(2)/n-heptane(3) exists as a vapor at 101.33 kPa and 100°C with mole fractions $z_1 = 0.45$, $z_2 = 0.30$, $z_3 = 0.25$. The system is slowly cooled at constant pressure until it is completely condensed into a water phase and a hydrocarbon phase. Assuming that the two liquid phases are immiscible, that the vapor phase is an ideal gas, and that the hydrocarbons obey Raoult's law, determine:

(*a*) The dew-point temperature of the mixture and the composition of the first condensate.

(*b*) The temperature at which the second liquid phase first appears and its initial composition.

(*c*) The bubble-point temperature and the composition of the last bubble of vapor.

See the Problems section of Chapt. 12 for vapor pressure equations.

14.15. Work the preceding problem for a system composition of $z_1 = 0.32$, $z_2 = 0.45$, $z_3 = 0.23$.

14.16. The Case I behavior for SLE (Sec. 14.4) has an analog for VLE. Develop the analogy.

14.17. An assertion with respect to Case II behavior for SLE (Sec. 14.4) was that the condition $z_i \gamma_i^s = 1$ corresponds to complete immiscibility for all species in the solid state. Prove this.

14.18. Use results of Sec. 14.4 to develop the following (approximate) rules of thumb:

(*a*) The solubility of a solid in a liquid solvent increases with increasing T.

(*b*) The solubility of a solid in a liquid solvent is independent of the identity of the solvent species.

(*c*) Of two solids with roughly the same heat of fusion, that solid with the lower melting point is the more soluble in a given liquid solvent at a given T.

(*d*) Of two solids with similar melting points, that solid with the smaller heat of fusion is the more soluble in a given liquid solvent at a given T.

14.19. Estimate the solubility of naphthalene(1) in carbon dioxide(2) at a temperature of 80°C at pressures up to 300 bar. Use the procedure described in Sec. 14.5, with $l_{12} = 0.088$. Compare the results with those shown by Fig. 14.15. Discuss any differences. $P_1^{\text{sat}} = 0.0102$ bar.

14.20. Estimate the solubility of naphthalene(1) in nitrogen(2) at a temperature of 35°C at pressures up to 300 bar. Use the procedure described in Sec. 14.5, with $l_{12} = 0$. Compare the results with those shown by Fig. 14.15 for the naphthalene/CO_2 system at 35°C with $l_{12} = 0$. Discuss any differences.

14.21. The qualitative features of SVE at high pressures shown by Fig. 14.15 are determined by the equation of state for the gas. To what extent can these features be represented by the two-term virial equation in pressure, Eq. (3.31)?

14.22. The UNILAN equation for pure-species adsorption is

$$n = \frac{m}{2s} \ln \left(\frac{c + Pe^s}{c + Pe^{-s}} \right)$$

where m, s, and c are positive empirical constants.

(*a*) Show that the UNILAN equation reduces to the Langmuir isotherm for $s = 0$. (*Hint*: Apply l'Hôpital's rule.)

(*b*) Show that Henry's constant k for the UNILAN equation is

$$k(\text{UNILAN}) = \frac{m}{cs} \sinh s$$

(*c*) Examine the *detailed* behavior of the UNILAN equation at zero pressure ($P \rightarrow 0$, $n \rightarrow 0$).

14.23. In Example 14.8, Henry's constant for adsorption k, identified as the intercept on a plot of n/P vs. n, was found from a polynomial curve-fit of n/P vs. n. An alternative procedure is based on a plot of $\ln(P/n)$ vs. n. Suppose that the adsorbate equation of state is a power series in n: $z = 1 + Bn + Cn^2 + \dots$ Show how from a plot (or a polynomial curve-fit) of $\ln(P/n)$ vs. n one can extract values of k and B. [*Hint*: Start with Eq. (14.49).]

14.24. It was assumed in the development of Eq. (14.49) that the gas phase is *ideal*, with $Z = 1$. Suppose for a *real* gas phase that $Z = Z(T, P)$. Determine the analogous expression to Eq. (14.49) appropriate for a real (nonideal) gas phase. [*Hint*: Start with Eq. (14.45).]

14.25. Use results reported in Example 14.8 to prepare plots of Π vs. n and z vs. n for ethylene adsorbed on a carbon molecular sieve. Discuss the plots.

14.26. Suppose that the adsorbate equation of state is given by $z = (1 - bn)^{-1}$, where b is a constant. Find the implied adsorption isotherm, and show under what conditions it reduces to the Langmuir isotherm.

14.27. Suppose that the adsorbate equation of state is given by $z = 1 + \beta n$, where β is a function of T only. Find the implied adsorption isotherm, and show under what conditions it reduces to the Langmuir isotherm.

14.28. Derive the result given in step 3 of the procedure for predicting adsorption equilibria by ideal-adsorbed-solution theory at the end of Sec. 14.6.

CHAPTER 15

CHEMICAL-REACTION EQUILIBRIA

The transformation of raw materials into products of greater value by means of chemical reaction is a major industry, and a vast array of commercial products is obtained by chemical synthesis. Sulfuric acid, ammonia, ethylene, propylene, phosphoric acid, chlorine, nitric acid, urea, benzene, methanol, ethanol, and ethylene glycol are examples of chemicals produced in the United States in billions of kilograms each year. These in turn are used in the large-scale manufacture of fibers, paints, detergents, plastics, rubber, paper, fertilizers, insecticides, etc. Clearly, the chemical engineer must be familiar with chemical-reactor design and operation.

Both the rate and the equilibrium conversion of a chemical reaction depend on the temperature, pressure, and composition of reactants. Consider, for example, the oxidation of sulfur dioxide to sulfur trioxide. A catalyst is required if a reasonable reaction rate is to be attained. With a vanadium pentoxide catalyst the rate becomes appreciable at about 300°C and continues to increase at higher temperatures. On the basis of rate alone, one would operate the reactor at the highest practical temperature. However, the equilibrium conversion to sulfur trioxide falls as temperature rises, decreasing from about 90 percent at 520°C to 50 percent at about 680°C. These values represent maximum possible conversions regardless of catalyst or reaction rate. The evident conclusion is that both equilibrium and rate must be considered in the exploitation of chemical reactions for commercial purposes. Although reaction *rates* are not susceptible to thermodynamic treatment, equilibrium conversions are. Therefore, the purpose of this chapter is to determine the effect of temperature, pressure, and initial composition on the equilibrium conversions of chemical reactions.

Many industrial reactions are not carried to equilibrium; reactor design is then based primarily on reaction rate. However, the choice of operating conditions may still be influenced by equilibrium considerations. Moreover, the equilibrium

559

conversion of a reaction provides a goal by which to measure improvements in a process. Similarly, it may determine whether or not an experimental investigation of a new process is worthwhile. For example, if thermodynamic analysis indicates that a yield of only 20 percent is possible at equilibrium and if a 50 percent yield is necessary for the process to be economically attractive, there is no purpose to an experimental study. On the other hand, if the equilibrium yield is 80 percent, an experimental program to determine the reaction rate for various conditions of operation (catalyst, temperature, pressure, etc.) may be warranted.

Reaction stoichiometry is treated in Sec. 15.1, and reaction equilibrium, in Sec. 15.2. The equilibrium constant is introduced in Sec. 15.3, and its temperature dependence and evaluation are considered in Secs. 15.4 and 15.5. The connection between the equilibrium constant and composition is developed in Sec. 15.6. The calculation of equilibrium conversions for single reactions is taken up in Sec. 15.7. In Sec. 15.8, the phase rule is reconsidered; finally, multireaction equilibrium is treated in Sec. 15.9.[1]

15.1 The Reaction Coordinate

The general chemical reaction of Sec. 4.7 is written here as

$$|\nu_1|A_1 + |\nu_2|A_2 + \cdots \rightarrow |\nu_3|A_3 + |\nu_4|A_4 + \cdots \tag{15.1}$$

where the $|\nu_i|$ are stoichiometric coefficients and the A_i stand for chemical formulas. The ν_i themselves are called stoichiometric numbers, and we recall the sign convention that makes them positive for products and negative for reactants. Thus for the reaction

$$CH_4 + H_2O \rightarrow CO + 3H_2$$

the stoichiometric numbers are

$$\nu_{CH_4} = -1 \qquad \nu_{H_2O} = -1 \qquad \nu_{CO} = 1 \qquad \nu_{H_2} = 3$$

The stoichiometric number for any inert species is zero.

For the reaction represented by Eq. (15.1), the *changes* in the numbers of moles of the species present are in direct proportion to the stoichiometric numbers. Thus for the preceding reaction, if 0.5 mol of CH_4 disappears by reaction, 0.5 mol of H_2O must also disappear; simultaneously 0.5 mol of CO and 1.5 mol of H_2 are formed. Applying this principle to a differential amount of reaction, we can write

$$\frac{dn_2}{\nu_2} = \frac{dn_1}{\nu_1} \qquad \frac{dn_3}{\nu_3} = \frac{dn_1}{\nu_1} \qquad \text{etc.}$$

The list continues to include all species. Comparison of these equations shows that

$$\frac{dn_1}{\nu_1} = \frac{dn_2}{\nu_2} = \frac{dn_3}{\nu_3} = \frac{dn_4}{\nu_4} = \cdots$$

---

[1]For a comprehensive treatment of chemical-reaction equilibria, see W. R. Smith and R. W. Missen, *Chemical Reaction Equilibrium Analysis*, John Wiley & Sons, New York, 1982.

All terms being equal, they can be identified collectively with a single quantity representing an amount of reaction. Thus a *definition* of $d\varepsilon$ is provided by the equation

$$\frac{dn_1}{\nu_1} = \frac{dn_2}{\nu_2} = \frac{dn_3}{\nu_3} = \frac{dn_4}{\nu_4} = \cdots \equiv d\varepsilon \qquad (15.2)$$

The general relation between a differential change dn_i in the number of moles of a reacting species and $d\varepsilon$ is therefore

$$dn_i = \nu_i \, d\varepsilon \qquad (i = 1, 2, \ldots, N) \qquad (15.3)$$

This new variable ε, called the *reaction coordinate*, characterizes the extent or degree to which a reaction has taken place.[2] Equations (15.2) and (15.3) define *changes* in ε with respect to changes in the numbers of moles of the reacting species. The definition of ε itself is completed for each application by the specification that it be *zero* for the initial state of the system prior to reaction. Thus, integration of Eq. (15.3) from an initial unreacted state where $\varepsilon = 0$ and $n_i = n_{i_0}$ to a state reached after an arbitrary amount of reaction gives

$$\int_{n_{i_0}}^{n_i} dn_i = \nu_i \int_0^\varepsilon d\varepsilon$$

or

$$n_i = n_{i_0} + \nu_i \varepsilon \qquad (i = 1, 2, \ldots, N) \qquad (15.4)$$

Summation over all species yields

$$n = \sum_i n_i = \sum_i n_{i_0} + \varepsilon \sum_i \nu_i$$

or

$$n = n_0 + \nu \varepsilon$$

where

$$n \equiv \sum_i n_i \qquad n_0 \equiv \sum_i n_{i_0} \qquad \nu \equiv \sum_i \nu_i$$

Thus the mole fractions y_i of the species present are related to ε by

$$y_i = \frac{n_i}{n} = \frac{n_{i_0} + \nu_i \varepsilon}{n_0 + \nu \varepsilon} \qquad (15.5)$$

Application of this equation is illustrated in the following examples.

---

[2] The reaction coordinate ε has been given various other names, such as: degree of advancement, degree of reaction, extent of reaction, and progress variable.

Example 15.1   For a system in which the following reaction occurs,

$$CH_4 + H_2O \rightarrow CO + 3H_2$$

assume there are present initially 2 mol CH_4, 1 mol H_2O, 1 mol CO, and 4 mol H_2. Determine expressions for the mole fractions y_i as functions of ε.

SOLUTION   For the given reaction,

$$\nu = \sum_i \nu_i = -1 - 1 + 1 + 3 = 2$$

For the given numbers of moles of species initially present,

$$n_0 = \sum_i n_{i_0} = 2 + 1 + 1 + 4 = 8$$

Equation (15.5) now yields

$$y_{CH_4} = \frac{2 - \varepsilon}{8 + 2\varepsilon} \qquad y_{H_2O} = \frac{1 - \varepsilon}{8 + 2\varepsilon}$$

$$y_{CO} = \frac{1 + \varepsilon}{8 + 2\varepsilon} \qquad y_{H_2} = \frac{4 + 3\varepsilon}{8 + 2\varepsilon}$$

The mole fractions of the species in the reacting mixture are seen to be functions of the single variable ε.

Example 15.2   Consider a vessel which initially contains only n_0 moles of water vapor. If decomposition occurs according to the reaction

$$H_2O \rightarrow H_2 + \tfrac{1}{2}O_2$$

find expressions which relate the number of moles and the mole fraction of each chemical species to the reaction coordinate ε.

SOLUTION   For the given reaction, $\nu = -1 + 1 + \frac{1}{2} = \frac{1}{2}$. Application of Eqs. (15.4) and (15.5) yields

$$n_{H_2O} = n_0 - \varepsilon \qquad y_{H_2O} = \frac{n_0 - \varepsilon}{n_0 + \frac{1}{2}\varepsilon}$$

$$n_{H_2} = \varepsilon \qquad y_{H_2} = \frac{\varepsilon}{n_0 + \frac{1}{2}\varepsilon}$$

$$n_{O_2} = \tfrac{1}{2}\varepsilon \qquad y_{O_2} = \frac{\frac{1}{2}\varepsilon}{n_0 + \frac{1}{2}\varepsilon}$$

The fractional decomposition of water vapor is

$$\frac{n_0 - n_{H_2O}}{n_0} = \frac{n_0 - (n_0 - \varepsilon)}{n_0} = \frac{\varepsilon}{n_0}$$

Thus when $n_0 = 1$, ε can be identified with the fractional decomposition of the water vapor.

Since the ν_i are pure numbers without units, Eq. (15.3) shows that ε must be expressed in moles. This allows one to speak of a *mole of reaction*, meaning that ε has changed by a unit amount, i.e., by one mole. When $\Delta\varepsilon = 1$ mol, the reaction proceeds to such an extent that the change in mole number of each reactant and product is equal to its stoichiometric number.

When two or more independent reactions proceed simultaneously, we let subscript j be the reaction index, and associate a separate reaction coordinate ε_j with each reaction. The stoichiometric numbers are doubly subscripted to identify their association with both a species and a reaction. Thus $\nu_{i,j}$ designates the stoichiometric number of species i in reaction j. Since the number of moles of a species n_i may change because of several reactions, the general equation analogous to Eq. (15.3) includes a sum:

$$dn_i = \sum_j \nu_{i,j}\, d\varepsilon_j \qquad (i = 1, 2, \dots, N)$$

Integration from $n_i = n_{i_0}$ and $\varepsilon_j = 0$ to arbitrary n_i and ε_j gives

$$n_i = n_{i_0} + \sum_j \nu_{i,j}\varepsilon_j \qquad (i = 1, 2, \dots, N) \qquad (15.6)$$

Summing over all species yields

$$n = \sum_i n_{i_0} + \sum_i \sum_j \nu_{i,j}\varepsilon_j$$

This may also be written

$$n = n_0 + \sum_j \left(\sum_i \nu_{i,j} \right) \varepsilon_j$$

Analogous to the definition of ν for a single reaction, we here adopt the definition

$$\nu_j \equiv \sum_i \nu_{i,j}$$

Then

$$n = n_0 + \sum_j \nu_j \varepsilon_j$$

Combination of this equation with Eq. (15.6) gives the mole fraction:

$$\boxed{y_i = \frac{n_{i_0} + \sum_j \nu_{i,j}\varepsilon_j}{n_0 + \sum_j \nu_j \varepsilon_j} \qquad (i = 1, 2, \dots, N)} \qquad (15.7)$$

Example 15.3 Consider a system in which the following reactions occur:

$$CH_4 + H_2O \rightarrow CO + 3H_2 \quad (1)$$

$$CH_4 + 2H_2O \rightarrow CO_2 + 4H_2 \quad (2)$$

where the numbers (1) and (2) indicate the value of j, the reaction index. If there are present initially 2 mol CH_4 and 3 mol H_2O, determine expressions for the y_i as functions of ε_1 and ε_2.

SOLUTION The stoichiometric numbers $\nu_{i,j}$ can be arrayed as follows:

$i =$	CH_4	H_2O	CO	CO_2	H_2	
j						ν_j
1	-1	-1	1	0	3	2
2	-1	-2	0	1	4	2

Application of Eq. (15.7) now gives

$$y_{CH_4} = \frac{2 - \varepsilon_1 - \varepsilon_2}{5 + 2\varepsilon_1 + 2\varepsilon_2}$$

$$y_{H_2O} = \frac{3 - \varepsilon_1 - 2\varepsilon_2}{5 + 2\varepsilon_1 + 2\varepsilon_2}$$

$$y_{CO} = \frac{\varepsilon_1}{5 + 2\varepsilon_1 + 2\varepsilon_2}$$

$$y_{CO_2} = \frac{\varepsilon_2}{5 + 2\varepsilon_1 + 2\varepsilon_2}$$

$$y_{H_2} = \frac{3\varepsilon_1 + 4\varepsilon_2}{5 + 2\varepsilon_1 + 2\varepsilon_2}$$

The composition of the system is a function of the independent variables ε_1 and ε_2.

15.2 Application of Equilibrium Criteria to Chemical Reactions

In Sec. 14.1 it is shown that the total Gibbs energy of a closed system at constant T and P must decrease during an irreversible process and that the condition for equilibrium is reached when G^t attains its minimum value. At this equilibrium state,

$$(dG^t)_{T,P} = 0 \qquad\qquad (14.4)$$

Thus if a mixture of chemical species is not in chemical equilibrium, any reaction that occurs at constant T and P must lead to a decrease in the total Gibbs energy of the system. The significance of this for a single chemical reaction is seen in Fig. 15.1, which shows a schematic diagram of G^t vs. ε, the reaction coordinate. Since ε is the single variable that characterizes the progress of the reaction, and therefore the composition of the system, the total Gibbs energy at constant T and P is determined by ε. The arrows along the curve in Fig. 15.1 indicate the directions of changes in $(G^t)_{T,P}$ that are possible on account of reaction. The reaction coordinate has its equilibrium value ε_e at the minimum of the curve. The meaning of Eq. (14.4) is that differential displacements of the chemical reaction can occur at the equilibrium state without causing changes in the total Gibbs energy of the system.

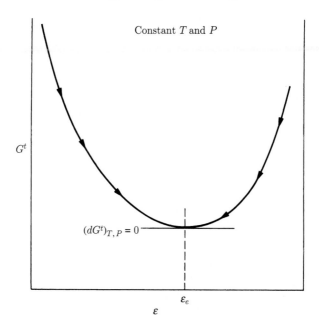

Figure 15.1: The total Gibbs energy in relation to the reaction coordinate.

Figure 15.1 indicates the two distinctive features of the equilibrium state for given T and P: (1) The total Gibbs energy G^t is a minimum; (2) its differential is zero. Each of these may serve as a criterion of equilibrium. Thus, we may write an expression for G^t as a function of ε and seek the value of ε which minimizes G^t, or we may differentiate the expression, equate it to zero, and solve for ε. The latter procedure is almost always used for single reactions (Fig. 15.1), and leads to the method of equilibrium constants, as described in the following sections. It may also be extended to multiple reactions, but in this case the direct minimization of G^t is often more convenient, and is considered in Sec. 15.9.

Although the equilibrium expressions are *developed* for closed systems at constant T and P, they are not restricted in *application* to systems that are actually closed and reach equilibrium states along paths of constant T and P. Once an equilibrium state is reached, no further changes occur, and the system continues to exist in this state at fixed T and P. How this state was *actually* attained does not matter. Once it is known that an equilibrium state exists at given T and P, the criteria apply.

15.3 The Standard Gibbs Energy Change and the Equilibrium Constant

Equation (10.2), the fundamental property relation for single-phase systems, provides an expression for the total differential of the Gibbs energy:

$$d(nG) = (nV)dP - (nS)dT + \sum_i \mu_i \, dn_i \qquad (10.2)$$

If changes in the mole numbers n_i occur as the result of a single chemical reaction in a closed system, then by Eq. (15.3) each dn_i may be replaced by the product $\nu_i d\varepsilon$. Equation (10.2) then becomes

$$d(nG) = (nV)dP - (nS)dT + \sum_i \nu_i \mu_i \, d\varepsilon$$

Since nG is a state function, the right-hand side of this equation is an exact differential expression; it follows that

$$\sum_i \nu_i \mu_i = \left[\frac{\partial (nG)}{\partial \varepsilon} \right]_{T,P} = \left[\frac{\partial (G^t)}{\partial \varepsilon} \right]_{T,P}$$

Thus the quantity $\sum_i \nu_i \mu_i$ represents, in general, the rate of change of the total Gibbs energy of the system with the reaction coordinate at constant T and P. Figure 15.1 shows that this quantity is zero at the equilibrium state. Therefore the criterion of chemical-reaction equilibrium may be written

$$\sum_i \nu_i \mu_i = 0 \qquad (15.8)$$

We recall the definition of the fugacity of a species in solution:

$$\mu_i = \Gamma_i(T) + RT \ln \hat{f}_i \qquad (10.42)$$

Moreover, we can write Eq. (10.30) for pure species i in its *standard state*[3] at the same temperature:

$$G_i^\circ = \Gamma_i(T) + RT \ln f_i^\circ$$

The difference between these two equations is then

$$\mu_i - G_i^\circ = RT \ln \frac{\hat{f}_i}{f_i^\circ} \qquad (15.9)$$

The ratio $\hat{f}_i / f_i^\circ$ is called the *activity* $\hat{a}_i$ of species i in solution. Thus by definition,

$$\hat{a}_i \equiv \frac{\hat{f}_i}{f_i^\circ} \qquad (15.10)$$

and the preceding equation becomes

$$\mu_i = G_i^\circ + RT \ln \hat{a}_i \qquad (15.11)$$

---

[3]Standard states are introduced and discussed in Sec. 4.4.

Combining Eq. (15.8) with Eq. (15.11) to eliminate μ_i gives for the equilibrium state of a chemical reaction

$$\sum_i \nu_i(G_i^\circ + RT \ln \hat{a}_i) = 0$$

or

$$\sum_i \nu_i G_i^\circ + RT \sum_i \ln(\hat{a}_i)^{\nu_i} = 0$$

or

$$\ln \prod_i (\hat{a}_i)^{\nu_i} = \frac{-\sum_i \nu_i G_i^\circ}{RT} \tag{15.12}$$

where $\prod_i$ signifies the product over all species i. In exponential form, Eq. (15.12) becomes

$$\boxed{\prod_i (\hat{a}_i)^{\nu_i} = \exp \frac{-\sum_i \nu_i G_i^\circ}{RT} \equiv K} \tag{15.13}$$

Included in this equation is the definition of K. Since G_i° is a property of pure species i in its standard state at fixed pressure, it depends only on temperature. Equation (15.13) shows that K is also a function of temperature only. In spite of its dependence on temperature, K is called the equilibrium *constant* for the reaction. Equation (15.12) may now be written

$$\boxed{-RT \ln K = \sum_i \nu_i G_i^\circ \equiv \Delta G^\circ} \tag{15.14}$$

The final term ΔG° is the conventional way of representing the quantity $\sum_i \nu_i G_i^\circ$. It is called the *standard Gibbs energy change of reaction.*

The activities $\hat{a}_i$ in Eq. (15.13) provide the connection between the *equilibrium* state of interest and the *standard* states of the individual species, for which data are presumed available, as discussed in Sec. 15.5. The standard states are arbitrary, but must always be at the equilibrium temperature T. The standard states selected need not be the same for all species taking part in a reaction. However, for a *particular* species the standard state represented by G_i° must be the same state as for the fugacity f_i° upon which the activity $\hat{a}_i$ is based.

The function $\Delta G^\circ \equiv \sum_i \nu_i G_i^\circ$ in Eq. (15.14) is the difference between the Gibbs energies of the products and reactants (weighted by their stoichiometric coefficients) when each is in its standard state as a pure substance at the system temperature and at a fixed pressure. Thus the value of ΔG° is fixed for a given reaction once the temperature is established, and is independent of the equilibrium pressure and composition. Other *standard property changes of reaction* are similarly defined. Thus, for the general property M, we write

$$\Delta M^\circ = \sum_i \nu_i M_i^\circ$$

In accord with this, ΔH° is defined by Eq. (4.20) and ΔC_P° by Eq. (4.22). For the standard entropy change of reaction ΔM° becomes ΔS°. These quantities are all functions of temperature only for a given reaction, and are related to one another by equations analogous to property relations for pure species.

As an example we develop the relation between the standard heat of reaction and the standard Gibbs-energy change of reaction. Equation (6.31) written for species i in its standard state becomes

$$H_i^\circ = -RT^2 \frac{d(G_i^\circ/RT)}{dT}$$

Total derivatives are appropriate here because the properties in the standard state are functions of temperature only. Multiplication of both sides of this equation by ν_i and summation over all species gives

$$\sum_i \nu_i H_i^\circ = -RT^2 \frac{d(\sum_i \nu_i G_i^\circ/RT)}{dT}$$

In view of the definitions of Eqs. (4.20) and (15.14), this may be written

$$\Delta H^\circ = -RT^2 \frac{d(\Delta G^\circ/RT)}{dT} \tag{15.15}$$

15.4   Effect of Temperature on the Equilibrium Constant

Since the standard-state temperature is that of the equilibrium mixture, the standard property changes of reaction, such as ΔG° and ΔH°, vary with the equilibrium temperature. The dependence of ΔG° on T is given by Eq. (15.15), which may be rewritten as

$$\frac{d(\Delta G^\circ/RT)}{dT} = \frac{-\Delta H^\circ}{RT^2}$$

According to Eq. (15.14),

$$\frac{\Delta G^\circ}{RT} = -\ln K$$

Therefore

$$\boxed{\frac{d\ln K}{dT} = \frac{\Delta H^\circ}{RT^2}} \tag{15.16}$$

Equation (15.16) gives the effect of temperature on the equilibrium constant, and hence on the equilibrium conversion. If ΔH° is negative, i.e., if the reaction is exothermic, the equilibrium constant decreases as the temperature increases. Conversely, K increases with T for an endothermic reaction.

If ΔH°, the standard enthalpy change (heat) of reaction, is assumed independent of T, integration of Eq. (15.16) from a particular temperature T_1 to an arbitrary temperature T leads to the simple result,

$$\ln \frac{K}{K_1} = -\frac{\Delta H^\circ}{R}\left(\frac{1}{T} - \frac{1}{T_1}\right) \tag{15.17}$$

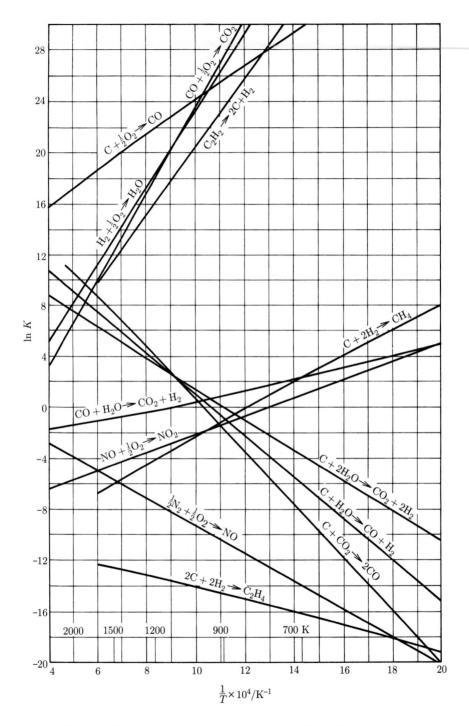

Figure 15.2: Equilibrium constants as a function of temperature.

This approximate equation implies that a plot of $\ln K$ vs. the reciprocal of absolute temperature is a straight line. Figure 15.2, a plot of $\ln K$ vs. $1/T$ for a number of common reactions, illustrates this near linearity. Thus, Eq. (15.17) provides a reasonably accurate relation for the interpolation and extrapolation of equilibrium-constant data.

A convenient starting point for *rigorous* development of the effect of temperature on the equilibrium constant is the general relation

$$\Delta G^\circ = \Delta H^\circ - T \Delta S^\circ \tag{15.18}$$

This equation follows from the definition of the Gibbs energy, $G \equiv H - TS$, applied to each species of a chemical reaction in its standard state at temperature T; thus

$$G_i^\circ = H_i^\circ - T S_i^\circ$$

Multiplication by ν_i and summation over all species gives

$$\sum_i \nu_i G_i^\circ = \sum_i \nu_i H_i^\circ - T \sum_i \nu_i S_i^\circ$$

Equation (15.18) follows immediately from the definition of a standard property change of reaction.

The standard heat of reaction is related to temperature by Eq. (4.24):

$$\Delta H^\circ = \Delta H_0^\circ + R \int_{T_0}^{T} \frac{\Delta C_P^\circ}{R} dT \tag{4.24}$$

The temperature dependence of the standard entropy change of reaction is developed similarly. Equation (6.21) is written for the standard-state entropy of species i at the constant standard-state pressure P°:

$$dS_i^\circ = C_{P_i}^\circ \frac{dT}{T}$$

Multiplying by ν_i, summing over all species, and invoking the definition of a standard property change of reaction transforms this equation into

$$d\Delta S^\circ = \Delta C_P^\circ \frac{dT}{T}$$

Integration gives

$$\Delta S^\circ = \Delta S_0^\circ + R \int_{T_0}^{T} \frac{\Delta C_P^\circ}{R} \frac{dT}{T} \tag{15.19}$$

where ΔS° and ΔS_0° are standard entropy changes of reaction at temperature T and at reference temperature T_0 respectively. Equations (15.18), (4.24), and (15.19) are combined to yield

$$\Delta G^\circ = \Delta H_0^\circ + R \int_{T_0}^{T} \frac{\Delta C_P^\circ}{R} dT - T \Delta S_0^\circ - RT \int_{T_0}^{T} \frac{\Delta C_P^\circ}{R} \frac{dT}{T}$$

However,

$$\Delta S_0^\circ = \frac{\Delta H_0^\circ - \Delta G_0^\circ}{T_0}$$

whence

$$\Delta G^\circ = \Delta H_0^\circ - \frac{T}{T_0}(\Delta H_0^\circ - \Delta G_0^\circ) + R\int_{T_0}^{T}\frac{\Delta C_P^\circ}{R}dT - RT\int_{T_0}^{T}\frac{\Delta C_P^\circ}{R}\frac{dT}{T}$$

Finally, division by RT yields

$$\frac{\Delta G^\circ}{RT} = \frac{\Delta G_0^\circ - \Delta H_0^\circ}{RT_0} + \frac{\Delta H_0^\circ}{RT} + \frac{1}{T}\int_{T_0}^{T}\frac{\Delta C_P^\circ}{R}dT - \int_{T_0}^{T}\frac{\Delta C_P^\circ}{R}\frac{dT}{T} \qquad (15.20)$$

We recall that by Eq. (15.14), $\ln K = -\Delta G^\circ/RT$.

When the temperature dependence of the heat capacity of each species is given by Eq. (4.4), the first integral on the right-hand side of Eq. (15.20) is given by Eq. (4.25), which for computational purposes we have named

$$\mathtt{IDCPH(T0,T;DA,DB,DC,DD)}$$

Similarly, the second integral is given by the analog of Eq. (5.15):

$$\int_{T_0}^{T}\frac{\Delta C_P^\circ}{R}\frac{dT}{T} = \Delta A\ln\tau + \left[\Delta B\,T_0 + \left(\Delta C\,T_0^2 + \frac{\Delta D}{\tau^2 T_0^2}\right)\left(\frac{\tau+1}{2}\right)\right](\tau-1) \quad (15.21)$$

where

$$\tau \equiv \frac{T}{T_0}$$

The integral is evaluated by a function of exactly the same form as given by Eq. (5.15), and the same computer program therefore serves for evaluation of either integral. The only difference is in the name of the function. For computational purposes the name used here is

$$\mathtt{IDCPS(T0,T;DA,DB,DC,DD)}$$

Thus $\Delta G^\circ/RT(= -\ln K)$ as given by Eq. (15.20) is readily calculated at any temperature from the standard heat of reaction and the standard Gibbs-energy change of reaction at a reference temperature (usually 298.15 K), and from two functions which can be evaluated by standard computational procedures.

15.5 Evaluation of Equilibrium Constants

Values of ΔG° for many *formation reactions* are tabulated in standard references.[4] The reported values of ΔG_f° are not measured experimentally, but are calculated

---

[4]For example, "TRC Thermodynamic Tables–Hydrocarbons" and "TRC Thermodynamic Tables–Non-hydrocarbons," serial publications of the Thermodynamics Research Center, Texas A & M Univ. System, College Station, Texas; "The NBS Tables of Chemical Thermodynamic Properties," *J. Physical and Chemical Reference Data*, vol. 11, supp. 2, 1982.

by Eq. (15.18). The determination of ΔS_f° may be based on the third law of thermodynamics, discussed in Sec. 5.8. Combination of values from Eq. (5.20) for the absolute entropies of the species taking part in the reaction gives the value of ΔS_f°. Entropies (and heat capacities) are also commonly determined from statistical calculations based on spectroscopic data.[5]

We list values of $\Delta G_{f_{298}}^\circ$ for a limited number of chemical compounds in Table C.4 of App. C. These are for a temperature of 298.15 K, as are the values of $\Delta H_{f_{298}}^\circ$ listed in the same table. Values of ΔG° for other reactions are calculated from values for formation reactions in exactly the same way that ΔH° values for other reactions are determined from values for formation reactions (Sec. 4.5). In more extensive compilations of data, values of ΔG_f° and ΔH_f° are given for a wide range of temperatures, rather than just at 298.15 K. Where data are lacking, methods of estimation are available; these are reviewed by Reid, Prausnitz, and Poling.[6]

Example 15.4 Calculate the equilibrium constant for the vapor-phase hydration of ethylene at 145 and at 320°C from data given in App. C.

SOLUTION   We first determine values for ΔA, ΔB, ΔC, and ΔD. For the reaction

$$C_2H_4(g) + H_2O(g) \rightarrow C_2H_5OH(g)$$

the meaning of Δ is indicated by

$$\Delta = (C_2H_5OH) - (C_2H_4) - (H_2O)$$

Thus, from the heat-capacity data of Table C.1:

$$\Delta A = 3.518 - 1.424 - 3.470 = -1.376$$

$$\Delta B = (20.001 - 14.394 - 1.450) \times 10^{-3} = 4.157 \times 10^{-3}$$

$$\Delta C = (-6.002 + 4.392 - 0.000) \times 10^{-6} = -1.610 \times 10^{-6}$$

$$\Delta D = (-0.000 - 0.000 - 0.121) \times 10^5 = -0.121 \times 10^5$$

In addition, we require values of ΔH_{298}° and ΔG_{298}° at 298.15 K for the hydration reaction. These are found from the heat-of-formation and Gibbs-energy-of-formation data of Table C.4:

$$\Delta H_{298}^\circ = -235{,}100 - 52{,}510 - (-241{,}818) = -45{,}792 \text{ J mol}^{-1}$$

and

$$\Delta G_{298}^\circ = -168{,}490 - 68{,}460 - (-228{,}572) = -8{,}378 \text{ J mol}^{-1}$$

For $T = 145 + 273.15 = 418.15$ K, the values of the integrals in Eq. (15.20) are given by

IDCPH(298.15,418.15;-1.376,4.157E-3,-1.610E-6,-0.121E+5) ≡ -23.121

---

[5]K. S. Pitzer, *Thermodynamics*, 3d ed., chap. 5, McGraw-Hill, New York, 1995.

[6]R. C. Reid, J. M. Prausnitz, and B. E. Poling, *The Properties of Gases and Liquids*, 4th ed., chap. 6, McGraw-Hill, New York, 1987.

IDCPS(298.15,418.15;-1.376,4.157E-3,-1.610E-6,-0.121E+5) $\equiv$ -0.06924

Substitution of values into Eq. (15.20) for a reference temperature of 298.15 gives

$$\frac{\Delta G^{\circ}_{418}}{RT} = \frac{-8,378 + 45,792}{(8.314)(298.15)} + \frac{-45,792}{(8.314)(418.15)} + \frac{-23.121}{418.15} + 0.06924 = 1.936$$

For $T = 320 + 273.15 = 593.15$ K,

IDCPH(298.15,593.15;-1.376,4.157E-3,-1.610E-6,-0.121E+5) $\equiv$ 22.632

IDCPS(298.15,593.15;-1.376,4.157E-3,-1.610E-6,-0.121E+5) $\equiv$ 0.01731

and

$$\frac{\Delta G^{\circ}_{593}}{RT} = \frac{-8,378 + 45,792}{(8.314)(298.15)} + \frac{-45,792}{(8.314)(593.15)} + \frac{22.632}{593.15} - 0.01731 = 5.829$$

Then

at 418.15 K: $\ln K = -1.936$ and $K = 14.43 \times 10^{-2}$

at 593.15 K: $\ln K = -5.829$ and $K = 2.94 \times 10^{-3}$

15.6 Relation of Equilibrium Constants to Composition

Gas-Phase Reactions

The standard state for a gas is the ideal-gas state of the pure gas at the standard-state pressure P° of 1 bar. Since the fugacity of an ideal gas is equal to its pressure, $f_i^{\circ} = P^{\circ}$ for each species i. Thus for gas-phase reactions $\hat{a}_i = \hat{f}_i/f_i^{\circ} = \hat{f}_i/P^{\circ}$, and Eq. (15.13) becomes

$$K = \prod_i \left(\frac{\hat{f}_i}{P^{\circ}} \right)^{\nu_i} \tag{15.22}$$

The equilibrium constant K is a function of temperature only. However, Eq. (15.22) relates K to fugacities of the reacting species as they exist in the real equilibrium mixture. These fugacities reflect the nonidealities of the equilibrium mixture and are functions of temperature, pressure, and composition. This means that for a fixed temperature the composition at equilibrium must change with pressure in such a way that $\prod_i (\hat{f}_i/P^{\circ})^{\nu_i}$ remains constant.

The fugacity is related to the fugacity coefficient by Eq. (10.47), here written

$$\hat{f}_i = \hat{\phi}_i y_i P$$

Substitution of this equation into Eq. (15.22) provides an equilibrium expression displaying the pressure and the composition:

$$\boxed{\prod_i (y_i \hat{\phi}_i)^{\nu_i} = \left(\frac{P}{P^{\circ}} \right)^{-\nu} K} \tag{15.23}$$

where $\nu \equiv \sum_i \nu_i$ and P° is the standard-state pressure of 1 bar, *expressed in the same units used for P*. The y_i's may be eliminated in favor of the equilibrium value of the reaction coordinate ε_e. Then, for a fixed temperature Eq. (15.23) relates ε_e to P. In principle, specification of the pressure allows solution for ε_e. However, the problem may be complicated by the dependence of the $\hat{\phi}_i$'s on composition, i.e., on ε_e. The methods of Sec. 10.7 and 13.4 can be applied to the calculation of $\hat{\phi}_i$ values, for example, by Eq. (10.69) or (13.38). Because of the complexity of the calculations, an iterative procedure, initiated by setting $\hat{\phi}_i = 1$ and formulated for computer solution, is indicated. Once an initial set of y_i's is calculated, the $\hat{\phi}_i$'s are determined, and the procedure is repeated to convergence.

If the assumption that the equilibrium mixture is an *ideal solution* is justified, then each $\hat{\phi}_i$ becomes ϕ_i, the fugacity coefficient of pure species i at T and P [Eq. (10.85)]. In this case, Eq. (15.23) becomes

$$\prod_i (y_i \phi_i)^{\nu_i} = \left(\frac{P}{P^\circ}\right)^{-\nu} K \qquad (15.24)$$

The ϕ_i's for each pure species can be evaluated from a generalized correlation once the equilibrium T and P are specified.

For pressures sufficiently low or temperatures sufficiently high, the equilibrium mixture behaves essentially as an ideal gas. In this event, each $\hat{\phi}_i = 1$, and Eq. (15.23) reduces to

$$\prod_i (y_i)^{\nu_i} = \left(\frac{P}{P^\circ}\right)^{-\nu} K \qquad (15.25)$$

In this equation the temperature-, pressure-, and composition-dependent terms are distinct and separate, and solution for any one of ε_e, T, or P, given the other two, is straightforward.

Although Eq. (15.25) holds only for an ideal-gas reaction, we can base some conclusions on it that are true in general.

1. According to Eq. (15.16), the effect of temperature on the equilibrium constant K is determined by the sign of ΔH°. Thus when ΔH° is positive, i.e., when the standard reaction is *endothermic*, an increase in T results in an increase in K. Equation (15.25) shows that an increase in K at constant P results in an increase in $\prod_i (y_i)^{\nu_i}$; this implies a shift of the reaction to the right and an increase in ε_e. Conversely, when ΔH° is negative, i.e., when the standard reaction is *exothermic*, an increase in T causes a decrease in K and a decrease in $\prod_i (y_i)^{\nu_i}$ at constant P. This implies a shift of the reaction to the left and a decrease in ε_e.

2. If the total stoichiometric number ν ($\equiv \sum_i \nu_i$) is negative, Eq. (15.25) shows that an increase in P at constant T causes an increase in $\prod_i (y_i)^{\nu_i}$, implying a shift of the reaction to the right and an increase in ε_e. If ν is positive,

an increase in P at constant T causes a decrease in $\prod_i (y_i)^{\nu_i}$, a shift of the reaction to the left, and a decrease in ε_e.

Liquid-Phase Reactions

For a reaction occurring in the liquid phase, we return to Eq. (15.13), which relates K to activities:

$$K = \prod_i (\hat{a})^{\nu_i} \tag{15.26}$$

The most common standard state for liquids is the state of the pure liquid at the system temperature and at 1 bar. The activities are then given by

$$\hat{a}_i = \frac{\hat{f}_i}{f_i^\circ}$$

where f_i° is the fugacity of pure liquid i at the temperature of the system and at 1 bar.

According to Eq. (10.89), which defines the activity coefficient,

$$\hat{f}_i = \gamma_i x_i f_i$$

where f_i is the fugacity of pure liquid i at the temperature *and pressure* of the equilibrium mixture. The activity can now be expressed as

$$\hat{a}_i = \frac{\gamma_i x_i f_i}{f_i^\circ} = \gamma_i x_i \left(\frac{f_i}{f_i^\circ} \right) \tag{15.27}$$

Since the fugacities of liquids are weak functions of pressure, the ratio f_i/f_i° is often taken as unity. However, it is readily evaluated. We write Eq. (10.30) twice, first for pure liquid i at temperature T and pressure P, and second for pure liquid i at the same temperature but at the standard-state pressure of P°. Taking the difference between these two equations gives

$$G_i - G_i^\circ = RT \ln \frac{f_i}{f_i^\circ}$$

Integration of Eq. (6.10) at constant temperature T for the change of state of pure liquid i from P° to P yields

$$G_i - G_i^\circ = \int_{P^\circ}^{P} V_i \, dP$$

Combining these two equations, we get

$$RT \ln \frac{f_i}{f_i^\circ} = \int_{P^\circ}^{P} V_i \, dP$$

Since V_i changes little with pressure for liquids (and solids), integration from P° to P gives to an excellent approximation

$$\ln \frac{f_i}{f_i^\circ} = \frac{V_i(P - P^\circ)}{RT}$$

Equation (15.26) may now be written

$$K = \left[\prod_i (x_i \gamma_i)^{\nu_i} \right] \exp \left[\frac{(P - P^\circ)}{RT} \sum_i (\nu_i V_i) \right] \tag{15.28}$$

Except for high pressures, the exponential term is close to unity and may be omitted. In this case,

$$K = \prod_i (x_i \gamma_i)^{\nu_i} \tag{15.29}$$

and the only problem is determination of the activity coefficients. An equation such as the Wilson equation [Eq. (11.23)] or the UNIFAC method can in principle be applied, and the compositions can be found from Eq. (15.29) by a complex iterative computer program. However, the relative ease of experimental investigation for liquid mixtures has worked against the application of Eq. (15.29).

If the equilibrium mixture is an ideal solution, then all the γ_i's are unity, and Eq. (15.29) becomes

$$K = \prod_i (x_i)^{\nu_i} \tag{15.30}$$

This simple relation is known as the *law of mass action*. Since liquids often form nonideal solutions, Eq. (15.30) can be expected in most instances to yield poor results.

For species known to be present in high concentration, the equation $\hat{a}_i = x_i$ is usually nearly correct, because the Lewis/Randall rule [Eq. (10.84)] always becomes valid for a species as its concentration approaches $x_i = 1$, as discussed in Sec. 11.1. For species at low concentration in aqueous solution, a different procedure has been widely adopted, because in this case the equality of $\hat{a}_i$ and x_i is usually far from correct. The method is based on the use of a fictitious or hypothetical standard state for the solute, taken as the state that would exist if the solute obeyed Henry's law up to a *molality* m of unity. In this application, Henry's law is expressed as

$$\hat{f}_i = k_i m_i \tag{15.31}$$

and it is always valid for a species whose concentration approaches zero. This hypothetical state is illustrated in Fig. 15.3. The dashed line drawn tangent to the curve at the origin represents Henry's law, and is valid in the case shown to a molality much less than unity. However, one can calculate the properties the solute would have if it obeyed Henry's law to a concentration of 1 molal, and this hypothetical state often serves as a convenient standard state for solutes.

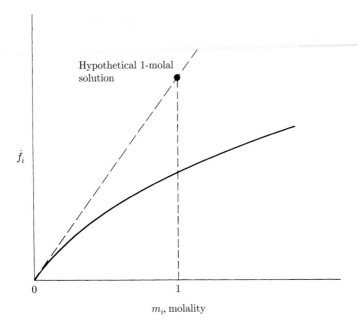

Figure 15.3: Standard state for dilute aqueous solutions.

The standard-state fugacity is

$$\hat{f}_i^\circ = k_i m_i^\circ = k_i \times 1 = k_i$$

Hence, for any species at a concentration low enough for Henry's law to hold,

$$\hat{f}_i = k_i m_i = \hat{f}_i^\circ m_i$$

and

$$\hat{a}_i = \frac{\hat{f}_i}{\hat{f}_i^\circ} = m_i \tag{15.32}$$

The advantage of this standard state is that it provides a very simple relation between activity and concentration for cases in which Henry's law is at least approximately valid. Its range does not commonly extend to a concentration of 1 molal. In the rare case where it does, the standard state is a real state of the solute. This standard state is useful only where ΔG° data are available for the standard state of a 1-molal solution, for otherwise the equilibrium constant cannot be evaluated by Eq. (15.14).

15.7 Equilibrium Conversions for Single Reactions

Suppose a single reaction occurs in a *homogeneous* system, and suppose the equilibrium constant is known. In this event, the calculation of the phase composition at equilibrium is straightforward if the phase is assumed an ideal gas [Eq. (15.25)] or an ideal solution [Eq. (15.24) or (15.30)]. When an assumption of ideality is not reasonable, the problem is still tractable for gas-phase reactions through application of an equation of state and solution by computer. For *heterogeneous* systems, where more than one phase is present, the problem is more complicated and requires the superposition of the criterion for phase equilibrium developed in Sec. 10.6. At equilibrium, there can be no tendency for change to occur, either by mass transfer between phases or by chemical reaction. We present in what follows, mainly by example, the procedures in use for equilibrium calculations, first, for single-phase reactions, and second, for heterogeneous reactions.

Single-Phase Reactions

The following examples illustrate application of the equations developed in the preceding section.

> **Example 15.5**   The water-gas-shift reaction
>
> $$CO(g) + H_2O(g) \rightarrow CO_2(g) + H_2(g)$$
>
> is carried out under the different sets of conditions described below. Calculate the fraction of steam reacted in each case. Assume the mixture behaves as an ideal gas.
>
> (*a*) The reactants consist of 1 mol of H_2O vapor and 1 mol of CO. The temperature is 1100 K and the pressure is 1 bar.
>
> (*b*) Same as (*a*) except that the pressure is 10 bar.
>
> (*c*) Same as (*a*) except that 2 mol of N_2 is included in the reactants.
>
> (*d*) The reactants are 2 mol of H_2O and 1 mol of CO. Other conditions are the same as in (*a*).
>
> (*e*) The reactants are 1 mol of H_2O and 2 mol of CO. Other conditions are the same as in (*a*).
>
> (*f*) The initial mixture consists of 1 mol of H_2O, 1 mol of CO, and 1 mol of CO_2. Other conditions are the same as in (*a*).
>
> (*g*) Same as (*a*) except that the temperature is 1650 K.

SOLUTION   (*a*) For the given reaction at 1100 K, $10^4/T = 9.05$, and Fig. 15.2 provides the value, $\ln K = 0$ or $K = 1$. For this reaction $\nu = \sum_i \nu_i = 1 + 1 - 1 - 1 = 0$. Since the reaction mixture is an ideal gas, Eq. (15.25) applies, and here becomes

$$\frac{y_{H_2} y_{CO_2}}{y_{CO} y_{H_2O}} = K = 1 \qquad\qquad (A)$$

By Eq. (15.5), we have

$$y_{CO} = \frac{1 - \varepsilon_e}{2} \qquad y_{H_2O} = \frac{1 - \varepsilon_e}{2}$$

$$y_{CO_2} = \frac{\varepsilon_e}{2} \qquad y_{H_2} = \frac{\varepsilon_e}{2}$$

Substitution of these values into Eq. (A) gives

$$\frac{\varepsilon_e^2}{(1 - \varepsilon_e)^2} = 1 \qquad \text{or} \qquad \varepsilon_e = 0.5$$

Therefore the fraction of the steam that reacts is 0.5.

(b) Since $\nu = 0$, the increase in pressure has no effect on the ideal-gas reaction, and ε_e is still 0.5.

(c) The N_2 does not take part in the reaction, and serves only as a diluent. It does increase the initial number of moles n_0 from 2 to 4, and the mole fractions are all reduced by a factor of 2. However, Eq. (A) is unchanged and reduces to the same expression as before. Therefore, ε_e is again 0.5.

(d) In this case the mole fractions at equilibrium are

$$y_{CO} = \frac{1 - \varepsilon_e}{3} \qquad y_{H_2O} = \frac{2 - \varepsilon_e}{3}$$

$$y_{CO_2} = \frac{\varepsilon_e}{3} \qquad y_{H_2} = \frac{\varepsilon_e}{3}$$

and Eq. (A) becomes

$$\frac{\varepsilon_e^2}{(1 - \varepsilon_e)(2 - \varepsilon_e)} = 1 \qquad \text{or} \qquad \varepsilon_e = 0.667$$

The fraction of steam that reacts is then $0.667/2 = 0.333$.

(e) Here the expressions for y_{CO} and y_{H_2O} are interchanged, but this leaves the equilibrium equation the same as in (d). Therefore $\varepsilon_e = 0.667$, and the fraction of steam that reacts is 0.667.

(f) In this case Eq. (A) becomes

$$\frac{\varepsilon_e(1 + \varepsilon_e)}{(1 - \varepsilon_e)^2} = 1 \qquad \text{or} \qquad \varepsilon_e = 0.333$$

The fraction of steam reacted is 0.333.

(g) At 1650 K, $10^4/T = 6.06$, and from Fig. 15.2 we have $\ln K = -1.15$ and $K = 0.316$. Therefore Eq. (A) becomes

$$\frac{\varepsilon_e^2}{(1 - \varepsilon_e)^2} = 0.316 \qquad \text{or} \qquad \varepsilon_e = 0.36$$

Since the reaction is exothermic, the conversion decreases with increasing temperature.

Example 15.6 Estimate the maximum conversion of ethylene to ethanol by vapor-phase hydration at 250°C and 35 bars for an initial steam-to-ethylene ratio of 5.

SOLUTION The calculation of K for this reaction is treated in Example 15.4. For a temperature of 250°C or 523.15 K the calculation yields

$$K = 10.02 \times 10^{-3}$$

The appropriate equilibrium expression is Eq. (15.23). This equation requires evaluation of the fugacity coefficients of the species present in the equilibrium mixture. This may be accomplished with Eq. (10.69). However, the calculations involve iteration, because the fugacity coefficients are functions of composition. For purposes of illustration, we carry out only the first iteration, based on the assumption that the reaction mixture is an ideal solution. In this case Eq. (15.23) reduces to Eq. (15.24), which requires fugacity coefficients of the *pure* gases of the reacting mixture at the equilibrium T and P. Since $\nu = \sum_i \nu_i = -1$, this equation becomes

$$\frac{y_{EtOH}\phi_{EtOH}}{y_{C_2H_4}\phi_{C_2H_4}y_{H_2O}\phi_{H_2O}} = \left(\frac{P}{P^\circ}\right)(10.02 \times 10^{-3}) \tag{A}$$

Computations based on Eq. (10.64) in conjunction with Eqs. (3.50) and (3.51) provide values represented by

$$\text{PHIB(TR,PR,OMEGA)} \equiv \phi_i$$

The results of these calculations are summarized in the following table:

	T_c/K	P_c/bar	ω_i	T_{r_i}	P_{r_i}	B^0	B^1	ϕ_i
C_2H_4	282.3	50.40	0.087	1.853	0.694	-0.074	0.126	0.977
H_2O	647.1	220.55	0.345	0.808	0.159	-0.511	-0.281	0.887
EtOH	513.9	61.48	0.645	1.018	0.569	-0.327	-0.021	0.827

The critical data and ω_i's are from App. B. The temperature and pressure in all cases are 523.15 K and 35 bar. Substitution of values for the ϕ_i's and for (P/P°) into Eq. (A) gives

$$\frac{y_{EtOH}}{y_{C_2H_4}y_{H_2O}} = \frac{(0.977)(0.887)}{(0.827)}(35)(10.02 \times 10^{-3}) = 0.367 \tag{B}$$

By Eq. (15.5),

$$y_{C_2H_4} = \frac{1 - \varepsilon_e}{6 - \varepsilon_e} \qquad y_{H_2O} = \frac{5 - \varepsilon_e}{6 - \varepsilon_e} \qquad y_{EtOH} = \frac{\varepsilon_e}{6 - \varepsilon_e}$$

Substituting these into Eq. (B) gives

$$\frac{\varepsilon_e(6 - \varepsilon_e)}{(5 - \varepsilon_e)(1 - \varepsilon_e)} = 0.367$$

This reduces to

$$\varepsilon_e^2 - 6.000\varepsilon_e + 1.342 = 0$$

Solving this quadratic equation gives

$$\varepsilon_e = 0.233$$

for the smaller root. Since the larger root is greater than unity, it does not represent a physically possible result. The maximum conversion of ethylene to ethanol under the stated conditions is therefore 23.3 percent.

In this reaction, increasing the temperature decreases K and hence the conversion. Increasing the pressure increases the conversion. Equilibrium considerations therefore suggest that the operating pressure be as high as possible (limited by condensation), and the temperature as low as possible. However, even with the best catalyst known, the minimum temperature for a reasonable reaction rate is about $150°C$. This is an instance where both equilibrium and reaction rate influence the commercialization of a reaction process.

The equilibrium conversion is a function of temperature, pressure, and the steam-to-ethylene ratio in the feed. The effects of all three variables are shown in Fig. 15.4. The curves in this figure come from calculations just like those illustrated in this example, except that a less precise relation for K as a function of T was used.

Example 15.7 In a laboratory investigation, acetylene is catalytically hydrogenated to ethylene at $1,120°C$ and 1 bar. If the feed is an equimolar mixture of acetylene and hydrogen, what is the composition of the product stream at equilibrium?

SOLUTION The required reaction is obtained by addition of the two formation reactions written as follows:

$$C_2H_2 \rightarrow 2C + H_2 \quad (1)$$

$$2C + 2H_2 \rightarrow C_2H_4 \quad (2)$$

The sum of reactions (1) and (2) is the hydrogenation reaction

$$C_2H_2 + H_2 \rightarrow C_2H_4$$

Also

$$\Delta G° = \Delta G_1° + \Delta G_2°$$

By Eq. (15.14),

$$-RT \ln K = -RT \ln K_1 - RT \ln K_2$$

or

$$K = K_1 K_2$$

Data for both reactions (1) and (2) are given by Fig. 15.2. For $1,120°C$ [1,393 K], $10^4/T = 7.18$, the following values are read from the graph:

$$\ln K_1 = 12.9 \qquad K_1 = 4.0 \times 10^5$$

$$\ln K_2 = -12.9 \qquad K_2 = 2.5 \times 10^{-6}$$

Therefore

$$K = K_1 K_2 = 1.0$$

At this elevated temperature and for a pressure of 1 bar, we can safely assume ideal gases. Application of Eq. (15.25) leads to the expression

$$\frac{y_{C_2H_4}}{y_{H_2} y_{C_2H_2}} = 1$$

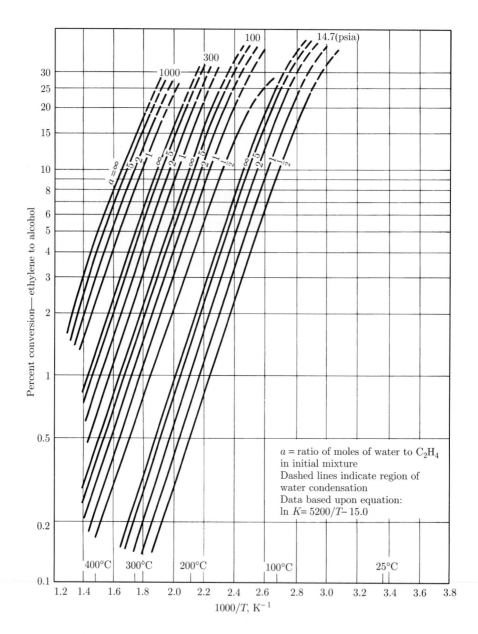

Figure 15.4: Equilibrium conversion of ethylene to ethyl alcohol in the vapor phase.

On the basis of one mole initially of each reactant, Eq. (15.5) gives

$$y_{H_2} = y_{C_2H_2} = \frac{1 - \varepsilon_e}{2 - \varepsilon_e} \qquad \text{and} \qquad y_{C_2H_4} = \frac{\varepsilon_e}{2 - \varepsilon_e}$$

Therefore

$$\frac{\varepsilon_e(2 - \varepsilon_e)}{(1 - \varepsilon_e)^2} = 1$$

The smaller root of this quadratic expression (the larger is > 1) is

$$\varepsilon_e = 0.293$$

The equilibrium composition of the product gas is then

$$y_{H_2} = y_{C_2H_2} = \frac{1 - 0.293}{2 - 0.293} = 0.414$$

$$y_{C_2H_4} = \frac{0.293}{2 - 0.293} = 0.172$$

Example 15.8 Acetic acid is esterified in the liquid phase with ethanol at $100°C$ and atmospheric pressure to produce ethyl acetate and water according to the reaction

$$CH_3COOH(l) + C_2H_5OH(l) \rightarrow CH_3COOC_2H_5(l) + H_2O(l)$$

If initially there is one mole each of acetic acid and ethanol, estimate the mole fraction of ethyl acetate in the reacting mixture at equilibrium.

SOLUTION Data for $\Delta H^\circ_{f_{298}}$ and $\Delta G^\circ_{f_{298}}$ are given for liquid acetic acid, ethanol, and water in Table C.4. For liquid ethyl acetate, the corresponding values are (Thermodynamics Research Center, Texas A & M Univ. System, College Station, Texas):

$$\Delta H^\circ_{f_{298}} = -480,000 \text{ J} \qquad \text{and} \qquad \Delta G^\circ_{f_{298}} = -332,200 \text{ J}$$

The values of ΔH°_{298} and ΔG°_{298} for the reaction are therefore

$$\Delta H^\circ_{298} = -480,000 - 285,830 + 484,500 + 277,690 = -3,640 \text{ J}$$

$$\Delta G^\circ_{298} = -332,200 - 237,130 + 389,900 + 174,780 = -4,650 \text{ J}$$

By Eq. (15.14),

$$\ln K_{298} = \frac{-\Delta G^\circ_{298}}{RT} = \frac{4,650}{(8.314)(298.15)} = 1.876$$

$$K_{298} = 6.527$$

For the small temperature change from 298.15 to 373.15 K, Eq. (15.17) is adequate for estimation of K. Thus

$$\ln \frac{K_{373}}{K_{298}} = \frac{-\Delta H^\circ_{298}}{R} \left(\frac{1}{373.15} - \frac{1}{298.15} \right)$$

or

$$\ln \frac{K_{373}}{6.527} = \frac{3,640}{8.314} \left(\frac{1}{373.15} - \frac{1}{298.15} \right) = -0.295$$

$$K_{373} = (6.527)(0.744) = 4.859$$

For the given reaction, Eq. (15.5), with x replacing y, yields

$$x_{\text{AcH}} = x_{\text{EtOH}} = \frac{1 - \varepsilon_e}{2}$$

$$x_{\text{EtAc}} = x_{\text{H}_2\text{O}} = \frac{\varepsilon_e}{2}$$

Since the pressure is low, Eq. (15.29) is applicable. In the absence of data for the activity coefficients in this complex system, we assume that the reacting species form an ideal solution. In this case Eq. (15.30) is employed, giving

$$K = \frac{x_{\text{EtAc}} x_{\text{H}_2\text{O}}}{x_{\text{AcH}} x_{\text{EtOH}}}$$

Thus

$$4.859 = \left(\frac{\varepsilon_e}{1 - \varepsilon_e}\right)^2$$

from which

$$\varepsilon_e = 0.6879$$

and

$$x_{\text{EtAc}} = 0.6879/2 = 0.344$$

This result is in good agreement with experiment, which yields a value of about 0.33. The assumption of an ideal solution evidently introduces little error, probably because of cancellation effects.

Example 15.9 The gas-phase oxidation of SO_2 to SO_3 is carried out at a pressure of 1 bar with 20% excess air in an adiabatic reactor. Assuming that the reactants enter at 25°C and that equilibrium is attained at the exit, determine the composition and temperature of the product stream from the reactor.

SOLUTION   The reaction is

$$SO_2 + \tfrac{1}{2}O_2 \rightarrow SO_3$$

for which

$$\Delta H^\circ_{298} = -98{,}890 \text{ J mol}^{-1}$$

$$\Delta G^\circ_{298} = -70{,}866 \text{ J mol}^{-1}$$

On the basis of one mole of SO_2 entering the reactor,

$$\text{Moles } O_2 \text{ entering} = (0.5)(1.2) = 0.6$$
$$\text{Moles } N_2 \text{ entering} = (0.6)(79/21) = 2.257$$

The amount of each species in the product stream is found by application of Eq. (15.4):

$$\text{Moles } SO_2 = 1 - \varepsilon_e$$
$$\text{Moles } O_2 = 0.6 - 0.5\varepsilon_e$$
$$\text{Moles } SO_3 = \varepsilon_e$$
$$\text{Moles } N_2 = 2.257$$

$$\overline{\text{Total moles} = 3.857 - 0.5\varepsilon_e}$$

Two equations must be written if we are to solve for both ε_e and the temperature. They are an energy balance and an equilibrium equation. For the energy balance, we proceed as in Example 4.7:

$$\Delta H_{298}^{\circ} \varepsilon_e + \Delta H_P^{\circ} = \Delta H = 0 \qquad (A)$$

where all enthalpies are on the basis of 1 mol SO_2 entering the reactor. The enthalpy change of the products as they are heated from 298.15 K to T is given by

$$\Delta H_P^{\circ} = \langle C_P^{\circ} \rangle_H (T - 298.15) \qquad (B)$$

where $\langle C_P^{\circ} \rangle_H$ is defined as the *total* heat capacity of the product stream:

$$\langle C_P^{\circ} \rangle_H \equiv \sum_i n_i \langle C_{P_i}^{\circ} \rangle_H$$

Data from Table C.1 provide $\langle C_{P_i}^{\circ} \rangle_H / R$ values:

SO_2:   MCPH(298.15,T;5.699,0.801E-3,0.0,-1.015E+5)
O_2:   MCPH(298.15,T;3.639,0.506E-3,0.0,-0.227E+5)
SO_3:   MCPH(298.15,T;8.060,1.056E-3,0.0,-2.028E+5)
N_2:   MCPH(298.15,T;3.280,0.593E-3,0.0,0.040E+5)

Equations (A) and (B) combine to yield

$$\Delta H_{298}^{\circ} \varepsilon_e + \langle C_P^{\circ} \rangle_H (T - 298.15) = 0$$

Solution for T gives

$$T = \frac{-\Delta H_{298}^{\circ} \varepsilon_e}{\langle C_P^{\circ} \rangle_H} + 298.15 \qquad (C)$$

At the conditions of temperature and pressure of the equilibrium state, the assumption of ideal gases is fully justified, and the equilibrium constant is therefore given by Eq. (15.25), which here becomes

$$K = \left(\frac{\varepsilon_e}{1 - \varepsilon_e} \right) \left(\frac{3.857 - 0.5\varepsilon_e}{0.6 - 0.5\varepsilon_e} \right)^{0.5} \qquad (D)$$

Since $-\ln K = \Delta G^{\circ}/RT$, Eq. (15.20) can be written

$$-\ln K = \frac{\Delta G_0^{\circ} - \Delta H_0^{\circ}}{RT_0} + \frac{\Delta H_0^{\circ}}{RT} + \frac{1}{T} \int_{T_0}^{T} \frac{\Delta C_P^{\circ}}{R} dT - \int_{T_0}^{T} \frac{\Delta C_P^{\circ}}{R} \frac{dT}{T}$$

Substitution of numerical values yields

$$\ln K = -11.3054 + \frac{11,894.4}{T} + \text{IDCPS} - \tfrac{1}{T}*\text{IDCPH} \qquad (E)$$

where

$$\text{IDCPS} \equiv \text{IDCPS}(298.15,T;0.5415,0.002E-3,0.0,-0.8995E+5)$$

$$\text{IDCPH} \equiv \text{IDCPH}(298.15,T;0.5415,0.002E-3,0.0,-0.8995E+5)$$

These expressions for the computed values of the integrals show parameters ΔA, ΔB, ΔC, and ΔD as evaluated from data of Table C.1.

An iteration scheme for solution of these equations for ε_e and T that converges fairly rapidly is as follows:

1. Assume a starting value for T.

2. Evaluate IDCPH and IDCPS at this value of T.

3. Solve Eq. (E) for K and Eq. (D) for ε_e, probably by trial.

4. Evaluate $\langle C_P^\circ \rangle_H$ and solve Eq. (C) for T.

5. Find a new value of T as the arithmetic mean of the value just calculated and the initial value; return to step 2.

This scheme converges on the values $\varepsilon_e = 0.77$ and $T = 855.7$ K. For the product stream,

$$y_{SO_2} = \frac{1 - 0.77}{3.857 - (0.5)(0.77)} = \frac{0.23}{3.472} = 0.0662$$

$$y_{O_2} = \frac{0.6 - (0.5)(0.77)}{3.472} = \frac{0.215}{3.472} = 0.0619$$

$$y_{SO_3} = \frac{0.77}{3.472} = 0.2218$$

$$y_{N_2} = \frac{2.257}{3.472} = 0.6501$$

Reactions in Heterogeneous Systems

When liquid and gas phases are both present in an equilibrium mixture of reacting species, Eq. (10.44), a criterion of vapor/liquid equilibrium, must be satisfied along with the equation of chemical-reaction equilibrium. There is considerable choice in the method of treatment of such cases. For example, consider a reaction of gas A and water B to form an aqueous solution C. The reaction may be assumed to occur entirely in the gas phase with simultaneous transfer of material between phases to maintain phase equilibrium. In this case, the equilibrium constant is evaluated from ΔG° data based on standard states for the species as gases, i.e., the ideal-gas states at 1 bar and the reaction temperature. On the other hand, the reaction may be assumed to occur in the liquid phase, in which case ΔG° is based on standard states for the species as liquids. Alternatively, the reaction may be written

$$A(g) + B(l) \rightarrow C(aq)$$

in which case the ΔG° value is for mixed standard states: C as a solute in an ideal 1-molal aqueous solution, B as a pure liquid at 1 bar, and A as a pure ideal gas at 1 bar. For this choice of standard states, the equilibrium constant as given by Eq. (15.13) becomes

$$K = \frac{\hat{a}_C}{\hat{a}_B \hat{a}_A} = \frac{m_C}{(\gamma_B x_B)(\hat{f}_A / P^\circ)}$$

The last term arises from Eq. (15.32) applied to species C, Eq. (15.27) applied to B with $f_B / f_B^\circ = 1$, and the fact that $\hat{a}_A = \hat{f}_A / P^\circ$ for species A in the gas phase. Since K depends on the standard states, this value of K is not the same as that obtained when the standard state for each species is chosen as the ideal-gas state at 1 bar. However, all methods theoretically lead to the same equilibrium composition, provided Henry's law as applied to species C in solution is valid. In practice, a particular choice of standard states may simplify calculations or yield more accurate results, because it makes better use of the available data. The nature of the calculations required for heterogeneous reactions is illustrated in the following example.

> **Example 15.10** Estimate the compositions of the liquid and vapor phases when ethylene reacts with water to form ethanol at 200°C and 34.5 bar, conditions which assure the presence of both liquid and vapor phases. The reaction vessel is maintained at 34.5 bar by connection to a source of ethylene at this pressure. Assume no other reactions to occur.
>
> SOLUTION   According to the phase rule (see Sec. 15.8), the system has two degrees of freedom. Specification of both the temperature and the pressure leaves no other degrees of freedom, and fixes the intensive state of the system, independent of the initial amounts of reactants. Therefore, material-balance equations do not enter into the solution of this problem, and we can make no use of equations that relate compositions to the reaction coordinate. Instead, phase-equilibrium relations must be employed to provide a sufficient number of equations to allow solution for the unknown compositions.
>
> The most convenient approach to this problem is to regard the chemical reaction as occurring in the vapor phase. Thus
>
> $$C_2H_4(g) + H_2O(g) \rightarrow C_2H_5OH(g)$$
>
> and the standard states are those of the pure ideal gases at 1 bar. For these standard states, the equilibrium expression is Eq. (15.22), which in this case becomes
>
> $$K = \frac{\hat{f}_{\text{EtOH}}}{\hat{f}_{C_2H_4}\hat{f}_{H_2O}} P^\circ \qquad (A)$$
>
> where the standard-state pressure P° is 1 bar (expressed in appropriate units). A general expression for $\ln K$ as a function of T is provided by the results of Example 15.4. For 200°C [473.15 K], this equation yields
>
> $$\ln K = -3.473 \qquad K = 0.0310$$
>
> The task now is to incorporate the phase-equilibrium equations,
>
> $$\hat{f}_i^v = \hat{f}_i^l$$
>
> into Eq. (A) and to relate the fugacities to the compositions in such a way that the equations can be readily solved. Equation (A) may be written
>
> $$K = \frac{\hat{f}_{\text{EtOH}}^v}{\hat{f}_{C_2H_4}^v \hat{f}_{H_2O}^v} P^\circ = \frac{\hat{f}_{\text{EtOH}}^l}{\hat{f}_{C_2H_4}^l \hat{f}_{H_2O}^l} P^\circ \qquad (B)$$
>
> The liquid-phase fugacities are related to activity coefficients by Eq. (10.89):

$$\hat{f}_i^l = x_i \gamma_i f_i^l \tag{C}$$

and the vapor-phase fugacity is related to the fugacity coefficient by Eq. (10.47):

$$\hat{f}_i^v = y_i \hat{\phi}_i P \tag{D}$$

Elimination of the fugacities in Eq. (B) by Eqs. (C) and (D) gives

$$K = \frac{x_{\text{EtOH}} \gamma_{\text{EtOH}} f_{\text{EtOH}}^l P^\circ}{(y_{\text{C}_2\text{H}_4} \hat{\phi}_{\text{C}_2\text{H}_4} P)(x_{\text{H}_2\text{O}} \gamma_{\text{H}_2\text{O}} f_{\text{H}_2\text{O}}^l)} \tag{E}$$

The fugacity f_i^l is for pure liquid i at the temperature and pressure of the system. However, pressure has small effect on the fugacity of a liquid, and to a good approximation we can write

$$f_i^l = f_i^{\text{sat}}$$

and therefore by Eqs. (10.38) and (10.39),

$$f_i^l = \phi_i^{\text{sat}} P_i^{\text{sat}} \tag{F}$$

In this equation ϕ_i^{sat} is the fugacity coefficient of pure saturated i (either liquid or vapor) evaluated at the temperature of the system and at P_i^{sat}, the vapor pressure of pure species i. The assumption that the vapor phase is an ideal solution allows substitution of $\phi_{\text{C}_2\text{H}_4}$ for $\hat{\phi}_{\text{C}_2\text{H}_4}$, where $\phi_{\text{C}_2\text{H}_4}$ is the fugacity coefficient of pure ethylene at the system T and P. With this substitution and that of Eq. (F), Eq. (E) becomes

$$K = \frac{x_{\text{EtOH}} \gamma_{\text{EtOH}} \phi_{\text{EtOH}}^{\text{sat}} P_{\text{EtOH}}^{\text{sat}} P^\circ}{(y_{\text{C}_2\text{H}_4} \phi_{\text{C}_2\text{H}_4} P)(x_{\text{H}_2\text{O}} \gamma_{\text{H}_2\text{O}} \phi_{\text{H}_2\text{O}}^{\text{sat}} P_{\text{H}_2\text{O}}^{\text{sat}})} \tag{G}$$

where the standard-state pressure P° is 1 bar, expressed in the units used for pressure.

In addition to Eq. (G) the following expressions can be written. Since $\sum_i y_i = 1$,

$$y_{\text{C}_2\text{H}_4} = 1 - y_{\text{EtOH}} - y_{\text{H}_2\text{O}} \tag{H}$$

We can eliminate y_{EtOH} and $y_{\text{H}_2\text{O}}$ in this equation in favor of x_{EtOH} and $x_{\text{H}_2\text{O}}$ by the vapor/liquid equilibrium relation:

$$\hat{f}_i^v = \hat{f}_i^l$$

Combining this with Eqs. (C), (D), and (F), we obtain

$$y_i = \frac{\gamma_i x_i \phi_i^{\text{sat}} P_i^{\text{sat}}}{\phi_i P} \tag{I}$$

where ϕ_i replaces $\hat{\phi}_i$ because of the assumption that the vapor phase is an ideal solution. Equations (H) and (I) yield

$$y_{\text{C}_2\text{H}_4} = 1 - \frac{x_{\text{EtOH}} \gamma_{\text{EtOH}} \phi_{\text{EtOH}}^{\text{sat}} P_{\text{EtOH}}^{\text{sat}}}{\phi_{\text{EtOH}} P} - \frac{x_{\text{H}_2\text{O}} \gamma_{\text{H}_2\text{O}} \phi_{\text{H}_2\text{O}}^{\text{sat}} P_{\text{H}_2\text{O}}^{\text{sat}}}{\phi_{\text{H}_2\text{O}} P} \tag{J}$$

Since ethylene is far more volatile than ethanol or water, we assume that $x_{\text{C}_2\text{H}_4} = 0$. Then

$$x_{\text{H}_2\text{O}} = 1 - x_{\text{EtOH}} \tag{K}$$

Equations (G), (J), and (K) form the basis for solution of the problem. The three primary variables in these equations are x_{H_2O}, x_{EtOH}, and $y_{C_2H_4}$, and all other quantities are either given or are determined from correlations of data. The values of P_i^{sat} are

$$P_{H_2O}^{\text{sat}} = 15.55 \qquad P_{EtOH}^{\text{sat}} = 30.22 \text{ bar}$$

The quantities ϕ_i^{sat} and ϕ_i are found from the generalized correlation represented by Eq. (10.64) with B^0 and B^1 given by Eqs. (3.50) and (3.51). Computed results are represented by `PHIB(TR,PR,OMEGA)`. With $T = 473.15$ K, $P = 34.5$ bar, and critical data and the ω's from App. B, the computations provide the following values:

	T_c/K	P_c/bar	ω_i	T_{r_i}	P_{r_i}	$P_{r_i}^{\text{sat}}$	B^0	B^1	ϕ_i	ϕ_i^{sat}
EtOH	513.9	61.48	0.645	0.921	0.561	0.492	-0.399	-0.104	0.753	0.780
H_2O	647.1	220.55	0.345	0.731	0.156	0.071	-0.613	-0.502	0.846	0.926
C_2H_4	282.3	50.40	0.087	1.676	0.685	$\cdots$	-0.102	0.119	0.963	$\cdots$

Substitution of all values so far determined into Eqs. (G), (J), and (K) reduces these three equations to the following:

$$K = \frac{0.0493 x_{EtOH} \gamma_{EtOH}}{y_{C_2H_4} x_{H_2O} \gamma_{H_2O}} \qquad (L)$$

$$y_{C_2H_4} = 1 - 0.907 x_{EtOH} \gamma_{EtOH} - 0.493 x_{H_2O} \gamma_{H_2O} \qquad (M)$$

$$x_{H_2O} = 1 - x_{EtOH} \qquad (K)$$

The only remaining undetermined thermodynamic properties are γ_{H_2O} and γ_{EtOH}. Because of the highly nonideal behavior of a liquid solution of ethanol and water, these must be determined from experimental data. The required data, found from VLE measurements, are given by Otsuki and Williams.[7] From their results for the ethanol/water system one can estimate values of γ_{H_2O} and γ_{EtOH} at 200°C. (Pressure has little effect on the activity coefficients of liquids.)

A procedure for solution of the foregoing three equations is as follows.

1. Assume a value for x_{EtOH} and calculate x_{H_2O} by Eq. (K).

2. Determine γ_{H_2O} and γ_{EtOH} from data in the reference cited.

3. Calculate $y_{C_2H_4}$ by Eq. (M).

4. Calculate K by Eq. (L) and compare with the value of 0.0310 determined from standard-reaction data.

5. If the two values agree, the assumed value of x_{EtOH} is correct. If they do not agree, assume a new value of x_{EtOH} and repeat the procedure.

---

[7]H. Otsuki and F. C. Williams, *Chem. Engr. Progr. Symp. Series No. 6*, vol. 49, pp. 55–67, 1953.

If we take $x_{EtOH} = 0.06$, then by Eq. (K), $x_{H_2O} = 0.94$, and from the reference cited,

$$\gamma_{EtOH} = 3.34 \qquad \text{and} \qquad \gamma_{H_2O} = 1.00$$

By Eq. (M),

$$y_{C_2H_4} = 1 - (0.907)(3.34)(0.06) - (0.493)(1.00)(0.94) = 0.355$$

The value of K given by Eq. (L) is then

$$K = \frac{(0.0493)(0.06)(3.34)}{(0.355)(0.94)(1.00)} = 0.0296$$

This result is in close enough agreement with the value (0.0310) found from standard-reaction data that we can take $x_{EtOH} = 0.06$ and $x_{H_2O} = 0.94$ as the liquid-phase compositions. The remaining vapor-phase compositions ($y_{C_2H_4}$ has already been determined as 0.356) are found by solution of Eq. (I) for y_{H_2O} or y_{EtOH}. All results are summarized in the following table.

	x_i	y_i
EtOH	0.060	0.180
H_2O	0.940	0.464
C_2H_4	0.000	0.356
	$\sum_i x_i = 1.000$	$\sum_i y_i = 1.000$

These results are probably reasonable estimates of actual values, provided no other reactions take place.

15.8   Phase Rule & Duhem's Theorem for Reacting Systems

The phase rule (applicable to intensive properties) as discussed in Secs. 2.8 and 12.2 for nonreacting systems of π phases and N chemical species is

$$F = 2 - \pi + N$$

It must be modified for application to systems in which chemical reactions occur. The phase-rule variables are unchanged: temperature, pressure, and $N - 1$ mole fractions in each phase. The total number of these variables is $2 + (N - 1)(\pi)$. The same phase-equilibrium equations apply as before, and they number $(\pi - 1)(N)$. However, Eq. (15.8) provides for each independent reaction an additional relation that must be satisfied at equilibrium. Since the μ_i's are functions of temperature, pressure, and the phase compositions, Eq. (15.8) represents a relation connecting phase-rule variables. If there are r independent chemical reactions at equilibrium within the system, then there is a total of $(\pi - 1)(N) + r$ independent equations relating the phase-rule variables. Taking the difference between the number of variables and the number of equations, we obtain

$$F = [2 + (N - 1)(\pi)] - [(\pi - 1)(N) + r]$$

or

$$F = 2 - \pi + N - r \qquad (15.33)$$

This is the phase rule for reacting systems.

The only remaining problem for application is to determine the number of independent chemical reactions. This can be done systematically as follows.

1. Write chemical equations for the formation, from the *constituent elements*, of each chemcial compound considered present in the system.

2. Combine these equations so as to eliminate from them all elements not considered present *as elements* in the system. A systematic procedure is to select one equation and combine it with each of the others of the set to eliminate a particular element. Then the process is repeated to eliminate another element from the new set of equations. This is done for each element eliminated [see Example 15.11(d)], and usually reduces the set by one equation for each element eliminated. However, the simultaneous elimination of two or more elements may occur.

The set of r equations resulting from this reduction procedure is a complete set of independent reactions for the N species considered present in the system. However, more than one such set is possible, depending on how the reduction procedure is carried out, but all sets number r and are equivalent. The reduction procedure also ensures the following relation:

$r \geq$ number of compounds present in the system
$\quad$ − number of constituent elements *not* present *as elements*

The phase-equilibrium and chemical-reaction-equilibrium equations are the only ones considered in the foregoing treatment as interrelating the phase-rule variables. However, in certain situations *special constraints* may be placed on the system that allow additional equations to be written over and above those considered in the development of Eq. (15.33). If the number of equations resulting from special constraints is s, then Eq. (15.33) must be modified to take account of these s additional equations. The still more general form of the phase rule that results is

$$F = 2 - \pi + N - r - s \qquad (15.34)$$

Example 15.11 shows how Eqs. (15.33) and (15.34) may be applied to specific systems.

Example 15.11 Determine the number of degrees of freedom F for each of the following systems.

(a) A system of two miscible nonreacting species which exists as an azeotrope in vapor/liquid equilibrium.

(b) A system prepared by partially decomposing $CaCO_3$ into an evacuated space.

(c) A system prepared by partially decomposing NH_4Cl into an evacuated space.

(d) A system consisting of the gases CO, CO_2, H_2, H_2O, and CH_4 in chemical equilibrium.

SOLUTION   (a) The system consists of two nonreacting species in two phases. If there were no azeotrope, we would apply Eq. (15.33):

$$F = 2 - \pi + N - r = 2 - 2 + 2 - 0 = 2$$

This is the usual result for binary VLE. However, a special constraint is imposed on the system; it is an azeotrope. This provides an equation, $x_1 = y_1$, not considered in the development of Eq. (15.33). Thus, we apply Eq. (15.34) with $s = 1$. The result is that $F = 1$. If the system is to be an azeotrope, then just one phase-rule variable—T, P, or x_1 $(= y_1)$—may be arbitrarily specified.

(b) Here there is a single chemical reaction:

$$CaCO_3(s) \rightarrow CaO(s) + CO_2(g)$$

and $r = 1$. There are three chemical species and three phases—solid $CaCO_3$, solid CaO, and gaseous CO_2. One might think a special constraint has been imposed by the requirement that the system be prepared in a special way—by decomposing $CaCO_3$. This is not the case, because no equation connecting the phase-rule variables can be written as a result of this requirement. Therefore

$$F = 2 - \pi + N - r - s = 2 - 3 + 3 - 1 - 0 = 1$$

and there is a single degree of freedom. This is the reason that $CaCO_3$ exerts a fixed decomposition pressure at fixed T.

(c) The chemical reaction here is

$$NH_4Cl(s) \rightarrow NH_3(g) + HCl(g)$$

Three species, but only two phases, are present in this case, solid NH_4Cl and a gas mixture of NH_3 and HCl. In addition, there is a special constraint, because the requirement that the system be formed by the decomposition of NH_4Cl means that the gas phase is equimolar in NH_3 and HCl. Thus a special equation, $y_{NH_3} = y_{HCl}$ $(= 0.5)$, connecting the phase-rule variables can be written. Application of Eq. (15.34) gives

$$F = 2 - \pi + N - r - s = 2 - 2 + 3 - 1 - 1 = 1$$

and the system has but one degree of freedom. This result is the same as that for part (b), and it is a matter of experience that NH_4Cl has a given decomposition pressure at a given temperature. This conclusion is reached quite differently in the two cases.

(d) This system contains five species, all in a single gas phase. There are no special constraints. Only r remains to be determined. The formation reactions for the compounds present are

$$C + \tfrac{1}{2}O_2 \rightarrow CO \qquad (A)$$

$$C + O_2 \rightarrow CO_2 \qquad (B)$$

$$H_2 + \tfrac{1}{2}O_2 \rightarrow H_2O \qquad (C)$$

$$C + 2H_2 \rightarrow CH_4 \qquad (D)$$

Systematic elimination of C and O_2, the elements not present in the system, leads to two equations. One such pair of equations is obtained in the following way. We eliminate C from this set of equations by combining Eq. (B), first with Eq. (A) and then with Eq. (D). The two resulting reactions are

$$\text{From } (B) \text{ and } (A): \quad CO + \tfrac{1}{2}O_2 \rightarrow CO_2 \qquad (E)$$

$$\text{From } (B) \text{ and } (D): \quad CH_4 + O_2 \rightarrow 2H_2 + CO_2 \qquad (F)$$

Equations (C), (E), and (F) are the new set, and we now eliminate O_2 by combining Eq. (C), first with Eq. (E) and then with Eq. (F). This gives

$$\text{From } (C) \text{ and } (E): \quad CO_2 + H_2 \rightarrow CO + H_2O \qquad (G)$$

$$\text{From } (C) \text{ and } (F): \quad CH_4 + 2H_2O \rightarrow CO_2 + 4H_2 \qquad (H)$$

Equations (G) and (H) are an independent set and indicate that $r = 2$. The use of different elimination procedures produces other pairs of equations, but always just two equations.

Application of Eq. (15.34) yields

$$F = 2 - \pi + N - r - s = 2 - 1 + 5 - 2 - 0 = 4$$

This result means that one is free to specify four phase-rule variables, for example, T, P, and two mole fractions, in an equilibrium mixture of these five chemical species, provided that nothing else is arbitrarily set. In other words, there can be no special constraints, such as the specification that the system be prepared from given amounts of CH_4 and H_2O. This imposes special constraints through material balances that reduce the degrees of freedom to two. (Duhem's theorem; see the following paragraphs.)

Duhem's theorem states that, for any closed system formed initially from given masses of particular chemical species, the equilibrium state is *completely determined* (extensive as well as intensive properties) by specification of any two independent variables. This theorem was developed in Sec. 12.2 for nonreacting systems. It was shown there that the difference between the number of independent variables that completely determine the state of the system and the number of independent equations that can be written connecting these variables is

$$[2 + (N - 1)(\pi) + \pi] - [(\pi - 1)(N) + N] = 2$$

If chemical reactions occur, then we must introduce a new variable, the reaction coordinate ε_j for each independent reaction, in order to formulate the material-balance equations. Furthermore, we are able to write a new equilibrium relation [Eq. (15.8)] for each independent reaction. Therefore, when chemical-reaction equilibrium is superimposed on phase equilibrium, r new variables appear and r new equations can be written. The difference between the number of variables and number of equations therefore is unchanged, and Duhem's theorem as originally stated holds for reacting systems as well as for nonreacting systems.

Most chemical-reaction equilibrium problems are so posed that it is Duhem's theorem that makes them determinate. The usual problem is to find the composition of a system that reaches equilibrium from an initial state of *fixed amounts of of reacting species* when the *two* variables T and P are specified.

15.9   Multireaction Equilibria

When the equilibrium state in a reacting system depends on two or more independent chemical reactions, the equilibrium composition can be found by a direct extension of the methods developed for single reactions. One first determines a set of independent reactions as discussed in Sec. 15.8. With each independent reaction there is associated a reaction coordinate in accord with the treatment of Sec. 15.1. In addition, a separate equilibrium constant is evaluated for each reaction, and Eq. (15.13) becomes

$$K_j = \prod_i (\hat{a}_i)^{\nu_{i,j}} \qquad (15.35)$$

where j is the reaction index. For a gas-phase reaction Eq. (15.35) takes the form

$$K_j = \prod_i \left(\frac{\hat{f}_i}{P^\circ} \right)^{\nu_{i,j}} \qquad (15.36)$$

If the equilibrium mixture is an ideal gas, we may write

$$\prod_i (y_i)^{\nu_{i,j}} = \left(\frac{P}{P^\circ} \right)^{-\nu_j} K_j \qquad (15.37)$$

For r independent reactions there are r separate equations of this kind, and the y_i's can be eliminated by Eq. (15.7) in favor of the r reaction coordinates ε_j. The set of equations is then solved simultaneously for the r reaction coordinates, as illustrated in the following example.

> **Example 15.12**   A bed of coal (assume pure carbon) in a coal gasifier is fed with steam and air and produces a gas stream containing H_2, CO, O_2, H_2O, CO_2, and N_2. If the feed to the gasifier consists of 1 mol of steam and 2.38 mol of air, calculate the equilibrium composition of the gas stream at $P = 20$ bar for temperatures of 1,000, 1,100, 1,200, 1,300, 1,400, and 1,500 K. The following data are available:

T/K	$\Delta G_f^\circ/\text{J mol}^{-1}$		
	H_2O	CO	CO_2
1,000	$-192,420$	$-200,240$	$-395,790$
1,100	$-187,000$	$-209,110$	$-395,960$
1,200	$-181,380$	$-217,830$	$-396,020$
1,300	$-175,720$	$-226,530$	$-396,080$
1,400	$-170,020$	$-235,130$	$-396,130$
1,500	$-164,310$	$-243,740$	$-396,160$

SOLUTION   The feed stream to the coal bed consists of 1 mol of steam and 2.38 mol of air, containing

$$O_2: (0.21)(2.38) = 0.5 \text{ mol}$$

$$N_2: (0.79)(2.38) = 1.88 \text{ mol}$$

The species present at equilibrium are C, H_2, O_2, N_2, H_2O, CO, and CO_2. The formation reactions for the compounds present are

$$H_2 + \tfrac{1}{2}O_2 \rightarrow H_2O \quad (1)$$

$$C + \tfrac{1}{2}O_2 \rightarrow CO \quad (2)$$

$$C + O_2 \rightarrow CO_2 \quad (3)$$

Since the elements hydrogen, oxygen, and carbon are themselves presumed present in the system, this set of three independent reactions is a complete set.

All species are present as gases except carbon, which is present as a pure solid phase. In the equilibrium expression, Eq. (15.35), the activity of the pure carbon is $\hat{a}_C = a_C = f_C/f_C^\circ$. The fugacity ratio is the fugacity of carbon at 20 bar divided by the fugacity of carbon at 1 bar. Since the effect of pressure on the fugacity of a solid is very small, negligible error is introduced by the assumption that this ratio is unity. The activity of the carbon is then $\hat{a}_C = 1$, and it may be omitted from the equilibrium expression. With the assumption that the remaining species are ideal gases, Eq. (15.37) is written for the gas phase only, and it provides the following equilibrium expressions for reactions (1) through (3):

$$K_1 = \frac{y_{H_2O}}{y_{O_2}^{1/2} y_{H_2}} \left(\frac{P}{P^\circ}\right)^{-1/2}$$

$$K_2 = \frac{y_{CO}}{y_{O_2}^{1/2}} \left(\frac{P}{P^\circ}\right)^{1/2}$$

$$K_3 = \frac{y_{CO_2}}{y_{O_2}}$$

The reaction coordinates for the three reactions are designated ε_1, ε_2, and ε_3, and they are here taken to be the equilibrium values. For the initial state,

$n_{H_2} = n_{CO} = n_{CO_2} = 0$, $n_{H_2O} = 1$, $n_{O_2} = 0.5$, and $n_{N_2} = 1.88$. Moreover, since only the gas-phase species are considered, $\nu_1 = -\frac{1}{2}$, $\nu_2 = \frac{1}{2}$, and $\nu_3 = 0$. Applying Eq. (15.7) to each species gives

$$y_{H_2} = \frac{-\varepsilon_1}{3.38 + (\varepsilon_2 - \varepsilon_1)/2} \qquad y_{CO} = \frac{\varepsilon_2}{3.38 + (\varepsilon_2 - \varepsilon_1)/2}$$

$$y_{O_2} = \frac{\frac{1}{2}(1 - \varepsilon_1 - \varepsilon_2) - \varepsilon_3}{3.38 + (\varepsilon_2 - \varepsilon_1)/2} \qquad y_{H_2O} = \frac{1 + \varepsilon_1}{3.38 + (\varepsilon_2 - \varepsilon_1)/2}$$

$$y_{CO_2} = \frac{\varepsilon_3}{3.38 + (\varepsilon_2 - \varepsilon_1)/2} \qquad y_{N_2} = \frac{1.88}{3.38 + (\varepsilon_2 - \varepsilon_1)/2}$$

Substitution of these expressions for y_i into the equilibrium equations gives

$$K_1 = \frac{(1 + \varepsilon_1)(2n)^{1/2}(P/P^\circ)^{-1/2}}{(1 - \varepsilon_1 - \varepsilon_2 - 2\varepsilon_3)^{1/2}(-\varepsilon_1)}$$

$$K_2 = \frac{\sqrt{2}\varepsilon_2(P/P^\circ)^{1/2}}{(1 - \varepsilon_1 - \varepsilon_2 - 2\varepsilon_3)^{1/2}n^{1/2}}$$

$$K_3 = \frac{2\varepsilon_3}{(1 - \varepsilon_1 - \varepsilon_2 - 2\varepsilon_3)}$$

where

$$n \equiv 3.38 + \frac{\varepsilon_2 - \varepsilon_1}{2}$$

Numerical values for the K_i calculated by Eq. (15.14) are found to be very large. For example, at 1,500 K,

$$\ln K_1 = \frac{-\Delta G_1^\circ}{RT} = \frac{164,310}{(8.314)(1,500)} = 13.2 \qquad K_1 \sim 10^6$$

$$\ln K_2 = \frac{-\Delta G_2^\circ}{RT} = \frac{243,740}{(8.314)(1,500)} = 19.6 \qquad K_2 \sim 10^8$$

$$\ln K_3 = \frac{-\Delta G_3^\circ}{RT} = \frac{396,160}{(8.314)(1,500)} = 31.8 \qquad K_3 \sim 10^{14}$$

With K_i's so large, the quantity $1 - \varepsilon_1 - \varepsilon_2 - 2\varepsilon_3$ in the denominator of each equilibrium equation must be nearly zero. This means that the mole fraction of oxygen in the equilibrium mixture is very small. For practical purposes, no oxygen is present.

We therefore reformulate the problem by eliminating O_2 from the formation reactions. For this, we combine Eq. (1), first with Eq. (2), and then with Eq. (3). This provides the two equations

$$C + CO_2 \rightarrow 2CO \qquad (a)$$

$$H_2O + C \rightarrow H_2 + CO \qquad (b)$$

The corresponding equilibrium equations are

$$K_a = \frac{y_{CO}^2}{y_{CO_2}} \left(\frac{P}{P^\circ}\right)$$

and

$$K_b = \frac{y_{H_2} y_{CO}}{y_{H_2O}} \left(\frac{P}{P^\circ}\right)$$

The input stream is specified to contain 1 mol H_2, 0.5 mol O_2, and 1.88 mol N_2. Since O_2 has been eliminated from the set of reaction equations, we replace the 0.5 mol of O_2 in the feed by 0.5 mol of CO_2. The presumption is that this amount of CO_2 has been formed by prior reaction of the 0.5 mol O_2 with carbon. Thus the equivalent feed stream contains 1 mol H_2, 0.5 mol CO_2, and 1.88 mol N_2, and application of Eq. (15.7) to Eqs. (a) and (b) gives

$$y_{H_2} = \frac{\varepsilon_b}{3.38 + \varepsilon_a + \varepsilon_b}$$

$$y_{CO} = \frac{2\varepsilon_a + \varepsilon_b}{3.38 + \varepsilon_a + \varepsilon_b}$$

$$y_{H_2O} = \frac{1 - \varepsilon_b}{3.38 + \varepsilon_a + \varepsilon_b}$$

$$y_{CO_2} = \frac{0.5 - \varepsilon_a}{3.38 + \varepsilon_a + \varepsilon_b}$$

$$y_{N_2} = \frac{1.88}{3.38 + \varepsilon_a + \varepsilon_b}$$

Since values of y_i must lie between zero and unity, we see from the first and third of these expressions that

$$0 \leq \varepsilon_b \leq 1$$

and from the second and fourth that

$$-0.5 \leq \varepsilon_a \leq 0.5$$

Combining the expressions for the y_i with the equilibrium equations, we get

$$K_a = \frac{(2\varepsilon_a + \varepsilon_b)^2}{(0.5 - \varepsilon_a)(3.38 + \varepsilon_a + \varepsilon_b)} \left(\frac{P}{P^\circ}\right) \tag{A}$$

and

$$K_b = \frac{\varepsilon_b(2\varepsilon_a + \varepsilon_b)}{(1 - \varepsilon_b)(3.38 + \varepsilon_a + \varepsilon_b)} \left(\frac{P}{P^\circ}\right) \tag{B}$$

For reaction (a) at 1,000 K,

$$\Delta G^\circ_{1000} = 2(-200,240) - (-395,790) = -4,690$$

and by Eq. (15.14)

$$\ln K_a = \frac{4,690}{(8.314)(1,000)} = 0.5641 \qquad K_a = 1.758$$

Similarly, for reaction (b),

$$\Delta G^{\circ}_{1000} = (-200{,}240) - (-192{,}420) = -7{,}820$$

and

$$\ln K_b = \frac{7{,}820}{(8.314)(1{,}000)} = 0.9406 \qquad K_b = 2.561$$

Equations (A) and (B) with these values for K_a and K_b and with $(P/P^{\circ}) = 20$ constitute two nonlinear equations in unknowns ε_a and ε_b. An *ad hoc* iteration scheme can be devised for their solution, but Newton's method for solving an array of nonlinear algebraic equations is attractive. It is described and applied to this example in App. H. Moreover, the Mathcad® program for solving these equations is included in App. D.2. The results of calculations for all temperatures are shown in the following table.

T/K	K_a	K_b	ε_a	ε_b
1,000	1.758	2.561	−0.0506	0.5336
1,100	11.405	11.219	0.1210	0.7124
1,200	53.155	38.609	0.3168	0.8551
1,300	194.430	110.064	0.4301	0.9357
1,400	584.85	268.76	0.4739	0.9713
1,500	1,514.12	583.58	0.4896	0.9863

Values for the mole fractions y_i of the species in the equilibrium mixture are calculated by the equations already given. The results of all such calculations appear in the following table and are shown graphically in Fig. 15.5.

T/K	y_{H_2}	y_{CO}	y_{H_2O}	y_{CO_2}	y_{N_2}
1,000	0.138	0.112	0.121	0.143	0.486
1,100	0.169	0.226	0.068	0.090	0.447
1,200	0.188	0.327	0.032	0.040	0.413
1,300	0.197	0.378	0.014	0.015	0.396
1,400	0.201	0.398	0.006	0.005	0.390
1,500	0.203	0.405	0.003	0.002	0.387

At the higher temperatures the values of ε_a and ε_b are approaching their upper limiting values of 0.5 and 1.0, indicating that reactions (a) and (b) are proceeding nearly to completion. In this limit, which is approached even more closely at still higher temperatures, the mole fractions of CO_2 and H_2O approach zero, and for the product species,

$$y_{H_2} = \frac{1}{3.38 + 0.5 + 1.0} = 0.205$$

$$y_{CO} = \frac{1+1}{3.38 + 0.5 + 1.0} = 0.410$$

$$y_{N_2} = \frac{1.88}{3.38 + 0.5 + 1.0} = 0.385$$

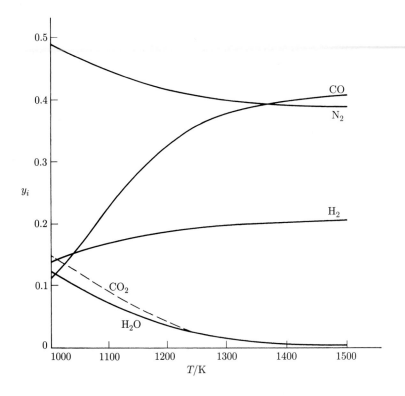

Figure 15.5: Equilibrium compositions of the product gases in Example 15.12.

In this example we have assumed a sufficient depth for the coal bed that equilibrium is approached by the gases while they are in contact with the incandescent carbon. This need not be the case; if oxygen and steam are supplied at too high a rate, the reactions may not attain equilibrium or may reach equilibrium after they have left the coal bed. In this event, carbon is not present at equilibrium, and the problem must again be reformulated.

Although the Eqs. (A) and (B) of the preceding example are readily solved, the method of equilibrium constants does not lend itself to standardization so as to allow a *general* program to be written for computer solution. An alternative criterion of equilibrium, mentioned in Sec. 15.2, is based on the fact that at equilibrium the total Gibbs energy of the system has its minimum value, as illustrated for a single reaction in Fig. 15.1. Applied to multiple reactions, this criterion is the basis for a general scheme of computer solution.

The total Gibbs energy of a single-phase system is given by Eq. (10.2), which shows that

$$(G^t)_{T,P} = g(n_1, n_2, n_3, \ldots, n_N)$$

The problem is to find the set of n_i's which minimizes G^t for specified T and P, subject to the constraints of the material balances. The standard solution to this type of problem is based on the method of Lagrange's undetermined multipliers. The procedure for gas-phase reactions is described as follows.

1. The first step is to formulate the constraining equations, i.e., the material balances. Although reacting molecular species are not conserved in a closed system, the total number of atoms of each *element* is constant. Let subscript k identify a particular atom. Then define A_k as the total number of atomic masses of the kth element in the system, as determined by the initial constitution of the system. Further, let a_{ik} be the number of atoms of the kth element present in each molecule of chemical species i. The material balance on each element k may then be written

$$\sum_i n_i a_{ik} = A_k \qquad (k = 1, 2, \ldots, w) \qquad (15.38)$$

or

$$\sum_i n_i a_{ik} - A_k = 0 \qquad (k = 1, 2, \ldots, w)$$

where w is the total number of elements comprising the system.

2. Next, we introduce the Lagrange multipliers λ_k, one for each element, by multiplying each element balance by its λ_k:

$$\lambda_k \left(\sum_i n_i a_{ik} - A_k \right) = 0 \qquad (k = 1, 2, \ldots, w)$$

These equations are summed over k, giving

$$\sum_k \lambda_k \left(\sum_i n_i a_{ik} - A_k \right) = 0$$

3. Then a new function F is formed by addition of this last sum to G^t. Thus,

$$F = G^t + \sum_k \lambda_k \left(\sum_i n_i a_{ik} - A_k \right)$$

This new function is identical with G^t, because the summation term is zero. However, the partial derivatives of F and G^t with respect to n_i are different, because the function F incorporates the constraints of the material balances.

4. The minimum value of both F and G^t occurs when the partial derivatives of F with respect to n_i are zero. Therefore, we set the expression for these derivatives equal to zero:

$$\left(\frac{\partial F}{\partial n_i} \right)_{T,P,n_j} = \left(\frac{\partial G^t}{\partial n_i} \right)_{T,P,n_j} + \sum_k \lambda_k a_{ik} = 0$$

Since the first term on the right is the definition of the chemical potential [see Eq. (10.1)], this equation can be written

$$\mu_i + \sum_k \lambda_k a_{ik} = 0 \qquad (i = 1, 2, \ldots, N) \qquad (15.39)$$

However, the chemical potential is given by Eq. (15.11):

$$\mu_i = G_i^\circ + RT \ln \hat{a}_i$$

For gas-phase reactions and standard states as the pure ideal gases at 1 bar [or 1(atm)], this becomes

$$\mu_i = G_i^\circ + RT \ln(\hat{f}_i/P^\circ)$$

If G_i° is arbitrarily set equal to zero for all *elements* in their standard states, then for compounds $G_i^\circ = \Delta G_{f_i}^\circ$, the standard Gibbs-energy change of formation for species i. In addition, the fugacity is eliminated in favor of the fugacity coefficient by Eq. (10.47), $\hat{f}_i = y_i \hat{\phi}_i P$. With these substitutions, the equation for μ_i becomes

$$\mu_i = \Delta G_{f_i}^\circ + RT \ln(y_i \hat{\phi}_i P/P^\circ)$$

Combination with Eq. (15.39) gives

$$\Delta G_{f_i}^\circ + RT \ln(y_i \hat{\phi}_i P/P^\circ) + \sum_k \lambda_k a_{ik} = 0 \qquad (i = 1, 2, \ldots, N) \qquad (15.40)$$

Again we note that P° is 1 bar, expressed in the units used for pressure. If species i is an element, $\Delta G_{f_i}^\circ$ is zero.

Equation (15.40) represents N equilibrium equations, one for each chemical species, and Eq. (15.38) represents w material-balance equations, one for each element—a total of $N + w$ equations. The unknowns in these equations are the n_i's (note that $y_i = n_i / \sum_i n_i$), of which there are N, and the λ_k's, of which there are w—a total of $N + w$ unknowns. Thus the number of equations is sufficient for the determination of all unknowns.

The foregoing discussion has presumed that the $\hat{\phi}_i$'s are known. If the phase is an ideal gas, then each $\hat{\phi}_i$ is unity. If the phase is an ideal solution, each $\hat{\phi}_i$ becomes ϕ_i, and can at least be estimated. For real gases, each $\hat{\phi}_i$ is a function of the y_i's, the quantities being calculated. Thus an iterative procedure is indicated. The calculations are initiated with each $\hat{\phi}_i$ set equal to unity. Solution of the equations then provides a preliminary set of y_i's. For low pressures or high temperatures this result is usually adequate. Where it is not satisfactory, an equation of state is used together with the calculated y_i's to give a new and more nearly correct set of $\hat{\phi}_i$'s for use in Eq. (15.40). Then a new set of y_i's is determined. The process is repeated until successive iterations produce no significant change in the y_i's. All calculations are well suited to computer solution, including the calculation of the $\hat{\phi}_i$'s by equations such as Eq. (10.69) or (13.38).

In the procedure just described, the question of what chemical reactions are involved never enters directly into any of the equations. However, the choice of a set of species is entirely equivalent to the choice of a set of independent reactions among the species. In any event, a set of species or an equivalent set of independent reactions must always be assumed, and different assumptions produce different results.

Example 15.13 Calculate the equilibrium compositions at 1,000 K and 1 bar of a gas-phase system containing the species CH_4, H_2O, CO, CO_2, and H_2. In the initial unreacted state there are present 2 mol of CH_4 and 3 mol of H_2O. Values of $\Delta G^\circ_{f_i}$ at 1,000 K are

$$\Delta G^\circ_{f_{CH_4}} = 19{,}720 \text{ J mol}^{-1}$$

$$\Delta G^\circ_{f_{H_2O}} = -192{,}420 \text{ J mol}^{-1}$$

$$\Delta G^\circ_{f_{CO}} = -200{,}240 \text{ J mol}^{-1}$$

$$\Delta G^\circ_{f_{CO_2}} = -395{,}790 \text{ J mol}^{-1}$$

SOLUTION   The required values of A_k are determined from the initial numbers of moles, and the values of a_{ik} come directly from the chemical formulas of the species. These are shown in the accompanying table.

		Element k	
	Carbon	Oxygen	Hydrogen
	A_k = no. of atomic masses of k in the system		
	$A_C = 2$	$A_O = 3$	$A_H = 14$
Species i	a_{ik} = no. of atoms of k per molecule of i		
CH_4	$a_{CH_4,C} = 1$	$a_{CH_4,O} = 0$	$a_{CH_4,H} = 4$
H_2O	$a_{H_2O,C} = 0$	$a_{H_2O,O} = 1$	$a_{H_2O,H} = 2$
CO	$a_{CO,C} = 1$	$a_{CO,O} = 1$	$a_{CO,H} = 0$
CO_2	$a_{CO_2,C} = 1$	$a_{CO_2,O} = 2$	$a_{CO_2,H} = 0$
H_2	$a_{H_2,C} = 0$	$a_{H_2,O} = 0$	$a_{H_2,H} = 2$

At 1 bar and 1,000 K the assumption of ideal gases is justified, and the $\hat{\phi}_i$'s are all unity. Since $P = 1$ bar, $P/P^\circ = 1$, and Eq. (15.40) is written

$$\frac{\Delta G^\circ_{f_i}}{RT} + \ln \frac{n_i}{\sum_i n_i} + \sum_k \frac{\lambda_k}{RT} a_{ik} = 0$$

The five equations for the five species then become

$$\text{CH}_4: \quad \frac{19,720}{RT} + \ln\frac{n_{\text{CH}_4}}{\sum_i n_i} + \frac{\lambda_\text{C}}{RT} + \frac{4\lambda_\text{H}}{RT} = 0$$

$$\text{H}_2\text{O}: \quad \frac{-192,420}{RT} + \ln\frac{n_{\text{H}_2\text{O}}}{\sum_i n_i} + \frac{2\lambda_\text{H}}{RT} + \frac{\lambda_\text{O}}{RT} = 0$$

$$\text{CO}: \quad \frac{-200,240}{RT} + \ln\frac{n_{\text{CO}}}{\sum_i n_i} + \frac{\lambda_\text{C}}{RT} + \frac{\lambda_\text{O}}{RT} = 0$$

$$\text{CO}_2: \quad \frac{-395,790}{RT} + \ln\frac{n_{\text{CO}_2}}{\sum_i n_i} + \frac{\lambda_\text{C}}{RT} + \frac{2\lambda_\text{O}}{RT} = 0$$

$$\text{H}_2: \quad \ln\frac{n_{\text{H}_2}}{\sum_i n_i} + \frac{2\lambda_\text{H}}{RT} = 0$$

The three material-balance equations [Eq. (15.38)] are

$$\text{C}: \quad n_{\text{CH}_4} + n_{\text{CO}} + n_{\text{CO}_2} = 2$$

$$\text{H}: \quad 4n_{\text{CH}_4} + 2n_{\text{H}_2\text{O}} + 2n_{\text{H}_2} = 14$$

$$\text{O}: \quad n_{\text{H}_2\text{O}} + n_{\text{CO}} + 2n_{\text{CO}_2} = 3$$

Simultaneous computer solution of these eight equations,[8] with

$$RT = 8{,}314 \text{ J mol}^{-1}$$

and

$$\sum_i n_i = n_{\text{CH}_4} + n_{\text{H}_2\text{O}} + n_{\text{CO}} + n_{\text{CO}_2} + n_{\text{H}_2}$$

produces the following results ($y_i = n_i/\sum_i n_i$):

$$y_{\text{CH}_4} = 0.0196 \qquad \frac{\lambda_\text{C}}{RT} = 0.7635$$

$$y_{\text{H}_2\text{O}} = 0.0980$$

$$y_{\text{CO}} = 0.1743 \qquad \frac{\lambda_\text{O}}{RT} = 25.068$$

$$y_{\text{CO}_2} = 0.0371$$

$$y_{\text{H}_2} = 0.6710 \qquad \frac{\lambda_\text{H}}{RT} = 0.1994$$

$$\sum_i y_i = 1.0000$$

---

[8] The Mathcad® formulation of this problem is given in App. D.2.

The values of λ_k/RT are of no significance, but are included for the sake of completeness.

PROBLEMS

15.1. Develop expressions for the mole fractions of reacting species as functions of the reaction coordinate for:

(a) A system initially containing 2 mol NH_3 and 5 mol O_2 and undergoing the reaction
$$4NH_3(g) + 5O_2(g) \rightarrow 4NO(g) + 6H_2O(g)$$

(b) A system initially containing 3 mol H_2S and 5 mol O_2 and undergoing the reaction:
$$2H_2S(g) + 3O_2(g) \rightarrow 2H_2O(g) + 2SO_2(g)$$

(c) A system initially containing 3 mol NO_2, 4 mol NH_3, and 1 mol N_2 and undergoing the reaction:
$$6NO_2(g) + 8NH_3(g) \rightarrow 7N_2(g) + 12H_2O(g)$$

15.2. A system initially containing 2 mol CO_2, 5 mol H_2, and 1 mol CO undergoes the following reactions:
$$CO_2(g) + 3H_2(g) \rightarrow CH_3OH(g) + H_2O(g)$$
$$CO_2(g) + H_2(g) \rightarrow CO(g) + H_2O(g)$$
Develop expressions for the mole fractions of the reacting species as functions of the reaction coordinates for the two reactions.

15.3. A system initially containing 2 mol C_2H_4 and 3 mol O_2 undergoes the following reactions:
$$C_2H_4(g) + \tfrac{1}{2}O_2(g) \rightarrow \langle(CH_2)_2\rangle O(g)$$
$$C_2H_4(g) + 3O_2(g) \rightarrow 2CO_2(g) + 2H_2O(g)$$
Develop expressions for the mole fractions of the reacting species as functions of the reaction coordinates for the two reactions.

15.4. Consider the water-gas-shift reaction,
$$H_2(g) + CO_2(g) \rightarrow H_2O(g) + CO(g)$$

At high temperatures and low to moderate pressures the reacting species form an ideal-gas mixture. Application of the summability equation to Eq. (10.26) yields
$$G = \sum_i y_i G_i + RT \sum_i y_i \ln y_i$$

If the Gibbs energies of the elements in their standard states are set equal to zero, $G_i = \Delta G_{f_i}^{\circ}$ for each species, and then

$$G = \sum_i y_i \Delta G^\circ_{f_i} + RT \sum_i y_i \ln y_i \qquad (A)$$

With the understanding that T and P are constant, we may write the equilibrium criterion of Eq. (14.4) for this reacting system as

$$dG^t = d(nG) = n\,dG + G\,dn = 0$$

or

$$n\frac{dG}{d\varepsilon} + G\frac{dn}{d\varepsilon} = 0$$

But for the water-gas-shift reaction, $dn/d\varepsilon = 0$. The equilibrium criterion therefore becomes

$$\frac{dG}{d\varepsilon} = 0 \qquad (B)$$

Once the y_i are eliminated in favor of ε, Eq. (A) relates G to ε. Data for $\Delta G^\circ_{f_i}$ for the compounds of interest are given with Example 15.12. For a temperature of 1,000 K (the reaction is unaffected by P) and for a feed of 1 mol H_2 and 1 mol CO_2,

(a) Determine the equilibrium value of ε by application of Eq. (B).

(b) Plot G vs. ε, indicating the location of the equilibrium value of ε determined in (a).

15.5. Rework Prob. 15.4 for a temperature of 1,100 K.

15.6. Rework Prob. 15.4 for a temperature of 1,200 K.

15.7. Rework Prob. 15.4 for a temperature of 1,300 K.

15.8. Verify the answer to Prob. 15.4, part (a), by the method of equilibrium constants.

15.9. Verify the answer to Prob. 15.5, part (a), by the method of equilibrium constants.

15.10. Verify the answer to Prob. 15.6, part (a), by the method of equilibrium constants.

15.11. Verify the answer to Prob. 15.7, part (a), by the method of equilibrium constants.

15.12. Develop a general equation for the standard Gibbs energy change of reaction ΔG° as a function of temperature for one of the reactions given in parts (a), (f), (i), (n), (r), (t), (u), (x), and (y) of Prob. 4.18.

15.13. For ideal gases, exact mathematical expressions can be developed for the effect of T and P on ε_e. For conciseness we let $\prod_i (y_i)^{\nu_i} \equiv K_y$. Then we can write the mathematical relations:

$$\left(\frac{\partial \varepsilon_e}{\partial T}\right)_P = \left(\frac{\partial K_y}{\partial T}\right)_P \frac{d\varepsilon_e}{dK_y} \qquad \text{and} \qquad \left(\frac{\partial \varepsilon_e}{\partial P}\right)_T = \left(\frac{\partial K_y}{\partial P}\right)_T \frac{d\varepsilon_e}{dK_y}$$

Using Eqs. (15.25) and (15.16), show that

(a) $\left(\dfrac{\partial \varepsilon_e}{\partial T}\right)_P = \dfrac{K_y}{RT^2}\dfrac{d\varepsilon_e}{dK_y}\Delta H^\circ$

(b) $\left(\dfrac{\partial \varepsilon_e}{\partial P}\right)_T = \dfrac{K_y}{P}\dfrac{d\varepsilon_e}{dK_y}(-\nu)$

(c) $d\varepsilon_e/dK_y$ is always positive. (*Note*: It is equally valid and perhaps easier to show that the reciprocal is positive.)

15.14. For the ammonia synthesis reaction written

$$\tfrac{1}{2}N_2(g) + \tfrac{3}{2}H_2(g) \to NH_3(g)$$

with 0.5 mol N_2 and 1.5 mol H_2 as the initial amounts of reactants and with the assumption that the equilibrium mixture is an ideal gas, show that

$$\varepsilon_e = 1 - \left(1 + 1.299K\,\frac{P}{P^\circ}\right)^{-1/2}$$

15.15. Peter, Paul, and Mary, members of a thermodynamics class, are asked to find the equilibrium composition at a particular T and P and for given initial amounts of reactants for the following gas-phase reaction:

$$2NH_3 + 3NO \to 3H_2O + \tfrac{5}{2}N_2 \qquad\qquad (A)$$

Each solves the problem correctly in a different way. Mary bases her solution on reaction (A) as written. Paul, who prefers whole numbers, multiplies reaction (A) by 2:

$$4NH_3 + 6NO \to 6H_2O + 5N_2 \qquad\qquad (B)$$

Peter, who usually does things backward, deals with the reaction:

$$3H_2O + \tfrac{5}{2}N_2 \to 2NH_3 + 3NO \qquad\qquad (C)$$

Write the chemical-equilibrium equations for the three reactions, indicate how the equilibrium constants are related, and show why Peter, Paul, and Mary all obtain the same result.

15.16. The following reaction reaches equilibrium at 500°C and 2 bar:

$$4HCl(g) + O_2(g) \to 2H_2O(g) + 2Cl_2(g)$$

If the system initially contains 5 mol HCl for each mole of oxygen, what is the composition of the system at equilibrium? Assume ideal gases.

15.17. The following reaction reaches equilibrium at 650°C and atmospheric pressure:

$$N_2(g) + C_2H_2(g) \to 2HCN(g)$$

If the system initially is an equimolar mixture of nitrogen and acetylene, what is the composition of the system at equilibrium? What would be the effect of doubling the pressure? Assume ideal gases.

15.18. The following reaction reaches equilibrium at 350°C and 3 bar:

$$CH_3CHO(g) + H_2(g) \to C_2H_5OH(g)$$

If the system initially contains 1.5 mol H_2 for each mole of acetaldehyde, what is the composition of the system at equilibrium? What would be the effect of reducing the pressure to 1 bar? Assume ideal gases.

15.19. The following reaction reaches equilibrium at $650°C$ and atmospheric pressure:

$$C_6H_5CH{:}CH_2(g) + H_2(g) \rightarrow C_6H_5{.}C_2H_5(g)$$

If the system initially contains 1.5 mol H_2 for each mole of styrene, what is the composition of the system at equilibrium? Assume ideal gases.

15.20. The gas stream from a sulfur burner is composed of 15-mole-% SO_2, 20-mole-% O_2, and 65-mole-% N_2. This gas stream at 1 bar and $480°C$ enters a catalytic converter, where the SO_2 is further oxidized to SO_3. Assuming that the reaction reaches equilibrium, how much heat must be removed from the converter to maintain isothermal conditions? Base your answer on 1 mol of entering gas.

15.21. For the cracking reaction,

$$C_3H_8(g) \rightarrow C_2H_4(g) + CH_4(g)$$

the equilibrium conversion is negligible at 300 K, but becomes appreciable at temperatures above 500 K. For a pressure of 1 bar, determine

(*a*) The fractional conversion of propane at 625 K.

(*b*) The temperature at which the fractional conversion is 85%.

15.22. Ethylene is produced by the dehydrogenation of ethane. If the feed includes 0.5 mol of steam (an inert diluent) per mole of ethane and if the reaction reaches equilibrium at 1,100 K and 1 bar, what is the composition of the product gas on a water-free basis?

15.23. The production of 1,3-butadiene can be carried out by the dehydrogenation of 1-butene:

$$C_2H_5CH{:}CH_2(g) \rightarrow CH_2{:}CHCH{:}CH_2(g) + H_2(g)$$

Side reactions are supressed by the introduction of steam. If equilibrium is attained at 950 K and 1 bar and if the reactor product contains 10-mole-% 1,3-butadiene, determine

(*a*) The mole fractions of the other species in the product gas.

(*b*) The mole fraction of steam required in the feed.

15.24. The production of 1,3-butadiene can be carried out by the dehydrogenation of *n*-butane:

$$C_4H_{10}(g) \rightarrow CH_2{:}CHCH{:}CH_2(g) + 2H_2(g)$$

Side reactions are supressed by the introduction of steam. If equilibrium is attained at 925 K and 1 bar and if the reactor product contains 12-mole-% 1,3-butadiene, determine

(*a*) The mole fractions of the other species in the product gas.

(*b*) The mole fraction of steam required in the feed.

15.25. For the ammonia synthesis reaction,

$$\tfrac{1}{2}N_2(g) + \tfrac{3}{2}H_2(g) \rightarrow NH_3(g)$$

the equilibrium conversion to ammonia is large at 300 K, but decreases rapidly with increasing T. However, reaction rates become appreciable only at higher temperatures. For a feed mixture of hydrogen and nitrogen in the stoichiometric proportions,

(a) Determine the mole fraction of ammonia in the equilibrium mixture at 1 bar and 300 K.

(b) At what temperature does the equilibrium mole fraction of ammonia decrease to 0.50 for a pressure of 1 bar?

(c) At what temperature does the equilibrium mole fraction of ammonia decrease to 0.50 for a pressure of 100 bar assuming the equilibrium mixture an ideal gas?

(d) At what temperature does the equilibrium mole fraction of ammonia decrease to 0.50 for a pressure of 100 bar, assuming the equilibrium mixture an ideal solution of gases?

15.26. For the methanol synthesis reaction,

$$CO(g) + 2H_2(g) \rightarrow CH_3OH(g)$$

the equilibrium conversion to methanol is large at 300 K, but decreases rapidly with increasing T. However, reaction rates become appreciable only at higher temperatures. For a feed mixture of carbon monoxide and hydrogen in the stoichiometric proportions,

(a) Determine the mole fraction of methanol in the equilibrium mixture at 1 bar and 300 K.

(b) At what temperature does the equilibrium mole fraction of methanol decrease to 0.50 for a pressure of 1 bar?

(c) At what temperature does the equilibrium mole fraction of methanol decrease to 0.50 for a pressure of 100 bar, assuming the equilibrium mixture an ideal gas?

(d) At what temperature does the equilibrium mole fraction of methanol decrease to 0.50 for a pressure of 100 bar, assuming the equilibrium mixture an ideal solution of gases?

15.27. Limestone ($CaCO_3$) decomposes upon heating to yield quicklime (CaO) and carbon dioxide. At what temperature does limestone exert a decomposition pressure of 1(atm)?

15.28. Ammonium chloride [$NH_4Cl(s)$] decomposes upon heating to yield a gas mixture of ammonia and hydrochloric acid. At what temperature does ammonium chloride exert a decomposition pressure of 1.5 bar? For $NH_4Cl(s)$, $\Delta H^\circ_{f_{298}} = -314,430$ J and $\Delta G^\circ_{f_{298}} = -202,870$ J.

15.29. A chemically reactive system contains the following species in the gas phase: NH_3, NO, NO_2, O_2, and H_2O. Determine a complete set of independent reactions for this system. How many degrees of freedom does the system have?

15.30. The relative compositions of the pollutants NO and NO_2 in air are governed by the reaction,

$$NO + \tfrac{1}{2}O_2 \rightarrow NO_2$$

For air containing 21-mole-% O_2 at 25°C and 1.0133 bar, what is the concentration of NO in parts per million if the total concentration of the two nitrogen oxides is 5 ppm?

15.31. Consider the gas-phase oxidation of ethylene to ethylene oxide at a pressure of 1 bar with 25% excess air. If the reactants enter the process at 25°C, if the reaction proceeds adiabatically to equilibrium, and if there are no side reactions, determine the composition and temperature of the product stream from the reactor.

15.32. Carbon black is produced by the decomposition of methane:

$$CH_4(g) \rightarrow C(s) + 2H_2(g)$$

For equilibrium at 650°C and 1 bar,

(*a*) What is the gas-phase composition if pure methane enters the reactor, and what fraction of the methane decomposes?

(*b*) Repeat part (*a*) if the feed is an equimolar mixture of methane and nitrogen.

15.33. Consider the reactions,

$$\tfrac{1}{2}N_2(g) + \tfrac{1}{2}O_2(g) \rightarrow NO(g)$$

$$\tfrac{1}{2}N_2(g) + O_2(g) \rightarrow NO_2(g)$$

If these reactions come to equilibrium after combustion in an internal-combustion engine at 2,000 K and 200 bar, estimate the mole fractions of NO and NO_2 present for mole fractions of nitrogen and oxygen in the combustion products of 0.70 and 0.05.

15.34. Oil refineries frequently have both H_2S and SO_2 to dispose of. The following reaction suggests a means of getting rid of both at once:

$$2H_2S(g) + SO_2(g) \rightarrow 3S(s) + 2H_2O(g)$$

For reactants in the stoichiometric proportion, estimate the percent conversion of each reactant if the reaction comes to equilibrium at 450°C and 8 bar.

15.35. The species N_2O_4 and NO_2 as gases attain rapid equilibrium by the reaction:

$$N_2O_4 \rightarrow 2NO_2$$

(*a*) For $T = 350$ K and $P = 5$ bar, calculate the mole fractions of these species in the equilibrium mixture. Assume ideal gases.

(*b*) If an equilibrium mixture of N_2O_4 and NO_2 at the conditions of part (*a*) flows through a throttle valve to a pressure of 1 bar and through a heat exchanger that restores its initial temperature, how much heat must be exchanged, assuming chemical equilibrium is again attained in the final state? Base the answer on an amount of mixture equivalent to 1 mol of N_2O_4, i.e., as though all the NO_2 were present as N_2O_4.

15.36. The following isomerization reaction occurs in the *liquid* phase:

$$A \rightarrow B$$

where A and B are miscible liquids for which

$$G^E/RT = 0.1\, x_A x_B$$

If $\Delta G^\circ_{298} = -1,000$ J, what is the equilibrium composition of the mixture at 25°C? How much error is introduced if one assumes that A and B form an ideal solution?

15.37. The feed gas to a methanol synthesis reactor is composed of 75-mole-% H_2, 15-mole-% CO, 5-mole-% CO_2, and 5-mole-% N_2. The system comes to equilibrium at 550 K and 100 bar with respect to the following reactions:

$$2H_2(g) + CO(g) \rightarrow CH_3OH(g)$$

$$H_2(g) + CO_2(g) \rightarrow CO(g) + H_2O(g)$$

Assuming ideal gases, determine the composition of the equilibrium mixture.

15.38. Hydrogen gas is produced by the reaction of steam with "water gas," an equimolar mixture of H_2 and CO obtained by the reaction of steam with coal. A stream of "water gas" mixed with steam is passed over a catalyst to convert CO to CO_2 by the reaction:

$$H_2O(g) + CO(g) \rightarrow H_2(g) + CO_2(g)$$

Subsequently, unreacted water is condensed and carbon dioxide is absorbed, leaving a product that is mostly hydrogen. The equilibrium conditions are 1 bar and 800 K.

(a) Would there be any advantage to carrying out the reaction at pressures above 1 bar?

(b) If the equilibrium temperature were raised, would the conversion of CO be increased?

(c) For the given equilibrium conditions, determine the molar ratio of steam to "water gas" (H_2 + CO) required to produce a *product* gas containing only 2-mole-% CO after cooling to $20°C$, where the unreacted H_2O has been virtually all condensed out.

(d) Is there any danger that solid carbon will form at the equilibrium conditions by the reaction

$$2CO(g) \rightarrow CO_2(g) + C(s)$$

15.39. One method for the manufacture of "synthesis gas" is the catalytic reforming of methane with steam:

$$CH_4(g) + H_2O(g) \rightarrow CO(g) + 3H_2(g)$$

The only other reaction considered is

$$CO(g) + H_2O(g) \rightarrow CO_2(g) + H_2(g)$$

Assume equilibrium is attained for both reactions at 1 bar and 1,300 K.

(a) Would it be better to carry out the reaction at pressures above 1 bar?

(b) Would it be better to carry out the reaction at temperatures below 1,300 K?

(c) Estimate the molar ratio of hydrogen to carbon monoxide in the synthesis gas if the feed consists of an equimolar mixture of steam and methane.

(d) Repeat part (c) for a steam to methane mole ratio in the feed of 2.

(e) How could the feed composition be altered to yield a lower ratio of hydrogen to carbon monoxide in the synthesis gas than is obtained in part (c)?

(*f*) Is there any danger that carbon will deposit by the reaction $2CO \rightarrow C + CO_2$ under conditions of part (*c*)? Part (*d*)? If so, how could the feed be altered to prevent carbon deposition?

15.40. Set up the equations required for solution of Example 15.13 by the method of equilibrium constants. Verify that your equations yield the same equilibrium compositions as given in the example.

15.41. Ethylene oxide as a vapor and water as liquid, both at 25°C and 101.33 kPa, react to form an aqueous solution of ethylene glycol (1,2-ethanediol) at the same conditions:

$$\langle (CH_2)_2 \rangle O + H_2O \rightarrow CH_2OH.CH_2OH$$

If the initial molar ratio of ethylene oxide to water is 3.0, estimate the equilibrium conversion of ethylene oxide to ethylene glycol.

At equilibrium the system consists of liquid and vapor in equilibrium, and the intensive state of the system is fixed by the specification of T and P. Therefore, one must first determine the phase compositions, independent of the ratio of reactants. These results may then be applied in the material-balance equations to find the equilibrium conversion.

Choose as standard states for water and ethylene glycol the pure liquids at 1 bar and for ethylene oxide the pure ideal gas at 1 bar. Assume that the Lewis/Randall rule applies to the water in the liquid phase and that the vapor phase is an ideal gas. The partial pressure of ethylene oxide over the liquid phase is given by

$$p_i/\text{kPa} = 415\, x_i$$

The vapor pressure of ethylene glycol at 25°C is so low that its concentration in the vapor phase is negligible.

CHAPTER 16

THERMODYNAMIC ANALYSIS OF PROCESSES

The object of this chapter is the evaluation of processes from the thermodynamic point of view. No new fundamental ideas are needed; a combination of the first and second laws provides the basis. Hence, the chapter affords a review of thermodynamic principles.

Real irreversible processes are amenable to thermodynamic analysis. The goal of such an analysis is to determine how efficiently energy is used or produced and to show quantitatively the effect of inefficiencies in each step of a process. The cost of energy is of concern in any manufacturing operation, and the first step in any attempt to reduce energy requirements is to determine where and to what extent energy is wasted through process irreversibilities. The treatment here is limited to steady-state flow processes, because of their predominance in industrial practice.

16.1 Calculation of Ideal Work

In any steady-state flow process requiring work, there is an absolute minimum amount which must be expended to accomplish the desired change of state of the fluid flowing through the control volume. In a process producing work, there is an absolute maximum amount which may be accomplished as the result of a given change of state of the fluid flowing through the control volume. In either case, the limiting value obtains when the change of state associated with the process is accomplished *completely reversibly*. The implications of this requirement, listed in Chap. 7 in connection with Eq. (7.19), are repeated here:

1. The process is internally reversible within the control volume.

2. Heat transfer external to the control volume is reversible.

The second item means that heat transfer between the control volume and its surroundings must occur at the temperature of the surroundings. We presume that the control volume exists in surroundings that constitute a heat reservoir at a constant and uniform temperature denoted by T_σ. In many cases Carnot engines or heat pumps must be presumed present so as to provide for the reversible transfer of heat between temperatures at the control surface and the temperature T_σ of the surroundings. Since Carnot engines and heat pumps are cyclic, they undergo no net change of state.

For any completely reversible process, the entropy generation is zero, and Eq. (7.20), written for the uniform surroundings temperature T_σ, becomes

$$\dot{Q} = T_\sigma \Delta(S\dot{m})_{\text{fs}}$$

Substituting for $\dot{Q}$ in the energy balance given by Eq. (7.15), we get

$$\Delta\left[\left(H + \tfrac{1}{2}u^2 + zg\right)\dot{m}\right]_{\text{fs}} = T_\sigma \Delta(S\dot{m})_{\text{fs}} + \dot{W}_s(\text{rev})$$

where $\dot{W}_s(\text{rev})$ indicates that the shaft work is for a completely reversible process. We call this work the *ideal work*, $\dot{W}_{\text{ideal}}$. Thus

$$\dot{W}_{\text{ideal}} = \Delta\left[\left(H + \tfrac{1}{2}u^2 + zg\right)\dot{m}\right]_{\text{fs}} - T_\sigma \Delta(S\dot{m})_{\text{fs}} \qquad (16.1)$$

In most applications to chemical processes, the kinetic- and potential-energy terms are negligible compared with the others; in this event Eq. (16.1) is written

$$\boxed{\dot{W}_{\text{ideal}} = \Delta(H\dot{m})_{\text{fs}} - T_\sigma \Delta(S\dot{m})_{\text{fs}}} \qquad (16.2)$$

For the special case of a single stream flowing through the control volume, Eq. (16.2) becomes

$$\dot{W}_{\text{ideal}} = \dot{m}(\Delta H - T_\sigma \Delta S) \qquad (16.3)$$

Division by $\dot{m}$ puts this equation on a unit-mass basis

$$W_{\text{ideal}} = \Delta H - T_\sigma \Delta S \qquad (16.4)$$

A completely reversible processes is hypothetical, devised solely for determination of the ideal work associated with a given change of state. Its only connection with an actual process is that it brings about the same change of state as the actual process. Our objective is to compare the actual work of a process with the work of the hypothetical reversible process. No description is ever required of hypothetical processes devised for the calculation of ideal work. One need only realize that such processes may always be imagined. Nevertheless, an illustration of a hypothetical reversible process is given in Example 16.1.

Equations (16.1) through (16.4) give the work of completely reversible processes associated with given property changes in the flowing streams. When the same property changes occur in actual processes, the actual work $\dot{W}_s$ (or W_s) is given by an energy balance, and we can compare the actual work with the ideal work. When $\dot{W}_{\text{ideal}}$ (or W_{ideal}) is positive, it is the *minimum work required* to bring about a given change in the properties of the flowing streams, and is smaller than $\dot{W}_s$. In this case we define a thermodynamic efficiency η_t as the ratio of the ideal work to the actual work:

$$\eta_t(\text{work required}) = \frac{\dot{W}_{\text{ideal}}}{\dot{W}_s} \tag{16.5}$$

When $\dot{W}_{\text{ideal}}$ (or W_{ideal}) is negative, $|\dot{W}_{\text{ideal}}|$ is the *maximum work obtainable* from a given change in the properties of the flowing streams, and is larger than $|\dot{W}_s|$. In this case, the thermodynamic efficiency is defined as the ratio of the actual work to the ideal work:

$$\eta_t(\text{work produced}) = \frac{\dot{W}_s}{\dot{W}_{\text{ideal}}} \tag{16.6}$$

Example 16.1   What is the maximum work that can be obtained in a steady-state flow process from 1 mol of nitrogen (assumed an ideal gas) at 800 K and 50 bar? Take the temperature and pressure of the surroundings as 300 K and 1.0133 bar.

SOLUTION   The maximum possible work is obtained from any completely reversible process that reduces the nitrogen to the temperature and pressure of the surroundings, i.e., to 300 K and 1.0133 bar. (The maintenance of a final temperature or pressure below that of the surroundings would require work in an amount at least equal to any gain in work from the process as a result of the lower level.) The result is obtained directly by solution of Eq. (16.4), where ΔS and ΔH are the molar entropy and enthalpy changes of the nitrogen as its state is changed from 800 K and 50 bar to 300 K and 1.0133 bar. For an ideal gas, enthalpy is independent of pressure, and its change is given by

$$\Delta H = \int_{T_1}^{T_2} C_P^{ig}\, dT$$

The value of this integral is found from Eq. (4.7), and is represented by

8.314*ICPH(800,300;3.280,0.593E-3,0.0,0.040E+5) $\equiv -15{,}060$ J mol^{-1}

The parameters in the heat-capacity equation for nitrogen come from Table C.1. Similarly, the entropy change is found from Eq. (5.14), written here as

$$\Delta S = \int_{T_1}^{T_2} C_P^{ig}\, \frac{dT}{T} - R\ln\frac{P_2}{P_1}$$

The value of the integral is found from Eq. (5.15), and is represented by

8.314*ICPS(800,300;3.280,0.593E-3,0.0,0.040E+5) $\equiv -29.373$ J mol^{-1} K^{-1}

whence
$$\Delta S = -29.373 - 8.314\ln\frac{1.0133}{50} = 3.042 \text{ J mol}^{-1}\text{ K}^{-1}$$

With these values of ΔH and ΔS, Eq. (16.4) becomes

$$W_{\text{ideal}} = -15{,}060 - (300)(3.042) = -15{,}973 \text{ J mol}^{-1}$$

The significance of this simple calculation becomes evident when we consider in detail the steps of a specific reversible process designed to bring about the same change of state. Suppose the nitrogen is continuously changed to its final state at 1.0133 bar and $T_2 = T_\sigma = 300$ K by the following two-step process:

1. Reversible, adiabatic expansion (as in a turbine) from initial state P_1, T_1, H_1 to 1.0133 bar. Let the temperature at the end of this isentropic step be T'.

2. Cooling (or heating, if T' is less than T_2) to the final temperature T_2 at a constant pressure of 1.0133 bar.

For step 1, a steady-state flow process, the energy balance is

$$Q + W_s = \Delta H$$

or, since the process is adiabatic,

$$W_s = \Delta H = (H' - H_1)$$

where H' is the enthalpy at the intermediate state of T' and 1.0133 bar. For maximum work production, step 2 must also be reversible, with heat transferred reversibly to the surroundings at T_σ. These requirements are met by use of Carnot engines which receive heat from the nitrogen, produce work W_{Carnot}, and reject heat to the surroundings at T_σ. Since the temperature of the heat source, the nitrogen, decreases from T' to T_2, the expression for the work of the Carnot engines is written in differential form:

$$dW_{\text{Carnot}} = \frac{T - T_\sigma}{T}(dQ)$$

Note here that dQ is negative with reference to the nitrogen, which is taken as the system. Integration yields

$$W_{\text{Carnot}} = Q - T_\sigma \int_{T'}^{T_2} \frac{dQ}{T}$$

Quantity Q, the heat exchanged with the nitrogen, is equal to the enthalpy change $H_2 - H'$. The integral is the change in entropy of the nitrogen as it is cooled by the Carnot engines. Since step 1 occurs at constant entropy, the integral also represents ΔS for both steps. Hence

$$W_{\text{Carnot}} = (H_2 - H') - T_\sigma \Delta S$$

The sum of W_s and W_{Carnot} gives the ideal work; thus

$$\begin{aligned} W_{\text{ideal}} &= (H' - H_1) + (H_2 - H') - T_\sigma \Delta S \\ &= (H_2 - H_1) - T_\sigma \Delta S \\ &= \Delta H - T_\sigma \Delta S \end{aligned}$$

which is the same as Eq. (16.4).

This derivation makes clear the difference between W_s, the ideal shaft work of the turbine, and W_{ideal}. The ideal work includes not only the ideal shaft work, but also all work obtainable by the operation of heat engines for the reversible transfer of heat to the surroundings at T_σ.

Example 16.2 Rework Example 5.5, making use of the equation for ideal work.

SOLUTION   The procedure here is to calculate the maximum possible work W_{ideal} which can be obtained from 1 kg of steam in a flow process as it undergoes a change in state from saturated steam at 100°C to liquid water at 0°C. Now the problem reduces to the question of whether this amount of work is sufficient to operate a Carnot heat pump delivering 2,000 kJ as heat at 200°C and taking heat from the unlimited supply of cooling water at 0°C.

For the steam,

$$\Delta H = 0 - 2{,}676.0 = -2{,}676.0$$
$$\Delta S = 0 - 7.3554 = -7.3554$$

Neglecting kinetic- and potential-energy terms, we have by Eq. (16.4)

$$W_{ideal} = \Delta H - T_\sigma \Delta S = -2{,}676.0 - (273.15)(-7.3554) = -666.9 \text{ kJ kg}^{-1}$$

If this amount of work, numerically the maximum obtainable from the steam, is used to drive a Carnot heat pump operating between the temperatures of 0 and 200°C, the heat transferred at the higher temperature is

$$|Q| = |W|\frac{T}{T_\sigma - T} = (666.9)\left(\frac{200 + 273.15}{200 - 0}\right) = 1{,}577.7 \text{ kJ}$$

This is the maximum possible heat release at 200°C; it is less than the claimed value of 2,000 kJ. As in Example 5.5, we conclude that the process described is not possible.

Example 16.3 What is the thermodynamic efficiency of the compression process of Example 7.12 if $T_\sigma = 300$ K?

SOLUTION   Saturated steam at 100 kPa is compressed adiabatically to 300 kPa with a compressor efficiency of 0.75. From the results of Example 7.12, we have

$$\Delta H = 2{,}959.9 - 2{,}675.4 = 284.5 \text{ kJ kg}^{-1}$$
$$\Delta S = 7.5019 - 7.3598 = 0.1421 \text{ kJ kg}^{-1} \text{ K}^{-1}$$

and

$$W_s = 284.5 \text{ kJ kg}^{-1}$$

Application of Eq. (16.4) gives

$$W_{ideal} = \Delta H - T_\sigma \Delta S = 284.5 - (300)(0.1421) = 241.9 \text{ kJ kg}^{-1}$$

Then by Eq. (16.6),

$$\eta_t = \frac{W_{ideal}}{W_s} = \frac{241.9}{284.5} = 0.850$$

The *compressor* efficiency η, based on reversible compression to a final state where $S_2 = S_1$, is different from the *thermodynamic* efficiency η_t, which is based on reversible compression to the *actual* final state where $S_2 > S_1$.

16.2 Lost Work

Work that is wasted as the result of irreversibilities in a process is called *lost work*, W_{lost}, and is defined as the difference between the actual work of a process and the ideal work for the process. Thus by definition,

$$W_{\text{lost}} \equiv W_s - W_{\text{ideal}} \tag{16.7}$$

In terms of rates this is written

$$\dot{W}_{\text{lost}} \equiv \dot{W}_s - \dot{W}_{\text{ideal}} \tag{16.8}$$

The actual work rate comes from Eq. (7.15):

$$\dot{W}_s = \Delta\left[\left(H + \tfrac{1}{2}u^2 + zg\right)\dot{m}\right]_{\text{fs}} - \dot{Q}$$

The ideal work rate is given by Eq. (16.1):

$$\dot{W}_{\text{ideal}} = \Delta\left[\left(H + \tfrac{1}{2}u^2 + zg\right)\dot{m}\right]_{\text{fs}} - T_\sigma\Delta(S\dot{m})_{\text{fs}}$$

The difference between these two equations gives

$$\boxed{\dot{W}_{\text{lost}} = T_\sigma\Delta(S\dot{m})_{\text{fs}} - \dot{Q}} \tag{16.9}$$

Equation (7.20) may be written for the case of a single surroundings temperature T_σ:

$$\dot{S}_{G,\text{total}} = \Delta(S\dot{m})_{\text{fs}} - \frac{\dot{Q}}{T_\sigma} \tag{16.10}$$

Multiplication by T_σ gives

$$T_\sigma\dot{S}_{G,\text{total}} = T_\sigma\Delta(S\dot{m})_{\text{fs}} - \dot{Q}$$

Since the right-hand sides of this equation and of Eq. (16.9) are identical, it follows that

$$\boxed{\dot{W}_{\text{lost}} = T_\sigma\dot{S}_{G,\text{total}}} \tag{16.11}$$

Since the second law of thermodynamics requires that $\dot{S}_{G,\text{total}} \geq 0$, it follows that $\dot{W}_{\text{lost}} \geq 0$. When a process is completely reversible, the equality holds, and the lost work is zero. For irreversible processes the inequality holds, and the lost work, i.e., the energy that becomes unavailable for work, is positive. The engineering significance of this result is clear: The greater the irreversibility of a process, the greater the rate of entropy production and the greater the amount of energy that becomes unavailable for work. Thus every irreversibility carries with it a price.

For the special case of a single stream flowing through the control volume,

$$\dot{W}_{\text{lost}} = \dot{m}T_\sigma\Delta S - \dot{Q} \tag{16.12}$$

Division by $\dot{m}$ gives

$$W_{\text{lost}} = T_\sigma \Delta S - Q \qquad (16.13)$$

where the basis is now a unit amount of fluid flowing through the control volume. Similarly, for a single stream, Eq. (16.10) becomes

$$\dot{S}_{G,\text{total}} = \dot{m}\Delta S - \frac{\dot{Q}}{T_\sigma} \qquad (16.14)$$

Division by $\dot{m}$ provides an equation based on a unit amount of fluid flowing through the control volume:

$$S_{G,\text{total}} = \Delta S - \frac{Q}{T_\sigma} \qquad (16.15)$$

Equations (16.13) and (16.16) combine for a unit amount of fluid to give

$$W_{\text{lost}} = T_\sigma S_{G,\text{total}} \qquad (16.16)$$

Again, since $S_{G,\text{total}} \geq 0$, it follows that $W_{\text{lost}} \geq 0$.

> **Example 16.4**   What is the lost work associated with the compression process of Example (16.3)?
>
> SOLUTION   Since the compression process is adiabatic, Eq. (16.15) reduces to
>
> $$S_{G,\text{total}} = \Delta S$$
>
> and Eq. (16.16) becomes
>
> $$W_{\text{lost}} = T_\sigma \Delta S$$
>
> where ΔS is the entropy change of the steam as a result of compression. Taking this value from Example 16.3, we find
>
> $$W_{\text{lost}} = (300)(0.1421) = 42.6 \text{ kJ kg}^{-1}$$
>
> This result is also given by Eq. (16.7), where values are from Example 16.3:
>
> $$W_{\text{lost}} = W_s - W_{\text{ideal}} = 284.5 - 241.9 = 42.6 \text{ kJ kg}^{-1}$$

16.3   Thermodynamic Analysis of Steady-State Flow Processes

Many processes consist of a number of steps, and lost-work calculations are then made for each step separately. By Eq. (16.11),

$$\dot{W}_{\text{lost}} = T_\sigma \dot{S}_{G,\text{total}}$$

Summing over the steps of a process gives

$$\sum \dot{W}_{\text{lost}} = T_\sigma \sum \dot{S}_{G,\text{total}}$$

Dividing the former equation by the latter yields

$$\frac{\dot{W}_{\text{lost}}}{\sum \dot{W}_{\text{lost}}} = \frac{\dot{S}_{G,\text{total}}}{\sum \dot{S}_{G,\text{total}}}$$

Thus an analysis of the lost work, made by calculation of the fraction that each individual lost-work term represents of the total lost work, is the same as an analysis of the rate of entropy generation, made by expressing each individual entropy-generation term as a fraction of the sum of all entropy-generation terms.

An alternative to the lost-work or entropy-generation analysis is a work analysis. For this, we write Eq. (16.8) as

$$\sum \dot{W}_{\text{lost}} = \dot{W}_s - \dot{W}_{\text{ideal}} \tag{16.17}$$

For a work-requiring process, all of these work quantities are positive and $\dot{W}_s > \dot{W}_{\text{ideal}}$. We therefore write the preceding equation as

$$\boxed{\dot{W}_s = \dot{W}_{\text{ideal}} + \sum \dot{W}_{\text{lost}}} \tag{16.18}$$

A work analysis then expresses each of the individual work terms on the right as a fraction of $\dot{W}_s$.

For a work-producing process, $\dot{W}_s$ and $\dot{W}_{\text{ideal}}$ are negative, and $|\dot{W}_{\text{ideal}}| > |\dot{W}_s|$. Equation (16.17) is therefore best written

$$\boxed{|\dot{W}_{\text{ideal}}| = |\dot{W}_s| + \sum \dot{W}_{\text{lost}}} \tag{16.19}$$

A work analysis here expresses each of the individual work terms on the right as a fraction of $|\dot{W}_{\text{ideal}}|$. A work analysis cannot be carried out in the case where a process is so inefficient that $\dot{W}_{\text{ideal}}$ is negative, indicating that the process should produce work, but $\dot{W}_s$ is positive, indicating that the process in fact requires work. A lost-work or entropy-generation analysis is always possible.

Example 16.5 The operating conditions of a practical steam power plant are described in Example 8.1, parts (b) and (c). In addition, steam is generated in a furnace/boiler unit where methane is burned completely to CO_2 and H_2O with 25 percent excess air. The flue gas leaving the furnace has a temperature of 460 K, and $T_\sigma = 298.15$ K. Make a thermodynamic analysis of the power plant.

SOLUTION   A flow diagram of the power plant is shown in Fig. 16.1. The conditions and properties for key points in the steam cycle, taken from Example 8.1, are listed in the following table.

Point	State of steam	$t/°C$	P/kPa	$H/kJ\ kg^{-1}$	$S/kJ\ kg^{-1}K^{-1}$
1	Subcooled liquid	45.83	8,600	203.4	0.6580
2	Superheated vapor	500	8,600	3,391.6	6.6858
3	Wet vapor, $x = 0.9378$	45.83	10	2,436.0	7.6846
4	Saturated liquid	45.83	10	191.8	0.6493

Since the steam undergoes a cyclic process, the only changes that need be considered for calculation of the ideal work are those of the gases passing through the furnace. The reaction occurring is

$$CH_4 + 2O_2 \rightarrow CO_2 + 2H_2O$$

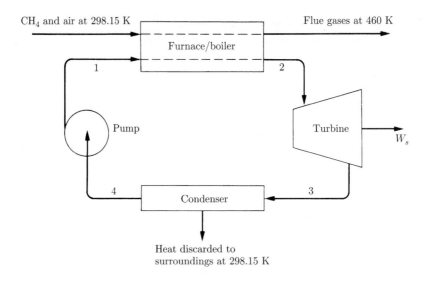

Figure 16.1: Power cycle of Example 16.5.

For this reaction, data from Table C.4 give

$$\Delta H^{\circ}_{298} = -393{,}509 + (2)(-241{,}818) - (-74{,}520) = -802{,}625 \text{ J}$$

$$\Delta G^{\circ}_{298} = -394{,}359 + (2)(-228{,}572) - (-50{,}460) = -801{,}043 \text{ J}$$

Moreover,

$$\Delta S^{\circ}_{298} = \frac{\Delta H^{\circ}_{298} - \Delta G^{\circ}_{298}}{298.15} = -5.306 \text{ J K}^{-1}$$

On the basis of 1 mol of methane burned with 25 percent excess air, the air entering the furnace contains

$$
\begin{aligned}
O_2: &\quad (2)(1.25) = 2.5 \text{ mol} \\
N_2: &\quad (2.5)(79/21) = 9.405 \text{ mol} \\
\hline
\text{Total:} &\quad 11.905 \text{ mol air}
\end{aligned}
$$

After complete combustion of the methane, the flue gas contains

CO_2:	1 mol	$y_{CO_2} = 0.0775$
H_2O:	2 mol	$y_{H_2O} = 0.1550$
O_2:	0.5 mol	$y_{O_2} = 0.0387$
N_2:	9.405 mol	$y_{N_2} = 0.7288$
Total:	12.905 mol flue gas	$\sum y_i = 1.0000$

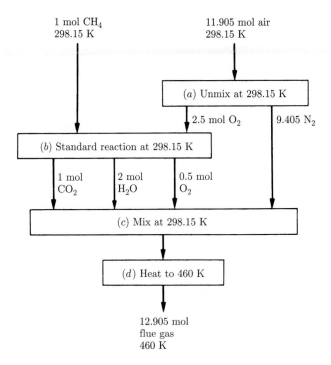

Figure 16.2: Calculation path for combustion process of Example 16.5.

The change of state that occurs in the furnace is from methane and air at atmospheric pressure and 298.15 K, the temperature of the surroundings, to flue gas at atmospheric pressure and 460 K. To calculate ΔH and ΔS for this change of state, we devise the path shown in Fig. 16.2. The assumption of ideal gases is reasonable here, and on this basis we calculate ΔH and ΔS for each of the four steps shown in Fig. 16.2.

Step a: For unmixing the entering air, Eqs. (11.37) and (11.35) with changes of sign give

$$\Delta H_a = 0$$
$$\Delta S_a = nR\sum_i y_i \ln y_i$$
$$= (11.905)(8.314)(0.21\ln 0.21 + 0.79\ln 0.79)$$
$$= -50.870 \text{ J K}^{-1}$$

Step b: For the standard reaction at 298.15 K,

$$\Delta H_b = \Delta H^{\circ}_{298} = -802{,}625 \text{ J}$$
$$\Delta S_b = \Delta S^{\circ}_{298} = -5.306 \text{ J K}^{-1}$$

Step c: For mixing to form the flue gas,

$$\Delta H_c = 0$$

$$\Delta S_c = -nR \sum_i y_i \ln y_i$$
$$= -(12.905)(8.314)(0.0775 \ln 0.0775 + 0.1550 \ln 0.1550$$
$$+ 0.0387 \ln 0.0387 + 0.7288 \ln 0.7288)$$
$$= 90.510 \text{ J K}^{-1}$$

Step d: For the heating step, the mean heat capacities between 298.15 and 460 K are calculated by Eqs. (4.8) and (5.17) with data from Table C.1. The results in J mol^{-1} K^{-1} are summarized as follows:

	$\langle C_P \rangle_H$	$\langle C_P \rangle_S$
CO_2	41.649	41.377
H_2O	34.153	34.106
N_2	29.381	29.360
O_2	30.473	30.405

We multiply each individual heat capacity by the number of moles of that species in the flue gas and sum over all species. This gives total mean heat capacities for the 12.905 mol of mixture:

$$\langle C_P^t \rangle_H = 401.520 \qquad \text{and} \qquad \langle C_P^t \rangle_S = 400.922 \text{ J K}^{-1}$$

Then

$$\Delta H_d = \langle C_P^t \rangle_H (T_2 - T_1) = (401.520)(460 - 298.15) = 64,986 \text{ J}$$

and

$$\Delta S_d = \langle C_P^t \rangle_S \ln \frac{T_2}{T_1} = 400.922 \ln \frac{460}{298.15} = 173.852 \text{ J K}^{-1}$$

For the total process on the basis of 1 mol CH_4 burned,

$$\Delta H = \sum \Delta H_i = 0 - 802,625 + 0 + 64,986 = -737,639 \text{ J}$$

or

$$\Delta H = -737.64 \text{ kJ}$$

and

$$\Delta S = \sum \Delta S_i = -50.870 - 5.306 + 90.510 + 173.852 = 208.186 \text{ J K}^{-1}$$

or

$$\Delta S = 0.2082 \text{ kJ K}^{-1}$$

The steam rate found in Example 8.1 is

$$\dot{m} = 84.75 \text{ kg s}^{-1}$$

An energy balance for the furnace/boiler unit, where heat is transferred from the combustion gases to the steam, allows calculation of the entering methane rate $\dot{n}_{CH_4}$:

$$(84.75)(3,391.6 - 203.4) + \dot{n}_{CH_4}(-737.64) = 0$$

whence

$$\dot{n}_{CH_4} = 366.30 \text{ mol s}^{-1}$$

The ideal work for the process, given by Eq. (16.3), is

$$\dot{W}_{\text{ideal}} = 366.30[-737.64 - (298.15)(0.2082)] = -292.94 \times 10^3 \text{ kJ s}^{-1}$$

or

$$\dot{W}_{\text{ideal}} = -292.94 \times 10^3 \text{ kW}$$

The rate of entropy generation in each of the four units of the power plant is calculated by Eq. (16.10), and the lost work is then given by Eq. (16.11).

Furnace/boiler: We have assumed no heat transfer from the furnace/boiler to the surroundings; therefore $\dot{Q} = 0$. The term $\Delta(S\dot{m})_{\text{fs}}$ is simply the sum of the entropy changes of the two streams multiplied by their rates:

$$\dot{S}_{G,\text{total}} = (366.30)(0.2082) + (84.75)(6.6858 - 0.6580) = 587.12 \text{ kJ s}^{-1} \text{ K}^{-1}$$

or

$$\dot{S}_{G,\text{total}} = 587.12 \text{ kW K}^{-1}$$

and

$$\dot{W}_{\text{lost}} = T_\sigma \dot{S}_{G,\text{total}} = (298.15)(587.12) = 175.05 \times 10^3 \text{ kW}$$

Turbine: For adiabatic operation,

$$\dot{S}_{G,\text{total}} = (84.75)(7.6846 - 6.6858) = 84.65 \text{ kW K}^{-1}$$

and

$$\dot{W}_{\text{lost}} = (298.15)(84.65) = 25.24 \times 10^3 \text{ kW}$$

Condenser: The condenser transfers heat from the condensing steam to the surroundings at 298.15 K in an amount determined in Example 8.1:

$$\dot{Q}(\text{condenser}) = -190.2 \times 10^3 \text{ kJ s}^{-1}$$

Thus

$$\dot{S}_{G,\text{total}} = (84.75)(0.6493 - 7.6846) + \frac{190,200}{298.15} = 41.69 \text{ kW K}^{-1}$$

and

$$\dot{W}_{\text{lost}} = (298.15)(41.69) = 12.32 \times 10^3 \text{ kW}$$

Pump: Since the pump operates adiabatically,

$$\dot{S}_{G,\text{total}} = (84.75)(0.6580 - 0.6493) = 0.74 \text{ kW K}^{-1}$$

and

$$\dot{W}_{\text{lost}} = 0.22 \times 10^3 \text{ kW}$$

The entropy-generation analysis is as follows:

	kW K^{-1}	Percent of $\sum \dot{S}_{G,\text{total}}$
$\dot{S}_{G,\text{total}}$(furnace/boiler)	587.12	82.2
$\dot{S}_{G,\text{total}}$(turbine)	84.65	11.9
$\dot{S}_{G,\text{total}}$(condenser)	41.69	5.8
$\dot{S}_{G,\text{total}}$(pump)	0.74	0.1
$\sum \dot{S}_{G,\text{total}}$	714.20	100.0

A work analysis is carried out in accord with Eq. (16.19):

$$|\dot{W}_{\text{ideal}}| = |\dot{W}_s| + \sum \dot{W}_{\text{lost}}$$

The results of this analysis are shown in the following table:

	kW	Percent of $|\dot{W}_{\text{ideal}}|$
$|\dot{W}_s|$ (from Example 8.1)	80.00×10^3	$27.3(=\eta_t)$
$\dot{W}_{\text{lost}}$(furnace/boiler)	175.05×10^3	59.8
$\dot{W}_{\text{lost}}$(turbine)	25.24×10^3	8.6
$\dot{W}_{\text{lost}}$(condenser)	12.43×10^3	4.2
$\dot{W}_{\text{lost}}$(pump)	0.22×10^3	0.1
$|\dot{W}_{\text{ideal}}|$	292.94×10^3	100.0

The thermodynamic efficiency of the power plant is 27.3 percent, and the major source of inefficiency is the furnace/boiler. The combustion process itself accounts for most of the entropy generation in this unit, and the remainder is the result of heat transfer across finite temperature differences.

Example 16.6 Methane is liquefied in a simple Linde system, as shown in Fig. 16.3. The methane enters the compressor at 1 bar and 300 K, and after compression to 60 bar is cooled back to 300 K. The product is saturated liquid methane at 1 bar. The unliquefied methane, also at 1 bar, is returned through a heat exchanger where it is heated to 295 K by the high-pressure methane. A heat leak into the heat exchanger of 5 kJ is assumed for each kilogram of methane entering the compressor. Heat leaks to other parts of the liquefier are assumed negligible. Make a thermodynamic analysis of the process for a surroundings temperature of $T_\sigma = 300$ K.

SOLUTION   Methane compression from 1 to 60 bar is assumed to be carried out in a three-stage machine with inter- and after-cooling to 300 K and a compressor efficiency of 75 percent. The actual work of this compression is estimated as 1,000 kJ per kilogram of methane. The fraction of the methane that is liquefied z is calculated by an energy balance:

$$H_4 z + H_6(1 - z) - H_2 = Q$$

where Q is the heat leak from the surroundings. Solution for z gives

$$z = \frac{H_6 - H_2 - Q}{H_6 - H_4} = \frac{1{,}188.9 - 1{,}140.0 - 5}{1{,}188.9 - 285.4} = 0.0486$$

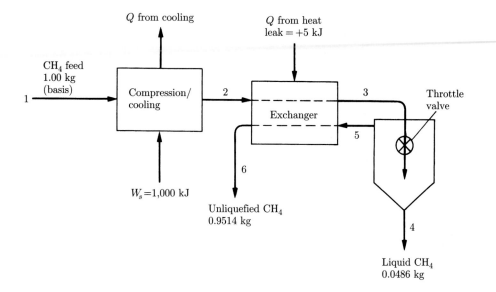

Figure 16.3: Linde liquefaction system for Example 16.6.

This result may be compared with the value of 0.0541 obtained in Example 9.3 for the same operating conditions, but no heat leak. The properties at the various key points of the process, given in the accompanying table, are either available as data or are calculated by standard methods. Data are from Perry and Green.[1] The basis of all calculations is 1 kg of methane entering the process, and all rates are expressed on this basis.

Point	State of the CH$_4$	T/K	P/bar	H/kJ kg^{-1}	S/kJ kg^{-1}K^{-1}
1	Superheated vapor	300.0	1	1,199.8	11.629
2	Superheated vapor	300.0	60	1,140.0	9.359
3	Superheated vapor	207.1	60	772.0	7.798
4	Saturated liquid	111.5	1	285.4	4.962
5	Saturated vapor	111.5	1	796.9	9.523
6	Superheated vapor	295.0	1	1,188.9	11.589

The ideal work depends on the overall changes in the methane passing through the liquefier. Application of Eq. (16.2) gives

$$\dot{W}_{\text{ideal}} = \Delta(H\dot{m})_{\text{fs}} - T_\sigma \Delta(S\dot{m})_{\text{fs}}$$
$$= [(0.0486)(285.4) + (0.9514)(1,188.9) - 1,199.8]$$
$$- (300)[(0.0486)(4.962) + (0.9514)(11.589) - 11.629]$$
$$= 53.8 \text{ kJ}$$

[1] R. H. Perry and D. Green, *Perry's Chemical Engineers' Handbook*, p. 3-203, McGraw-Hill, New York, 1984.

The rate of entropy generation and the lost work for each of the individual steps of the process are calculated by Eqs. (16.10) and (16.11).

The heat transfer for the compression/cooling step is calculated by an energy balance:

$$\dot{Q} = \Delta H - \dot{W}_s = (H_2 - H_1) - \dot{W}_s$$
$$= (1,140.0 - 1,199.8) - 1,000 = -1,059.8 \text{ kJ}$$

Then

$$\dot{S}_{G,\text{total}}(\text{compression/cooling}) = S_2 - S_1 - \frac{\dot{Q}}{T_\sigma}$$

from which

$$\dot{S}_{G,\text{total}}(\text{compression/cooling}) = 9.359 - 11.629 + \frac{1,059.8}{300}$$
$$= 1.2627 \text{ kJ kg}^{-1} \text{ K}^{-1}$$

and

$$\dot{W}_{\text{lost}}(\text{compression/cooling}) = (300)(1.2627) = 378.8 \text{ kJ kg}^{-1}$$

For the exchanger, with $\dot{Q}$ equal to the heat leak,

$$\dot{S}_{G,\text{total}}(\text{exchanger}) = (S_6 - S_5)(1 - z) + (S_3 - S_2)(1) - \frac{\dot{Q}}{T_\sigma}$$

Hence

$$\dot{S}_{G,\text{total}}(\text{exchanger}) = (11.589 - 9.523)(0.9514) + (7.798 - 9.359) - \frac{5}{300}$$
$$= 0.3879 \text{ kJ kg}^{-1} \text{ K}^{-1}$$

and

$$\dot{W}_{\text{lost}}(\text{exchanger}) = (300)(0.3879) = 116.4 \text{ kJ kg}^{-1}$$

For the throttle and separator, assuming adiabatic operation,

$$\dot{S}_{G,\text{total}}(\text{throttle}) = S_4 z + S_5(1 - z) - S_3$$
$$= (4.962)(0.0486) + (9.523)(0.9514) - 7.798$$
$$= 1.5033 \text{ kJ kg}^{-1} \text{ K}^{-1}$$

and

$$\dot{W}_{\text{lost}}(\text{throttle}) = (300)(1.5033) = 451.0 \text{ kJ kg}^{-1}$$

Analysis of the process with respect to entropy generation is shown in the following table:

	kJ kg^{-1} K^{-1}	Percent of $\sum \dot{S}_{G,\text{total}}$
$\dot{S}_{G,\text{total}}(\text{compression/cooling})$	1.2627	40.0
$\dot{S}_{G,\text{total}}(\text{exchanger})$	0.3879	12.3
$\dot{S}_{G,\text{total}}(\text{throttle})$	1.5033	47.7
$\sum \dot{S}_{G,\text{total}}$	3.1539	100.0

The work analysis, based on Eq. (16.18),

$$\dot{W}_s = \dot{W}_{\text{ideal}} + \sum \dot{W}_{\text{lost}}$$

is shown in the following table:

	kW kg^{-1}	Percent of $\dot{W}_s$
$\dot{W}_{\text{ideal}}$	53.8	5.4($=\eta_t$)
$\dot{W}_{\text{lost}}$(compression/cooling)	378.8	37.9
$\dot{W}_{\text{lost}}$(exchanger)	116.4	11.6
$\dot{W}_{\text{lost}}$(throttle)	451.0	45.1
$\dot{W}_s$	1,000.0	100.0

The largest loss occurs in the throttling step. Elimination of this highly irre-versible process in favor of a turbine results in a considerable increase in efficiency.

From the standpoint of energy conservation, the thermodynamic efficiency of a process should be as high as possible, and the entropy generation or lost work as low as possible. The final design depends largely on economic considerations, and the cost of energy is an important factor. The thermodynamic analysis of a specific process shows the locations of the major inefficiencies, and hence the pieces of equipment or steps in the process that could be altered or replaced to advantage. However, this sort of analysis gives no hint as to the nature of the changes that might be made. It merely shows that the present design is wasteful of energy and that there is room for improvement. One function of the chemical engineer is to try to devise a better process and to use ingenuity to keep operating costs, as well as capital expenditures, low. Each newly devised process may, of course, be analyzed to determine what improvement has been made.

PROBLEMS

16.1. Determine the maximum amount of work that can be obtained in a flow process from 1 kg of steam at 3,000 kPa and 450°C for surrounding conditions of 300 K and 101.33 kPa.

16.2. Liquid water at 325 K and 8,000 kPa flows into a boiler at the rate of 10 kg s^{-1} and is vaporized, producing saturated vapor at 8,000 kPa. What is the maximum fraction of the heat added to the water in the boiler that can be converted into work in a process whose product is water at initial conditions, if $T_\sigma = 300$ K? What happens to the rest of the heat? What is the rate of entropy change in the surroundings as a result of the work-producing process? In the system? Total?

16.3. Suppose the heat added to the water in the boiler in Prob. 16.2 comes from a furnace at a temperature of 600°C. What is the total rate of entropy generation as a result of the heating process? What is $\dot{W}_{\text{lost}}$?

16.4. What is the ideal-work rate for the expansion process of Example 7.10? What is the thermodynamic efficiency of the process? What is the rate of entropy generation $\dot{S}_{G,\text{total}}$? What is $\dot{W}_{\text{lost}}$? Take $T_\sigma = 300$ K.

16.5. What is the ideal work for the compression process of Example 7.13? What is the thermodynamic efficiency of the process? What is $S_{G,\text{total}}$? What is W_{lost}? Take $T_\sigma = 293.15$ K.

16.6. What is the ideal work for the pumping process of Example 7.14? What is the thermodynamic efficiency of the process? What is $S_{G,\text{total}}$? What is W_{lost}? Take $T_\sigma = 300$ K.

16.7. What is the ideal work for the separation of an equimolar mixture of methane and ethane at $175°$C and 3 bar in a steady-flow process into product streams of the pure gases at $35°$C and 1 bar if $T_\sigma = 300$ K?

16.8. What is the work required for the separation of air (21-mole-% oxygen and 79-mole-% nitrogen) at $25°$C and 1 bar in a steady-flow process into product streams of pure oxygen and nitrogen, also at $25°$C and 1 bar, if the thermodynamic efficiency of the process is 5 percent and if $T_\sigma = 300$ K?

16.9. An ideal gas at 2,500 kPa is throttled adiabatically to 150 kPa at the rate of 20 mol s^{-1}. Determine $\dot{S}_{G,\text{total}}$ and $\dot{W}_{\text{lost}}$ if $T_\sigma = 300$ K.

16.10. A refrigeration system cools a brine solution from $25°$C to $-15°$C at the rate of 20 kg s^{-1}. Heat is discarded to the atmosphere at a temperatue of $30°$C. What is the power requirement if the thermodynamic efficiency of the system is 0.27? The specific heat of the brine is 3.5 kJ kg^{-1} °C^{-1}.

16.11. An ice plant produces 0.5 kg s^{-1} of flake ice at $0°$C from water at $20°$C (T_σ) in a continuous process. If the latent heat of fusion of water is 333.4 kJ kg^{-1} and if the thermodynamic efficiency of the process is 32%, what is the power requirement of the plant?

16.12. Exhaust gas at $400°$C and 1 bar from internal-combustion engines flows at the rate of 125 mol s^{-1} into a waste-heat boiler where saturated steam is generated at a pressure of 1,200 kPa. Water enters the boiler at $20°$C (T_σ), and the exhaust gases are cooled to within $10°$C of the steam temperature. The heat capacity of the exhaust gases is $C_P/R = 3.34 + 1.12 \times 10^{-3}T$, where T is in kelvins. The steam flows into an adiabatic turbine from which it exhausts at a pressure of 25 kPa. If the turbine efficiency η is 72%,

(a) What is $\dot{W}_s$, the power output of the turbine?

(b) What is the thermodynamic efficiency of the boiler/turbine combination?

(c) Determine $\dot{S}_{G,\text{total}}$ for the boiler and for the turbine.

(d) Express $\dot{W}_{\text{lost}}$(boiler) and $\dot{W}_{\text{lost}}$(turbine) as fractions of $|\dot{W}_{\text{ideal}}|$, the ideal work of the process.

16.13. Consider the direct heat transfer from a heat reservoir at T_1 to another heat reservoir at temperature T_2, where $T_1 > T_2 > T_\sigma$. It is not obvious why the lost work of this process should depend on T_σ, the temperature of the surroundings, because the surroundings are not involved in the actual heat-transfer process. Through appropriate use of the Carnot-engine formula, show for the transfer of an amount of heat equal to $|Q|$ that

$$W_{\text{lost}} = T_\sigma |Q| \frac{T_1 - T_2}{T_1 T_2} = T_\sigma S_{G,\text{total}}$$

16.14. An inventor has developed a complicated process for making heat continuously available at an elevated temperature. Saturated steam at 100°C is the only source of energy. Assuming that there is plenty of cooling water available at 0°C, what is the maximum temperature level at which heat in the amount of 2,000 kJ can be made available for each kilogram of steam flowing through the process?

16.15. A plant takes in water at 70(°F), cools it to 32(°F), and freezes it at this temperature, producing $1(lb_m)(s)^{-1}$ of ice. Heat rejection is at 70(°F). The heat of fusion of water is $143.3(Btu)(lb_m)^{-1}$.

(a) What is $\dot{W}_{ideal}$ for the process?

(b) What is the power requirement of a single Carnot heat pump operating between 32 and 70(°F)? What is the thermodynamic efficiency of this process? What is its irreversible feature?

(c) What is the power requirement if an ideal tetrafluoroethane vapor-compression refrigeration cycle is used? *Ideal* here implies isentropic compression, infinite cooling-water rate in the condenser, and minimum heat-transfer driving forces in evaporator and condenser of 0(°F). What is the thermodynamic efficiency of this process? What are its irreversible features?

(d) What is the power requirement of a tetrafluoroethane vapor-compression cycle for which the compressor efficiency is 75%, the minimum temperature differences in evaporator and condenser are 8(°F), and the temperature rise of the cooling water in the condenser is 20(°F)? Make a thermodynamic analysis of this process.

16.16. Consider a steady-flow process in which the following gas-phase reaction takes place: $CO + \frac{1}{2}O_2 \rightarrow CO_2$. The surroundings are at 300 K.

(a) What is W_{ideal} when the reactants enter the process as pure carbon monoxide and as air containing the stoichiometric amount of oxygen, both at 25°C and 1 bar, and the products of complete combustion leave the process at the same conditions?

(b) The overall process is exactly the same as in (a). However, we now specify that the CO is burned in an adiabatic reactor at 1 bar. What is W_{ideal} for the process of cooling the flue gases to 25°C? What is the irreversible feature of the overall process? What is its thermodynamic efficiency? What has increased in entropy? By how much?

16.17. A chemical plant has saturated steam available at 2,700 kPa, but because of a process change has little use for steam at this pressure. Rather, steam at 1,000 kPa is required. Also available is saturated exhaust steam at 275 kPa. A suggestion is that the 275-kPa steam be compressed to 1,000 kPa by using all the work of expansion of the 2,700-kPa steam to 1,000 kPa. The two streams at 1,000 kPa would then be mixed. Determine the rates at which steam at each initial pressure must be supplied to provide enough steam at 1,000 kPa so that upon condensation to saturated liquid heat in the amount of 300 kJ s^{-1} is released,

(a) If the process is carried out in a completely reversible manner.

(b) If the higher-pressure steam expands in a turbine of 78% efficiency and the lower-pressure steam is compressed in a machine of 75% efficiency. Make a thermodynamic analysis of this process.

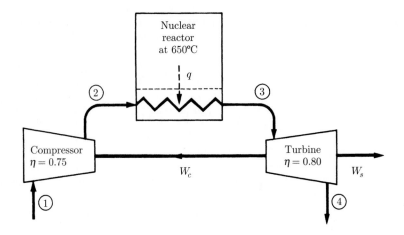

Figure P16.22

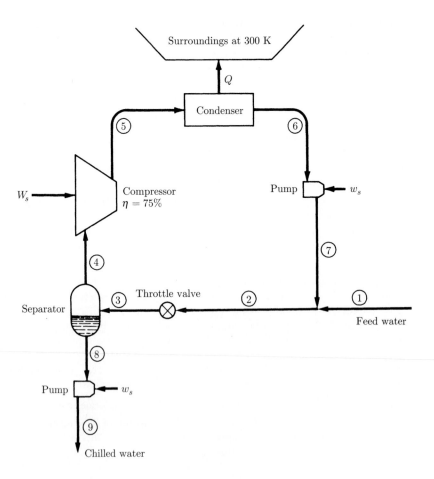

Figure P16.24

16.18. Make a thermodynamic analysis of the refrigeration cycle of Example 9.1(c).

16.19. Make a thermodynamic analysis of the refrigeration cycle described in one of the parts of Prob. 9.3. Assume that the refrigeration effect maintains a heat reservoir at a temperature 10(°F) above the evaporation temperature and that T_σ is 10(°F) below the condensation temperature.

16.20. Make a thermodynamic analysis of the refrigeration cycle described in the first paragraph of Prob. 9.6. Assume that the refrigeration effect maintains a heat reservoir at a temperature 10(°F) above the evaporation temperature and that T_σ is 10(°F) below the condensation temperature.

16.21. A colloidal solution enters a single-effect evaporator at 100°C. Water is vaporized from the solution, producing a more concentrated solution and 0.5 kg s^{-1} of steam at 100°C. This steam is compressed and sent to the heating coils of the evaporator to supply the heat required for its operation. For a minimum heat-transfer driving force across the evaporator coils of 10°C, for a compressor efficiency of 75%, and for adiabatic operation, what is the state of the stream leaving the heating coils of the evaporator? For a surroundings temperature of 300 K, make a thermodynamic analysis of the process.

16.22. An elementary nuclear-powered gas-turbine power plant operates as shown in Fig. P16.22. Air entering at point 1 is compressed adiabatically to point 2, heated at constant pressure between points 2 and 3, and expanded adiabatically from point 3 to point 4. Specified conditions are:

$$\begin{aligned}
&\text{Point 1:} \quad t = 30°\text{C},\ P = 1(\text{atm})\\
&\text{Point 2:} \quad P = 5(\text{atm})\\
&\text{Point 3:} \quad t = 575°\text{C},\ P = 5(\text{atm})\\
&\text{Point 4:} \quad P = 1(\text{atm})
\end{aligned}$$

The work to drive the compressor W_c comes from the turbine, and the additional work of the turbine W_s is the net work output of the power plant. The compressor and turbine efficiencies are given on the figure. Assume air an ideal gas for which $C_P = (7/2)R$. Including the nuclear reactor as part of the system and treating it as a heat reservoir at 650°C, make a thermodynamic analysis of the process. $T_\sigma = 293.15$ K.

16.23. Make a thermodynamic analysis of the process described in Prob. 8.7. Take the surroundings temperature as 80(°F).

16.24. Figure P16.24 shows a process that accomplishes the chilling of 0.5 kg s^{-1} of water from 26 to 4°C. The water acts as its own refrigerant by means of a recycle loop. The compressor (vacuum pump) maintains a suction pressure at point 4 such that the saturation temperature in the separator is 4°C, and discharges at point 5 to a pressure of 6 kPa. The compressor operates adiabatically with an efficiency of 75%. The condenser discharges saturated liquid water at 6 kPa. Make a thermodynamic analysis of the process, considering it to consist of the following parts:

(a) Points 6 and 1 to points 4 and 9.

(b) Point 4 to point 5.

(c) Point 5 to point 6.

Apart from the condenser, the process may be assumed adiabatic.

APPENDIX A

CONVERSION FACTORS AND VALUES OF THE GAS CONSTANT

Because standard reference books contain data in diverse units, we include Tables A.1 and A.2 to aid the conversion of values from one set of units to another. Those units having no connection with the SI system are enclosed in parentheses. The following definitions are noted:

$$\text{(ft)} \equiv \text{U.S. defined foot} \equiv 3.048 \times 10^{-1} \text{ m}$$
$$\text{(in)} \equiv \text{U.S. defined inch} \equiv 2.54 \times 10^{-2} \text{ m}$$
$$\text{(lb}_\text{m}) \equiv \text{U.S. defined pound } \textit{mass} \text{ (avoirdupois)}$$
$$\equiv 4.5359237 \times 10^{-1} \text{ kg}$$
$$\text{(lb}_\text{f}) \equiv \text{force to accelerate 1 (lb}_\text{m}) \text{ by 32.1740 (ft) s}^{-2}$$
$$\text{(atm)} \equiv \text{standard atmospheric pressure} \equiv 101{,}325 \text{ Pa}$$
$$\text{(psia)} \equiv \text{pounds } \textit{force} \text{ per square inch absolute pressure}$$
$$\text{(torr)} \equiv \text{pressure exerted by 1 mm mercury at 0}°\text{C and standard gravity}$$
$$\text{(cal)} \equiv \text{thermochemical calorie}$$
$$\text{(Btu)} \equiv \text{international steam table British thermal unit}$$
$$\text{(lb mole)} \equiv \text{mass in pounds } \textit{mass} \text{ with numerical value equal to the molar mass}$$
$$\text{(R)} \equiv \text{absolute temperature in Rankines}$$

The conversion factors of Table A.1 are referred to a single basic or derived unit of the SI system. Conversions between other pairs of units for a given quantity are made as in the following example:

$$1 \text{ bar} = 0.986923(\text{atm}) = 750.061(\text{torr})$$

thus

$$1(\text{atm}) = \frac{750.061}{0.986923} = 760.00(\text{torr})$$

Table A.1: Conversion factors

Quantity	Conversion
Length	$1 \text{ m} = 100 \text{ cm}$ $= 3.28084(\text{ft}) = 39.3701(\text{in})$
Mass	$1 \text{ kg} = 10^3 \text{ g}$ $= 2.20462(\text{lb}_\text{m})$
Force	$1 \text{ N} = 1 \text{ kg m s}^{-2}$ $= 10^5(\text{dyne})$ $= 0.224809(\text{lb}_\text{f})$
Pressure	$1 \text{ bar} = 10^5 \text{ kg m}^{-1} \text{ s}^{-2} = 10^5 \text{ N m}^{-2}$ $= 10^5 \text{ Pa} = 10^2 \text{ kPa}$ $= 10^6(\text{dyne}) \text{ cm}^{-2}$ $= 0.986923(\text{atm})$ $= 14.5038(\text{psia})$ $= 750.061(\text{torr})$
Volume	$1 \text{ m}^3 = 10^6 \text{ cm}^3$ $= 35.3147(\text{ft})^3$
Density	$1 \text{ g cm}^{-3} = 10^3 \text{ kg m}^{-3}$ $= 62.4278(\text{lb}_\text{m})(\text{ft})^{-3}$
Energy	$1 \text{ J} = 1 \text{ kg m}^2 \text{ s}^{-2} = 1 \text{ N m}$ $= 1 \text{ m}^3 \text{ Pa} = 10^{-5} \text{ m}^3 \text{ bar} = 10 \text{ cm}^3 \text{ bar}$ $= 9.86923 \text{ cm}^3(\text{atm})$ $= 10^7(\text{dyne}) \text{ cm} = 10^7(\text{erg})$ $= 0.239006(\text{cal})$ $= 5.12197 \times 10^{-3}(\text{ft})^3(\text{psia}) = 0.737562(\text{ft})(\text{lb}_\text{f})$ $= 9.47831 \times 10^{-4}(\text{Btu})$
Power	$1 \text{ kW} = 10^3 \text{ W} = 10^3 \text{ kg m}^2 \text{ s}^{-3} = 10^3 \text{ J s}^{-1}$ $= 239.006(\text{cal}) \text{ s}^{-1}$ $= 737.562(\text{ft})(\text{lb}_\text{f}) \text{ s}^{-1}$ $= 0.947831(\text{Btu}) \text{ s}^{-1}$ $= 1.34102(\text{hp})$

Table A.2: Values of the universal gas constant

$R = 8.314 \text{ J mol}^{-1} \text{ K}^{-1} = 8.314 \text{ m}^3 \text{ Pa mol}^{-1} \text{ K}^{-1}$
$= 83.14 \text{ cm}^3 \text{ bar mol}^{-1} \text{ K}^{-1} = 8{,}314 \text{ cm}^3 \text{ kPa mol}^{-1} \text{ K}^{-1}$
$= 82.06 \text{ cm}^3(\text{atm}) \text{ mol}^{-1} \text{ K}^{-1} = 62{,}356 \text{ cm}^3(\text{torr}) \text{ mol}^{-1} \text{ K}^{-1}$
$= 1.987(\text{cal}) \text{ mol}^{-1} \text{ K}^{-1} = 1.986(\text{Btu})(\text{lb mole})^{-1}(\text{R})^{-1}$
$= 0.7302(\text{ft})^3(\text{atm})(\text{lb mol})^{-1}(\text{R})^{-1} = 10.73(\text{ft})^3(\text{psia})(\text{lb mol})^{-1}(\text{R})^{-1}$
$= 1{,}545(\text{ft})(\text{lb}_\text{f})(\text{lb mol})^{-1}(\text{R})^{-1}$

APPENDIX B

PROPERTIES OF PURE SPECIES

Listed here for various chemical species are values for the molar mass (molecular weight), acentric factor ω, critical temperature T_c, critical pressure P_c, critical compressibility factor Z_c, critical molar volume V_c, and normal boiling point T_n. Abstracted from Project 801, DIPPR®, Design Institute for Physical Property Data of the American Institute of Chemical Engineers, they are reproduced with permission. The full data compilation is published by T. E. Daubert, R. P. Danner, H. M. Sibul, and C. C. Stebbins, *Physical and Thermodynamic Properties of Pure Chemicals: Data Compilation*, Taylor & Francis, Bristol, PA, 1,405 chemicals, extant 1995. Included are values for 26 physical constants and regressed values of parameters in equations for the temperature dependence of 13 thermodynamic and transport properties.

Electronic versions by the same authors include:

- *DIPPR® Data Compilation of Pure Compound Properties*, ASCII Files, National Institute of Science and Technology, Standard Reference Data, Gaithersburg, MD, 1,458 chemicals, extant 1995.

- *DIPPR® Data Compilation, Student DIPPR Database*, PC-DOS Version, National Institute of Science and Technology, Standard Reference Data, Gaithersburg, MD, 100 common chemicals for teaching purposes, 1995.

Table B.1: Properties of Pure Species

	Molar mass	ω	T_c/K	P_c/bar	Z_c	V_c cm^3 mol^{-1}	T_n/K
Methane	16.043	0.012	190.6	45.99	0.286	98.6	111.4
Ethane	30.070	0.100	305.3	48.72	0.279	145.5	184.6
Propane	44.097	0.152	369.8	42.48	0.276	200.0	231.1
n-Butane	58.123	0.200	425.1	37.96	0.274	255.	272.7
n-Pentane	72.150	0.252	469.7	33.70	0.270	313.	309.2
n-Hexane	86.177	0.301	507.6	30.25	0.266	371.	341.9
n-Heptane	100.204	0.350	540.2	27.40	0.261	428.	371.6
n-Octane	114.231	0.400	568.7	24.90	0.256	486.	398.8
n-Nonane	128.258	0.444	594.6	22.90	0.252	544.	424.0
n-Decane	142.285	0.492	617.7	21.10	0.247	600.	447.3
Isobutane	58.123	0.181	408.1	36.48	0.282	262.7	261.4
Isooctane	114.231	0.302	544.0	25.68	0.266	468.	372.4
Cyclopentane	70.134	0.196	511.8	45.02	0.273	258.	322.4
Cyclohexane	84.161	0.210	553.6	40.73	0.273	308.	353.9
Methylcyclopentane	84.161	0.230	532.8	37.85	0.272	319.	345.0
Methylcyclohexane	98.188	0.235	572.2	34.71	0.269	368.	374.1
Ethylene	28.054	0.087	282.3	50.40	0.281	131.	169.4
Propylene	42.081	0.140	365.6	46.65	0.289	188.4	225.5
1-Butene	56.108	0.191	420.0	40.43	0.277	239.3	266.9
cis-2-Butene	56.108	0.205	435.6	42.43	0.273	233.8	276.9
$trans$-2-Butene	56.108	0.218	428.6	41.00	0.275	237.7	274.0
1-Hexene	84.161	0.280	504.0	31.40	0.265	354.	336.3
Isobutylene	56.108	0.194	417.9	40.00	0.275	238.9	266.3
1,3-Butadiene	54.092	0.190	425.2	42.77	0.267	220.4	268.7
Cyclohexene	82.145	0.212	560.4	43.50	0.272	291.	356.1
Acetylene	26.038	0.187	308.3	61.39	0.271	113.	189.4
Benzene	78.114	0.210	562.2	48.98	0.271	259.	353.2
Toluene	92.141	0.262	591.8	41.06	0.264	316.	383.8
Ethylbenzene	106.167	0.303	617.2	36.06	0.263	374.	409.4
Cumene	120.194	0.326	631.1	32.09	0.261	427.	425.6
o-Xylene	106.167	0.310	630.3	37.34	0.263	369.	417.6
m-Xylene	106.167	0.326	617.1	35.36	0.259	376.	412.3
p-Xylene	106.167	0.322	616.2	35.11	0.260	379.	411.5
Styrene	104.152	0.297	636.0	38.40	0.256	352.	418.3
Naphthalene	128.174	0.302	748.4	40.51	0.269	413.	
Biphenyl	154.211	0.365	789.3	38.50	0.295	502.	528.2
Formaldehyde	30.026	0.282	408.0	65.90	0.223	115.	154.1
Acetaldehyde	44.053	0.291	466.0	55.50	0.221	154.	294.0
Methyl acetate	74.079	0.331	506.6	47.50	0.257	228.	330.1
Ethyl acetate	88.106	0.366	523.3	38.80	0.255	286.	350.2
Acetone	58.080	0.307	508.2	47.01	0.233	209.	329.4
Methyl ethyl ketone	72.107	0.323	535.5	41.50	0.249	267.	352.8
Diethyl ether	74.123	0.281	466.7	36.40	0.263	280.	307.6
Methyl t-butyl ether	88.150	0.266	497.1	34.30	0.273	329.	328.4

Table B.1: Properties of Pure Species (Continued)

	Molar mass	ω	T_c/K	P_c/bar	Z_c	V_c cm^3 mol^{-1}	T_n/K
Methanol	32.042	0.564	512.6	80.97	0.224	118.	337.9
Ethanol	46.069	0.645	513.9	61.48	0.240	167.	351.4
1-Propanol	60.096	0.622	536.8	51.75	0.254	219.	370.4
1-Butanol	74.123	0.594	563.1	44.23	0.260	275.	390.8
1-Hexanol	102.177	0.579	611.4	35.10	0.263	381.	430.6
2-Propanol	60.096	0.668	508.3	47.62	0.248	220.	355.4
Phenol	94.113	0.444	694.3	61.30	0.243	229.	455.0
Ethylene glycol	62.068	0.487	719.7	77.00	0.246	191.0	470.5
Acetic acid	60.053	0.467	592.0	57.86	0.211	179.7	391.1
n-Butyric acid	88.106	0.681	615.7	40.64	0.232	291.7	436.4
Benzoic acid	122.123	0.603	751.0	44.70	0.246	344.	522.4
Acetonitrile	41.053	0.338	545.5	48.30	0.184	173.	354.8
Methylamine	31.057	0.281	430.1	74.60	0.321	154.	266.8
Ethylamine	45.084	0.285	456.2	56.20	0.307	207.	289.7
Nitromethane	61.040	0.348	588.2	63.10	0.223	173.	374.4
Carbon tetrachloride	153.822	0.193	556.4	45.60	0.272	276.	349.8
Chloroform	119.377	0.222	536.4	54.72	0.293	239.	334.3
Dichloromethane	84.932	0.199	510.0	60.80	0.265	185.	312.9
Methyl chloride	50.488	0.153	416.3	66.80	0.276	143.	249.1
Ethyl chloride	64.514	0.190	460.4	52.70	0.275	200.	285.4
Chlorobenzene	112.558	0.250	632.4	45.20	0.265	308.	404.9
Argon	39.948	0.000	150.9	48.98	0.291	74.6	87.3
Krypton	83.800	0.000	209.4	55.02	0.288	91.2	119.8
Xenon	165.03	0.000	289.7	58.40	0.286	118.0	165.0
Helium 4	4.003	−0.390	5.2	2.28	0.302	57.3	4.2
Hydrogen	2.016	−0.216	33.19	13.13	0.305	64.1	20.4
Oxygen	31.999	0.022	154.6	50.43	0.288	73.4	90.2
Nitrogen	28.014	0.038	126.2	34.00	0.289	89.2	77.3
Chlorine	70.905	0.069	417.2	77.10	0.265	124.	239.1
Carbon monoxide	28.010	0.048	132.9	34.99	0.299	93.4	81.7
Carbon dioxide	44.010	0.224	304.2	73.83	0.274	94.0	
Carbon disulfide	76.143	0.111	552.0	79.00	0.275	160.	319.4
Hydrogen sulfide	34.082	0.094	373.5	89.63	0.284	98.5	212.8
Sulfur dioxide	64.065	0.245	430.8	78.84	0.269	122.	263.1
Sulfur trioxide	80.064	0.424	490.9	82.10	0.255	127.	317.9
Nitric oxide(NO)	30.006	0.583	180.2	64.80	0.251	58.0	121.4
Nitrous oxide(N_2O)	44.013	0.141	309.6	72.45	0.274	97.4	184.7
Hydrogen chloride	36.461	0.132	324.7	83.10	0.249	81.	188.2
Hydrogen cyanide	27.026	0.410	456.7	53.90	0.197	139.	298.9
Water	18.015	0.345	647.1	220.55	0.229	55.9	373.2
Ammonia	17.031	0.253	405.7	112.80	0.242	72.5	239.7
Nitric acid	63.013	0.714	520.0	68.90	0.231	145.	356.2
Sulfuric acid	98.080	. . .	924.0	64.00	0.147	177.	610.0

APPENDIX C

HEAT CAPACITIES AND PROPERTY CHANGES OF FORMATION

Table C.1: Heat Capacities of Gases in the Ideal-Gas State[†]

Constants in equation $C_P^{ig}/R = A + BT + CT^2 + DT^{-2}$ T (kelvins) from 298 to T_{max}

Chemical species		T_{max}	A	$10^3 B$	$10^6 C$	$10^{-5} D$
Paraffins:						
Methane	CH_4	1500	1.702	9.081	−2.164	
Ethane	C_2H_6	1500	1.131	19.225	−5.561	
Propane	C_3H_8	1500	1.213	28.785	−8.824	
n-Butane	C_4H_{10}	1500	1.935	36.915	−11.402	
iso-Butane	C_4H_{10}	1500	1.677	37.853	−11.945	
n-Pentane	C_5H_{12}	1500	2.464	45.351	−14.111	
n-Hexane	C_6H_{14}	1500	3.025	53.722	−16.791	
n-Heptane	C_7H_{16}	1500	3.570	62.127	−19.486	
n-Octane	C_8H_{18}	1500	8.163	70.567	−22.208	
1-Alkenes:						
Ethylene	C_2H_4	1500	1.424	14.394	−4.392	
Propylene	C_3H_6	1500	1.637	22.706	−6.915	
1-Butene	C_4H_8	1500	1.967	31.630	−9.873	
1-Pentene	C_5H_{10}	1500	2.691	39.753	−12.447	
1-Hexene	C_6H_{12}	1500	3.220	48.189	−15.157	
1-Heptene	C_7H_{14}	1500	3.768	56.588	−17.847	
1-Octene	C_8H_{16}	1500	4.324	64.960	−20.521	
Miscellaneous organics:						
Acetaldehyde	C_2H_4O	1000	1.693	17.978	−6.158	
Acetylene	C_2H_2	1500	6.132	1.952	· · · · · ·	−1.299
Benzene	C_6H_6	1500	−0.206	39.064	−13.301	
1,3-Butadiene	C_4H_6	1500	2.734	26.786	−8.882	
Cyclohexane	C_6H_{12}	1500	−3.876	63.249	−20.928	
Ethanol	C_2H_6O	1500	3.518	20.001	−6.002	
Ethylbenzene	C_8H_{10}	1500	1.124	55.380	−18.476	
Ethylene oxide	C_2H_4O	1000	−0.385	23.463	−9.296	
Formaldehyde	CH_2O	1500	2.264	7.022	−1.877	
Methanol	CH_4O	1500	2.211	12.216	−3.450	
Toluene	C_7H_8	1500	0.290	47.052	−15.716	
Styrene	C_8H_8	1500	2.050	50.192	−16.662	
Miscellaneous inorganics:						
Air		2000	3.355	0.575	· · · · · ·	−0.016
Ammonia	NH_3	1800	3.578	3.020	· · · · · ·	−0.186
Bromine	Br_2	3000	4.493	0.056	· · · · · ·	−0.154
Carbon monoxide	CO	2500	3.376	0.557	· · · · · ·	−0.031
Carbon dioxide	CO_2	2000	5.457	1.045	· · · · · ·	−1.157
Carbon disulfide	CS_2	1800	6.311	0.805	· · · · · ·	−0.906
Chlorine	Cl_2	3000	4.442	0.089	· · · · · ·	−0.344
Hydrogen	H_2	3000	3.249	0.422	· · · · · ·	0.083
Hydrogen sulfide	H_2S	2300	3.931	1.490	· · · · · ·	−0.232
Hydrogen chloride	HCl	2000	3.156	0.623	· · · · · ·	0.151
Hydrogen cyanide	HCN	2500	4.736	1.359	· · · · · ·	−0.725
Nitrogen	N_2	2000	3.280	0.593	· · · · · ·	0.040
Nitrous oxide	N_2O	2000	5.328	1.214	· · · · · ·	−0.928
Nitric oxide	NO	2000	3.387	0.629	· · · · · ·	0.014
Nitrogen dioxide	NO_2	2000	4.982	1.195	· · · · · ·	−0.792
Dinitrogen tetroxide	N_2O_4	2000	11.660	2.257	· · · · · ·	−2.787
Oxygen	O_2	2000	3.639	0.506	· · · · · ·	−0.227
Sulfur dioxide	SO_2	2000	5.699	0.801	· · · · · ·	−1.015
Sulfur trioxide	SO_3	2000	8.060	1.056	· · · · · ·	−2.028
Water	H_2O	2000	3.470	1.450	· · · · · ·	0.121

[†]Selected from H. M. Spencer, *Ind. Eng. Chem.*, vol. 40, pp. 2152–2154, 1948; K. K. Kelley, *U.S. Bur. Mines Bull. 584*, 1960; L. B. Pankratz, *U.S. Bur. Mines Bull. 672*, 1982.

Table C.2: Heat Capacities of Solids[†]

Constants for the equation $C_P/R = A + BT + DT^{-2}$
T (kelvins) from 298 K to T_{max}

Chemical species	T_{max}	A	$10^3\ B$	$10^{-5}\ D$
CaO	2000	6.104	0.443	−1.047
$CaCO_3$	1200	12.572	2.637	−3.120
$Ca(OH)_2$	700	9.597	5.435	
CaC_2	720	8.254	1.429	−1.042
$CaCl_2$	1055	8.646	1.530	−0.302
C(graphite)	2000	1.771	0.771	−0.867
Cu	1357	2.677	0.815	0.035
CuO	1400	5.780	0.973	−0.874
$Fe(\alpha)$	1043	−0.111	6.111	1.150
Fe_2O_3	960	11.812	9.697	−1.976
Fe_3O_4	850	9.594	27.112	0.409
FeS	411	2.612	13.286	
I_2	386.8	6.481	1.502	
NH_4Cl	458	5.939	16.105	
Na	371	1.988	4.688	
NaCl	1073	5.526	1.963	
NaOH	566	0.121	16.316	1.948
$NaHCO_3$	400	5.128	18.148	
S (rhombic)	368.3	4.114	−1.728	−0.783
SiO_2 (quartz)	847	4.871	5.365	−1.001

[†]Selected from K. K. Kelley, *U.S. Bur. Mines Bull. 584*, 1960; L. B. Pankratz, *U.S. Bur. Mines Bull. 672*, 1982.

Table C.3: Heat Capacities of Liquids[†]

Constants for the equation $C_P/R = A + BT + CT^2$
T from 273.15 to 373.15 K

Chemical species	A	$10^3\ B$	$10^6\ C$
Ammonia	22.626	−100.75	192.71
Aniline	15.819	29.03	−15.80
Benzene	−0.747	67.96	−37.78
1,3-Butadiene	22.711	−87.96	205.79
Carbon tetrachloride	21.155	−48.28	101.14
Chlorobenzene	11.278	32.86	−31.90
Chloroform	19.215	−42.89	83.01
Cyclohexane	−9.048	141.38	−161.62
Ethanol	33.866	−172.60	349.17
Ethylene oxide	21.039	−86.41	172.28
Methanol	13.431	−51.28	131.13
n-Propanol	41.653	−210.32	427.20
Sulfur trioxide	−2.930	137.08	−84.73
Toluene	15.133	6.79	16.35
Water	8.712	1.25	−0.18

[†]Based on correlations presented by J. W. Miller, Jr., G. R. Schorr, and C. L. Yaws, *Chem. Eng.*, vol. 83(23), p. 129, 1976.

Table C.4: Standard Enthalpies and Gibbs Energies of Formation at 298.15 K[†]

Joules per mole of the substance formed

Chemical species		State (Note 2)	$\Delta H^\circ_{f_{298}}$	$\Delta G^\circ_{f_{298}}$
Paraffins:				
Methane	CH_4	(g)	$-74,520$	$-50,460$
Ethane	C_2H_6	(g)	$-83,820$	$-31,855$
Propane	C_3H_8	(g)	$-104,680$	$-24,290$
n-Butane	C_4H_{10}	(g)	$-125,790$	$-16,570$
n-Pentane	C_5H_{12}	(g)	$-146,760$	$-8,650$
n-Hexane	C_6H_{14}	(g)	$-166,920$	150
n-Heptane	C_7H_{14}	(g)	$-187,780$	$8,260$
n-Octane	C_8H_{16}	(g)	$-208,750$	$16,260$
1-Alkenes:				
Ethylene	C_2H_4	(g)	$52,510$	$68,460$
Propylene	C_3H_6	(g)	$19,710$	$62,205$
1-Butene	C_4H_8	(g)	-540	$70,340$
1-Pentene	C_5H_{10}	(g)	$-21,280$	$78,410$
1-Hexene	C_6H_{12}	(g)	$-41,950$	$86,830$
1-Heptene	C_7H_{14}	(g)	$-62,760$	
Miscellaneous organics:				
Acetaldehyde	C_2H_4O	(g)	$-166,190$	$-128,860$
Acetic acid	$C_2H_4O_2$	(l)	$-484,500$	$-389,900$
Acetylene	C_2H_2	(g)	$227,480$	$209,970$
Benzene	C_6H_6	(g)	$82,930$	$129,665$
Benzene	C_6H_6	(l)	$49,080$	$124,520$
1,3-Butadiene	C_4H_6	(g)	$109,240$	$149,795$
Cyclohexane	C_6H_{12}	(g)	$-123,140$	$31,920$
Cyclohexane	C_6H_{12}	(l)	$-156,230$	$26,850$
1,2-Ethanediol	$C_2H_6O_2$	(l)	$-454,800$	$-323,080$
Ethanol	C_2H_6O	(g)	$-235,100$	$-168,490$
Ethanol	C_2H_6O	(l)	$-277,690$	$-174,780$
Ethylbenzene	C_8H_{10}	(g)	$29,920$	$130,890$
Ethylene oxide	C_2H_4O	(g)	$-52,630$	$-13,010$
Formaldehyde	CH_2O	(g)	$-108,570$	$-102,530$
Methanol	CH_4O	(g)	$-200,660$	$-161,960$
Methanol	CH_4O	(l)	$-238,660$	$-166,270$
Methylcyclohexane	C_7H_{14}	(g)	$-154,770$	$27,480$
Methylcyclohexane	C_7H_{14}	(l)	$-190,160$	$20,560$
Styrene	C_8H_8	(g)	$147,360$	$213,900$
Toluene	C_7H_8	(g)	$50,170$	$122,050$
Toluene	C_7H_8	(l)	$12,180$	$113,630$

Table C.4 (Continued)

Chemical species		State (Note 2)	ΔH°_{f298}	ΔG°_{f298}
Miscellaneous inorganics:				
Ammonia	NH_3	(g)	$-46,110$	$-16,450$
Ammonia	NH_3	(aq)		$-26,500$
Calcium carbide	CaC_2	(s)	$-59,800$	$-64,900$
Calcium carbonate	$CaCO_3$	(s)	$-1,206,920$	$-1,128,790$
Calcium chloride	$CaCl_2$	(s)	$-795,800$	$-748,100$
Calcium chloride	$CaCl_2$	(aq)		$-8,101,900$
Calcium chloride	$CaCl_2 \cdot 6H_2O$	(s)	$-2,607,900$	
Calcium hydroxide	$Ca(OH)_2$	(s)	$-986,090$	$-898,490$
Calcium hydroxide	$Ca(OH)_2$	(aq)		$-868,070$
Calcium oxide	CaO	(s)	$-635,090$	$-604,030$
Carbon dioxide	CO_2	(g)	$-393,509$	$-394,359$
Carbon monoxide	CO	(g)	$-110,525$	$-137,169$
Hydrochloric acid	HCl	(g)	$-92,307$	$-95,299$
Hydrogen cyanide	HCN	(g)	$135,100$	$124,700$
Hydrogen sulfide	H_2S	(g)	$-20,630$	$-33,560$
Iron oxide	FeO	(s)	$-272,000$	
Iron oxide(hematite)	Fe_2O_3	(s)	$-824,200$	$-742,200$
Iron oxide(magnetite)	Fe_3O_4	(s)	$-1,118,400$	$-1,015,400$
Iron sulfide(pyrite)	FeS_2	(s)	$-178,200$	$-166,900$
Lithium chloride	$LiCl$	(s)	$-408,610$	
Lithium chloride	$LiCl \cdot H_2O$	(s)	$-712,580$	
Lithium chloride	$LiCl \cdot 2H_2O$	(s)	$-1,012,650$	
Lithium chloride	$LiCl \cdot 3H_2O$	(s)	$-1,311,300$	
Nitric acid	HNO_3	(l)	$-174,100$	$-80,710$
Nitric acid	HNO_3	(aq)		$-111,250$
Nitrogen oxides	NO	(g)	$90,250$	$86,550$
	NO_2	(g)	$33,180$	$51,310$
	N_2O	(g)	$82,050$	$104,200$
	N_2O_4	(g)	$9,160$	$97,540$
Sodium carbonate	Na_2CO_3	(s)	$-1,130,680$	$-1,044,440$
Sodium carbonate	$Na_2CO_3 \cdot 10H_2O$	(s)	$-4,081,320$	
Sodium chloride	$NaCl$	(s)	$-411,153$	$-384,138$
Sodium chloride	$NaCl$	(aq)		$-393,133$
Sodium hydroxide	$NaOH$	(s)	$-425,609$	$-379,494$
Sodium hydroxide	$NaOH$	(aq)		$-419,150$
Sulfur dioxide	SO_2	(g)	$-296,830$	$-300,194$
Sulfur trioxide	SO_3	(g)	$-395,720$	$-371,060$
Sulfur trioxide	SO_3	(l)	$-441,040$	
Sulfuric acid	H_2SO_4	(l)	$-813,989$	$-690,003$
Sulfuric acid	H_2SO_4	(aq)		$-744,530$
Water	H_2O	(g)	$-241,818$	$-228,572$
Water	H_2O	(l)	$-285,830$	$-237,129$

[†]Taken from *TRC Thermodynamic Tables—Hydrocarbons*, Thermodynamics Research Center, Texas A & M Univ. System, College Station, Texas; "The NBS Tables of Chemical Thermodynamic Properties," *J. Physical and Chemical Reference Data*, vol. 11, supp. 2, 1982.

Notes

1. The standard Gibbs energy of formation ΔG°_{f298} is the change in the Gibbs energy when 1 mol of the listed compound is formed from its elements with each substance in its standard state at 298.15 K (25°C).

2. Standard states: (a) Gases (g): the pure ideal gas at 1 bar and 25°C. (b) Liquids (l) and solids (s): the pure substance at 1 bar and 25°C. (c) Solutes in aqueous solution (aq): The hypothetical ideal 1-molal solution of the solute in water at 1 bar and 25°C.

APPENDIX D

REPRESENTATIVE COMPUTER PROGRAMS

D.1 Defined Functions

By Eq. (4.8),

$$\texttt{MCPH} \equiv \frac{\langle C_P \rangle_H}{R} = A + \frac{B}{2}T_0(\tau + 1) + \frac{C}{3}T_0^2(\tau^2 + \tau + 1) + \frac{D}{\tau T_0^2}$$

from which,

$$\texttt{ICPH} \equiv \int_{T_0}^{T} \frac{C_P}{R} dT = \texttt{MCPH} * (T - T_0)$$

By Eq. (5.17),

$$\texttt{MCPS} \equiv \frac{\langle C_P^{ig} \rangle_S}{R} = A + \left[BT_0 + \left(CT_0^2 + \frac{D}{\tau^2 T_0^2} \right) \left(\frac{\tau + 1}{2} \right) \right] \left(\frac{\tau - 1}{\ln \tau} \right)$$

from which

$$\texttt{ICPS} \equiv \int_{T_0}^{T} \frac{C_P^{ig}}{R} \frac{dT}{T} = \texttt{MCPS} * \ln \tau$$

where

$$\tau \equiv \frac{T}{T_0}$$

Maple®

```
tau:=(T0,T)->T/T0:
H2:=(T0,T,B)->(B/2)*T0*(tau(T0,T)+1):
H3:=(T0,T,C)->(C/3)*T0^2*(1+tau(T0,T)*(1+tau(T0,T))):
H4:=(T0,T,D)->D/(tau(T0,T)*T0^2):
S2:=(T0,T,C,D)->C*T0^2+D/(tau(T0,T)*tau(T0,T)*T0*T0):
S3:=(T0,T)->(tau(T0,T)+1)/2:
S4:=(T0,T)->(tau(T0,T)-1)/ln(tau(T0,T)):


MCPH:=(T0,T,A,B,C,D)->A+H2(T0,T,B)+H3(T0,T,C)+H4(T0,T,D):

ICPH:=(T0,T,A,B,C,D)->MCPH(T0,T,A,B,C,D)*(T-T0):

MCPS:=(T0,T,A,B,C,D)->A+(B*T0+S2(T0,T,C,D)*S3(T0,T))*S4(T0,T):

ICPS:=(T0,T,A,B,C,D)->MCPS(T0,T,A,B,C,D)*ln(tau(T0,T)):
```

Mathcad®

$$\tau(T_0,T):=\frac{T}{T_0}$$

$$H_2(T_0,T,B):=\frac{B}{2}\cdot T_0\cdot(\tau(T_0,T)+1)$$

$$H_3(T_0,T,C):=\frac{C}{3}\cdot T_0^2\cdot((\tau(T_0,T)^2+\tau(T_0,T)+1)$$

$$H_4(T_0,T,D):=\frac{D}{tau(T_0,T)}\cdot T_0^2$$

$$S_2(T_0,T,C,D):=C\cdot T_0^2+\frac{D}{\tau(T_0,T)^2\cdot T_0^2}$$

$$S_3(T_0,T):=\frac{\tau(T_0,T)+1}{2}$$

$$S_4(T_0,T):=\frac{\tau(T_0,T)-1}{\ln(\tau(T_0,T))}$$

$$MCPH(T_0,T,A,B,C,D):=A+H_2(T_0,T,B)+H_3(T_0,T,C)+H_4(T_0,T,D)$$

$$ICPH(T_0,T,A,B,C,D):=MCPH(T_0,T,A,B,C,D)\cdot(T-T_0)$$

$$MCPS(T_0,T,A,B,C,D):=A+(B\cdot T_0+S_2(T_0,T,C,D)\cdot S_3(T_0,T))\cdot S_4(T_0,T)$$

$$ICPS(T_0,T,A,B,C,D):=MCPS(T_0,T,A,B,C,D)\cdot\ln(\tau(T_0,T))$$

By Eq. (6.62) and (6.63),

$$\text{HRB} \equiv \frac{H^R}{RT_c} = P_r \left[B^0 - T_r \frac{dB^0}{dT_r} + \omega \left(B^1 - T_r \frac{dB^1}{dT_r} \right) \right]$$

and

$$\text{SRB} \equiv \frac{S^R}{R} = -P_r \left(\frac{dB^0}{dT_r} + \omega \frac{dB^1}{dT_r} \right)$$

By Eq. (10.64),

$$\text{PHIB} \equiv \phi = \exp \left[\frac{P_r}{T_r} (B^0 + \omega B^1) \right]$$

Maple®

```
B0:=(TR)->0.083-0.422/TR^1.6:
DB0:=(TR)->0.675/TR^2.6:
B1:=(TR)->0.139-0.172/TR^4.2:
DB1:=(TR)->0.722/TR^5.2:
HRB:=(TR,PR,omega)->PR*(B0(TR)-TR*DB0(TR)+omega*(B1(TR)
    -TR*DB1(TR))):
SRB:=(TR,PR,omega)->-PR*(DB0(TR)+omega*DB1(TR)):
PHIB:=(TR,PR,omega)->exp((PR/TR)*(B0(TR)+omega*B1(TR))):
```

Mathcad®

$$B_0(T_r) := 0.083 - \frac{0.422}{T_r^{1.6}}$$

$$DB_0(T_r) := \frac{0.675}{T_r^{2.6}}$$

$$B_1(T_r) := 0.139 - \frac{0.172}{T_r^{4.2}}$$

$$DB_1(T_r) := \frac{0.722}{T_r^{5.2}}$$

$$\text{HRB}(T_r, P_r, \omega) := P_r \cdot (B_0(T_r) - T_r \cdot DB_0(T_r) + \omega \cdot (B_1(T_r) - T_r \cdot DB_1(T_r)))$$

$$\text{SRB}(T_r, P_r, \omega) := -P_r \cdot (DB_0(T_r) + \omega \cdot DB_1(T_r))$$

$$\text{PHIB}(T_r, P_r, \omega) := \exp \left[\frac{P_r}{T_r} \cdot (B_0(T_r) + \omega \cdot B_1(T_r)) \right]$$

D.2 Solution of Example Problems by Mathcad$^{®}$

Example 3.7 — Molar volumes by the Redlich/Kwong equation.

Given: R≡ 83.14

P:=13.76   T:=333.15   a:=$1.5641 \cdot 10^8$ b:=44.891

Initial guess: V:=2000

Solve block: GIVEN

$$P = \frac{R \cdot T}{V - b} - \frac{a}{T^{0.5} \cdot V \cdot (V + b)} \qquad V \geq b$$

$$FIND(V) = 1.713 \cdot 10^3$$

Change the initial guess to V:=45 and the answer is 71.34.

Example 12.2 — Dewpoint & bubblepoint calculations.

The problem formulation is the same for all of its parts:

Antoine vapor-pressure equations:

A1:=16.6780   B1:=3640.20   C1:=53.54

A2:=16.2887   B2:=3816.44   C2:=46.13

$$PSAT_1 := \exp\left(A1 - \frac{B1}{T - C1} \right) \qquad PSAT_2 := \exp\left(A2 - \frac{B2}{T - C2} \right)$$

Wilson-equation parameters: R≡ 1.987

a_{12}:=437.98 a_{21}:=1238.00 V_1:=76.92 V_2:=18.05

Activity-coefficient equations:

$$\Lambda_{12}(T) := \frac{V_2}{V_1} \cdot \exp\left(\frac{-a_{12}}{R \cdot T} \right) \qquad \Lambda_{21}(T) := \frac{V_1}{V_2} \cdot \exp\left(\frac{-a_{21}}{R \cdot T} \right)$$

$\gamma_1(x_1, x_2, T) :=$

$$\exp\left[-\ln(x_1 + x_2 \cdot \Lambda_{12}(T)) + x_2 \cdot \left(\frac{\Lambda_{12}(T)}{x_1 + x_2 \cdot \Lambda_{12}(T)} - \frac{\Lambda_{21}(T)}{x_2 + x_1 \cdot \Lambda_{21}(T)} \right) \right]$$

$\gamma_2(x_1, x_2, T) :=$

$$\exp\left[-\ln(x_2 + x_1 \cdot \Lambda_{21}(T)) - x_1 \cdot \left(\frac{\Lambda_{12}(T)}{x_1 + x_2 \cdot \Lambda_{12}(T)} - \frac{\Lambda_{21}(T)}{x_2 + x_1 \cdot \Lambda_{21}(T)} \right) \right]$$

(a) BUBL P Calculation:

Given: $x_1 := 0.25$ $x_2 := 1-x_1$ $T := 353.15$

Initial guesses: $y_1 := 0.30$ $y_2 := 1-y_1$ $P := 100$

Solve block: GIVEN

$y_1 \cdot P = x_1 \cdot \gamma_1(x_1, x_2, T) \cdot PSAT_1(T)$ $y_2 \cdot P = x_2 \cdot \gamma_2(x_1, x_2, T) \cdot PSAT_2(T)$

$x_1 + x_2 = 1$ $y_1 + y_2 = 1$

$$FIND(y_1, y_2, P) = \begin{pmatrix} 0.538 \\ 0.462 \\ 91.47 \end{pmatrix}$$

(b) DEW P Calculation:

Given: $y_1 := 0.60$ $y_2 := 1-y_1$ $T := 353.15$

Initial guesses: $x_1 := 0.50$ $x_2 := 1-x_1$ $P := 100$

Solve block: GIVEN

$y_1 \cdot P = x_1 \cdot \gamma_1(x_1, x_2, T) \cdot PSAT_1(T)$ $y_2 \cdot P = x_2 \cdot \gamma_2(x_1, x_2, T) \cdot PSAT_2(T)$

$x_1 + x_2 = 1$ $y_1 + y_2 = 1$

$$FIND(x_1, x_2, P) = \begin{pmatrix} 0.449 \\ 0.551 \\ 96.71 \end{pmatrix}$$

(c) BUBL T Calculation:

Given: $x_1 := 0.85$ $x_2 := 1-x_1$ $P := 101.33$

Initial guesses: $y_1 := 0.30$ $y_2 := 1-y_1$ $T := 300$

Solve block: GIVEN

$y_1 \cdot P = x_1 \cdot \gamma_1(x_1, x_2, T) \cdot PSAT_1(T)$ $y_2 \cdot P = x_2 \cdot \gamma_2(x_1, x_2, T) \cdot PSAT_2(T)$

$x_1 + x_2 = 1$ $y_1 + y_2 = 1$

$$FIND(y_1, y_2, T) = \begin{pmatrix} 0.815 \\ 0.185 \\ 353.85 \end{pmatrix}$$

(d) DEW T Calculation:

Given: $y_1 := 0.40$ $y_2 := 1 - y_1$ $P := 101.33$

Initial guesses: $x_1 := 0.50$ $x_2 := 1 - x_1$ $T := 300$

Solve block: GIVEN

$y_1 \cdot P = x_1 \cdot \gamma_1(x_1, x_2, T) \cdot \text{PSAT}_1(T)$ $y_2 \cdot P = x_2 \cdot \gamma_2(x_1, x_2, T) \cdot \text{PSAT}_2(T)$

$x_1 + x_2 = 1$ $y_1 + y_2 = 1$

$$\text{FIND}(x_1, x_2, T) = \begin{pmatrix} 0.064 \\ 0.936 \\ 360.62 \end{pmatrix}$$

(e) Azeotrope Calculation:

Given: $T := 353.15$

Initial guesses: $x_1 := 0.50$ $x_2 := 1 - x_1$ $P := 100$

                       $y_1 := 0.50$ $y_2 := 1 - y_1$

Solve block: GIVEN

$y_1 \cdot P = x_1 \cdot \gamma_1(x_1, x_2, T) \cdot \text{PSAT}_1(T)$ $y_2 \cdot P = x_2 \cdot \gamma_2(x_1, x_2, T) \cdot \text{PSAT}_2(T)$

$x_1 + x_2 = 1$ $y_1 + y_2 = 1$

$x_1 = y_1$ $x_2 = y_2$

$$\text{FIND}(x_1, x_2, y_1, y_2, P) = \begin{pmatrix} 0.717 \\ 0.283 \\ 0.717 \\ 0.283 \\ 99.83 \end{pmatrix}$$

Example 15.12 — Solution of two reaction-equilibrium equations.

Given: $K_a := 1.758$ $K_b := 2.561$

Initial guesses: $\varepsilon_a := 0.1$ $\varepsilon_b := 0.7$

Solve block:　　　GIVEN

$$K_a = \frac{(2 \cdot \varepsilon_a + \varepsilon_b)^2}{(0.5 - \varepsilon_a) \cdot (3.38 + \varepsilon_a + \varepsilon_b)} \cdot 20 \qquad K_b = \frac{\varepsilon_b (2 \cdot \varepsilon_a + \varepsilon_b)}{(1 - \varepsilon_b) \cdot (3.38 + \varepsilon_a + \varepsilon_b)} \cdot 20$$

$$0.5 \geq \varepsilon_a \geq -0.5 \qquad\qquad 0 \leq \varepsilon_b \leq 1$$

$$\mathtt{FIND}(\varepsilon_a, \varepsilon_b) = \begin{pmatrix} -0.0506 \\ 0.5336 \end{pmatrix}$$

Example 15.13 — Reaction equilibrium by minimizing the Gibbs energy.

In the following, define: $\Lambda_i \equiv \lambda_i / RT$ and $RT \equiv R \times T = 8314$

Definition:　　　　　$RT \equiv 8314$

Initial guesses:　　$\Lambda_C := 1$　　$\Lambda_H := 1$　　$\Lambda_0 := 1$　　$n := 1$

$y_{CH4} := 0.01$　　$y_{H2O} := 0.01$　　$y_{CO} := 0.01$　　$y_{CO2} := 0.01$　　$y_{H2} := 0.96$

The 9 unknowns (above) require 9 equations (below):

Solve block:　　　GIVEN

$$y_{CH4} + y_{CO} + y_{CO2} = \frac{2}{n} \qquad\qquad 4 \cdot y_{CH4} + 2 \cdot y_{H2O} + 2 \cdot y_{H2} = \frac{14}{n}$$

$$y_{H2O} + y_{CO} + 2 \cdot y_{CO2} = \frac{3}{n} \qquad\qquad y_{CH4} + y_{H2O} + y_{CO} + y_{CO2} + y_{H2} = 1$$

$$\frac{19720}{RT} + \ln(y_{CH4}) + \Lambda_C + 4 \cdot \Lambda_H = 0 \qquad\qquad -\frac{192420}{RT} + \ln(y_{H2O}) + 2 \cdot \Lambda_H + \Lambda_0 = 0$$

$$-\frac{200240}{RT} + \ln(y_{CO}) + \Lambda_C + \Lambda_0 = 0 \qquad\qquad -\frac{395790}{RT} + \ln(y_{CO2}) + \Lambda_C + 2 \cdot \Lambda_0 = 0$$

$$\ln(y_{H2}) + 2 \cdot \Lambda_H = 0$$

$$0 \leq y_{CH4} \leq 1 \quad 0 \leq y_{H2O} \leq 1 \quad 0 \leq y_{CO} \leq 1 \quad 0 \leq y_{CO2} \leq 1 \quad 0 \leq y_{H2} \leq 1$$

$$\mathtt{FIND}(y_{CH4}, y_{H2O}, y_{CO}, y_{CO2}, y_{H2}, \Lambda_C, \Lambda_H, \Lambda_0, n) = \begin{pmatrix} 0.0196 \\ 0.0980 \\ 0.1743 \\ 0.0371 \\ 0.6710 \\ 0.7635 \\ 0.1994 \\ 25.068 \\ 8.6608 \end{pmatrix}$$

APPENDIX E

THE LEE/KESLER GENERALIZED-CORRELATION TABLES

The Lee/Kesler tables are adapted and published by permission from "A Generalized Thermodynamic Correlation Based on Three-Parameter Corresponding States," by Byung Ik Lee and Michael G. Kesler, *AIChE J.*, **21**, 510–527 (1975). The numbers printed in italic type are liquid-phase properties.

Tables E.1 – E.4   Correlation for the compressibility factor.

Tables E.5 – E.8   Correlation for the residual enthalpy.

Tables E.9 – E.12   Correlation for the residual entropy.

Tables E.13 – E.16   Correlation for the fugacity coefficient.

Table E.1: Values of Z^0

$P_r =$	0.0100	0.0500	0.1000	0.2000	0.4000	0.6000	0.8000	1.0000
T_r								
0.30	*0.0029*	*0.0145*	*0.0290*	*0.0579*	*0.1158*	*0.1737*	*0.2315*	0.2892
0.35	*0.0026*	*0.0130*	*0.0261*	*0.0522*	*0.1043*	*0.1564*	*0.2084*	0.2604
0.40	*0.0024*	*0.0119*	*0.0239*	*0.0477*	*0.0953*	*0.1429*	*0.1904*	0.2379
0.45	*0.0022*	*0.0110*	*0.0221*	*0.0442*	*0.0882*	*0.1322*	*0.1762*	0.2200
0.50	*0.0021*	*0.0103*	*0.0207*	*0.0413*	*0.0825*	*0.1236*	*0.1647*	0.2056
0.55	0.9804	*0.0098*	*0.0195*	*0.0390*	*0.0778*	*0.1166*	*0.1553*	0.1939
0.60	0.9849	*0.0093*	*0.0186*	*0.0371*	*0.0741*	*0.1109*	*0.1476*	0.1842
0.65	0.9881	0.9377	*0.0178*	*0.0356*	*0.0710*	*0.1063*	*0.1415*	0.1765
0.70	0.9904	0.9504	0.8958	*0.0344*	*0.0687*	*0.1027*	*0.1366*	0.1703
0.75	0.9922	0.9598	0.9165	*0.0336*	*0.0670*	*0.1001*	*0.1330*	0.1656
0.80	0.9935	0.9669	0.9319	0.8539	*0.0661*	*0.0985*	*0.1307*	0.1626
0.85	0.9946	0.9725	0.9436	0.8810	*0.0661*	*0.0983*	*0.1301*	0.1614
0.90	0.9954	0.9768	0.9528	0.9015	0.7800	*0.1006*	*0.1321*	0.1630
0.93	0.9959	0.9790	0.9573	0.9115	0.8059	0.6635	*0.1359*	0.1664
0.95	0.9961	0.9803	0.9600	0.9174	0.8206	0.6967	*0.1410*	0.1705
0.97	0.9963	0.9815	0.9625	0.9227	0.8338	0.7240	0.5580	0.1779
0.98	0.9965	0.9821	0.9637	0.9253	0.8398	0.7360	0.5887	0.1844
0.99	0.9966	0.9826	0.9648	0.9277	0.8455	0.7471	0.6138	0.1959
1.00	0.9967	0.9832	0.9659	0.9300	0.8509	0.7574	0.6355	0.2901
1.01	0.9968	0.9837	0.9669	0.9322	0.8561	0.7671	0.6542	0.4648
1.02	0.9969	0.9842	0.9679	0.9343	0.8610	0.7761	0.6710	0.5146
1.05	0.9971	0.9855	0.9707	0.9401	0.8743	0.8002	0.7130	0.6026
1.10	0.9975	0.9874	0.9747	0.9485	0.8930	0.8323	0.7649	0.6880
1.15	0.9978	0.9891	0.9780	0.9554	0.9081	0.8576	0.8032	0.7443
1.20	0.9981	0.9904	0.9808	0.9611	0.9205	0.8779	0.8330	0.7858
1.30	0.9985	0.9926	0.9852	0.9702	0.9396	0.9083	0.8764	0.8438
1.40	0.9988	0.9942	0.9884	0.9768	0.9534	0.9298	0.9062	0.8827
1.50	0.9991	0.9954	0.9909	0.9818	0.9636	0.9456	0.9278	0.9103
1.60	0.9993	0.9964	0.9928	0.9856	0.9714	0.9575	0.9439	0.9308
1.70	0.9994	0.9971	0.9943	0.9886	0.9775	0.9667	0.9563	0.9463
1.80	0.9995	0.9977	0.9955	0.9910	0.9823	0.9739	0.9659	0.9583
1.90	0.9996	0.9982	0.9964	0.9929	0.9861	0.9796	0.9735	0.9678
2.00	0.9997	0.9986	0.9972	0.9944	0.9892	0.9842	0.9796	0.9754
2.20	0.9998	0.9992	0.9983	0.9967	0.9937	0.9910	0.9886	0.9865
2.40	0.9999	0.9996	0.9991	0.9983	0.9969	0.9957	0.9948	0.9941
2.60	1.0000	0.9998	0.9997	0.9994	0.9991	0.9990	0.9990	0.9993
2.80	1.0000	1.0000	1.0001	1.0002	1.0007	1.0013	1.0021	1.0031
3.00	1.0000	1.0002	1.0004	1.0008	1.0018	1.0030	1.0043	1.0057
3.50	1.0001	1.0004	1.0008	1.0017	1.0035	1.0055	1.0075	1.0097
4.00	1.0001	1.0005	1.0010	1.0021	1.0043	1.0066	1.0090	1.0115

Table E.2: Values of Z^1

$P_r =$	0.0100	0.0500	0.1000	0.2000	0.4000	0.6000	0.8000	1.0000
T_r								
0.30	*−0.0008*	*−0.0040*	*−0.0081*	*−0.0161*	*−0.0323*	*−0.0484*	*−0.0645*	−0.0806
0.35	*−0.0009*	*−0.0046*	*−0.0093*	*−0.0185*	*−0.0370*	*−0.0554*	*−0.0738*	−0.0921
0.40	*−0.0010*	*−0.0048*	*−0.0095*	*−0.0190*	*−0.0380*	*−0.0570*	*−0.0758*	−0.0946
0.45	*−0.0009*	*−0.0047*	*−0.0094*	*−0.0187*	*−0.0374*	*−0.0560*	*−0.0745*	−0.0929
0.50	*−0.0009*	*−0.0045*	*−0.0090*	*−0.0181*	*−0.0360*	*−0.0539*	*−0.0716*	−0.0893
0.55	−0.0314	*−0.0043*	*−0.0086*	*−0.0172*	*−0.0343*	*−0.0513*	*−0.0682*	−0.0849
0.60	−0.0205	*−0.0041*	*−0.0082*	*−0.0164*	*−0.0326*	*−0.0487*	*−0.0646*	−0.0803
0.65	−0.0137	−0.0772	*−0.0078*	*−0.0156*	*−0.0309*	*−0.0461*	*−0.0611*	−0.0759
0.70	−0.0093	−0.0507	−0.1161	*−0.0148*	*−0.0294*	*−0.0438*	*−0.0579*	−0.0718
0.75	−0.0064	−0.0339	−0.0744	−0.0143	*−0.0282*	*−0.0417*	*−0.0550*	−0.0681
0.80	−0.0044	−0.0228	−0.0487	−0.1160	*−0.0272*	*−0.0401*	*−0.0526*	−0.0648
0.85	−0.0029	−0.0152	−0.0319	−0.0715	*−0.0268*	*−0.0391*	*−0.0509*	−0.0622
0.90	−0.0019	−0.0099	−0.0205	−0.0442	−0.1118	*−0.0396*	*−0.0503*	−0.0604
0.93	−0.0015	−0.0075	−0.0154	−0.0326	−0.0763	−0.1662	*−0.0514*	−0.0602
0.95	−0.0012	−0.0062	−0.0126	−0.0262	−0.0589	−0.1110	*−0.0540*	−0.0607
0.97	−0.0010	−0.0050	−0.0101	−0.0208	−0.0450	−0.0770	−0.1647	−0.0623
0.98	−0.0009	−0.0044	−0.0090	−0.0184	−0.0390	−0.0641	−0.1100	−0.0641
0.99	−0.0008	−0.0039	−0.0079	−0.0161	−0.0335	−0.0531	−0.0796	−0.0680
1.00	−0.0007	−0.0034	−0.0069	−0.0140	−0.0285	−0.0435	−0.0588	−0.0879
1.01	−0.0006	−0.0030	−0.0060	−0.0120	−0.0240	−0.0351	−0.0429	−0.0223
1.02	−0.0005	−0.0026	−0.0051	−0.0102	−0.0198	−0.0277	−0.0303	−0.0062
1.05	−0.0003	−0.0015	−0.0029	−0.0054	−0.0092	−0.0097	−0.0032	0.0220
1.10	0.0000	0.0000	0.0001	0.0007	0.0038	0.0106	0.0236	0.0476
1.15	0.0002	0.0011	0.0023	0.0052	0.0127	0.0237	0.0396	0.0625
1.20	0.0004	0.0019	0.0039	0.0084	0.0190	0.0326	0.0499	0.0719
1.30	0.0006	0.0030	0.0061	0.0125	0.0267	0.0429	0.0612	0.0819
1.40	0.0007	0.0036	0.0072	0.0147	0.0306	0.0477	0.0661	0.0857
1.50	0.0008	0.0039	0.0078	0.0158	0.0323	0.0497	0.0677	0.0864
1.60	0.0008	0.0040	0.0080	0.0162	0.0330	0.0501	0.0677	0.0855
1.70	0.0008	0.0040	0.0081	0.0163	0.0329	0.0497	0.0667	0.0838
1.80	0.0008	0.0040	0.0081	0.0162	0.0325	0.0488	0.0652	0.0814
1.90	0.0008	0.0040	0.0079	0.0159	0.0318	0.0477	0.0635	0.0792
2.00	0.0008	0.0039	0.0078	0.0155	0.0310	0.0464	0.0617	0.0767
2.20	0.0007	0.0037	0.0074	0.0147	0.0293	0.0437	0.0579	0.0719
2.40	0.0007	0.0035	0.0070	0.0139	0.0276	0.0411	0.0544	0.0675
2.60	0.0007	0.0033	0.0066	0.0131	0.0260	0.0387	0.0512	0.0634
2.80	0.0006	0.0031	0.0062	0.0124	0.0245	0.0365	0.0483	0.0598
3.00	0.0006	0.0029	0.0059	0.0117	0.0232	0.0345	0.0456	0.0565
3.50	0.0005	0.0026	0.0052	0.0103	0.0204	0.0303	0.0401	0.0497
4.00	0.0005	0.0023	0.0046	0.0091	0.0182	0.0270	0.0357	0.0443

Table E.3: Values of Z^0

$P_r =$	1.0000	1.2000	1.5000	2.0000	3.0000	5.0000	7.0000	10.000
T_r								
0.30	0.2892	0.3479	0.4335	0.5775	0.8648	1.4366	2.0048	2.8507
0.35	0.2604	0.3123	0.3901	0.5195	0.7775	1.2902	1.7987	2.5539
0.40	0.2379	0.2853	0.3563	0.4744	0.7095	1.1758	1.6373	2.3211
0.45	0.2200	0.2638	0.3294	0.4384	0.6551	1.0841	1.5077	2.1338
0.50	0.2056	0.2465	0.3077	0.4092	0.6110	1.0094	1.4017	1.9801
0.55	0.1939	0.2323	0.2899	0.3853	0.5747	0.9475	1.3137	1.8520
0.60	0.1842	0.2207	0.2753	0.3657	0.5446	0.8959	1.2398	1.7440
0.65	0.1765	0.2113	0.2634	0.3495	0.5197	0.8526	1.1773	1.6519
0.70	0.1703	0.2038	0.2538	0.3364	0.4991	0.8161	1.1341	1.5729
0.75	0.1656	0.1981	0.2464	0.3260	0.4823	0.7854	1.0787	1.5047
0.80	0.1626	0.1942	0.2411	0.3182	0.4690	0.7598	1.0400	1.4456
0.85	0.1614	0.1924	0.2382	0.3132	0.4591	0.7388	1.0071	1.3943
0.90	0.1630	0.1935	0.2383	0.3114	0.4527	0.7220	0.9793	1.3496
0.93	0.1664	0.1963	0.2405	0.3122	0.4507	0.7138	0.9648	1.3257
0.95	0.1705	0.1998	0.2432	0.3138	0.4501	0.7092	0.9561	1.3108
0.97	0.1779	0.2055	0.2474	0.3164	0.4504	0.7052	0.9480	1.2968
0.98	0.1844	0.2097	0.2503	0.3182	0.4508	0.7035	0.9442	1.2901
0.99	0.1959	0.2154	0.2538	0.3204	0.4514	0.7018	0.9406	1.2835
1.00	0.2901	0.2237	0.2583	0.3229	0.4522	0.7004	0.9372	1.2772
1.01	0.4648	0.2370	0.2640	0.3260	0.4533	0.6991	0.9339	1.2710
1.02	0.5146	0.2629	0.2715	0.3297	0.4547	0.6980	0.9307	1.2650
1.05	0.6026	0.4437	0.3131	0.3452	0.4604	0.6956	0.9222	1.2481
1.10	0.6880	0.5984	0.4580	0.3953	0.4770	0.6950	0.9110	1.2232
1.15	0.7443	0.6803	0.5798	0.4760	0.5042	0.6987	0.9033	1.2021
1.20	0.7858	0.7363	0.6605	0.5605	0.5425	0.7069	0.8990	1.1844
1.30	0.8438	0.8111	0.7624	0.6908	0.6344	0.7358	0.8998	1.1580
1.40	0.8827	0.8595	0.8256	0.7753	0.7202	0.7761	0.9112	1.1419
1.50	0.9103	0.8933	0.8689	0.8328	0.7887	0.8200	0.9297	1.1339
1.60	0.9308	0.9180	0.9000	0.8738	0.8410	0.8617	0.9518	1.1320
1.70	0.9463	0.9367	0.9234	0.9043	0.8809	0.8984	0.9745	1.1343
1.80	0.9583	0.9511	0.9413	0.9275	0.9118	0.9297	0.9961	1.1391
1.90	0.9678	0.9624	0.9552	0.9456	0.9359	0.9557	1.0157	1.1452
2.00	0.9754	0.9715	0.9664	0.9599	0.9550	0.9772	1.0328	1.1516
2.20	0.9856	0.9847	0.9826	0.9806	0.9827	1.0094	1.0600	1.1635
2.40	0.9941	0.9936	0.9935	0.9945	1.0011	1.0313	1.0793	1.1728
2.60	0.9993	0.9998	1.0010	1.0040	1.0137	1.0463	1.0926	1.1792
2.80	1.0031	1.0042	1.0063	1.0106	1.0223	1.0565	1.1016	1.1830
3.00	1.0057	1.0074	1.0101	1.0153	1.0284	1.0635	1.1075	1.1848
3.50	1.0097	1.0120	1.0156	1.0221	1.0368	1.0723	1.1138	1.1834
4.00	1.0115	1.0140	1.0179	1.0249	1.0401	1.0747	1.1136	1.1773

Table E.4: Values of Z^1

$P_r =$	1.0000	1.2000	1.5000	2.0000	3.0000	5.0000	7.0000	10.000
T_r								
0.30	−0.0806	−0.0966	−0.1207	−0.1608	−0.2407	−0.3996	−0.5572	−0.7915
0.35	−0.0921	−0.1105	−0.1379	−0.1834	−0.2738	−0.4523	−0.6279	−0.8863
0.40	−0.0946	−0.1134	−0.1414	−0.1879	−0.2799	−0.4603	−0.6365	−0.8936
0.45	−0.0929	−0.1113	−0.1387	−0.1840	−0.2734	−0.4475	−0.6162	−0.8608
0.50	−0.0893	−0.1069	−0.1330	−0.1762	−0.2611	−0.4253	−0.5831	−0.8099
0.55	−0.0849	−0.1015	−0.1263	−0.1669	−0.2465	−0.3991	−0.5446	−0.7521
0.60	−0.0803	−0.0960	−0.1192	−0.1572	−0.2312	−0.3718	−0.5047	−0.6928
0.65	−0.0759	−0.0906	−0.1122	−0.1476	−0.2160	−0.3447	−0.4653	−0.6346
0.70	−0.0718	−0.0855	−0.1057	−0.1385	−0.2013	−0.3184	−0.4270	−0.5785
0.75	−0.0681	−0.0808	−0.0996	−0.1298	−0.1872	−0.2929	−0.3901	−0.5250
0.80	−0.0648	−0.0767	−0.0940	−0.1217	−0.1736	−0.2682	−0.3545	−0.4740
0.85	−0.0622	−0.0731	−0.0888	−0.1138	−0.1602	−0.2439	−0.3201	−0.4254
0.90	−0.0604	−0.0701	−0.0840	−0.1059	−0.1463	−0.2195	−0.2862	−0.3788
0.93	−0.0602	−0.0687	−0.0810	−0.1007	−0.1374	−0.2045	−0.2661	−0.3516
0.95	−0.0607	−0.0678	−0.0788	−0.0967	−0.1310	−0.1943	−0.2526	−0.3339
0.97	−0.0623	−0.0669	−0.0759	−0.0921	−0.1240	−0.1837	−0.2391	−0.3163
0.98	−0.0641	−0.0661	−0.0740	−0.0893	−0.1202	−0.1783	−0.2322	−0.3075
0.99	−0.0680	−0.0646	−0.0715	−0.0861	−0.1162	−0.1728	−0.2254	−0.2989
1.00	−0.0879	−0.0609	−0.0678	−0.0824	−0.1118	−0.1672	−0.2185	−0.2902
1.01	−0.0223	−0.0473	−0.0621	−0.0778	−0.1072	−0.1615	−0.2116	−0.2816
1.02	−0.0062	−0.0227	−0.0524	−0.0722	−0.1021	−0.1556	−0.2047	−0.2731
1.05	0.0220	0.1059	0.0451	−0.0432	−0.0838	−0.1370	−0.1835	−0.2476
1.10	0.0476	0.0897	0.1630	0.0698	−0.0373	−0.1021	−0.1469	−0.2056
1.15	0.0625	0.0943	0.1548	0.1667	0.0332	−0.0611	−0.1084	−0.1642
1.20	0.0719	0.0991	0.1477	0.1990	0.1095	−0.0141	−0.0678	−0.1231
1.30	0.0819	0.1048	0.1420	0.1991	0.2079	0.0875	0.0176	−0.0423
1.40	0.0857	0.1063	0.1383	0.1894	0.2397	0.1737	0.1008	0.0350
1.50	0.0854	0.1055	0.1345	0.1806	0.2433	0.2309	0.1717	0.1058
1.60	0.0855	0.1035	0.1303	0.1729	0.2381	0.2631	0.2255	0.1673
1.70	0.0838	0.1008	0.1259	0.1658	0.2305	0.2788	0.2628	0.2179
1.80	0.0816	0.0978	0.1216	0.1593	0.2224	0.2846	0.2871	0.2576
1.90	0.0792	0.0947	0.1173	0.1532	0.2144	0.2848	0.3017	0.2876
2.00	0.0767	0.0916	0.1133	0.1476	0.2069	0.2819	0.3097	0.3096
2.20	0.0719	0.0857	0.1057	0.1374	0.1932	0.2720	0.3135	0.3355
2.40	0.0675	0.0803	0.0989	0.1285	0.1812	0.2602	0.3089	0.3459
2.60	0.0634	0.0754	0.0929	0.1207	0.1706	0.2484	0.3009	0.3475
2.80	0.0598	0.0711	0.0876	0.1138	0.1613	0.2372	0.2915	0.3443
3.00	0.0535	0.0672	0.0828	0.1076	0.1529	0.2268	0.2817	0.3385
3.50	0.0497	0.0591	0.0728	0.0949	0.1356	0.2042	0.2584	0.3194
4.00	0.0443	0.0527	0.0651	0.0849	0.1219	0.1857	0.2378	0.2994

Table E.5: Values of $(H^R)^0/RT_c$

$P_r =$	0.0100	0.0500	0.1000	0.2000	0.4000	0.6000	0.8000	1.0000
T_r								
0.30	−6.045	−6.043	−6.040	−6.034	−6.022	−6.011	−5.999	−5.987
0.35	−5.906	−5.904	−5.901	−5.895	−5.882	−5.870	−5.858	−5.845
0.40	−5.763	−5.761	−5.757	−5.751	−5.738	−5.726	−5.713	−5.700
0.45	−5.615	−5.612	−5.609	−5.603	−5.590	−5.577	−5.564	−5.551
0.50	−5.465	−5.463	−5.459	−5.453	−5.440	−5.427	−5.414	−5.401
0.55	−0.032	−5.312	−5.309	−5.303	−5.290	−5.278	−5.265	−5.252
0.60	−0.027	−5.162	−5.159	−5.153	−5.141	−5.129	−5.116	−5.104
0.65	−0.023	−0.118	−5.008	−5.002	−4.991	−4.980	−4.968	−4.956
0.70	−0.020	−0.101	−0.213	−4.848	−4.838	−4.828	−4.818	−4.808
0.75	−0.017	−0.088	−0.183	−4.687	−4.679	−4.672	−4.664	−4.655
0.80	−0.015	−0.078	−0.160	−0.345	−4.507	−4.504	−4.499	−4.494
0.85	−0.014	−0.069	−0.141	−0.300	−4.309	−4.313	−4.316	−4.316
0.90	−0.012	−0.062	−0.126	−0.264	−0.596	−4.074	−4.094	−4.108
0.93	−0.011	−0.058	−0.118	−0.246	−0.545	−0.960	−3.920	−3.953
0.95	−0.011	−0.056	−0.113	−0.235	−0.516	−0.885	−3.763	−3.825
0.97	−0.011	−0.054	−0.109	−0.225	−0.490	−0.824	−1.356	−3.658
0.98	−0.010	−0.053	−0.107	−0.221	−0.478	−0.797	−1.273	−3.544
0.99	−0.010	−0.052	−0.105	−0.216	−0.466	−0.773	−1.206	−3.376
1.00	−0.010	−0.051	−0.103	−0.212	−0.455	−0.750	−1.151	−2.584
1.01	−0.010	−0.050	−0.101	−0.208	−0.445	−0.721	−1.102	−1.796
1.02	−0.010	−0.049	−0.099	−0.203	−0.434	−0.708	−1.060	−1.627
1.05	−0.009	−0.046	−0.094	−0.192	−0.407	−0.654	−0.955	−1.359
1.10	−0.008	−0.042	−0.086	−0.175	−0.367	−0.581	−0.827	−1.120
1.15	−0.008	−0.039	−0.079	−0.160	−0.334	−0.523	−0.732	−0.968
1.20	−0.007	−0.036	−0.073	−0.148	−0.305	−0.474	−0.657	−0.857
1.30	−0.006	−0.031	−0.063	−0.127	−0.259	−0.399	−0.545	−0.698
1.40	−0.005	−0.027	−0.055	−0.110	−0.224	−0.341	−0.463	−0.588
1.50	−0.005	−0.024	−0.048	−0.097	−0.196	−0.297	−0.400	−0.505
1.60	−0.004	−0.021	−0.043	−0.086	−0.173	−0.261	−0.350	−0.440
1.70	−0.004	−0.019	−0.038	−0.076	−0.153	−0.231	−0.309	−0.387
1.80	−0.003	−0.017	−0.034	−0.068	−0.137	−0.206	−0.275	−0.344
1.90	−0.003	−0.015	−0.031	−0.062	−0.123	−0.185	−0.246	−0.307
2.00	−0.003	−0.014	−0.028	−0.056	−0.111	−0.167	−0.222	−0.276
2.20	−0.002	−0.012	−0.023	−0.046	−0.092	−0.137	−0.182	−0.226
2.40	−0.002	−0.010	−0.019	−0.038	−0.076	−0.114	−0.150	−0.187
2.60	−0.002	−0.008	−0.016	−0.032	−0.064	−0.095	−0.125	−0.155
2.80	−0.001	−0.007	−0.014	−0.027	−0.054	−0.080	−0.105	−0.130
3.00	−0.001	−0.006	−0.011	−0.023	−0.045	−0.067	−0.088	−0.109
3.50	−0.001	−0.004	−0.007	−0.015	−0.029	−0.043	−0.056	−0.069
4.00	−0.000	−0.002	−0.005	−0.009	−0.017	−0.026	−0.033	−0.041

Table E.6: Values of $(H^R)^1/RT_c$

$P_r =$	0.0100	0.0500	0.1000	0.2000	0.4000	0.6000	0.8000	1.0000
T_r								
0.30	-11.098	-11.096	-11.095	-11.091	-11.083	-11.076	-11.069	-11.062
0.35	-10.656	-10.655	-10.654	-10.653	-10.650	-10.646	-10.643	-10.640
0.40	-10.121	-10.121	-10.121	-10.120	-10.121	-10.121	-10.121	-10.121
0.45	-9.515	-9.515	-9.516	-9.517	-9.519	-9.521	-9.523	-9.525
0.50	-8.868	-8.869	-8.870	-8.872	-8.876	-8.880	-8.884	-8.888
0.55	-0.080	-8.211	-8.212	-8.215	-8.221	-8.226	-8.232	-8.238
0.60	-0.059	-7.568	-7.570	-7.573	-7.579	-7.585	-7.591	-7.596
0.65	-0.045	-0.247	-6.949	-6.952	-6.959	-6.966	-6.973	-6.980
0.70	-0.034	-0.185	-0.415	-6.360	-6.367	-6.373	-6.381	-6.388
0.75	-0.027	-0.142	-0.306	-5.796	-5.802	-5.809	-5.816	-5.824
0.80	-0.021	-0.110	-0.234	-0.542	-5.266	-5.271	-5.278	-5.285
0.85	-0.017	-0.087	-0.182	-0.401	-4.753	-4.754	-4.758	-4.763
0.90	-0.014	-0.070	-0.144	-0.308	-0.751	-4.254	-4.248	-4.249
0.93	-0.012	-0.061	-0.126	-0.265	-0.612	-1.236	-3.942	-3.934
0.95	-0.011	-0.056	-0.115	-0.241	-0.542	-0.994	-3.737	-3.712
0.97	-0.010	-0.052	-0.105	-0.219	-0.483	-0.837	-1.616	-3.470
0.98	-0.010	-0.050	-0.101	-0.209	-0.457	-0.776	-1.324	-3.332
0.99	-0.009	-0.048	-0.097	-0.200	-0.433	-0.722	-1.154	-3.164
1.00	-0.009	-0.046	-0.093	-0.191	-0.410	-0.675	-1.034	-2.471
1.01	-0.009	-0.044	-0.089	-0.183	-0.389	-0.632	-0.940	-1.375
1.02	-0.008	-0.042	-0.085	-0.175	-0.370	-0.594	-0.863	-1.180
1.05	-0.007	-0.037	-0.075	-0.153	-0.318	-0.498	-0.691	-0.877
1.10	-0.006	-0.030	-0.061	-0.123	-0.251	-0.381	-0.507	-0.617
1.15	-0.005	-0.025	-0.050	-0.099	-0.199	-0.296	-0.385	-0.459
1.20	-0.004	-0.020	-0.040	-0.080	-0.158	-0.232	-0.297	-0.349
1.30	-0.003	-0.013	-0.026	-0.052	-0.100	-0.142	-0.177	-0.203
1.40	-0.002	-0.008	-0.016	-0.032	-0.060	-0.083	-0.100	-0.111
1.50	-0.001	-0.005	-0.009	-0.018	-0.032	-0.042	-0.048	-0.049
1.60	-0.000	-0.002	-0.004	-0.007	-0.012	-0.013	-0.011	-0.005
1.70	-0.000	-0.000	-0.000	-0.000	0.003	0.009	0.017	0.027
1.80	0.000	0.001	0.003	0.006	0.015	0.025	0.037	0.051
1.90	0.001	0.003	0.005	0.011	0.023	0.037	0.053	0.070
2.00	0.001	0.003	0.007	0.015	0.030	0.047	0.065	0.085
2.20	0.001	0.005	0.010	0.020	0.040	0.062	0.083	0.106
2.40	0.001	0.006	0.012	0.023	0.047	0.071	0.095	0.120
2.60	0.001	0.006	0.013	0.026	0.052	0.078	0.104	0.130
2.80	0.001	0.007	0.014	0.028	0.055	0.082	0.110	0.137
3.00	0.001	0.007	0.014	0.029	0.058	0.086	0.114	0.142
3.50	0.002	0.008	0.016	0.031	0.062	0.092	0.122	0.152
4.00	0.002	0.008	0.016	0.032	0.064	0.096	0.127	0.158

Table E.7: Values of $(H^R)^0/RT_c$

$P_r =$	1.0000	1.2000	1.5000	2.0000	3.0000	5.0000	7.0000	10.000
T_r								
0.30	−5.987	−5.975	−5.957	−5.927	−5.868	−5.748	−5.628	−5.446
0.35	−5.845	−5.833	−5.814	−5.783	−5.721	−5.595	−5.469	−5.278
0.40	−5.700	−5.687	−5.668	−5.636	−5.572	−5.442	−5.311	−5.113
0.45	−5.551	−5.538	−5.519	−5.486	−5.421	−5.288	−5.154	−5.950
0.50	−5.401	−5.388	−5.369	−5.336	−5.279	−5.135	−4.999	−4.791
0.55	−5.252	−5.239	−5.220	−5.187	−5.121	−4.986	−4.849	−4.638
0.60	−5.104	−5.091	−5.073	−5.041	−4.976	−4.842	−4.794	−4.492
0.65	−4.956	−4.949	−4.927	−4.896	−4.833	−4.702	−4.565	−4.353
0.70	−4.808	−4.797	−4.781	−4.752	−4.693	−4.566	−4.432	−4.221
0.75	−4.655	−4.646	−4.632	−4.607	−4.554	−4.434	−4.393	−4.095
0.80	−4.494	−4.488	−4.478	−4.459	−4.413	−4.303	−4.178	−3.974
0.85	−4.316	−4.316	−4.312	−4.302	−4.269	−4.173	−4.056	−3.857
0.90	−4.108	−4.118	−4.127	−4.132	−4.119	−4.043	−3.935	−3.744
0.93	−3.953	−3.976	−4.000	−4.020	−4.024	−3.963	−3.863	−3.678
0.95	−3.825	−3.865	−3.904	−3.940	−3.958	−3.910	−3.815	−3.634
0.97	−3.658	−3.732	−3.796	−3.853	−3.890	−3.856	−3.767	−3.591
0.98	−3.544	−3.652	−3.736	−3.806	−3.854	−3.829	−3.743	−3.569
0.99	−3.376	−3.558	−3.670	−3.758	−3.818	−3.801	−3.719	−3.548
1.00	−2.584	−3.441	−3.598	−3.706	−3.782	−3.774	−3.695	−3.526
1.01	−1.796	−3.283	−3.516	−3.652	−3.744	−3.746	−3.671	−3.505
1.02	−1.627	−3.039	−3.422	−3.595	−3.705	−3.718	−3.647	−3.484
1.05	−1.359	−2.034	−3.030	−3.398	−3.583	−3.632	−3.575	−3.420
1.10	−1.120	−1.487	−2.203	−2.965	−3.353	−3.484	−3.453	−3.315
1.15	−0.968	−1.239	−1.719	−2.479	−3.091	−3.329	−3.329	−3.211
1.20	−0.857	−1.076	−1.443	−2.079	−2.801	−3.166	−3.202	−3.107
1.30	−0.698	−0.860	−1.116	−1.560	−2.274	−2.825	−2.942	−2.899
1.40	−0.588	−0.716	−0.915	−1.253	−1.857	−2.486	−2.679	−2.692
1.50	−0.505	−0.611	−0.774	−1.046	−1.549	−2.175	−2.421	−2.486
1.60	−0.440	−0.531	−0.667	−0.894	−1.318	−1.904	−2.177	−2.285
1.70	−0.387	−0.446	−0.583	−0.777	−1.139	−1.672	−1.953	−2.091
1.80	−0.344	−0.413	−0.515	−0.683	−0.996	−1.476	−1.751	−1.908
1.90	−0.307	−0.368	−0.458	−0.606	−0.880	−1.309	−1.571	−1.736
2.00	−0.276	−0.330	−0.411	−0.541	−0.782	−1.167	−1.411	−1.577
2.20	−0.226	−0.269	−0.334	−0.437	−0.629	−0.937	−1.143	−1.295
2.40	−0.187	−0.222	−0.275	−0.359	−0.513	−0.761	−0.929	−1.058
2.60	−0.155	−0.185	−0.228	−0.297	−0.422	−0.621	−0.756	−0.858
2.80	−0.130	−0.154	−0.190	−0.246	−0.348	−0.508	−0.614	−0.689
3.00	−0.109	−0.129	−0.159	−0.205	−0.288	−0.415	−0.495	−0.545
3.50	−0.069	−0.081	−0.099	−0.127	−0.174	−0.239	−0.270	−0.264
4.00	−0.041	−0.048	−0.058	−0.072	−0.095	−0.116	−0.110	−0.061

Table E.8: Values of $(H^R)^1/RT_c$

$P_r =$	1.0000	1.2000	1.5000	2.0000	3.0000	5.0000	7.0000	10.000
T_r								
0.30	−11.062	−11.055	−11.044	−11.027	−10.992	−10.935	−10.872	−10.781
0.35	−10.640	−10.637	−10.632	−10.624	−10.609	−10.581	−10.554	−10.529
0.40	−10.121	−10.121	−10.121	−10.122	−10.123	−10.128	−10.135	−10.150
0.45	−9.525	−9.527	−9.531	−9.537	−9.549	−9.576	−9.611	−9.663
0.50	−8.888	−8.892	−8.899	−8.909	−8.932	−8.978	−9.030	−9.111
0.55	−8.238	−8.243	−8.252	−8.267	−8.298	−8.360	−8.425	−8.531
0.60	−7.596	−7.603	−7.614	−7.632	−7.669	−7.745	−7.824	−7.950
0.65	−6.980	−6.987	−6.997	−7.017	−7.059	−7.147	−7.239	−7.381
0.70	−6.388	−6.395	−6.407	−6.429	−6.475	−6.574	−6.677	−6.837
0.75	−5.824	−5.832	−5.845	−5.868	−5.918	−6.027	−6.142	−6.318
0.80	−5.285	−5.293	−5.306	−5.330	−5.385	−5.506	−5.632	−5.824
0.85	−4.763	−4.771	−4.784	−4.810	−4.872	−5.000	−5.149	−5.358
0.90	−4.249	−4.255	−4.268	−4.298	−4.371	−4.530	−4.688	−4.916
0.93	−3.934	−3.937	−3.951	−3.987	−4.073	−4.251	−4.422	−4.662
0.95	−3.712	−3.713	−3.730	−3.773	−3.873	−4.068	−4.248	−4.497
0.97	−3.470	−3.467	−3.492	−3.551	−3.670	−3.885	−4.077	−4.336
0.98	−3.332	−3.327	−3.363	−3.434	−3.568	−3.795	−3.992	−4.257
0.99	−3.164	−3.164	−3.223	−3.313	−3.464	−3.705	−3.909	−4.178
1.00	−2.471	−2.952	−3.065	−3.186	−3.358	−3.615	−3.825	−4.100
1.01	−1.375	−2.595	−2.880	−3.051	−3.251	−3.525	−3.742	−4.023
1.02	−1.180	−1.723	−2.650	−2.906	−3.142	−3.435	−3.661	−3.947
1.05	−0.877	−0.878	−1.496	−2.381	−2.800	−3.167	−3.418	−3.722
1.10	−0.617	−0.673	−0.617	−1.261	−2.167	−2.720	−3.023	−3.362
1.15	−0.459	−0.503	−0.487	−0.604	−1.497	−2.275	−2.641	−3.019
1.20	−0.349	−0.381	−0.381	−0.361	−0.934	−1.840	−2.273	−2.692
1.30	−0.203	−0.218	−0.218	−0.178	−0.300	−1.066	−1.592	−2.086
1.40	−0.111	−0.115	−0.128	−0.070	−0.044	−0.504	−1.012	−1.547
1.50	−0.049	−0.046	−0.032	0.008	0.078	−0.142	−0.556	−1.080
1.60	−0.005	0.004	0.023	0.065	0.151	0.082	−0.217	−0.689
1.70	0.027	0.040	0.063	0.109	0.202	0.223	0.028	−0.369
1.80	0.051	0.067	0.094	0.143	0.241	0.317	0.203	−0.112
1.90	0.070	0.088	0.117	0.169	0.271	0.381	0.330	0.092
2.00	0.085	0.105	0.136	0.190	0.295	0.428	0.424	0.255
2.20	0.106	0.128	0.163	0.221	0.331	0.493	0.551	0.489
2.40	0.120	0.144	0.181	0.242	0.356	0.535	0.631	0.645
2.60	0.130	0.156	0.194	0.257	0.376	0.567	0.687	0.754
2.80	0.137	0.164	0.204	0.269	0.391	0.591	0.729	0.836
3.00	0.142	0.170	0.211	0.278	0.403	0.611	0.763	0.899
3.50	0.152	0.181	0.224	0.294	0.425	0.650	0.827	1.015
4.00	0.158	0.188	0.233	0.306	0.442	0.680	0.874	1.097

Table E.9: Values of $(S^R)^0/R$

$P_r =$	0.0100	0.0500	0.1000	0.2000	0.4000	0.6000	0.8000	1.0000
T_r								
0.30	−11.614	−10.008	−9.319	−8.635	−7.961	−7.574	−7.304	−7.099
0.35	−11.185	−9.579	−8.890	−8.205	−7.529	−7.140	−6.869	−6.663
0.40	−10.802	−9.196	−8.506	−7.821	−7.144	−6.755	−6.483	−6.275
0.45	−10.453	−8.847	−8.157	−7.472	−6.794	−6.404	−6.132	−5.924
0.50	−10.137	−8.531	−7.841	−7.156	−6.479	−6.089	−5.816	−5.608
0.55	−0.038	−8.245	−7.555	−6.870	−6.193	−5.803	−5.531	−5.324
0.60	−0.029	−7.983	−7.294	−6.610	−5.933	−5.544	−5.273	−5.066
0.65	−0.023	−0.122	−7.052	−6.368	−5.694	−5.306	−5.036	−4.830
0.70	−0.018	−0.096	−0.206	−6.140	−5.467	−5.082	−4.814	−4.610
0.75	−0.015	−0.078	−0.164	−5.917	−5.248	−4.866	−4.600	−4.399
0.80	−0.013	−0.064	−0.134	−0.294	−5.026	−4.694	−4.388	−4.191
0.85	−0.011	−0.054	−0.111	−0.239	−4.785	−4.418	−4.166	−3.976
0.90	−0.009	−0.046	−0.094	−0.199	−0.463	−4.145	−3.912	−3.738
0.93	−0.008	−0.042	−0.085	−0.179	−0.408	−0.750	−3.723	−3.569
0.95	−0.008	−0.039	−0.080	−0.168	−0.377	−0.671	−3.556	−3.433
0.97	−0.007	−0.037	−0.075	−0.157	−0.350	−0.607	−1.056	−3.259
0.98	−0.007	−0.036	−0.073	−0.153	−0.337	−0.580	−0.971	−3.142
0.99	−0.007	−0.035	−0.071	−0.148	−0.326	−0.555	−0.903	−2.972
1.00	−0.007	−0.034	−0.069	−0.144	−0.315	−0.532	−0.847	−2.178
1.01	−0.007	−0.033	−0.067	−0.139	−0.304	−0.510	−0.799	−1.391
1.02	−0.006	−0.032	−0.065	−0.135	−0.294	−0.491	−0.757	−1.225
1.05	−0.006	−0.030	−0.060	−0.124	−0.267	−0.439	−0.656	−0.965
1.10	−0.005	−0.026	−0.053	−0.108	−0.230	−0.371	−0.537	−0.742
1.15	−0.005	−0.023	−0.047	−0.096	−0.201	−0.319	−0.452	−0.607
1.20	−0.004	−0.021	−0.042	−0.085	−0.177	−0.277	−0.389	−0.512
1.30	−0.003	−0.017	−0.033	−0.068	−0.140	−0.217	−0.298	−0.385
1.40	−0.003	−0.014	−0.027	−0.056	−0.114	−0.174	−0.237	−0.303
1.50	−0.002	−0.011	−0.023	−0.046	−0.094	−0.143	−0.194	−0.246
1.60	−0.002	−0.010	−0.019	−0.039	−0.079	−0.120	−0.162	−0.204
1.70	−0.002	−0.008	−0.017	−0.033	−0.067	−0.102	−0.137	−0.172
1.80	−0.001	−0.007	−0.014	−0.029	−0.058	−0.088	−0.117	−0.147
1.90	−0.001	−0.006	−0.013	−0.025	−0.051	−0.076	−0.102	−0.127
2.00	−0.001	−0.006	−0.011	−0.022	−0.044	−0.067	−0.089	−0.111
2.20	−0.001	−0.004	−0.009	−0.018	−0.035	−0.053	−0.070	−0.087
2.40	−0.001	−0.004	−0.007	−0.014	−0.028	−0.042	−0.056	−0.070
2.60	−0.001	−0.003	−0.006	−0.012	−0.023	−0.035	−0.046	−0.058
2.80	−0.000	−0.002	−0.005	−0.010	−0.020	−0.029	−0.039	−0.048
3.00	−0.000	−0.002	−0.004	−0.008	−0.017	−0.025	−0.033	−0.041
3.50	−0.000	−0.001	−0.003	−0.006	−0.012	−0.017	−0.023	−0.029
4.00	−0.000	−0.001	−0.002	−0.004	−0.009	−0.013	−0.017	−0.021

Table E.10: Values of $(S^R)^1/R$

$P_r =$	0.0100	0.0500	0.1000	0.2000	0.4000	0.6000	0.8000	1.0000
T_r								
0.30	−16.782	−16.774	−16.764	−16.744	−16.705	−16.665	−16.626	−16.586
0.35	−15.413	−15.408	−15.401	−15.387	−15.359	−15.333	−15.305	−15.278
0.40	−13.990	−13.986	−13.981	−13.972	−13.953	−13.934	−13.915	−13.896
0.45	−12.564	−12.561	−12.558	−12.551	−12.537	−12.523	−12.509	−12.496
0.50	−11.202	−11.200	−11.197	−11.092	−11.082	−11.172	−11.162	−11.153
0.55	−0.115	−9.948	−9.946	−9.942	−9.935	−9.928	−9.921	−9.914
0.60	−0.078	−8.828	−8.826	−8.823	−8.817	−8.811	−8.806	−8.799
0.65	−0.055	−0.309	−7.832	−7.829	−7.824	−7.819	−7.815	−7.510
0.70	−0.040	−0.216	−0.491	−6.951	−6.945	−6.941	−6.937	−6.933
0.75	−0.029	−0.156	−0.340	−6.173	−6.167	−6.162	−6.158	−6.155
0.80	−0.022	−0.116	−0.246	−0.578	−5.475	−5.468	−5.462	−5.458
0.85	−0.017	−0.088	−0.183	−0.400	−4.853	−4.841	−4.832	−4.826
0.90	−0.013	−0.068	−0.140	−0.301	−0.744	−4.269	−4.249	−4.238
0.93	−0.011	−0.058	−0.120	−0.254	−0.593	−1.219	−3.914	−3.894
0.95	−0.010	−0.053	−0.109	−0.228	−0.517	−0.961	−3.697	−3.658
0.97	−0.010	−0.048	−0.099	−0.206	−0.456	−0.797	−1.570	−3.406
0.98	−0.009	−0.046	−0.094	−0.196	−0.429	−0.734	−1.270	−3.264
0.99	−0.009	−0.044	−0.090	−0.186	−0.405	−0.680	−1.098	−3.093
1.00	−0.008	−0.042	−0.086	−0.177	−0.382	−0.632	−0.977	−2.399
1.01	−0.008	−0.040	−0.082	−0.169	−0.361	−0.590	−0.883	−1.306
1.02	−0.008	−0.039	−0.078	−0.161	−0.342	−0.552	−0.807	−1.113
1.05	−0.007	−0.034	−0.069	−0.140	−0.292	−0.460	−0.642	−0.820
1.10	−0.005	−0.028	−0.055	−0.112	−0.229	−0.350	−0.470	−0.577
1.15	−0.005	−0.023	−0.045	−0.091	−0.183	−0.275	−0.361	−0.437
1.20	−0.004	−0.019	−0.037	−0.075	−0.149	−0.220	−0.286	−0.343
1.30	−0.003	−0.013	−0.026	−0.052	−0.102	−0.148	−0.190	−0.226
1.40	−0.002	−0.010	−0.019	−0.037	−0.072	−0.104	−0.133	−0.158
1.50	−0.001	−0.007	−0.014	−0.027	−0.053	−0.076	−0.097	−0.115
1.60	−0.001	−0.005	−0.011	−0.021	−0.040	−0.057	−0.073	−0.086
1.70	−0.001	−0.004	−0.008	−0.016	−0.031	−0.044	−0.056	−0.067
1.80	−0.001	−0.003	−0.006	−0.013	−0.024	−0.035	−0.044	−0.053
1.90	−0.001	−0.003	−0.005	−0.010	−0.019	−0.028	−0.036	−0.043
2.00	−0.000	−0.002	−0.004	−0.008	−0.016	−0.023	−0.029	−0.035
2.20	−0.000	−0.001	−0.003	−0.006	−0.011	−0.016	−0.021	−0.025
2.40	−0.000	−0.001	−0.002	−0.004	−0.008	−0.012	−0.015	−0.019
2.60	−0.000	−0.001	−0.002	−0.003	−0.006	−0.009	−0.012	−0.015
2.80	−0.000	−0.001	−0.001	−0.003	−0.005	−0.008	−0.010	−0.012
3.00	−0.000	−0.001	−0.001	−0.002	−0.004	−0.006	−0.008	−0.010
3.50	−0.000	−0.000	−0.001	−0.001	−0.003	−0.004	−0.006	−0.007
4.00	−0.000	−0.000	−0.001	−0.001	−0.002	−0.003	−0.005	−0.006

Table E.11: Values of $(S^R)^0/R$

$P_r =$	1.0000	1.2000	1.5000	2.0000	3.0000	5.0000	7.0000	10.000
T_r								
0.30	−7.099	−6.935	−6.740	−6.497	−6.180	−5.847	−5.683	−5.578
0.35	−6.663	−6.497	−6.299	−6.052	−5.728	−5.376	−5.194	−5.060
0.40	−6.275	−6.109	−5.909	−5.660	−5.330	−4.967	−4.772	−4.619
0.45	−5.924	−5.757	−5.557	−5.306	−4.974	−4.603	−4.401	−4.234
0.50	−5.608	−5.441	−5.240	−4.989	−4.656	−4.282	−4.074	−3.899
0.55	−5.324	−5.157	−4.956	−4.706	−4.373	−3.998	−3.788	−3.607
0.60	−5.066	−4.900	−4.700	−4.451	−4.120	−3.747	−3.537	−3.353
0.65	−4.830	−4.665	−4.467	−4.220	−3.892	−3.523	−3.315	−3.131
0.70	−4.610	−4.446	−4.250	−4.007	−3.684	−3.322	−3.117	−2.935
0.75	−4.399	−4.238	−4.045	−3.807	−3.491	−3.138	−2.939	−2.761
0.80	−4.191	−4.034	−3.846	−3.615	−3.310	−2.970	−2.777	−2.605
0.85	−3.976	−3.825	−3.646	−3.425	−3.135	−2.812	−2.629	−2.463
0.90	−3.738	−3.599	−3.434	−3.231	−2.964	−2.663	−2.491	−2.334
0.93	−3.569	−3.444	−3.295	−3.108	−2.860	−2.577	−2.412	−2.262
0.95	−3.433	−3.326	−3.193	−3.023	−2.790	−2.520	−2.362	−2.215
0.97	−3.259	−3.188	−3.081	−2.932	−2.719	−2.463	−2.312	−2.170
0.98	−3.142	−3.106	−3.019	−2.884	−2.682	−2.436	−2.287	−2.148
0.99	−2.972	−3.010	−2.953	−2.835	−2.646	−2.408	−2.263	−2.126
1.00	−2.178	−2.893	−2.879	−2.784	−2.609	−2.380	−2.239	−2.105
1.01	−1.391	−2.736	−2.798	−2.730	−2.571	−2.352	−2.215	−2.083
1.02	−1.225	−2.495	−2.706	−2.673	−2.533	−2.325	−2.191	−2.062
1.05	−0.965	−1.523	−2.328	−2.483	−2.415	−2.242	−2.121	−2.001
1.10	−0.742	−1.012	−1.557	−2.081	−2.202	−2.104	−2.007	−1.903
1.15	−0.607	−0.790	−1.126	−1.649	−1.968	−1.966	−1.897	−1.810
1.20	−0.512	−0.651	−0.890	−1.308	−1.727	−1.827	−1.789	−1.722
1.30	−0.385	−0.478	−0.628	−0.891	−1.299	−1.554	−1.581	−1.556
1.40	−0.303	−0.375	−0.478	−0.663	−0.990	−1.303	−1.386	−1.402
1.50	−0.246	−0.299	−0.381	−0.520	−0.777	−1.088	−1.208	−1.260
1.60	−0.204	−0.247	−0.312	−0.421	−0.628	−0.913	−1.050	−1.130
1.70	−0.172	−0.208	−0.261	−0.350	−0.519	−0.773	−0.915	−1.013
1.80	−0.147	−0.177	−0.222	−0.296	−0.438	−0.661	−0.799	−0.908
1.90	−0.127	−0.153	−0.191	−0.255	−0.375	−0.570	−0.702	−0.815
2.00	−0.111	−0.134	−0.167	−0.221	−0.625	−0.497	−0.620	−0.733
2.20	−0.087	−0.105	−0.130	−0.172	−0.251	−0.388	−0.492	−0.599
2.40	−0.070	−0.084	−0.104	−0.138	−0.201	−0.311	−0.399	−0.496
2.60	−0.058	−0.069	−0.086	−0.113	−0.164	−0.255	−0.329	−0.416
2.80	−0.048	−0.058	−0.072	−0.094	−0.137	−0.213	−0.277	−0.353
3.00	−0.041	−0.049	−0.061	−0.080	−0.116	−0.181	−0.236	−0.303
3.50	−0.029	−0.034	−0.042	−0.056	−0.081	−0.126	−0.166	−0.216
4.00	−0.021	−0.025	−0.031	−0.041	−0.059	−0.093	−0.123	−0.162

Table E.12: Values of $(S^R)^1/R$

$P_r =$	1.0000	1.2000	1.5000	2.0000	3.0000	5.0000	7.0000	10.000
T_r								
0.30	−16.586	−16.547	−16.488	−16.390	−16.195	−15.837	−15.468	−14.925
0.35	−15.278	−15.251	−15.211	−15.144	−15.011	−14.751	−14.496	−14.153
0.40	−13.896	−13.877	−13.849	−13.803	−13.714	−13.541	−13.376	−13.144
0.45	−12.496	−12.482	−12.462	−12.430	−12.367	−12.248	−12.145	−11.999
0.50	−11.153	−11.143	−11.129	−11.107	−11.063	−10.985	−10.920	−10.836
0.55	−9.914	−9.907	−9.897	−9.882	−9.853	−9.806	−9.769	−9.732
0.60	−8.799	−8.794	−8.787	−8.777	−8.760	−8.736	−8.723	−8.720
0.65	−7.810	−7.807	−7.801	−7.794	−7.784	−7.779	−7.785	−7.811
0.70	−6.933	−6.930	−6.926	−6.922	−6.919	−6.929	−6.952	−7.002
0.75	−6.155	−6.152	−6.149	−6.147	−6.149	−6.174	−6.213	−6.285
0.80	−5.458	−5.455	−5.453	−5.452	−5.461	−5.501	−5.555	−5.648
0.85	−4.826	−4.822	−4.820	−4.822	−4.839	−4.898	−4.969	−5.082
0.90	−4.238	−4.232	−4.230	−4.236	−4.267	−4.351	−4.442	−4.578
0.93	−3.894	−3.885	−3.884	−3.896	−3.941	−4.046	−4.151	−4.300
0.95	−3.658	−3.647	−3.648	−3.669	−3.728	−3.851	−3.966	−4.125
0.97	−3.406	−3.391	−3.401	−3.437	−3.517	−3.661	−3.788	−3.957
0.98	−3.264	−3.247	−3.268	−3.318	−3.412	−3.569	−3.701	−3.875
0.99	−3.093	−3.082	−3.126	−3.195	−3.306	−3.477	−3.616	−3.796
1.00	−2.399	−2.868	−2.967	−3.067	−3.200	−3.387	−3.532	−3.717
1.01	−1.306	−2.513	−2.784	−2.933	−3.094	−3.297	−3.450	−3.640
1.02	−1.113	−1.655	−2.557	−2.790	−2.986	−3.209	−3.369	−3.565
1.05	−0.820	−0.831	−1.443	−2.283	−2.655	−2.949	−3.134	−3.348
1.10	−0.577	−0.640	−0.618	−1.241	−2.067	−2.534	−2.767	−3.013
1.15	−0.437	−0.489	−0.502	−0.654	−1.471	−2.138	−2.428	−2.708
1.20	−0.343	−0.385	−0.412	−0.447	−0.991	−1.767	−2.115	−2.430
1.30	−0.226	−0.254	−0.282	−0.300	−0.481	−1.147	−1.569	−1.944
1.40	−0.158	−0.178	−0.200	−0.220	−0.290	−0.730	−1.138	−1.544
1.50	−0.115	−0.130	−0.147	−0.166	−0.206	−0.479	−0.823	−1.222
1.60	−0.086	−0.098	−0.112	−0.129	−0.159	−0.334	−0.604	−0.969
1.70	−0.067	−0.076	−0.087	−0.102	−0.127	−0.248	−0.456	−0.775
1.80	−0.053	−0.060	−0.070	−0.083	−0.105	−0.195	−0.355	−0.628
1.90	−0.043	−0.049	−0.057	−0.069	−0.089	−0.160	−0.286	−0.518
2.00	−0.035	−0.040	−0.048	−0.058	−0.077	−0.136	−0.238	−0.434
2.20	−0.025	−0.029	−0.035	−0.043	−0.060	−0.105	−0.178	−0.322
2.40	−0.019	−0.022	−0.027	−0.034	−0.048	−0.086	−0.143	−0.254
2.60	−0.015	−0.018	−0.021	−0.028	−0.041	−0.074	−0.120	−0.210
2.80	−0.012	−0.014	−0.018	−0.023	−0.025	−0.065	−0.104	−0.180
3.00	−0.010	−0.012	−0.015	−0.020	−0.031	−0.058	−0.093	−0.158
3.50	−0.007	−0.009	−0.011	−0.015	−0.024	−0.046	−0.073	−0.122
4.00	−0.006	−0.007	−0.009	−0.012	−0.020	−0.038	−0.060	−0.100

Table E.13: Values of ϕ^0

$P_r =$	0.0100	0.0500	0.1000	0.2000	0.4000	0.6000	0.8000	1.0000
T_r								
0.30	*0.0002*	*0.0000*	*0.0000*	*0.0000*	*0.0000*	*0.0000*	*0.0000*	0.0000
0.35	*0.0034*	*0.0007*	*0.0003*	*0.0002*	*0.0001*	*0.0001*	*0.0001*	0.0000
0.40	*0.0272*	*0.0055*	*0.0028*	*0.0014*	*0.0007*	*0.0005*	*0.0004*	0.0003
0.45	*0.1321*	*0.0266*	*0.0135*	*0.0069*	*0.0036*	*0.0025*	*0.0020*	0.0016
0.50	*0.4529*	*0.0912*	*0.0461*	*0.0235*	*0.0122*	*0.0085*	*0.0067*	0.0055
0.55	0.9817	*0.2432*	*0.1227*	*0.0625*	*0.0325*	*0.0225*	*0.0176*	0.0146
0.60	0.9840	*0.5383*	*0.2716*	*0.1384*	*0.0718*	*0.0497*	*0.0386*	0.0321
0.65	0.9886	0.9419	*0.5212*	*0.2655*	*0.1374*	*0.0948*	*0.0738*	0.0611
0.70	0.9908	0.9528	0.9057	0.4560	*0.2360*	*0.1626*	*0.1262*	0.1045
0.75	0.9931	0.9616	0.9226	0.7178	*0.3715*	*0.2559*	*0.1982*	0.1641
0.80	0.9931	0.9683	0.9354	0.8730	*0.5445*	*0.3750*	*0.2904*	0.2404
0.85	0.9954	0.9727	0.9462	0.8933	*0.7534*	*0.5188*	*0.4018*	0.3319
0.90	0.9954	0.9772	0.9550	0.9099	0.8204	*0.6823*	*0.5297*	0.4375
0.93	0.9954	0.9795	0.9594	0.9183	0.8375	0.7551	*0.6109*	0.5058
0.95	0.9954	0.9817	0.9616	0.9226	0.8472	0.7709	*0.6668*	0.5521
0.97	0.9954	0.9817	0.9638	0.9268	0.8570	0.7852	0.7112	0.5984
0.98	0.9954	0.9817	0.9638	0.9290	0.8610	0.7925	0.7211	0.6223
0.99	0.9977	0.9840	0.9661	0.9311	0.8650	0.7980	0.7295	0.6442
1.00	0.9977	0.9840	0.9661	0.9333	0.8690	0.8035	0.7379	0.6668
1.01	0.9977	0.9840	0.9683	0.9354	0.8730	0.8110	0.7464	0.6792
1.02	0.9977	0.9840	0.9683	0.9376	0.8770	0.8166	0.7551	0.6902
1.05	0.9977	0.9863	0.9705	0.9441	0.8872	0.8318	0.7762	0.7194
1.10	0.9977	0.9886	0.9750	0.9506	0.9016	0.8531	0.8072	0.7586
1.15	0.9977	0.9886	0.9795	0.9572	0.9141	0.8730	0.8318	0.7907
1.20	0.9977	0.9908	0.9817	0.9616	0.9247	0.8892	0.8531	0.8166
1.30	0.9977	0.9931	0.9863	0.9705	0.9419	0.9141	0.8872	0.8590
1.40	0.9977	0.9931	0.9886	0.9772	0.9550	0.9333	0.9120	0.8892
1.50	1.0000	0.9954	0.9908	0.9817	0.9638	0.9462	0.9290	0.9141
1.60	1.0000	0.9954	0.9931	0.9863	0.9727	0.9572	0.9441	0.9311
1.70	1.0000	0.9977	0.9954	0.9886	0.9772	0.9661	0.9550	0.9462
1.80	1.0000	0.9977	0.9954	0.9908	0.9817	0.9727	0.9661	0.9572
1.90	1.0000	0.9977	0.9954	0.9931	0.9863	0.9795	0.9727	0.9661
2.00	1.0000	0.9977	0.9977	0.9954	0.9886	0.9840	0.9795	0.9727
2.20	1.0000	1.0000	0.9977	0.9977	0.9931	0.9908	0.9886	0.9840
2.40	1.0000	1.0000	1.0000	0.9977	0.9977	0.9954	0.9931	0.9931
2.60	1.0000	1.0000	1.0000	1.0000	1.0000	0.9977	0.9977	0.9977
2.80	1.0000	1.0000	1.0000	1.0000	1.0000	1.0000	1.0023	1.0023
3.00	1.0000	1.0000	1.0000	1.0000	1.0023	1.0023	1.0046	1.0046
3.50	1.0000	1.0000	1.0000	1.0023	1.0023	1.0046	1.0069	1.0093
4.00	1.0000	1.0000	1.0000	1.0023	1.0046	1.0069	1.0093	1.0116

Table E.14: Values of ϕ^1

$P_r =$	0.0100	0.0500	0.1000	0.2000	0.4000	0.6000	0.8000	1.0000
T_r								
0.30	*0.0000*	*0.0000*	*0.0000*	*0.0000*	*0.0000*	*0.0000*	*0.0000*	0.0000
0.35	*0.0000*	*0.0000*	*0.0000*	*0.0000*	*0.0000*	*0.0000*	*0.0000*	0.0000
0.40	*0.0000*	*0.0000*	*0.0000*	*0.0000*	*0.0000*	*0.0000*	*0.0000*	0.0000
0.45	*0.0002*	*0.0002*	*0.0002*	*0.0002*	*0.0002*	*0.0002*	*0.0002*	0.0002
0.50	*0.0014*	*0.0014*	*0.0014*	*0.0014*	*0.0014*	*0.0014*	*0.0013*	0.0013
0.55	0.9705	*0.0069*	*0.0068*	*0.0068*	*0.0066*	*0.0065*	*0.0064*	0.0063
0.60	0.9795	*0.0227*	*0.0226*	*0.0223*	*0.0220*	*0.0216*	*0.0213*	0.0210
0.65	0.9863	0.9311	*0.0572*	*0.0568*	*0.0559*	*0.0551*	*0.0543*	0.0535
0.70	0.9908	0.9528	0.9036	*0.1182*	*0.1163*	*0.1147*	*0.1131*	0.1116
0.75	0.9931	0.9683	0.9332	*0.2112*	*0.2078*	*0.2050*	*0.2022*	0.1994
0.80	0.9954	0.9772	0.9550	0.9057	*0.3302*	*0.3257*	*0.3212*	0.3168
0.85	0.9977	0.9863	0.9705	0.9375	*0.4774*	*0.4708*	*0.4654*	0.4590
0.90	0.9977	0.9908	0.9795	0.9594	0.9141	*0.6323*	*0.6250*	0.6165
0.93	0.9977	0.9931	0.9840	0.9705	0.9354	0.8953	*0.7227*	0.7144
0.95	0.9977	0.9931	0.9885	0.9750	0.9484	0.9183	*0.7888*	0.7797
0.97	1.0000	0.9954	0.9908	0.9795	0.9594	0.9354	0.9078	0.8413
0.98	1.0000	0.9954	0.9908	0.9817	0.9638	0.9440	0.9225	0.8729
0.99	1.0000	0.9954	0.9931	0.9840	0.9683	0.9528	0.9332	0.9036
1.00	1.0000	0.9977	0.9931	0.9863	0.9727	0.9594	0.9440	0.9311
1.01	1.0000	0.9977	0.9931	0.9885	0.9772	0.9638	0.9528	0.9462
1.02	1.0000	0.9977	0.9954	0.9908	0.9795	0.9705	0.9616	0.9572
1.05	1.0000	0.9977	0.9977	0.9954	0.9885	0.9863	0.9840	0.9840
1.10	1.0000	1.0000	1.0000	1.0000	1.0023	1.0046	1.0093	1.0163
1.15	1.0000	1.0000	1.0023	1.0046	1.0116	1.0186	1.0257	1.0375
1.20	1.0000	1.0023	1.0046	1.0069	1.0163	1.0280	1.0399	1.0544
1.30	1.0000	1.0023	1.0069	1.0116	1.0257	1.0399	1.0544	1.0716
1.40	1.0000	1.0046	1.0069	1.0139	1.0304	1.0471	1.0642	1.0815
1.50	1.0000	1.0046	1.0069	1.0163	1.0328	1.0496	1.0666	1.0865
1.60	1.0000	1.0046	1.0069	1.0163	1.0328	1.0496	1.0691	1.0865
1.70	1.0000	1.0046	1.0093	1.0163	1.0328	1.0496	1.0691	1.0865
1.80	1.0000	1.0046	1.0069	1.0163	1.0328	1.0496	1.0666	1.0840
1.90	1.0000	1.0046	1.0069	1.0163	1.0328	1.0496	1.0666	1.0815
2.00	1.0000	1.0046	1.0069	1.0163	1.0304	1.0471	1.0642	1.0815
2.20	1.0000	1.0046	1.0069	1.0139	1.0304	1.0447	1.0593	1.0765
2.40	1.0000	1.0046	1.0069	1.0139	1.0280	1.0423	1.0568	1.0716
2.60	1.0000	1.0023	1.0069	1.0139	1.0257	1.0399	1.0544	1.0666
2.80	1.0000	1.0023	1.0069	1.0116	1.0257	1.0375	1.0496	1.0642
3.00	1.0000	1.0023	1.0069	1.0116	1.0233	1.0352	1.0471	1.0593
3.50	1.0000	1.0023	1.0046	1.0023	1.0209	1.0304	1.0423	1.0520
4.00	1.0000	1.0023	1.0046	1.0093	1.0186	1.0280	1.0375	1.0471

Table E.15: Values of ϕ^0

$P_r =$	1.0000	1.2000	1.5000	2.0000	3.0000	5.0000	7.0000	10.000
T_r								
0.30	0.0000	0.0000	0.0000	0.0000	0.0000	0.0000	0.0000	0.0000
0.35	0.0000	0.0000	0.0000	0.0000	0.0000	0.0000	0.0000	0.0000
0.40	0.0003	0.0003	0.0003	0.0002	0.0002	0.0002	0.0002	0.0003
0.45	0.0016	0.0014	0.0012	0.0010	0.0008	0.0008	0.0009	0.0012
0.50	0.0055	0.0048	0.0041	0.0034	0.0028	0.0025	0.0027	0.0034
0.55	0.0146	0.0127	0.0107	0.0089	0.0072	0.0063	0.0066	0.0080
0.60	0.0321	0.0277	0.0234	0.0193	0.0154	0.0132	0.0135	0.0160
0.65	0.0611	0.0527	0.0445	0.0364	0.0289	0.0244	0.0245	0.0282
0.70	0.1045	0.0902	0.0759	0.0619	0.0488	0.0406	0.0402	0.0453
0.75	0.1641	0.1413	0.1188	0.0966	0.0757	0.0625	0.0610	0.0673
0.80	0.2404	0.2065	0.1738	0.1409	0.1102	0.0899	0.0867	0.0942
0.85	0.3319	0.2858	0.2399	0.1945	0.1517	0.1227	0.1175	0.1256
0.90	0.4375	0.3767	0.3162	0.2564	0.1995	0.1607	0.1524	0.1611
0.93	0.5058	0.4355	0.3656	0.2972	0.2307	0.1854	0.1754	0.1841
0.95	0.5521	0.4764	0.3999	0.3251	0.2523	0.2028	0.1910	0.2000
0.97	0.5984	0.5164	0.4345	0.3532	0.2748	0.2203	0.2075	0.2163
0.98	0.6223	0.5370	0.4529	0.3681	0.2864	0.2296	0.2158	0.2244
0.99	0.6442	0.5572	0.4699	0.3828	0.2978	0.2388	0.2244	0.2328
1.00	0.6668	0.5781	0.4875	0.3972	0.3097	0.2483	0.2328	0.2415
1.01	0.6792	0.5970	0.5047	0.4121	0.3214	0.2576	0.2415	0.2500
1.02	0.6902	0.6166	0.5224	0.4266	0.3334	0.2673	0.2506	0.2582
1.05	0.7194	0.6607	0.5728	0.4710	0.3690	0.2958	0.2773	0.2844
1.10	0.7586	0.7112	0.6412	0.5408	0.4285	0.3451	0.3228	0.3296
1.15	0.7907	0.7499	0.6918	0.6026	0.4875	0.3954	0.3690	0.3750
1.20	0.8166	0.7834	0.7328	0.6546	0.5420	0.4446	0.4150	0.4198
1.30	0.8590	0.8318	0.7943	0.7345	0.6383	0.5383	0.5058	0.5093
1.40	0.8892	0.8690	0.8395	0.7925	0.7145	0.6237	0.5902	0.5943
1.50	0.9141	0.8974	0.8730	0.8375	0.7745	0.6966	0.6668	0.6714
1.60	0.9311	0.9183	0.8995	0.8710	0.8222	0.7586	0.7328	0.7430
1.70	0.9462	0.9354	0.9204	0.8995	0.8610	0.8091	0.7907	0.8054
1.80	0.9572	0.9484	0.9376	0.9204	0.8913	0.8531	0.8414	0.8590
1.90	0.9661	0.9594	0.9506	0.9376	0.9162	0.8872	0.8831	0.9057
2.00	0.9727	0.9683	0.9616	0.9528	0.9354	0.9183	0.9183	0.9462
2.20	0.9840	0.9817	0.9795	0.9727	0.9661	0.9616	0.9727	1.0093
2.40	0.9931	0.9908	0.9908	0.9886	0.9863	0.9931	1.0116	1.0568
2.60	0.9977	0.9977	0.9977	0.9977	1.0023	1.0162	1.0399	1.0889
2.80	1.0023	1.0023	1.0046	1.0069	1.0116	1.0328	1.0593	1.1117
3.00	1.0046	1.0069	1.0069	1.0116	1.0209	1.0423	1.0740	1.1298
3.50	1.0093	1.0116	1.0139	1.0186	1.0304	1.0593	1.0914	1.1508
4.00	1.0116	1.0139	1.0162	1.0233	1.0375	1.0666	1.0990	1.1588

Table E.16: Values of ϕ^1

$P_r =$	1.0000	1.2000	1.5000	2.0000	3.0000	5.0000	7.0000	10.000
T_r								
0.30	0.0000	0.0000	0.0000	0.0000	0.0000	0.0000	0.0000	0.0000
0.35	0.0000	0.0000	0.0000	0.0000	0.0000	0.0000	0.0000	0.0000
0.40	0.0000	0.0000	0.0000	0.0000	0.0000	0.0000	0.0000	0.0000
0.45	0.0002	0.0002	0.0002	0.0002	0.0001	0.0001	0.0001	0.0001
0.50	0.0013	0.0013	0.0013	0.0012	0.0011	0.0009	0.0008	0.0006
0.55	0.0063	0.0062	0.0061	0.0058	0.0053	0.0045	0.0039	0.0031
0.60	0.0210	0.0207	0.0202	0.0194	0.0179	0.0154	0.0133	0.0108
0.65	0.0536	0.0527	0.0516	0.0497	0.0461	0.0401	0.0350	0.0289
0.70	0.1117	0.1102	0.1079	0.1040	0.0970	0.0851	0.0752	0.0629
0.75	0.1995	0.1972	0.1932	0.1871	0.1754	0.1552	0.1387	0.1178
0.80	0.3170	0.3133	0.3076	0.2978	0.2812	0.2512	0.2265	0.1954
0.85	0.4592	0.4539	0.4457	0.4325	0.4093	0.3698	0.3365	0.2951
0.90	0.6166	0.6095	0.5998	0.5834	0.5546	0.5058	0.4645	0.4130
0.93	0.7145	0.7063	0.6950	0.6761	0.6457	0.5916	0.5470	0.4898
0.95	0.7798	0.7691	0.7568	0.7379	0.7063	0.6501	0.6026	0.5432
0.97	0.8414	0.8318	0.8185	0.7998	0.7656	0.7096	0.6607	0.5984
0.98	0.8730	0.8630	0.8492	0.8298	0.7962	0.7379	0.6887	0.6266
0.99	0.9036	0.8913	0.8790	0.8590	0.8241	0.7674	0.7178	0.6546
1.00	0.9311	0.9204	0.9078	0.8872	0.8531	0.7962	0.7464	0.6823
1.01	0.9462	0.9462	0.9333	0.9162	0.8831	0.8241	0.7745	0.7096
1.02	0.9572	0.9661	0.9594	0.9419	0.9099	0.8531	0.8035	0.7379
1.05	0.9840	0.9954	1.0186	1.0162	0.9886	0.9354	0.8872	0.8222
1.10	1.0162	1.0280	1.0593	1.0990	1.1015	1.0617	1.0186	0.9572
1.15	1.0375	1.0520	1.0814	1.1376	1.1858	1.1722	1.1403	1.0864
1.20	1.0544	1.0691	1.0990	1.1588	1.2388	1.2647	1.2474	1.2050
1.30	1.0715	1.0914	1.1194	1.1776	1.2853	1.3868	1.4125	1.4061
1.40	1.0814	1.0990	1.1298	1.1858	1.2942	1.4488	1.5171	1.5524
1.50	1.0864	1.1041	1.1350	1.1858	1.2942	1.4689	1.5740	1.6520
1.60	1.0864	1.1041	1.1350	1.1858	1.2883	1.4689	1.5996	1.7140
1.70	1.0864	1.1041	1.1324	1.1803	1.2794	1.4622	1.6033	1.7458
1.80	1.0839	1.1015	1.1298	1.1749	1.2706	1.4488	1.5959	1.7620
1.90	1.0814	1.0990	1.1272	1.1695	1.2618	1.4355	1.5849	1.7620
2.00	1.0814	1.0965	1.1220	1.1641	1.2503	1.4191	1.5704	1.7539
2.20	1.0765	1.0914	1.1143	1.1535	1.2331	1.3900	1.5346	1.7219
2.40	1.0715	1.0864	1.1066	1.1429	1.2190	1.3614	1.4997	1.6866
2.60	1.0666	1.0814	1.1015	1.1350	1.2023	1.3397	1.4689	1.6482
2.80	1.0641	1.0765	1.0940	1.1272	1.1912	1.3183	1.4388	1.6144
3.00	1.0593	1.0715	1.0889	1.1194	1.1803	1.3002	1.4158	1.5813
3.50	1.0520	1.0617	1.0789	1.1041	1.1561	1.2618	1.3614	1.5101
4.00	1.0471	1.0544	1.0691	1.0914	1.1403	1.2303	1.3213	1.4555

APPENDIX F

STEAM TABLES

All tables are generated by computer from programs based on "The 1976 IFC[1] Formulation for Industrial Use: A Formulation of the Thermodynamic Properties of Ordinary Water Substance," as published in the *ASME Steam Tables*, 4th ed., App. I, pp. 11–29, The Am. Soc. Mech. Engrs., New York, 1979.

[1]International Formulation Committee.

Table F.1: Saturated Steam, SI units

V = SPECIFIC VOLUME   cm³ g⁻¹
U = SPECIFIC INTERNAL ENERGY   kJ kg⁻¹
H = SPECIFIC ENTHALPY   kJ kg⁻¹
S = SPECIFIC ENTROPY   kJ kg⁻¹ K⁻¹

t °C	T K	P kPa	SPECIFIC VOLUME V			INTERNAL ENERGY U			ENTHALPY H			ENTROPY S		
			sat. liq.	evap.	sat. vap.	sat. liq.	evap.	sat. vap.	sat. liq.	evap.	sat. vap.	sat. liq.	evap.	sat. vap.
0	273.15	0.611	1.000	206300.	206300.	-0.04	2375.7	2375.6	-0.04	2501.7	2501.6	0.0000	9.1578	9.1578
0.01	273.16	0.611	1.000	206200.	206200.	0.00	2375.6	2375.6	0.00	2501.6	2501.6	0.0000	9.1575	9.1575
1	274.15	0.657	1.000	192600.	192600.	4.17	2372.7	2376.9	4.17	2499.2	2503.4	0.0153	9.1158	9.1311
2	275.15	0.705	1.000	179900.	179900.	8.39	2369.9	2378.3	8.39	2496.8	2505.2	0.0306	9.0741	9.1047
3	276.15	0.757	1.000	168200.	168200.	12.60	2367.1	2379.7	12.60	2494.5	2507.1	0.0459	9.0326	9.0785
4	277.15	0.813	1.000	157300.	157300.	16.80	2364.3	2381.1	16.80	2492.1	2508.9	0.0611	8.9915	9.0526
5	278.15	0.872	1.000	147200.	147200.	21.01	2361.4	2382.4	21.01	2489.7	2510.7	0.0762	8.9507	9.0269
6	279.15	0.935	1.000	137800.	137800.	25.21	2358.6	2383.8	25.21	2487.4	2512.6	0.0913	8.9102	9.0014
7	280.15	1.001	1.000	129100.	129100.	29.41	2355.8	2385.2	29.41	2485.0	2514.4	0.1063	8.8699	8.9762
8	281.15	1.072	1.000	121000.	121000.	33.60	2353.0	2386.6	33.60	2482.6	2516.2	0.1213	8.8300	8.9513
9	282.15	1.147	1.000	113400.	113400.	37.80	2350.1	2387.9	37.80	2480.3	2518.1	0.1362	8.7903	8.9265
10	283.15	1.227	1.000	106400.	106400.	41.99	2347.3	2389.3	41.99	2477.9	2519.9	0.1510	8.7510	8.9020
11	284.15	1.312	1.000	99910.	99910.	46.18	2344.5	2390.7	46.19	2475.5	2521.7	0.1658	8.7119	8.8776
12	285.15	1.401	1.000	93840.	93840.	50.38	2341.7	2392.1	50.38	2473.2	2523.6	0.1805	8.6731	8.8536
13	286.15	1.497	1.001	88180.	88180.	54.56	2338.9	2393.4	54.57	2470.8	2525.4	0.1952	8.6345	8.8297
14	287.15	1.597	1.001	82900.	82900.	58.75	2336.1	2394.8	58.75	2468.5	2527.2	0.2098	8.5963	8.8060
15	288.15	1.704	1.001	77980.	77980.	62.94	2333.2	2396.2	62.94	2466.1	2529.1	0.2243	8.5582	8.7826
16	289.15	1.817	1.001	73380.	73380.	67.12	2330.4	2397.6	67.13	2463.8	2530.9	0.2388	8.5205	8.7593
17	290.15	1.936	1.001	69090.	69090.	71.31	2327.6	2398.9	71.31	2461.4	2532.7	0.2533	8.4830	8.7363
18	291.15	2.062	1.001	65090.	65090.	75.49	2324.8	2400.3	75.50	2459.0	2534.5	0.2677	8.4458	8.7135
19	292.15	2.196	1.002	61340.	61340.	79.68	2322.0	2401.7	79.68	2456.7	2536.4	0.2820	8.4088	8.6908
20	293.15	2.337	1.002	57840.	57840.	83.86	2319.2	2403.0	83.86	2454.3	2538.2	0.2963	8.3721	8.6684
21	294.15	2.485	1.002	54560.	54560.	88.04	2316.4	2404.4	88.04	2452.0	2540.0	0.3105	8.3356	8.6462
22	295.15	2.642	1.002	51490.	51490.	92.22	2313.6	2405.8	92.23	2449.6	2541.8	0.3247	8.2994	8.6241
23	296.15	2.808	1.002	48620.	48620.	96.40	2310.7	2407.1	96.41	2447.2	2543.6	0.3389	8.2634	8.6023
24	297.15	2.982	1.003	45920.	45930.	100.6	2307.9	2408.5	100.6	2444.9	2545.5	0.3530	8.2277	8.5806
25	298.15	3.166	1.003	43400.	43400.	104.8	2305.1	2409.9	104.8	2442.5	2547.3	0.3670	8.1922	8.5592
26	299.15	3.360	1.003	41030.	41030.	108.9	2302.3	2411.2	108.9	2440.2	2549.1	0.3810	8.1569	8.5379
27	300.15	3.564	1.003	38810.	38810.	113.1	2299.5	2412.6	113.1	2437.8	2550.9	0.3949	8.1218	8.5168
28	301.15	3.778	1.004	36730.	36730.	117.3	2296.7	2414.0	117.3	2435.4	2552.7	0.4088	8.0870	8.4959
29	302.15	4.004	1.004	34770.	34770.	121.5	2293.8	2415.3	121.5	2433.1	2554.5	0.4227	8.0524	8.4751

No.	T	P	v_f	v_g	v_{fg}	u_f	u_{fg}	u_g	h_f	h_{fg}	h_g	s_f	s_{fg}	s_g
30	303.15	4.241	1.004	32930.	32930.	125.7	2291.0	2416.7	125.7	2430.7	2556.4	0.4365	8.0180	8.4546
31	304.15	4.491	1.005	31200.	31200.	129.8	2288.2	2418.0	129.8	2428.3	2558.2	0.4503	7.9839	8.4342
32	305.15	4.753	1.005	29570.	29570.	134.0	2285.4	2419.4	134.0	2425.9	2560.0	0.4640	7.9500	8.4140
33	306.15	5.029	1.005	28040.	28040.	138.2	2282.6	2420.8	138.2	2423.6	2561.8	0.4777	7.9163	8.3939
34	307.15	5.318	1.006	26600.	26600.	142.4	2279.7	2422.1	142.4	2421.2	2563.6	0.4913	7.8828	8.3740
35	308.15	5.622	1.006	25240.	25240.	146.6	2276.9	2423.5	146.6	2418.8	2565.4	0.5049	7.8495	8.3543
36	309.15	5.940	1.006	23970.	23970.	150.7	2274.1	2424.8	150.7	2416.4	2567.2	0.5184	7.8164	8.3348
37	310.15	6.274	1.007	22760.	22760.	154.9	2271.3	2426.2	154.9	2414.1	2569.0	0.5319	7.7835	8.3154
38	311.15	6.624	1.007	21630.	21630.	159.1	2268.4	2427.5	159.1	2411.7	2570.8	0.5453	7.7509	8.2962
39	312.15	6.991	1.007	20560.	20560.	163.3	2265.6	2428.9	163.3	2409.3	2572.6	0.5588	7.7184	8.2772
40	313.15	7.375	1.008	19550.	19550.	167.4	2262.8	2430.2	167.5	2406.9	2574.4	0.5721	7.6861	8.2583
41	314.15	7.777	1.008	18590.	18590.	171.6	2259.9	2431.6	171.6	2404.5	2576.2	0.5854	7.6541	8.2395
42	315.15	8.198	1.009	17690.	17690.	175.8	2257.1	2432.9	175.8	2402.1	2577.9	0.5987	7.6222	8.2209
43	316.15	8.639	1.009	16840.	16840.	180.0	2254.3	2434.2	180.0	2399.7	2579.7	0.6120	7.5905	8.2025
44	317.15	9.100	1.009	16040.	16040.	184.2	2251.4	2435.6	184.2	2397.3	2581.5	0.6252	7.5590	8.1842
45	318.15	9.582	1.010	15280.	15280.	188.3	2248.6	2436.9	188.4	2394.9	2583.3	0.6383	7.5277	8.1661
46	319.15	10.09	1.010	14560.	14560.	192.5	2245.7	2438.3	192.5	2392.5	2585.1	0.6514	7.4966	8.1481
47	320.15	10.61	1.011	13880.	13880.	196.7	2242.9	2439.6	196.7	2390.1	2586.9	0.6645	7.4657	8.1302
48	321.15	11.16	1.011	13230.	13230.	200.9	2240.0	2440.9	200.9	2387.7	2588.6	0.6776	7.4350	8.1125
49	322.15	11.74	1.012	12620.	12620.	205.1	2237.2	2442.3	205.1	2385.3	2590.4	0.6906	7.4044	8.0950
50	323.15	12.34	1.012	12050.	12040.	209.2	2234.3	2443.6	209.3	2382.9	2592.2	0.7035	7.3741	8.0776
51	324.15	12.96	1.013	11500.	11500.	213.4	2231.5	2444.9	213.4	2380.5	2593.9	0.7164	7.3439	8.0603
52	325.15	13.61	1.013	10980.	10980.	217.6	2228.6	2446.2	217.6	2378.1	2595.7	0.7293	7.3138	8.0432
53	326.15	14.29	1.014	10490.	10490.	221.8	2225.8	2447.6	221.8	2375.7	2597.5	0.7422	7.2840	8.0262
54	327.15	15.00	1.014	10020.	10020.	226.0	2222.9	2448.9	226.0	2373.2	2599.2	0.7550	7.2543	8.0093
55	328.15	15.74	1.015	9578.9	9577.9	230.2	2220.0	2450.2	230.2	2370.8	2601.0	0.7677	7.2248	7.9925
56	329.15	16.51	1.015	9158.7	9157.7	234.3	2217.2	2451.5	234.4	2368.4	2602.7	0.7804	7.1955	7.9759
57	330.15	17.31	1.016	8759.8	8758.7	238.5	2214.3	2452.8	238.5	2365.9	2604.5	0.7931	7.1663	7.9595
58	331.15	18.15	1.016	8380.8	8379.8	242.7	2211.4	2454.1	242.7	2363.5	2606.2	0.8058	7.1373	7.9431
59	332.15	19.02	1.017	8020.8	8019.7	246.9	2208.6	2455.4	246.9	2361.1	2608.0	0.8184	7.1085	7.9269
60	333.15	19.92	1.017	7678.5	7677.5	251.1	2205.7	2456.8	251.1	2358.6	2609.7	0.8310	7.0798	7.9108
61	334.15	20.86	1.018	7353.2	7352.1	255.3	2202.8	2458.1	255.3	2356.2	2611.4	0.8435	7.0513	7.8948
62	335.15	21.84	1.018	7043.7	7042.7	259.4	2199.9	2459.4	259.5	2353.7	2613.2	0.8560	7.0230	7.8790
63	336.15	22.86	1.019	6749.3	6748.2	263.6	2197.0	2460.7	263.6	2351.3	2614.9	0.8685	6.9948	7.8633
64	337.15	23.91	1.019	6469.0	6468.0	267.8	2194.1	2462.0	267.8	2348.8	2616.6	0.8809	6.9667	7.8477
65	338.15	25.01	1.020	6202.3	6201.3	272.0	2191.2	2463.2	272.0	2346.3	2618.4	0.8933	6.9388	7.8322
66	339.15	26.15	1.020	5948.2	5947.1	276.2	2188.3	2464.5	276.2	2343.9	2620.1	0.9057	6.9111	7.8168
67	340.15	27.33	1.021	5706.2	5705.2	280.4	2185.4	2465.8	280.4	2341.4	2621.8	0.9180	6.8835	7.8015
68	341.15	28.56	1.022	5475.6	5474.6	284.6	2182.5	2467.1	284.6	2338.9	2623.5	0.9303	6.8561	7.7864
69	342.15	29.84	1.022	5255.8	5254.8	288.8	2179.6	2468.4	288.8	2336.4	2625.2	0.9426	6.8288	7.7714
70	343.15	31.16	1.023	5046.3	5045.2	292.9	2176.7	2469.7	293.0	2334.0	2626.9	0.9548	6.8017	7.7565
71	344.15	32.53	1.023	4846.4	4845.4	297.1	2173.8	2470.9	297.2	2331.5	2628.6	0.9670	6.7747	7.7417
72	345.15	33.96	1.024	4655.7	4654.7	301.3	2170.9	2472.2	301.4	2329.0	2630.3	0.9792	6.7478	7.7270
73	346.15	35.43	1.025	4473.7	4472.7	305.5	2168.0	2473.5	305.5	2326.5	2632.0	0.9913	6.7211	7.7124
74	347.15	36.96	1.025	4300.0	4299.0	309.7	2165.1	2474.8	309.7	2324.0	2633.7	1.0034	6.6945	7.6979

Table F.1: Saturated Steam, SI units (Continued)

t °C	T K	P kPa	SPECIFIC VOLUME V			INTERNAL ENERGY U			ENTHALPY H			ENTROPY S		
			sat. liq.	evap.	sat. vap.	sat. liq.	evap.	sat. vap.	sat. liq.	evap.	sat. vap.	sat. liq.	evap.	sat. vap.
75	348.15	38.55	1.026	4133.1	4134.1	313.9	2162.1	2476.0	313.9	2321.5	2635.4	1.0154	6.6681	7.6835
76	349.15	40.19	1.027	3974.6	3975.7	318.1	2159.2	2477.3	318.1	2318.9	2637.1	1.0275	6.6418	7.6693
77	350.15	41.89	1.027	3823.3	3824.3	322.3	2156.3	2478.5	322.3	2316.4	2638.7	1.0395	6.6156	7.6551
78	351.15	43.65	1.028	3678.6	3679.6	326.5	2153.3	2479.8	326.5	2313.9	2640.4	1.0514	6.5896	7.6410
79	352.15	45.47	1.029	3540.3	3541.3	330.7	2150.4	2481.1	330.7	2311.4	2642.1	1.0634	6.5637	7.6271
80	353.15	47.36	1.029	3408.1	3409.1	334.9	2147.4	2482.3	334.9	2308.8	2643.8	1.0753	6.5380	7.6132
81	354.15	49.31	1.030	3281.6	3282.6	339.1	2144.5	2483.5	339.1	2306.3	2645.4	1.0871	6.5123	7.5995
82	355.15	51.33	1.031	3160.6	3161.6	343.3	2141.5	2484.8	343.3	2303.8	2647.1	1.0990	6.4868	7.5858
83	356.15	53.42	1.031	3044.8	3045.8	347.5	2138.6	2486.0	347.5	2301.2	2648.7	1.1108	6.4615	7.5722
84	357.15	55.57	1.032	2933.9	2935.0	351.7	2135.6	2487.3	351.7	2298.6	2650.4	1.1225	6.4362	7.5587
85	358.15	57.80	1.033	2827.8	2828.8	355.9	2132.6	2488.5	355.9	2296.1	2652.0	1.1343	6.4111	7.5454
86	359.15	60.11	1.033	2726.1	2727.2	360.1	2129.7	2489.7	360.1	2293.5	2653.6	1.1460	6.3861	7.5321
87	360.15	62.49	1.034	2628.8	2629.8	364.3	2126.7	2490.9	364.3	2290.9	2655.3	1.1577	6.3612	7.5189
88	361.15	64.95	1.035	2535.5	2536.5	368.5	2123.7	2492.2	368.5	2288.4	2656.9	1.1693	6.3365	7.5058
89	362.15	67.49	1.035	2446.0	2447.0	372.7	2120.7	2493.4	372.7	2285.8	2658.5	1.1809	6.3119	7.4928
90	363.15	70.11	1.036	2360.3	2361.3	376.9	2117.7	2494.6	376.9	2283.2	2660.1	1.1925	6.2873	7.4799
91	364.15	72.81	1.037	2278.0	2279.1	381.1	2114.7	2495.8	381.1	2280.6	2661.7	1.2041	6.2629	7.4670
92	365.15	75.61	1.038	2199.2	2200.2	385.3	2111.7	2497.0	385.4	2278.0	2663.4	1.2156	6.2387	7.4543
93	366.15	78.49	1.038	2123.5	2124.5	389.5	2108.7	2498.2	389.6	2275.4	2665.0	1.2271	6.2145	7.4416
94	367.15	81.46	1.039	2050.9	2051.9	393.7	2105.7	2499.4	393.8	2272.8	2666.6	1.2386	6.1905	7.4291
95	368.15	84.53	1.040	1981.2	1982.2	397.9	2102.7	2500.6	398.0	2270.2	2668.1	1.2501	6.1665	7.4166
96	369.15	87.69	1.041	1914.3	1915.3	402.1	2099.7	2501.8	402.2	2267.5	2669.7	1.2615	6.1427	7.4042
97	370.15	90.94	1.041	1850.0	1851.0	406.3	2096.6	2503.0	406.4	2264.9	2671.3	1.2729	6.1190	7.3919
98	371.15	94.30	1.042	1788.3	1789.3	410.5	2093.6	2504.1	410.6	2262.2	2672.9	1.2842	6.0954	7.3796
99	372.15	97.76	1.043	1729.0	1730.0	414.7	2090.6	2505.3	414.8	2259.6	2674.4	1.2956	6.0719	7.3675
100	373.15	101.33	1.044	1672.0	1673.0	419.0	2087.5	2506.5	419.1	2256.9	2676.0	1.3069	6.0485	7.3554
102	375.15	108.78	1.045	1564.5	1565.5	427.4	2081.4	2508.8	427.5	2251.6	2679.1	1.3294	6.0021	7.3315
104	377.15	116.68	1.047	1465.1	1466.2	435.8	2075.3	2511.1	435.9	2246.3	2682.2	1.3518	5.9560	7.3078
106	379.15	125.04	1.049	1373.1	1374.2	444.3	2069.2	2513.4	444.4	2240.9	2685.3	1.3742	5.9104	7.2845
108	381.15	133.90	1.050	1287.9	1288.9	452.7	2063.0	2515.7	452.9	2235.4	2688.3	1.3964	5.8651	7.2615
110	383.15	143.27	1.052	1208.9	1209.9	461.2	2056.8	2518.0	461.3	2230.0	2691.3	1.4185	5.8203	7.2388
112	385.15	153.16	1.054	1135.6	1136.6	469.6	2050.6	2520.2	469.8	2224.5	2694.3	1.4405	5.7758	7.2164
114	387.15	163.62	1.055	1067.5	1068.5	478.1	2044.3	2522.4	478.3	2219.0	2697.2	1.4624	5.7318	7.1942
116	389.15	174.65	1.057	1004.2	1005.2	486.6	2038.1	2524.6	486.7	2213.4	2700.2	1.4842	5.6881	7.1723
118	391.15	186.28	1.059	945.3	946.3	495.0	2031.8	2526.8	495.2	2207.9	2703.1	1.5060	5.6447	7.1507
120	393.15	198.54	1.061	890.5	891.5	503.5	2025.4	2529.0	503.7	2202.2	2706.0	1.5276	5.6017	7.1293
122	395.15	211.45	1.062	839.4	840.5	512.0	2019.1	2531.1	512.2	2196.6	2708.8	1.5491	5.5590	7.1082
124	397.15	225.04	1.064	791.8	792.8	520.5	2012.7	2533.2	520.7	2190.9	2711.6	1.5706	5.5167	7.0873
126	399.15	239.33	1.066	747.3	748.4	529.0	2006.3	2535.3	529.2	2185.2	2714.4	1.5919	5.4747	7.0666
128	401.15	254.35	1.068	705.8	706.9	537.5	1999.9	2537.4	537.8	2179.4	2717.2	1.6132	5.4330	7.0462

130	403.15	270.13	1.070	667.1	668.1	546.0	1993.4	2539.4	546.3	2173.6	2719.9	1.6344	5.3917	7.0261
132	405.15	286.70	1.072	630.8	631.9	554.5	1986.9	2541.4	554.8	2167.8	2722.6	1.6555	5.3507	7.0061
134	407.15	304.07	1.074	596.9	598.0	563.1	1980.4	2543.4	563.4	2161.9	2725.3	1.6765	5.3099	6.9864
136	409.15	322.29	1.076	565.1	566.2	571.5	1973.8	2545.4	572.0	2155.9	2727.9	1.6974	5.2695	6.9669
138	411.15	341.38	1.078	535.3	536.4	580.2	1967.2	2547.4	580.5	2150.0	2730.5	1.7182	5.2293	6.9475
140	413.15	361.38	1.080	507.4	508.5	588.7	1960.6	2549.3	589.3	2144.0	2733.1	1.7390	5.1894	6.9284
142	415.15	382.31	1.082	481.2	482.3	597.3	1953.9	2551.2	597.7	2137.9	2735.6	1.7597	5.1499	6.9095
144	417.15	404.20	1.084	456.6	457.7	605.9	1947.2	2553.1	606.3	2131.8	2738.1	1.7803	5.1105	6.8908
146	419.15	427.09	1.086	433.5	434.6	614.4	1940.5	2554.9	614.9	2125.7	2740.6	1.8008	5.0715	6.8723
148	421.15	451.01	1.089	411.8	412.9	623.0	1933.7	2556.5	623.5	2119.5	2743.0	1.8213	5.0327	6.8539
150	423.15	476.00	1.091	391.4	392.4	631.6	1926.9	2558.6	632.1	2113.2	2745.4	1.8416	4.9941	6.8358
152	425.15	502.08	1.093	372.1	373.2	640.2	1920.1	2560.3	640.8	2106.9	2747.7	1.8619	4.9558	6.8178
154	427.15	529.29	1.095	354.0	355.1	648.9	1913.2	2562.1	649.4	2100.6	2750.0	1.8822	4.9178	6.8000
156	429.15	557.67	1.098	336.9	338.0	657.5	1906.3	2563.8	658.1	2094.2	2752.3	1.9023	4.8800	6.7823
158	431.15	587.25	1.100	320.8	321.9	666.1	1899.3	2565.5	666.8	2087.7	2754.5	1.9224	4.8424	6.7648
160	433.15	618.06	1.102	305.7	306.8	674.8	1892.3	2567.1	675.5	2081.3	2756.7	1.9425	4.8050	6.7475
162	435.15	650.16	1.105	291.3	292.4	683.5	1885.3	2568.8	684.2	2074.7	2758.9	1.9624	4.7679	6.7303
164	437.15	683.56	1.107	277.8	278.9	692.1	1878.2	2570.4	692.9	2068.1	2761.0	1.9823	4.7309	6.7133
166	439.15	718.31	1.109	265.0	266.1	700.8	1871.1	2571.9	701.6	2061.4	2763.1	2.0022	4.6942	6.6964
168	441.15	754.45	1.112	252.9	254.0	709.5	1863.9	2573.4	710.4	2054.7	2765.1	2.0219	4.6577	6.6796
170	443.15	792.02	1.114	241.4	242.6	718.2	1856.7	2574.9	719.1	2047.9	2767.1	2.0416	4.6214	6.6630
172	445.15	831.06	1.117	230.6	231.7	727.0	1849.5	2576.4	727.9	2041.1	2769.0	2.0613	4.5853	6.6465
174	447.15	871.60	1.120	220.3	221.5	735.7	1842.2	2577.8	736.7	2034.2	2770.9	2.0809	4.5493	6.6302
176	449.15	913.68	1.122	210.6	211.7	744.4	1834.8	2579.3	745.5	2027.3	2772.7	2.1004	4.5136	6.6140
178	451.15	957.36	1.125	201.4	202.5	753.2	1827.4	2580.6	754.3	2020.2	2774.5	2.1199	4.4780	6.5979
180	453.15	1002.7	1.128	192.7	193.8	762.0	1820.0	2581.9	763.1	2013.1	2776.3	2.1393	4.4426	6.5819
182	455.15	1049.6	1.130	184.4	185.5	770.8	1812.5	2583.2	772.0	2006.0	2778.0	2.1587	4.4074	6.5660
184	457.15	1098.3	1.133	176.5	177.6	779.6	1804.9	2584.5	780.8	1998.8	2779.6	2.1780	4.3723	6.5503
186	459.15	1148.8	1.136	169.0	170.2	788.4	1797.3	2585.7	789.7	1991.5	2781.2	2.1972	4.3374	6.5346
188	461.15	1201.0	1.139	161.9	163.1	797.2	1789.7	2586.9	798.6	1984.2	2782.8	2.2164	4.3026	6.5191
190	463.15	1255.1	1.142	155.2	156.3	806.1	1782.0	2588.1	807.5	1976.7	2784.3	2.2356	4.2680	6.5036
192	465.15	1311.1	1.144	148.8	149.9	814.9	1774.2	2589.2	816.5	1969.3	2785.7	2.2547	4.2336	6.4883
194	467.15	1369.0	1.147	142.6	143.8	823.8	1766.4	2590.2	825.4	1961.7	2787.1	2.2738	4.1993	6.4730
196	469.15	1428.9	1.150	136.8	138.0	832.7	1758.6	2591.3	834.4	1954.1	2788.4	2.2928	4.1651	6.4578
198	471.15	1490.9	1.153	131.3	132.4	841.6	1750.6	2592.3	843.4	1946.4	2789.7	2.3117	4.1310	6.4428
200	473.15	1554.9	1.156	126.0	127.2	850.6	1742.6	2593.2	852.4	1938.6	2790.9	2.3307	4.0971	6.4278
202	475.15	1621.0	1.160	121.0	122.1	859.5	1734.6	2594.1	861.4	1930.7	2792.1	2.3495	4.0633	6.4128
204	477.15	1689.3	1.163	116.2	117.3	868.5	1726.5	2595.0	870.5	1922.8	2793.2	2.3684	4.0296	6.3980
206	479.15	1759.8	1.166	111.6	112.8	877.5	1718.3	2595.8	879.5	1914.7	2794.3	2.3872	3.9961	6.3832
208	481.15	1832.6	1.169	107.2	108.4	886.5	1710.1	2596.6	888.6	1906.6	2795.3	2.4059	3.9626	6.3686
210	483.15	1907.7	1.173	103.1	104.2	895.5	1701.8	2597.3	897.7	1898.5	2796.2	2.4247	3.9293	6.3539
212	485.15	1985.2	1.176	99.09	100.2	904.5	1693.5	2598.0	906.9	1890.2	2797.1	2.4434	3.8960	6.3394
214	487.15	2065.1	1.179	95.28	96.46	913.6	1685.1	2598.7	916.0	1881.8	2797.9	2.4620	3.8629	6.3249
216	489.15	2147.5	1.183	91.65	92.83	922.7	1676.6	2599.3	925.2	1873.4	2798.6	2.4806	3.8298	6.3104
218	491.15	2232.4	1.186	88.17	89.36	931.8	1668.0	2599.8	934.4	1864.9	2799.3	2.4992	3.7968	6.2960

Table F.1: Saturated Steam, SI units (Continued)

t °C	T K	P kPa	SPECIFIC VOLUME V			INTERNAL ENERGY U			ENTHALPY H			ENTROPY S		
			sat. liq.	evap.	sat. vap.	sat. liq.	evap.	sat. vap.	sat. liq.	evap.	sat. vap.	sat. liq.	evap.	sat. vap.
220	493.15	2319.8	1.190	84.85	86.04	940.9	1659.4	2600.3	943.7	1856.2	2799.9	2.5178	3.7639	6.2817
222	495.15	2409.9	1.194	81.67	82.86	950.1	1650.7	2600.8	952.9	1847.5	2800.5	2.5363	3.7311	6.2674
224	497.15	2502.7	1.197	78.62	79.82	959.2	1642.0	2601.2	962.2	1838.7	2800.8	2.5548	3.6984	6.2532
226	499.15	2598.2	1.201	75.71	76.91	968.4	1633.1	2601.5	971.5	1829.8	2801.4	2.5733	3.6657	6.2390
228	501.15	2696.5	1.205	72.92	74.12	977.6	1624.2	2601.8	980.9	1820.8	2801.7	2.5917	3.6331	6.2249
230	503.15	2797.6	1.209	70.24	71.45	986.9	1615.2	2602.1	990.3	1811.7	2802.0	2.6102	3.6006	6.2107
232	505.15	2901.6	1.213	67.68	68.89	996.2	1606.1	2602.3	999.7	1802.5	2802.2	2.6286	3.5681	6.1967
234	507.15	3008.0	1.217	65.22	66.43	1005.4	1597.0	2602.4	1009.1	1793.2	2802.3	2.6470	3.5356	6.1826
236	509.15	3118.6	1.221	62.86	64.08	1014.8	1587.7	2602.4	1018.6	1783.8	2802.3	2.6653	3.5033	6.1686
238	511.15	3231.7	1.225	60.60	61.82	1024.1	1578.4	2602.5	1028.1	1774.2	2802.3	2.6837	3.4709	6.1546
240	513.15	3347.8	1.229	58.43	59.65	1033.5	1569.0	2602.5	1037.6	1764.6	2802.2	2.7020	3.4386	6.1406
242	515.15	3467.2	1.233	56.34	57.57	1042.9	1559.5	2602.4	1047.2	1754.9	2802.1	2.7203	3.4063	6.1266
244	517.15	3589.8	1.238	54.34	55.58	1052.3	1549.9	2602.2	1056.8	1745.0	2801.8	2.7386	3.3740	6.1127
246	519.15	3715.7	1.242	52.41	53.66	1061.8	1540.2	2602.0	1066.4	1735.0	2801.4	2.7569	3.3418	6.0987
248	521.15	3844.9	1.247	50.56	51.81	1071.3	1530.5	2601.8	1076.1	1724.9	2801.0	2.7752	3.3096	6.0848
250	523.15	3977.6	1.251	48.79	50.04	1080.8	1520.6	2601.4	1085.8	1714.7	2800.4	2.7935	3.2773	6.0708
252	525.15	4113.7	1.256	47.08	48.33	1090.4	1510.6	2601.0	1095.5	1704.3	2799.8	2.8118	3.2451	6.0569
254	527.15	4253.4	1.261	45.43	46.69	1100.0	1500.5	2600.5	1105.3	1693.8	2799.1	2.8300	3.2129	6.0429
256	529.15	4396.5	1.266	43.85	45.11	1109.6	1490.4	2600.0	1115.2	1683.2	2798.3	2.8483	3.1807	6.0290
258	531.15	4543.7	1.271	42.33	43.60	1119.3	1480.1	2599.3	1125.0	1672.4	2797.4	2.8666	3.1484	6.0150
260	533.15	4694.3	1.276	40.86	42.13	1129.0	1469.7	2598.6	1134.9	1661.5	2796.4	2.8848	3.1161	6.0010
262	535.15	4848.8	1.281	39.44	40.73	1138.7	1459.2	2597.8	1144.9	1650.4	2795.3	2.9031	3.0838	5.9869
264	537.15	5007.1	1.286	38.08	39.37	1148.5	1448.5	2597.0	1154.9	1639.2	2794.1	2.9214	3.0515	5.9729
266	539.15	5169.3	1.291	36.77	38.06	1158.3	1437.8	2596.1	1165.0	1627.8	2792.8	2.9397	3.0191	5.9588
268	541.15	5335.5	1.297	35.51	36.80	1168.2	1426.9	2595.0	1175.1	1616.3	2791.4	2.9580	2.9866	5.9446
270	543.15	5505.8	1.303	34.29	35.59	1178.1	1415.8	2593.9	1185.2	1604.6	2789.9	2.9763	2.9541	5.9304
272	545.15	5680.2	1.308	33.11	34.42	1188.0	1404.7	2592.8	1195.4	1592.8	2788.2	2.9947	2.9215	5.9162
274	547.15	5858.7	1.314	31.97	33.29	1198.0	1393.4	2591.4	1205.7	1580.8	2786.5	3.0131	2.8889	5.9019
276	549.15	6041.5	1.320	30.88	32.20	1208.0	1382.0	2590.1	1216.0	1568.5	2784.6	3.0314	2.8561	5.8876
278	551.15	6228.7	1.326	29.82	31.14	1218.1	1370.4	2588.6	1226.4	1556.2	2782.6	3.0499	2.8233	5.8731
280	553.15	6420.2	1.332	28.79	30.13	1228.3	1358.7	2587.0	1236.8	1543.6	2780.4	3.0683	2.7903	5.8586
282	555.15	6616.1	1.339	27.81	29.14	1238.5	1346.8	2585.3	1247.3	1530.8	2778.1	3.0868	2.7573	5.8440
284	557.15	6816.6	1.345	26.85	28.20	1248.7	1334.8	2583.5	1257.9	1517.6	2775.5	3.1053	2.7241	5.8294
286	559.15	7021.8	1.352	25.93	27.28	1259.0	1322.6	2581.6	1268.5	1504.6	2773.2	3.1238	2.6908	5.8146
288	561.15	7231.5	1.359	25.03	26.39	1269.4	1310.2	2579.6	1279.2	1491.2	2770.5	3.1424	2.6573	5.7997
290	563.15	7446.1	1.366	24.17	25.54	1279.8	1297.7	2577.5	1290.0	1477.6	2767.6	3.1611	2.6237	5.7848
292	565.15	7665.6	1.373	23.33	24.71	1290.3	1284.9	2575.3	1300.9	1463.8	2764.6	3.1798	2.5899	5.7697
294	567.15	7889.7	1.381	22.52	23.90	1300.9	1272.0	2572.9	1311.8	1449.7	2761.5	3.1985	2.5560	5.7545
296	569.15	8118.9	1.388	21.74	23.13	1311.5	1258.9	2570.4	1322.8	1435.4	2758.2	3.2173	2.5218	5.7392
298	571.15	8353.2	1.396	20.98	22.38	1322.2	1245.6	2567.8	1333.9	1420.8	2754.7	3.2362	2.4875	5.7237

300	573.15	8592.7	1.404	20.24	21.65	1333.0	1232.0	2565.0	1345.1	1406.0	2751.0	3.2552	2.4529	5.7081
302	575.15	8837.4	1.412	19.53	20.94	1343.8	1218.3	2562.1	1356.3	1390.9	2747.2	3.2742	2.4182	5.6924
304	577.15	9087.3	1.421	18.84	20.26	1354.8	1204.3	2559.0	1367.7	1375.5	2743.2	3.2933	2.3832	5.6765
306	579.15	9342.7	1.430	18.17	19.60	1365.8	1190.1	2555.9	1379.1	1359.8	2739.0	3.3125	2.3479	5.6604
308	581.15	9603.6	1.439	17.52	18.96	1376.9	1175.6	2552.5	1390.7	1343.9	2734.6	3.3318	2.3124	5.6442
310	583.15	9870.0	1.448	16.89	18.33	1388.1	1161.0	2549.1	1402.4	1327.6	2730.0	3.3512	2.2766	5.6278
312	585.15	10142.1	1.458	16.27	17.73	1399.4	1146.0	2545.4	1414.2	1311.0	2725.2	3.3707	2.2404	5.6111
314	587.15	10420.0	1.468	15.68	17.14	1410.8	1130.8	2541.6	1426.1	1294.1	2720.2	3.3903	2.2040	5.5943
316	589.15	10703.0	1.478	15.09	16.57	1422.3	1115.2	2537.5	1438.1	1276.8	2714.9	3.4101	2.1672	5.5772
318	591.15	10993.4	1.488	14.53	16.02	1433.9	1099.4	2533.3	1450.3	1259.1	2709.4	3.4300	2.1300	5.5599
320	593.15	11289.1	1.500	13.98	15.48	1445.7	1083.2	2528.9	1462.6	1241.1	2703.7	3.4500	2.0923	5.5423
322	595.15	11591.0	1.511	13.44	14.96	1457.5	1066.7	2524.3	1475.1	1222.6	2697.6	3.4702	2.0542	5.5244
324	597.15	11899.2	1.523	12.92	14.45	1469.5	1049.9	2519.4	1487.7	1203.6	2691.3	3.4906	2.0156	5.5062
326	599.15	12213.7	1.535	12.41	13.95	1481.7	1032.6	2514.3	1500.4	1184.2	2684.6	3.5111	1.9764	5.4876
328	601.15	12534.8	1.548	11.91	13.46	1494.0	1014.8	2508.8	1513.4	1164.2	2677.6	3.5319	1.9367	5.4685
330	603.15	12862.5	1.561	11.43	12.99	1506.4	996.7	2503.1	1526.5	1143.6	2670.2	3.5528	1.8962	5.4490
332	605.15	13197.0	1.575	10.95	12.53	1519.1	978.0	2497.0	1539.9	1122.5	2662.3	3.5740	1.8550	5.4290
334	607.15	13538.3	1.590	10.49	12.08	1531.9	958.7	2490.6	1553.4	1100.7	2654.1	3.5955	1.8129	5.4084
336	609.15	13886.7	1.606	10.03	11.63	1544.9	938.9	2483.7	1567.2	1078.1	2645.3	3.6172	1.7700	5.3872
338	611.15	14242.3	1.622	9.58	11.20	1558.1	918.4	2476.4	1581.2	1054.8	2636.0	3.6392	1.7261	5.3653
340	613.15	14605.2	1.639	9.14	10.78	1571.5	897.2	2468.7	1595.5	1030.7	2626.2	3.6616	1.6811	5.3427
342	615.15	14975.5	1.657	8.71	10.37	1585.2	875.2	2460.5	1610.0	1005.7	2615.7	3.6844	1.6350	5.3194
344	617.15	15353.5	1.676	8.286	9.962	1599.2	852.5	2451.7	1624.9	979.7	2604.7	3.7075	1.5877	5.2952
346	619.15	15739.3	1.696	7.870	9.566	1613.5	828.9	2442.4	1640.2	952.8	2593.0	3.7311	1.5391	5.2702
348	621.15	16133.1	1.718	7.461	9.178	1628.1	804.5	2432.6	1655.8	924.8	2580.7	3.7553	1.4891	5.2444
350	623.15	16535.1	1.741	7.058	8.799	1643.0	779.2	2422.2	1671.8	895.9	2567.7	3.7801	1.4375	5.2177
352	625.15	16945.5	1.766	6.654	8.420	1659.4	751.5	2410.8	1689.3	864.2	2553.5	3.8071	1.3822	5.1893
354	627.15	17364.4	1.794	6.252	8.045	1676.3	722.4	2398.7	1707.5	830.9	2538.4	3.8349	1.3247	5.1596
356	629.15	17792.2	1.824	5.850	7.674	1693.4	692.2	2385.6	1725.9	796.2	2522.1	3.8629	1.2654	5.1283
358	631.15	18229.0	1.858	5.448	7.306	1710.8	660.5	2371.4	1744.7	759.9	2504.6	3.8915	1.2037	5.0953
360	633.15	18675.1	1.896	5.044	6.940	1728.8	627.1	2355.8	1764.2	721.3	2485.4	3.9210	1.1390	5.0600
361	634.15	18901.7	1.917	4.840	6.757	1738.0	609.5	2347.5	1774.2	701.0	2475.2	3.9362	1.1052	5.0414
362	635.15	19130.7	1.939	4.634	6.573	1747.5	591.2	2338.7	1784.6	679.8	2464.4	3.9518	1.0702	5.0220
363	636.15	19362.1	1.963	4.425	6.388	1757.3	572.1	2329.3	1795.3	657.8	2453.0	3.9679	1.0338	5.0017
364	637.15	19596.1	1.988	4.213	6.201	1767.4	552.0	2319.4	1806.4	634.6	2440.9	3.9846	0.9958	4.9804
365	638.15	19832.6	2.016	3.996	6.012	1778.0	530.8	2308.8	1818.0	610.0	2428.0	4.0021	0.9558	4.9579
366	639.15	20071.6	2.046	3.772	5.819	1789.1	508.2	2297.3	1830.2	583.9	2414.1	4.0205	0.9134	4.9339
367	640.15	20313.2	2.080	3.540	5.621	1801.0	483.8	2284.8	1843.2	555.7	2399.0	4.0401	0.8680	4.9081
368	641.15	20557.5	2.118	3.298	5.416	1813.8	457.5	2271.1	1857.3	525.1	2382.4	4.0613	0.8189	4.8801
369	642.15	20804.4	2.162	3.039	5.201	1827.8	427.9	2255.7	1872.8	491.1	2363.9	4.0846	0.7647	4.8492
370	643.15	21054.0	2.214	2.759	4.973	1843.6	394.5	2238.1	1890.2	452.6	2342.8	4.1108	0.7036	4.8144
371	644.15	21306.4	2.278	2.446	4.723	1862.0	355.3	2217.3	1910.5	407.4	2317.9	4.1414	0.6324	4.7738
372	645.15	21561.6	2.364	2.075	4.439	1884.6	306.6	2191.2	1935.6	351.4	2287.0	4.1794	0.5446	4.7240
373	646.15	21819.7	2.496	1.588	4.084	1916.0	238.9	2154.9	1970.5	273.5	2244.0	4.2325	0.4233	4.6559
374	647.15	22080.5	2.843	0.623	3.466	1983.9	95.7	2079.2	2046.7	109.5	2156.2	4.3493	0.1692	4.5185
374.15	647.30	22120.0	3.170	0.000	3.170	2037.3	0.0	2037.3	2107.4	0.0	2107.4	4.4429	0.0000	4.4429

Table F.2: Superheated Steam, SI units

TEMPERATURE: $t\,°C$
(TEMPERATURE: T kelvins)

P/kPa (t^{sat}/°C)		sat. liq.	sat. vap.	75 (348.15)	100 (373.15)	125 (398.15)	150 (423.15)	175 (448.15)	200 (473.15)	225 (498.15)	250 (523.15)
1 (6.98)	V	1.000	129200.	160640.	172180.	183720.	195270.	206810.	218350.	229890.	241430.
	U	29.334	2385.2	2480.8	2516.4	2552.3	2588.5	2624.9	2661.7	2698.8	2736.3
	H	29.335	2514.4	2641.5	2688.6	2736.0	2783.7	2831.7	2880.1	2928.7	2977.7
	S	0.1060	8.9767	9.3828	9.5136	9.6365	9.7527	9.8629	9.9679	10.0681	10.1641
10 (45.83)	V	1.010	14670.	16030.	17190.	18350.	19510.	20660.	21820.	22980.	24130.
	U	191.822	2438.0	2479.7	2515.6	2551.6	2588.0	2624.5	2661.4	2698.6	2736.1
	H	191.832	2584.8	2640.0	2687.5	2735.2	2783.1	2831.2	2879.6	2928.4	2977.4
	S	0.6493	8.1511	8.3168	8.4486	8.5722	8.6888	8.7994	8.9045	9.0049	9.1010
20 (60.09)	V	1.017	7649.8	8000.0	8584.7	9167.1	9748.0	10320.	10900.	11480.	12060.
	U	251.432	2456.9	2478.4	2514.6	2550.9	2587.4	2624.1	2661.0	2698.3	2735.8
	H	251.453	2609.9	2638.4	2686.3	2734.2	2782.3	2830.6	2879.2	2928.0	2977.1
	S	0.8321	7.9094	7.9933	8.1261	8.2504	8.3676	8.4785	8.5839	8.6844	8.7806
30 (69.12)	V	1.022	5229.3	5322.0	5714.4	6104.6	6493.2	6880.8	7267.5	7653.8	8039.7
	U	289.271	2468.6	2477.1	2513.6	2550.2	2586.8	2623.6	2660.7	2698.0	2735.6
	H	289.302	2625.4	2636.8	2685.1	2733.3	2781.6	2830.0	2878.7	2927.6	2976.8
	S	0.9441	7.7695	7.8024	7.9363	8.0614	8.1791	8.2903	8.3960	8.4967	8.5930
40 (75.89)	V	1.027	3993.4	......	4279.2	4573.3	4865.8	5157.2	5447.8	5738.0	6027.7
	U	317.609	2477.1	......	2512.6	2549.4	2586.2	2623.2	2660.3	2697.7	2735.4
	H	317.650	2636.9	......	2683.8	2732.3	2780.9	2829.5	2878.2	2927.2	2976.5
	S	1.0261	7.6709	......	7.8009	7.9268	8.0450	8.1566	8.2624	8.3633	8.4598
50 (81.35)	V	1.030	3240.2	......	3418.1	3654.5	3889.3	4123.0	4356.0	4588.5	4820.5
	U	340.513	2484.0	......	2511.7	2548.6	2585.6	2622.7	2659.9	2697.4	2735.1
	H	340.564	2646.0	......	2682.6	2731.4	2780.1	2828.9	2877.7	2926.8	2976.1
	S	1.0912	7.5947	......	7.6953	7.8219	7.9406	8.0526	8.1587	8.2598	8.3564
75 (91.79)	V	1.037	2216.9	......	2269.8	2429.4	2587.3	2744.2	2900.2	3055.8	3210.9
	U	384.374	2496.7	......	2509.2	2546.7	2584.2	2621.6	2659.0	2696.7	2734.5
	H	384.451	2663.0	......	2679.4	2728.9	2778.2	2827.4	2876.6	2925.8	2975.5
	S	1.2131	7.4570	......	7.5014	7.6300	7.7500	7.8629	7.9697	8.0712	8.1681
100 (99.63)	V	1.043	1693.7	......	1695.5	1816.7	1936.3	2054.7	2172.3	2289.4	2406.1
	U	417.406	2506.1	......	2506.6	2544.8	2582.7	2620.4	2658.1	2695.9	2733.9
	H	417.511	2675.4	......	2676.2	2726.5	2776.3	2825.9	2875.4	2924.9	2974.5
	S	1.3027	7.3598	......	7.3618	7.4923	7.6137	7.7275	7.8349	7.9369	8.0342

P, kPa (t, °C)		Sat. liq.	Sat. vap.								
101.325 (100.00)	V	1.044	1673.0	·····	1673.0	1792.7	1910.7	2027.7	2143.8	2259.3	2374.5
	U	418.959	2506.5	·····	2506.5	2544.7	2582.6	2620.4	2658.1	2695.9	2733.9
	H	419.064	2676.0	·····	2676.0	2726.4	2776.2	2825.8	2875.3	2924.8	2974.5
	S	1.3069	7.3554	·····	7.3554	7.4860	7.6075	7.7213	7.8288	7.9308	8.0280
125 (105.99)	V	1.049	1374.6	·····	·····	1449.1	1545.6	1641.0	1735.6	1829.6	1923.2
	U	444.224	2513.4	·····	·····	2542.9	2581.2	2619.3	2657.2	2695.2	2733.3
	H	444.356	2685.2	·····	·····	2724.0	2774.4	2824.4	2874.2	2923.9	2973.7
	S	1.3740	7.2847	·····	·····	7.3844	7.5072	7.6219	7.7300	7.8324	7.9300
150 (111.37)	V	1.053	1159.0	·····	·····	1204.0	1285.2	1365.2	1444.4	1523.0	1601.3
	U	466.968	2519.5	·····	·····	2540.9	2579.7	2618.1	2656.3	2694.4	2732.7
	H	467.126	2693.4	·····	·····	2721.5	2772.5	2822.9	2872.9	2922.9	2972.9
	S	1.4336	7.2234	·····	·····	7.2953	7.4194	7.5352	7.6439	7.7468	7.8447
175 (116.06)	V	1.057	1003.34	·····	·····	1028.8	1099.1	1168.2	1236.4	1304.1	1371.3
	U	486.815	2524.7	·····	·····	2538.9	2578.2	2616.9	2655.3	2693.7	2732.1
	H	487.000	2700.3	·····	·····	2719.0	2770.5	2821.3	2871.7	2921.9	2972.0
	S	1.4849	7.1716	·····	·····	7.2191	7.3447	7.4614	7.5708	7.6741	7.7724
200 (120.23)	V	1.061	885.44	·····	·····	897.47	959.54	1020.4	1080.4	1139.8	1198.9
	U	504.489	2529.2	·····	·····	2536.9	2576.6	2615.7	2654.4	2692.9	2731.4
	H	504.701	2706.3	·····	·····	2716.4	2768.5	2819.8	2870.5	2920.9	2971.2
	S	1.5301	7.1268	·····	·····	7.1523	7.2794	7.3971	7.5072	7.6110	7.7096
225 (123.99)	V	1.064	792.97	·····	·····	795.25	850.97	905.44	959.06	1012.1	1064.7
	U	520.465	2533.2	·····	·····	2534.8	2575.1	2614.5	2653.5	2692.1	2730.8
	H	520.705	2711.6	·····	·····	2713.8	2766.5	2818.2	2869.3	2919.9	2970.4
	S	1.5705	7.0873	·····	·····	7.0928	7.2213	7.3400	7.4508	7.5551	7.6540
250 (127.43)	V	1.068	718.44	·····	·····	·····	764.09	813.47	861.98	909.91	957.41
	U	535.077	2536.8	·····	·····	·····	2573.5	2613.3	2652.5	2691.4	2730.2
	H	535.343	2716.4	·····	·····	·····	2764.5	2816.7	2868.0	2918.9	2969.6
	S	1.6071	7.0520	·····	·····	·····	7.1689	7.2886	7.4001	7.5050	7.6042
275 (130.60)	V	1.071	657.04	·····	·····	·····	693.00	738.21	782.55	826.29	869.61
	U	548.564	2540.0	·····	·····	·····	2571.9	2612.1	2651.6	2690.7	2729.6
	H	548.858	2720.7	·····	·····	·····	2762.5	2815.1	2866.8	2917.9	2968.7
	S	1.6407	7.0201	·····	·····	·····	7.1211	7.2419	7.3541	7.4594	7.5590
300 (133.54)	V	1.073	605.56	·····	·····	·····	633.74	675.49	716.35	756.60	796.44
	U	561.107	2543.0	·····	·····	·····	2570.3	2610.8	2650.6	2689.9	2729.0
	H	561.429	2724.7	·····	·····	·····	2760.4	2813.5	2865.5	2916.9	2967.9
	S	1.6716	6.9909	·····	·····	·····	7.0771	7.1990	7.3119	7.4177	7.5176

Table F.2: Superheated Steam, SI units (Continued)

TEMPERATURE: t °C
(TEMPERATURE: T kelvins)

P/kPa (t^{sat}/°C)		sat. liq.	sat. vap.	300 (573.15)	350 (623.15)	400 (673.15)	450 (723.15)	500 (773.15)	550 (823.15)	600 (873.15)	650 (923.15)
1 (6.98)	V	1.000	129200.	264500.	287580.	310660.	333730.	356810.	379880.	402960.	426040.
	U	29.334	2385.2	2812.3	2889.9	2969.1	3049.9	3132.4	3216.7	3302.6	3390.3
	H	29.335	2514.4	3076.8	3177.5	3279.7	3383.6	3489.2	3596.5	3705.6	3816.4
	S	0.1060	8.9767	10.3450	10.5133	10.6711	10.8200	10.9612	11.0957	11.2243	11.3476
10 (45.83)	V	1.010	14670.	26440.	28750.	31060.	33370.	35670.	37980.	40290.	42600.
	U	191.822	2438.0	2812.2	2889.8	2969.0	3049.8	3132.3	3216.6	3302.6	3390.3
	H	191.832	2584.8	3076.6	3177.3	3279.6	3383.5	3489.1	3596.5	3705.5	3816.3
	S	0.6493	8.1511	9.2820	9.4504	9.6083	9.7572	9.8984	10.0329	10.1616	10.2849
20 (60.09)	V	1.017	7649.8	13210.	14370.	15520.	16680.	17830.	18990.	20140.	21300.
	U	251.432	2456.9	2812.0	2889.6	2968.9	3049.7	3132.3	3216.5	3302.5	3390.2
	H	251.453	2609.9	3076.4	3177.1	3279.4	3383.4	3489.0	3596.4	3705.4	3816.2
	S	0.8321	7.9094	8.9618	9.1303	9.2882	9.4372	9.5784	9.7130	9.8416	9.9650
30 (69.12)	V	1.022	5229.3	8810.8	9581.2	10350.	11120.	11890.	12660.	13430.	14190.
	U	289.271	2468.6	2811.8	2889.5	2968.7	3049.6	3132.2	3216.5	3302.5	3390.2
	H	289.302	2625.4	3076.1	3176.9	3279.3	3383.3	3488.9	3596.3	3705.4	3816.2
	S	0.9441	7.7695	8.7744	8.9430	9.1010	9.2499	9.3912	9.5257	9.6544	9.7778
40 (75.89)	V	1.027	3993.4	6606.5	7184.6	7762.5	8340.1	8917.6	9494.9	10070.	10640.
	U	317.609	2477.1	2811.6	2889.4	2968.6	3049.5	3132.1	3216.4	3302.4	3390.1
	H	317.650	2636.9	3075.9	3176.8	3279.1	3383.1	3488.8	3596.2	3705.3	3816.1
	S	1.0261	7.6709	8.6413	8.8100	8.9680	9.1170	9.2583	9.3929	9.5216	9.6450
50 (81.35)	V	1.030	3240.2	5283.9	5746.7	6209.1	6671.4	7133.5	7595.5	8057.4	8519.2
	U	340.513	2484.0	2811.5	2889.2	2968.5	3049.4	3132.0	3216.3	3302.3	3390.1
	H	340.564	2646.0	3075.7	3176.6	3279.0	3383.0	3488.7	3596.1	3705.2	3816.0
	S	1.0912	7.5947	8.5380	8.7068	8.8649	9.0139	9.1552	9.2898	9.4185	9.5419
75 (91.79)	V	1.037	2216.9	3520.5	3829.4	4138.0	4446.4	4754.7	5062.8	5370.9	5678.9
	U	384.374	2496.7	2811.0	2888.9	2968.2	3049.2	3131.8	3216.1	3302.2	3389.9
	H	384.451	2663.0	3075.1	3176.1	3278.6	3382.7	3488.4	3595.8	3705.0	3815.9
	S	1.2131	7.4570	8.3502	8.5191	8.6773	8.8265	8.9678	9.1025	9.2312	9.3546
100 (99.63)	V	1.043	1693.7	2638.7	2870.8	3102.5	3334.0	3565.3	3796.5	4027.7	4258.8
	U	417.406	2506.1	2810.6	2888.6	2968.0	3049.0	3131.6	3216.0	3302.0	3389.8
	H	417.511	2675.4	3074.5	3175.6	3278.2	3382.4	3488.1	3595.6	3704.8	3815.7
	S	1.3027	7.3598	8.2166	8.3858	8.5442	8.6934	8.8348	8.9695	9.0982	9.2217

p (kPa) (t_{sat})		Sat.									
101.325 (100.00)	V	1.044	1673.0	2604.2	2833.2	3061.9	3290.3	3518.7	3746.9	3975.0	4203.1
	U	418.959	2506.5	2810.6	2888.5	2968.0	3048.0	3131.6	3215.6	3302.0	3389.8
	H	419.064	2676.0	3074.4	3175.6	3278.2	3382.3	3488.1	3595.6	3704.8	3815.7
	S	1.3069	7.3554	8.2105	8.3797	8.5381	8.6873	8.8287	8.9634	9.0922	9.2156
125 (105.99)	V	1.049	1374.6	2109.7	2295.6	2481.2	2666.5	2851.7	3036.8	3221.8	3406.7
	U	444.224	2513.4	2810.2	2888.2	2967.7	3048.7	3131.4	3215.8	3301.9	3389.7
	H	444.356	2685.2	3073.9	3175.2	3277.8	3382.0	3487.9	3595.4	3704.6	3815.5
	S	1.3740	7.2847	8.1129	8.2823	8.4408	8.5901	8.7316	8.8663	8.9951	9.1186
150 (111.37)	V	1.053	1159.0	1757.0	1912.2	2066.9	2221.5	2375.9	2530.2	2684.5	2838.6
	U	466.968	2519.5	2809.7	2887.9	2967.4	3048.5	3131.6	3215.6	3301.7	3389.5
	H	467.126	2693.4	3073.3	3174.7	3277.5	3381.6	3487.6	3595.1	3704.2	3815.3
	S	1.4336	7.2234	8.0280	8.1976	8.3562	8.5056	8.6472	8.7819	8.9108	9.0343
175 (116.06)	V	1.057	1003.34	1505.1	1638.3	1771.1	1903.7	2036.1	2168.4	2300.7	2432.9
	U	486.815	2524.7	2809.3	2887.5	2967.1	3048.3	3131.0	3215.4	3301.6	3389.4
	H	487.000	2700.3	3072.7	3174.2	3277.1	3381.4	3487.3	3594.9	3704.0	3815.1
	S	1.4849	7.1716	7.9561	8.1259	8.2847	8.4341	8.5758	8.7106	8.8394	8.9630
200 (120.23)	V	1.061	885.44	1316.2	1432.8	1549.2	1665.3	1781.2	1897.1	2012.9	2128.6
	U	504.489	2529.2	2808.8	2887.2	2966.9	3048.0	3130.8	3215.3	3301.4	3389.2
	H	504.701	2706.3	3072.1	3173.8	3276.7	3381.1	3487.0	3594.7	3704.0	3815.0
	S	1.5301	7.1268	7.8937	8.0638	8.2226	8.3722	8.5139	8.6487	8.7776	8.9012
225 (123.99)	V	1.064	792.97	1169.2	1273.1	1376.6	1479.9	1583.0	1686.0	1789.0	1891.9
	U	520.465	2533.2	2808.4	2886.9	2966.6	3047.8	3130.6	3215.1	3301.3	3389.1
	H	520.705	2711.6	3071.5	3173.3	3276.3	3380.8	3486.8	3594.4	3703.6	3814.8
	S	1.5705	7.0873	7.8385	8.0088	8.1679	8.3175	8.4593	8.5942	8.7231	8.8467
250 (127.43)	V	1.068	718.44	1051.6	1145.2	1238.5	1331.5	1424.4	1517.2	1609.9	1702.5
	U	535.077	2536.8	2808.0	2886.5	2966.3	3047.6	3130.4	3214.9	3301.1	3389.0
	H	535.343	2716.4	3070.9	3172.8	3275.9	3380.4	3486.5	3594.2	3703.6	3814.6
	S	1.6071	7.0520	7.7891	7.9597	8.1188	8.2686	8.4104	8.5453	8.6743	8.7980
275 (130.60)	V	1.071	657.04	955.45	1040.7	1125.5	1210.2	1294.7	1379.0	1463.3	1547.6
	U	548.564	2540.0	2807.5	2886.2	2966.0	3047.3	3130.2	3214.7	3300.9	3388.8
	H	548.858	2720.7	3070.3	3172.4	3275.5	3380.1	3486.2	3594.0	3703.4	3814.4
	S	1.6407	7.0201	7.7444	7.9151	8.0744	8.2243	8.3661	8.5011	8.6301	8.7538
300 (133.54)	V	1.073	605.56	875.29	953.52	1031.4	1109.0	1186.5	1263.9	1341.2	1418.5
	U	561.107	2543.0	2807.1	2885.8	2965.8	3047.1	3130.0	3214.5	3300.8	3388.7
	H	561.429	2724.7	3069.7	3171.9	3275.2	3379.7	3486.0	3593.7	3703.2	3814.2
	S	1.6716	6.9909	7.7034	7.8744	8.0338	8.1838	8.3257	8.4608	8.5898	8.7135

Table F.2: Superheated Steam, SI units (Continued)

TEMPERATURE: $t\,°C$
(TEMPERATURE: T kelvins)

P/kPa ($t^{sat}/°C$)		sat. liq.	sat. vap.	150 (423.15)	175 (448.15)	200 (473.15)	220 (493.15)	240 (513.15)	260 (533.15)	280 (553.15)	300 (573.15)
325 (136.29)	V	1.076	561.75	583.58	622.41	660.33	690.22	719.81	749.18	778.39	807.47
	U	572.847	2545.7	2568.7	2609.6	2649.6	2681.2	2712.7	2744.0	2775.3	2806.6
	H	573.197	2728.3	2758.4	2811.9	2864.2	2905.6	2946.6	2987.5	3028.2	3069.0
	S	1.7004	6.9640	7.0363	7.1592	7.2729	7.3585	7.4400	7.5181	7.5933	7.6657
350 (138.87)	V	1.079	524.00	540.58	576.90	612.31	640.18	667.75	695.09	722.27	749.33
	U	583.892	2548.2	2567.1	2608.3	2648.6	2680.4	2712.0	2743.4	2774.8	2806.2
	H	584.270	2731.6	2756.3	2810.3	2863.0	2904.5	2945.7	2986.7	3027.6	3068.4
	S	1.7273	6.9392	6.9982	7.1222	7.2366	7.3226	7.4045	7.4828	7.5581	7.6307
375 (141.31)	V	1.081	491.13	503.29	537.46	570.69	596.81	622.62	648.22	673.64	698.94
	U	594.332	2550.6	2565.4	2607.1	2647.7	2679.6	2711.3	2742.8	2774.3	2805.7
	H	594.737	2734.7	2754.1	2808.6	2861.7	2903.4	2944.8	2985.9	3026.9	3067.8
	S	1.7526	6.9160	6.9624	7.0875	7.2027	7.2891	7.3713	7.4499	7.5254	7.5981
400 (143.62)	V	1.084	462.22	470.66	502.93	534.26	558.85	583.14	607.20	631.09	654.85
	U	604.237	2552.7	2563.7	2605.8	2646.7	2678.8	2710.6	2742.2	2773.7	2805.3
	H	604.670	2737.6	2752.7	2807.0	2860.4	2902.3	2943.9	2985.1	3026.2	3067.2
	S	1.7764	6.8943	6.9285	7.0548	7.1708	7.2576	7.3402	7.4190	7.4947	7.5675
425 (145.82)	V	1.086	436.61	441.85	472.47	502.12	525.36	548.30	571.01	593.54	615.95
	U	613.667	2554.8	2562.0	2604.5	2645.7	2678.0	2709.9	2741.6	2773.2	2804.8
	H	614.128	2740.3	2749.8	2805.3	2859.1	2901.2	2942.9	2984.3	3025.5	3066.6
	S	1.7990	6.8739	6.8965	7.0239	7.1407	7.2280	7.3108	7.3899	7.4657	7.5388
450 (147.92)	V	1.088	413.75	416.24	445.38	473.55	495.59	517.33	538.83	560.17	581.37
	U	622.672	2556.7	2560.3	2603.2	2644.7	2677.1	2709.2	2741.0	2772.7	2804.4
	H	623.162	2742.9	2747.7	2803.7	2857.8	2900.2	2942.0	2983.5	3024.8	3066.0
	S	1.8204	6.8547	6.8660	6.9946	7.1121	7.1999	7.2831	7.3624	7.4384	7.5116
475 (149.92)	V	1.091	393.22	393.31	421.14	447.97	468.95	489.62	510.05	530.30	550.43
	U	631.294	2558.5	2558.6	2601.9	2643.7	2676.3	2708.5	2740.4	2772.2	2803.9
	H	631.812	2745.3	2745.5	2802.0	2856.5	2899.1	2941.1	2982.7	3024.1	3065.4
	S	1.8408	6.8365	6.8369	6.9667	7.0850	7.1732	7.2567	7.3363	7.4125	7.4858
500 (151.84)	V	1.093	374.68	· · · · · · ·	399.31	424.96	444.97	464.67	484.14	503.43	522.58
	U	639.569	2560.2	· · · · · · ·	2600.6	2642.7	2675.5	2707.8	2739.8	2771.7	2803.5
	H	640.116	2747.5	· · · · · · ·	2800.3	2855.1	2898.0	2940.1	2981.9	3023.4	3064.8
	S	1.8604	6.8192		6.9400	7.0592	7.1478	7.2317	7.3115	7.3879	7.4614

P (T sat)		Sat. liq.	Sat. vap.								
525 (153.69)	V	1.095	357.84	· · ·	379.56	404.13	423.28	442.11	460.70	479.11	497.38
	U	647.528	2561.8	· · ·	2599.3	2641.6	2674.6	2707.1	2739.2	2771.2	2803.0
	H	648.103	2749.7	· · ·	2798.6	2853.8	2896.8	2939.2	2981.1	3022.7	3064.1
	S	1.8790	6.8027	· · ·	6.9145	7.0345	7.1236	7.2078	7.2879	7.3645	7.4381
550 (155.47)	V	1.097	342.48	· · ·	361.60	385.19	403.55	421.59	439.38	457.00	474.48
	U	655.199	2563.3	· · ·	2598.0	2640.6	2673.8	2706.4	2738.6	2770.6	2802.6
	H	655.802	2751.7	· · ·	2796.8	2852.5	2895.7	2938.3	2980.3	3022.0	3063.5
	S	1.8970	6.7870	· · ·	6.8900	7.0108	7.1004	7.1849	7.2653	7.3421	7.4158
575 (157.18)	V	1.099	328.41	· · ·	345.20	367.90	385.54	402.85	419.92	436.81	453.56
	U	662.603	2564.8	· · ·	2596.6	2639.6	2672.9	2705.7	2738.0	2770.1	2802.1
	H	663.235	2753.6	· · ·	2795.1	2851.1	2894.6	2937.3	2979.5	3021.3	3062.9
	S	1.9142	6.7720	· · ·	6.8664	6.9880	7.0781	7.1630	7.2436	7.3206	7.3945
600 (158.84)	V	1.101	315.47	· · ·	330.16	352.04	369.03	385.68	402.08	418.31	434.39
	U	669.762	2566.2	· · ·	2595.3	2638.5	2672.1	2705.0	2737.4	2769.6	2801.6
	H	670.423	2755.5	· · ·	2793.3	2849.7	2893.5	2936.4	2978.7	3020.6	3062.3
	S	1.9308	6.7575	· · ·	6.8437	6.9662	7.0567	7.1419	7.2228	7.3000	7.3740
625 (160.44)	V	1.103	303.54	· · ·	316.31	337.45	353.83	369.87	385.67	401.28	416.75
	U	676.695	2567.5	· · ·	2593.9	2637.5	2671.2	2704.2	2736.8	2769.1	2801.2
	H	677.384	2757.2	· · ·	2791.6	2848.4	2892.3	2935.4	2977.8	3019.9	3061.7
	S	1.9469	6.7437	· · ·	6.8217	6.9451	7.0361	7.1217	7.2028	7.2802	7.3544
650 (161.99)	V	1.105	292.49	· · ·	303.53	323.98	339.80	355.29	370.52	385.56	400.47
	U	683.417	2568.7	· · ·	2592.5	2636.4	2670.3	2703.5	2736.2	2768.5	2800.7
	H	684.135	2758.9	· · ·	2789.8	2847.0	2891.2	2934.4	2977.0	3019.2	3061.0
	S	1.9623	6.7304	· · ·	6.8004	6.9247	7.0162	7.1021	7.1835	7.2611	7.3355
675 (163.49)	V	1.106	282.23	· · ·	291.69	311.51	326.81	341.78	356.49	371.01	385.39
	U	689.943	2570.5	· · ·	2591.1	2635.4	2669.5	2702.8	2735.6	2768.0	2800.3
	H	690.689	2760.5	· · ·	2788.0	2845.6	2890.1	2933.5	2976.2	3018.5	3060.4
	S	1.9773	6.7176	· · ·	6.7798	6.9050	6.9970	7.0833	7.1650	7.2428	7.3173
700 (164.96)	V	1.108	272.68	· · ·	280.69	299.92	314.75	329.23	343.46	357.50	371.39
	U	696.285	2571.1	· · ·	2589.7	2634.3	2668.6	2702.1	2735.0	2767.5	2799.8
	H	697.061	2762.0	· · ·	2786.2	2844.2	2888.9	2932.5	2975.4	3017.7	3059.8
	S	1.9918	6.7052	· · ·	6.7598	6.8859	6.9784	7.0651	7.1470	7.2250	7.2997
725 (166.38)	V	1.110	263.77	· · ·	270.45	289.13	303.51	317.55	331.33	344.92	358.36
	U	702.457	2572.2	· · ·	2588.3	2633.2	2667.7	2701.3	2734.3	2767.0	2799.1
	H	703.261	2763.4	· · ·	2784.4	2842.8	2887.7	2931.5	2974.6	3017.0	3059.1
	S	2.0059	6.6932	· · ·	6.7404	6.8673	6.9604	7.0474	7.1296	7.2078	7.2827

Table F.2: Superheated Steam, SI units (Continued)

| | TEMPERATURE: $t\,^{\circ}C$ | | | | | | | | | |
| | (TEMPERATURE: T kelvins) | | | | | | | | | |

P/kPa ($t^{sat}/^{\circ}C$)		sat. liq.	sat. vap.	325 (598.15)	350 (623.15)	400 (673.15)	450 (723.15)	500 (773.15)	550 (823.15)	600 (873.15)	650 (923.15)
325 (136.29)	V	1.076	561.75	843.68	879.78	951.73	1023.5	1095.0	1166.5	1237.9	1309.2
	U	572.847	2545.4	2845.9	2885.5	2965.5	3046.9	3129.8	3214.4	3300.6	3388.6
	H	573.197	2728.3	3120.1	3171.4	3274.8	3379.5	3485.7	3593.5	3702.9	3814.1
	S	1.7004	6.9640	7.7530	7.8369	7.9965	8.1465	8.2885	8.4236	8.5527	8.6764
350 (138.87)	V	1.079	524.00	783.01	816.57	883.45	950.11	1016.6	1083.0	1149.3	1215.6
	U	583.892	2548.2	2845.6	2885.1	2965.2	3046.6	3129.6	3214.2	3300.5	3388.4
	H	584.270	2731.6	3119.6	3170.9	3274.4	3379.2	3485.4	3593.3	3702.7	3813.9
	S	1.7273	6.9392	7.7181	7.8022	7.9619	8.1120	8.2540	8.3892	8.5183	8.6421
375 (141.31)	V	1.081	491.13	730.42	761.79	824.28	886.54	948.66	1010.7	1072.6	1134.5
	U	594.332	2550.6	2845.2	2884.8	2964.9	3046.4	3129.4	3214.0	3300.3	3388.3
	H	594.737	2734.7	3119.1	3170.5	3274.0	3378.8	3485.1	3593.0	3702.5	3813.7
	S	1.7526	6.9160	7.6856	7.7698	7.9296	8.0798	8.2219	8.3571	8.4863	8.6101
400 (143.62)	V	1.084	462.22	684.41	713.85	772.50	830.92	889.19	947.35	1005.4	1063.4
	U	604.237	2552.7	2844.8	2884.5	2964.6	3046.2	3129.2	3213.8	3300.2	3388.2
	H	604.670	2737.6	3118.5	3170.0	3273.6	3378.5	3484.9	3592.8	3702.3	3813.5
	S	1.7764	6.8943	7.6552	7.7395	7.8994	8.0497	8.1919	8.3271	8.4563	8.5802
425 (145.82)	V	1.086	436.61	643.81	671.56	726.81	781.84	836.72	891.49	946.17	1000.8
	U	613.667	2554.8	2844.4	2884.1	2964.4	3045.9	3129.0	3213.7	3300.0	3388.0
	H	614.128	2740.3	3118.0	3169.5	3273.3	3378.2	3484.6	3592.5	3702.1	3813.4
	S	1.7990	6.8739	7.6265	7.7109	7.8710	8.0214	8.1636	8.2989	8.4282	8.5520
450 (147.92)	V	1.088	413.75	607.73	633.97	686.20	738.21	790.07	841.83	893.50	945.10
	U	622.672	2556.7	2844.0	2883.8	2964.1	3045.7	3128.8	3213.5	3299.8	3387.9
	H	623.162	2742.9	3117.5	3169.1	3272.9	3377.9	3484.3	3592.3	3701.9	3813.2
	S	1.8204	6.8547	7.5995	7.6840	7.8442	7.9947	8.1370	8.2723	8.4016	8.5255
475 (149.92)	V	1.091	393.22	575.44	600.33	649.87	699.18	748.34	797.40	846.37	895.27
	U	631.294	2558.5	2843.4	2883.4	2963.6	3045.4	3128.6	3213.3	3299.7	3387.7
	H	631.812	2745.3	3116.9	3168.6	3272.5	3377.6	3484.0	3592.1	3701.7	3813.0
	S	1.8408	6.8365	7.5739	7.6585	7.8189	7.9694	8.1118	8.2472	8.3765	8.5004
500 (151.84)	V	1.093	374.68	546.38	570.05	617.16	664.05	710.78	757.41	803.95	850.42
	U	639.569	2560.2	2843.2	2883.1	2963.5	3045.2	3128.4	3213.1	3299.5	3387.6
	H	640.116	2747.5	3116.4	3168.1	3272.1	3377.2	3483.8	3591.8	3701.5	3812.8
	S	1.8604	6.8192	7.5496	7.6343	7.7948	7.9454	8.0879	8.2233	8.3526	8.4766

P (Sat. temp)											
525 (153.69)	V	1.095	357.84	520.08	542.66	587.58	632.26	676.80	721.23	765.57	809.85
	U	647.528	2561.8	2842.8	2882.7	2963.2	3045.0	3128.2	3213.0	3299.4	3387.5
	H	648.103	2749.7	3115.9	3167.6	3271.7	3376.9	3483.5	3591.6	3701.3	3812.6
	S	1.8790	6.8027	7.5264	7.6112	7.7719	7.9226	8.0651	8.2006	8.3299	8.4539
550 (155.47)	V	1.097	342.48	496.18	517.76	560.68	603.37	645.91	688.34	730.68	772.96
	U	655.199	2563.3	2842.4	2882.4	2963.0	3044.7	3128.0	3212.8	3299.2	3387.3
	H	655.802	2751.7	3115.3	3167.2	3271.3	3376.6	3483.2	3591.4	3701.1	3812.5
	S	1.8970	6.7870	7.5043	7.5892	7.7500	7.9008	8.0433	8.1789	8.3083	8.4323
575 (157.18)	V	1.099	328.41	474.36	495.03	536.12	576.98	617.70	658.30	698.83	739.28
	U	662.603	2564.8	2842.0	2882.1	2962.7	3044.5	3127.8	3212.6	3299.1	3387.2
	H	663.235	2753.6	3114.8	3166.7	3271.0	3376.3	3482.9	3591.1	3700.9	3812.3
	S	1.9142	6.7720	7.4831	7.5681	7.7290	7.8799	8.0226	8.1581	8.2876	8.4116
600 (158.84)	V	1.101	315.47	454.35	474.19	513.61	552.80	591.84	630.78	669.63	708.41
	U	669.762	2566.2	2841.6	2881.7	2962.4	3044.3	3127.6	3212.4	3298.9	3387.1
	H	670.423	2755.5	3114.3	3166.2	3270.6	3376.0	3482.7	3590.9	3700.7	3812.1
	S	1.9308	6.7575	7.4628	7.5479	7.7090	7.8600	8.0027	8.1383	8.2678	8.3919
625 (160.44)	V	1.103	303.54	435.94	455.01	492.89	530.55	568.05	605.45	642.76	680.01
	U	676.695	2567.5	2841.2	2881.4	2962.1	3044.0	3127.4	3212.2	3298.8	3386.9
	H	677.384	2757.2	3113.7	3165.7	3270.2	3375.6	3482.4	3590.7	3700.5	3811.9
	S	1.9469	6.7437	7.4433	7.5285	7.6897	7.8408	7.9836	8.1192	8.2488	8.3729
650 (161.99)	V	1.105	292.49	418.95	437.31	473.78	510.01	546.10	582.07	617.96	653.79
	U	683.417	2568.7	2840.9	2881.0	2961.8	3043.8	3127.2	3212.1	3298.6	3386.8
	H	684.135	2758.9	3113.2	3165.3	3269.8	3375.3	3482.1	3590.4	3700.3	3811.8
	S	1.9623	6.7304	7.4245	7.5099	7.6712	7.8224	7.9652	8.1009	8.2305	8.3546
675 (163.49)	V	1.106	282.23	403.22	420.92	456.07	491.00	525.77	560.43	595.00	629.51
	U	689.943	2570.0	2840.5	2880.7	2961.6	3043.6	3127.0	3211.9	3298.5	3386.7
	H	690.689	2760.5	3112.6	3164.8	3269.4	3375.0	3481.8	3590.2	3700.1	3811.6
	S	1.9773	6.7176	7.4064	7.4919	7.6534	7.8046	7.9475	8.0833	8.2129	8.3371
700 (164.96)	V	1.108	272.68	388.61	405.71	439.64	473.34	506.89	540.33	573.68	606.97
	U	696.285	2571.1	2840.1	2880.3	2961.3	3043.3	3126.8	3211.7	3298.3	3386.5
	H	697.061	2762.0	3112.1	3164.3	3269.0	3374.7	3481.6	3589.9	3699.9	3811.4
	S	1.9918	6.7052	7.3890	7.4745	7.6362	7.7875	7.9305	8.0663	8.1959	8.3201
725 (166.38)	V	1.110	263.77	375.01	391.54	424.33	456.90	489.31	521.61	553.83	585.99
	U	702.457	2572.2	2839.7	2880.0	2961.0	3043.1	3126.6	3211.5	3298.1	3386.4
	H	703.261	2763.4	3111.5	3163.8	3268.7	3374.3	3481.3	3589.7	3699.7	3811.2
	S	2.0059	6.6932	7.3721	7.4578	7.6196	7.7710	7.9140	8.0499	8.1796	8.3038

Table F.2: Superheated Steam, SI units (Continued)

TEMPERATURE: t °C
(TEMPERATURE: T kelvins)

P/kPa (t^{sat}/°C)		sat. liq.	sat. vap.	175 (448.15)	200 (473.15)	220 (493.15)	240 (513.15)	260 (533.15)	280 (553.15)	300 (573.15)	325 (598.15)
750 (167.76)	V	1.112	255.43	260.88	279.05	293.03	306.65	320.01	333.17	346.19	362.32
	U	708.467	2573.3	2586.9	2632.1	2666.8	2700.6	2733.7	2766.4	2798.9	2839.3
	H	709.301	2764.8	2782.5	2841.4	2886.6	2930.6	2973.7	3016.3	3058.5	3111.0
	S	2.0195	6.6817	6.7215	6.8494	6.9429	7.0303	7.1128	7.1912	7.2662	7.3558
775 (169.10)	V	1.113	247.61	251.93	269.63	283.22	296.45	309.41	322.19	334.81	350.44
	U	714.326	2574.3	2585.4	2631.0	2665.9	2699.8	2733.1	2765.9	2798.4	2838.9
	H	715.189	2766.2	2780.7	2840.0	2885.4	2929.6	2972.9	3015.6	3057.9	3110.5
	S	2.0328	6.6705	6.7031	6.8319	6.9259	7.0137	7.0965	7.1751	7.2502	7.3400
800 (170.41)	V	1.115	240.26	243.53	260.79	274.02	286.88	299.48	311.89	324.14	339.31
	U	720.043	2575.5	2584.0	2629.9	2665.0	2699.1	2732.5	2765.4	2797.9	2838.5
	H	720.935	2767.5	2778.8	2838.6	2884.2	2928.6	2972.1	3014.9	3057.3	3109.9
	S	2.0457	6.6596	6.6851	6.8148	6.9094	6.9976	7.0807	7.1595	7.2348	7.3247
825 (171.69)	V	1.117	233.34	235.64	252.48	265.37	277.90	290.15	302.21	314.12	328.85
	U	725.625	2576.2	2582.5	2628.8	2664.1	2698.4	2731.8	2764.8	2797.5	2838.1
	H	726.547	2768.7	2776.9	2837.1	2883.1	2927.6	2971.2	3014.1	3056.6	3109.4
	S	2.0583	6.6491	6.6675	6.7982	6.8933	6.9819	7.0653	7.1443	7.2197	7.3098
850 (172.94)	V	1.118	226.81	228.21	244.66	257.24	269.44	281.37	293.10	304.68	319.00
	U	731.080	2577.1	2581.1	2627.7	2663.2	2697.6	2731.2	2764.3	2797.0	2837.7
	H	732.031	2769.9	2775.1	2835.7	2881.9	2926.6	2970.4	3013.4	3056.0	3108.8
	S	2.0705	6.6388	6.6504	6.7820	6.8777	6.9666	7.0503	7.1295	7.2051	7.2954
875 (174.16)	V	1.120	220.65	221.20	237.29	249.56	261.46	273.09	284.51	295.79	309.72
	U	736.415	2578.0	2579.6	2626.6	2662.3	2696.8	2730.6	2763.7	2796.5	2837.3
	H	737.394	2771.0	2773.1	2834.2	2880.7	2925.6	2969.5	3012.7	3055.3	3108.3
	S	2.0825	6.6289	6.6336	6.7662	6.8624	6.9518	7.0357	7.1152	7.1909	7.2813
900 (175.36)	V	1.121	214.81	. . . . .	230.32	242.31	253.93	265.27	276.40	287.39	300.96
	U	741.635	2578.8	. . . . .	2625.5	2661.4	2696.1	2729.9	2763.2	2796.1	2836.9
	H	742.644	2772.1	. . . . .	2832.7	2879.5	2924.6	2968.7	3012.0	3054.7	3107.7
	S	2.0941	6.6192		6.7508	6.8475	6.9373	7.0215	7.1012	7.1771	7.2676
925 (176.53)	V	1.123	209.28	. . . . .	223.73	235.46	246.80	257.87	268.73	279.44	292.66
	U	746.746	2579.6	. . . . .	2624.3	2660.5	2695.3	2729.3	2762.6	2795.6	2836.5
	H	747.784	2773.2	. . . . .	2831.3	2878.3	2923.6	2967.8	3011.2	3054.1	3107.2
	S	2.1055	6.6097		6.7357	6.8329	6.9231	7.0076	7.0875	7.1636	7.2543

Abs. Press. kPa (Sat. Temp.)		Sat.								
950 (177.67)	V	1.124	204.03	. . . . .	217.48	228.96	240.05	250.86	261.46	271.91	284.81
	U	751.754	2580.4	. . . . .	2623.2	2659.5	2694.6	2728.7	2762.1	2795.1	2836.0
	H	752.822	2774.2	. . . . .	2829.8	2877.0	2922.6	2967.0	3010.5	3053.4	3106.6
	S	2.1166	6.6005	. . . . .	6.7209	6.8187	6.9093	6.9941	7.0742	7.1505	7.2413
975 (178.79)	V	1.126	199.04	. . . . .	211.55	222.79	233.64	244.20	254.56	264.76	277.35
	U	756.663	2581.1	. . . . .	2622.0	2658.6	2693.8	2728.0	2761.5	2794.6	2835.6
	H	757.761	2775.2	. . . . .	2828.3	2875.8	2921.6	2966.1	3009.7	3052.8	3106.1
	S	2.1275	6.5916	. . . . .	6.7064	6.8048	6.8958	6.9809	7.0612	7.1377	7.2286
1000 (179.88)	V	1.127	194.29	. . . . .	205.92	216.93	227.55	237.89	248.01	257.98	270.27
	U	761.478	2581.9	. . . . .	2620.9	2657.7	2693.0	2727.4	2761.0	2794.2	2835.2
	H	762.605	2776.2	. . . . .	2826.8	2874.6	2920.6	2965.2	3009.0	3052.1	3105.5
	S	2.1382	6.5828	. . . . .	6.6922	6.7911	6.8825	6.9680	7.0485	7.1251	7.2163
1050 (182.02)	V	1.130	185.45	. . . . .	195.45	206.04	216.24	226.15	235.84	245.37	257.12
	U	770.843	2583.3	. . . . .	2618.5	2655.8	2691.5	2726.1	2759.9	2793.2	2834.4
	H	772.029	2778.0	. . . . .	2823.8	2872.1	2918.5	2963.5	3007.5	3050.8	3104.4
	S	2.1588	6.5659	. . . . .	6.6645	6.7647	6.8569	6.9430	7.0240	7.1009	7.1924
1100 (184.07)	V	1.133	177.38	. . . . .	185.92	196.14	205.96	215.47	224.77	233.91	245.16
	U	779.878	2584.5	. . . . .	2616.2	2653.9	2689.9	2724.7	2758.8	2792.2	2833.6
	H	781.124	2779.7	. . . . .	2820.7	2869.6	2916.4	2961.8	3006.0	3049.6	3103.3
	S	2.1786	6.5497	. . . . .	6.6379	6.7392	6.8323	6.9190	7.0005	7.0778	7.1695
1150 (186.05)	V	1.136	169.99	. . . . .	177.22	187.10	196.56	205.73	214.67	223.44	234.25
	U	788.611	2585.8	. . . . .	2613.8	2651.9	2688.3	2723.4	2757.7	2791.3	2832.8
	H	789.917	2781.3	. . . . .	2817.6	2867.1	2914.4	2960.0	3004.5	3048.2	3102.2
	S	2.1977	6.5342	. . . . .	6.6122	6.7147	6.8086	6.8959	6.9779	7.0556	7.1476
1200 (187.96)	V	1.139	163.20	. . . . .	169.23	178.80	187.95	196.79	205.40	213.85	224.24
	U	797.064	2586.9	. . . . .	2611.3	2650.0	2686.7	2722.1	2756.5	2790.3	2832.0
	H	798.430	2782.7	. . . . .	2814.4	2864.5	2912.2	2958.2	3003.0	3046.9	3101.0
	S	2.2161	6.5194	. . . . .	6.5872	6.6909	6.7858	6.8738	6.9562	7.0342	7.1266
1250 (189.81)	V	1.141	156.93	. . . . .	161.88	171.17	180.02	188.56	196.88	205.02	215.03
	U	805.259	2588.0	. . . . .	2608.9	2648.0	2685.1	2720.8	2755.4	2789.3	2831.1
	H	806.685	2784.1	. . . . .	2811.2	2861.9	2910.1	2956.5	3001.5	3045.6	3099.9
	S	2.2338	6.5050	. . . . .	6.5630	6.6680	6.7637	6.8523	6.9353	7.0136	7.1064
1300 (191.61)	V	1.144	151.13	. . . . .	155.09	164.11	172.70	180.97	189.01	196.87	206.53
	U	813.213	2589.0	. . . . .	2606.4	2646.0	2683.5	2719.4	2754.3	2788.4	2830.3
	H	814.700	2785.4	. . . . .	2808.0	2859.3	2908.0	2954.7	3000.0	3044.3	3098.8
	S	2.2510	6.4913	. . . . .	6.5394	6.6457	6.7424	6.8316	6.9151	6.9938	7.0869

Table F.2: Superheated Steam, SI units (Continued)

TEMPERATURE: $t\,°C$
(TEMPERATURE: T kelvins)

P/kPa ($t^{sat}/°C$)		sat. liq.	sat. vap.	350 (623.15)	375 (648.15)	400 (673.15)	450 (723.15)	500 (773.15)	550 (833.15)	600 (873.15)	650 (923.15)
750	V	1.112	255.43	378.31	394.22	410.05	441.55	472.90	504.15	535.30	566.40
(167.76)	U	708.467	2573.3	2879.6	2920.1	2960.7	3042.9	3126.3	3211.4	3298.0	3386.2
	H	709.301	2764.8	3163.4	3215.7	3268.3	3374.0	3481.0	3589.5	3699.5	3811.0
	S	2.0195	6.6817	7.4416	7.5240	7.6035	7.7550	7.8981	8.0340	8.1637	8.2880
775	V	1.113	247.61	365.94	381.35	396.69	427.20	457.56	487.81	517.97	548.07
(169.10)	U	714.326	2574.3	2879.3	2919.8	2960.4	3042.6	3126.1	3211.2	3297.8	3386.1
	H	715.189	2766.2	3162.9	3215.3	3267.9	3373.7	3480.8	3589.2	3699.3	3810.9
	S	2.0328	6.6705	7.4259	7.5084	7.5880	7.7396	7.8827	8.0187	8.1484	8.2727
800	V	1.115	240.26	354.34	369.29	384.16	413.74	443.17	472.49	501.72	530.89
(170.41)	U	720.043	2575.3	2878.9	2919.5	2960.2	3042.4	3125.9	3211.0	3297.7	3386.0
	H	720.935	2767.5	3162.4	3214.9	3267.5	3373.4	3480.5	3589.0	3699.1	3810.7
	S	2.0457	6.6596	7.4107	7.4932	7.5729	7.7246	7.8678	8.0038	8.1336	8.2579
825	V	1.117	233.34	343.45	357.96	372.39	401.10	429.65	458.10	486.46	514.76
(171.69)	U	725.625	2576.2	2878.6	2919.1	2959.9	3042.2	3125.7	3210.8	3297.5	3385.8
	H	726.547	2768.7	3161.9	3214.5	3267.1	3373.1	3480.2	3588.8	3698.8	3810.5
	S	2.0583	6.6491	7.3959	7.4786	7.5583	7.7101	7.8533	7.9894	8.1192	8.2436
850	V	1.118	226.81	333.20	347.29	361.31	389.20	416.93	444.56	472.09	499.57
(172.94)	U	731.080	2577.1	2878.2	2918.8	2959.6	3041.9	3125.5	3210.7	3297.4	3385.7
	H	732.031	2769.9	3161.4	3214.0	3266.7	3372.7	3479.9	3588.5	3698.6	3810.3
	S	2.0705	6.6388	7.3815	7.4643	7.5441	7.6960	7.8393	7.9754	8.1053	8.2296
875	V	1.120	220.65	323.53	337.24	350.87	377.98	404.94	431.79	458.55	485.25
(174.16)	U	736.415	2578.0	2877.9	2918.5	2959.3	3041.7	3125.3	3210.5	3297.2	3385.6
	H	737.394	2771.0	3161.0	3213.6	3266.3	3372.4	3479.7	3588.3	3698.4	3810.2
	S	2.0825	6.6289	7.3676	7.4504	7.5303	7.6823	7.8257	7.9618	8.0917	8.2161
900	V	1.121	214.81	314.40	327.74	341.01	367.39	393.61	419.73	445.76	471.72
(175.36)	U	741.635	2578.8	2877.5	2918.2	2959.0	3041.4	3125.1	3210.3	3297.1	3385.4
	H	742.644	2772.1	3160.5	3213.2	3266.0	3372.1	3479.4	3588.1	3698.2	3810.0
	S	2.0941	6.6192	7.3540	7.4370	7.5169	7.6689	7.8124	7.9486	8.0785	8.2030
925	V	1.123	209.28	305.76	318.75	331.68	357.36	382.90	408.32	433.66	458.93
(176.53)	U	746.746	2579.6	2877.2	2917.9	2958.8	3041.2	3124.9	3210.1	3296.9	3385.3
	H	747.784	2773.2	3160.0	3212.7	3265.6	3371.8	3479.1	3587.8	3698.0	3809.8
	S	2.1055	6.6097	7.3408	7.4238	7.5038	7.6560	7.7995	7.9357	8.0657	8.1902

P (kPa) (T_sat)		Sat.									
950 (177.67)	V	1.124	204.03	297.57	310.24	322.84	347.87	372.74	397.51	422.19	446.81
	U	751.754	2580.4	2876.8	2917.6	2958.5	3041.0	3124.7	3209.9	3296.7	3385.1
	H	752.822	2774.2	3159.5	3212.3	3265.2	3371.5	3478.8	3587.6	3697.8	3809.6
	S	2.1166	6.6005	7.3279	7.4110	7.4911	7.6433	7.7869	7.9232	8.0532	8.1777
975 (178.79)	V	1.126	199.04	289.81	302.17	314.45	338.86	363.11	387.26	411.32	435.31
	U	756.663	2581.1	2876.5	2917.3	2958.2	3040.7	3124.5	3209.8	3296.6	3385.0
	H	757.761	2775.2	3159.0	3211.9	3264.8	3371.1	3478.6	3587.3	3697.6	3809.4
	S	2.1275	6.5916	7.3154	7.3986	7.4787	7.6310	7.7747	7.9110	8.0410	8.1656
1000 (179.88)	V	1.127	194.29	282.43	294.50	306.49	330.30	353.96	377.52	400.98	424.38
	U	761.478	2581.9	2876.1	2917.0	2957.9	3040.5	3124.3	3209.6	3296.4	3384.9
	H	762.605	2776.2	3158.5	3211.5	3264.4	3370.8	3478.3	3587.1	3697.4	3809.3
	S	2.1382	6.5828	7.3031	7.3864	7.4665	7.6190	7.7627	7.8991	8.0292	8.1537
1050 (182.02)	V	1.130	185.45	268.74	280.25	291.69	314.41	336.97	359.43	381.79	404.10
	U	770.843	2583.3	2875.4	2916.3	2957.4	3040.0	3123.9	3209.2	3296.1	3384.6
	H	772.029	2778.0	3157.6	3210.6	3263.6	3370.2	3477.7	3586.6	3697.0	3808.9
	S	2.1588	6.5659	7.2795	7.3629	7.4432	7.5958	7.7397	7.8762	8.0063	8.1309
1100 (184.07)	V	1.133	177.38	256.28	267.30	278.24	299.96	321.53	342.98	364.35	385.65
	U	779.878	2584.5	2874.7	2915.7	2956.8	3039.6	3123.5	3208.9	3295.8	3384.3
	H	781.124	2779.7	3156.6	3209.7	3262.9	3369.5	3477.2	3586.2	3696.6	3808.5
	S	2.1786	6.5497	7.2569	7.3405	7.4209	7.5737	7.7177	7.8543	7.9845	8.1092
1150 (186.05)	V	1.136	169.99	244.91	255.47	265.96	286.77	307.42	327.97	348.42	368.81
	U	788.611	2585.8	2874.0	2915.1	2956.2	3039.1	3123.1	3208.5	3295.5	3384.1
	H	789.917	2781.3	3155.6	3208.9	3262.1	3368.9	3476.6	3585.7	3696.2	3808.2
	S	2.1977	6.5342	7.2352	7.3190	7.3995	7.5525	7.6966	7.8333	7.9636	8.0883
1200 (187.96)	V	1.139	163.20	234.49	244.63	254.70	274.68	294.50	314.20	333.82	353.38
	U	797.064	2586.9	2873.3	2914.4	2955.7	3038.6	3122.7	3208.2	3295.2	3383.8
	H	798.430	2782.7	3154.6	3208.0	3261.3	3368.2	3476.1	3585.2	3695.8	3807.8
	S	2.2161	6.5194	7.2144	7.2983	7.3790	7.5323	7.6765	7.8132	7.9436	8.0684
1250 (189.81)	V	1.141	156.93	224.90	234.66	244.35	263.55	282.60	301.54	320.39	339.18
	U	805.259	2588.0	2872.5	2913.8	2955.1	3038.1	3122.3	3207.8	3294.9	3383.5
	H	806.685	2784.1	3153.7	3207.1	3260.5	3367.6	3475.5	3584.7	3695.4	3807.5
	S	2.2338	6.5050	7.1944	7.2785	7.3593	7.5128	7.6571	7.7940	7.9244	8.0493
1300 (191.61)	V	1.144	151.13	216.05	225.46	234.79	253.28	271.62	289.85	307.99	326.07
	U	813.213	2589.0	2871.8	2913.2	2954.5	3037.7	3121.9	3207.5	3294.6	3383.2
	H	814.700	2785.4	3152.7	3206.3	3259.7	3366.9	3475.0	3584.3	3695.0	3807.1
	S	2.2510	6.4913	7.1751	7.2594	7.3404	7.4940	7.6385	7.7754	7.9060	8.0309

Table F.2: Superheated Steam, SI units (Continued)

TEMPERATURE: $t\,°C$
(TEMPERATURE: T kelvins)

P/kPa ($t^{sat}/°C$)		sat. liq.	sat. vap.	200 (473.15)	225 (498.15)	250 (523.15)	275 (548.15)	300 (573.15)	325 (598.15)	350 (623.15)	375 (648.15)
1350 (193.35)	V	1.146	145.74	148.79	159.70	169.96	179.79	189.33	198.66	207.85	216.93
	U	820.944	2589.9	2603.9	2653.6	2700.1	2744.4	2787.4	2829.5	2871.1	2912.5
	H	822.491	2786.6	2804.7	2869.2	2929.5	2987.1	3043.0	3097.7	3151.7	3205.4
	S	2.2676	6.4780	6.5165	6.6493	6.7675	6.8750	6.9746	7.0681	7.1566	7.2410
1400 (195.04)	V	1.149	140.72	142.94	153.57	163.55	173.08	182.32	191.35	200.24	209.02
	U	828.465	2590.8	2601.3	2651.7	2698.6	2743.2	2786.4	2828.6	2870.4	2911.9
	H	830.074	2787.8	2801.4	2866.7	2927.6	2985.5	3041.6	3096.5	3150.7	3204.5
	S	2.2837	6.4651	6.4941	6.6285	6.7477	6.8560	6.9561	7.0499	7.1386	7.2233
1450 (196.69)	V	1.151	136.04	137.48	147.86	157.57	166.83	175.79	184.54	193.15	201.65
	U	835.791	2591.6	2598.7	2649.7	2697.1	2742.0	2785.4	2827.8	2869.7	2911.3
	H	837.460	2788.9	2798.1	2864.1	2925.5	2983.9	3040.3	3095.4	3149.7	3203.6
	S	2.2993	6.4526	6.4722	6.6082	6.7286	6.8376	6.9381	7.0322	7.1212	7.2061
1500 (198.29)	V	1.154	131.66	132.38	142.53	151.99	161.00	169.70	178.19	186.53	194.77
	U	842.933	2592.4	2596.1	2647.7	2695.5	2740.8	2784.4	2826.9	2868.9	2910.6
	H	844.663	2789.9	2794.7	2861.5	2923.5	2982.3	3038.9	3094.2	3148.7	3202.8
	S	2.3145	6.4406	6.4508	6.5885	6.7099	6.8196	6.9207	7.0152	7.1044	7.1894
1550 (199.85)	V	1.156	127.55	127.61	137.54	146.77	155.54	164.00	172.25	180.34	188.33
	U	849.901	2593.2	2593.5	2645.8	2694.0	2739.5	2783.4	2826.1	2868.2	2910.0
	H	851.694	2790.8	2791.3	2858.9	2921.5	2980.6	3037.6	3093.1	3147.7	3201.9
	S	2.3292	6.4289	6.4298	6.5692	6.6917	6.8022	6.9038	6.9986	7.0081	7.1733
1600 (201.37)	V	1.159	123.69	......	132.85	141.87	150.42	158.66	166.68	174.54	182.30
	U	856.707	2593.8	......	2643.7	2692.4	2738.3	2782.4	2825.2	2867.5	2909.3
	H	858.561	2791.7	......	2856.3	2919.4	2979.0	3036.2	3091.9	3146.7	3201.0
	S	2.3436	6.4175	......	6.5503	6.6740	6.7852	6.8873	6.9825	7.0723	7.1577
1650 (202.86)	V	1.161	120.05	......	128.45	137.27	145.61	153.64	161.44	169.09	176.63
	U	863.359	2594.5	......	2641.7	2690.9	2737.1	2781.3	2824.4	2866.7	2908.7
	H	865.275	2792.6	......	2853.6	2917.4	2977.3	3034.8	3090.8	3145.7	3200.1
	S	2.3576	6.4065	......	6.5319	6.6567	6.7687	6.8713	6.9669	7.0569	7.1425
1700 (204.31)	V	1.163	116.62	......	124.31	132.94	141.09	148.91	156.51	163.96	171.30
	U	869.866	2595.1	......	2639.6	2689.3	2735.8	2780.3	2823.5	2866.0	2908.0
	H	871.843	2793.4	......	2851.0	2915.3	2975.6	3033.5	3089.6	3144.7	3199.2
	S	2.3713	6.3957	......	6.5138	6.6398	6.7526	6.8557	6.9516	7.0419	7.1277

P, kPa (T_sat)		Sat. Liquid	Sat. Vapor								
1750 (205.72)	V	1.166	113.38	.....	120.39	128.85	136.82	144.45	151.87	159.12	166.27
	U	876.234	2595.7	.....	2637.6	2687.7	2734.5	2779.3	2822.7	2865.3	2907.4
	H	878.274	2794.1	.....	2848.2	2913.2	2974.0	3032.1	3088.4	3143.7	3198.4
	S	2.3846	6.3853	.....	6.4961	6.6233	6.7368	6.8405	6.9368	7.0273	7.1133
1800 (207.11)	V	1.168	110.32	.....	116.69	124.99	132.78	140.24	147.48	154.55	161.51
	U	882.472	2596.3	.....	2635.5	2686.1	2733.3	2778.2	2821.8	2864.5	2906.7
	H	884.574	2794.8	.....	2845.5	2911.0	2972.3	3030.7	3087.3	3142.7	3197.5
	S	2.3976	6.3751	.....	6.4787	6.6071	6.7214	6.8257	6.9223	7.0131	7.0993
1850 (208.47)	V	1.170	107.41	.....	113.19	121.33	128.96	136.26	143.33	150.23	157.02
	U	888.585	2596.8	.....	2633.3	2684.4	2732.0	2777.2	2820.9	2863.8	2906.1
	H	890.750	2795.5	.....	2842.8	2908.9	2970.6	3029.3	3086.1	3141.7	3196.6
	S	2.4103	6.3651	.....	6.4616	6.5912	6.7064	6.8112	6.9082	6.9993	7.0856
1900 (209.80)	V	1.172	104.65	.....	109.87	117.87	125.35	132.49	139.39	146.14	152.76
	U	894.580	2597.3	.....	2631.2	2682.8	2730.7	2776.2	2820.1	2863.0	2905.4
	H	896.807	2796.1	.....	2840.0	2906.2	2968.8	3027.9	3084.9	3140.7	3195.7
	S	2.4228	6.3554	.....	6.4448	6.5757	6.6917	6.7970	6.8944	6.9857	7.0723
1950 (211.10)	V	1.174	102.031	.....	106.72	114.58	121.91	128.90	135.66	142.25	148.72
	U	900.461	2597.7	.....	2629.0	2681.1	2729.4	2775.1	2819.2	2862.3	2904.8
	H	902.752	2796.7	.....	2837.1	2904.6	2967.1	3026.5	3083.7	3139.7	3194.8
	S	2.4349	6.3459	.....	6.4283	6.5604	6.6772	6.7831	6.8809	6.9725	7.0593
2000 (212.37)	V	1.177	99.536	.....	103.72	111.45	118.65	125.50	132.11	138.56	144.89
	U	906.236	2598.2	.....	2626.9	2679.5	2728.1	2774.0	2818.3	2861.5	2904.1
	H	908.589	2797.2	.....	2834.3	2902.4	2965.4	3025.0	3082.5	3138.6	3193.9
	S	2.4469	6.3366	.....	6.4120	6.5454	6.6631	6.7696	6.8677	6.9596	7.0466
2100 (214.85)	V	1.181	94.890	.....	98.147	105.64	112.59	119.18	125.53	131.70	137.76
	U	917.479	2598.9	.....	2622.4	2676.1	2725.4	2771.9	2816.5	2860.0	2902.8
	H	919.959	2798.2	.....	2828.5	2897.9	2961.9	3022.2	3080.1	3136.6	3192.1
	S	2.4700	6.3187	.....	6.3802	6.5162	6.6356	6.7432	6.8422	6.9347	7.0220
2200 (217.24)	V	1.185	90.652	.....	93.067	100.35	107.07	113.43	119.53	125.47	131.28
	U	928.346	2599.6	.....	2617.9	2672.7	2722.7	2769.7	2814.7	2858.5	2901.5
	H	930.953	2799.1	.....	2822.7	2893.4	2958.3	3019.3	3077.7	3134.5	3190.3
	S	2.4922	6.3015	.....	6.3492	6.4879	6.6091	6.7179	6.8177	6.9107	6.9985
2300 (219.55)	V	1.189	86.769	.....	88.420	95.513	102.03	108.18	114.06	119.77	125.36
	U	938.866	2600.2	.....	2613.3	2669.2	2720.1	2767.6	2812.9	2857.0	2900.2
	H	941.601	2799.8	.....	2816.7	2888.9	2954.7	3016.4	3075.3	3132.4	3188.5
	S	2.5136	6.2849	.....	6.3190	6.4605	6.5835	6.6935	6.7941	6.8877	6.9759

Table F.2: Superheated Steam, SI units (Continued)

TEMPERATURE: t °C
(TEMPERATURE: T kelvins)

P/kPa (t^{sat}/°C)		sat. liq.	sat. vap.	400 (673.15)	425 (698.15)	450 (723.15)	475 (748.15)	500 (773.15)	550 (823.15)	600 (873.15)	650 (923.15)
1350 (193.35)	V	1.146	145.74	225.94	234.88	243.78	252.63	261.46	279.03	296.51	313.93
	U	820.944	2589.9	2953.9	2995.5	3037.2	3079.2	3121.5	3207.1	3294.3	3383.0
	H	822.491	2786.6	3259.0	3312.6	3366.3	3420.2	3474.4	3583.8	3694.5	3806.8
	S	2.2676	6.4780	7.3221	7.4003	7.4759	7.5493	7.6205	7.7576	7.8882	8.0132
1400 (195.04)	V	1.149	140.72	217.72	226.35	234.95	243.50	252.02	268.98	285.85	302.66
	U	828.465	2590.8	2953.4	2994.9	3036.7	3078.7	3121.1	3206.8	3293.9	3382.7
	H	830.074	2787.8	3258.2	3311.8	3365.6	3419.6	3473.9	3583.3	3694.1	3806.4
	S	2.2837	6.4651	7.3045	7.3828	7.4585	7.5319	7.6032	7.7404	7.8710	7.9961
1450 (196.69)	V	1.151	136.04	210.06	218.42	226.72	234.99	243.23	259.62	275.93	292.16
	U	835.791	2591.6	2952.8	2994.4	3036.2	3078.3	3120.7	3206.4	3293.6	3382.4
	H	837.460	2788.9	3257.4	3311.1	3365.0	3419.0	3473.3	3582.9	3693.7	3806.1
	S	2.2993	6.4526	7.2874	7.3658	7.4416	7.5151	7.5865	7.7237	7.8545	7.9796
1500 (198.29)	V	1.154	131.66	202.92	211.01	219.05	227.06	235.03	250.89	266.66	282.37
	U	842.933	2592.4	2952.2	2993.9	3035.8	3077.9	3120.3	3206.0	3293.3	3382.1
	H	844.663	2789.9	3256.6	3310.4	3364.3	3418.4	3472.8	3582.4	3693.3	3805.7
	S	2.3145	6.4406	7.2709	7.3494	7.4253	7.4989	7.5703	7.7077	7.8385	7.9636
1550 (199.85)	V	1.156	127.55	196.24	204.08	211.87	219.63	227.35	242.77	258.00	273.21
	U	849.901	2593.2	2951.7	2993.4	3035.3	3077.4	3119.8	3205.7	3293.0	3381.9
	H	851.694	2790.8	3255.8	3309.7	3363.7	3417.8	3472.2	3581.9	3692.9	3805.3
	S	2.3292	6.4289	7.2550	7.3336	7.4095	7.4832	7.5547	7.6921	7.8230	7.9482
1600 (201.37)	V	1.159	123.69	189.97	197.58	205.15	212.67	220.16	235.06	249.87	264.62
	U	856.707	2593.8	2951.1	2992.9	3034.8	3077.0	3119.4	3205.3	3292.7	3381.6
	H	858.561	2791.7	3255.0	3309.0	3363.0	3417.2	3471.7	3581.4	3692.5	3805.0
	S	2.3436	6.4175	7.2394	7.3182	7.3942	7.4679	7.5395	7.6770	7.8080	7.9333
1650 (202.86)	V	1.161	120.05	184.09	191.48	198.82	206.13	213.40	227.86	242.24	256.55
	U	863.359	2594.5	2950.5	2992.3	3034.3	3076.5	3119.0	3205.0	3292.4	3381.3
	H	865.275	2792.6	3254.2	3308.3	3362.4	3416.7	3471.1	3581.0	3692.1	3804.6
	S	2.3576	6.4065	7.2244	7.3032	7.3794	7.4531	7.5248	7.6624	7.7934	7.9188
1700 (204.31)	V	1.163	116.62	178.55	185.74	192.87	199.97	207.04	221.09	235.06	248.96
	U	869.866	2595.1	2949.8	2991.8	3033.9	3076.1	3118.6	3204.6	3292.1	3381.0
	H	871.843	2793.4	3253.5	3307.6	3361.7	3416.1	3470.6	3580.5	3691.7	3804.3
	S	2.3713	6.3957	7.2098	7.2887	7.3649	7.4388	7.5105	7.6482	7.7793	7.9047

689

P (Tsat)											
1750 (205.72)	V	1.166	113.38	173.32	180.32	187.26	194.17	201.04	214.71	228.28	241.80
	U	876.234	2595.7	2949.3	2991.3	3033.4	3075.7	3118.2	3204.3	3291.8	3380.8
	H	878.274	2794.1	3252.7	3306.9	3361.1	3415.5	3470.0	3580.0	3691.3	3803.9
	S	2.3846	6.3853	7.1955	7.2746	7.3509	7.4248	7.4965	7.6344	7.7656	7.8910
1800 (207.11)	V	1.168	110.32	168.39	175.20	181.97	188.69	195.38	208.68	221.89	235.03
	U	882.472	2596.3	2948.8	2990.8	3032.9	3075.2	3117.8	3203.9	3291.5	3380.5
	H	884.574	2794.8	3251.9	3306.1	3360.4	3414.9	3469.5	3579.5	3690.9	3803.6
	S	2.3976	6.3751	7.1816	7.2608	7.3372	7.4112	7.4830	7.6209	7.7522	7.8777
1850 (208.47)	V	1.170	107.41	163.73	170.37	176.96	183.50	190.02	202.97	215.84	228.64
	U	888.585	2596.8	2948.2	2990.3	3032.4	3074.8	3117.4	3203.6	3291.1	3380.2
	H	890.750	2795.5	3251.1	3305.4	3359.8	3414.3	3468.9	3579.1	3690.4	3803.2
	S	2.4103	6.3651	7.1681	7.2474	7.3239	7.3980	7.4698	7.6079	7.7392	7.8648
1900 (209.80)	V	1.172	104.65	159.30	165.78	172.21	178.59	184.94	197.57	210.11	222.58
	U	894.580	2597.3	2947.6	2989.7	3031.9	3074.3	3117.0	3203.2	3290.8	3380.0
	H	896.807	2796.1	3250.3	3304.7	3359.1	3413.7	3468.4	3578.6	3690.0	3802.8
	S	2.4228	6.3554	7.1550	7.2344	7.3109	7.3851	7.4570	7.5951	7.7265	7.8522
1950 (211.10)	V	1.174	102.031	155.11	161.43	167.70	173.93	180.13	192.44	204.67	216.83
	U	900.461	2597.7	2947.0	2989.2	3031.5	3073.9	3116.6	3202.9	3290.5	3379.7
	H	902.752	2796.7	3249.5	3304.0	3358.5	3413.1	3467.8	3578.1	3689.6	3802.5
	S	2.4349	6.3459	7.1421	7.2216	7.2983	7.3725	7.4445	7.5827	7.7142	7.8399
2000 (212.37)	V	1.177	99.536	151.13	157.30	163.42	169.51	175.55	187.57	199.50	211.36
	U	906.236	2598.2	2946.4	2988.7	3031.0	3073.5	3116.2	3202.5	3290.2	3379.4
	H	908.589	2797.2	3248.7	3303.3	3357.8	3412.5	3467.3	3577.6	3689.2	3802.1
	S	2.4469	6.3366	7.1296	7.2092	7.2859	7.3602	7.4323	7.5706	7.7022	7.8279
2100 (214.85)	V	1.181	94.890	143.73	149.63	155.48	161.28	167.06	178.53	189.91	201.22
	U	917.479	2598.9	2945.3	2987.6	3030.0	3072.6	3115.3	3201.8	3289.2	3378.9
	H	919.959	2798.2	3247.1	3301.8	3356.5	3411.3	3466.2	3576.7	3688.4	3801.4
	S	2.4700	6.3187	7.1053	7.1851	7.2621	7.3365	7.4087	7.5472	7.6789	7.8048
2200 (217.24)	V	1.185	90.652	137.00	142.65	148.25	153.81	159.34	170.30	181.19	192.00
	U	928.346	2599.6	2944.1	2986.6	3029.1	3071.7	3114.5	3201.1	3289.0	3378.3
	H	930.953	2799.1	3245.5	3300.4	3355.2	3410.1	3465.1	3575.7	3687.6	3800.7
	S	2.4922	6.3015	7.0821	7.1621	7.2393	7.3139	7.3862	7.5249	7.6568	7.7827
2300 (219.55)	V	1.189	86.769	130.85	136.28	141.65	146.99	152.28	162.80	173.22	183.58
	U	938.866	2600.2	2942.9	2985.5	3028.1	3070.8	3113.7	3200.4	3288.3	3377.8
	H	941.601	2799.8	3243.9	3299.0	3353.9	3408.9	3464.0	3574.8	3686.7	3800.0
	S	2.5136	6.2849	7.0598	7.1401	7.2174	7.2922	7.3646	7.5035	7.6355	7.7616

Table F.2: Superheated Steam, SI units (Continued)

TEMPERATURE: t °C
(TEMPERATURE: T kelvins)

P/kPa (t^{sat}/°C)		sat. liq.	sat. vap.	225 (498.15)	250 (523.15)	275 (548.15)	300 (573.15)	325 (598.15)	350 (623.15)	375 (648.15)	400 (673.15)
2400 (221.78)	V	1.193	83.199	84.149	91.075	97.411	103.36	109.05	114.55	119.93	125.22
	U	949.066	2600.7	2608.6	2665.6	2717.3	2765.4	2811.1	2855.4	2898.8	2941.7
	H	951.929	2800.4	2810.6	2884.2	2951.1	3013.4	3072.8	3130.4	3186.7	3242.3
	S	2.5343	6.2690	6.2894	6.4338	6.5586	6.6699	6.7714	6.8656	6.9542	7.0384
2500 (223.94)	V	1.197	79.905	80.210	86.985	93.154	98.925	104.43	109.75	114.94	120.04
	U	958.969	2601.2	2603.8	2662.0	2714.5	2763.1	2809.3	2853.9	2897.5	2940.6
	H	961.962	2800.9	2804.3	2879.5	2947.4	3010.4	3070.4	3128.2	3184.8	3240.7
	S	2.5543	6.2536	6.2604	6.4077	6.5345	6.6470	6.7494	6.8442	6.9333	7.0178
2600 (226.04)	V	1.201	76.856	. . . . . .	83.205	89.220	94.830	100.17	105.32	110.33	115.26
	U	968.597	2601.5	. . . . . .	2658.4	2711.7	2760.9	2807.4	2852.3	2896.1	2939.4
	H	971.720	2801.4	. . . . . .	2874.7	2943.6	3007.4	3067.9	3126.1	3183.0	3239.0
	S	2.5736	6.2387		6.3823	6.5110	6.6249	6.7281	6.8236	6.9131	6.9979
2700 (228.07)	V	1.205	74.025	. . . . . .	79.698	85.575	91.036	96.218	101.21	106.07	110.83
	U	977.968	2601.8	. . . . . .	2654.7	2708.8	2758.6	2805.6	2850.7	2894.8	2938.2
	H	981.222	2801.7	. . . . . .	2869.9	2939.8	3004.4	3065.4	3124.0	3181.2	3237.4
	S	2.5924	6.2244		6.3575	6.4882	6.6034	6.7075	6.8036	6.8935	6.9787
2800 (230.05)	V	1.209	71.389	. . . . . .	76.437	82.187	87.510	92.550	97.395	102.10	106.71
	U	987.100	2602.1	. . . . . .	2650.9	2705.9	2756.3	2803.7	2849.2	2893.4	2937.0
	H	990.485	2802.0	. . . . . .	2864.9	2936.0	3001.3	3062.8	3121.9	3179.5	3235.8
	S	2.6106	6.2104		6.3331	6.4659	6.5824	6.6875	6.7842	6.8746	6.9601
2900 (231.97)	V	1.213	68.928		73.395	79.029	84.226	89.133	93.843	98.414	102.88
	U	996.008	2602.3		2647.1	2702.9	2754.0	2801.8	2847.6	2892.0	2935.8
	H	999.524	2802.2		2859.9	2932.1	2998.2	3060.3	3119.7	3177.4	3234.1
	S	2.6283	6.1969		6.3092	6.4441	6.5621	6.6681	6.7654	6.8563	6.9421
3000 (233.84)	V	1.216	66.626		70.551	76.078	81.159	85.943	90.526	94.969	99.310
	U	1004.7	2602.4		2643.2	2700.0	2751.6	2799.9	2846.0	2890.7	2934.6
	H	1008.4	2802.3		2854.8	2928.2	2995.1	3057.7	3117.5	3175.6	3232.5
	S	2.6455	6.1837		6.2857	6.4228	6.5422	6.6491	6.7471	6.8385	6.9246
3100 (235.67)	V	1.220	64.467		67.885	73.315	78.287	82.958	87.423	91.745	95.965
	U	1013.2	2602.5		2639.2	2697.0	2749.2	2797.9	2844.3	2889.3	2933.4
	H	1017.0	2802.3		2849.6	2924.2	2991.9	3055.1	3115.4	3173.7	3230.8
	S	2.6623	6.1709		6.2626	6.4019	6.5227	6.6307	6.7294	6.8212	6.9077

P (T sat)											
3200 (237.45)	V	1.224	62.439	· · ·	65.380	70.721	75.593	80.158	84.513	88.723	92.829
	U	1021.5	2602.5	· · ·	2635.2	2693.9	2746.8	2796.0	2842.7	2887.9	2932.1
	H	1025.4	2802.3	· · ·	2844.4	2920.2	2988.7	3052.5	3113.2	3171.8	3229.2
	S	2.6786	6.1585	· · ·	6.2398	6.3815	6.5037	6.6127	6.7120	6.8043	6.8912
3300 (239.18)	V	1.227	60.529	· · ·	63.021	68.282	73.061	77.526	81.778	85.883	89.883
	U	1029.7	2602.5	· · ·	2631.1	2690.8	2744.4	2794.0	2841.1	2886.5	2930.9
	H	1033.7	2802.3	· · ·	2839.0	2916.1	2985.5	3049.9	3110.9	3169.9	3227.5
	S	2.6945	6.1463	· · ·	6.2173	6.3614	6.4851	6.5951	6.6952	6.7879	6.8752
3400 (240.88)	V	1.231	58.728	· · ·	60.796	65.982	70.675	75.048	79.204	83.210	87.110
	U	1037.6	2602.5	· · ·	2626.9	2687.7	2741.9	2792.0	2839.4	2885.1	2929.7
	H	1041.8	2802.1	· · ·	2833.6	2912.0	2982.2	3047.2	3108.7	3168.0	3225.9
	S	2.7101	6.1344	· · ·	6.1951	6.3416	6.4669	6.5779	6.6787	6.7719	6.8595
3500 (242.54)	V	1.235	57.025	· · ·	58.693	63.812	68.424	72.710	76.776	80.689	84.494
	U	1045.4	2602.4	· · ·	2622.7	2684.5	2739.5	2790.0	2837.8	2883.7	2928.4
	H	1049.8	2802.0	· · ·	2828.1	2907.8	2979.0	3044.5	3106.5	3166.1	3224.2
	S	2.7253	6.1228	· · ·	6.1732	6.3221	6.4491	6.5611	6.6626	6.7563	6.8443
3600 (244.16)	V	1.238	55.415	· · ·	56.702	61.759	66.297	70.501	74.482	78.308	82.024
	U	1053.1	2602.2	· · ·	2618.4	2681.3	2737.0	2788.0	2836.1	2882.3	2927.2
	H	1057.6	2801.7	· · ·	2822.5	2903.6	2975.6	3041.8	3104.2	3164.2	3222.5
	S	2.7401	6.1115	· · ·	6.1514	6.3030	6.4315	6.5446	6.6468	6.7411	6.8294
3700 (245.75)	V	1.242	53.888	· · ·	54.812	59.814	64.282	68.410	72.311	76.055	79.687
	U	1060.6	2602.1	· · ·	2614.0	2678.0	2734.4	2786.0	2834.4	2880.8	2926.0
	H	1065.2	2801.4	· · ·	2816.8	2899.3	2972.3	3039.1	3102.0	3162.2	3220.8
	S	2.7547	6.1004	· · ·	6.1299	6.2841	6.4143	6.5284	6.6314	6.7262	6.8149
3800 (247.31)	V	1.245	52.438	· · ·	53.017	57.968	62.372	66.429	70.254	73.920	77.473
	U	1068.0	2601.9	· · ·	2609.5	2674.7	2731.9	2783.9	2832.7	2879.4	2924.7
	H	1072.7	2801.1	· · ·	2811.0	2895.0	2968.9	3036.4	3099.7	3160.3	3219.1
	S	2.7689	6.0896	· · ·	6.1085	6.2654	6.3973	6.5126	6.6163	6.7117	6.8007
3900 (248.84)	V	1.249	51.061	· · ·	51.308	56.215	60.558	64.547	68.302	71.894	75.372
	U	1075.3	2601.6	· · ·	2605.0	2671.4	2729.3	2781.9	2831.0	2877.9	2923.5
	H	1080.1	2800.8	· · ·	2805.1	2890.6	2965.5	3033.6	3097.4	3158.3	3217.4
	S	2.7828	6.0789	· · ·	6.0872	6.2470	6.3806	6.4970	6.6015	6.6974	6.7868
4000 (250.33)	V	1.252	49.749	· · ·	· · ·	54.546	58.833	62.759	66.446	69.969	73.376
	U	1082.4	2601.3	· · ·	· · ·	2668.0	2726.7	2779.8	2829.3	2876.5	2922.2
	H	1087.4	2800.3	· · ·	· · ·	2886.1	2962.0	3030.8	3095.1	3156.4	3215.7
	S	2.7965	6.0685	· · ·	· · ·	6.2288	6.3642	6.4817	6.5870	6.6834	6.7733

Table F.2: Superheated Steam, SI units (Continued)

TEMPERATURE: $t\,°C$
(TEMPERATURE: T kelvins)

P/kPa (t^{sat}/°C)		sat. liq.	sat. vap.	425 (698.15)	450 (723.15)	475 (748.15)	500 (773.15)	525 (798.15)	550 (823.15)	600 (873.15)	650 (923.15)
2400 (221.78)	V	1.193	83.199	130.44	135.61	140.73	145.82	150.88	155.91	165.92	175.86
	U	949.066	2600.7	2984.5	3027.1	3069.9	3112.9	3156.1	3199.6	3287.7	3377.2
	H	951.929	2800.4	3297.5	3352.6	3407.7	3462.9	3518.2	3573.8	3685.9	3799.3
	S	2.5343	6.2690	7.1189	7.1964	7.2713	7.3439	7.4144	7.4830	7.6152	7.7414
2500 (223.94)	V	1.197	79.905	125.07	130.04	134.97	139.87	144.74	149.58	159.21	168.76
	U	958.969	2601.2	2983.4	3026.2	3069.0	3112.1	3155.4	3198.9	3287.1	3376.7
	H	961.962	2800.9	3296.1	3351.3	3406.5	3461.7	3517.2	3572.9	3685.1	3798.6
	S	2.5543	6.2536	7.0986	7.1763	7.2513	7.3240	7.3946	7.4633	7.5956	7.7220
2600 (226.04)	V	1.201	76.856	120.11	124.91	129.66	134.38	139.07	143.74	153.01	162.21
	U	968.597	2601.5	2982.3	3025.2	3068.1	3111.2	3154.6	3198.2	3286.5	3376.1
	H	971.720	2801.4	3294.6	3349.9	3405.3	3460.6	3516.2	3571.9	3684.3	3797.9
	S	2.5736	6.2387	7.0789	7.1568	7.2320	7.3048	7.3755	7.4443	7.5768	7.7033
2700 (228.07)	V	1.205	74.025	115.52	120.15	124.74	129.30	133.82	138.33	147.27	156.14
	U	977.968	2601.8	2981.2	3024.2	3067.2	3110.4	3153.8	3197.5	3285.8	3375.6
	H	981.222	2801.7	3293.1	3348.6	3404.0	3459.5	3515.2	3571.0	3683.5	3797.1
	S	2.5924	6.2244	7.0600	7.1381	7.2134	7.2863	7.3571	7.4260	7.5587	7.6853
2800 (230.05)	V	1.209	71.389	111.25	115.74	120.17	124.58	128.95	133.30	141.94	150.50
	U	987.100	2602.1	2980.2	3023.2	3066.3	3109.6	3153.1	3196.8	3285.2	3375.0
	H	990.485	2802.0	3291.7	3347.3	3402.8	3458.4	3514.1	3570.1	3682.6	3796.4
	S	2.6106	6.2104	7.0416	7.1199	7.1954	7.2685	7.3394	7.4084	7.5412	7.6679
2900 (231.97)	V	1.213	68.928	107.28	111.62	115.92	120.18	124.42	128.62	136.97	145.26
	U	996.008	2602.3	2979.1	3022.3	3065.5	3108.8	3152.3	3196.1	3284.6	3374.5
	H	999.524	2802.2	3290.2	3346.0	3401.6	3457.3	3513.1	3569.1	3681.8	3795.7
	S	2.6283	6.1969	7.0239	7.1024	7.1780	7.2512	7.3222	7.3913	7.5243	7.6511
3000 (233.84)	V	1.216	66.626	103.58	107.79	111.95	116.08	120.18	124.26	132.34	140.36
	U	1004.7	2602.4	2978.0	3021.3	3064.6	3107.9	3151.4	3195.4	3284.0	3373.9
	H	1008.4	2802.3	3288.7	3344.6	3400.4	3456.2	3512.1	3568.1	3681.0	3795.0
	S	2.6455	6.1837	7.0067	7.0854	7.1612	7.2345	7.3056	7.3748	7.5079	7.6349
3100 (235.67)	V	1.220	64.467	100.11	104.20	108.24	112.24	116.22	120.17	128.01	135.78
	U	1013.2	2602.5	2976.9	3020.3	3063.7	3107.1	3150.8	3194.7	3283.3	3373.4
	H	1017.0	2802.3	3287.3	3343.3	3399.2	3455.1	3511.0	3567.2	3680.2	3794.3
	S	2.6623	6.1709	6.9900	7.0689	7.1448	7.2183	7.2895	7.3588	7.4920	7.6191

(Sat. temp)											
3200 (237.45)	V	1.224	62.439	96.859	100.83	104.76	108.65	112.51	116.34	123.95	131.48
	U	1021.5	2602.5	2975.9	3019.3	3062.8	3106.3	3150.0	3193.9	3282.7	3372.8
	H	1025.4	2802.3	3285.8	3342.0	3398.0	3454.0	3510.0	3566.2	3679.3	3793.6
	S	2.6786	6.1585	6.9738	7.0528	7.1290	7.2026	7.2739	7.3433	7.4767	7.6039
3300 (239.18)	V	1.227	60.529	93.805	97.668	101.49	105.27	109.02	112.74	120.13	127.45
	U	1029.7	2602.5	2974.8	3018.3	3061.9	3105.5	3149.2	3193.2	3282.1	3372.3
	H	1033.7	2802.3	3284.3	3340.6	3396.8	3452.8	3509.0	3565.3	3678.5	3792.9
	S	2.6945	6.1463	6.9580	7.0373	7.1136	7.1873	7.2588	7.3282	7.4618	7.5891
3400 (240.88)	V	1.231	58.728	90.930	94.692	98.408	102.09	105.74	109.36	116.54	123.65
	U	1037.6	2602.5	2973.7	3017.4	3061.0	3104.6	3148.4	3192.5	3281.5	3371.7
	H	1041.8	2802.1	3282.8	3339.3	3395.5	3451.7	3507.9	3564.3	3677.7	3792.1
	S	2.7101	6.1344	6.9426	7.0221	7.0986	7.1724	7.2440	7.3136	7.4473	7.5747
3500 (242.54)	V	1.235	57.025	88.220	91.886	95.505	99.088	102.64	106.17	113.15	120.07
	U	1045.4	2602.4	2972.6	3016.4	3060.1	3103.8	3147.7	3191.8	3280.8	3371.2
	H	1049.8	2802.0	3281.3	3338.0	3394.3	3450.6	3506.9	3563.4	3676.9	3791.4
	S	2.7253	6.1228	6.9277	7.0074	7.0840	7.1580	7.2297	7.2993	7.4332	7.5607
3600 (244.16)	V	1.238	55.415	85.660	89.236	92.764	96.255	99.716	103.15	109.96	116.69
	U	1053.1	2602.2	2971.5	3015.4	3059.2	3103.0	3146.9	3191.1	3280.2	3370.6
	H	1057.6	2801.7	3279.8	3336.6	3393.1	3449.5	3505.9	3562.4	3676.1	3790.7
	S	2.7401	6.1115	6.9131	6.9930	7.0698	7.1439	7.2157	7.2854	7.4195	7.5471
3700 (245.75)	V	1.242	53.888	83.238	86.728	90.171	93.576	96.950	100.30	106.93	113.49
	U	1060.6	2602.1	2970.4	3014.4	3058.2	3102.1	3146.1	3190.4	3279.6	3370.1
	H	1065.2	2801.4	3278.4	3335.3	3391.9	3448.4	3504.9	3561.5	3675.2	3790.0
	S	2.7547	6.1004	6.8989	6.9790	7.0559	7.1302	7.2021	7.2719	7.4061	7.5339
3800 (247.31)	V	1.245	52.438	80.944	84.353	87.714	91.038	94.330	97.596	104.06	110.46
	U	1068.0	2601.9	2969.3	3013.4	3057.3	3101.3	3145.4	3189.6	3279.0	3369.5
	H	1072.7	2801.1	3276.8	3333.9	3390.7	3447.2	3503.8	3560.5	3674.4	3789.3
	S	2.7689	6.0896	6.8849	6.9653	7.0424	7.1168	7.1888	7.2587	7.3931	7.5210
3900 (248.84)	V	1.249	51.061	78.767	82.099	85.383	88.629	91.844	95.033	101.35	107.59
	U	1075.3	2601.6	2968.2	3012.4	3056.4	3100.5	3144.6	3188.9	3278.3	3369.0
	H	1080.1	2800.8	3275.3	3332.6	3389.4	3446.1	3502.8	3559.5	3673.6	3788.6
	S	2.7828	6.0789	6.8713	6.9519	7.0292	7.1037	7.1759	7.2459	7.3804	7.5084
4000 (250.33)	V	1.252	49.749	76.698	79.958	83.169	86.341	89.483	92.598	98.763	104.86
	U	1082.4	2601.3	2967.0	3011.4	3055.5	3099.6	3143.8	3188.2	3277.7	3368.4
	H	1087.4	2800.3	3273.8	3331.2	3388.2	3445.0	3501.7	3558.6	3672.8	3787.9
	S	2.7965	6.0685	6.8581	6.9388	7.0163	7.0909	7.1632	7.2333	7.3680	7.4961

Table F.2: Superheated Steam, SI units (Continued)

TEMPERATURE: $t\,°C$
(TEMPERATURE: T kelvins)

P/kPa (t^{sat}/°C)		sat. liq.	sat. vap.	260 (533.15)	275 (548.15)	300 (573.15)	325 (598.15)	350 (623.15)	375 (648.15)	400 (673.15)	425 (698.15)
4100 (251.80)	V	1.256	48.500	50.150	52.955	57.191	61.057	64.680	68.137	71.476	74.730
	U	1089.4	2601.0	2624.6	2664.5	2724.0	2777.7	2827.6	2875.0	2920.9	2965.9
	H	1094.6	2799.9	2830.3	2881.6	2958.5	3028.0	3092.8	3154.4	3214.0	3272.3
	S	2.8099	6.0583	6.1157	6.2107	6.3480	6.4667	6.5727	6.6697	6.7600	6.8450
4200 (253.24)	V	1.259	47.307	48.654	51.438	55.625	59.435	62.998	66.392	69.667	72.856
	U	1096.3	2600.7	2620.4	2661.0	2721.4	2775.6	2825.8	2873.6	2919.7	2964.8
	H	1101.6	2799.4	2824.8	2877.1	2955.0	3025.2	3090.4	3152.4	3212.3	3270.8
	S	2.8231	6.0482	6.0962	6.1929	6.3320	6.4519	6.5587	6.6563	6.7469	6.8323
4300 (254.66)	V	1.262	46.168	47.223	49.988	54.130	57.887	61.393	64.728	67.942	71.069
	U	1103.1	2600.3	2616.2	2657.5	2718.7	2773.4	2824.1	2872.1	2918.4	2963.7
	H	1108.5	2798.9	2819.2	2872.4	2951.4	3022.3	3088.1	3150.4	3210.5	3269.3
	S	2.8360	6.0383	6.0768	6.1752	6.3162	6.4373	6.5450	6.6431	6.7341	6.8198
4400 (256.05)	V	1.266	45.079	45.853	48.601	52.702	56.409	59.861	63.139	66.295	69.363
	U	1109.8	2599.9	2611.8	2653.9	2716.0	2771.3	2822.3	2870.6	2917.1	2962.5
	H	1115.4	2798.3	2813.6	2867.8	2947.8	3019.5	3085.7	3148.4	3208.8	3267.7
	S	2.8487	6.0286	6.0575	6.1577	6.3006	6.4230	6.5315	6.6301	6.7216	6.8076
4500 (257.41)	V	1.269	44.037	44.540	47.273	51.336	54.996	58.396	61.620	64.721	67.732
	U	1116.4	2599.5	2607.4	2650.3	2713.2	2769.1	2820.5	2869.1	2915.8	2961.4
	H	1122.1	2797.7	2807.9	2863.0	2944.2	3016.6	3083.3	3146.4	3207.1	3266.2
	S	2.8612	6.0191	6.0382	6.1403	6.2852	6.4088	6.5182	6.6174	6.7093	6.7955
4600 (258.75)	V	1.272	43.038	43.278	46.000	50.027	53.643	56.994	60.167	63.215	66.172
	U	1122.9	2599.1	2602.9	2646.6	2710.4	2766.9	2818.7	2867.6	2914.5	2960.3
	H	1128.8	2797.0	2802.0	2858.2	2940.5	3013.7	3080.9	3144.4	3205.3	3264.7
	S	2.8735	6.0097	6.0190	6.1230	6.2700	6.3949	6.5050	6.6049	6.6972	6.7838
4700 (260.07)	V	1.276	42.081	· · · · · ·	44.778	48.772	52.346	55.651	58.775	61.773	64.679
	U	1129.3	2598.6	· · · · · ·	2642.9	2707.6	2764.7	2816.9	2866.1	2913.2	2959.1
	H	1135.3	2796.4	· · · · · ·	2853.3	2936.8	3010.7	3078.5	3142.3	3203.6	3263.1
	S	2.8855	6.0004	· · · · · ·	6.1058	6.2549	6.3811	6.4921	6.5926	6.6853	6.7722
4800 (261.37)	V	1.279	41.161	· · · · · ·	43.604	47.569	51.103	54.364	57.441	60.390	63.247
	U	1135.8	2598.1	· · · · · ·	2639.1	2704.8	2762.5	2815.1	2864.6	2911.9	2958.0
	H	1141.8	2795.7	· · · · · ·	2848.4	2933.1	3007.8	3076.1	3140.3	3201.8	3261.6
	S	2.8974	5.9913	· · · · · ·	6.0887	6.2399	6.3675	6.4794	6.5805	6.6736	6.7608

4900 (262.65)	V	1.282	40.278	.....	42.475	46.412	49.909	53.128	56.161	59.064	61.874
	U	1141.9	2597.6	.....	2635.2	2701.9	2760.2	2813.3	2863.0	2910.6	2956.9
	H	1148.2	2794.9	.....	2843.3	2929.3	3004.8	3073.6	3138.2	3200.0	3260.0
	S	2.9091	5.9823	.....	6.0717	6.2252	6.3541	6.4669	6.5685	6.6621	6.7496
5000 (263.91)	V	1.286	39.429	.....	41.388	45.301	48.762	51.941	54.932	57.791	60.555
	U	1148.0	2597.0	.....	2631.3	2699.0	2758.0	2811.5	2861.5	2909.3	2955.7
	H	1154.5	2794.2	.....	2838.2	2925.5	3001.8	3071.2	3136.2	3198.3	3258.5
	S	2.9206	5.9735	.....	6.0547	6.2105	6.3408	6.4545	6.5568	6.6508	6.7386
5100 (265.15)	V	1.289	38.611	.....	40.340	44.231	47.660	50.801	53.750	56.567	59.288
	U	1154.1	2596.5	.....	2627.3	2696.1	2755.7	2809.6	2860.0	2908.0	2954.5
	H	1160.7	2793.4	.....	2833.1	2921.7	2998.7	3068.7	3134.1	3196.5	3256.9
	S	2.9319	5.9648	.....	6.0378	6.1960	6.3277	6.4423	6.5452	6.6396	6.7278
5200 (266.37)	V	1.292	37.824	.....	39.330	43.201	46.599	49.703	52.614	55.390	58.070
	U	1160.1	2595.9	.....	2623.3	2693.1	2753.4	2807.8	2858.4	2906.7	2953.4
	H	1166.8	2792.6	.....	2827.8	2917.8	2995.7	3066.2	3132.0	3194.7	3255.4
	S	2.9431	5.9561	.....	6.0210	6.1815	6.3147	6.4302	6.5338	6.6287	6.7172
5300 (267.58)	V	1.296	37.066	.....	38.354	42.209	45.577	48.647	51.520	54.257	56.897
	U	1166.1	2595.3	.....	2619.2	2690.1	2751.0	2805.9	2856.9	2905.3	2952.2
	H	1172.9	2791.7	.....	2822.5	2913.8	2992.6	3063.7	3129.9	3192.9	3253.8
	S	2.9541	5.9476	.....	6.0041	6.1672	6.3018	6.4183	6.5225	6.6179	6.7067
5400 (268.76)	V	1.299	36.334	.....	37.411	41.251	44.591	47.628	50.466	53.166	55.768
	U	1171.9	2594.6	.....	2615.0	2687.1	2748.7	2804.0	2855.3	2904.0	2951.1
	H	1178.9	2790.8	.....	2817.0	2909.8	2989.5	3061.2	3127.8	3191.1	3252.2
	S	2.9650	5.9392	.....	5.9873	6.1530	6.2891	6.4066	6.5114	6.6072	6.6963
5500 (269.93)	V	1.302	35.628	.....	36.499	40.327	43.641	46.647	49.450	52.115	54.679
	U	1177.7	2594.0	.....	2610.8	2684.0	2746.3	2802.1	2853.7	2902.7	2949.9
	H	1184.9	2789.9	.....	2811.5	2905.8	2986.4	3058.7	3125.7	3189.3	3250.6
	S	2.9757	5.9309	.....	5.9705	6.1388	6.2765	6.3949	6.5004	6.5967	6.6862
5600 (271.09)	V	1.306	34.946	.....	35.617	39.434	42.724	45.700	48.470	51.100	53.630
	U	1183.5	2593.3	.....	2606.5	2680.9	2744.0	2800.2	2852.1	2901.3	2948.7
	H	1190.8	2789.0	.....	2805.9	2901.7	2983.2	3056.1	3123.6	3187.5	3249.0
	S	2.9863	5.9227	.....	5.9537	6.1248	6.2640	6.3834	6.4896	6.5863	6.6761
5700 (272.22)	V	1.309	34.288	.....	34.761	38.571	41.838	44.785	47.525	50.121	52.617
	U	1189.1	2592.6	.....	2602.1	2677.8	2741.6	2798.3	2850.5	2899.9	2947.5
	H	1196.6	2788.0	.....	2800.2	2897.6	2980.0	3053.5	3121.4	3185.6	3247.5
	S	2.9968	5.9146	.....	5.9369	6.1108	6.2516	6.3720	6.4789	6.5761	6.6663

Table F.2: Superheated Steam, SI units (Continued)

TEMPERATURE: t °C
(TEMPERATURE: T kelvins)

P/kPa (t^{sat}/°C)		sat. liq.	sat. vap.	450 (723.15)	475 (748.15)	500 (773.15)	525 (798.15)	550 (823.15)	575 (848.15)	600 (873.15)	650 (923.15)
4100 (251.80)	V	1.256	48.500	77.921	81.062	84.165	87.236	90.281	93.303	96.306	102.26
	U	1089.4	2601.0	3010.4	3054.6	3098.8	3143.0	3187.5	3232.1	3277.1	3367.9
	H	1094.6	2799.9	3329.9	3387.0	3443.9	3500.7	3557.6	3614.7	3671.9	3787.1
	S	2.8099	6.0583	6.9260	7.0037	7.0785	7.1508	7.2210	7.2893	7.3558	7.4842
4200 (253.24)	V	1.259	47.307	75.981	79.056	82.092	85.097	88.075	91.030	93.966	99.787
	U	1096.3	2600.7	3009.4	3053.7	3097.9	3142.3	3186.8	3231.5	3276.5	3367.3
	H	1101.6	2799.4	3328.5	3385.7	3442.7	3499.7	3556.7	3613.8	3671.1	3786.4
	S	2.8231	6.0482	6.9135	6.9913	7.0662	7.1387	7.2090	7.2774	7.3440	7.4724
4300 (254.66)	V	1.262	46.168	74.131	77.143	80.116	83.057	85.971	88.863	91.735	97.428
	U	1103.1	2600.3	3008.4	3052.8	3097.1	3141.5	3186.0	3230.8	3275.8	3366.8
	H	1108.5	2798.9	3327.1	3384.5	3441.6	3498.6	3555.7	3612.9	3670.3	3785.7
	S	2.8360	6.0383	6.9012	6.9792	7.0543	7.1269	7.1973	7.2658	7.3324	7.4610
4400 (256.05)	V	1.266	45.079	72.365	75.317	78.229	81.110	83.963	86.794	89.605	95.177
	U	1109.8	2599.9	3007.4	3051.9	3096.3	3140.7	3185.3	3230.1	3275.2	3366.2
	H	1115.4	2798.3	3325.8	3383.3	3440.5	3497.6	3554.7	3612.0	3669.5	3785.0
	S	2.8487	6.0286	6.8892	6.9674	7.0426	7.1153	7.1858	7.2544	7.3211	7.4498
4500 (257.41)	V	1.269	44.037	70.677	73.572	76.427	79.249	82.044	84.817	87.570	93.025
	U	1116.4	2599.5	3006.3	3050.9	3095.4	3139.9	3184.6	3229.5	3274.6	3365.7
	H	1122.1	2797.7	3324.4	3382.0	3439.3	3496.6	3553.8	3611.1	3668.6	3784.3
	S	2.8612	6.0191	6.8774	6.9558	7.0311	7.1040	7.1746	7.2432	7.3100	7.4388
4600 (258.75)	V	1.272	43.038	69.063	71.903	74.702	77.469	80.209	82.926	85.623	90.967
	U	1122.9	2599.1	3005.3	3050.0	3094.6	3139.2	3183.9	3228.8	3273.9	3365.1
	H	1128.8	2797.0	3323.0	3380.8	3438.2	3495.5	3552.8	3610.2	3667.8	3783.6
	S	2.8735	6.0097	6.8659	6.9444	7.0199	7.0928	7.1636	7.2323	7.2991	7.4281
4700 (260.07)	V	1.276	42.081	67.517	70.304	73.051	75.765	78.452	81.116	83.760	88.997
	U	1129.3	2598.6	3004.3	3049.1	3093.7	3138.4	3183.1	3228.1	3273.3	3364.6
	H	1135.3	2796.4	3321.6	3379.5	3437.1	3494.5	3551.9	3609.3	3667.0	3782.9
	S	2.8855	6.0004	6.8545	6.9332	7.0089	7.0819	7.1527	7.2215	7.2885	7.4176
4800 (261.37)	V	1.279	41.161	66.036	68.773	71.469	74.132	76.768	79.381	81.973	87.109
	U	1135.6	2598.1	3003.3	3048.2	3092.9	3137.6	3182.4	3227.4	3272.7	3364.0
	H	1141.8	2795.7	3320.3	3378.3	3435.9	3493.4	3550.9	3608.5	3666.2	3782.1
	S	2.8974	5.9913	6.8434	6.9223	6.9981	7.0712	7.1422	7.2110	7.2781	7.4072

4900 (262.65)	V	1.282	40.278	64.615	67.303	69.951	72.565	75.152	77.716	80.260	85.298
	U	1141.9	2597.6	3002.3	3047.2	3092.0	3136.8	3181.7	3226.8	3272.0	3363.5
	H	1148.2	2794.9	3318.9	3377.2	3434.8	3492.4	3549.9	3607.6	3665.3	3781.4
	S	2.9091	5.9823	6.8324	6.9115	6.9874	7.0607	7.1318	7.2007	7.2678	7.3971
5000 (263.91)	V	1.286	39.429	63.250	65.893	68.494	71.061	73.602	76.119	78.616	83.559
	U	1148.0	2597.0	3001.2	3046.3	3091.2	3136.0	3181.0	3226.1	3271.4	3362.9
	H	1154.5	2794.2	3317.5	3375.8	3433.7	3491.3	3549.0	3606.7	3664.5	3780.7
	S	2.9206	5.9735	6.8217	6.9009	6.9770	7.0504	7.1215	7.1906	7.2578	7.3872
5100 (265.15)	V	1.289	38.611	61.940	64.537	67.094	69.616	72.112	74.584	77.035	81.888
	U	1154.1	2596.5	3000.2	3045.4	3090.3	3135.3	3180.2	3225.4	3270.8	3362.4
	H	1160.7	2793.4	3316.1	3374.5	3432.5	3490.3	3548.0	3605.8	3663.7	3780.0
	S	2.9319	5.9648	6.8111	6.8905	6.9668	7.0403	7.1115	7.1807	7.2479	7.3775
5200 (266.37)	V	1.292	37.824	60.679	63.234	65.747	68.227	70.679	73.108	75.516	80.282
	U	1160.1	2595.9	2999.2	3044.5	3089.5	3134.5	3179.5	3224.7	3270.2	3361.8
	H	1166.8	2792.6	3314.7	3373.3	3431.4	3489.3	3547.1	3604.9	3662.8	3779.3
	S	2.9431	5.9561	6.8007	6.8803	6.9567	7.0304	7.1017	7.1709	7.2382	7.3679
5300 (267.58)	V	1.296	37.066	59.466	61.980	64.452	66.890	69.300	71.687	74.054	78.736
	U	1166.1	2595.3	2998.2	3043.5	3088.6	3133.7	3178.8	3224.1	3269.5	3361.3
	H	1172.9	2791.7	3313.3	3372.0	3430.2	3488.2	3546.1	3604.0	3662.0	3778.6
	S	2.9541	5.9476	6.7905	6.8703	6.9468	7.0206	7.0920	7.1613	7.2287	7.3585
5400 (268.76)	V	1.299	36.334	58.297	60.772	63.204	65.603	67.973	70.320	72.646	77.248
	U	1171.9	2594.6	2997.1	3042.6	3087.8	3132.9	3178.1	3223.4	3268.9	3360.7
	H	1178.9	2790.8	3311.9	3370.8	3429.1	3487.2	3545.1	3603.1	3661.2	3777.8
	S	2.9650	5.9392	6.7804	6.8604	6.9371	7.0110	7.0825	7.1519	7.2194	7.3493
5500 (269.93)	V	1.302	35.628	57.171	59.608	62.002	64.362	66.694	69.002	71.289	75.814
	U	1177.7	2594.0	2996.1	3041.7	3086.9	3132.1	3177.3	3222.7	3268.3	3360.2
	H	1184.9	2789.9	3310.5	3369.5	3427.9	3486.1	3544.2	3602.2	3660.4	3777.1
	S	2.9757	5.9309	6.7705	6.8507	6.9275	7.0015	7.0731	7.1426	7.2102	7.3402
5600 (271.09)	V	1.306	34.946	56.085	58.486	60.843	63.165	65.460	67.731	69.981	74.431
	U	1183.5	2593.3	2995.0	3040.7	3086.1	3131.3	3176.6	3222.0	3267.6	3359.6
	H	1190.8	2789.0	3309.1	3368.2	3426.8	3485.1	3543.2	3601.3	3659.5	3776.4
	S	2.9863	5.9227	6.7607	6.8411	6.9181	6.9922	7.0639	7.1335	7.2011	7.3313
5700 (272.22)	V	1.309	34.288	55.038	57.403	59.724	62.011	64.270	66.504	68.719	73.096
	U	1189.1	2592.6	2994.0	3039.8	3085.2	3130.5	3175.9	3221.3	3267.0	3359.1
	H	1196.6	2788.0	3307.7	3367.0	3425.6	3484.0	3542.2	3600.4	3658.7	3775.7
	S	2.9968	5.9146	6.7511	6.8316	6.9088	6.9831	7.0549	7.1245	7.1923	7.3226

Table F.2: Superheated Steam, SI units (Continued)

TEMPERATURE: $t°C$
(TEMPERATURE: T kelvins)

P/kPa ($t^{sat}/°C$)		sat. liq.	sat. vap.	280 (553.15)	290 (563.15)	300 (573.15)	325 (598.15)	350 (623.15)	375 (648.15)	400 (673.15)	425 (698.15)
5800 (273.35)	V	1.312	33.651	34.756	36.301	37.736	40.982	43.902	46.611	49.176	51.638
	U	1194.7	2591.9	2614.4	2645.7	2674.6	2739.1	2796.3	2848.9	2898.6	2946.4
	H	1202.3	2787.0	2816.0	2856.3	2893.5	2976.8	3051.0	3119.3	3183.8	3245.9
	S	3.0071	5.9066	5.9592	6.0314	6.0969	6.2393	6.3608	6.4683	6.5660	6.6565
5900 (274.46)	V	1.315	33.034	33.953	35.497	36.928	40.154	43.048	45.728	48.262	50.693
	U	1200.3	2591.1	2610.2	2642.1	2671.4	2736.7	2794.4	2847.3	2897.2	2945.2
	H	1208.0	2786.0	2810.5	2851.5	2889.3	2973.6	3048.4	3117.1	3182.0	3244.3
	S	3.0172	5.8986	5.9431	6.0166	6.0830	6.2272	6.3496	6.4578	6.5560	6.6469
6000 (275.55)	V	1.319	32.438	33.173	34.718	36.145	39.353	42.222	44.874	47.379	49.779
	U	1205.8	2590.4	2605.9	2638.4	2668.1	2734.2	2792.4	2845.7	2895.8	2944.0
	H	1213.7	2785.0	2804.9	2846.7	2885.0	2970.4	3045.8	3115.0	3180.1	3242.6
	S	3.0273	5.8908	5.9270	6.0017	6.0692	6.2151	6.3386	6.4475	6.5462	6.6374
6100 (276.63)	V	1.322	31.860	32.415	33.962	35.386	38.577	41.422	44.048	46.524	48.895
	U	1211.2	2589.6	2601.5	2634.6	2664.8	2731.7	2790.4	2844.1	2894.5	2942.8
	H	1219.3	2783.9	2799.3	2841.8	2880.7	2967.1	3043.1	3112.8	3178.3	3241.0
	S	3.0372	5.8830	5.9108	5.9869	6.0555	6.2031	6.3277	6.4373	6.5364	6.6280
6200 (277.70)	V	1.325	31.300	31.679	33.227	34.650	37.825	40.648	43.248	45.697	48.039
	U	1216.6	2588.8	2597.1	2630.8	2661.5	2729.2	2788.5	2842.4	2893.1	2941.6
	H	1224.8	2782.9	2793.5	2836.8	2876.3	2963.8	3040.5	3110.6	3176.4	3239.4
	S	3.0471	5.8753	5.8946	5.9721	6.0418	6.1911	6.3168	6.4272	6.5268	6.6188
6300 (278.75)	V	1.328	30.757	30.962	32.514	33.935	37.097	39.898	42.473	44.895	47.210
	U	1221.9	2588.0	2592.6	2626.9	2658.1	2726.7	2786.5	2840.8	2891.7	2940.4
	H	1230.3	2781.8	2787.6	2831.7	2871.9	2960.4	3037.8	3108.4	3174.5	3237.8
	S	3.0568	5.8677	5.8783	5.9573	6.0281	6.1793	6.3061	6.4172	6.5173	6.6096
6400 (279.79)	V	1.332	30.230	30.265	31.821	33.241	36.390	39.170	41.722	44.119	46.407
	U	1227.2	2587.2	2587.2	2623.0	2654.7	2724.2	2784.4	2839.1	2890.3	2939.2
	H	1235.7	2780.6	2781.6	2826.6	2867.5	2957.1	3035.1	3106.2	3172.7	3236.2
	S	3.0664	5.8601	5.8619	5.9425	6.0144	6.1675	6.2955	6.4072	6.5079	6.6006
6500 (280.82)	V	1.335	29.719	· · · · · · ·	31.146	32.567	35.704	38.465	40.994	43.366	45.629
	U	1232.5	2586.3	· · · · · · ·	2619.0	2651.2	2721.6	2782.4	2837.5	2888.9	2938.0
	H	1241.1	2779.5	· · · · · · ·	2821.4	2862.9	2953.7	3032.4	3103.9	3170.8	3234.5
	S	3.0759	5.8527	· · · · · · ·	5.9277	6.0008	6.1558	6.2849	6.3974	6.4986	6.5917

P (kPa) (Tsat °C)		Sat. liq		Sat. vap								
6600 (281.84)	V	1.338	....	29.223	....	30.490	31.911	35.038	37.781	40.287	42.636	44.874
	U	1237.6	....	2585.5	....	2614.9	2647.7	2719.0	2780.4	2835.8	2887.5	2936.7
	H	1246.5	....	2778.3	....	2816.1	2858.4	2950.2	3029.7	3101.7	3168.9	3232.9
	S	3.0853	....	5.8452	....	5.9129	5.9872	6.1442	6.2744	6.3877	6.4894	6.5828
6700 (282.84)	V	1.342	....	28.741	....	29.850	31.273	34.391	37.116	39.601	41.927	44.141
	U	1242.8	....	2584.6	....	2610.8	2644.2	2716.4	2778.3	2834.1	2886.1	2935.5
	H	1251.8	....	2777.1	....	2810.8	2853.7	2946.8	3027.0	3099.5	3167.0	3231.3
	S	3.0946	....	5.8379	....	5.8980	5.9736	6.1326	6.2640	6.3781	6.4803	6.5741
6800 (283.84)	V	1.345	....	28.272	....	29.226	30.652	33.762	36.470	38.935	41.239	43.430
	U	1247.9	....	2583.7	....	2606.6	2640.6	2713.7	2776.2	2832.4	2884.7	2934.3
	H	1257.0	....	2775.9	....	2805.3	2849.0	2943.3	3024.2	3097.2	3165.1	3229.6
	S	3.1038	....	5.8306	....	5.8830	5.9599	6.1211	6.2537	6.3686	6.4713	6.5655
7000 (285.79)	V	1.351	....	27.373	....	28.024	29.457	32.556	35.233	37.660	39.922	42.068
	U	1258.0	....	2581.8	....	2597.9	2633.2	2708.4	2772.1	2829.0	2881.8	2931.8
	H	1267.4	....	2773.5	....	2794.1	2839.4	2936.3	3018.7	3092.7	3161.2	3226.3
	S	3.1219	....	5.8162	....	5.8530	5.9327	6.0982	6.2333	6.3497	6.4536	6.5485
7200 (287.70)	V	1.358	....	26.522	....	26.878	28.321	31.413	34.063	36.454	38.676	40.781
	U	1267.9	....	2579.9	....	2589.0	2625.6	2702.9	2767.8	2825.6	2878.9	2929.4
	H	1277.6	....	2770.9	....	2782.5	2829.5	2929.1	3013.1	3088.1	3157.4	3223.0
	S	3.1397	....	5.8020	....	5.8226	5.9054	6.0755	6.2132	6.3312	6.4362	6.5319
7400 (289.57)	V	1.364	....	25.715	....	25.781	27.238	30.328	32.954	35.312	37.497	39.564
	U	1277.6	....	2578.0	....	2579.7	2617.8	2697.3	2763.5	2822.1	2876.0	2926.9
	H	1287.6	....	2768.3	....	2770.5	2819.3	2921.8	3007.4	3083.4	3153.5	3219.6
	S	3.1571	....	5.7880	....	5.7919	5.8779	6.0530	6.1933	6.3130	6.4190	6.5156
7600 (291.41)	V	1.371	....	24.949	....	....	26.204	29.297	31.901	34.229	36.380	38.409
	U	1287.2	....	2575.9	....	....	2609.7	2691.7	2759.2	2818.6	2873.1	2924.3
	H	1297.6	....	2765.5	....	....	2808.8	2914.3	3001.6	3078.7	3149.6	3216.3
	S	3.1742	....	5.7742	....	....	5.8503	6.0306	6.1737	6.2950	6.4022	6.4996
7800 (293.21)	V	1.378	....	24.220	....	....	25.214	28.315	30.900	33.200	35.319	37.314
	U	1296.7	....	2573.8	....	....	2601.3	2685.9	2754.8	2815.1	2870.1	2921.8
	H	1307.4	....	2762.8	....	....	2798.0	2906.7	2995.8	3074.0	3145.6	3212.9
	S	3.1911	....	5.7605	....	....	5.8224	6.0082	6.1542	6.2773	6.3857	6.4839
8000 (294.97)	V	1.384	....	23.525	....	....	24.264	27.378	29.948	32.222	34.310	36.273
	U	1306.0	....	2571.7	....	....	2592.7	2679.9	2750.3	2811.5	2867.1	2919.3
	H	1317.1	....	2759.9	....	....	2786.8	2899.0	2989.9	3069.2	3141.6	3209.5
	S	3.2076	....	5.7471	....	....	5.7942	5.9860	6.1349	6.2599	6.3694	6.4684

Table F.2: Superheated Steam, SI units (Continued)

TEMPERATURE: $t\,^\circ$C
(TEMPERATURE: T kelvins)

P/kPa ($t^{sat}/^\circ$C)		sat. liq.	sat. vap.	450 (723.15)	475 (748.15)	500 (773.15)	525 (798.15)	550 (823.15)	575 (848.15)	600 (873.15)	650 (923.15)
5800	V	1.312	33.651	54.026	56.357	58.644	60.896	63.120	65.320	67.500	71.807
(273.35)	U	1194.7	2591.9	2992.9	3038.8	3084.4	3129.8	3175.2	3220.7	3266.4	3358.5
	H	1202.3	2787.0	3306.3	3365.7	3424.5	3483.0	3541.2	3599.5	3657.9	3775.0
	S	3.0071	5.9066	6.7416	6.8223	6.8996	6.9740	7.0460	7.1157	7.1835	7.3139
5900	V	1.315	33.034	53.048	55.346	57.600	59.819	62.010	64.176	66.322	70.563
(274.46)	U	1200.3	2591.1	2991.9	3037.9	3083.5	3129.0	3174.4	3220.0	3265.7	3357.9
	H	1208.0	2786.0	3304.9	3364.4	3423.3	3481.9	3540.3	3598.6	3657.0	3774.3
	S	3.0172	5.8986	6.7322	6.8132	6.8906	6.9652	7.0372	7.1070	7.1749	7.3054
6000	V	1.319	32.438	52.103	54.369	56.592	58.778	60.937	63.071	65.184	69.359
(275.55)	U	1205.8	2590.4	2990.8	3036.9	3082.6	3128.2	3173.7	3219.3	3265.1	3357.4
	H	1213.7	2785.0	3303.5	3363.2	3422.2	3480.8	3539.3	3597.7	3656.2	3773.5
	S	3.0273	5.8908	6.7230	6.8041	6.8818	6.9564	7.0285	7.0985	7.1664	7.2971
6100	V	1.322	31.860	51.189	53.424	55.616	57.771	59.898	62.001	64.083	68.196
(276.63)	U	1211.2	2589.6	2989.8	3036.0	3081.8	3127.4	3173.0	3218.6	3264.5	3356.8
	H	1219.3	2783.9	3302.0	3361.9	3421.0	3479.8	3538.3	3596.8	3655.4	3772.8
	S	3.0372	5.8830	6.7139	6.7952	6.8730	6.9478	7.0200	7.0900	7.1581	7.2889
6200	V	1.325	31.300	50.304	52.510	54.671	56.797	58.894	60.966	63.018	67.069
(277.70)	U	1216.6	2588.8	2988.7	3035.0	3080.9	3126.6	3172.2	3218.0	3263.8	3356.3
	H	1224.8	2782.9	3300.6	3360.6	3419.9	3478.7	3537.4	3595.9	3654.5	3772.1
	S	3.0471	5.8753	6.7049	6.7864	6.8644	6.9393	7.0116	7.0817	7.1498	7.2808
6300	V	1.328	30.757	49.447	51.624	53.757	55.853	57.921	59.964	61.986	65.979
(278.75)	U	1221.9	2588.0	2987.7	3034.1	3080.1	3125.8	3171.5	3217.3	3263.2	3355.7
	H	1230.3	2781.8	3299.2	3359.3	3418.7	3477.7	3536.4	3595.0	3653.7	3771.4
	S	3.0568	5.8677	6.6960	6.7778	6.8559	6.9309	7.0034	7.0735	7.1417	7.2728
6400	V	1.332	30.230	48.617	50.767	52.871	54.939	56.978	58.993	60.987	64.922
(279.79)	U	1227.2	2587.2	2986.6	3033.1	3079.2	3125.0	3170.8	3216.6	3262.6	3355.2
	H	1235.7	2780.6	3297.7	3358.0	3417.6	3476.6	3535.4	3594.1	3652.9	3770.7
	S	3.0664	5.8601	6.6872	6.7692	6.8475	6.9226	6.9952	7.0655	7.1337	7.2649
6500	V	1.335	29.719	47.812	49.935	52.012	54.053	56.065	58.052	60.018	63.898
(280.82)	U	1232.5	2586.3	2985.5	3032.2	3078.4	3124.2	3170.0	3215.9	3261.9	3354.6
	H	1241.1	2779.5	3296.3	3356.8	3416.4	3475.4	3534.4	3593.2	3652.1	3770.0
	S	3.0759	5.8527	6.6786	6.7608	6.8392	6.9145	6.9871	7.0575	7.1258	7.2572

P (Tsat)		Sat.	Sat.								
6600 (281.84)	V	1.338	29.223	47.031	49.129	51.180	53.194	55.179	57.139	59.079	62.905
	U	1237.6	2585.5	2984.5	3031.2	3077.4	3123.4	3169.3	3215.2	3261.3	3354.1
	H	1246.5	2778.3	3294.9	3355.5	3415.2	3474.5	3533.5	3592.3	3651.2	3769.2
	S	3.0853	5.8452	6.6700	6.7524	6.8310	6.9064	6.9792	7.0497	7.1181	7.2495
6700 (282.84)	V	1.342	28.741	46.274	48.346	50.372	52.361	54.320	56.254	58.168	61.942
	U	1242.8	2584.6	2983.4	3030.3	3076.6	3122.6	3168.6	3214.5	3260.7	3353.5
	H	1251.8	2777.1	3293.4	3354.2	3414.1	3473.4	3532.5	3591.4	3650.4	3768.5
	S	3.0946	5.8379	6.6616	6.7442	6.8229	6.8985	6.9714	7.0419	7.1104	7.2420
6800 (283.84)	V	1.345	28.272	45.539	47.587	49.588	51.552	53.486	55.395	57.283	61.007
	U	1247.9	2583.7	2982.3	3029.3	3075.7	3121.8	3167.8	3213.9	3260.0	3353.0
	H	1257.0	2775.9	3292.0	3352.9	3412.9	3472.4	3531.5	3590.5	3649.6	3767.8
	S	3.1038	5.8306	6.6532	6.7361	6.8150	6.8907	6.9636	7.0343	7.1028	7.2345
7000 (285.79)	V	1.351	27.373	44.131	46.133	48.086	50.003	51.889	53.750	55.590	59.217
	U	1258.0	2581.8	2980.1	3027.4	3074.0	3120.2	3166.3	3212.5	3258.8	3351.9
	H	1267.4	2773.5	3289.1	3350.3	3410.6	3470.2	3529.6	3588.7	3647.9	3766.4
	S	3.1219	5.8162	6.6368	6.7201	6.7993	6.8753	6.9485	7.0193	7.0880	7.2200
7200 (287.70)	V	1.358	26.522	42.802	44.759	46.668	48.540	50.381	52.197	53.991	57.527
	U	1267.9	2579.9	2978.0	3025.4	3072.2	3118.6	3164.9	3211.1	3257.5	3350.7
	H	1277.6	2770.9	3286.1	3347.7	3408.2	3468.1	3527.6	3586.9	3646.2	3764.9
	S	3.1397	5.8020	6.6208	6.7044	6.7840	6.8602	6.9337	7.0047	7.0735	7.2058
7400 (289.57)	V	1.364	25.715	41.544	43.460	45.327	47.156	48.954	50.727	52.478	55.928
	U	1277.6	2578.0	2975.8	3023.5	3070.4	3117.0	3163.4	3209.8	3256.2	3349.6
	H	1287.7	2768.3	3283.2	3345.1	3405.9	3466.0	3525.7	3585.1	3644.5	3763.5
	S	3.1571	5.7880	6.6050	6.6892	6.7691	6.8456	6.9192	6.9904	7.0594	7.1919
7600 (291.41)	V	1.371	24.949	40.351	42.228	44.056	45.845	47.603	49.335	51.045	54.413
	U	1287.2	2575.9	2973.6	3021.5	3068.7	3115.4	3161.9	3208.4	3254.9	3348.5
	H	1297.6	2765.5	3280.3	3342.5	3403.5	3463.8	3523.7	3583.3	3642.9	3762.1
	S	3.1742	5.7742	6.5896	6.6742	6.7545	6.8312	6.9051	6.9765	7.0457	7.1784
7800 (293.21)	V	1.378	24.220	39.220	41.060	42.850	44.601	46.320	48.014	49.686	52.976
	U	1296.7	2573.8	2971.4	3019.6	3066.9	3113.8	3160.4	3207.0	3253.7	3347.4
	H	1307.4	2762.8	3277.3	3339.8	3401.1	3461.7	3521.7	3581.5	3641.2	3760.6
	S	3.1911	5.7605	6.5745	6.6596	6.7402	6.8172	6.8913	6.9629	7.0322	7.1652
8000 (294.97)	V	1.384	23.525	38.145	39.950	41.704	43.419	45.102	46.759	48.394	51.611
	U	1306.0	2571.7	2969.2	3017.6	3065.1	3112.2	3158.9	3205.6	3252.4	3346.3
	H	1317.1	2759.9	3274.3	3337.2	3398.8	3459.5	3519.7	3579.7	3639.5	3759.2
	S	3.2076	5.7471	6.5597	6.6452	6.7262	6.8035	6.8778	6.9496	7.0191	7.1523

Table F.2: Superheated Steam, SI units (Continued)

TEMPERATURE: $t\,^{\circ}C$
(TEMPERATURE: T kelvins)

P/kPa (t^{sat}/°C)		sat. liq.	sat. vap.	300 (573.15)	320 (593.15)	340 (613.15)	360 (633.15)	380 (653.15)	400 (673.15)	425 (698.15)	450 (723.15)
8200 (296.70)	V	1.391	22.863	23.350	25.916	28.064	29.968	31.715	33.350	35.282	37.121
	U	1315.2	2569.5	2583.7	2657.7	2718.5	2771.5	2819.5	2864.1	2916.7	2966.9
	H	1326.6	2757.0	2775.2	2870.2	2948.6	3017.2	3079.5	3137.6	3206.0	3271.3
	S	3.2239	5.7338	5.7656	5.9288	6.0588	6.1689	6.2659	6.3534	6.4532	6.5452
8400 (298.39)	V	1.398	22.231	22.469	25.058	27.203	29.094	30.821	32.435	34.337	36.147
	U	1324.3	2567.2	2574.4	2651.1	2713.4	2767.3	2816.0	2861.1	2914.1	2964.7
	H	1336.1	2754.6	2763.1	2861.6	2941.9	3011.7	3074.8	3133.5	3202.6	3268.3
	S	3.2399	5.7207	5.7366	5.9056	6.0388	6.1509	6.2491	6.3376	6.4383	6.5309
8600 (300.06)	V	1.404	21.627	. . . . . .	24.236	26.380	28.258	29.968	31.561	33.437	35.217
	U	1333.3	2564.9	. . . . . .	2644.3	2708.1	2763.1	2812.4	2858.0	2911.5	2962.4
	H	1345.4	2750.9	. . . . . .	2852.7	2935.0	3006.1	3070.1	3129.4	3199.1	3265.3
	S	3.2557	5.7076	. . . . . .	5.8823	6.0189	6.1330	6.2326	6.3220	6.4236	6.5168
8800 (301.70)	V	1.411	21.049	. . . . . .	23.446	25.592	27.459	29.153	30.727	32.576	34.329
	U	1342.2	2562.6	. . . . . .	2637.3	2702.8	2758.8	2808.8	2854.9	2908.9	2960.1
	H	1354.6	2747.8	. . . . . .	2843.6	2928.0	3000.4	3065.3	3125.3	3195.6	3262.2
	S	3.2713	5.6948	. . . . . .	5.8590	5.9990	6.1152	6.2162	6.3067	6.4092	6.5030
9000 (303.31)	V	1.418	20.495	. . . . . .	22.685	24.836	26.694	28.372	29.929	31.754	33.480
	U	1351.0	2560.1	. . . . . .	2630.1	2697.4	2754.4	2805.2	2851.8	2906.3	2957.8
	H	1363.7	2744.6	. . . . . .	2834.3	2920.9	2994.7	3060.5	3121.2	3192.0	3259.2
	S	3.2867	5.6820	. . . . . .	5.8355	5.9792	6.0976	6.2000	6.2915	6.3949	6.4894
9200 (304.89)	V	1.425	19.964	. . . . . .	21.952	24.110	25.961	27.625	29.165	30.966	32.668
	U	1359.7	2557.7	. . . . . .	2622.9	2691.9	2750.0	2801.5	2848.7	2903.6	2955.5
	H	1372.8	2741.3	. . . . . .	2824.7	2913.7	2988.9	3055.7	3117.0	3188.5	3256.1
	S	3.3018	5.6694	. . . . . .	5.8118	5.9594	6.0801	6.1840	6.2765	6.3808	6.4760
9400 (306.44)	V	1.432	19.455	. . . . . .	21.245	23.412	25.257	26.909	28.433	30.212	31.891
	U	1368.2	2555.2	. . . . . .	2615.1	2686.3	2745.6	2797.8	2845.5	2900.9	2953.2
	H	1381.7	2738.0	. . . . . .	2814.8	2906.3	2983.0	3050.7	3112.8	3184.9	3253.0
	S	3.3168	5.6568	. . . . . .	5.7879	5.9397	6.0627	6.1681	6.2617	6.3669	6.4628
9600 (307.97)	V	1.439	18.965	. . . . . .	20.561	22.740	24.581	26.221	27.731	29.489	31.145
	U	1376.7	2552.6	. . . . . .	2607.3	2680.5	2741.0	2794.1	2842.3	2898.2	2950.9
	H	1390.6	2734.7	. . . . . .	2804.7	2898.8	2977.0	3045.8	3108.5	3181.3	3249.9
	S	3.3315	5.6444	. . . . . .	5.7637	5.9199	6.0454	6.1524	6.2470	6.3532	6.4498

(P) (Tsat)											
9800 (309.48)	V	1.446	18.494	......	19.899	22.093	23.931	25.561	27.056	28.795	30.429
	U	1385.2	2550.0	......	2599.2	2674.7	2736.4	2790.3	2839.1	2895.5	2948.6
	H	1399.3	2731.2	......	2794.3	2891.2	2971.0	3040.8	3104.2	3177.7	3246.8
	S	3.3461	5.6321	......	5.7393	5.9001	6.0282	6.1368	6.2325	6.3397	6.4369
10000 (310.96)	V	1.453	18.041	......	19.256	21.468	23.305	24.926	26.408	28.128	29.742
	U	1393.5	2547.3	......	2590.9	2668.7	2731.8	2786.4	2835.8	2892.8	2946.2
	H	1408.0	2727.7	......	2783.5	2883.4	2964.8	3035.7	3099.9	3174.1	3243.6
	S	3.3605	5.6198	......	5.7145	5.8803	6.0110	6.1213	6.2182	6.3264	6.4243
10200 (312.42)	V	1.460	17.605	......	18.632	20.865	22.702	24.315	25.785	27.487	29.081
	U	1401.8	2544.6	......	2582.3	2662.6	2727.0	2782.6	2832.6	2890.0	2943.9
	H	1416.7	2724.2	......	2772.3	2875.4	2958.6	3030.6	3095.6	3170.4	3240.5
	S	3.3748	5.6076	......	5.6894	5.8604	5.9940	6.1059	6.2040	6.3131	6.4118
10400 (313.86)	V	1.467	17.184	......	18.024	20.282	22.121	23.726	25.185	26.870	28.446
	U	1410.0	2541.8	......	2573.4	2656.3	2722.2	2778.7	2829.3	2887.3	2941.5
	H	1425.2	2720.6	......	2760.8	2867.2	2952.3	3025.4	3091.2	3166.7	3237.3
	S	3.3889	5.5955	......	5.6638	5.8404	5.9769	6.0907	6.1899	6.3001	6.3994
10600 (315.27)	V	1.474	16.778	......	17.432	19.717	21.560	23.159	24.607	26.276	27.834
	U	1418.1	2539.0	......	2564.1	2649.9	2717.4	2774.7	2825.9	2884.5	2939.1
	H	1433.7	2716.9	......	2748.9	2858.9	2945.9	3020.2	3086.8	3163.0	3234.1
	S	3.4029	5.5835	......	5.6376	5.8203	5.9599	6.0755	6.1759	6.2872	6.3872
10800 (316.67)	V	1.481	16.385	......	16.852	19.170	21.018	22.612	24.050	25.703	27.245
	U	1426.2	2536.2	......	2554.5	2643.4	2712.4	2770.7	2822.6	2881.7	2936.7
	H	1442.2	2713.1	......	2736.5	2850.4	2939.4	3014.9	3082.3	3159.3	3230.9
	S	3.4167	5.5715	......	5.6109	5.8000	5.9429	6.0604	6.1621	6.2744	6.3752
11000 (318.05)	V	1.489	16.006	......	16.285	18.639	20.494	22.083	23.512	25.151	26.676
	U	1434.2	2533.2	......	2544.4	2636.7	2707.4	2766.7	2819.2	2878.9	2934.3
	H	1450.6	2709.3	......	2723.5	2841.7	2932.8	3009.6	3077.8	3155.5	3227.7
	S	3.4304	5.5595	......	5.5835	5.7797	5.9259	6.0454	6.1483	6.2617	6.3633
11200 (319.40)	V	1.496	15.639	......	15.726	18.124	19.987	21.573	22.993	24.619	26.128
	U	1442.1	2530.3	......	2533.8	2629.8	2702.2	2762.6	2815.8	2876.0	2931.8
	H	1458.9	2705.4	......	2710.0	2832.8	2926.1	3004.2	3073.3	3151.7	3224.5
	S	3.4440	5.5476	......	5.5553	5.7591	5.9090	6.0305	6.1347	6.2491	6.3515
11400 (320.74)	V	1.504	15.284	......	......	17.622	19.495	21.079	22.492	24.104	25.599
	U	1450.0	2527.2	......	......	2622.7	2697.0	2758.4	2812.3	2873.1	2929.4
	H	1467.2	2701.5	......	......	2823.6	2919.3	2998.7	3068.7	3147.9	3221.2
	S	3.4575	5.5357	......	......	5.7383	5.8920	6.0156	6.1211	6.2367	6.3399

Table F.2: Superheated Steam, SI units (Continued)

TEMPERATURE: t °C
(TEMPERATURE: T kelvins)

P/kPa (t^{sat}/°C)		sat. liq.	sat. vap.	475 (748.15)	500 (773.15)	525 (798.15)	550 (823.15)	575 (848.15)	600 (873.15)	625 (898.15)	650 (923.15)
8200 (296.70)	V	1.391	22.863	38.893	40.614	42.295	43.943	45.566	47.166	48.747	50.313
	U	1315.2	2569.5	3015.6	3063.3	3110.5	3157.4	3204.3	3251.1	3298.1	3345.2
	H	1326.6	2757.0	3334.5	3396.4	3457.3	3517.8	3577.9	3637.9	3697.8	3757.7
	S	3.2239	5.7338	6.6311	6.7124	6.7900	6.8646	6.9365	7.0062	7.0739	7.1397
8400 (298.39)	V	1.398	22.231	37.887	39.576	41.224	42.839	44.429	45.996	47.544	49.076
	U	1324.3	2567.2	3013.6	3061.6	3108.9	3155.9	3202.9	3249.8	3296.9	3344.1
	H	1336.1	2754.0	3331.9	3394.0	3455.2	3515.8	3576.1	3636.2	3696.2	3756.3
	S	3.2399	5.7207	6.6173	6.6990	6.7769	6.8516	6.9238	6.9936	7.0614	7.1274
8600 (300.06)	V	1.404	21.627	36.928	38.586	40.202	41.787	43.345	44.880	46.397	47.897
	U	1333.3	2564.9	3011.6	3059.8	3107.3	3154.4	3201.5	3248.5	3295.7	3342.9
	H	1345.4	2750.9	3329.2	3391.6	3453.0	3513.8	3574.3	3634.5	3694.7	3754.9
	S	3.2557	5.7076	6.6037	6.6858	6.7639	6.8390	6.9113	6.9813	7.0492	7.1153
8800 (301.70)	V	1.411	21.049	36.011	37.640	39.228	40.782	42.310	43.815	45.301	46.771
	U	1342.2	2562.6	3009.6	3058.0	3105.6	3152.9	3200.1	3247.2	3294.5	3341.8
	H	1354.6	2747.8	3326.5	3389.2	3450.8	3511.8	3572.4	3632.8	3693.1	3753.4
	S	3.2713	5.6948	6.5904	6.6728	6.7513	6.8265	6.8990	6.9692	7.0373	7.1035
9000 (303.31)	V	1.418	20.495	35.136	36.737	38.296	39.822	41.321	42.798	44.255	45.695
	U	1351.0	2560.1	3007.6	3056.1	3104.0	3151.4	3198.7	3246.0	3293.3	3340.7
	H	1363.7	2744.6	3323.8	3386.8	3448.7	3509.8	3570.6	3631.1	3691.6	3752.0
	S	3.2867	5.6820	6.5773	6.6600	6.7388	6.8143	6.8870	6.9574	7.0256	7.0919
9200 (304.89)	V	1.425	19.964	34.298	35.872	37.405	38.904	40.375	41.824	43.254	44.667
	U	1359.7	2557.7	3005.6	3054.3	3102.3	3149.9	3197.3	3244.7	3292.1	3339.6
	H	1372.8	2741.3	3321.1	3384.4	3446.5	3507.8	3568.8	3629.5	3690.0	3750.5
	S	3.3018	5.6694	6.5644	6.6475	6.7266	6.8023	6.8752	6.9457	7.0141	7.0806
9400 (306.44)	V	1.432	19.455	33.495	35.045	36.552	38.024	39.470	40.892	42.295	43.682
	U	1368.2	2555.2	3003.5	3052.5	3100.7	3148.4	3195.9	3243.4	3290.9	3338.5
	H	1381.7	2738.0	3318.4	3381.9	3444.3	3505.9	3566.9	3627.8	3688.4	3749.1
	S	3.3168	5.6568	6.5517	6.6352	6.7146	6.7906	6.8637	6.9343	7.0029	7.0695
9600 (307.97)	V	1.439	18.965	32.726	34.252	35.734	37.182	38.602	39.999	41.377	42.738
	U	1376.7	2552.6	3001.5	3050.7	3099.0	3146.9	3194.5	3242.1	3289.7	3337.4
	H	1390.6	2734.7	3315.6	3379.5	3442.1	3503.9	3565.1	3626.1	3686.9	3747.6
	S	3.3315	5.6444	6.5392	6.6231	6.7028	6.6790	6.8523	6.9231	6.9918	7.0585

P (Tsat)											
9800 (309.48)	V	1.446	18.494	31.988	33.491	34.949	36.373	37.769	39.142	40.496	41.832
	U	1385.2	2550.0	2999.4	3048.8	3097.4	3145.4	3193.1	3240.8	3288.5	3336.2
	H	1399.3	2731.2	3312.9	3377.0	3439.9	3501.9	3563.3	3624.4	3685.3	3746.2
	S	3.3461	5.6321	6.5268	6.6112	6.6912	6.7676	6.8411	6.9121	6.9810	7.0478
10000 (310.96)	V	1.453	18.041	31.280	32.760	34.196	35.597	36.970	38.320	39.650	40.963
	U	1393.5	2547.3	2997.4	3047.0	3095.7	3143.9	3191.7	3239.5	3287.3	3335.1
	H	1408.0	2727.7	3310.1	3374.6	3437.6	3499.8	3561.4	3622.7	3683.8	3744.7
	S	3.3605	5.6198	6.5147	6.5994	6.6797	6.7564	6.8302	6.9013	6.9703	7.0373
10200 (312.42)	V	1.460	17.605	30.599	32.058	33.472	34.851	36.202	37.530	38.837	40.128
	U	1401.8	2544.6	2995.3	3045.2	3094.0	3142.3	3190.3	3238.2	3286.1	3334.0
	H	1416.7	2724.2	3307.4	3372.1	3435.5	3497.8	3559.6	3621.0	3682.2	3743.3
	S	3.3748	5.6076	6.5027	6.5879	6.6685	6.7454	6.8194	6.8907	6.9598	7.0269
10400 (313.86)	V	1.467	17.184	29.943	31.382	32.776	34.134	35.464	36.770	38.056	39.325
	U	1410.0	2541.8	2993.2	3043.3	3092.4	3140.8	3188.9	3236.9	3284.8	3332.9
	H	1425.2	2720.6	3304.6	3369.7	3433.2	3495.8	3557.8	3619.3	3680.6	3741.8
	S	3.3889	5.5955	6.4909	6.5765	6.6574	6.7346	6.8087	6.8803	6.9495	7.0167
10600 (315.27)	V	1.474	16.778	29.313	30.732	32.106	33.444	34.753	36.039	37.304	38.552
	U	1418.1	2539.0	2991.1	3041.4	3090.7	3139.3	3187.5	3235.6	3283.6	3331.7
	H	1433.7	2716.9	3301.8	3367.2	3431.0	3493.8	3555.9	3617.6	3679.1	3740.4
	S	3.4029	5.5835	6.4793	6.5652	6.6465	6.7239	6.7983	6.8700	6.9394	7.0067
10800 (316.67)	V	1.481	16.385	28.706	30.106	31.461	32.779	34.069	35.335	36.580	37.808
	U	1426.2	2536.4	2989.0	3039.6	3089.0	3137.8	3186.1	3234.3	3282.4	3330.6
	H	1442.2	2713.1	3299.0	3364.7	3428.8	3491.8	3554.1	3615.9	3677.5	3738.9
	S	3.4167	5.5715	6.4678	6.5542	6.6357	6.7134	6.7880	6.8599	6.9294	6.9969
11000 (318.05)	V	1.489	16.006	28.120	29.503	30.839	32.139	33.410	34.656	35.882	37.091
	U	1434.2	2533.0	2986.9	3037.7	3087.3	3136.2	3184.7	3233.0	3281.2	3329.5
	H	1450.6	2709.3	3296.2	3362.2	3426.5	3489.7	3552.2	3614.2	3675.9	3737.5
	S	3.4304	5.5595	6.4564	6.5432	6.6251	6.7031	6.7779	6.8499	6.9196	6.9872
11200 (319.40)	V	1.496	15.639	27.555	28.921	30.240	31.521	32.774	34.002	35.210	36.400
	U	1442.1	2530.3	2984.8	3035.8	3085.6	3134.7	3183.3	3231.7	3280.0	3328.4
	H	1458.9	2705.4	3293.4	3359.7	3424.3	3487.7	3550.4	3612.5	3674.4	3736.0
	S	3.4440	5.5476	6.4452	6.5324	6.6147	6.6929	6.7679	6.8401	6.9099	6.9777
11400 (320.74)	V	1.504	15.284	27.010	28.359	29.661	30.925	32.160	33.370	34.560	35.733
	U	1450.0	2527.2	2982.6	3033.9	3083.9	3133.1	3181.9	3230.4	3278.8	3327.2
	H	1467.2	2701.5	3290.5	3357.2	3422.1	3485.7	3548.5	3610.8	3672.8	3734.6
	S	3.4575	5.5357	6.4341	6.5218	6.6043	6.6828	6.7580	6.8304	6.9004	6.9683

Table F.3: Saturated Steam, English units

V = SPECIFIC VOLUME $(ft)^3 (lb_m)^{-1}$
U = SPECIFIC INTERNAL ENERGY $(Btu) (lb_m)^{-1}$
H = SPECIFIC ENTHALPY $(Btu) (lb_m)^{-1}$
S = SPECIFIC ENTROPY $(Btu) (lb_m)^{-1} R^{-1}$

t °F	P (psia)	SPECIFIC VOLUME V			INTERNAL ENERGY U			ENTHALPY H			ENTROPY S		
		sat. liq.	evap.	sat. vap.	sat. liq.	evap.	sat. vap.	sat. liq.	evap.	sat. vap.	sat. liq.	evap.	sat. vap.
32	0.0886	0.01602	3304.6	3304.6	-0.02	1021.3	1021.3	-0.02	1075.5	1075.5	0.0	2.1873	2.1873
34	0.0960	0.01602	3061.9	3061.9	2.00	1020.0	1022.0	2.00	1074.4	1076.4	0.0041	2.1762	2.1802
36	0.1040	0.01602	2839.0	2839.0	4.01	1018.6	1022.6	4.01	1073.2	1077.2	0.0081	2.1651	2.1732
38	0.1125	0.01602	2634.1	2634.2	6.02	1017.3	1023.3	6.02	1072.1	1078.1	0.0122	2.1541	2.1663
40	0.1216	0.01602	2445.8	2445.8	8.03	1015.9	1023.9	8.03	1071.0	1079.0	0.0162	2.1432	2.1594
42	0.1314	0.01602	2272.4	2272.4	10.03	1014.6	1024.6	10.03	1069.8	1079.9	0.0202	2.1325	2.1527
44	0.1419	0.01602	2112.8	2112.8	12.04	1013.2	1025.2	12.04	1068.7	1080.7	0.0242	2.1217	2.1459
46	0.1531	0.01602	1965.7	1965.7	14.05	1011.9	1025.9	14.05	1067.6	1081.6	0.0282	2.1111	2.1393
48	0.1651	0.01602	1830.0	1830.0	16.05	1010.5	1026.6	16.05	1066.4	1082.5	0.0321	2.1006	2.1327
50	0.1780	0.01602	1704.8	1704.8	18.05	1009.2	1027.2	18.05	1065.3	1083.4	0.0361	2.0901	2.1262
52	0.1916	0.01602	1589.2	1589.2	20.06	1007.9	1027.9	20.06	1064.2	1084.2	0.0400	2.0798	2.1197
54	0.2063	0.01603	1482.4	1482.4	22.06	1006.5	1028.5	22.06	1063.1	1085.1	0.0439	2.0695	2.1134
56	0.2218	0.01603	1383.6	1383.6	24.06	1005.1	1029.2	24.06	1061.9	1086.0	0.0478	2.0593	2.1070
58	0.2384	0.01603	1292.2	1292.2	26.06	1003.8	1029.8	26.06	1060.8	1086.9	0.0516	2.0491	2.1008
60	0.2561	0.01603	1207.6	1207.6	28.06	1002.4	1030.5	28.06	1059.7	1087.7	0.0555	2.0391	2.0946
62	0.2749	0.01604	1129.2	1129.2	30.06	1001.1	1031.2	30.06	1058.5	1088.6	0.0593	2.0291	2.0885
64	0.2950	0.01604	1056.5	1056.5	32.06	999.8	1031.8	32.06	1057.4	1089.5	0.0632	2.0192	2.0824
66	0.3163	0.01604	989.0	989.1	34.06	998.4	1032.5	34.06	1056.3	1090.4	0.0670	2.0094	2.0764
68	0.3389	0.01605	926.5	926.5	36.05	997.1	1033.1	36.05	1055.2	1091.2	0.0708	1.9996	2.0704
70	0.3629	0.01605	868.3	868.4	38.05	995.7	1033.8	38.05	1054.0	1092.1	0.0745	1.9900	2.0645
72	0.3884	0.01605	814.3	814.3	40.05	994.4	1034.4	40.05	1052.9	1093.0	0.0783	1.9804	2.0587
74	0.4155	0.01606	764.1	764.1	42.05	993.0	1035.1	42.05	1051.8	1093.8	0.0821	1.9708	2.0529
76	0.4442	0.01606	717.4	717.4	44.04	991.7	1035.7	44.04	1050.7	1094.7	0.0858	1.9614	2.0472
78	0.4746	0.01607	673.8	673.9	46.04	990.3	1036.4	46.04	1049.5	1095.6	0.0895	1.9520	2.0415
80	0.5068	0.01607	633.3	633.3	48.03	989.0	1037.0	48.04	1048.4	1096.4	0.0932	1.9426	2.0359

2.0303	1.9334	0.0969	1097.3	1047.3	50.03	1037.7	987.7	50.03	595.6	595.5	0.01608	0.5409	82
2.0248	1.9242	0.1006	1098.2	1046.1	52.03	1038.3	986.3	52.03	560.3	560.3	0.01608	0.5770	84
2.0193	1.9151	0.1043	1099.0	1045.0	54.03	1039.0	985.0	54.02	527.5	527.5	0.01609	0.6152	86
2.0139	1.9060	0.1079	1099.9	1043.9	56.02	1039.6	983.6	56.02	496.8	496.8	0.01609	0.6555	88
2.0086	1.8970	0.1115	1100.8	1042.7	58.02	1040.3	982.3	58.02	468.1	468.1	0.01610	0.6981	90
2.0033	1.8881	0.1152	1101.6	1041.6	60.01	1040.9	980.9	60.01	441.3	441.3	0.01610	0.7431	92
1.9980	1.8792	0.1188	1102.5	1040.5	62.01	1041.6	979.6	62.01	416.3	416.3	0.01611	0.7906	94
1.9928	1.8704	0.1224	1103.3	1039.3	64.01	1042.2	978.2	64.00	392.8	392.9	0.01612	0.8407	96
1.9876	1.8617	0.1260	1104.2	1038.2	66.00	1042.9	976.9	66.00	370.9	370.9	0.01612	0.8936	98
1.9825	1.8530	0.1295	1105.1	1037.1	68.00	1043.5	975.5	68.00	350.4	350.4	0.01613	0.9492	100
1.9775	1.8444	0.1331	1105.9	1035.9	70.00	1044.2	974.2	69.99	331.1	331.1	0.01614	1.0079	102
1.9725	1.8358	0.1366	1106.8	1034.8	71.99	1044.8	972.8	71.99	313.1	313.1	0.01614	1.0697	104
1.9675	1.8273	0.1402	1107.6	1033.6	73.99	1045.4	971.5	73.98	296.2	296.2	0.01615	1.1347	106
1.9626	1.8188	0.1437	1108.5	1032.5	75.98	1046.1	970.1	75.98	280.3	280.3	0.01616	1.2030	108
1.9577	1.8105	0.1472	1109.3	1031.4	77.98	1046.7	968.8	77.98	265.4	265.4	0.01617	1.275	110
1.9528	1.8021	0.1507	1110.2	1030.2	79.98	1047.4	967.4	79.97	251.4	251.4	0.01617	1.351	112
1.9480	1.7938	0.1542	1111.0	1029.1	81.97	1048.0	966.0	81.97	238.2	238.2	0.01618	1.430	114
1.9433	1.7856	0.1577	1111.9	1027.9	83.97	1048.6	964.7	83.97	225.9	225.8	0.01619	1.513	116
1.9386	1.7774	0.1611	1112.7	1026.8	85.97	1049.3	963.3	85.96	214.2	214.2	0.01620	1.601	118
1.9339	1.7693	0.1646	1113.6	1025.6	87.97	1049.9	962.0	87.96	203.26	203.25	0.01620	1.693	120
1.9293	1.7613	0.1680	1114.4	1024.5	89.96	1050.6	960.6	89.96	192.95	192.94	0.01621	1.789	122
1.9247	1.7533	0.1715	1115.3	1023.3	91.96	1051.2	959.2	91.96	183.24	183.23	0.01622	1.890	124
1.9202	1.7453	0.1749	1116.1	1022.2	93.96	1051.8	957.9	93.95	174.09	174.08	0.01623	1.996	126
1.9157	1.7374	0.1783	1117.0	1021.0	95.96	1052.4	956.5	95.95	165.47	165.45	0.01624	2.107	128
1.9112	1.7295	0.1817	1117.8	1019.8	97.96	1053.1	955.1	97.95	157.33	157.32	0.01625	2.223	130
1.9068	1.7217	0.1851	1118.6	1018.7	99.95	1053.7	953.8	99.95	149.66	149.64	0.01626	2.345	132
1.9024	1.7140	0.1884	1119.5	1017.5	101.95	1054.3	952.4	101.94	142.41	142.40	0.01626	2.472	134
1.8980	1.7063	0.1918	1120.3	1016.4	103.95	1055.0	951.0	103.94	135.57	135.55	0.01627	2.605	136
1.8937	1.6986	0.1951	1121.1	1015.2	105.95	1055.6	949.6	105.94	129.11	129.09	0.01628	2.744	138
1.8895	1.6910	0.1985	1122.0	1014.0	107.95	1056.2	948.3	107.94	123.00	122.98	0.01629	2.889	140
1.8852	1.6834	0.2018	1122.8	1012.9	109.95	1056.8	946.9	109.94	117.22	117.21	0.01630	3.041	142
1.8810	1.6759	0.2051	1123.6	1011.7	111.95	1057.5	945.5	111.94	111.76	111.74	0.01631	3.200	144
1.8769	1.6684	0.2084	1124.5	1010.5	113.95	1058.1	944.1	113.94	106.59	106.58	0.01632	3.365	146
1.8727	1.6610	0.2117	1125.3	1009.3	115.95	1058.7	942.8	115.94	101.70	101.68	0.01633	3.538	148
1.8686	1.6536	0.2150	1126.1	1008.2	117.95	1059.3	941.4	117.94	97.07	97.05	0.01634	3.718	150
1.8646	1.6463	0.2183	1126.9	1007.0	119.95	1059.9	940.0	119.94	92.68	92.66	0.01635	3.906	152
1.8606	1.6390	0.2216	1127.7	1005.8	121.95	1060.5	938.6	121.94	88.52	88.50	0.01636	4.102	154
1.8566	1.6318	0.2248	1128.6	1004.6	123.95	1061.2	937.2	123.94	84.57	84.56	0.01637	4.307	156
1.8526	1.6245	0.2281	1129.4	1003.4	125.96	1061.8	935.8	125.94	80.83	80.82	0.01638	4.520	158
1.8487	1.6174	0.2313	1130.2	1002.2	127.96	1062.4	934.4	127.94	77.29	77.27	0.01640	4.741	160

Table F.3: Saturated Steam, English units (Continued)

t °F	P (psia)	SPECIFIC VOLUME V sat. liq.	evap.	sat. vap.	INTERNAL ENERGY U sat. liq.	evap.	sat. vap.	ENTHALPY H sat. liq.	evap.	sat. vap.	ENTROPY S sat. liq.	evap.	sat. vap.
162	4.972	0.01641	73.90	73.92	129.95	933.0	1063.0	129.96	1001.0	1131.0	0.2345	1.6103	1.8448
164	5.212	0.01642	70.70	70.72	131.95	931.6	1063.6	131.96	999.8	1131.8	0.2377	1.6032	1.8409
166	5.462	0.01643	67.67	67.68	133.95	930.2	1064.2	133.97	998.6	1132.6	0.2409	1.5961	1.8371
168	5.722	0.01644	64.78	64.80	135.95	928.8	1064.8	135.97	997.4	1133.4	0.2441	1.5892	1.8333
170	5.993	0.01645	62.04	62.06	137.96	927.4	1065.4	137.97	996.2	1134.2	0.2473	1.5822	1.8295
172	6.274	0.01646	59.43	59.45	139.96	926.0	1066.0	139.98	995.0	1135.0	0.2505	1.5753	1.8258
174	6.566	0.01647	56.95	56.97	141.96	924.6	1066.6	141.98	993.8	1135.8	0.2537	1.5684	1.8221
176	6.869	0.01649	54.59	54.61	143.97	923.2	1067.2	143.99	992.6	1136.6	0.2568	1.5616	1.8184
178	7.184	0.01650	52.35	52.36	145.97	921.8	1067.8	145.99	991.4	1137.4	0.2600	1.5548	1.8147
180	7.511	0.01651	50.21	50.22	147.98	920.4	1068.4	148.00	990.2	1138.2	0.2631	1.5480	1.8111
182	7.850	0.01652	48.17	48.19	149.98	919.0	1069.0	150.01	989.0	1139.0	0.2662	1.5413	1.8075
184	8.203	0.01653	46.23	46.25	151.99	917.6	1069.6	152.01	987.8	1139.8	0.2694	1.5346	1.8040
186	8.568	0.01655	44.38	44.40	153.99	916.2	1070.2	154.02	986.5	1140.5	0.2725	1.5279	1.8004
188	8.947	0.01656	42.62	42.64	156.00	914.7	1070.7	156.03	985.3	1141.3	0.2756	1.5213	1.7969
190	9.340	0.01657	40.94	40.96	158.01	913.3	1071.3	158.04	984.1	1142.1	0.2787	1.5148	1.7934
192	9.747	0.01658	39.34	39.35	160.02	911.9	1071.9	160.05	982.8	1142.9	0.2818	1.5082	1.7900
194	10.168	0.01660	37.81	37.82	162.02	910.5	1072.5	162.05	981.6	1143.7	0.2848	1.5017	1.7865
196	10.605	0.01661	36.35	36.36	164.03	909.0	1073.1	164.06	980.4	1144.4	0.2879	1.4952	1.7831
198	11.058	0.01662	34.95	34.97	166.04	907.6	1073.6	166.08	979.1	1145.2	0.2910	1.4888	1.7798
200	11.526	0.01664	33.62	33.64	168.05	906.2	1074.2	168.00	977.9	1146.0	0.2940	1.4824	1.7764
202	12.011	0.01665	32.35	32.37	170.06	904.7	1074.8	170.10	976.6	1146.7	0.2971	1.4760	1.7731
204	12.512	0.01666	31.13	31.15	172.07	903.3	1075.3	172.11	975.4	1147.5	0.3001	1.4697	1.7698
206	13.031	0.01668	29.97	29.99	174.08	901.8	1075.9	174.12	974.1	1148.2	0.3031	1.4634	1.7665
208	13.568	0.01669	28.86	28.88	176.09	900.4	1076.5	176.14	972.8	1149.0	0.3061	1.4571	1.7632
210	14.123	0.01670	27.80	27.82	178.11	898.9	1077.0	178.15	971.6	1149.7	0.3091	1.4509	1.7600
212	14.696	0.01672	26.78	26.80	180.12	897.5	1077.6	180.17	970.3	1150.5	0.3121	1.4447	1.7568
215	15.592	0.01674	25.34	25.36	183.14	895.3	1078.4	183.19	968.4	1151.6	0.3166	1.4354	1.7520
220	17.186	0.01678	23.13	23.15	188.18	891.6	1079.8	188.22	965.2	1153.4	0.3241	1.4201	1.7442
225	18.912	0.01681	21.15	21.17	193.22	888.0	1081.2	193.28	962.0	1155.3	0.3315	1.4051	1.7365
230	20.78	0.01685	19.364	19.381	198.27	884.3	1082.5	198.33	958.7	1157.1	0.3388	1.3902	1.7290
235	22.79	0.01689	17.756	17.773	203.32	880.5	1083.9	203.39	955.4	1158.8	0.3461	1.3754	1.7215
240	24.97	0.01693	16.304	16.321	208.37	876.8	1085.2	208.45	952.1	1160.6	0.3533	1.3609	1.7142
245	27.31	0.01697	14.991	15.008	213.43	873.1	1086.5	213.52	948.8	1162.3	0.3606	1.3465	1.7070
250	29.82	0.01701	13.802	13.819	218.50	869.3	1087.8	218.59	945.4	1164.0	0.3677	1.3323	1.7000
255	32.53	0.01705	12.724	12.741	223.57	865.5	1089.0	223.67	942.1	1165.7	0.3748	1.3182	1.6930

260	35.43	0.01709	11.745	11.762	228.64	861.6	1090.3	228.76	938.6	1167.4	0.3819	1.3043	1.6862
265	38.53	0.01713	10.854	10.871	233.73	857.8	1091.5	233.85	935.2	1169.0	0.3890	1.2905	1.6795
270	41.86	0.01717	10.042	10.060	238.82	853.9	1092.7	238.95	931.7	1170.6	0.3960	1.2769	1.6729
275	45.41	0.01722	9.302	9.320	243.91	850.0	1093.9	244.06	928.2	1172.2	0.4029	1.2634	1.6663
280	49.20	0.01726	8.627	8.644	249.01	846.1	1095.1	249.17	924.6	1173.8	0.4098	1.2501	1.6599
285	53.24	0.01731	8.009	8.026	254.12	842.1	1096.2	254.29	921.0	1175.3	0.4167	1.2368	1.6536
290	57.55	0.01736	7.443	7.460	259.24	838.1	1097.4	259.43	917.4	1176.8	0.4236	1.2238	1.6473
295	62.13	0.01740	6.924	6.942	264.37	834.1	1098.5	264.57	913.7	1178.3	0.4304	1.2108	1.6412
300	67.01	0.01745	6.448	6.466	269.50	830.1	1099.6	269.71	910.0	1179.7	0.4372	1.1979	1.6351
305	72.18	0.01750	6.011	6.028	274.64	826.0	1100.6	274.87	906.3	1181.1	0.4439	1.1852	1.6291
310	77.67	0.01755	5.608	5.626	279.79	821.9	1101.7	280.04	902.5	1182.5	0.4506	1.1726	1.6232
315	83.48	0.01760	5.238	5.255	284.94	817.7	1102.7	285.21	898.7	1183.9	0.4573	1.1601	1.6174
320	89.64	0.01766	4.896	4.914	290.11	813.6	1103.7	290.40	894.8	1185.2	0.4640	1.1477	1.6116
325	96.16	0.01771	4.581	4.598	295.28	809.4	1104.6	295.60	890.9	1186.5	0.4706	1.1354	1.6059
330	103.05	0.01776	4.289	4.307	300.47	805.1	1105.6	300.81	886.9	1187.7	0.4772	1.1231	1.6003
335	110.32	0.01782	4.020	4.037	305.66	800.8	1106.5	306.03	882.9	1188.9	0.4837	1.1110	1.5947
340	117.99	0.01787	3.770	3.788	310.87	796.5	1107.4	311.26	878.8	1190.1	0.4902	1.0990	1.5892
345	126.08	0.01793	3.539	3.556	316.08	792.2	1108.2	316.50	874.7	1191.2	0.4967	1.0871	1.5838
350	134.60	0.01799	3.324	3.342	321.31	787.8	1109.1	321.76	870.6	1192.3	0.5032	1.0752	1.5784
355	143.57	0.01805	3.124	3.143	326.55	783.3	1109.9	327.03	866.3	1193.4	0.5097	1.0634	1.5731
360	153.01	0.01811	2.939	2.957	331.79	778.9	1110.7	332.31	862.1	1194.4	0.5161	1.0517	1.5678
365	162.93	0.01817	2.767	2.785	337.05	774.3	1111.4	337.60	857.8	1195.4	0.5225	1.0401	1.5626
370	173.34	0.01823	2.606	2.624	342.33	769.8	1112.1	342.91	853.4	1196.3	0.5289	1.0286	1.5575
375	184.27	0.01830	2.457	2.475	347.61	765.2	1112.8	348.24	849.0	1197.2	0.5352	1.0171	1.5523
380	195.73	0.01836	2.317	2.335	352.91	760.5	1113.5	353.58	844.5	1198.0	0.5416	1.0057	1.5473
385	207.74	0.01843	2.187	2.205	358.22	755.9	1114.1	358.93	839.9	1198.8	0.5479	0.9944	1.5422
390	220.32	0.01850	2.065	2.083	363.55	751.1	1114.7	364.30	835.3	1199.6	0.5542	0.9831	1.5372
395	233.49	0.01857	1.9510	1.9695	368.89	746.3	1115.2	369.69	830.6	1200.3	0.5604	0.9718	1.5323
400	247.26	0.01864	1.8444	1.8630	374.24	741.5	1115.7	375.09	825.9	1201.0	0.5667	0.9607	1.5274
405	261.65	0.01871	1.7445	1.7633	379.61	736.6	1116.2	380.52	821.1	1201.6	0.5729	0.9496	1.5225
410	276.69	0.01878	1.6510	1.6697	384.99	731.7	1116.7	385.96	816.2	1202.1	0.5791	0.9385	1.5176
415	292.40	0.01886	1.5632	1.5820	390.40	726.7	1117.1	391.42	811.2	1202.7	0.5853	0.9275	1.5128
420	308.78	0.01894	1.4808	1.4997	395.81	721.6	1117.4	396.90	806.2	1203.1	0.5915	0.9165	1.5080
425	325.87	0.01901	1.4033	1.4224	401.25	716.5	1117.8	402.40	801.1	1203.5	0.5977	0.9055	1.5032
430	343.67	0.01909	1.3306	1.3496	406.70	711.3	1118.0	407.92	796.0	1203.9	0.6038	0.8946	1.4985
435	362.23	0.01918	1.2621	1.2812	412.18	706.1	1118.3	413.46	790.7	1204.2	0.6100	0.8838	1.4937
440	381.54	0.01926	1.1976	1.2169	417.67	700.8	1118.5	419.03	785.4	1204.4	0.6161	0.8729	1.4890
445	401.64	0.01934	1.1369	1.1562	423.18	695.5	1118.7	424.62	780.0	1204.6	0.6222	0.8621	1.4843
450	422.55	0.01943	1.0796	1.0991	428.71	690.1	1118.8	430.23	774.5	1204.7	0.6283	0.8514	1.4797
455	444.28	0.0195	1.0256	1.0451	434.27	684.6	1118.9	435.87	768.9	1204.8	0.6344	0.8406	1.4750

Table F.3: Saturated Steam, English units (Continued)

t °F	P (psia)	SPECIFIC VOLUME V			INTERNAL ENERGY U			ENTHALPY H			ENTROPY S		
		sat. liq.	evap.	sat. vap.	sat. liq.	evap.	sat. vap.	sat. liq.	evap.	sat. vap.	sat. liq.	evap.	sat. vap.
460	466.87	0.0196	0.9746	0.9942	439.84	679.0	1118.9	441.54	763.2	1204.8	0.6405	0.8299	1.4704
465	490.32	0.0197	0.9265	0.9462	445.44	673.4	1118.9	447.23	757.5	1204.7	0.6466	0.8192	1.4657
470	514.67	0.0198	0.8810	0.9008	451.06	667.7	1118.8	452.95	751.6	1204.6	0.6527	0.8084	1.4611
475	539.94	0.0199	0.8379	0.8578	456.71	662.0	1118.7	458.70	745.7	1204.4	0.6587	0.7977	1.4565
480	566.15	0.0200	0.7972	0.8172	462.39	656.1	1118.5	464.48	739.6	1204.1	0.6648	0.7871	1.4518
485	593.32	0.0201	0.7586	0.7787	468.09	650.2	1118.3	470.29	733.5	1203.8	0.6708	0.7764	1.4472
490	621.48	0.0202	0.7220	0.7422	473.82	644.2	1118.0	476.14	727.2	1203.3	0.6769	0.7657	1.4426
495	650.65	0.0203	0.6874	0.7077	479.57	638.0	1117.6	482.02	720.8	1202.8	0.6830	0.7550	1.4380
500	680.86	0.0204	0.6545	0.6749	485.36	631.8	1117.2	487.94	714.3	1202.2	0.6890	0.7443	1.4333
505	712.12	0.0205	0.6233	0.6438	491.2	625.6	1116.7	493.9	707.7	1201.6	0.6951	0.7336	1.4286
510	744.47	0.0207	0.5936	0.6143	497.0	619.2	1116.2	499.9	700.9	1200.8	0.7012	0.7228	1.4240
515	777.93	0.0208	0.5654	0.5862	502.9	612.7	1115.6	505.9	694.1	1200.0	0.7072	0.7120	1.4193
520	812.53	0.0209	0.5386	0.5596	508.8	606.1	1114.9	512.0	687.0	1199.0	0.7133	0.7013	1.4146
525	848.28	0.0210	0.5131	0.5342	514.8	599.3	1114.2	518.1	679.9	1198.0	0.7194	0.6904	1.4098
530	885.23	0.0212	0.4889	0.5100	520.8	592.5	1113.3	524.3	672.6	1196.9	0.7255	0.6796	1.4051
535	923.39	0.0213	0.4657	0.4870	526.9	585.6	1112.4	530.5	665.1	1195.6	0.7316	0.6686	1.4003
540	962.79	0.0215	0.4437	0.4651	532.9	578.5	1111.4	536.8	657.5	1194.3	0.7378	0.6577	1.3954
545	1003.5	0.0216	0.4226	0.4442	539.1	571.2	1110.3	543.1	649.7	1192.8	0.7439	0.6467	1.3906
550	1045.4	0.0218	0.4026	0.4243	545.3	563.9	1109.1	549.5	641.8	1191.2	0.7501	0.6356	1.3856
555	1088.7	0.0219	0.3834	0.4053	551.5	556.4	1107.9	555.9	633.6	1189.5	0.7562	0.6244	1.3807
560	1133.4	0.0221	0.3651	0.3871	557.8	548.7	1106.5	562.4	625.3	1187.7	0.7625	0.6132	1.3757
565	1179.4	0.0222	0.3475	0.3698	564.1	540.9	1105.0	569.0	616.8	1185.7	0.7687	0.6019	1.3706
570	1226.9	0.0224	0.3308	0.3532	570.5	532.9	1103.4	575.6	608.0	1183.6	0.7750	0.5905	1.3654
575	1275.8	0.0226	0.3147	0.3373	577.0	524.8	1101.7	582.3	599.1	1181.4	0.7813	0.5790	1.3602
580	1326.2	0.0228	0.2994	0.3222	583.5	516.4	1099.9	589.1	589.9	1179.0	0.7876	0.5673	1.3550
585	1378.1	0.0230	0.2846	0.3076	590.1	507.9	1098.0	596.0	580.4	1176.4	0.7940	0.5556	1.3496
590	1431.5	0.0232	0.2705	0.2937	596.8	499.1	1095.9	602.9	570.8	1173.7	0.8004	0.5437	1.3442
595	1486.6	0.0234	0.2569	0.2803	603.5	490.2	1093.7	610.0	560.8	1170.8	0.8069	0.5317	1.3386
600	1543.2	0.0236	0.2438	0.2675	610.4	481.0	1091.3	617.1	550.6	1167.7	0.8134	0.5196	1.3330
605	1601.5	0.0239	0.2313	0.2551	617.3	471.5	1088.8	624.4	540.0	1164.4	0.8200	0.5072	1.3273
610	1661.6	0.0241	0.2191	0.2433	624.4	461.8	1086.1	631.8	529.2	1160.9	0.8267	0.4947	1.3214
615	1723.3	0.0244	0.2075	0.2318	631.5	451.8	1083.3	639.3	517.9	1157.2	0.8334	0.4819	1.3154
620	1786.9	0.0247	0.1961	0.2208	638.8	441.4	1080.2	646.9	506.3	1153.2	0.8403	0.4689	1.3092
625	1852.2	0.0250	0.1852	0.2102	646.2	430.7	1076.8	654.7	494.2	1148.9	0.8472	0.4556	1.3028
630	1919.5	0.0253	0.1746	0.1999	653.7	419.5	1073.2	662.7	481.6	1144.2	0.8542	0.4419	1.2962

635	1988.7	0.0256	0.1643	0.1899	661.4	407.9	1069.3	670.8	468.4	1139.2	0.8614	0.4279	1.2893
640	2059.9	0.0259	0.1543	0.1802	669.2	395.8	1065.0	679.1	454.6	1133.7	0.8686	0.4134	1.2821
645	2133.1	0.0263	0.1445	0.1708	677.3	383.1	1060.4	687.7	440.2	1127.8	0.8761	0.3985	1.2746
650	2208.4	0.0267	0.1350	0.1617	685.5	369.8	1055.3	696.4	425.0	1121.4	0.8837	0.3830	1.2667
655	2285.9	0.0272	0.1257	0.1529	694.0	355.8	1049.8	705.5	409.0	1114.5	0.8915	0.3670	1.2584
660	2365.7	0.0277	0.1166	0.1443	702.8	341.0	1043.9	714.9	392.1	1107.0	0.8995	0.3502	1.2498
662	2398.2	0.0279	0.1131	0.1409	706.4	335.0	1041.4	718.7	385.2	1103.9	0.9029	0.3433	1.2462
664	2431.1	0.0281	0.1095	0.1376	710.2	328.5	1038.7	722.9	377.7	1100.6	0.9064	0.3361	1.2425
666	2464.4	0.0283	0.1059	0.1342	714.2	321.7	1035.9	727.1	370.0	1097.1	0.9100	0.3286	1.2387
668	2498.1	0.0286	0.1023	0.1309	718.3	314.8	1033.0	731.5	362.1	1093.5	0.9137	0.3210	1.2347
670	2532.2	0.0288	0.0987	0.1275	722.3	307.7	1030.0	735.8	354.0	1089.8	0.9174	0.3133	1.2307
672	2566.6	0.0291	0.0951	0.1242	726.4	300.5	1026.9	740.2	345.7	1085.9	0.9211	0.3054	1.2266
674	2601.5	0.0294	0.0916	0.1210	730.5	293.1	1023.6	744.7	337.2	1081.9	0.9249	0.2974	1.2223
676	2636.8	0.0297	0.0880	0.1177	734.7	285.5	1020.2	749.2	328.5	1077.6	0.9287	0.2892	1.2179
678	2672.5	0.0300	0.0844	0.1144	738.9	277.7	1016.6	753.8	319.4	1073.2	0.9326	0.2807	1.2133
680	2708.6	0.0304	0.0808	0.1112	743.2	269.6	1012.8	758.5	310.1	1068.5	0.9365	0.2720	1.2086
682	2745.1	0.0307	0.0772	0.1079	747.7	261.2	1008.8	763.3	300.4	1063.6	0.9406	0.2631	1.2036
684	2782.1	0.0311	0.0735	0.1046	752.2	252.4	1004.6	768.2	290.2	1058.4	0.9447	0.2537	1.1984
686	2819.5	0.0316	0.0698	0.1013	756.9	243.1	1000.0	773.4	279.5	1052.9	0.9490	0.2439	1.1930
688	2857.4	0.0320	0.0659	0.0980	761.8	233.3	995.2	778.8	268.2	1047.0	0.9535	0.2337	1.1872
690	2895.7	0.0326	0.0620	0.0946	767.0	222.9	989.9	784.5	256.1	1040.6	0.9583	0.2227	1.1810
692	2934.5	0.0331	0.0580	0.0911	772.5	211.6	984.1	790.5	243.1	1033.6	0.9634	0.2110	1.1744
694	2973.7	0.0338	0.0537	0.0875	778.5	199.2	977.7	797.1	228.8	1025.9	0.9689	0.1983	1.1671
696	3013.4	0.0345	0.0492	0.0837	785.1	185.4	970.5	804.4	212.8	1017.2	0.9749	0.1841	1.1591
698	3053.6	0.0355	0.0442	0.0797	792.6	169.6	962.2	812.6	194.6	1007.2	0.9818	0.1681	1.1499
700	3094.3	0.0366	0.0386	0.0752	801.5	150.7	952.1	822.4	172.7	995.2	0.9901	0.1490	1.1390
702	3135.5	0.0382	0.0317	0.0700	812.8	126.3	939.1	835.0	144.7	979.7	1.0006	0.1246	1.1252
704	3177.2	0.0411	0.0219	0.0630	830.1	89.1	919.2	854.2	102.0	956.2	1.0169	0.0876	1.1046
705.47	3208.2	0.0508	0.0000	0.0508	875.9	-0.0	875.9	906.0	-0.0	906.0	1.0612	0.0000	1.0612

Table F.4: Superheated Steam, English units

TEMPERATURE: t °F

P/(psia) (t^{sat}/°F)		sat. liq.	sat. vap.	200	250	300	350	400	450	500
1 (101.74)	V	0.0161	333.60	392.5	422.4	452.3	482.1	511.9	541.7	571.5
	U	69.73	1044.1	1077.5	1094.7	1112.0	1129.5	1147.1	1164.9	1182.8
	H	69.73	1105.8	1150.2	1172.9	1195.7	1218.7	1241.8	1265.1	1288.6
	S	0.1326	1.9781	2.0509	2.0841	2.1152	2.1445	2.1722	2.1985	2.2237
5 (162.24)	V	0.0164	73.532	78.14	84.21	90.24	96.25	102.2	108.2	114.2
	U	130.18	1063.1	1076.3	1093.8	1111.3	1128.9	1146.7	1164.5	1182.6
	H	130.20	1131.1	1148.6	1171.7	1194.8	1218.0	1241.3	1264.7	1288.2
	S	0.2349	1.8443	1.8716	1.9054	1.9369	1.9664	1.9943	2.0208	2.0460
10 (193.21)	V	0.0166	38.420	38.84	41.93	44.98	48.02	51.03	54.04	57.04
	U	161.23	1072.3	1074.7	1092.6	1110.4	1128.3	1146.1	1164.1	1182.2
	H	161.26	1143.3	1146.6	1170.2	1193.7	1217.1	1240.6	1264.1	1287.8
	S	0.2836	1.7879	1.7928	1.8273	1.8593	1.8892	1.9173	1.9439	1.9692
14.696 (212.00)	V	0.0167	26.799	. . . . .	28.42	30.52	32.60	34.67	36.72	38.77
	U	180.12	1077.6	. . . . .	1091.5	1109.6	1127.6	1145.7	1163.7	1181.9
	H	180.17	1150.5	. . . . .	1168.8	1192.6	1216.3	1239.9	1263.6	1287.4
	S	0.3121	1.7568	. . . . .	1.7833	1.8158	1.8460	1.8743	1.9010	1.9265
15 (213.03)	V	0.0167	26.290	. . . . .	27.84	29.90	31.94	33.96	35.98	37.98
	U	181.16	1077.9	. . . . .	1091.4	1109.5	1127.6	1145.6	1163.7	1181.9
	H	181.21	1150.9	. . . . .	1168.7	1192.5	1216.2	1239.9	1263.6	1287.3
	S	0.3137	1.7552	. . . . .	1.7809	1.8134	1.8436	1.8720	1.8988	1.9242
20 (227.96)	V	0.0168	20.087	. . . . .	20.79	22.36	23.90	25.43	26.95	28.46
	U	196.21	1082.0	. . . . .	1090.3	1108.6	1126.9	1145.1	1163.3	1181.6
	H	196.27	1156.3	. . . . .	1167.1	1191.4	1215.4	1239.2	1263.0	1286.9
	S	0.3358	1.7320	. . . . .	1.7475	1.7805	1.8111	1.8397	1.8666	1.8921
25 (240.07)	V	0.0169	16.301	. . . . .	16.56	17.83	19.08	20.31	21.53	22.74
	U	208.44	1085.2	. . . . .	1089.0	1107.7	1126.2	1144.6	1162.9	1181.2
	H	208.52	1160.6	. . . . .	1165.6	1190.2	1214.5	1238.5	1262.5	1286.4
	S	0.3535	1.7141	. . . . .	1.7212	1.7547	1.7856	1.8145	1.8415	1.8672
30 (250.34)	V	0.0170	13.744	. . . . .	. . . . .	14.81	15.86	16.89	17.91	18.93
	U	218.84	1087.9	. . . . .	. . . . .	1106.8	1125.5	1144.0	1162.5	1180.9
	H	218.93	1164.1	. . . . .	. . . . .	1189.0	1213.6	1237.8	1261.9	1286.0
	S	0.3682	1.6995	. . . . .	. . . . .	1.7334	1.7647	1.7937	1.8210	1.8467

Abs. Press. (Sat. Temp.)		(Sat. Liq.)	(Sat. Vap.)						
35 (259.29)	V	0.0171	11.896	...	12.65	13.56	14.45	15.33	16.21
	U	227.92	1090.1	...	1105.9	1124.8	1143.5	1162.0	1180.5
	H	228.03	1167.1	...	1187.8	1212.7	1237.1	1261.3	1285.5
	S	0.3809	1.6872	...	1.7152	1.7468	1.7761	1.8035	1.8294
40 (267.25)	V	0.0172	10.497	...	11.04	11.84	12.62	13.40	14.16
	U	236.02	1092.1	...	1104.9	1124.1	1142.9	1161.6	1180.2
	H	236.14	1169.8	...	1186.6	1211.7	1236.4	1260.8	1285.0
	S	0.3921	1.6765	...	1.6992	1.7312	1.7608	1.7883	1.8143
45 (274.44)	V	0.0172	9.399	...	9.777	10.50	11.20	11.89	12.58
	U	243.34	1093.8	...	1104.0	1123.4	1142.4	1161.2	1179.8
	H	243.49	1172.0	...	1185.4	1210.8	1235.7	1260.2	1284.6
	S	0.4021	1.6671	...	1.6849	1.7173	1.7471	1.7749	1.8009
50 (281.01)	V	0.0173	8.514	...	8.769	9.424	10.06	10.69	11.31
	U	250.05	1095.3	...	1103.0	1122.7	1141.8	1160.7	1179.5
	H	250.21	1174.1	...	1184.1	1209.9	1234.9	1259.6	1284.1
	S	0.4112	1.6586	...	1.6720	1.7048	1.7349	1.7628	1.7890
55 (287.08)	V	0.0173	7.785	...	7.945	8.546	9.130	9.702	10.27
	U	256.25	1096.7	...	1102.0	1121.9	1141.3	1160.3	1179.1
	H	256.43	1175.9	...	1182.8	1208.9	1234.2	1259.1	1283.6
	S	0.4196	1.6510	...	1.6601	1.6934	1.7237	1.7518	1.7781
60 (292.71)	V	0.0174	7.174	...	7.257	7.815	8.354	8.881	9.400
	U	262.02	1098.0	...	1101.0	1121.2	1140.7	1159.9	1178.8
	H	262.21	1177.6	...	1181.6	1208.0	1233.5	1258.5	1283.2
	S	0.4273	1.6440	...	1.6492	1.6829	1.7134	1.7417	1.7681
65 (297.98)	V	0.0174	6.653	...	6.675	7.195	7.697	8.186	8.667
	U	267.42	1099.1	...	1100.0	1120.4	1140.2	1159.4	1178.4
	H	267.63	1179.1	...	1180.3	1207.0	1232.7	1257.9	1282.7
	S	0.4344	1.6375	...	1.6390	1.6731	1.7040	1.7324	1.7589
70 (302.93)	V	0.0175	6.205	...	...	6.664	7.133	7.590	8.039
	U	272.51	1100.2	...	...	1119.7	1139.6	1159.0	1178.1
	H	272.74	1180.6	...	...	1206.0	1232.0	1257.3	1282.2
	S	0.4411	1.6316	...	...	1.6640	1.6951	1.7237	1.7504
75 (307.61)	V	0.0175	5.814	...	...	6.204	6.645	7.074	7.494
	U	277.32	1101.2	...	...	1118.9	1139.0	1158.5	1177.7
	H	277.56	1181.9	...	...	1205.0	1231.2	1256.7	1281.7
	S	0.4474	1.6260	...	...	1.6554	1.6868	1.7156	1.7424

Table F.4: Superheated Steam, English units (Continued)

TEMPERATURE: $t\,^{\circ}F$

P/(psia) ($t^{sat}/^{\circ}F$)		sat. liq.	sat. vap.	600	700	800	900	1000	1100	1200
1 (101.74)	V	0.0161	333.60	631.1	690.7	750.3	809.9	869.5	929.0	988.6
	U	69.73	1044.1	1219.3	1256.7	1294.9	1334.0	1374.0	1414.9	1456.7
	H	69.73	1105.8	1336.1	1384.5	1433.7	1483.8	1534.9	1586.8	1639.7
	S	0.1326	1.9781	2.2708	2.3144	2.3551	2.3934	2.4296	2.4640	2.4969
5 (162.24)	V	0.0164	73.532	126.1	138.1	150.0	161.9	173.9	185.8	197.7
	U	130.18	1063.1	1219.2	1256.5	1294.8	1333.9	1373.9	1414.8	1456.7
	H	130.20	1131.1	1335.9	1384.3	1433.6	1483.7	1534.7	1586.7	1639.6
	S	0.2349	1.8443	2.0932	2.1369	2.1776	2.2159	2.2521	2.2866	2.3194
10 (193.21)	V	0.0166	38.420	63.03	69.00	74.98	80.94	86.91	92.87	98.84
	U	161.23	1072.3	1218.9	1256.4	1294.6	1333.7	1373.8	1414.7	1456.6
	H	161.26	1143.3	1335.5	1384.0	1433.4	1483.5	1534.6	1586.6	1639.5
	S	0.2836	1.7879	2.0166	2.0603	2.1011	2.1394	2.1757	2.2101	2.2430
14.696 (212.00)	V	0.0167	26.799	42.86	46.93	51.00	55.06	59.13	63.19	67.25
	U	180.12	1077.6	1218.7	1256.2	1294.5	1333.6	1373.7	1414.6	1456.5
	H	180.17	1150.5	1335.2	1383.8	1433.2	1483.4	1534.5	1586.6	1639.4
	S	0.3121	1.7568	1.9739	2.0177	2.0585	2.0969	2.1331	2.1676	2.2005
15 (213.03)	V	0.0167	26.290	41.99	45.98	49.96	53.95	57.93	61.90	65.88
	U	181.16	1077.9	1218.7	1256.2	1294.5	1333.6	1373.7	1414.6	1456.5
	H	181.21	1150.9	1335.2	1383.8	1433.2	1483.4	1534.5	1586.5	1639.4
	S	0.3137	1.7552	1.9717	2.0155	2.0563	2.0946	2.1309	2.1653	2.1982
20 (227.96)	V	0.0168	20.087	31.47	34.46	37.46	40.45	43.43	46.42	49.40
	U	196.21	1082.0	1218.4	1256.0	1294.3	1333.5	1373.6	1414.5	1456.4
	H	196.27	1156.3	1334.9	1383.5	1432.9	1483.2	1534.3	1586.3	1639.3
	S	0.3358	1.7320	1.9397	1.9836	2.0244	2.0628	2.0991	2.1336	2.1665
25 (240.07)	V	0.0169	16.301	25.15	27.56	29.95	32.35	34.74	37.13	39.52
	U	208.44	1085.2	1218.2	1255.8	1294.2	1333.4	1373.5	1414.4	1456.3
	H	208.52	1160.6	1334.6	1383.3	1432.7	1483.0	1534.2	1586.2	1639.2
	S	0.3535	1.7141	1.9149	1.9588	1.9997	2.0381	2.0744	2.1089	2.1418
30 (250.34)	V	0.0170	13.744	20.95	22.95	24.95	26.95	28.94	30.94	32.93
	U	218.84	1087.9	1218.0	1255.6	1294.0	1333.2	1373.3	1414.3	1456.3
	H	218.93	1164.1	1334.2	1383.0	1432.5	1482.8	1534.0	1586.1	1639.0
	S	0.3682	1.6995	1.8946	1.9386	1.9795	2.0179	2.0543	2.0888	2.1217

Abs. Press. Lb/Sq In. (Sat. Temp)		Sat.								
35 (259.29)	V	0.0171	11.896	17.94	19.66	21.38	23.09	24.80	26.51	28.22
	U	227.92	1090.1	1217.7	1255.4	1293.9	1333.1	1373.2	1414.3	1456.2
	H	228.03	1167.1	1333.9	1382.1	1432.3	1482.7	1533.9	1586.0	1638.9
	S	0.3809	1.6872	1.8774	1.9214	1.9624	2.0009	2.0372	2.0717	2.1046
40 (267.25)	V	0.0172	10.497	15.68	17.19	18.70	20.20	21.70	23.19	24.69
	U	236.02	1092.1	1217.5	1255.3	1293.7	1333.0	1373.1	1414.2	1456.1
	H	236.14	1169.8	1333.6	1382.5	1432.1	1482.5	1533.7	1585.8	1638.8
	S	0.3921	1.6765	1.8624	1.9065	1.9476	1.9860	2.0224	2.0569	2.0899
45 (274.44)	V	0.0172	9.399	13.93	15.28	16.61	17.95	19.28	20.61	21.94
	U	243.34	1093.8	1217.2	1255.1	1293.6	1332.9	1373.0	1414.1	1456.0
	H	243.49	1172.0	1333.3	1382.3	1431.9	1482.3	1533.6	1585.7	1638.7
	S	0.4021	1.6671	1.8492	1.8934	1.9345	1.9730	2.0093	2.0439	2.0768
50 (281.01)	V	0.0173	8.514	12.53	13.74	14.95	16.15	17.35	18.55	19.75
	U	250.05	1095.3	1217.0	1254.9	1293.4	1332.7	1372.9	1414.0	1455.9
	H	250.21	1174.1	1332.9	1382.0	1431.7	1482.2	1533.4	1585.6	1638.6
	S	0.4112	1.6586	1.8374	1.8816	1.9227	1.9613	1.9977	2.0322	2.0652
55 (287.08)	V	0.0173	7.785	11.38	12.48	13.58	14.68	15.77	16.86	17.95
	U	256.25	1096.7	1216.8	1254.7	1293.3	1332.6	1372.8	1413.9	1455.8
	H	256.43	1175.9	1332.6	1381.8	1431.5	1482.0	1533.3	1585.5	1638.5
	S	0.4196	1.6510	1.8266	1.8710	1.9121	1.9507	1.9871	2.0216	2.0546
60 (292.71)	V	0.0174	7.174	10.42	11.44	12.45	13.45	14.45	15.45	16.45
	U	262.02	1098.0	1216.5	1254.5	1293.1	1332.5	1372.7	1413.8	1455.8
	H	262.21	1177.6	1332.2	1381.5	1431.3	1481.8	1533.2	1585.3	1638.4
	S	0.4273	1.6440	1.8168	1.8612	1.9024	1.9410	1.9774	2.0120	2.0450
65 (297.98)	V	0.0174	6.653	9.615	10.55	11.48	12.41	13.34	14.26	15.18
	U	267.42	1099.1	1216.3	1254.3	1293.0	1332.4	1372.6	1413.7	1455.7
	H	267.63	1179.1	1331.9	1381.3	1431.1	1481.6	1533.0	1585.2	1638.3
	S	0.4344	1.6375	1.8077	1.8522	1.8935	1.9321	1.9685	2.0031	2.0361
70 (302.93)	V	0.0175	6.205	8.922	9.793	10.66	11.52	12.38	13.24	14.10
	U	272.51	1100.2	1216.0	1254.1	1292.8	1332.2	1372.5	1413.6	1455.6
	H	272.74	1180.6	1331.6	1381.0	1430.9	1481.5	1532.9	1585.1	1638.2
	S	0.4411	1.6316	1.7993	1.8439	1.8852	1.9238	1.9603	1.9949	2.0279
75 (307.61)	V	0.0175	5.814	8.320	9.135	9.945	10.75	11.55	12.35	13.15
	U	277.32	1101.2	1215.8	1254.0	1292.7	1332.1	1372.4	1413.5	1455.5
	H	277.56	1181.9	1331.3	1380.7	1430.7	1481.3	1532.7	1585.0	1638.1
	S	0.4474	1.6260	1.7915	1.8361	1.8774	1.9161	1.9526	1.9872	2.0202

Table F.4: Superheated Steam, English units (Continued)

TEMPERATURE: $t °F$

P/(psia) (t^{sat}/°F)		sat. liq.	sat. vap.	340	360	380	400	420	450	500
80 (312.04)	V	0.0176	5.471	5.715	5.885	6.053	6.218	6.381	6.622	7.018
	U	281.89	1102.1	1114.0	1122.3	1130.4	1138.4	1146.3	1158.1	1177.4
	H	282.15	1183.1	1198.6	1209.4	1220.0	1230.5	1240.8	1256.1	1281.3
	S	0.4534	1.6208	1.6405	1.6539	1.6667	1.6790	1.6909	1.7080	1.7349
85 (316.26)	V	0.0176	5.167	5.364	5.525	5.684	5.840	5.995	6.223	6.597
	U	286.24	1102.9	1113.1	1121.5	1129.7	1137.8	1145.8	1157.6	1177.0
	H	286.52	1184.2	1197.5	1208.4	1219.1	1229.7	1240.1	1255.5	1280.8
	S	0.4590	1.6159	1.6328	1.6463	1.6592	1.6716	1.6836	1.7008	1.7279
90 (320.28)	V	0.0177	4.895	5.051	5.205	5.356	5.505	5.652	5.869	6.223
	U	290.40	1103.7	1112.3	1120.8	1129.1	1137.2	1145.3	1157.2	1176.7
	H	290.69	1185.3	1196.4	1207.5	1218.3	1228.9	1239.4	1254.9	1280.3
	S	0.4643	1.6113	1.6254	1.6391	1.6521	1.6646	1.6767	1.6940	1.7212
95 (324.13)	V	0.0177	4.651	4.771	4.919	5.063	5.205	5.345	5.551	5.889
	U	294.38	1104.5	1111.4	1120.0	1128.4	1136.6	1144.7	1156.7	1176.3
	H	294.70	1186.2	1195.3	1206.5	1217.4	1228.1	1238.7	1254.3	1279.8
	S	0.4694	1.6069	1.6184	1.6322	1.6453	1.6580	1.6701	1.6876	1.7149
100 (327.82)	V	0.0177	4.431	4.519	4.660	4.799	4.935	5.068	5.266	5.588
	U	298.21	1105.2	1110.6	1119.2	1127.7	1136.0	1144.2	1156.3	1175.9
	H	298.54	1187.2	1194.2	1205.5	1216.5	1227.4	1238.0	1253.7	1279.3
	S	0.4743	1.6027	1.6116	1.6255	1.6389	1.6516	1.6638	1.6814	1.7088
105 (331.37)	V	0.0178	4.231	4.291	4.427	4.560	4.690	4.818	5.007	5.315
	U	301.89	1105.8	1109.7	1118.5	1127.0	1135.4	1143.7	1155.8	1175.6
	H	302.24	1188.0	1193.1	1204.5	1215.6	1226.6	1237.3	1253.1	1278.8
	S	0.4790	1.5988	1.6051	1.6192	1.6326	1.6455	1.6578	1.6755	1.7031
110 (334.79)	V	0.0178	4.048	4.083	4.214	4.343	4.468	4.591	4.772	5.068
	U	305.44	1106.5	1108.8	1117.7	1126.4	1134.8	1143.1	1155.3	1175.2
	H	305.80	1188.9	1191.9	1203.5	1214.7	1225.8	1236.6	1252.5	1278.3
	S	0.4834	1.5950	1.5988	1.6131	1.6267	1.6396	1.6521	1.6698	1.6975
115 (338.08)	V	0.0179	3.881	3.894	4.020	4.144	4.265	4.383	4.558	4.841
	U	308.87	1107.0	1107.9	1116.9	1125.7	1134.2	1142.6	1154.8	1174.8
	H	309.25	1189.6	1190.8	1202.5	1213.8	1225.0	1235.8	1251.8	1277.9
	S	0.4877	1.5913	1.5928	1.6072	1.6209	1.6340	1.6465	1.6644	1.6922

P (Sat. Temp)		Sat.							
120 (341.27)	V	0.0179	3.728	3.842	3.962	4.079	4.193	4.361	4.634
	U	312.19	1107.6	1116.1	1124.9	1133.6	1142.0	1154.4	1174.5
	H	312.58	1190.4	1201.4	1212.9	1224.1	1235.1	1251.2	1277.4
	S	0.4919	1.5879	1.6015	1.6154	1.6286	1.6412	1.6592	1.6872
125 (344.35)	V	0.0179	3.586	3.679	3.794	3.907	4.018	4.180	4.443
	U	315.40	1108.1	1115.3	1124.2	1132.9	1141.4	1153.9	1174.1
	H	315.82	1191.1	1200.4	1212.0	1223.3	1234.4	1250.6	1276.9
	S	0.4959	1.5845	1.5960	1.6100	1.6233	1.6360	1.6541	1.6823
130 (347.33)	V	0.0180	3.454	3.527	3.639	3.749	3.856	4.013	4.267
	U	318.52	1108.6	1114.5	1123.5	1132.3	1140.9	1153.4	1173.7
	H	318.95	1191.7	1199.4	1211.1	1222.5	1233.6	1249.9	1276.4
	S	0.4998	1.5813	1.5907	1.6048	1.6182	1.6310	1.6493	1.6775
135 (350.23)	V	0.0180	3.332	3.387	3.496	3.602	3.706	3.858	4.104
	U	321.55	1109.1	1113.7	1122.8	1131.7	1140.3	1152.9	1173.3
	H	322.00	1192.4	1198.3	1210.1	1221.6	1232.9	1249.3	1275.8
	S	0.5035	1.5782	1.5855	1.5997	1.6133	1.6262	1.6446	1.6730
140 (353.04)	V	0.0180	3.219	3.257	3.363	3.466	3.567	3.714	3.953
	U	324.49	1109.6	1112.9	1122.1	1131.0	1139.7	1152.4	1172.9
	H	324.96	1193.0	1197.2	1209.2	1220.8	1232.1	1248.7	1275.3
	S	0.5071	1.5752	1.5804	1.5948	1.6085	1.6215	1.6400	1.6686
145 (355.77)	V	0.0181	3.113	3.135	3.239	3.339	3.437	3.580	3.812
	U	327.36	1110.0	1112.0	1121.3	1130.4	1139.1	1151.9	1172.6
	H	327.84	1193.5	1196.1	1208.2	1220.0	1231.4	1248.0	1274.8
	S	0.5107	1.5723	1.5755	1.5901	1.6039	1.6170	1.6356	1.6643
150 (358.43)	V	0.0181	3.014	3.022	3.123	3.221	3.316	3.455	3.680
	U	330.15	1110.4	1111.2	1120.6	1129.7	1138.6	1151.4	1172.2
	H	330.65	1194.1	1195.1	1207.3	1219.1	1230.6	1247.4	1274.3
	S	0.5141	1.5695	1.5707	1.5854	1.5993	1.6126	1.6313	1.6602
155 (361.02)	V	0.0181	2.921		3.014	3.110	3.203	3.339	3.557
	U	332.87	1110.8		1119.8	1129.0	1138.0	1150.9	1171.8
	H	333.39	1194.6		1206.3	1218.2	1229.8	1246.7	1273.8
	S	0.5174	1.5668		1.5809	1.5949	1.6083	1.6271	1.6561
160 (363.55)	V	0.0182	2.834		2.913	3.006	3.097	3.229	3.441
	U	335.53	1111.2		1119.1	1128.4	1137.4	1150.4	1171.4
	H	336.07	1195.1		1205.3	1217.4	1229.1	1246.0	1273.3
	S	0.5206	1.5641		1.5764	1.5906	1.6041	1.6231	1.6522

Table F.4: Superheated Steam, English units (Continued)

TEMPERATURE: $t\,°F$

$P/(\text{psia})$ $(t^{\text{sat}}/°F)$		sat. liq.	sat. vap.	600	700	800	900	1000	1100	1200
80 (312.04)	V	0.0176	5.471	7.794	8.560	9.319	10.08	10.83	11.58	12.33
	U	281.89	1102.1	1215.5	1253.8	1292.5	1332.0	1372.3	1413.4	1455.4
	H	282.15	1183.1	1330.9	1380.5	1430.5	1481.1	1532.6	1584.9	1638.0
	S	0.4534	1.6208	1.7842	1.8289	1.8702	1.9089	1.9454	1.9800	2.0131
85 (316.26)	V	0.0176	5.167	7.330	8.052	8.768	9.480	10.19	10.90	11.60
	U	286.24	1102.9	1215.3	1253.6	1292.4	1331.9	1372.2	1413.3	1455.4
	H	286.52	1184.2	1330.6	1380.2	1430.3	1481.0	1532.4	1584.7	1637.9
	S	0.4590	1.6159	1.7772	1.8220	1.8634	1.9021	1.9386	1.9733	2.0063
90 (320.28)	V	0.0177	4.895	6.917	7.600	8.277	8.950	9.621	10.29	10.96
	U	290.40	1103.7	1215.0	1253.4	1292.2	1331.7	1372.0	1413.2	1455.3
	H	290.69	1185.3	1330.2	1380.0	1430.1	1480.8	1532.3	1584.6	1637.8
	S	0.4643	1.6113	1.7707	1.8156	1.8570	1.8957	1.9323	1.9669	2.0000
95 (324.13)	V	0.0177	4.651	6.548	7.196	7.838	8.477	9.113	9.747	10.38
	U	294.38	1104.5	1214.8	1253.2	1292.1	1331.6	1371.9	1413.1	1455.2
	H	294.70	1186.2	1329.9	1379.7	1429.9	1480.6	1532.1	1584.5	1637.7
	S	0.4694	1.6069	1.7645	1.8094	1.8509	1.8897	1.9262	1.9609	1.9940
100 (327.82)	V	0.0177	4.431	6.216	6.833	7.443	8.050	8.655	9.258	9.860
	U	298.21	1105.2	1214.5	1253.0	1291.9	1331.5	1371.8	1413.0	1455.1
	H	298.54	1187.2	1329.6	1379.5	1429.7	1480.4	1532.0	1584.4	1637.6
	S	0.4743	1.6027	1.7586	1.8036	1.8451	1.8839	1.9205	1.9552	1.9883
105 (331.37)	V	0.0178	4.231	5.915	6.504	7.086	7.665	8.241	8.816	9.389
	U	301.89	1105.8	1214.3	1252.8	1291.8	1331.3	1371.7	1412.9	1455.0
	H	302.24	1188.0	1329.2	1379.2	1429.4	1480.3	1531.8	1584.2	1637.5
	S	0.4790	1.5988	1.7530	1.7981	1.8396	1.8785	1.9151	1.9498	1.9828
110 (334.79)	V	0.0178	4.048	5.642	6.205	6.761	7.314	7.865	8.413	8.961
	U	305.44	1106.5	1214.0	1252.7	1291.6	1331.2	1371.6	1412.8	1455.0
	H	305.80	1188.9	1328.9	1378.9	1429.2	1480.1	1531.7	1584.1	1637.4
	S	0.4834	1.5950	1.7476	1.7928	1.8344	1.8732	1.9099	1.9446	1.9777
115 (338.08)	V	0.0179	3.881	5.392	5.932	6.465	6.994	7.521	8.046	8.570
	U	308.87	1107.0	1213.8	1252.5	1291.5	1331.1	1371.5	1412.8	1454.9
	H	309.25	1189.6	1328.6	1378.7	1429.0	1479.9	1531.6	1584.0	1637.2
	S	0.4877	1.5913	1.7425	1.7877	1.8294	1.8682	1.9049	1.9396	1.9727

P (T sat)		Sat								
120 (341.27)	V	0.0179	3.728	5.164	5.681	6.193	6.701	7.206	7.710	8.212
	U	312.19	1107.6	1213.5	1252.3	1291.3	1331.0	1371.4	1412.7	1454.8
	H	312.58	1190.4	1328.2	1378.4	1428.8	1479.8	1531.4	1583.9	1637.1
	S	0.4919	1.5879	1.7376	1.7829	1.8246	1.8635	1.9001	1.9349	1.9680
125 (344.35)	V	0.0179	3.586	4.953	5.451	5.943	6.431	6.916	7.400	7.882
	U	315.40	1108.1	1213.3	1252.1	1291.2	1330.8	1371.3	1412.6	1454.7
	H	315.82	1191.1	1327.9	1378.2	1428.6	1479.6	1531.3	1583.7	1637.0
	S	0.4959	1.5845	1.7328	1.7782	1.8199	1.8589	1.8955	1.9303	1.9634
130 (347.33)	V	0.0180	3.454	4.759	5.238	5.712	6.181	6.649	7.114	7.578
	U	318.52	1108.6	1213.0	1251.9	1291.0	1330.7	1371.2	1412.5	1454.6
	H	318.95	1191.7	1327.5	1377.9	1428.4	1479.4	1531.1	1583.6	1636.9
	S	0.4998	1.5813	1.7283	1.7737	1.8155	1.8545	1.8911	1.9259	1.9591
135 (350.23)	V	0.0180	3.332	4.579	5.042	5.498	5.951	6.401	6.849	7.296
	U	321.55	1109.1	1212.8	1251.7	1290.9	1330.6	1371.1	1412.4	1454.5
	H	322.00	1192.4	1327.2	1377.7	1428.2	1479.2	1531.0	1583.5	1636.8
	S	0.5035	1.5782	1.7239	1.7694	1.8112	1.8502	1.8869	1.9217	1.9548
140 (353.04)	V	0.0180	3.219	4.412	4.859	5.299	5.736	6.171	6.604	7.035
	U	324.49	1109.6	1212.5	1251.5	1290.7	1330.5	1371.0	1412.3	1454.5
	H	324.96	1193.0	1326.8	1377.4	1428.0	1479.1	1530.8	1583.4	1636.7
	S	0.5071	1.5752	1.7196	1.7652	1.8071	1.8461	1.8828	1.9176	1.9508
145 (355.77)	V	0.0181	3.113	4.256	4.689	5.115	5.537	5.957	6.375	6.791
	U	327.36	1110.0	1212.3	1251.3	1290.6	1330.3	1370.9	1412.2	1454.4
	H	327.84	1193.5	1326.5	1377.1	1427.8	1478.9	1530.7	1583.2	1636.6
	S	0.5107	1.5723	1.7155	1.7612	1.8031	1.8421	1.8789	1.9137	1.9469
150 (358.43)	V	0.0181	3.014	4.111	4.530	4.942	5.351	5.757	6.161	6.564
	U	330.15	1110.4	1212.0	1251.1	1290.4	1330.2	1370.7	1412.1	1454.3
	H	330.65	1194.1	1326.1	1376.9	1427.6	1478.7	1530.5	1583.1	1636.5
	S	0.5141	1.5695	1.7115	1.7573	1.7992	1.8383	1.8751	1.9099	1.9431
155 (361.02)	V	0.0181	2.921	3.975	4.381	4.781	5.177	5.570	5.961	6.352
	U	332.87	1110.8	1211.8	1251.0	1290.3	1330.1	1370.6	1412.0	1454.2
	H	333.39	1194.6	1325.8	1376.6	1427.4	1478.6	1530.4	1583.0	1636.4
	S	0.5174	1.5668	1.7077	1.7535	1.7955	1.8346	1.8714	1.9062	1.9394
160 (363.55)	V	0.0182	2.834	3.848	4.242	4.629	5.013	5.395	5.774	6.152
	U	335.53	1111.2	1211.5	1250.8	1290.1	1330.0	1370.5	1411.9	1454.1
	H	336.07	1195.1	1325.4	1376.4	1427.2	1478.4	1530.3	1582.9	1636.3
	S	0.5206	1.5641	1.7039	1.7499	1.7919	1.8310	1.8678	1.9027	1.9359

Table F.4: Superheated Steam, English units (Continued)

TEMPERATURE: $t\,°F$

P/(psia) (t^{sat}/°F)		sat. liq.	sat. vap.	400	420	440	460	480	500	550
165 (366.02)	V	0.0182	2.751	2.908	2.997	3.083	3.168	3.251	3.333	3.533
	U	338.12	1111.6	1127.7	1136.8	1145.6	1154.2	1162.7	1171.0	1191.3
	H	338.68	1195.6	1216.5	1228.3	1239.7	1251.0	1261.9	1272.8	1299.2
	S	0.5238	1.5616	1.5864	1.6000	1.6129	1.6252	1.6370	1.6484	1.6753
170 (368.42)	V	0.0182	2.674	2.816	2.903	2.987	3.070	3.151	3.231	3.425
	U	340.66	1111.9	1127.0	1136.2	1145.1	1153.7	1162.3	1170.6	1191.0
	H	341.24	1196.0	1215.6	1227.5	1239.0	1250.3	1261.4	1272.2	1298.8
	S	0.5269	1.5591	1.5823	1.5960	1.6090	1.6214	1.6333	1.6447	1.6717
175 (370.77)	V	0.0182	2.601	2.729	2.814	2.897	2.977	3.056	3.134	3.324
	U	343.15	1112.2	1126.3	1135.6	1144.5	1153.3	1161.8	1170.2	1190.7
	H	343.74	1196.4	1214.7	1226.7	1238.3	1249.7	1260.8	1271.7	1298.4
	S	0.5299	1.5567	1.5783	1.5921	1.6051	1.6176	1.6296	1.6411	1.6682
180 (373.08)	V	0.0183	2.531	2.647	2.730	2.811	2.890	2.967	3.043	3.229
	U	345.58	1112.5	1125.6	1134.9	1144.0	1152.8	1161.4	1169.8	1190.4
	H	346.19	1196.9	1213.8	1225.9	1237.6	1249.0	1260.2	1271.2	1297.9
	S	0.5328	1.5543	1.5743	1.5882	1.6014	1.6140	1.6260	1.6376	1.6647
184 (375.33)	V	0.0183	2.465	2.570	2.651	2.730	2.807	2.883	2.957	3.138
	U	347.96	1112.8	1124.9	1134.3	1143.4	1152.3	1160.9	1169.4	1190.1
	H	348.58	1197.2	1212.9	1225.1	1236.9	1248.4	1259.6	1270.7	1297.5
	S	0.5356	1.5520	1.5705	1.5845	1.5978	1.6104	1.6225	1.6341	1.6614
190 (377.53)	V	0.0183	2.403	2.496	2.576	2.654	2.729	2.803	2.876	3.052
	U	350.29	1113.1	1124.2	1133.7	1142.9	1151.8	1160.5	1169.0	1189.8
	H	350.94	1197.6	1212.0	1224.3	1236.2	1247.7	1259.0	1270.1	1297.1
	S	0.5384	1.5498	1.5667	1.5808	1.5942	1.6069	1.6191	1.6307	1.6581
195 (379.69)	V	0.0184	2.344	2.426	2.505	2.581	2.655	2.727	2.798	2.971
	U	352.58	1113.4	1123.5	1133.1	1142.3	1151.3	1160.0	1168.6	1189.4
	H	353.24	1198.0	1211.1	1223.4	1235.4	1247.1	1258.4	1269.6	1296.6
	S	0.5412	1.5476	1.5630	1.5772	1.5907	1.6035	1.6157	1.6274	1.6549
200 (381.80)	V	0.0184	2.287	2.360	2.437	2.511	2.584	2.655	2.725	2.894
	U	354.82	1113.7	1122.8	1132.4	1141.7	1150.8	1159.6	1168.2	1189.1
	H	355.51	1198.3	1210.1	1222.6	1234.7	1246.4	1257.9	1269.0	1296.2
	S	0.5438	1.5454	1.5593	1.5737	1.5872	1.6001	1.6124	1.6242	1.6518

Abs. Press. Lb/Sq In. (Sat. Temp)		Sat.								
205 (383.88)	V	0.0184	2.233	2.297	2.372	2.446	2.517	2.587	2.655	2.820
	U	357.03	1113.9	1122.1	1131.8	1141.2	1150.3	1159.1	1167.8	1188.8
	H	357.73	1198.7	1209.2	1221.8	1234.0	1245.8	1257.3	1268.5	1295.8
	S	0.5465	1.5434	1.5557	1.5702	1.5839	1.5969	1.6092	1.6211	1.6488
210 (385.92)	V	0.0184	2.182	2.236	2.311	2.383	2.453	2.521	2.588	2.750
	U	359.20	1114.2	1121.3	1131.2	1140.6	1149.8	1158.7	1167.4	1188.5
	H	359.91	1199.0	1208.2	1221.0	1233.2	1245.1	1256.7	1268.0	1295.3
	S	0.5490	1.5413	1.5522	1.5668	1.5806	1.5936	1.6061	1.6180	1.6458
215 (387.91)	V	0.0185	2.133	2.179	2.252	2.323	2.392	2.459	2.524	2.684
	U	361.32	1114.4	1120.6	1130.5	1140.0	1149.3	1158.2	1167.0	1188.1
	H	362.06	1199.3	1207.3	1220.1	1232.5	1244.4	1256.0	1267.4	1294.9
	S	0.5515	1.5393	1.5487	1.5634	1.5773	1.5905	1.6030	1.6149	1.6429
220 (389.88)	V	0.0185	2.086	2.124	2.196	2.266	2.333	2.399	2.464	2.620
	U	363.41	1114.6	1119.9	1129.9	1139.5	1148.7	1157.8	1166.6	1187.8
	H	364.17	1199.6	1206.3	1219.3	1231.7	1243.7	1255.4	1266.9	1294.5
	S	0.5540	1.5374	1.5453	1.5601	1.5741	1.5873	1.5999	1.6120	1.6400
225 (391.80)	V	0.0185	2.041	2.071	2.143	2.211	2.278	2.342	2.406	2.559
	U	365.47	1114.9	1119.1	1129.2	1138.3	1148.2	1157.3	1166.1	1187.5
	H	366.24	1199.9	1205.4	1218.4	1230.9	1243.1	1254.8	1266.3	1294.0
	S	0.5564	1.5354	1.5419	1.5569	1.5710	1.5843	1.5969	1.6090	1.6372
230 (393.70)	V	0.0185	1.9985	2.021	2.091	2.159	2.224	2.288	2.350	2.501
	U	367.49	1115.1	1118.4	1128.5	1138.3	1147.7	1156.8	1165.7	1187.2
	H	368.28	1200.1	1204.4	1217.5	1230.2	1242.4	1254.2	1265.7	1293.6
	S	0.5588	1.5336	1.5385	1.5537	1.5679	1.5813	1.5940	1.6062	1.6344
235 (395.56)	V	0.0186	1.9573	1.973	2.042	2.109	2.173	2.236	2.297	2.445
	U	369.48	1115.3	1117.6	1127.9	1137.7	1147.2	1156.4	1165.3	1186.8
	H	370.29	1200.4	1203.4	1216.7	1229.4	1241.7	1253.6	1265.2	1293.1
	S	0.5611	1.5317	1.5353	1.5505	1.5648	1.5783	1.5911	1.6033	1.6317
240 (397.39)	V	0.0186	1.9177	1.927	1.995	2.061	2.124	2.186	2.246	2.391
	U	371.45	1115.5	1116.8	1127.2	1137.1	1146.6	1155.9	1164.9	1186.5
	H	372.27	1200.6	1202.4	1215.8	1228.6	1241.0	1253.0	1264.6	1292.7
	S	0.5634	1.5299	1.5320	1.5474	1.5618	1.5754	1.5883	1.6006	1.6291
245 (399.19)	V	0.0186	1.8797	1.882	1.950	2.015	2.077	2.138	2.197	2.340
	U	373.38	1115.6	1116.1	1126.5	1136.5	1146.1	1155.4	1164.4	1186.2
	H	374.22	1200.9	1201.4	1214.9	1227.8	1240.3	1252.3	1264.1	1292.3
	S	0.5657	1.5281	1.5288	1.5443	1.5588	1.5725	1.5855	1.5978	1.6265

Table F.4: Superheated Steam, English units (Continued)

TEMPERATURE: $t\,°F$

$P/$(psia) ($t^{sat}/°F$)		sat. liq.	sat. vap.	600	700	800	900	1000	1100	1200
165 (366.02)	V	0.0182	2.751	3.728	4.111	4.487	4.860	5.230	5.598	5.965
	U	338.12	1111.6	1211.3	1250.6	1289.9	1329.8	1370.4	1411.8	1454.1
	H	338.68	1195.6	1325.1	1376.1	1427.0	1478.2	1530.1	1582.7	1636.2
	S	0.5238	1.5616	1.7003	1.7463	1.7884	1.8275	1.8643	1.8992	1.9324
170 (368.42)	V	0.0182	2.674	3.616	3.988	4.354	4.715	5.075	5.432	5.789
	U	340.66	1111.9	1211.0	1250.4	1289.8	1329.7	1370.3	1411.7	1454.0
	H	341.24	1196.0	1324.7	1375.8	1426.8	1478.0	1530.0	1582.6	1636.1
	S	0.5269	1.5591	1.6968	1.7428	1.7850	1.8241	1.8610	1.8959	1.9291
175 (370.77)	V	0.0182	2.601	3.510	3.872	4.227	4.579	4.929	5.276	5.623
	U	343.15	1112.2	1210.7	1250.2	1289.6	1329.6	1370.2	1411.6	1453.9
	H	343.74	1196.4	1324.4	1375.6	1426.5	1477.9	1529.8	1582.5	1636.0
	S	0.5299	1.5567	1.6933	1.7395	1.7816	1.8208	1.8577	1.8926	1.9258
180 (373.08)	V	0.0183	2.531	3.409	3.762	4.108	4.451	4.791	5.129	5.466
	U	345.58	1112.5	1210.5	1250.0	1289.5	1329.4	1370.1	1411.5	1453.8
	H	346.19	1196.9	1324.0	1375.3	1426.3	1477.7	1529.7	1582.4	1635.9
	S	0.5328	1.5543	1.6900	1.7362	1.7784	1.8176	1.8545	1.8894	1.9227
185 (375.33)	V	0.0183	2.465	3.314	3.658	3.996	4.329	4.660	4.989	5.317
	U	347.96	1112.8	1210.2	1249.8	1289.3	1329.3	1370.0	1411.4	1453.7
	H	348.58	1197.2	1323.7	1375.1	1426.1	1477.5	1529.5	1582.3	1635.8
	S	0.5356	1.5520	1.6867	1.7330	1.7753	1.8145	1.8514	1.8864	1.9196
190 (377.53)	V	0.0183	2.403	3.225	3.560	3.889	4.214	4.536	4.857	5.177
	U	350.29	1113.1	1209.9	1249.6	1289.2	1329.2	1369.9	1411.3	1453.7
	H	350.94	1197.6	1323.3	1374.8	1425.9	1477.4	1529.4	1582.1	1635.7
	S	0.5384	1.5498	1.6835	1.7299	1.7722	1.8115	1.8484	1.8834	1.9166
195 (379.69)	V	0.0184	2.344	3.139	3.467	3.788	4.105	4.419	4.732	5.043
	U	352.58	1113.4	1209.7	1249.4	1289.0	1329.1	1369.8	1411.3	1453.6
	H	353.24	1198.0	1323.0	1374.5	1425.7	1477.2	1529.2	1582.0	1635.6
	S	0.5412	1.5476	1.6804	1.7269	1.7692	1.8085	1.8455	1.8804	1.9137
200 (381.80)	V	0.0184	2.287	3.058	3.378	3.691	4.001	4.308	4.613	4.916
	U	354.82	1113.7	1209.4	1249.2	1288.9	1328.9	1369.7	1411.2	1453.5
	H	355.51	1198.3	1322.6	1374.3	1425.5	1477.0	1529.1	1581.9	1635.4
	S	0.5438	1.5454	1.6773	1.7239	1.7663	1.8057	1.8426	1.8776	1.9109

Abs. press., lb/in² (sat. temp.)		Sat.								
205 (383.88)	V	0.0184	2.233	2.981	3.294	3.600	3.902	4.202	4.499	4.796
	U	357.03	1113.9	1209.2	1249.0	1288.7	1328.8	1369.6	1411.1	1453.4
	H	357.73	1198.7	1322.3	1374.0	1425.3	1476.8	1528.9	1581.8	1635.3
	S	0.5465	1.5434	1.6744	1.7210	1.7635	1.8028	1.8398	1.8748	1.9081
210 (385.92)	V	0.0184	2.182	2.908	3.214	3.513	3.808	4.101	4.392	4.681
	U	359.20	1114.2	1208.9	1248.8	1288.6	1328.7	1369.4	1411.0	1453.3
	H	359.91	1199.0	1321.9	1373.7	1425.1	1476.7	1528.8	1581.6	1635.2
	S	0.5490	1.5413	1.6715	1.7182	1.7607	1.8001	1.8371	1.8721	1.9054
215 (387.91)	V	0.0185	2.133	2.838	3.137	3.430	3.718	4.004	4.289	4.572
	U	361.32	1114.4	1208.6	1248.7	1288.4	1328.6	1369.3	1410.9	1453.2
	H	362.06	1199.3	1321.5	1373.5	1424.9	1476.5	1528.7	1581.5	1635.1
	S	0.5515	1.5393	1.6686	1.7155	1.7580	1.7974	1.8344	1.8694	1.9028
220 (389.88)	V	0.0185	2.086	2.771	3.064	3.350	3.633	3.912	4.190	4.467
	U	363.41	1114.6	1208.4	1248.5	1288.3	1328.4	1369.2	1410.8	1453.2
	H	364.17	1199.6	1321.2	1373.2	1424.7	1476.3	1528.5	1581.4	1635.0
	S	0.5540	1.5374	1.6658	1.7128	1.7553	1.7948	1.8318	1.8668	1.9002
225 (391.80)	V	0.0185	2.041	2.707	2.994	3.275	3.551	3.825	4.097	4.367
	U	365.47	1114.9	1208.1	1248.3	1288.1	1328.3	1369.1	1410.7	1453.1
	H	366.24	1199.9	1320.8	1372.9	1424.5	1476.1	1528.4	1581.3	1634.9
	S	0.5564	1.5354	1.6631	1.7101	1.7527	1.7922	1.8293	1.8643	1.8977
230 (393.70)	V	0.0185	1.9984	2.646	2.928	3.202	3.473	3.741	4.007	4.272
	U	367.49	1115.1	1207.8	1248.1	1288.0	1328.2	1369.0	1410.6	1453.0
	H	368.28	1200.1	1320.4	1372.7	1424.2	1476.0	1528.2	1581.1	1634.8
	S	0.5588	1.5336	1.6604	1.7075	1.7502	1.7897	1.8268	1.8618	1.8952
235 (395.56)	V	0.0186	1.9573	2.588	2.864	3.133	3.398	3.660	3.921	4.180
	U	369.48	1115.3	1207.6	1247.9	1287.8	1328.0	1368.9	1410.5	1452.9
	H	370.29	1200.4	1320.1	1372.4	1424.0	1475.8	1528.1	1581.0	1634.7
	S	0.5611	1.5317	1.6578	1.7050	1.7477	1.7872	1.8243	1.8594	1.8928
240 (397.39)	V	0.0186	1.9177	2.532	2.802	3.066	3.326	3.583	3.839	4.093
	U	371.45	1115.5	1207.3	1247.7	1287.7	1327.9	1368.8	1410.4	1452.8
	H	372.27	1200.6	1319.7	1372.1	1423.8	1475.6	1527.9	1580.9	1634.6
	S	0.5634	1.5299	1.6552	1.7025	1.7452	1.7848	1.8219	1.8570	1.8904
245 (399.19)	V	0.0186	1.8797	2.478	2.744	3.002	3.257	3.509	3.760	4.009
	U	373.38	1115.6	1207.0	1247.5	1287.5	1327.8	1368.7	1410.3	1452.8
	H	374.22	1200.9	1319.4	1371.9	1423.6	1475.5	1527.8	1580.8	1634.5
	S	0.5657	1.5281	1.6527	1.7000	1.7428	1.7824	1.8196	1.8547	1.8881

Table F.4: Superheated Steam, English units (Continued)

TEMPERATURE: t °F

P/(psia) (t^{sat}/°F)		sat. liq.	sat. vap.	420	440	460	480	500	520	550
250 (400.97)	V	0.0187	1.8432	1.907	1.970	2.032	2.092	2.150	2.207	2.291
	U	375.28	1115.8	1125.8	1135.9	1145.6	1154.9	1164.0	1172.9	1185.8
	H	376.14	1201.1	1214.0	1227.1	1239.6	1251.7	1263.5	1275.0	1291.8
	S	0.5679	1.5264	1.5413	1.5559	1.5697	1.5827	1.5951	1.6070	1.6239
255 (402.72)	V	0.0187	1.8080	1.865	1.928	1.989	2.048	2.105	2.161	2.244
	U	377.15	1116.0	1125.1	1135.3	1145.0	1154.5	1163.6	1172.5	1185.5
	H	378.04	1201.3	1213.1	1226.3	1238.9	1251.1	1262.9	1274.5	1291.4
	S	0.5701	1.5247	1.5383	1.5530	1.5669	1.5800	1.5925	1.6044	1.6214
260 (404.44)	V	0.0187	1.7742	1.825	1.887	1.947	2.005	2.062	2.117	2.198
	U	379.00	1116.2	1124.5	1134.7	1144.5	1154.0	1163.1	1172.1	1185.1
	H	379.90	1201.5	1212.2	1225.5	1238.2	1250.4	1262.4	1274.0	1290.9
	S	0.5722	1.5230	1.5353	1.5502	1.5642	1.5774	1.5899	1.6019	1.6189
265 (406.13)	V	0.0187	1.7416	1.786	1.848	1.907	1.964	2.020	2.075	2.154
	U	380.83	1116.3	1123.8	1134.1	1144.0	1153.5	1162.7	1171.7	1184.8
	H	381.74	1201.7	1211.3	1224.7	1237.5	1249.8	1261.8	1273.4	1290.4
	S	0.5743	1.5214	1.5324	1.5474	1.5614	1.5747	1.5873	1.5993	1.6165
270 (407.80)	V	0.0188	1.7101	1.749	1.810	1.868	1.925	1.980	2.034	2.112
	U	382.62	1116.5	1123.1	1133.5	1143.4	1153.0	1162.3	1171.3	1184.5
	H	383.56	1201.9	1210.4	1223.9	1236.7	1249.2	1261.2	1272.9	1290.0
	S	0.5764	1.5197	1.5295	1.5446	1.5588	1.5721	1.5848	1.5969	1.6140
275 (409.45)	V	0.0188	1.6798	1.713	1.773	1.831	1.887	1.941	1.994	2.071
	U	384.40	1116.6	1122.3	1132.8	1142.9	1152.5	1161.8	1170.9	1184.1
	H	385.35	1202.1	1209.5	1223.1	1236.0	1248.5	1260.6	1272.4	1289.5
	S	0.5784	1.5181	1.5266	1.5419	1.5561	1.5696	1.5823	1.5944	1.6117
280 (411.07)	V	0.0188	1.6505	1.678	1.738	1.795	1.850	1.904	1.956	2.032
	U	386.15	1116.7	1121.6	1132.2	1142.3	1152.0	1161.4	1170.5	1183.8
	H	387.12	1202.3	1208.6	1222.2	1235.3	1247.9	1260.0	1271.9	1289.1
	S	0.5805	1.5166	1.5238	1.5391	1.5535	1.5670	1.5798	1.5920	1.6093
285 (412.67)	V	0.0188	1.6222	1.645	1.704	1.760	1.815	1.868	1.919	1.994
	U	387.88	1116.9	1120.9	1131.6	1141.7	1151.5	1160.9	1170.1	1183.4
	H	388.87	1202.4	1207.6	1221.4	1234.6	1247.2	1259.4	1271.3	1288.6
	S	0.5824	1.5150	1.5210	1.5365	1.5509	1.5645	1.5774	1.5897	1.6070

Press. (Sat. Temp.)		Sat.								
290 (414.25)	V	0.0188	1.5948	1.612	1.671	1.727	1.780	1.833	1.884	1.958
	U	389.59	1117.0	1120.9	1130.9	1141.2	1151.0	1160.5	1169.7	1183.1
	H	390.60	1202.6	1206.7	1220.6	1233.8	1246.6	1258.9	1270.8	1288.1
	S	0.5844	1.5135	1.5182	1.5338	1.5484	1.5621	1.5750	1.5873	1.6048
295 (415.81)	V	0.0189	1.5684	1.581	1.639	1.694	1.747	1.799	1.849	1.922
	U	391.27	1117.1	1119.5	1130.3	1140.6	1150.5	1160.0	1169.3	1182.7
	H	392.30	1202.7	1205.8	1219.7	1233.1	1245.9	1258.3	1270.2	1287.7
	S	0.5863	1.5120	1.5155	1.5312	1.5458	1.5596	1.5726	1.5850	1.6025
300 (417.35)	V	0.0189	1.5427	1.551	1.608	1.663	1.715	1.766	1.816	1.888
	U	392.94	1117.2	1118.7	1129.6	1140.0	1150.0	1159.6	1168.9	1182.4
	H	393.99	1202.9	1204.8	1218.9	1232.3	1245.2	1257.7	1269.7	1287.2
	S	0.5882	1.5105	1.5127	1.5286	1.5433	1.5572	1.5703	1.5827	1.6003
310 (420.36)	V	0.0189	1.4939	......	1.549	1.603	1.655	1.704	1.753	1.823
	U	396.21	1117.5	......	1128.3	1138.7	1149.0	1158.7	1168.1	1181.7
	H	397.30	1203.2	......	1217.2	1230.8	1243.9	1256.5	1268.6	1286.3
	S	0.5920	1.5076	......	1.5234	1.5384	1.5525	1.5657	1.5782	1.5960
320 (423.31)	V	0.0190	1.4480	......	1.494	1.547	1.597	1.646	1.694	1.762
	U	399.41	1117.7	......	1127.0	1137.7	1147.9	1157.8	1167.2	1181.0
	H	400.53	1203.4	......	1215.5	1229.3	1242.5	1255.2	1267.5	1285.3
	S	0.5956	1.5048	......	1.5184	1.5336	1.5478	1.5612	1.5739	1.5918
330 (426.18)	V	0.0190	1.4048	......	1.442	1.494	1.544	1.591	1.638	1.705
	U	402.53	1117.8	......	1125.7	1136.6	1146.9	1156.8	1166.4	1180.2
	H	403.70	1203.6	......	1213.8	1227.8	1241.2	1254.0	1266.4	1284.4
	S	0.5991	1.5021	......	1.5134	1.5289	1.5433	1.5568	1.5696	1.5876
340 (428.98)	V	0.0191	1.3640	......	1.393	1.444	1.493	1.540	1.585	1.651
	U	405.60	1118.0	......	1124.3	1135.4	1145.8	1155.9	1165.6	1179.5
	H	406.80	1203.8	......	1212.0	1226.2	1239.8	1252.8	1265.3	1283.4
	S	0.6026	1.4994	......	1.5086	1.5242	1.5388	1.5525	1.5654	1.5836
350 (431.73)	V	0.0191	1.3255	......	1.347	1.397	1.445	1.491	1.536	1.600
	U	408.59	1118.1	......	1123.0	1134.2	1144.8	1154.9	1164.7	1178.8
	H	409.83	1204.0	......	1210.2	1224.7	1238.4	1251.5	1264.2	1282.4
	S	0.6059	1.4968	......	1.5038	1.5197	1.5344	1.5483	1.5613	1.5797
360 (434.41)	V	0.0192	1.2891	......	1.303	1.353	1.400	1.445	1.489	1.552
	U	411.53	1118.3	......	1121.6	1132.9	1143.7	1154.0	1163.9	1178.1
	H	412.81	1204.1	......	1208.4	1223.1	1237.0	1250.3	1263.1	1281.5
	S	0.6092	1.4943	......	1.4990	1.5152	1.5301	1.5441	1.5573	1.5758

Table F.4: Superheated Steam, English units (Continued)

TEMPERATURE: $t\,^{\circ}\text{F}$

P/(psia) ($t^{\text{sat}}/^{\circ}\text{F}$)		sat. liq.	sat. vap.	600	700	800	900	1000	1100	1200
250 (400.97)	V	0.0187	1.8432	2.426	2.687	2.941	3.191	3.438	3.684	3.928
	U	375.28	1115.8	1206.7	1247.3	1287.3	1327.7	1368.6	1410.2	1452.7
	H	376.14	1201.1	1319.0	1371.6	1423.4	1475.3	1527.6	1580.6	1634.4
	S	0.5679	1.5264	1.6502	1.6976	1.7405	1.7801	1.8173	1.8524	1.8858
255 (402.72)	V	0.0187	1.8080	2.377	2.633	2.882	3.127	3.370	3.611	3.850
	U	377.15	1116.0	1206.5	1247.1	1287.2	1327.5	1368.5	1410.1	1452.6
	H	378.04	1201.3	1318.6	1371.3	1423.2	1475.1	1527.5	1580.5	1634.3
	S	0.5701	1.5247	1.6477	1.6953	1.7382	1.7778	1.8150	1.8502	1.8836
260 (404.44)	V	0.0187	1.7742	2.329	2.581	2.826	3.066	3.304	3.541	3.776
	U	379.00	1116.2	1206.2	1246.9	1287.0	1327.4	1368.4	1410.0	1452.5
	H	379.90	1201.5	1318.2	1371.1	1423.0	1474.9	1527.3	1580.4	1634.2
	S	0.5722	1.5230	1.6453	1.6930	1.7359	1.7756	1.8128	1.8480	1.8814
265 (406.13)	V	0.0187	1.7416	2.283	2.531	2.771	3.007	3.241	3.473	3.704
	U	380.83	1116.3	1205.9	1246.7	1286.9	1327.3	1368.2	1409.9	1452.4
	H	381.74	1201.7	1317.9	1370.8	1422.8	1474.8	1527.2	1580.3	1634.1
	S	0.5743	1.5214	1.6430	1.6907	1.7337	1.7734	1.8106	1.8458	1.8792
270 (407.80)	V	0.0188	1.7101	2.239	2.482	2.719	2.951	3.181	3.408	3.635
	U	382.62	1116.5	1205.6	1246.5	1286.7	1327.2	1368.1	1409.8	1452.3
	H	383.56	1201.9	1317.5	1370.5	1422.6	1474.6	1527.0	1580.1	1634.0
	S	0.5764	1.5197	1.6406	1.6885	1.7315	1.7713	1.8085	1.8437	1.8771
275 (409.45)	V	0.0188	1.6798	2.196	2.436	2.668	2.896	3.122	3.346	3.568
	U	384.40	1116.6	1205.4	1246.3	1286.6	1327.0	1368.0	1409.8	1452.3
	H	385.35	1202.1	1317.1	1370.3	1422.4	1474.4	1526.9	1580.0	1633.9
	S	0.5784	1.5181	1.6384	1.6863	1.7294	1.7691	1.8064	1.8416	1.8750
280 (411.07)	V	0.0188	1.6505	2.155	2.391	2.619	2.844	3.066	3.286	3.504
	U	386.15	1116.7	1205.1	1246.1	1286.4	1326.9	1367.9	1409.7	1452.2
	H	387.12	1202.3	1316.8	1370.0	1422.1	1474.2	1526.8	1579.9	1633.8
	S	0.5805	1.5166	1.6361	1.6841	1.7273	1.7671	1.8043	1.8395	1.8730
285 (412.67)	V	0.0188	1.6222	2.115	2.348	2.572	2.793	3.011	3.227	3.442
	U	387.88	1116.9	1204.8	1245.9	1286.3	1326.8	1367.8	1409.6	1452.1
	H	388.87	1202.4	1316.4	1369.7	1421.9	1474.1	1526.6	1579.8	1633.6
	S	0.5824	1.5150	1.6339	1.6820	1.7252	1.7650	1.8023	1.8375	1.8710

Temp °F (Sat. temp)										
290 (414.25)	ν	0.0188	1.5948	2.077	2.306	2.527	2.744	2.958	3.171	3.382
	U	389.59	1117.0	1204.5	1245.7	1286.1	1326.6	1367.7	1409.5	1452.0
	H	390.60	1202.6	1316.0	1369.5	1421.7	1473.9	1526.5	1579.6	1633.5
	S	0.5844	1.5135	1.6317	1.6799	1.7232	1.7630	1.8003	1.8356	1.8690
295 (415.81)	ν	0.0189	1.5684	2.040	2.265	2.483	2.697	2.908	3.117	3.325
	U	391.27	1117.1	1204.3	1245.5	1286.0	1326.5	1367.6	1409.4	1451.9
	H	392.30	1202.7	1315.6	1369.2	1421.5	1473.7	1526.3	1579.5	1633.4
	S	0.5863	1.5120	1.6295	1.6779	1.7211	1.7610	1.7984	1.8336	1.8671
300 (417.35)	ν	0.0189	1.5427	2.004	2.226	2.441	2.651	2.859	3.064	3.269
	U	392.94	1117.2	1204.0	1245.3	1285.8	1326.4	1367.5	1409.3	1451.9
	H	393.99	1202.9	1315.2	1368.9	1421.3	1473.6	1526.2	1579.4	1633.3
	S	0.5882	1.5105	1.6274	1.6758	1.7192	1.7591	1.7964	1.8317	1.8652
310 (420.36)	ν	0.0189	1.4939	1.936	2.152	2.360	2.564	2.765	2.964	3.162
	U	396.21	1117.5	1203.4	1244.9	1285.5	1326.1	1367.3	1409.1	1451.7
	H	397.30	1203.2	1314.5	1368.4	1420.9	1473.2	1525.9	1579.2	1633.1
	S	0.5920	1.5076	1.6233	1.6719	1.7153	1.7553	1.7927	1.8280	1.8615
320 (423.31)	ν	0.0190	1.4480	1.873	2.082	2.284	2.482	2.677	2.871	3.063
	U	399.41	1117.7	1202.8	1244.5	1285.2	1325.9	1367.0	1408.9	1451.5
	H	400.53	1203.4	1313.7	1367.8	1420.4	1472.9	1525.6	1578.9	1632.9
	S	0.5956	1.5048	1.6192	1.6680	1.7116	1.7516	1.7890	1.8243	1.8579
330 (426.18)	ν	0.0190	1.4048	1.813	2.017	2.213	2.405	2.595	2.783	2.969
	U	402.53	1117.8	1202.3	1244.1	1284.9	1325.6	1366.8	1408.7	1451.4
	H	403.70	1203.6	1313.0	1367.3	1420.0	1472.5	1525.3	1578.7	1632.7
	S	0.5991	1.5021	1.6153	1.6643	1.7079	1.7480	1.7855	1.8208	1.8544
340 (428.98)	ν	0.0191	1.3640	1.756	1.955	2.146	2.333	2.518	2.700	2.881
	U	405.60	1118.1	1201.7	1243.7	1284.6	1325.4	1366.6	1408.5	1451.2
	H	406.80	1203.8	1312.2	1366.7	1419.6	1472.2	1525.0	1578.4	1632.5
	S	0.6026	1.4994	1.6114	1.6606	1.7044	1.7445	1.7820	1.8174	1.8510
350 (431.73)	ν	0.0191	1.3255	1.703	1.897	2.083	2.265	2.444	2.622	2.798
	U	408.59	1118.1	1201.1	1243.3	1284.2	1325.1	1366.4	1408.3	1451.0
	H	409.83	1204.0	1311.4	1366.2	1419.2	1471.8	1524.7	1578.2	1632.3
	S	0.6059	1.4968	1.6077	1.6571	1.7009	1.7411	1.7787	1.8141	1.8477
360 (434.41)	ν	0.0192	1.2891	1.652	1.842	2.024	2.201	2.375	2.548	2.720
	U	411.53	1118.3	1200.5	1242.9	1283.9	1324.8	1366.2	1408.2	1450.9
	H	412.81	1204.1	1310.6	1365.6	1418.7	1471.5	1524.4	1577.9	1632.1
	S	0.6092	1.4943	1.6040	1.6536	1.6976	1.7379	1.7754	1.8109	1.8445

Table F.4: Superheated Steam, English units (Continued)

TEMPERATURE: $t\,°F$

P/(psia) ($t^{sat}/°F$)		sat. liq.	sat. vap.	460	480	500	520	540	560	580
370 (437.04)	V	0.0192	1.2546	1.311	1.357	1.402	1.445	1.486	1.527	1.566
	U	414.41	1118.4	1131.7	1142.6	1153.0	1163.0	1172.6	1182.0	1191.0
	H	415.73	1204.3	1221.4	1235.5	1249.0	1261.9	1274.4	1286.5	1298.3
	S	0.6125	1.4918	1.5107	1.5259	1.5401	1.5534	1.5660	1.5780	1.5894
380 (439.61)	V	0.0193	1.2218	1.271	1.317	1.361	1.403	1.444	1.483	1.522
	U	417.24	1118.5	1130.4	1141.5	1152.1	1162.1	1171.8	1181.2	1190.4
	H	418.59	1204.4	1219.8	1234.1	1247.7	1260.8	1273.3	1285.5	1297.4
	S	0.6156	1.4894	1.5063	1.5217	1.5360	1.5495	1.5622	1.5743	1.5858
390 (442.13)	V	0.0193	1.1906	1.233	1.278	1.321	1.363	1.403	1.442	1.480
	U	420.01	1118.6	1129.2	1140.4	1151.0	1161.2	1171.0	1180.5	1189.7
	H	421.40	1204.5	1218.2	1232.6	1246.4	1259.6	1272.3	1284.6	1296.5
	S	0.6187	1.4870	1.5020	1.5176	1.5321	1.5457	1.5585	1.5707	1.5823
400 (444.60)	V	0.0193	1.1610	1.197	1.242	1.284	1.325	1.364	1.403	1.440
	U	422.74	1118.7	1127.9	1139.3	1150.0	1160.3	1170.2	1179.8	1189.1
	H	424.17	1204.6	1216.5	1231.2	1245.1	1258.4	1271.2	1283.6	1295.7
	S	0.6217	1.4847	1.4978	1.5136	1.5282	1.5420	1.5549	1.5672	1.5789
410 (447.02)	V	0.0194	1.1327	1.163	1.207	1.249	1.289	1.328	1.365	1.402
	U	425.41	1118.7	1126.6	1138.1	1149.0	1159.4	1169.4	1179.1	1188.4
	H	426.88	1204.7	1214.8	1229.7	1243.8	1257.2	1270.2	1282.7	1294.8
	S	0.6247	1.4825	1.4936	1.5096	1.5244	1.5383	1.5514	1.5637	1.5755
420 (449.40)	V	0.0194	1.1057	1.130	1.173	1.215	1.254	1.293	1.330	1.366
	U	428.05	1118.8	1125.3	1137.0	1148.0	1158.5	1168.6	1178.3	1187.8
	H	429.56	1204.7	1213.1	1228.2	1242.4	1256.0	1269.1	1281.7	1293.9
	S	0.6276	1.4802	1.4894	1.5056	1.5206	1.5347	1.5479	1.5603	1.5722
430 (451.74)	V	0.0195	1.0800	1.099	1.142	1.183	1.222	1.259	1.296	1.331
	U	430.64	1118.8	1123.9	1135.8	1147.0	1157.6	1167.8	1177.6	1187.1
	H	432.19	1204.8	1211.4	1226.6	1241.1	1254.8	1268.0	1280.7	1293.0
	S	0.6304	1.4781	1.4853	1.5017	1.5169	1.5311	1.5444	1.5570	1.5689
440 (454.03)	V	0.0195	1.0554	1.069	1.111	1.152	1.190	1.227	1.263	1.298
	U	433.19	1118.8	1122.6	1134.6	1145.9	1156.7	1166.9	1176.9	1186.4
	H	434.77	1204.8	1209.6	1225.1	1239.7	1253.6	1266.9	1279.7	1292.1
	S	0.6332	1.4759	1.4812	1.4979	1.5132	1.5276	1.5410	1.5537	1.5657

P (Tsat)		Sat. liquid	Sat. vapor							
450	V	0.0195	1.0318	1.040	1.082	1.122	1.160	1.197	1.232	1.266
(456.28)	U	435.69	1118.9	1121.2	1133.4	1144.9	1155.8	1166.1	1176.1	1185.7
	H	437.32	1204.8	1207.8	1223.5	1238.3	1252.4	1265.8	1278.7	1291.2
	S	0.6360	1.4738	1.4771	1.4940	1.5096	1.5241	1.5377	1.5505	1.5626
460	V	0.0196	1.0092	1.012	1.054	1.094	1.132	1.168	1.203	1.236
(458.50)	U	438.17	1118.9	1119.8	1132.2	1143.8	1154.8	1165.3	1175.4	1185.1
	H	439.83	1204.8	1206.0	1222.0	1236.9	1251.1	1264.7	1277.7	1290.3
	S	0.6387	1.4718	1.4731	1.4903	1.5060	1.5207	1.5344	1.5473	1.5595
470	V	0.0196	0.9876	......	1.028	1.067	1.104	1.140	1.174	1.207
(460.68)	U	440.60	1118.9	......	1131.0	1142.8	1153.9	1164.5	1174.6	1184.4
	H	442.31	1204.8	......	1220.4	1235.5	1249.9	1263.6	1276.7	1289.4
	S	0.6413	1.4697	......	1.4865	1.5025	1.5173	1.5311	1.5441	1.5564
480	V	0.0197	0.9668	......	1.002	1.041	1.078	1.113	1.147	1.180
(462.82)	U	443.00	1118.9	......	1129.8	1141.7	1152.9	1163.6	1173.8	1183.7
	H	444.75	1204.8	......	1218.8	1234.1	1248.6	1262.4	1275.7	1288.5
	S	0.6439	1.4677	......	1.4828	1.4990	1.5139	1.5279	1.5410	1.5534
490	V	0.0197	0.9468	......	0.9774	1.016	1.052	1.087	1.121	1.153
(464.93)	U	445.36	1118.9	......	1128.5	1140.6	1151.9	1162.7	1173.1	1183.0
	H	447.15	1204.7	......	1217.1	1232.7	1247.4	1261.3	1274.7	1287.5
	S	0.6465	1.4658	......	1.4791	1.4955	1.5106	1.5247	1.5380	1.5504
500	V	0.0197	0.9276	......	0.9537	0.9919	1.028	1.062	1.095	1.127
(467.01)	U	447.70	1118.8	......	1127.2	1139.5	1151.0	1161.9	1172.3	1182.3
	H	449.52	1204.7	......	1215.5	1231.2	1246.1	1260.2	1273.6	1286.6
	S	0.6490	1.4639	......	1.4755	1.4921	1.5074	1.5216	1.5349	1.5475
510	V	0.0198	0.9091	......	0.9310	0.9688	1.005	1.039	1.071	1.103
(469.05)	U	450.00	1118.8	......	1126.0	1138.4	1150.0	1161.0	1171.5	1181.6
	H	451.87	1204.6	......	1213.8	1229.8	1244.8	1259.0	1272.6	1285.7
	S	0.6515	1.4620	......	1.4718	1.4886	1.5041	1.5185	1.5319	1.5446
520	V	0.0198	0.8914	......	0.9090	0.9466	0.9820	1.016	1.048	1.079
(471.07)	U	452.27	1118.8	......	1124.7	1137.2	1149.0	1160.1	1170.7	1180.9
	H	454.18	1204.5	......	1212.1	1228.3	1243.5	1257.8	1271.5	1284.7
	S	0.6539	1.4601	......	1.4682	1.4853	1.5009	1.5154	1.5290	1.5418
530	V	0.0199	0.8742	......	0.8878	0.9252	0.9603	0.9937	1.026	1.056
(473.05)	U	454.51	1118.7	......	1123.4	1136.1	1148.0	1159.2	1169.9	1180.1
	H	456.46	1204.5	......	1210.4	1226.8	1242.2	1256.7	1270.5	1283.8
	S	0.6564	1.4583	......	1.4646	1.4819	1.4977	1.5124	1.5261	1.5390

Table F.4: Superheated Steam, English units (Continued)

TEMPERATURE: t °F

P/(psia) (t^{sat}/°F)		sat. liq.	sat. vap.	600	700	800	900	1000	1100	1200
370 (437.04)	V	0.0192	1.2546	1.605	1.790	1.967	2.140	2.310	2.478	2.645
	U	414.41	1118.4	1199.9	1242.5	1283.6	1324.6	1366.0	1408.0	1450.7
	H	415.73	1204.3	1309.8	1365.1	1418.3	1471.1	1524.1	1577.7	1631.8
	S	0.6125	1.4918	1.6004	1.6503	1.6943	1.7346	1.7723	1.8077	1.8414
380 (439.61)	V	0.0193	1.2218	1.560	1.741	1.914	2.082	2.248	2.412	2.575
	U	417.24	1118.5	1199.3	1242.1	1283.3	1324.3	1365.7	1407.8	1450.6
	H	418.59	1204.4	1309.0	1364.5	1417.9	1470.8	1523.8	1577.4	1631.6
	S	0.6156	1.4894	1.5969	1.6470	1.6911	1.7315	1.7692	1.8047	1.8384
390 (442.13)	V	0.0193	1.1906	1.517	1.694	1.863	2.028	2.190	2.350	2.508
	U	420.01	1118.6	1198.8	1241.7	1283.0	1324.1	1365.5	1407.6	1450.4
	H	421.40	1204.5	1308.2	1364.0	1417.5	1470.4	1523.5	1577.2	1631.4
	S	0.6187	1.4870	1.5935	1.6437	1.6880	1.7285	1.7662	1.8017	1.8354
400 (444.60)	V	0.0193	1.1610	1.476	1.650	1.815	1.976	2.134	2.290	2.445
	U	422.74	1118.7	1198.2	1241.3	1282.7	1323.8	1365.3	1407.4	1450.2
	H	424.17	1204.6	1307.4	1363.4	1417.0	1470.1	1523.3	1576.9	1631.2
	S	0.6217	1.4847	1.5901	1.6406	1.6850	1.7255	1.7632	1.7988	1.8325
410 (447.02)	V	0.0194	1.1327	1.438	1.608	1.769	1.926	2.081	2.233	2.385
	U	425.41	1118.7	1197.6	1240.8	1282.4	1323.6	1365.1	1407.2	1450.1
	H	426.89	1204.7	1306.6	1362.8	1416.6	1469.7	1523.0	1576.7	1631.0
	S	0.6247	1.4825	1.5868	1.6375	1.6820	1.7226	1.7603	1.7959	1.8297
420 (449.40)	V	0.0194	1.1057	1.401	1.568	1.726	1.879	2.030	2.180	2.327
	U	428.05	1118.8	1196.9	1240.4	1282.0	1323.3	1364.9	1407.0	1449.9
	H	429.56	1204.7	1305.8	1362.3	1416.2	1469.4	1522.7	1576.4	1630.8
	S	0.6276	1.4802	1.5835	1.6345	1.6791	1.7197	1.7575	1.7932	1.8269
430 (451.74)	V	0.0195	1.0800	1.366	1.529	1.684	1.835	1.982	2.128	2.273
	U	430.64	1118.8	1196.3	1240.0	1281.7	1323.0	1364.6	1406.8	1449.7
	H	432.19	1204.8	1305.0	1361.7	1415.7	1469.0	1522.4	1576.2	1630.6
	S	0.6304	1.4781	1.5804	1.6315	1.6762	1.7169	1.7548	1.7904	1.8242
440 (454.03)	V	0.0195	1.0554	1.332	1.493	1.644	1.792	1.936	2.079	2.220
	U	433.19	1118.8	1195.7	1239.6	1281.4	1322.8	1364.4	1406.6	1449.6
	H	434.77	1204.8	1304.2	1361.1	1415.3	1468.7	1522.1	1575.9	1630.4
	S	0.6332	1.4759	1.5772	1.6286	1.6734	1.7142	1.7521	1.7878	1.8216

Abs. Press. Lb/Sq In. (Sat. Temp.)		Sat.								
450 (456.28)	V	0.0195	1.0318	1.300	1.458	1.607	1.751	1.892	2.032	2.170
	U	435.69	1118.8	1195.1	1239.2	1281.1	1322.5	1364.2	1406.5	1449.4
	H	437.32	1204.8	1303.3	1360.6	1414.9	1468.3	1521.8	1575.7	1630.1
	S	0.6360	1.4738	1.5742	1.6258	1.6707	1.7115	1.7495	1.7852	1.8190
460 (458.50)	V	0.0196	1.0092	1.269	1.424	1.570	1.712	1.850	1.987	2.123
	U	438.17	1118.9	1194.5	1238.8	1280.8	1322.3	1364.0	1406.3	1449.3
	H	439.83	1204.8	1302.5	1360.0	1414.4	1468.0	1521.5	1575.4	1629.9
	S	0.6387	1.4718	1.5711	1.6230	1.6680	1.7089	1.7469	1.7826	1.8165
470 (460.68)	V	0.0196	0.9875	1.240	1.392	1.536	1.674	1.810	1.944	2.077
	U	440.60	1118.8	1193.9	1238.3	1280.4	1322.0	1363.8	1406.1	1449.1
	H	442.31	1204.8	1301.7	1359.4	1414.0	1467.6	1521.2	1575.2	1629.7
	S	0.6413	1.4697	1.5681	1.6202	1.6654	1.7064	1.7444	1.7802	1.8141
480 (462.82)	V	0.0197	0.9668	1.211	1.361	1.502	1.638	1.772	1.903	2.033
	U	443.00	1118.9	1193.2	1237.9	1280.1	1321.7	1363.5	1405.9	1448.9
	H	444.75	1204.8	1300.8	1358.8	1413.6	1467.3	1520.9	1574.9	1629.5
	S	0.6439	1.4677	1.5652	1.6176	1.6628	1.7038	1.7419	1.7777	1.8116
490 (464.93)	V	0.0197	0.9468	1.184	1.332	1.470	1.604	1.735	1.864	1.991
	U	445.36	1118.9	1192.6	1237.5	1279.8	1321.5	1363.3	1405.7	1448.8
	H	447.15	1204.7	1300.0	1358.3	1413.1	1466.9	1520.6	1574.7	1629.3
	S	0.6465	1.4658	1.5623	1.6149	1.6603	1.7014	1.7395	1.7753	1.8093
500 (467.01)	V	0.0197	0.9276	1.158	1.304	1.440	1.571	1.699	1.826	1.951
	U	447.70	1118.8	1192.0	1237.1	1279.5	1321.2	1363.1	1405.5	1448.6
	H	449.52	1204.7	1299.1	1357.7	1412.7	1466.6	1520.3	1574.4	1629.1
	S	0.6490	1.4639	1.5595	1.6123	1.6578	1.6990	1.7371	1.7730	1.8069
510 (469.05)	V	0.0198	0.9091	1.133	1.277	1.410	1.539	1.665	1.789	1.912
	U	450.00	1118.8	1191.3	1236.6	1279.2	1321.0	1362.9	1405.3	1448.4
	H	451.87	1204.6	1298.3	1357.1	1412.2	1466.2	1520.0	1574.2	1628.9
	S	0.6515	1.4620	1.5567	1.6097	1.6554	1.6966	1.7348	1.7707	1.8047
520 (471.07)	V	0.0198	0.8914	1.109	1.250	1.382	1.509	1.632	1.754	1.875
	U	452.27	1118.8	1190.7	1236.2	1278.8	1320.7	1362.7	1405.1	1448.3
	H	454.18	1204.5	1297.4	1356.5	1411.8	1465.9	1519.7	1573.9	1628.7
	S	0.6539	1.4601	1.5539	1.6072	1.6530	1.6943	1.7325	1.7684	1.8024
530 (473.05)	V	0.0199	0.8742	1.086	1.225	1.355	1.479	1.601	1.720	1.839
	U	454.51	1118.7	1190.0	1235.9	1278.5	1320.4	1362.4	1404.9	1448.1
	H	456.46	1204.5	1296.5	1355.9	1411.4	1465.5	1519.4	1573.7	1628.4
	S	0.6564	1.4583	1.5512	1.6047	1.6506	1.6920	1.7302	1.7662	1.8002

Table F.4: Superheated Steam, English units (Continued)

TEMPERATURE: t °F

P/(psia) (t^{sat}/°F)		sat. liq.	sat. vap.	500	520	540	560	580	600	650
540 (475.01)	V	0.0199	0.8577	0.9045	0.9394	0.9725	1.004	1.035	1.064	1.134
	U	456.72	1118.7	1134.9	1147.0	1158.3	1169.1	1179.4	1189.4	1213.0
	H	458.71	1204.4	1225.3	1240.8	1255.5	1269.4	1282.8	1295.7	1326.3
	S	0.6587	1.4565	1.4786	1.4946	1.5094	1.5232	1.5362	1.5484	1.5767
550 (476.94)	V	0.0199	0.8418	0.8846	0.9192	0.9520	0.9833	1.013	1.042	1.112
	U	458.91	1118.6	1133.8	1145.9	1157.4	1168.3	1178.7	1188.7	1212.4
	H	460.94	1204.3	1223.8	1239.5	1254.3	1268.4	1281.8	1294.8	1325.6
	S	0.6611	1.4547	1.4753	1.4915	1.5064	1.5203	1.5334	1.5458	1.5742
560 (478.84)	V	0.0200	0.8264	0.8653	0.8997	0.9322	0.9632	0.9930	1.022	1.090
	U	461.07	1118.5	1132.6	1144.9	1156.5	1167.5	1178.0	1188.0	1211.9
	H	463.14	1204.2	1222.2	1238.1	1253.1	1267.3	1280.9	1293.9	1324.9
	S	0.6634	1.4529	1.4720	1.4884	1.5035	1.5175	1.5307	1.5431	1.5717
570 (480.72)	V	0.0200	0.8115	0.8467	0.8808	0.9131	0.9438	0.9733	1.002	1.069
	U	463.20	1118.5	1131.4	1143.9	1155.6	1166.6	1177.2	1187.4	1211.4
	H	465.32	1204.1	1220.7	1236.8	1251.9	1266.2	1279.9	1293.0	1324.2
	S	0.6657	1.4512	1.4687	1.4853	1.5005	1.5147	1.5280	1.5405	1.5693
580 (482.57)	V	0.0201	0.7971	0.8287	0.8626	0.8946	0.9251	0.9542	0.9824	1.049
	U	465.31	1118.4	1130.2	1142.8	1154.6	1165.8	1176.5	1186.7	1210.8
	H	467.47	1203.9	1219.1	1235.4	1250.7	1265.1	1278.9	1292.1	1323.4
	S	0.6679	1.4495	1.4654	1.4822	1.4976	1.5120	1.5254	1.5380	1.5668
590 (484.40)	V	0.0201	0.7832	0.8112	0.8450	0.8768	0.9069	0.9358	0.9637	1.030
	U	467.40	1118.3	1129.0	1141.7	1153.7	1165.0	1175.7	1186.0	1210.3
	H	469.59	1203.8	1217.5	1234.0	1249.4	1264.0	1277.9	1291.2	1322.7
	S	0.6701	1.4478	1.4622	1.4792	1.4948	1.5092	1.5227	1.5354	1.5645
600 (486.20)	V	0.0201	0.7697	0.7944	0.8279	0.8595	0.8894	0.9180	0.9456	1.011
	U	469.46	1118.2	1127.7	1140.7	1152.8	1164.1	1175.0	1185.3	1209.8
	H	471.70	1203.7	1215.9	1232.6	1248.2	1262.9	1276.9	1290.3	1322.0
	S	0.6723	1.4461	1.4590	1.4762	1.4919	1.5065	1.5201	1.5329	1.5621
610 (487.98)	V	0.0202	0.7567	0.7780	0.8114	0.8427	0.8724	0.9008	0.9281	0.9927
	U	471.50	1118.1	1126.5	1139.6	1151.8	1163.3	1174.2	1184.7	1209.2
	H	473.78	1203.5	1214.3	1231.2	1246.9	1261.8	1275.9	1289.4	1321.3
	S	0.6745	1.4445	1.4558	1.4732	1.4891	1.5038	1.5175	1.5304	1.5598

Temp (Sat.)		col1	col2	col3	col4	col5	col6	col7	col8	col9
620 (489.74)	V	0.0202	0.7441	0.7621	0.7954	0.8265	0.8560	0.8841	0.9112	0.9751
	U	473.52	1118.0	1125.2	1138.5	1150.8	1162.4	1173.5	1184.0	1208.7
	H	475.84	1203.4	1212.7	1229.7	1245.7	1260.7	1274.9	1288.5	1320.5
	S	0.6766	1.4428	1.4526	1.4702	1.4863	1.5011	1.5150	1.5279	1.5575
630 (491.48)	V	0.0202	0.7318	0.7467	0.7798	0.8108	0.8401	0.8680	0.8948	0.9580
	U	475.52	1117.9	1123.9	1137.4	1149.9	1161.6	1172.7	1183.3	1208.1
	H	477.88	1203.2	1211.0	1228.3	1244.4	1259.5	1273.9	1287.6	1319.8
	S	0.6787	1.4412	1.4494	1.4672	1.4835	1.4985	1.5124	1.5255	1.5552
640 (493.19)	V	0.0203	0.7200	0.7318	0.7648	0.7956	0.8246	0.8523	0.8788	0.9415
	U	477.49	1117.8	1122.7	1136.3	1148.9	1160.7	1171.9	1182.6	1207.6
	H	479.89	1203.0	1209.3	1226.8	1243.1	1258.4	1272.8	1286.7	1319.1
	S	0.6808	1.4396	1.4462	1.4643	1.4807	1.4959	1.5099	1.5231	1.5530
650 (494.89)	V	0.0203	0.7084	0.7173	0.7501	0.7808	0.8096	0.8371	0.8634	0.9254
	U	479.45	1117.6	1121.3	1135.1	1147.9	1159.8	1171.1	1181.9	1207.0
	H	481.89	1202.8	1207.6	1225.4	1241.8	1257.2	1271.8	1285.7	1318.3
	S	0.6828	1.4381	1.4430	1.4614	1.4780	1.4932	1.5074	1.5207	1.5507
660 (496.57)	V	0.0204	0.6972	0.7031	0.7359	0.7664	0.7951	0.8224	0.8485	0.9098
	U	481.38	1117.5	1120.0	1134.0	1146.9	1159.0	1170.3	1181.2	1206.5
	H	483.87	1202.7	1205.9	1223.9	1240.5	1256.1	1270.8	1284.8	1317.6
	S	0.6849	1.4365	1.4399	1.4584	1.4752	1.4907	1.5049	1.5183	1.5485
670 (498.22)	V	0.0204	0.6864	0.6894	0.7221	0.7525	0.7810	0.8080	0.8339	0.8947
	U	483.30	1117.4	1118.7	1132.8	1145.9	1158.1	1169.6	1180.5	1205.9
	H	485.83	1202.5	1204.2	1222.4	1239.2	1254.9	1269.7	1283.9	1316.8
	S	0.6869	1.4350	1.4367	1.4555	1.4725	1.4881	1.5025	1.5159	1.5463
680 (499.86)	V	0.0204	0.6758	0.6760	0.7087	0.7389	0.7673	0.7941	0.8198	0.8801
	U	485.20	1117.2	1117.3	1131.7	1144.9	1157.2	1168.8	1179.8	1205.3
	H	487.77	1202.3	1202.4	1220.8	1237.9	1253.7	1268.7	1282.9	1316.1
	S	0.6889	1.4334	1.4336	1.4526	1.4698	1.4855	1.5000	1.5136	1.5442
690 (501.48)	V	0.0205	0.6655	......	0.6956	0.7257	0.7539	0.7806	0.8061	0.8658
	U	487.08	1117.1	......	1130.5	1143.9	1156.3	1168.0	1179.0	1204.8
	H	489.70	1202.1	......	1219.3	1236.5	1252.5	1267.6	1282.0	1315.3
	S	0.6908	1.4319	......	1.4497	1.4671	1.4830	1.4976	1.5113	1.5421
700 (503.08)	V	0.0205	0.6556	......	0.6829	0.7129	0.7409	0.7675	0.7928	0.8520
	U	488.95	1116.9	......	1129.3	1142.8	1155.4	1167.1	1178.3	1204.2
	H	491.60	1201.8	......	1217.8	1235.2	1251.3	1266.6	1281.0	1314.6
	S	0.6928	1.4304	......	1.4468	1.4644	1.4805	1.4952	1.5090	1.5399

Table F.4: Superheated Steam, English units (Continued)

TEMPERATURE: $t\,°F$

P/(psia) (t^{sat}/°F)		sat. liq.	sat. vap.	700	750	800	900	1000	1100	1200
540 (475.01)	V	0.0199	0.8577	1.201	1.266	1.328	1.451	1.570	1.688	1.804
	U	456.72	1118.7	1235.3	1257.0	1278.2	1320.2	1362.2	1404.8	1447.9
	H	458.71	1204.4	1355.3	1383.4	1410.9	1465.1	1519.1	1573.4	1628.2
	S	0.6587	1.4565	1.6023	1.6260	1.6483	1.6897	1.7280	1.7640	1.7981
550 (476.94)	V	0.0199	0.8418	1.178	1.241	1.303	1.424	1.541	1.657	1.771
	U	458.91	1118.6	1234.9	1256.6	1277.9	1319.9	1362.0	1404.6	1447.8
	H	460.94	1204.3	1354.7	1382.9	1410.5	1464.8	1518.9	1573.2	1628.0
	S	0.6611	1.4547	1.5999	1.6237	1.6460	1.6875	1.7259	1.7619	1.7959
560 (478.84)	V	0.0200	0.8264	1.155	1.218	1.279	1.397	1.513	1.627	1.739
	U	461.07	1118.5	1234.4	1256.2	1277.5	1319.6	1361.8	1404.4	1447.6
	H	463.14	1204.2	1354.2	1382.4	1410.0	1464.4	1518.6	1572.9	1627.8
	S	0.6634	1.4529	1.5975	1.6214	1.6438	1.6853	1.7237	1.7598	1.7939
570 (480.72)	V	0.0200	0.8115	1.133	1.195	1.255	1.372	1.486	1.597	1.708
	U	463.20	1118.5	1234.0	1255.8	1277.2	1319.4	1361.6	1404.2	1447.5
	H	465.32	1204.1	1353.6	1381.9	1409.6	1464.1	1518.3	1572.7	1627.6
	S	0.6657	1.4512	1.5952	1.6191	1.6415	1.6832	1.7216	1.7577	1.7918
580 (482.57)	V	0.0201	0.7971	1.112	1.173	1.232	1.347	1.459	1.569	1.678
	U	465.31	1118.4	1233.6	1255.5	1276.9	1319.1	1361.3	1404.0	1447.3
	H	467.47	1203.9	1353.0	1381.4	1409.2	1463.7	1518.0	1572.4	1627.4
	S	0.6679	1.4495	1.5929	1.6169	1.6394	1.6811	1.7196	1.7556	1.7898
590 (484.40)	V	0.0201	0.7832	1.092	1.152	1.210	1.324	1.434	1.542	1.649
	U	467.40	1118.3	1233.1	1255.1	1276.5	1318.9	1361.1	1403.8	1447.1
	H	469.59	1203.8	1352.4	1380.9	1408.7	1463.4	1517.7	1572.2	1627.2
	S	0.6701	1.4478	1.5906	1.6147	1.6372	1.6790	1.7175	1.7536	1.7878
600 (486.20)	V	0.0201	0.7697	1.073	1.132	1.189	1.301	1.409	1.516	1.621
	U	469.46	1118.2	1232.7	1254.7	1276.2	1318.6	1360.9	1403.6	1447.0
	H	471.70	1203.7	1351.8	1380.4	1408.3	1463.0	1517.4	1571.9	1627.0
	S	0.6723	1.4461	1.5884	1.6125	1.6351	1.6769	1.7155	1.7517	1.7859
610 (487.98)	V	0.0202	0.7567	1.054	1.112	1.169	1.279	1.386	1.491	1.594
	U	471.50	1118.1	1232.2	1254.3	1275.9	1318.3	1360.7	1403.4	1446.8
	H	473.78	1203.5	1351.2	1379.9	1407.8	1462.7	1517.1	1571.7	1626.7
	S	0.6745	1.4445	1.5861	1.6104	1.6330	1.6749	1.7135	1.7497	1.7839

620 (489.74)	V	0.0202	0.7441	1.035	1.093	1.149	1.257	1.363	1.466	1.568
	U	473.52	1118.0	1231.8	1253.9	1275.6	1318.1	1360.5	1403.2	1446.6
	H	475.84	1203.4	1350.6	1379.3	1407.4	1462.3	1516.8	1571.4	1626.5
	S	0.6766	1.4428	1.5839	1.6082	1.6310	1.6729	1.7116	1.7478	1.7820
630 (491.48)	V	0.0202	0.7318	1.017	1.074	1.130	1.236	1.340	1.442	1.543
	U	475.52	1117.9	1231.3	1253.6	1275.2	1317.8	1360.2	1403.1	1446.5
	H	477.88	1203.2	1350.0	1378.8	1406.9	1461.9	1516.5	1571.2	1626.6
	S	0.6787	1.4412	1.5818	1.6062	1.6289	1.6710	1.7097	1.7459	1.7802
640 (493.19)	V	0.0203	0.7200	1.000	1.056	1.111	1.216	1.319	1.419	1.518
	U	477.49	1117.8	1230.9	1253.2	1274.9	1317.5	1360.0	1402.9	1446.3
	H	479.89	1203.0	1349.3	1378.3	1406.5	1461.6	1516.2	1570.9	1626.1
	S	0.6808	1.4396	1.5797	1.6041	1.6269	1.6690	1.7078	1.7441	1.7783
650 (494.89)	V	0.0203	0.7084	0.9835	1.039	1.093	1.197	1.298	1.397	1.494
	U	479.45	1117.6	1230.4	1252.8	1274.6	1317.3	1359.8	1402.7	1446.1
	H	481.89	1202.8	1348.7	1377.8	1406.0	1461.2	1515.9	1570.7	1625.9
	S	0.6828	1.4381	1.5775	1.6021	1.6249	1.6671	1.7059	1.7422	1.7765
660 (496.57)	V	0.0204	0.6972	0.9673	1.022	1.075	1.178	1.278	1.375	1.471
	U	481.38	1117.5	1230.0	1252.4	1274.2	1317.0	1359.6	1402.5	1446.0
	H	483.87	1202.7	1348.1	1377.3	1405.6	1460.9	1515.6	1570.4	1625.7
	S	0.6849	1.4365	1.5755	1.6001	1.6230	1.6652	1.7041	1.7404	1.7748
670 (498.22)	V	0.0204	0.6864	0.9516	1.006	1.058	1.160	1.258	1.354	1.449
	U	483.30	1117.4	1229.5	1252.0	1273.9	1316.7	1359.3	1402.3	1445.8
	H	485.83	1202.5	1347.5	1376.7	1405.1	1460.5	1515.3	1570.2	1625.5
	S	0.6869	1.4350	1.5734	1.5981	1.6211	1.6634	1.7023	1.7387	1.7730
680 (499.86)	V	0.0204	0.6758	0.9364	0.9900	1.042	1.142	1.239	1.334	1.427
	U	485.20	1117.2	1229.1	1251.6	1273.6	1316.5	1359.1	1402.1	1445.7
	H	487.77	1202.3	1346.9	1376.2	1404.7	1460.2	1515.0	1569.9	1625.3
	S	0.6889	1.4334	1.5714	1.5961	1.6192	1.6616	1.7005	1.7369	1.7713
690 (501.48)	V	0.0205	0.6655	0.9216	0.9746	1.026	1.125	1.220	1.314	1.406
	U	487.08	1117.1	1228.6	1251.3	1273.2	1316.2	1358.9	1401.9	1445.5
	H	489.70	1202.1	1346.3	1375.7	1404.2	1459.8	1514.7	1569.7	1625.0
	S	0.6908	1.4319	1.5693	1.5942	1.6173	1.6598	1.6987	1.7352	1.7696
700 (503.08)	V	0.0205	0.6556	0.9072	0.9596	1.010	1.108	1.202	1.295	1.386
	U	488.95	1116.9	1228.1	1250.9	1272.9	1315.9	1358.7	1401.7	1445.3
	H	491.60	1201.8	1345.6	1375.2	1403.7	1459.4	1514.4	1569.4	1624.8
	S	0.6928	1.4304	1.5673	1.5923	1.6154	1.6580	1.6970	1.7335	1.7679

Table F.4: Superheated Steam, English units (Continued)

TEMPERATURE: $t\,°F$

$P/(\text{psia})$ ($t^{\text{sat}}/°F$)		sat. liq.	sat. vap.	520	540	560	580	600	620	650
725 (507.01)	V	0.0206	0.6318	0.6525	0.6823	0.7100	0.7362	0.7610	0.7848	0.8190
	U	493.5	1116.5	1126.3	1140.2	1153.1	1165.1	1176.5	1187.3	1202.8
	H	496.3	1201.3	1213.8	1231.7	1248.3	1263.9	1278.6	1292.6	1312.6
	S	0.6975	1.4268	1.4396	1.4578	1.4742	1.4893	1.5033	1.5164	1.5347
750 (510.84)	V	0.0207	0.6095	0.6240	0.6536	0.6811	0.7069	0.7313	0.7547	0.7882
	U	498.0	1116.1	1123.1	1137.5	1150.7	1163.0	1174.6	1185.6	1201.3
	H	500.9	1200.7	1209.7	1228.2	1245.2	1261.1	1276.1	1290.4	1310.7
	S	0.7022	1.4232	1.4325	1.4511	1.4680	1.4835	1.4977	1.5111	1.5296
775 (514.57)	V	0.0208	0.5886	0.5971	0.6267	0.6539	0.6794	0.7035	0.7265	0.7594
	U	502.4	1115.6	1119.9	1134.7	1148.3	1160.9	1172.7	1183.9	1199.9
	H	505.4	1200.1	1205.6	1224.6	1242.1	1258.3	1273.6	1288.1	1308.8
	S	0.7067	1.4197	1.4253	1.4446	1.4619	1.4777	1.4923	1.5058	1.5247
800 (518.21)	V	0.0209	0.5690	0.5717	0.6013	0.6283	0.6536	0.6774	0.7000	0.7323
	U	506.7	1115.2	1116.6	1131.9	1145.9	1158.8	1170.8	1182.2	1198.4
	H	509.8	1199.4	1201.2	1220.9	1238.9	1255.5	1271.1	1285.9	1306.8
	S	0.7111	1.4163	1.4182	1.4381	1.4558	1.4720	1.4868	1.5007	1.5198
825 (521.76)	V	0.0210	0.5505	......	0.5773	0.6042	0.6293	0.6528	0.6751	0.7069
	U	510.9	1114.6	......	1129.0	1143.4	1156.6	1168.5	1180.5	1196.9
	H	514.1	1198.7	......	1217.1	1235.6	1252.6	1268.5	1283.6	1304.8
	S	0.7155	1.4129	......	1.4315	1.4498	1.4664	1.4815	1.4956	1.5150
850 (525.24)	V	0.0211	0.5330	......	0.5546	0.5815	0.6063	0.6296	0.6516	0.6829
	U	515.1	1114.1	......	1126.0	1140.8	1154.3	1166.9	1178.7	1195.3
	H	518.4	1198.0	......	1213.3	1232.2	1249.7	1265.9	1281.2	1302.8
	S	0.7197	1.4096	......	1.4250	1.4439	1.4608	1.4763	1.4906	1.5102
875 (528.63)	V	0.0211	0.5165	......	0.5330	0.5599	0.5846	0.6077	0.6294	0.6602
	U	519.2	1113.6	......	1123.0	1138.2	1152.0	1164.9	1176.9	1193.8
	H	522.6	1197.2	......	1209.3	1228.8	1246.7	1263.3	1278.8	1300.7
	S	0.7238	1.4064	......	1.4185	1.4379	1.4553	1.4711	1.4856	1.5056
900 (531.95)	V	0.0212	0.5009	......	0.5126	0.5394	0.5640	0.5869	0.6084	0.6388
	U	523.2	1113.0	......	1119.8	1135.5	1149.7	1162.8	1175.1	1192.2
	H	526.7	1196.4	......	1205.2	1225.3	1243.6	1260.6	1276.4	1298.6
	S	0.7279	1.4032	......	1.4120	1.4320	1.4498	1.4659	1.4807	1.5010

925 (535.21)	V	0.0213	0.4861	....	0.4930	0.5200	0.5445	0.5672	0.5885	0.6186
	U	527.1	1112.4	....	1116.5	1132.7	1147.3	1160.8	1173.2	1190.7
	H	530.8	1195.6	....	1200.9	1221.7	1240.5	1257.8	1274.0	1296.6
	S	0.7319	1.4001	....	1.4054	1.4260	1.4443	1.4608	1.4759	1.4965
950 (538.39)	V	0.0214	0.4721	....	0.4744	0.5014	0.5259	0.5485	0.5696	0.5993
	U	531.0	1111.7	....	1113.2	1129.9	1144.9	1158.6	1171.4	1189.1
	H	534.7	1194.7	....	1196.6	1218.0	1237.4	1255.1	1271.5	1294.4
	S	0.7358	1.3970	....	1.3988	1.4201	1.4389	1.4557	1.4711	1.4921
975 (541.52)	V	0.0215	0.4587	....	....	0.4837	0.5082	0.5307	0.5517	0.5810
	U	534.8	1111.1	....	....	1127.0	1142.4	1156.5	1169.5	1187.5
	H	538.7	1193.8	....	....	1214.3	1234.1	1252.2	1269.0	1292.3
	S	0.7396	1.3940	....	....	1.4142	1.4335	1.4507	1.4664	1.4877
1000 (544.58)	V	0.0216	0.4460	....	....	0.4668	0.4913	0.5137	0.5346	0.5636
	U	538.6	1110.4	....	....	1124.0	1139.9	1154.3	1167.5	1185.8
	H	542.6	1192.9	....	....	1210.4	1230.8	1249.3	1266.5	1290.1
	S	0.7434	1.3910	....	....	1.4082	1.4281	1.4457	1.4617	1.4833
1025 (547.58)	V	0.0217	0.4338	....	....	0.4506	0.4752	0.4975	0.5183	0.5471
	U	542.3	1109.7	....	....	1120.9	1137.3	1152.0	1165.6	1184.2
	H	546.4	1192.0	....	....	1206.4	1227.4	1246.4	1263.9	1287.9
	S	0.7471	1.3880	....	....	1.4022	1.4227	1.4407	1.4571	1.4791
1050 (550.53)	V	0.0218	0.4222	....	....	0.4350	0.4597	0.4821	0.5027	0.5312
	U	545.9	1109.0	....	....	1117.8	1134.7	1149.8	1163.6	1182.5
	H	550.1	1191.0	....	....	1202.3	1224.0	1243.4	1261.2	1285.7
	S	0.7507	1.3851	....	....	1.3962	1.4173	1.4358	1.4524	1.4748
1075 (553.43)	V	0.0219	0.4112	....	....	0.4200	0.4449	0.4673	0.4878	0.5161
	U	549.5	1108.3	....	....	1114.5	1131.9	1147.4	1161.5	1180.8
	H	553.9	1190.1	....	....	1198.1	1220.4	1240.4	1258.6	1283.5
	S	0.7543	1.3822	....	....	1.3901	1.4118	1.4308	1.4479	1.4706
1100 (556.28)	V	0.0220	0.4006	....	....	0.4056	0.4307	0.4531	0.4735	0.5017
	U	553.1	1107.5	....	....	1111.2	1129.1	1145.1	1159.5	1179.1
	H	557.5	1189.1	....	....	1193.7	1216.8	1237.3	1255.9	1281.2
	S	0.7578	1.3794	....	....	1.3840	1.4064	1.4259	1.4433	1.4664
1125 (559.07)	V	0.0220	0.3904	....	....	0.3917	0.4170	0.4394	0.4599	0.4879
	U	556.6	1106.8	....	....	1107.7	1126.3	1142.6	1157.4	1177.3
	H	561.2	1188.0	....	....	1189.2	1213.1	1234.1	1253.1	1278.9
	S	0.7613	1.3766	....	....	1.3778	1.4009	1.4210	1.4387	1.4623

Table F.4: Superheated Steam, English units (Continued)

TEMPERATURE: $t\,°F$

$P/(\text{psia})$ ($t^{\text{sat}}/°F$)		sat. liq.	sat. vap.	700	750	800	900	1000	1100	1200
725 (507.01)	V	0.0206	0.6318	0.8729	0.9240	0.9732	1.068	1.159	1.249	1.337
	U	493.5	1116.5	1227.0	1249.9	1272.0	1315.3	1358.1	1401.3	1444.9
	H	496.3	1201.3	1344.1	1373.8	1402.6	1458.5	1513.7	1568.8	1624.3
	S	0.6975	1.4268	1.5624	1.5876	1.6109	1.6536	1.6927	1.7293	1.7638
750 (510.84)	V	0.0207	0.6095	0.8409	0.8907	0.9386	1.031	1.119	1.206	1.292
	U	498.0	1116.1	1225.8	1248.9	1271.2	1314.6	1357.6	1400.8	1444.5
	H	500.9	1200.7	1342.5	1372.5	1401.5	1457.6	1512.9	1568.2	1623.8
	S	0.7022	1.4232	1.5577	1.5830	1.6065	1.6494	1.6886	1.7252	1.7598
775 (514.57)	V	0.0208	0.5886	0.8109	0.8595	0.9062	0.9957	1.082	1.166	1.249
	U	502.4	1115.6	1224.6	1247.9	1270.3	1313.9	1357.0	1400.3	1444.1
	H	505.4	1200.1	1340.9	1371.2	1400.3	1456.7	1512.2	1567.6	1623.2
	S	0.7067	1.4197	1.5530	1.5786	1.6022	1.6453	1.6846	1.7213	1.7559
800 (518.21)	V	0.0209	0.5690	0.7828	0.8303	0.8759	0.9631	1.047	1.129	1.209
	U	506.7	1115.2	1223.4	1246.9	1269.5	1313.2	1356.4	1399.8	1443.7
	H	509.8	1199.4	1339.3	1369.8	1399.1	1455.8	1511.4	1566.9	1622.7
	S	0.7111	1.4163	1.5484	1.5742	1.5980	1.6413	1.6807	1.7175	1.7522
825 (521.76)	V	0.0210	0.5505	0.7564	0.8029	0.8473	0.9323	1.014	1.094	1.172
	U	510.9	1114.6	1222.2	1245.9	1268.6	1312.6	1355.9	1399.3	1443.3
	H	514.1	1198.7	1337.7	1368.5	1398.0	1454.9	1510.7	1566.3	1622.2
	S	0.7155	1.4129	1.5440	1.5700	1.5939	1.6374	1.6770	1.7138	1.7485
850 (525.24)	V	0.0211	0.5330	0.7315	0.7770	0.8205	0.9034	0.9830	1.061	1.137
	U	515.1	1114.1	1221.0	1244.9	1267.7	1311.9	1355.3	1398.9	1442.9
	H	518.4	1198.0	1336.0	1367.1	1396.8	1454.0	1510.0	1565.7	1621.6
	S	0.7197	1.4096	1.5396	1.5658	1.5899	1.6336	1.6733	1.7102	1.7450
875 (528.63)	V	0.0211	0.5165	0.7080	0.7526	0.7952	0.8762	0.9538	1.029	1.103
	U	519.2	1113.6	1219.7	1243.9	1266.9	1311.2	1354.8	1398.4	1442.5
	H	522.6	1197.2	1334.4	1365.7	1395.6	1453.1	1509.2	1565.1	1621.1
	S	0.7238	1.4064	1.5353	1.5618	1.5860	1.6299	1.6697	1.7067	1.7416
900 (531.95)	V	0.0212	0.5009	0.6858	0.7296	0.7713	0.8504	0.9262	0.9998	1.072
	U	523.2	1113.0	1218.5	1242.8	1266.0	1310.5	1354.2	1397.9	1442.0
	H	526.7	1196.4	1332.7	1364.3	1394.4	1452.2	1508.5	1564.4	1620.6
	S	0.7279	1.4032	1.5311	1.5578	1.5822	1.6263	1.6662	1.7033	1.7382

P (Sat. Temp)										
925 (535.21)	V	0.0213	0.4861	0.6648	0.7078	0.7486	0.8261	0.9001	0.9719	1.042
	U	527.1	1112.4	1217.2	1241.8	1265.1	1309.8	1353.6	1397.4	1441.6
	H	530.8	1195.6	1331.0	1362.9	1393.2	1451.2	1507.7	1563.8	1620.0
	S	0.7319	1.4001	1.5269	1.5539	1.5784	1.6227	1.6628	1.7000	1.7349
950 (538.39)	V	0.0214	0.4721	0.6449	0.6871	0.7272	0.8030	0.8753	0.9455	1.014
	U	531.0	1111.7	1216.0	1240.7	1264.2	1309.1	1353.1	1397.0	1441.2
	H	534.7	1194.7	1329.3	1361.5	1392.0	1450.3	1507.0	1563.2	1619.5
	S	0.7358	1.3970	1.5228	1.5500	1.5748	1.6193	1.6595	1.6967	1.7317
975 (541.52)	V	0.0215	0.4587	0.6259	0.6675	0.7068	0.7811	0.8518	0.9204	0.9875
	U	534.8	1111.1	1214.7	1239.7	1263.3	1308.5	1352.5	1396.5	1440.8
	H	538.7	1193.8	1327.6	1360.1	1390.8	1449.4	1506.2	1562.5	1619.0
	S	0.7396	1.3940	1.5188	1.5463	1.5712	1.6159	1.6562	1.6936	1.7286
1000 (544.58)	V	0.0216	0.4460	0.6080	0.6489	0.6875	0.7603	0.8295	0.8966	0.9621
	U	538.6	1110.4	1213.4	1238.6	1262.4	1307.8	1351.9	1396.0	1440.4
	H	542.6	1192.9	1325.9	1358.7	1389.6	1448.5	1505.4	1561.9	1618.4
	S	0.7434	1.3910	1.5149	1.5426	1.5677	1.6126	1.6530	1.6905	1.7256
1025 (547.58)	V	0.0217	0.4338	0.5908	0.6311	0.6690	0.7405	0.8083	0.8739	0.9380
	U	542.3	1109.7	1212.1	1237.5	1261.5	1307.1	1351.4	1395.5	1440.0
	H	546.4	1192.0	1324.2	1357.3	1388.4	1447.5	1504.7	1561.3	1617.9
	S	0.7471	1.3880	1.5110	1.5389	1.5642	1.6094	1.6499	1.6874	1.7226
1050 (550.53)	V	0.0218	0.4222	0.5745	0.6142	0.6515	0.7216	0.7881	0.8524	0.9151
	U	545.9	1109.0	1210.8	1236.5	1260.6	1306.4	1350.8	1395.0	1439.6
	H	550.1	1191.0	1322.4	1355.8	1387.2	1446.6	1503.9	1560.7	1617.4
	S	0.7507	1.3851	1.5072	1.5354	1.5608	1.6062	1.6469	1.6845	1.7197
1075 (553.43)	V	0.0219	0.4112	0.5589	0.5981	0.6348	0.7037	0.7688	0.8318	0.8932
	U	549.5	1108.3	1209.4	1235.4	1259.7	1305.7	1350.2	1394.6	1439.2
	H	553.9	1190.1	1320.6	1354.4	1386.0	1445.7	1503.2	1560.0	1616.8
	S	0.7543	1.3822	1.5034	1.5319	1.5575	1.6031	1.6439	1.6816	1.7169
1100 (556.28)	V	0.0220	0.4006	0.5440	0.5826	0.6188	0.6865	0.7505	0.8121	0.8723
	U	553.1	1107.5	1208.1	1234.3	1258.8	1305.0	1349.7	1394.1	1438.7
	H	557.5	1189.1	1318.8	1352.9	1384.7	1444.7	1502.4	1559.4	1616.3
	S	0.7578	1.3794	1.4996	1.5284	1.5542	1.6000	1.6410	1.6787	1.7141
1125 (559.07)	V	0.0220	0.3904	0.5298	0.5679	0.6035	0.6701	0.7329	0.7934	0.8523
	U	556.6	1106.8	1206.7	1233.2	1257.8	1304.3	1349.1	1393.6	1438.3
	H	561.2	1188.0	1317.0	1351.4	1383.5	1443.8	1501.7	1558.8	1615.8
	S	0.7613	1.3766	1.4959	1.5250	1.5509	1.5970	1.6381	1.6759	1.7114

APPENDIX G

UNIFAC METHOD

The UNIQUAC equation[1] treats $g \equiv G^E/RT$ as comprised of two additive parts, a *combinatorial* term g^C to account for molecular size and shape differences, and a *residual* term g^R (not a residual property as defined in Sec. 6.2) to account for molecular interactions:

$$g \equiv g^C + g^R \qquad (G.1)$$

Function g^C contains pure-species parameters only, whereas function g^R incorporates two *binary* parameters for each pair of molecules. For a multicomponent system,

$$g^C = \sum_i x_i \ln \frac{\Phi_i}{x_i} + 5 \sum_i q_i x_i \ln \frac{\theta_i}{\Phi_i} \qquad (G.2)$$

and

$$g^R = -\sum_i q_i x_i \ln \left(\sum_j \theta_j \tau_{ji} \right) \qquad (G.3)$$

where

[1]D. S. Abrams and J. M. Prausnitz, *AIChE J.*, vol. 21, pp. 116–128, 1975.

$$\Phi_i \equiv \frac{x_i r_i}{\sum\limits_j x_j r_j} \tag{G.4}$$

and

$$\theta_i \equiv \frac{x_i q_i}{\sum\limits_j x_j q_j} \tag{G.5}$$

Subscript i identifies species, and j is a dummy index; all summations are over all species. Note that $\tau_{ji} \neq \tau_{ij}$; however, when $i = j$, then $\tau_{ii} = \tau_{jj} = 1$. In these equations r_i (a relative molecular volume) and q_i (a relative molecular surface area) are pure-species parameters. The influence of temperature on g enters through the interaction parameters τ_{ji} of Eq. (G.3), which are temperature dependent:

$$\tau_{ji} = \exp\frac{-(u_{ji} - u_{ii})}{RT} \tag{G.6}$$

Parameters for the UNIQUAC equation are therefore values of $(u_{ji} - u_{ii})$.

An expression for $\ln\gamma_i$ is found by application of Eq. (10.94) to the UNIQUAC equation for g [Eqs. (G.1) through (G.3)]. The result is given by the following equations:

$$\ln\gamma_i = \ln\gamma_i^C + \ln\gamma_i^R \tag{G.7}$$

$$\ln\gamma_i^C = 1 - J_i + \ln J_i - 5q_i \left(1 - \frac{J_i}{L_i} + \ln\frac{J_i}{L_i}\right) \tag{G.8}$$

and

$$\ln\gamma_i^R = q_i \left(1 - \ln s_i - \sum_j \theta_j \frac{\tau_{ij}}{s_j}\right) \tag{G.9}$$

where in addition to Eqs. (G.5) and (G.6)

$$J_i = \frac{r_i}{\sum\limits_j r_j x_j} \tag{G.10}$$

$$L_i = \frac{q_i}{\sum\limits_j q_j x_j} \tag{G.11}$$

$$s_i = \sum_l \theta_l \tau_{li} \tag{G.12}$$

Again subscript i identifies species, and j and l are dummy indices. All summations are over all species, and $\tau_{ij} = 1$ for $i = j$. Values for the parameters $(u_{ij} - u_{jj})$ are

Table G.1: UNIFAC–VLE subgroup parameters[†]

Main group	Subgroup	k	R_k	Q_k	Examples of molecules and their constituent groups	
1 "CH$_2$"	CH$_3$	1	0.9011	0.848	n-Butane:	2CH$_3$, 2CH$_2$
	CH$_2$	2	0.6744	0.540	Isobutane:	3CH$_3$, 1CH
	CH	3	0.4469	0.228	2,2-Dimethyl	
	C	4	0.2195	0.000	propane:	4CH$_3$, 1C
3 "ACH" (AC = aromatic carbon)	ACH	10	0.5313	0.400	Benzene:	6ACH
4 "ACCH$_2$"	ACCH$_3$	12	1.2663	0.968	Toluene:	5ACH, 1ACCH$_3$
	ACCH$_2$	13	1.0396	0.660	Ethylbenzene:	1CH$_3$, 5ACH, 1ACCH$_2$
5 "OH"	OH	15	1.0000	1.200	Ethanol:	1CH$_3$, 1CH$_2$, 1OH
7 "H$_2$O"	H$_2$O	17	0.9200	1.400	Water:	1H$_2$O
9 "CH$_2$CO"	CH$_3$CO	19	1.6724	1.488	Acetone:	1CH$_3$CO, 1CH$_3$
	CH$_2$CO	20	1.4457	1.180	3-Pentanone:	2CH$_3$, 1CH$_2$CO, 1CH$_2$
13 "CH$_2$O"	CH$_3$O	25	1.1450	1.088	Dimethyl ether:	1CH$_3$, 1CH$_3$O
	CH$_2$O	26	0.9183	0.780	Diethyl ether:	2CH$_3$, 1CH$_2$, 1CH$_2$O
	CH–O	27	0.6908	0.468	Diisopropyl ether:	4CH$_3$, 1CH, 1CH–O
15 "CNH"	CH$_3$NH	32	1.4337	1.244	Dimethylamine:	1CH$_3$, 1CH$_3$NH
	CH$_2$NH	33	1.2070	0.936	Diethylamine:	2CH$_3$, 1CH$_2$, 1CH$_2$NH
	CHNH	34	0.9795	0.624	Diisopropylamine:	4CH$_3$, 1CH, 1CHNH
19 "CCN"	CH$_3$CN	41	1.8701	1.724	Acetonitrile:	1CH$_3$CN
	CH$_2$CN	42	1.6434	1.416	Propionitrile:	1CH$_3$, 1CH$_2$CN

[†]H. K. Hansen, P. Rasmussen, Aa. Fredenslund, M. Schiller, and J. Gmehling, *IEC Research*, vol. 30, pp. 2352–2355, 1991.

found by regression of binary VLE data, and are given by Gmehling et al.[2]

The UNIFAC method for estimation of activity coefficients[3] depends on the concept that a liquid mixture may be considered a solution of the structural units from which the molecules are formed rather than a solution of the molecules themselves. These structural units are called *subgroups*, and a few of them are listed in the second column of Table G.1. A number, designated k, identifies each subgroup. The relative volume R_k and relative surface area Q_k are properties of the subgroups, and values are listed in columns 4 and 5 of Table G.1. Also shown (columns 6 and 7) are examples of the subgroup compositions of molecular species. When it is possible to construct a molecule from more than one set of subgroups, the

[2]J. Gmehling, U. Onken, and W. Arlt, *Vapor-Liquid Equilibrium Data Collection*, Chemistry Data Series, vol. I, parts 1–8, DECHEMA, Frankfurt/Main, 1974–1990.

[3]Aa. Fredenslund, R. L. Jones, and J. M. Prausnitz, *AIChE J.*, vol. 21, pp. 1086–1099, 1975.

set containing the least number of *different* subgroups is the correct set. The great advantage of the UNIFAC method is that a relatively small number of subgroups combine to form a very large number of molecules.

Activity coefficients depend not only on the subgroup properties R_k and Q_k, but also on interactions between subgroups. Here, similar subgroups are assigned to a main group, as shown in the first two columns of Table G.1. The designations of main groups, such as "CH$_2$", "ACH", etc., are descriptive only. All subgroups belonging to the same main group are considered identical with respect to group interactions. Therefore parameters characterizing group interactions are identified with pairs of *main* groups. Parameter values a_{mk} for a few such pairs are given in Table G.2.

The UNIFAC method is based on the UNIQUAC equation, for which the activity coefficients are given by Eq. (G.7). When applied to a solution of groups, Eqs. (G.8) and (G.9) are written:

$$\ln \gamma_i^C = 1 - J_i + \ln J_i - 5q_i \left(1 - \frac{J_i}{L_i} + \ln \frac{J_i}{L_i} \right) \tag{G.13}$$

and

$$\ln \gamma_i^R = q_i \left[1 - \sum_k \left(\theta_k \frac{\beta_{ik}}{s_k} - e_{ki} \ln \frac{\beta_{ik}}{s_k} \right) \right] \tag{G.14}$$

The quantities J_i and L_i are still given by Eqs. (G.10) and (G.11). In addition, the following definitions apply:

$$r_i = \sum_k \nu_k^{(i)} R_k \tag{G.15}$$

$$q_i = \sum_k \nu_k^{(i)} Q_k \tag{G.16}$$

$$e_{ki} = \frac{\nu_k^{(i)} Q_k}{q_i} \tag{G.17}$$

$$\beta_{ik} = \sum_m e_{mi} \tau_{mk} \tag{G.18}$$

$$\theta_k = \frac{\sum_i x_i q_i e_{ki}}{\sum_j x_j q_j} \tag{G.19}$$

$$s_k = \sum_m \theta_m \tau_{mk} \tag{G.20}$$

Table G.2: UNIFAC–VLE interaction parameters, a_{mk}, in kelvins[†]

	1	3	4	5	7	9	13	15	19
1 CH₂	0.00	61.13	76.50	986.50	1,318.00	476.40	251.50	255.70	597.00
3 ACH	−11.12	0.00	167.00	636.10	903.80	25.77	32.14	122.80	212.50
4 ACCH₂	−69.70	−146.80	0.00	803.20	5,695.00	−52.10	213.10	−49.29	6,096.00
5 OH	156.40	89.60	25.82	0.00	353.50	84.00	28.06	42.70	6.712
7 H₂O	300.00	362.30	377.60	−229.10	0.00	−195.40	540.50	168.00	112.60
9 CH₂CO	26.76	140.10	365.80	164.50	472.50	0.00	−103.60	−174.20	481.70
13 CH₂O	83.36	52.13	65.69	237.70	−314.70	191.10	0.00	251.50	−18.51
15 CNH	65.33	−22.31	223.00	−150.00	−448.20	394.60	−56.08	0.00	147.10
19 CCN	24.82	−22.97	−138.40	185.40	242.80	−287.50	38.81	−108.50	0.00

[†]H. K. Hansen, P. Rasmussen, Aa. Fredenslund, M. Schiller, and J. Gmehling, *IEC Research*, vol. 30, pp. 2352–2355, 1991.

$$\tau_{mk} = \exp \frac{-a_{mk}}{T} \qquad (G.21)$$

Subscript i identifies species, and j is a dummy index running over all species. Subscript k identifies subgroups, and m is a dummy index running over all subgroups. The quantity $\nu_k^{(i)}$ is the number of subgroups of type k in a molecule of species i. Values of the subgroup parameters R_k and Q_k and of the group interaction parameters a_{mk} come from tabulations in the literature. Tables G.1 and G.2 show a few parameter values; the number designations of the complete tables are retained.[4]

The equations for the UNIFAC method are presented here in a form convenient for computer programming. In the following example we run through a set of hand calculations to demonstrate their application.

Example G.1   For the binary system diethylamine(1)/n-heptane(2) at 308.15 K, find γ_1 and γ_2 when $x_1 = 0.4$ and $x_2 = 0.6$.

SOLUTION   The subgroups involved are indicated by the chemical formulas:

$$CH_3\text{--}CH_2NH\text{--}CH_2\text{--}CH_3(1)/CH_3\text{--}(CH_2)_5\text{--}CH_3(2)$$

The following table shows the subgroups, their identification numbers k, values of parameters R_k and Q_k (from Table G.1), and the numbers of each subgroup in each molecule:

	k	R_k	Q_k	$\nu_k^{(1)}$	$\nu_k^{(2)}$
CH_3	1	0.9011	0.848	2	2
CH_2	2	0.6744	0.540	1	5
CH_2NH	33	1.2070	0.936	1	0

By Eq. (G.15),

$$r_1 = (2)(0.9011) + (1)(0.6744) + (1)(1.2070) = 3.6836$$

Similarly,

$$r_2 = (2)(0.9011) + (5)(0.6744) = 5.1742$$

In like manner, by Eq. (G.16),

$$q_1 = 3.1720 \qquad \text{and} \qquad q_2 = 4.3960$$

The r_i and q_i values are molecular properties, independent of composition.

---

[4]H. K. Hansen, P. Rasmussen, Aa. Fredenslund, M. Schiller, and J. Gmehling, *IEC Research*, vol. 30, pp. 2352–2355, 1991.

Substituting known values into Eq. (G.17) generates the following table for e_{ki}:

e_{ki}		
k	$i = 1$	$i = 2$
1	0.5347	0.3858
2	0.1702	0.6142
33	0.2951	0.0000

The following interaction parameters are found from Table G.2:

$$a_{1,1} = a_{1,2} = a_{2,1} = a_{2,2} = a_{33,33} = 0 \text{ K}$$

$$a_{1,33} = a_{2,33} = 255.7 \text{ K}$$

$$a_{33,1} = a_{33,2} = 65.33 \text{ K}$$

Substitution of these values into Eq. (G.21) with $T = 308.15$ K gives

$$\tau_{1,1} = \tau_{1,2} = \tau_{2,1} = \tau_{2,2} = \tau_{33,33} = 1$$

$$\tau_{1,33} = \tau_{2,33} = 0.4361$$

$$\tau_{33,1} = \tau_{33,2} = 0.8090$$

Application of Eq. (G.18) leads to the values of β_{ik} in the following table:

β_{ik}		
i	$k = 1$	$k = 2$	$k = 33$
1	0.9436	0.9436	0.6024
2	1.0000	1.0000	0.4360

Substitution of these results into Eq. (G.19) yields:

$$\theta_1 = 0.4342 \qquad \theta_2 = 0.4700 \qquad \theta_{33} = 0.0958$$

and by Eq. (G.20),

$$s_1 = 0.9817 \qquad s_2 = 0.9817 \qquad s_{33} = 0.4901$$

The activity coefficients may now be calculated. By Eq. (G.13),

$$\ln \gamma_1^C = -0.0213 \qquad \text{and} \qquad \ln \gamma_2^C = -0.0076$$

and by Eq. (G.14),

$$\ln \gamma_1^R = 0.1463 \qquad \text{and} \qquad \ln \gamma_2^R = 0.0537$$

Finally, Eq. (G.7) gives

$$\gamma_1 = 1.133 \qquad \text{and} \qquad \gamma_2 = 1.047$$

NEWTON'S METHOD

Newton's method is a procedure for the numerical solution of algebraic equations, applicable to any number M of such equations expressed as functions of M variables.

Consider first a single equation $f(X) = 0$, in which $f(X)$ is a function of the single variable X. Our purpose is to find a root of this equation, i.e., the value of X for which the function is zero. A simple function is illustrated in Fig. H.1; it exhibits a single root at the point where the curve crosses the X-axis. When it is not possible to solve directly for the root,[1] a numerical procedure, such as Newton's method, is employed.

The application of Newton's method is illustrated in Fig. H.1. In the neighborhood of an arbitrary value $X = X_0$ the function $f(X)$ can be approximated by the tangent line drawn at $X = X_0$. The equation of the tangent line is given by the linear relation

$$g(X) = f(X_0) + \left[\frac{d f(X)}{dX} \right]_{X=X_0} (X - X_0)$$

where $g(X)$ is the value of the ordinate at X, as shown in Fig. H.1. The root of this equation is found by setting $g(X) = 0$ and solving for X; as indicated in Fig. H.1, the value is X_1. Since the actual function is not linear, this is not the root of $f(X)$. However, it lies closer to the root than does the starting value X_0. The function $f(X)$ is now approximated by a second line, drawn tangent to the curve at $X = X_1$, and the procedure is repeated, leading to a root for this linear approximation at X_2, a value still closer to the root of $f(X)$. This root can be approached as closely as desired by continued successive linear approximation of the original function. The general formula for iteration is

---

[1] For example, when $e^X + X^2 + 10 = 0$.

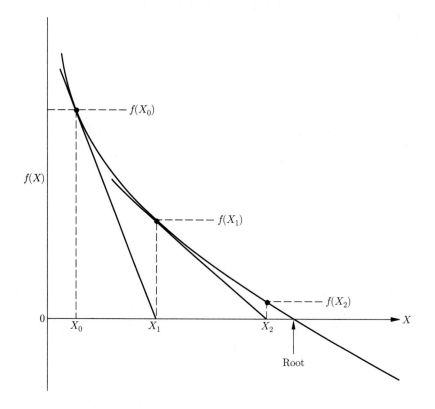

Figure H.1: Newton's method applied to a single function.

$$f(X_n) + \left[\frac{d\,f(X)}{dX}\right]_{X=X_n} \Delta X_n = 0 \qquad (H.1)$$

where

$$\Delta X_n \equiv X_{n+1} - X_n \qquad \text{or} \qquad X_{n+1} = X_n + \Delta X_n$$

Equation (H.1), written for successive iterations ($n = 0, 1, 2, \ldots$), produces successive values of ΔX_n and successive values of $f(X_n)$. The process starts with an initial value X_0 and continues until either ΔX_n or $f(X_n)$ approaches zero to within a preset tolerence.

Newton's method is readily extended to the solution of simultaneous equations. For the case of two equations in two unknowns, let $f_\mathrm{I} \equiv f_\mathrm{I}(X_\mathrm{I}, X_\mathrm{II})$ and $f_\mathrm{II} \equiv f_\mathrm{II}(X_\mathrm{I}, X_\mathrm{II})$ represent two functions, the values of which depend on the two variables X_I and X_II. Our purpose is to find the values of X_I and X_II for which

The values of the increments that satisfy these equations are:

$$\Delta\varepsilon_a = -0.0962 \qquad \text{and} \qquad \Delta\varepsilon_b = -0.1614$$

from which,

$$\varepsilon_a = 0.1 - 0.0962 = 0.0038 \qquad \text{and} \qquad \varepsilon_b = 0.7 - 0.1614 = 0.5386$$

These values are the basis for a second iteration, and the process continues, yielding results as follows:

n	ε_a	ε_b	$\Delta\varepsilon_a$	$\Delta\varepsilon_b$
0	0.1000	0.7000	−0.0962	−0.1614
1	0.0038	0.5386	−0.0472	−0.0094
2	−0.0434	0.5292	−0.0071	0.0043
3	−0.0505	0.5335	−0.0001	0.0001
4	−0.0506	0.5336	0.0000	0.0000

Convergence is clearly rapid. Moreover, any reasonable starting values lead to convergence on the same answers.

Convergence problems can arise with Newton's method when one or more of the functions exhibit extrema. This is illustrated for the case of a single equation in Fig. H.2. The function has two roots, at points A and B. If Newton's method is applied with a starting value of X smaller than a, a very small range of X values produces convergence on each root, but for most values it does not converge, and neither root is found. With a starting value of X between a and b, it converges on root A only if the value is sufficiently close to A. With a starting value of X to the right of b, it converges on root B. In cases such as this, a proper starting value can be found by trial, or by graphing the function to determine its behavior.

NAME INDEX

SUBJECT INDEX

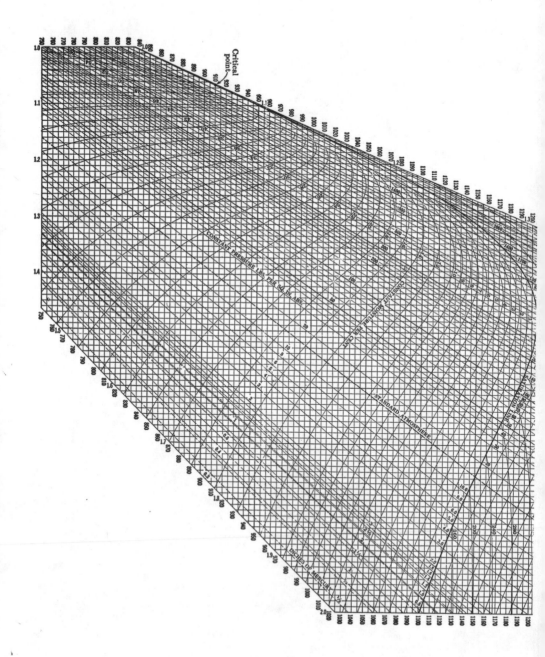